Applied Physics

Concepts into Practice

Gregory S. Romine
Indiana University/Purdue University
Indianapolis

Upper Saddle River, New Jersey
Columbus, Ohio

Library of Congress Cataloging-in-Publication Data

Romine, Gregory S.

  Applied physics: concepts into practice / Gregory S. Romine.
    p. cm.
  Includes index.
  ISBN 0-13-532466-1
  1. Physics. I. Title.

QC23 .R755 2001
530—dc21   00-036761

Vice President and Publisher: Dave Garza
Editor in Chief: Stephen Helba
Associate Editor: Katie E. Bradford
Production Editor: Louise N. Sette
Copy Editor: Maggie Diehl
Design Coordinator: Karrie Converse-Jones
Cover Designer: Tom Mack
Cover Art: © Kathy Hanley
Production Manager: Brian Fox
Marketing Manager: Chris Bracken

This book was set in Times Roman by York Graphic Services, Inc. It was printed and bound by Courier
Westford, Inc. The cover was printed by Phoenix Color Corp.

Photo Credits: p. 2, Richard Pasley, Stock Boston; p. 30, Eugene Gordon; p. 40, Gary S. Settles, Photo
Researchers, Inc.; p. 48, U. S. Navy Office of Information—East; p. 60, James D. Halderman; p. 74, Jose Carillo,
PhotoEdit; p. 77, Canadian Government Travel Bureau; p. 84, David Madison, Tony Stone Images; p. 94, U. S.
Army photo; p. 98, Ben Rose, The Image Bank; p. 120, Pearson Learning; p. 136, Ogust, The Image Works;
p. 150, Jet Propulsion Laboratory, NASA Headquarters; p. 169, AP/Wide World Photos; p. 176, Pearson
Education Corporate Digital Archives; p. 193, NASA Headquarters; p. 208, Phil Degginger, Tony Stone Images;
p. 217, Laima Druskis, Pearson Education/PH College; p. 229, Hulton Getty/Liaison Agency, Inc.; p. 236, Ford
Motor Company, Dearborn, MI; p. 243, Marc Anderson, Pearson Education/PH College; p. 264, Michael
Rosenfeld, Tony Stone Images; p. 269, Hulton Getty/Liaison Agency, Inc.; p. 282, Robert Pham, Pearson
Education/PH College; p. 291, Alan Band, Hulton Getty/Liaison Agency, Inc.; p. 304, Gary Beck, Albuquerque
Convention and Visitors Bureau; p. 328, Paul Seheult, Corbis; p. 334, Siede Preis, PhotoDisc, Inc.; p. 350,
PhotoDisc, Inc.; p. 360, Caterpillar Inc.; p. 363, Brownie Harris, The Stock Market; p. 374, Pearson Learning;
p. 390, Richard Megna, Fundamental Photographs; p. 414, Pearson Education/PH College; p. 448, Westinghouse
Electric; p. 476, E. E. Hertzog, U. S. Bureau of Reclamation; p. 486, Don Tremain, PhotoDisc, Inc.; p. 508,
Jeremy Woodhouse, PhotoDisc, Inc.; p. 534, National Institutes of Health; p. 543, National Radio Astronomy
Observatory; p. 566, Pearson Education/PH College; p. 577, National Institutes of Health; p. 596,
Runk/Schoenberger, Grant Heilman Photography, Inc.; p. 620, Biophoto Assoc., Photo Researchers, Inc.;
p. 640, NASA Headquarters; p. 646, Central Scientific Company; p. 652, U. S. Department of Energy; p. 664,
David Parker/Science Photo Library, Photo Researchers, Inc.

10 9 8 7 6 5 4 3 2 1

ISBN: 0-13-532466-1

Preface

Applied Physics: *Concepts into Practice* is intended for students enrolled in engineering technology, engineering, and medical degree programs. Students should have a basic knowledge of algebra, geometry, and trigonometry at the high school level. With this math background, students should be able to understand everything in the book. The text is intended for use in a two-semester sequence. For the first semester, Parts 1 and 2 cover mechanics and thermodynamics. For the second semester, Parts 3, 4, and 5 discuss electromagnetics, optics, and some modern physics.

Physics is the study of the forces in nature and how those forces interact with matter. Because these interactions can often be stated in a precise mathematical form, physics involves much use of mathematical modeling. The formulation of physical laws in a mathematical framework implies that a prerequisite of a serious study of physics is a certain fluency with mathematics. However, it is not necessary that students have such fluency before they begin the study of physics. Rather, the mathematical fluency can be developed along with the physical concepts.

In *Applied Physics*, mathematics is used as a language to describe the workings of nature, in contrast to being used as a number grinder where raw data are reduced to numerical answers that give no sense of perspective. Often in problem solving, physical concepts are used only to initially set up a problem, and then the problem becomes essentially one of mathematics. However, in this text, the internal consistency of the mathematics takes the problem to its final answer, which has a very real physical significance. Mathematical formalism is minimal, and a common-sense approach to mathematical methods is employed. The mathematics is "bootstrapped" along with the physics. By studying the text, students will naturally come to be able to "speak" mathematics with a certain degree of fluency.

In this way, I have gone beyond the mathematical limitations that an algebra-based text often imposes upon itself. I wrote *Applied Physics* because I could not find a textbook that integrated the physics and math well at the applied physics level. The mathematical approach for this kind of text is new. But in giving this book the mathematical continuity such a text so sorely needs, I have remained within the prerequisite student math background of only algebra, trigonometry, and geometry. With this background, many of the basic principles of higher math can be developed in a common sense way: on an as-needed basis when the physical concepts require it. The physics drives the math, not vice versa. A physicist, Sir Isaac Newton, invented the mathematical discipline known as *calculus* because the deeper mathematical perspective that calculus provided was necessary for a deeper physical understanding of the universe.

Role of Derivations

In my teaching experience, I've noticed that students have the most difficulty with problem solving in properly interpreting a problem. Identifying a problem's context is crucial to its solution. If students know where an equation comes from, they know what situation the equation describes and what the equation is capable of solving. They also know what a particular equation *cannot* do. Without the proper context, all equations are alike, and the students cannot see the forest for the trees. Therefore, derivations are given for most equations used in this text. By presenting the derivations, the text effectively demonstrates the use of mathematics as a language and modeling tool. Thus, a sense of perspective as to the physical concept can be attained.

Problem Solving

My advice to students is to do problems, work problems, and then do some more problems! In doing problems, students must struggle with the theory and in so doing gradually become familiar with it as they set up a problem, review it, and determine how to solve it. Answers are provided for *each* problem, giving students a goal to reach for. A great deal of material must be included in two semesters, and thus a fast pace is required. Therefore, the problems at the end of each chapter have been kept to a number that could reasonably be attempted during the time period allotted for each chapter. More difficult problems are indicated by an asterisk (*) before the problem number. There are very few "plug-and-chug" problems in this text. However, the book is sufficiently complete that students will not have to refer to an outside text to solve any of the problems.

Applied Physics gives students an opportunity to *understand* physics, in contrast to simply being exposed to it. Obtaining the correct answer for an exercise does not necessarily mean that the solver has any insight into the physical concept being investigated. For example, computers can crank numbers, but computers have absolutely no sense of perspective as to the nature of a given problem. Traveling the path to a correct answer is a significant learning experience for students. Their reward is a correct answer. The answers to the problems in *Applied Physics* are, in most cases, not nice round numbers because the world seldom operates in nice round numbers (take π, for instance). If the answer that a student obtains for a certain problem is essentially the same one given in the text, then the answer is probably correct. A student's answer may vary from the one given in the text because of round-off errors. Students shouldn't worry if their answers disagree slightly with the answers in the book. The *path* to an answer is what is important, not the number of decimal places to which the answer is carried out.

Units and Dimensional Analysis

Although I haven't devoted any space specifically to units and dimensional analysis, Appendix C includes a table of common symbols, dimensions, and units. Units are carried along in all of the examples. I feel that dimensional analysis is a contextual subject and thus can be taught most effectively by including appropriate units in each example. An answer for a given problem is usually not just a number, but a number that has some physical meaning associated with it. That physical meaning is usually assigned a name, such as a force in newtons or a mass in kilograms. In solving the problems, students are required to set up their own equations for a solution. The final equation should have the appropriate units. If, for example, the answer is energy in joules, the final equation should have dimensions of energy and not, say, momentum. If the final equation is dimensionally incorrect, then there is something wrong with the derivation.

Dimensional analysis is not a subject in its own right, but rather an aid to obtaining a proper perspective of a physical concept. Dimensional analysis also can be a big help in initially learning the language of mathematics. Often it is the only way that students know, in beginning their study of physics, whether or not the end equation will solve the problem. Dimensional analysis is ever-present in this book, but seldom directly focused upon.

Applying the Approach to the Real-World Classroom

Through several years of teaching at IUPUI, I have successfully taught introductory applied physics students using the approach of integrating physics and math—in both the classroom and the lab. The approach that math and physics go hand in hand is gaining ground. A nationwide workshop, called the *CPU (Constructing Physics Understanding) Project,* was recently conducted for high school teachers. The workshop was supported by the National Science Foundation and was funded by an Eisenhower Grant. In the workshop, laboratory-based elicitation exercises were used to develop the physical concepts being studied, and direct experience was used to debunk tightly held initial misconceptions about physics. In the force and motion studies, the laboratory equipment included motion carts and inclined planes as well as ultrasonic motion sensors interfaced to computers that, with the appropriate software, gave real-time plots of displacement, velocity, and acceleration versus time. Teachers in the role of students could obtain a direct correlation between the motion studied and the shape of a graph. Later teachers used these same graphs, coupled with the equation of a straight line and the ability to find the area inside simple geometric shapes such as a rectangle or a right triangle, to derive the four equations of motion. They learned that a continuum of ideas leads from the basic concept of motion to the mathematical representation of that motion on a graph to the derivation of equations describing that motion. The integration of math and physics and the continuum between the two is one of the guiding philosophies of this book.

Also in the workshop, the propagation of light was studied in optics. The equipment used was a small flashlight with a grain-of-wheat bulb as a point source of light. The light was shone on a small, solid square template, and the shadow was observed. Up to four flashlights were stacked one upon another, and the shadow went from a dark square through four steps to a fully illuminated screen. Next a bulb with a long vertical filament was substituted for the four stacked flashlights, and the shadow appeared as a dark shadow gradually tapering off to a fully illuminated screen. The student studying light would come to the conclusion that the long vertical filament was just like an infinite number of point sources stacked one on top of another with an infinitesimal space between each point source. The concepts of infinity and infinitesimals arise from purely physical phenomena. The physics forces on us a mathematical way of looking at the world, and without the math, much of the physics is lost. However, the math can be developed in such a way that it is constantly grounded to a real situation and the real utility of the mathematical method shown. This common sense approach works and can serve to defuse the math phobia that many students have.

Applying the Approach in the Laboratory

I have used this approach in student laboratories to good effect. I first give a short talk on the theory of the phenomenon or phenomena the students are going to explore, and I develop the equations they will use to describe the physical case under study. I then show how certain parameters in these equations (such as instantaneous velocity or force) can be gathered as data by the equipment, and I explain the equipment in terms of the concept under study. The data are then gathered and plotted (usually as a straight line) to show relationships visually. I am also able to show how the real world differs from the ideal by, for instance, pointing out that the graph the students actually obtained has a force of friction because of a nonzero y intercept. In the laboratory the physical concepts, the equipment, and the math all act together as an indivisible whole.

Features

Reviewers highlighted the writing style, examples, and problems sets as strengths of this text. The examples and exercises illustrate the application of physics theory.

StudyWizard e-tutorial CD-ROM—Packaged at the back of the text, this CD gives students another medium with which to learn physics. It includes review questions and vocabulary practice for each chapter and for Appendix A: Mathematics.

Writing Style—A friendly, casual writing style engages introductory physics students and makes physics less threatening for them.

Examples—Students will relate to and find interesting the real-world examples taken from everyday life.

Conceptual Problems—Multiple-choice and essay questions develop a mental feel for physics concepts without a focus on the math.

Exercises—Conveniently divided by chapter section for easy assignment of problems. An asterisk identifies more comprehensive exercises (solving these exercises requires an understanding of earlier material in the book).

Objectives—Along with the chapter-opening outline, these provide students with an overview of the chapter. Objectives are performance-based so that instructors can use them in curriculum development of activities and assessment materials. Questions in the Test Item File that accompanies the Romine text are tied to specific objectives to help instructors target their testing.

Key Concepts—These are bolded within the text, highlighted in the margin with their definition, and summarized at the end of the chapter.

Important Formulas—Formulas are highlighted within the chapter. End results of the derivations are summarized as numbered equations in a quick reference table at the end of the chapter. The table shows students the formulas that will be applied in the Exercises so they do not need to recall all of the derivations shown in the chapter and can focus on understanding the use of the end equations.

Chapter Summary—Highlights key points of the chapter and includes problem-solving and calculator use tips and dimensional analysis notes.

Math Appendix—This appendix supports the math in the text. Several chapter cross-references to the math appendix are provided to help students use the math in the text. The appendix covers trigonometric identities and rules for working with logarithms.

Also Available for Students

Lab Manual—Includes 27 experiments that begin with a Pre-Lab Activity to test students' readiness for the lab. Also includes a master Equipment List.

Companion Website at http://www.prenhall.com/romine—An additional learning resource, this provides scenarios and related multiple-choice questions for each of the five main parts of the text.

Supplements for Instructors

Instructors Resource and Solutions Manual—Includes answers to all Conceptual Problems and complete step-by-step solutions for all Exercises in the text. The answers and solutions have been thoroughly checked for accuracy. Also includes selected classroom demonstration ideas.

Lab Solutions Manual—Provides instructors with worked out solutions to lab assignments.

PowerPoint CD-ROM—Includes slides for all figures in the textbook and lab manual for instructors to use in classroom and lab presentations.

Test Item File—Provides both conceptual and mathematics-oriented multiple-choice questions and essay questions for each chapter. Questions are tied to specific Chapter Objectives in the textbook.

Prentice Hall Test Manager—This electronic version of the Test Item File can be used to create variations of exams; instructors can rearrange and edit provided questions, and can add their own questions to create unique exams.

Interactive Journey Through Physics CD-ROM—This salable product features physics animations and simulations.

Acknowledgments

This text has taken me roughly five years to complete. Over that period of time *something* has had to keep me motivated and productive through the initial writing, revisions, ancillary projects, and publishing. I want to thank the people at Prentice Hall for having faith in me and sticking with me through the long incubation of this book, especially Steve Helba for giving me the go-ahead for the project, and Katie Bradford, development editor, who cajoled, threatened, pleaded, promised, cursed, pushed, prodded, placated, and in general babysat me through at least three years of this process. I want to thank the reviewers whose constructive criticisms have given direction to the text and kept it from heretical utterings. These reviewers are:

Bobby K. Adams, College of the Albemarle;

Michael Crittenden, Genessee Community College;

William T. Divver, Jr., Spartanburg Technical College;

Bob Jackson, DeVry, Georgia;

Marv Johnson, Muskegon Community College;

Raymond P. Keegan, Heald College of San Francisco;

Tony Nelson, Oklahoma State University, Okmulgee;

Phyllis Salmons, Embry-Riddle Aeronautical University;

Lynn P. Thomson, Ricks College;

Clark H. Vangilder, DeVry, Phoenix;

Timothy R. Vierheller, The University of Akron, Wayne College; and

Michael W. Wolf, Ph.D., Embry-Riddle Aeronautical University.

Special thanks are due to Maggie Diehl, copy editor, and Louise Sette and Tricia Huhn, production editors, who helped bring the text, lab manual, and supplements to fruition. Also, I want to thank my colleague Steve Schuh, the master-distractor, who bought me some time by writing about a quarter of the exercises for the text, the objectives and summaries for each chapter, and for writing a large part of the Test Item File.

I especially want to thank my wife, Nevenka, who put up with my egotistical ravings and saw to it that I had the love and support that helps keep dreams alive. She has had faith in me and the book when faith was a hard commodity to come by. Without her, this project would have foundered long ago.

Gregory S. Romine

Brief Contents

Contents

Mechanics

PART I

This text is divided into five parts. Part I encompasses Chapters 1–12, or about half the text. Mechanics deals with motion and what causes particles to move (*forces*); it is the basis upon which most of the rest of physics is built.

Motion with a constant acceleration in a straight line is first developed. Then motion with a direction and vectors are included to investigate how objects move in a two-dimensional plane. The way a force acts on an object and the vector nature of forces are discussed next. From force flow the concepts of work and energy. We next look at a quantity intrinsic to a body: its momentum. A force can be redefined in terms of how momentum changes over time.

All of these ideas are reapplied to objects that rotate, and rotational motion, centripetal force, rotational energy, and angular momentum are developed. From rotating objects comes the idea of periodicity followed by development of wave motion and simple harmonic motion. The last three chapters in this section deal with applications of the basic concepts to fluids, materials, and simple machines.

This chapter is con-

cerned with uniform

acceleration in a

straight line. Drag

racing is a sport based

on acceleration over a

quarter-mile, straight-

line distance.

U
Uniformly Accelerated Motion in One Dimension

At the end of this chapter, the reader should be able to:

1. State the difference between *displacement, velocity,* and *acceleration.*

2. Explain the concept of a *differential* as a small change in a quantity.

3. Describe the difference between *instantaneous velocity* and *average velocity.*

4. State the four equations of uniformly accelerated motion and know when to use them.

5. Extract useful information, such as *displacement, velocity,* and *acceleration,* when given a graphical plot.

6. Demonstrate a much better understanding of how an object moves in space and time.

1.1   Definitions and Simple Equations of Motion

We live in a world governed by the laws of physics. Physics is all around us. Physics is the study of forces in nature and how those forces interact with matter. "Matter" includes us human beings. Why don't we notice physics in our everyday lives, except in special cases? We breathe, yet we seldom consciously notice. We can consciously stop breathing, and very soon the fact that we're not breathing forces its way to the forefront of our consciousness! Walking, for instance, is a very natural process, and we don't think about how we do it—we just do it. When we walk, we are constantly falling forward and catching ourselves by throwing a leg out in front to balance. If for some reason the leg doesn't get out in front of our center of gravity, we risk falling on our faces. We trip, and when that happens, our entire consciousness is centered on not falling on our faces! We don't notice the physics because we take it for granted, just as we take walking for granted. We don't notice until something goes wrong. To see the physics in our everyday life, we must make an effort to see it; and since it's all around us, we don't need a physics laboratory to see it.

Walking is a common phenomenon that involves motion in a gravitational field. If we walk from point A to point B, it takes a certain amount of time, and we cover a certain amount of distance. *This distance covered in a certain amount of time is **speed**.* (In the next chapter when we cover vectors, we'll see that a ***velocity** is a vector and thus has both magnitude **and** direction.* The magnitude of a velocity is its speed. When an object revolves in a circle at a constant rpm, we can say that it has a constant speed, but it would be wrong to say that it has a constant velocity because the direction in which the object moves is always changing.) If we run, we cover this same distance faster, or it takes less time. Running and walking are technically the same thing; they just differ in speed. However, when you go from a walk to a trot to a run, your speed is changing. You're *accelerating.*

Speed: The rate of change of distance over time. Its units are distance/time (m/s, km/h, ft/s, mph).

Velocity: Speed with the direction of motion given.

Acceleration: The rate of change of speed over time. Its units are distance/(time)^2 (m/s^2, ft/s^2).

Acceleration is another parameter of motion. There is another type of acceleration: that due to gravity. When something is dropped, it goes faster and faster until it strikes the ground. Maybe that's why we have an instinctive fear of heights. If we fall, the higher up we are, the harder (faster) we strike the ground. In walking (or running), we have the body in motion. Yet we can't walk without a gravitational field. It not only holds us to the earth so that we can walk across it, but it also allows us to move forward by continuously falling forward and catching ourselves. There's a lot of physics going on in just the simple act of walking.

However, before we can delve into the physics of motion, we must be able to describe it in a systematic way. We have already identified the basic parameters of motion above. They are distance and time. The concepts of velocity and acceleration are built around the parameters of distance and time. The definition of speed is the amount of distance covered, divided by the time it took. Mathematically, this takes the form

$$v = \frac{s}{t}$$

where v = speed, s = distance, and t = time. Two other equations can be derived from the first by algebraic manipulation.

$$s = vt \quad \text{and} \quad t = \frac{s}{v}$$

The first says that the distance covered is the velocity at which the distance was covered multiplied by the time it took to cover that distance. The second says that the time it took to cover a certain distance is that distance divided by the speed at which it was covered. We now have three forms of the equation defining speed: the mathematical definition of speed and solutions of that definition solving for distance and time.

We can proceed similarly with acceleration. By definition, *acceleration* is the change in velocity over the time it took the velocity to change.

$$a = \frac{\Delta v}{\Delta t}$$

(Equation 1.1)

The symbol Δ (delta) simply means "change"; Δv is a change in velocity, and Δt is a change in time. The magnitude of the change in velocity is found by subtracting the initial velocity from the final velocity, and the magnitude in the change in time is found by subtracting the initial time from the final time.

$$\Delta v = (v_f - v_i)$$

$$\Delta t = (t_f - t_i)$$

where the subscript *i* means "initial," and the subscript *f* means "final." If we start from rest as our initial condition, $t_i = 0$, and $v_i = 0$. The acceleration then becomes

$$a = \frac{v}{t}$$

where $v = v_f$, and $t = t_f$. But these assumptions are some of the ones that we unconsciously make when we begin to solve these problems. (We don't always start from an initial velocity or time of 0, and when this is the case, we must remember that the delta sign means to subtract the endpoints, or the initial value from the final value.) From the above with a little algebraic manipulation, we get a second equation for accelerated motion:

$$v = at$$

which in words states that the velocity an object achieves under uniform acceleration is equal to the acceleration multiplied by the time in which it accelerated.

It would seem that we have a system of equations for uniformly accelerated motion, but unfortunately it's not complete. What equation would we use if we wanted to know how far something traveled under uniform acceleration, or if we wanted to know its speed after it had traveled a certain distance under uniform acceleration? None of the above equations can give us these answers. We will now derive a complete set of equations for accelerated motion in one dimension. One dimension implies straight-line behavior. Since we're just starting our study of physics, we'll start with relatively simple situations and add complications *as needed*. After getting an idea as to what uniformly accelerated motion in one dimension is, we'll add the next complication in the next chapter, namely, vectors.

1.2   Velocity

Mathematics is the language of physics. Many of the concepts in physics are mathematically modeled. If you don't understand the mathematics, a good understanding of the physics is almost impossible. This doesn't mean that you must be a mathematician before beginning to read this text. It does mean that you must understand the necessary mathematics. In this text the mathematics (along with the physical concepts) will be bootstrapped into the material as needed. The mathematical prerequisites for this book are algebra, geometry, and some trigonometry. Along with the proper mathematical underpinning, most of the equations in this book will have their derivations given. Knowing an equation's origin allows you to place it in its proper context and to get a feel for where it can and can't be used. With this in mind, we can begin our derivation of the four equations of uniformly accelerated motion in one dimension. Let's start with the definition of acceleration.

$$a = \frac{\Delta V}{\Delta t}$$

As before, the Δ simply stands for change. The velocity is changing as the time increases. Using the Δ notation, we define

$$\Delta v = v_f - v_i \quad \text{and} \quad \Delta t = t_f - t_i$$

where the subscript f means "final" and the subscript i means "initial."

From the above equation for acceleration, we can derive another equation that will allow us to calculate the instantaneous velocity at any time t for a uniformly accelerating body. *If Δv is a change in velocity, and Δt is a change in time, then both are what is known as **differentials**.* When you subtract two numbers, you find their difference. A differential is just a small change. We can take the above equation for acceleration and rearrange it to get

Differential: A small change in some quantity, such as distance or time.

$$a\,\Delta t = \Delta v$$

The above equation has three terms: a constant term (a, acceleration which we assume doesn't change over time, in this chapter at least); and two differential terms, one with respect to time (Δt) and one with respect to velocity (Δv). Both of the differential terms are changing. Time is running, and the velocity is increasing or decreasing. To arrive at an equation for the instantaneous velocity at any time t, we must expand the differentials back into the notation that subtracts the endpoints. If we span a time from $t = 0$ to a time of $t = t$, then we also have a spread of velocities of v_0 (the initial velocity at time $t = 0$) to the final velocity v at time t. This gives us

$$a(t - 0) = (v - v_0)$$

Solving, we have $\qquad\qquad\qquad at = v - v_0$

Rearranging yields $\qquad\qquad\qquad v = v_0 + at$

We can see that if there had been an initial velocity (v_0) at time $t = 0$, $v = v_0$. Also, if we had started from rest ($v_0 = 0$), then $v = at$, which follows from the definition of acceleration.

Let's take a graphical look at the simple equation $v = at$. In Figure 1.1, we start at time $t = 0$ with a constant acceleration a_0. The vertical axis is acceleration, and the horizontal axis is time. As time runs, and if the acceleration doesn't change, the horizontal line a_0 will be plotted on the graph.

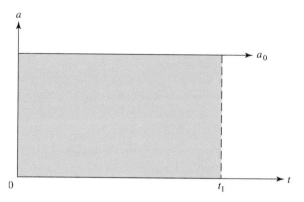

Figure 1.1 Area under a "Curve"

Now assume that we want to know the velocity of the accelerating object at the time $t = t_1$. We know that Figure 1.1 is a graph of the equation $v = at$. To calculate the velocity, we must multiply the acceleration a_0 by the time t_1 ($v_1 = a_0 t_1$). Look at the graph in Figure 1.1 again. Notice that the area of the filled-in box is also $a_0 t_1$. *In finding the instantaneous velocity from the acceleration and time, we calculate the area under the line.* In this case, the line is the straight horizontal line marked a_0.

We will now continue with our derivation of the equations of uniformly accelerated motion. We will use the same techniques as above, and we will start with the equation

$$v = v_0 + at$$

(Equation 1.2)

Equation (1.2) is the first of four formal equations for uniformly accelerated motion in one dimension.

1.3 Displacement

Just as we initially went from acceleration to velocity, we will now go from velocity to distance. In Equation (1.2), we have two variables: velocity (v) and time (t). These variables can change. We also have two constants: the initial velocity (v_0) and the acceleration (a). These constants, by definition, cannot change.

We know that a velocity is simply a change in displacement over a change in time. Mathematically, this is written as $v = \Delta x / \Delta t$, where x is the horizontal displacement. We substitute this into Equation (1.2) and get

$$\frac{\Delta x}{\Delta t} = v_0 + at$$

We again rearrange and get

$$\Delta x = v_0 \Delta t + at \Delta t$$

As before, we can easily expand the differential to the left of the equal sign. The differential immediately to the right of the equal sign is also as easily expanded because the constant term (v_0) doesn't change how we expand the differential. But what about the term at the far right where we have the mixed expression ($t \Delta t$)? Time (t) isn't constant, and Δt is the time interval over which t changes. The two terms interact. We need to look at two more graphs before we attempt to solve for our second equation of uniformly accelerated motion, one of which we've seen before.

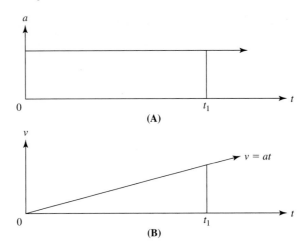

Figure 1.2 Plots of (A) Acceleration vs. Time and (B) Velocity vs. Time for the Same Constant Acceleration

Figure 1.2A is the acceleration-versus-time graph that was shown in Figure 1.1. Figure 1.2B simultaneously plots velocity versus time. If an object undergoes a constant acceleration, then its velocity is continuously increasing, and the object is also increasing its distance from its starting point. An acceleration is causing a changing velocity and a changing distance. Now look at Figure 1.2B, which is a graph of the velocity versus time. If the original acceleration is constant (Figure 1.2A), then the velocity will change linearly with time (Figure 1.2B). We know from Figure 1.2A that $v = at$, which is the equation of the diagonal line in Figure 1.2B. The area under the graph in Figure 1.2B will be velocity multiplied by time, or displacement (x). This area also has a triangular shape. The area of a triangle is one-half the base multiplied by the height. (The half comes from the fact that the triangle's area is half that of a rectangle with the same height and width.) Therefore, the displacement that an object undergoes can be calculated from Figure 1.2B as follows:

$$A = \left(\frac{1}{2}\right)vt \qquad v = at$$

$$A = \frac{1}{2}(at)t \qquad A = \frac{1}{2}at^2$$

where A is the area inside the triangle in Figure 1.2B.

We can now combine all of the elements from immediately above to obtain our next equation for uniformly accelerated motion.

$$x - x_0 = v_0(t - t_0) + \frac{1}{2}at^2$$

Since $t_0 = 0$,

(Equation 1.3)

$$x = x_0 + v_0 t + \frac{1}{2}at^2$$

Thus, we have the second of our equations that describe uniformly accelerated motion.

To recapitulate, the two equations are

(Equation 1.2)

$$v = v_0 + at$$

(Equation 1.3)   and

$$x = x_0 + v_0 t + \frac{1}{2}at^2$$

1.4 The Four Equations of Uniformly Accelerated Motion

Next, we want to obtain an equation without time (t) in it. We can use Equations (1.2) and (1.3) to do so. First we solve Equation (1.2) for time (t).

$$at = v_f - v_0$$

$$t = \frac{v_f - v_0}{a}$$

We then insert the above equation for time into Equation (1.3).

$$x = v_0 t + \frac{1}{2}at^2 + x_0$$

$$x = x_0 + v_0\left(\frac{v_f - v_0}{a}\right) + \frac{1}{2}a\left(\frac{v_f - v_0}{a}\right)^2$$

$$x = x_0 + \frac{v_0 v_f - v_0^2}{a} + \frac{1}{2}a\left(\frac{v_f^2 - 2v_f v_0 + v_0^2}{a^2}\right)$$

$$x = x_0 + \frac{v_0 v_f - v_0^2}{a} + \frac{v_f^2 - 2v_f v_0 + v_0^2}{2a}$$

$$2ax = 2ax_0 + 2v_0 v_f - 2v_0^2 + v_f^2 - 2v_f v_0 + v_0^2$$

$$2ax - 2ax_0 = v_f^2 - v_0^2$$

$$2a(x - x_0) = v_f^2 - v_0^2 \qquad \text{(Equation 1.4)}$$

The final variation is to find an equation describing uniformly accelerated motion that doesn't have the acceleration (a) in it. To do so, we use the concept of average velocity. *The distance traveled under uniform acceleration is equal to the **average velocity** over that distance multiplied by the time.* Mathematically,

Average velocity: The total distance traveled over the total time a trip takes.

$$x = v_{avg}t$$

where

$$v_{avg} = \frac{v_0 + v_f}{2}$$

The above equation is the simple average of two numbers, and it works only if the acceleration is constant. This gives us our fourth equation for uniformly accelerated motion.

$$x = \frac{v_0 + v_f}{2}t \qquad \text{(Equation 1.5)}$$

We now have our four equations for uniformly accelerated motion in one dimension.

$$v = v_0 + at \qquad \text{(Equation 1.2)}$$

$$x = x_0 + v_0 t + \frac{1}{2}at^2 \qquad \text{(Equation 1.3)}$$

$$2a(x - x_0) = v_f^2 - v_0^2 \qquad \text{(Equation 1.4)}$$

(Equation 1.5)

$$x = \frac{v_0 + v_f}{2}t$$

These four equations are the only ones you need to remember, because almost all of the exercises at the end of the chapter will revolve around the proper use of this system of equations.

1.5   Graphical Analysis and the Equations of Motion

For the rest of the chapter, we'll be learning how to use Equations (1.2) through (1.5), but remember how the use of simple equations interacted with visual graphical plots and areas to give us these equations. Also contemplate how such intuitive physical concepts of motion complement graphical analysis. These four equations describe quite well physically real situations.

To be able to start using the four equations, we must obtain some sort of sign convention. That is, we must be able to account for acceleration *and* deceleration, as well as both forward and reverse velocities. Sign conventions are usually arbitrary, which means that they don't have to be set up exactly the same way for each situation or problem. This allows you to take advantage of, for instance, a useful orientation for a given problem. A good example would be a body falling under the influence of gravity. A positive acceleration should point down because the speed of a falling body increases as the body accelerates downward. Conversely, a negative acceleration should point up, because when a ball is thrown up, it first slows down and stops for a moment before beginning to fall back to the earth. To get a better feel for that situation, look at the series of graphs in Figure 1.3. We'll start with some plots of velocity versus time. For all of these graphs, we'll define a positive acceleration as a line with a positive slope (increasing with time), and a negative acceleration (deceleration) as a line with a negative slope (decreasing with time). The sign conventions should become clear after you have taken a little time to study the examples below.

Figure 1.3A is a graph of the equation $v = v_0 + at$. The acceleration is positive, and it is the upper slanted line in the graph with the positive slope. The velocity is changing so that it increases with time. Figure 1.3A also has the information necessary to find the distance traveled (x) by finding the areas inside the right triangle and the rectangle. This information is the same as that provided by the equation $x = v_0 t + \frac{1}{2}at^2$, where $v_0 t$ is the area inside the lower, lightly shaded rectangle and $\frac{1}{2}at^2$ is the area inside the darkly shaded, right triangle. (Keep in mind that the area under a graph is related to some physical quantity, in this case displacement, x.) We can get both velocity and displacement information from the same graph, although it's plotted as velocity versus time instead of displacement versus time.

Now look at the graph in Figure 1.3B. We now have a graph of the equation $v_f = v_0 - at$, again with enough information from the plot to find the displacement, x. The displacement information corresponds to the equation $x = v_0 t - \frac{1}{2}at^2$. The acceleration is represented by the downward-sloping line that runs from v_0 at $t = 0$ on the axis (upper left) to v_f on the axis below the origin (lower right). The slope of that line is negative; therefore, the acceleration is also negative. The velocity gets smaller and smaller with time, is momentarily zero, and starts growing again in the reverse direction. For example, assume that we're standing at the side of a long, level air track and that a glider is moving to the right with a velocity v_0. (An air track is a straight, virtually frictionless

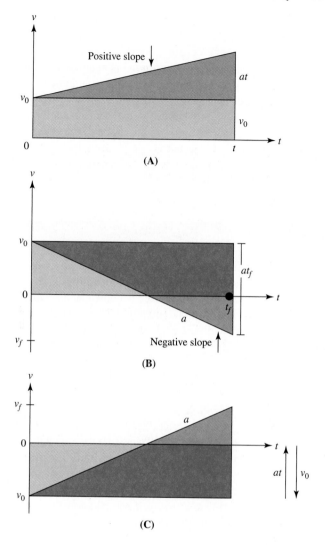

Figure 1.3 Plot of
(A) $v = v_0 + at$,
(B) $v_f = v_0 - at$, and
(C) $v_f = -v_0 + at$

surface; you'll probably get to work with one in the laboratory.) Also assume that an air jet is attached to the glider pointing in the direction opposite the glider's motion. If the air jet is switched on just as the glider passes you, it will provide a force that will act to slow the glider down, or decelerate it. If the air jet is left on long enough, the glider will eventually stop. If the jet is left on still longer, the glider will reverse its direction and will start moving back toward you again. Figure 1.3B is a graph of such motion in time. Note that v_f is a negative number on the graph because the air jet had been left on long enough so that the glider was moving back toward you again, or it had reversed its original direction. The two areas in the graph still represent the total displacement of the glider. The rectangle of height v_0 and base t_f ($v_0 t_f$) still represents the distance that the glider would have gone by time t_f if the original velocity hadn't changed. However, it was changing (decelerating), and the glider was slowing down. Thus, it wouldn't go as far, and that distance must be subtracted from rectangle $v_0 t_f$. What is subtracted is the large triangle $\frac{1}{2}at_f^2$. What is represented by the small triangle below the time axis is the distance covered by the glider after it has reversed direction.

Now look at Figure 1.3C. Notice that originally the acceleration is positive, and the velocity is negative. This situation is analogous to a glider on an air track moving to the left with the air jet also pointing to the left. The force from the jet will cause the glider to slow down its leftward movement, come to a stop, and start moving again to the right. The rectangle v_0t below the time axis is the distance the glider would have gone to the left without the deceleration. Since the glider is originally experiencing a deceleration, the dark gray triangle represents the fact that the glider doesn't go as far to the left (v_0t) as it would have if its velocity weren't changing. Figure 1.3C is a graph of the equation $v_f = -v_0 + at$. Using the area under the lines, the graph also represents the equation $x = -v_0t + \frac{1}{2}at^2$. (The area of the rectangle below the time axis is added to the area of the triangle to get the total displacement.)

1.6   Examples of Motion in One Dimension (Straight-Line Motion)

At this point we have developed concepts of motion and a mathematical framework (along with graphical analysis) that have allowed us to derive four equations for uniformly accelerated motion in one dimension. That part of classical physics is known as *mechanics* (which can be thought of as a study of forces and motion), and it rests largely on the concepts developed here. Indeed, these concepts will be carried throughout all of physics whenever and wherever motion is discussed. In your future studies of physics, keep this in mind: Physics is largely a cumulative process. You can't start in the middle and expect to understand what is going on, because some links in the logical chain will be missing. You must start at the beginning.

Now that we have developed a theory of motion, we need to ask, How well does it fit the real world? How useful to us are these four equations at this point? What situations in the real world correspond to accelerated motion? Two come to mind that are probably common to just about everyone: (1) the motion associated with riding in and driving a car and (2) the motion of a falling body. The first motion involves acceleration, deceleration (braking), relative velocity (passing and overtaking on the road), and distance traveled. The second situation is referred to as *free fall*. We all know that if we drop something, it falls toward the earth. The higher the elevation from which an object is dropped, the faster it is moving when it hits the ground. This implies that objects accelerate downward under the influence of gravity. We will now look at both situations in some detail.

Let's first look at the motion associated with driving a car. A car moves over the surface of the earth, and its motor provides the power for acceleration from rest to some cruising speed. Its brakes allow it to slow down and stop (deceleration), and it was designed to take us places (displacement). Figure 1.4 is the graph of a typical trip taken in an automobile. Notice that the graph is divided up into nine phases for the trip. In phase #1 the

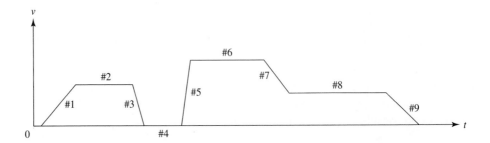

Figure 1.4 Plot of Velocity vs. Time for an Automobile Trip

car starts from rest and accelerates to the cruising speed of phase #2. In phase #2 the car cruises at speed v_2 for some time interval. In phase #3 the car brakes to a stop (possibly for a stoplight). In phase #4 the car waits at the stoplight until the light changes; then it accelerates away from the light to reach the cruising speed v_6. In phase #7 the car slows down (possibly because it has caught up with a slower car). In phase #8 the driver continues to follow the slower car. In phase #9 the driver brakes the car to a halt, having completed his or her journey. This graph describes the car's journey qualitatively, and if the velocity and time axes were calibrated, it would describe the journey quantitatively as well. Let's put numbers to such a graph in our first example.

EXAMPLE 1.1

Refer to Figure 1.5 for this example. (a) What are the values of acceleration for a_1 and a_3? (b) How much distance was covered during the trip? (c) What was the average speed for the trip?

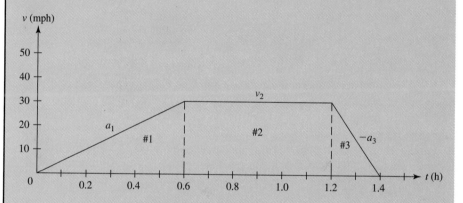

Figure 1.5 Plot of Velocity vs. Time for a Trip by Car

SOLUTION

(a) The value of a_1 is the slope of line a_1. We can also figure it using the mathematical definition of acceleration.

$$a = \frac{\Delta v}{\Delta t}$$

$$\Delta v = (v_f - v_i)$$

Since $v_f = 30$ mph, and $v_i = 0$ (both values taken from Figure 1.5),

$$\Delta v = 30 \text{ mph} - 0 \qquad \Delta v = 30 \text{ mph}$$

$$\Delta t = 0.6 \text{ h} - 0 \qquad \Delta t = 0.6 \text{ h}$$

$$a_1 = (30 \text{ mi/h})/0.6 \text{ h} \qquad a_1 = 50 \text{ mi/h}^2$$

The units for acceleration are distance/(time)^2. The units that you choose should be convenient for the problem at hand. But once chosen, they should be used consistently throughout the problem. Notice that the mathematical definition for acceleration ($a = \Delta v/\Delta t$) is also mathematically the same form as the equation for the slope of a line ($m = \Delta y/\Delta x$), or the rise over the run.

We can find a_3 the same way that we found a_1.

$$a = \frac{\Delta v}{\Delta t} \qquad\qquad \Delta v = (v_f - v_i)$$

But in this case, $v_f = 0$ $v_i = 30$ mph

and $\Delta t = 1.4$ h $- 1.2$ h $\Delta t = 0.2$ h

We can now calculate a_3.

$$a_3 = \frac{(v_f - v_i)}{\Delta t}$$

$$a_3 = \frac{(0 - 30 \text{ mph})}{0.2 \text{ h}}$$

$$a_3 = -150 \text{ mi/h}^2$$

Note that a_3 is negative, indicating braking, or a deceleration before coming to rest. A negative acceleration means that you're slowing down instead of speeding up.

(b) This distance can be calculated by determining the total area under the graph. Since the trip underwent three distinct phases, we can break the area up into the triangle under phase #1, the rectangle under phase #2, and the triangle under phase #3. Then we add the three areas together. By inspecting the graph, however, we can directly come up with an equation for the total distance covered.

$$s = \frac{1}{2}a_1 t_1^2 + v_2 t_2 + \frac{1}{2}a_3 t_3^2$$

Acceleration a_3 was negative due to braking, but the car covered some positive distance while coming to a stop. That's why the last term is added instead of subtracted. We can obtain t_1, t_2, and t_3 from the graph.

$$t_1 = 0.6 \text{ h} - 0 \quad \text{ or } \quad t_1 = 0.6 \text{ h}$$

$$t_2 = 1.2 \text{ h} - 0.6 \text{ h} \quad \text{ or } \quad t_2 = 0.6 \text{ h}$$

$$t_3 = 1.4 \text{ h} - 1.2 \text{ h} \quad \text{ or } \quad t_3 = 0.2 \text{ h}$$

Inserting all of the numbers yields

$$s = \frac{1}{2}(50 \text{ mi/h}^2)(0.6 \text{ h})^2 + (30 \text{ mph})(0.6 \text{ h}) + \frac{1}{2}(150 \text{ mi/h}^2)(0.2 \text{ h})^2$$

$$s = 26.4 \text{ mi}$$

This gives us a distance traveled of 30 mi.

(c) This speed is simply the distance traveled divided by the time the trip took.

$$v_{avg} = \frac{s}{t}$$

$$v_{avg} = 30 \text{ mi}/1.4 \text{ h}$$

$$v_{avg} = 21.4 \text{ mph}$$

Take care not to use an average velocity in any of the equations for uniformly accelerated motion, since all of the velocities in these equations are *instantaneous* velocities.

Of course, a graph need not accompany all problems, but in problem solving it's necessary to have at least some sort of mental picture of what is happening before you begin to work through the mathematics. Sometimes you can draw a very rudimentary picture; just the act of drawing the picture can order the problem in your mind so that you have a clear idea as to what is happening in the given situation. When aware of the situation, next find out what is being asked concerning the situation. Only after you know what the circumstances are and what is being asked about those circumstances can you begin to gather the resources necessary to solve the problem. In our next example, we'll follow this procedure.

EXAMPLE 1.2

You are traveling along a highway at night at a speed of 100 km/h when you spot an object directly in front of you in the road at the limit of your headlights. (a) If the maximum braking deceleration that your car can provide is 7 m/s², and if your headlights extend out to a range of 30 m, will you hit the object before coming to a stop? (b) How long will it take to stop?

SOLUTION
(a) We're talking about a panic stop here, in which the car will brake at its maximum ability until the car comes to rest. We know the initial and final velocities, the rate of deceleration, and the maximum distance we can go. If what is asked is, "Will we hit the object?" there are two ways to answer this question. Both will revolve around the same equation.

$$v_f^2 - v_i^2 = 2ax$$

This equation was chosen from the four concerning uniformly accelerated motion because we know all of the variable quantities: v_f, v_i, a, and x. To answer the question, let's solve for x, or the distance we'll go under maximum braking. We first set up the above equation to solve for x.

$$x = \frac{v_f^2 - v_i^2}{2a}$$

We can now solve for x, but all units must be consistent in the equation, which means that we must convert 100 km/h to meters per second (m/s).

$$\left(\frac{100 \text{ km}}{\text{h}}\right)\left(\frac{1000 \text{ m}}{\text{km}}\right)\left(\frac{1 \text{ h}}{3600 \text{ s}}\right) = 27.8 \text{ m/s}$$

Now we can substitute numbers into the equation.

$$x = \frac{0 - (27.8 \text{ m/s})^2}{2(-7 \text{ m/s}^2)}$$

or
$$x = 55.2 \text{ m}$$

We wouldn't be able to stop before hitting the object in the road. We would in effect outrun our headlights. It's easy to see that under a different set of conditions, such as rain with wet roads, the braking deceleration will be less, and you would need even more distance to stop before hitting something illuminated by the headlights.

(b) To solve this part, we have a choice of two equations from those available.

$$x = v_0 t - \frac{1}{2}at^2 \quad \text{or} \quad a = \frac{\Delta v}{\Delta t}$$

We're solving for time, and both equations should give us the same answer. However, let's not use the first one, since it is a quadratic equation (there's a t^2 in it!), and the math would be more difficult. Instead we'll use the second.

$$a = \frac{v_1 - v_i}{t}$$

Rearrange and solve for t.

$$t = \frac{v_f - v_i}{a}$$

and substitute the numbers.

$$t = \frac{0 - 27.8 \text{ m/s}}{-7 \text{ m/s}^2}$$

or
$$t = 3.97 \text{ s}$$

Note that often there is more than one way of solving a problem. Just because it's done one way here in the book doesn't mean that doing it another way is invalid. If the physics is good, then you'll get a correct answer.

If an automobile brakes continuously, it comes to rest and remains at rest. But if the deceleration doesn't end with braking, the object initially being decelerated will momentarily come to rest and then start accelerating in the opposite direction. This is what a spacecraft would experience using a retrorocket. The rocket engine attached to a spacecraft can provide a thrust that can be used for both acceleration and deceleration. If the rocket engine is pointing opposite the direction of the craft's velocity, then the engine's thrust will speed up the craft. If the engine is pointing in the same direction as the craft's velocity, then the engine's thrust will slow down and eventually halt the spacecraft. A longer burn of the spacecraft's engine will then cause the spacecraft to start accelerating in the opposite direction. Let's examine a similar situation in our next example.

EXAMPLE 1.3

Assume that you're in a stationary spacecraft and that another spacecraft is approaching you. As it passes you, you observe that it is moving at a relative velocity of 4000 m/s and is undergoing a deceleration of 50 m/s^2 (see Figure 1.6). (a) If the retrorocket fires continuously, how long will it be before the second spacecraft repasses you? (b) How long after the moving spacecraft passes the stationary craft the first time will it come to rest momentarily? (c) How fast will the second spacecraft be going when it repasses you?

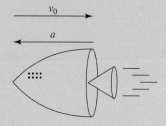

Figure 1.6 Retrorocket Decelerating Spaceship

SOLUTION

(a) We're solving for time, so we can eliminate one equation from the list of the four for uniformly accelerated motion [Equation (1.3)]. But that still leaves three. What else do we know? We know that $x = 0$, because when the second spacecraft repasses you, it really hasn't gone anywhere, has it? Its net displacement is zero! So from the remaining three equations, the list narrows down to only one $(x = v_0 t + \frac{1}{2}at^2)$, because only it contains the other quantities we know: v_0 and a.

$$x = v_0 t + \frac{1}{2}at^2$$

$$0 = v_0 t + \frac{1}{2}at^2$$

We now rearrange the equation to solve for t.

$$v_0 t = -\frac{1}{2}at^2$$

We can divide out t on both sides.

$$v_0 = -\frac{1}{2}at$$

Now set the equation up so that only t remains on the left side.

$$t = -\frac{2v_0}{a}$$

We can now substitute some numbers.

$$t = -2(4000 \text{ m/s})/(-50 \text{ m/s}^2)$$

$$t = 160 \text{ s}$$

It takes 160 s before the second craft repasses going the other way. Let's find out as much about this situation as we can from our toolbox of the four equations for uniformly accelerated motion.

(b) To solve this problem, we can use the equation $v_f = v_0 + at$. Since $v_f = 0$, the equation becomes

$$0 = v_0 + at$$

Rearranging yields $$t = -\frac{v_0}{a}$$

Substitute the numbers. $t = -(4000 \text{ m/s})/(-50 \text{ m/s}^2)$

$$t = 80 \text{ s}$$

Notice that this is half the time for the round trip. A certain symmetry becomes evident, and this symmetry can be used to your advantage in problem solving.

(c) If there's that certain symmetry to the problem, the second craft should repass us at 4000 m/s in the opposite direction. Let's see if it's true. To solve this problem rigorously, we must first calculate how far away the second craft was when it came to rest relative to us. We can use the equation $v_f^2 - v_i^2 = 2ax$.

$$v_f^2 - v_i^2 = 2ax$$

$$0 - v_i^2 = 2ax$$

Solve for x. $$x = -\frac{v_i^2}{2a}$$

Substitute numbers. $x = (-4000 \text{ m/s})^2/[(2)(-50 \text{ m/s}^2)]$

or $x = 160{,}000 \text{ m} = 160 \text{ km}$

Knowing how far away the spacecraft was when it momentarily came to rest, we can now calculate its speed when it repasses us. The same equation used above ($v_f^2 - v_i^2 = 2ax$) will do.

$$v_f^2 - v_i^2 = 2ax$$

$$v_f^2 - 0 = 2ax$$

$$v_f^2 = 2ax$$

$$v_f = \sqrt{2ax}$$

$$v_f = \sqrt{(2)(50 \text{ m/s}^2)(160,000 \text{ m})}$$

$$v_f = 4000 \text{ m/s}$$

It seems that when an acceleration is constantly acting in one direction, certain "tricks" can be used to help solve the problem. The first one we learned was that the time the spacecraft took to reach the spot where it momentarily came to rest relative to us was exactly half the round-trip time. The second trick was that it repassed us in the opposite direction at the same speed at which it first passed us.

1.7   The Acceleration of Gravity and Free Fall

We have discussed objects that are accelerated by some force associated with an engine of some sort, be it an internal combustion engine in a car or a rocket engine in a spacecraft. If you think about it, you will realize that acceleration is simply the result of an applied force. For the car and the spacecraft, the power provided by the engine provides the force to accelerate the vehicle. Braking is another force that comes from friction and provides a deceleration. We will examine the relationships among power, force, and acceleration in following chapters. What we are concentrating on in this chapter is how an object moves when accelerated.

When we drop an object and it falls to the earth, we are witnessing the behavior of an object in a gravitational field. The force of gravity between the object and the earth causes it to fall. The object is attracted to the earth by this force. Therefore, the object accelerates toward the earth. We notice that the higher the elevation from which we drop the object, the harder (faster) the object hits the ground. Thus, if the object is higher up, it will have more time to accelerate and will hit the earth with more speed and energy. Probably the most common example of accelerated motion is free fall. Close to the surface of the earth, the gravitational field is the same for all bodies. One of the implications of this fact is that all bodies (in a vacuum, to eliminate air friction) exhibit the same acceleration toward the earth, regardless of their weight. This finding, surprisingly, is relatively recent. Greek philosophers, notably Aristotle, assumed that the heavier an object, the greater its acceleration due to gravity. After all, a light feather drifts to earth while a rock rushes toward it at a great rate. This pronouncement wasn't questioned for almost two thousand years. In fact, a feather and a stone both fall at the same rate in a vacuum. However, a vacuum was hard to come by in ancient Greece. It wasn't until the late sixteenth century and the time of Galileo that it was proven that all objects near the surface of the earth fall with the same acceleration regardless of their mass. A lecture-hall demonstration of this principle can be performed with a feather and a small coin in a glass tube. At first the feather and the coin are let fall with the tube filled with air. Because of air resistance, the feather lags behind the coin and drifts down the tube, while the coin plummets to the bottom of the tube, hitting the bottom first. The tube, however, is connected to a vacuum pump, and the air can be pumped out. When the experiment is performed again with a vacuum in the tube instead of air, both feather and coin fall at the same rate, and both hit the bottom of the tube at the same instant, as shown in Figure 1.7.

However, the experiment does illustrate one of the limitations of using the equations for uniformly accelerated motion. We have assumed up to now that motion was frictionless. As was demonstrated in the feather-in-the-tube experiment, the resistance of the air can

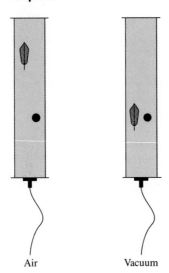

Figure 1.7 Coin and Feather Falling in a Tube with Air Inside (Left) and with a Vacuum Inside (Right)

greatly impede the acceleration of an object falling to earth. As a matter of fact, if an object falls from a great enough height, it will stop accelerating and will reach a terminal velocity. A parachute is a good example. We know that a parachute is pulled to the earth by the force of gravity, yet the parachuter floats gently to earth. The whole idea behind the large canopy is to provide for a lot of resistance to the air and to assure that the terminal velocity is as small as possible (and that therefore the fall is survivable). In short, the larger the canopy, the gentler the descent.

A characteristic acceleration is associated with gravity near the surface of the earth: 9.8 m/s² or 32 ft/s², depending on the system of units used. It is commonly referred to as g in equations, $1g$ being 9.8 m/s² or 32 ft/s². This acceleration acts in the same manner as the acceleration in Example 1.3 with the spacecraft firing its retrorocket. That is, it always points in the same direction (down), and it always has the same magnitude. We can therefore use the two tricks that we discovered in Example 1.3 in solving free-fall problems.

Look at Figure 1.8. At the left side of the figure, a ball is thrown into the air with an initial velocity v_0. It rises into the air, slowing as it goes, until it reaches its maximum height. At this point its velocity is zero. It then starts downward, accelerating as

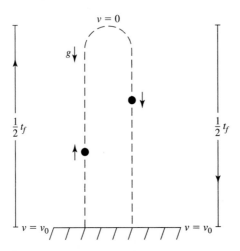

Figure 1.8 Object Thrown Vertically Upward

it goes until it strikes the ground with its initial velocity v_0. If t_f is the time of flight, then half of that time was spent with the ball going up, and the other half was spent with the ball going down. The acceleration of gravity points downward, and that accounts both for why the ball slows down and stops in the left upward part of its trajectory and for why it accelerates downward until it strikes the ground in the right downward part of its trajectory.

Let's solve a free-fall problem.

EXAMPLE 1.4

A ball is thrown into the air with an initial velocity of 20 ft/s. (a) How long is the ball in the air? (b) How high does the ball go? (c) At what time is the ball at a height of 3 ft?

SOLUTION

(a) We must choose an equation from our list of four that has time in it. There are two possible ways to solve this problem. We can choose Equation (1.2) and double that time for the time of flight, or we can choose Equation (1.3) and calculate the time of flight directly. Let's initially go with Equation (1.3).

$$y = y_0 + v_0 t + \frac{1}{2}at^2 \qquad y_0 = 0 \quad \text{and} \quad a = -g$$

(The sign in front of the gravitational acceleration is negative because the ball is initially fighting gravity.)

$$y = v_0 t - \frac{1}{2}gt^2$$

The ball starts and ends at the same place, so $y = 0$.

$$0 = v_0 t - \frac{1}{2}gt^2$$

$$\frac{1}{2}gt^2 = v_0 t$$

$$\frac{1}{2}gt = v_0$$

$$gt = 2v_0$$

$$t = \frac{2v_0}{g}$$

Insert numbers.

$$t = 2(20 \text{ ft/s})/(32 \text{ ft/s}^2)$$

$$t = 1.25 \text{ s}$$

This gives a flight time of 1.25 s. We can check this figure by determining the time using Equation (1.2) to find the time it takes the ball to rise until it is momentarily motionless.

$$v_f = v_0 + at \qquad v_f = 0 \quad \text{and} \quad a = -g$$

$$0 = v_0 - gt$$

$$gt = v_0$$

$$t = \frac{v_0}{g}$$

Insert numbers.

$$t = (20 \text{ ft/s})/(32 \text{ ft/s}^2)$$

$$t = 0.625 \text{ s}$$

Since this equation describes only the upward portion of the flight of the ball, we must double the above answer to get the time of flight.

$$t_f = 2t$$

$$t_f = 2(0.625 \text{ s})$$

$$t_f = 1.25 \text{ s}$$

The answer is the same as the one previously found.
(b) Both Equations (1.3) and (1.4) will give us the answer. We'll use Equation (1.4).

$$v_f^2 - v_0^2 = 2ay \qquad v_f = 0 \quad \text{and} \quad a = -g$$

$$0 - v_0^2 = -2gy$$

$$2gy = v_0^2$$

$$y = \frac{v_0^2}{(2g)}$$

$$y = (20 \text{ ft/s})^2/[2(32 \text{ ft/s}^2)]$$

$$y = 6.25 \text{ ft}$$

(c) There are two times when the ball is 3 ft from the ground: the first time on the way up and the second time on the way down. Again, there are two ways to answer this question. The first would be to look at the upward and downward parts of the trajectory separately, which would necessitate a two-part solution. The second way would be to use Equation (1.3). It's a quadratic and thus will yield two solutions. Since this can be done in a single step, we'll first solve Part (c) this way.

$$y = v_0 t - \frac{1}{2}gt^2$$

We're looking for t, so we solve the equation for t. But notice that t appears in the equation twice. It is a quadratic equation, which means that it has both t^1 (t to the first power) and t^2 (t to the second power, or t squared) in it. First we must put the above equation into the general form for a quadratic. That is, we must put it in descending order for powers of t and set the entire equation equal to zero.

See Appendix A, Section A.1, "The Quadratic Equation."

$$\frac{1}{2}gt^2 - v_0t + y = 0$$

$$gt^2 - 2v_0t + 2y = 0$$

The general form of the quadratic equation is as follows:

$$ax^2 + bx + c = 0$$

The solution of the quadratic equation is given next.

$$x = \frac{-b \pm \sqrt{b^2 - 4ac}}{2a}$$

Just as a, b, and c are the coefficients for x, so also are g, $-2v_0$, and $2y$ the coefficients for t. (A coefficient is the number in front of the variable. The variable in this equation is the time, t. So $a = g$, $b = -2v_0$, and $c = 2y$.) The above equation is our template for solving for t. When the proper coefficients for t are inserted, we have

$$t = \frac{2v_0 \pm \sqrt{(-2v_0)^2 - 4(g)(2y)}}{2(g)}$$

Now we can solve for t.

$$t = \frac{2v_0 \pm \sqrt{4v_0^2 - 8gy}}{2g}$$

$$t = \frac{2v_0 \pm 2\sqrt{v_0^2 - 2gy}}{2g}$$

Simplifying and rearranging yields

$$t = \frac{v_0 \pm \sqrt{v_0^2 - 2gy}}{g}$$

$$t = \frac{v_0}{g} \pm \frac{\sqrt{v_0^2 - 2gy}}{g}$$

Notice that we can expect two answers for t.

We can now insert our values and get some numbers: $v_0 = 20$ ft/s, $g = 32$ ft/s^2, and $y = 3$ ft.

$$t = \frac{20 \text{ ft/s}}{32 \text{ ft/s}^2} \pm \frac{\sqrt{(20 \text{ ft/s})^2 - 2(32 \text{ ft/s}^2)(3 \text{ ft})}}{32 \text{ ft/s}^2}$$

$$t = 0.625 \text{ s} \pm 0.45 \text{ s}$$

Notice that the time that is being added to and subtracted from (0.625 s) is the time it takes the ball to reach its maximum height. Since this marks the halfway point of the ball's trajectory, the ball will be at the height of 3 ft somewhat before and somewhat after this time. Therefore,

$$t = 0.175 \text{ s} \quad \text{and} \quad t = 1.075 \text{ s}$$

Since our time of flight is 1.25 s, we can see that the time at which the ball is at 3 ft during the downward part of its trajectory falls between the time at which it was at its maximum height and the time when it hit the ground again. This gives us a pretty good indication that the answers above are right. The self-consistency of the mathematics automatically gives us both answers directly, and these answers fit the schedule for the trajectory computed in Part (a). Again, look at the symmetry of the answer. You would expect the ball to arrive at the 3-ft height the same time before and the same time after the halfway point, or maximum height. It does. It goes through the 3-ft height 0.45 s before it reaches maximum height, and it again goes through the 3-ft mark 0.45 s after reaching maximum height.

We've now developed a way of describing motion mathematically. After the derivations and examples, you can see how intertwined the physics and the mathematics are. Mathematics is very much a descriptive language, and it will be so employed in this text. If you've never looked at mathematics in this way before, then think of studying physics also as studying a new language.

Summary

1. Displacement (x) is the distance between two points.
2. The average velocity is the total displacement divided by the total elapsed time to travel that distance.

$$v_{\text{avg}} = \frac{x}{t}$$

3. The instantaneous velocity is the velocity of an object at any given position along its path.

$$v = \frac{\Delta x}{\Delta t}$$

4. The slope of a displacement-versus-time graph is equal to the velocity at any given point in time ($\Delta x / \Delta t = v$).
5. The slope of a velocity-versus-time graph is equal to the acceleration at any given point in time ($\Delta v / \Delta t = a$).

Key Concepts

ACCELERATION: The rate of change of speed over time.

AVERAGE VELOCITY: The total distance traveled over the time a trip takes.

DIFFERENTIAL: A small change in some quantity, such as distance or time.

SPEED: The rate of change of distance over time.

VELOCITY: Speed with the direction of motion given.

Important Equations

(1.1) $a = \dfrac{\Delta v}{\Delta t}$ where $\Delta v = v_f - v_i$ and $\Delta t = t_f - t_i$ (Definition of acceleration)

(1.2) $v = v_0 + at$

(1.3) $x = x_0 + v_0 t + \dfrac{1}{2}at^2$

(1.4) $2a(x - x_0) = v_f^2 - v_0^2$

(1.5) $x = \dfrac{v_0 + v_f}{2}t$

(Four equations of uniformly accelerated motion)

Conceptual Problems

1. On a time-versus-distance graph, a horizontal straight line corresponds to motion at:

   a. zero speed

   b. constant speed

   c. increasing speed

   d. decreasing speed

2. On a time-versus-distance graph, a straight line sloping upward to the right corresponds to motion at:

   a. zero speed

   b. constant speed

   c. increasing speed

   d. decreasing speed

3. On a time-versus-distance graph, the motion of a car traveling along a straight road with a uniform acceleration of 2 m/s^2 would appear as a:

   a. horizontal straight line

   b. straight line sloping upward to the right

   c. straight line sloping downward to the right

   d. curved line whose upward slope to the right increases with time

4. The magnitude of the acceleration of a stone thrown upward is:

   a. greater than that of a stone thrown downward

   b. the same as that of a stone thrown downward

   c. smaller than that of a stone thrown downward

   d. zero when the stone momentarily comes to rest at the highest point in its motion

5. Can a rapidly moving object have the same acceleration as a slowly moving one? If so, give an example.

6. Must the path of an object undergoing an acceleration that is constant in both magnitude and direction be a straight line? If not, give an example.

7. Figure 1.9 shows the displacement-versus-time graph for nine different vehicles.

   a. Which vehicles are or have been moving in the forward direction?

   b. Which vehicles are or have been moving in reverse?

   c. Which vehicle has the highest constant velocity?

   d. Which vehicle has the highest constant velocity in the forward direction?

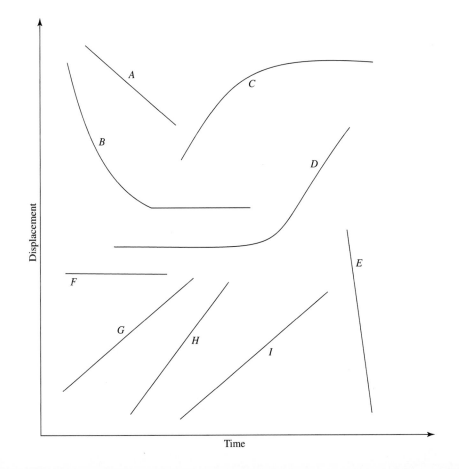

Figure 1.9 Displacement-vs.-Time Plots for Various Vehicles (Problem 7)

e. Which vehicle has the highest constant velocity in the backward direction?

f. Which vehicles have the same velocity?

g. Which vehicle has not moved at all?

h. Which vehicle has accelerated from rest to a constant velocity?

i. Which vehicle has been brought to a stop from an initial velocity in the forward direction?

j. Which vehicle has been brought to a stop from an initial velocity in the reverse direction?

8. A ball is thrown upward. At the top of its trajectory, what is its acceleration?

9. A man leaves his home running errands by going to the grocery and the drugstore; then he returns home. Is the displacement for the trip the same as the distance traveled? If not, why?

10. Two buses depart from Indianapolis. One is going to Chicago, and the other to New York. Both buses have the same speed of 60 mph. Do they have equal velocities? Explain.

11. What's wrong with this statement? "The car traveled around the track lap after lap at the same velocity." Explain.

12. Give an example of an object whose speed is zero for just an instant, but whose acceleration is nonzero.

13. An object moving with a constant acceleration is slowing down, but can it come to rest permanently if the acceleration remains constant?

14. A colleague shows you a position-versus-time graph in which the straight-line segments form a square whose sides are parallel to the position and time axes. If your colleague tells you that the data are from a real experiment, would you believe him or her, and why or why not?

15. The gas pedal of an automobile is often called the *accelerator.* What other automobile control can also be considered an accelerator?

16. Each second, a car moves one-half the remaining distance from its front bumper to a brick wall. Is there ever a collision between the car and the wall? Explain.

Exercises

In the following exercises, the time axis is always horizontal.

Velocity and Displacement

1. An airplane takes off at 10:00 A.M. and travels at 800 km/h until noon. At noon its speed is decreased to 600 km/h, and it holds this speed until it lands at 2:30 P.M. What is the airplane's average speed for the flight?

   Answer: 689 km/h

2. A car travels at 65 mph for 3 h; then it decreases its speed to 55 mph and maintains it for an hour. It decreases its speed to 35 mph and holds it there for a half hour. What is the car's average speed for the trip?

   Answer: 59.4 mph

*3. Someone attempting to catch up with a friend leaves in his car $\frac{1}{2}$ h after his friend has left. If his friend is traveling at 65 mph, how fast must the second car go to catch up with the first in 45 min?

   Answer: 108.3 mph

4. A BMW moving at 100 km/h is 40 km behind a VW moving in the same direction at 80 km/h. How far does the BMW go before it catches up with the VW?

   Answer: 200 km

5. A ball is rolling away from you at a constant speed of 3 ft/s. If you start trotting at a constant speed when it is 10 ft away, how fast must you go in order to intercept the ball just before it rolls into the street 80 ft from you?

   Answer: 3.43 ft/s

6. A car covers $\frac{1}{4}$ the distance to its destination at 30 mph, $\frac{1}{4}$ at 45 mph, and the remainder at 60 mph. Find the car's average speed for the entire distance.

   Answer: 45 mph

7. A man drives $\frac{3}{8}$ of the way home at 50 km/h. How fast must he drive the rest of the way to average 80 km/h for the entire trip?

   Answer: 125 km/h

8. Two race car drivers are racing around a banked circular track of length 4 km at a constant speed. They pass the start/finish line side by side, with one driver running at 180 km/h and the other at 200 km/h. How many laps will the faster driver have completed when he again catches the slower driver?

   Answer: 10 laps

Four Equations of Uniformly Accelerated Motion

9. A certain car is advertised as being capable of going from 0 to 60 mph in 10 s. (a) What is its acceleration? (b) How far does it go in the 10 s?

   Answer: (a) 8.8 ft/s^2
   (b) 440 ft

10. A AA-fuel dragster makes a run down a quarter-mile strip, posting an elapsed time of 7 s and a speed of 300 mph at the end of the strip. (a) What is the average speed for the fueler over the quarter mile? (b) What is the acceleration of the dragster? (This acceleration is substantially higher than *g*,

*Problems preceded by an asterisk (*) are more difficult.

that due to gravity.) (c) If the dragster has only ___ mi in which to stop, what must its lowest average deceleration be?

Answer: (a) 128.6 mph
(b) 53.9 ft/s²
(c) −36.7 ft/s²

11. A race car prepares to enter a sharp turn from a long, fast straightaway. It passes a turn marker indicating 300 ft at 140 mph and starts braking. It passes the next marker indicating 200 ft before the start of the turn $\frac{5}{8}$ s later. (a) How fast was the car going when it passed the 200-ft marker? (b) What was its braking deceleration?

Answer: (a) 78.2 mph
(b) −145 ft/s²

12. The time interval between a driver noticing a danger ahead and the application of the brakes is around 0.7 s. The maximum deceleration produced under braking for an average car might be 1g. Find the minimum distance required to stop if the car is traveling 100 km/h.

Answer: 58.9 m

13. When you are driving at night with your headlights at high beam, the headlights might reach ahead of the car 100 yd. Using the information from Exercise 12, how fast can a car go before it is moving too fast to stop when it spots a hazard at the extreme range of its headlight beam?

Answer: 80.5 mph

14. An elevator has an acceleration and a deceleration of magnitude 5.0 ft/s² and a maximum velocity of 20.0 ft/s. Find the time required for it to go from street level to the seventh floor, stopping at the seventh floor, assuming an average height of 15 ft per floor.

Answer: 9.25 s

*15. Two motocross bikes start from rest from the same corner of a square field of side with length s. The bikes arrive simultaneously at the corner diagonally opposite. One of the bikes travels along the diagonal with a constant acceleration a. The other bike moves along two edges of the square field with a constant speed v. What is the relationship between a and v? Assume that the bike that travels at a constant speed accelerates very quickly to v in a negligible amount of time.

Answer: $a = \dfrac{v^2}{\sqrt{2}s}$

*16. Three insects at the hub of a bicycle wheel start simultaneously crawling along different spokes toward the rim. They all arrive at the rim at the same time. One of the insects crawls with a constant speed v_1. The second starts from rest and crawls with a constant acceleration a. The third starts with an initial speed v_3 and decelerates with an acceleration a. (a) Show that $v_3 = 2v_1$. (b) Find the acceleration a.

Answer: (b) $a = \dfrac{2v_1^2}{R}$

Graphical Analysis and the Equations of Motion

17. The velocity-versus-time plot in Figure 1.10 is for the motion of a bumper car of the sort found in amusement parks. (a) Describe the motion of the car in the time interval shown. (b) What are the accelerations at 12 s and 18 s?

(c) How much distance was covered during the time interval plotted?

Answer: (b) $a_{12} = 0.67$ ft/s²; $a_{18} = -6.3$ ft/s²
(c) $s = 149.03$ ft

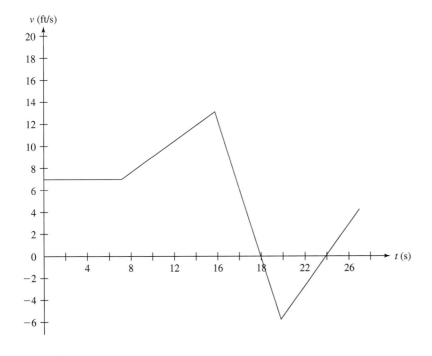

Figure 1.10 Velocity-vs.-Time Plot for Bumper Car Ride (Exercise 17)

18. If an object is dropped from rest, how much farther does it fall in the second second than in the first second?

   Answer: 3 times

19. How fast must the initial velocity of a ball be to reach a height of 10 m?

   Answer: 14 m/s

20. A ball dropped from the roof of a building takes 5.0 s to reach the street. How tall is the building?

   Answer: 400 ft

21. A stone is thrown vertically upward with a velocity of 15 m/s. (a) How long will it take the stone to reach the ground again? (b) What is the maximum height it reaches?

   Answer: (a) 3.06 s
   (b) 11.5 m

22. A stone is thrown vertically from a cliff 30 m high and reaches the ground 3.0 s later. What was the stone's initial velocity, and in what direction?

   Answer: 4.7 m/s, upward

23. The acceleration of gravity at the surface of Mars is 3.7 m/s^2. A stone is thrown vertically upward with a velocity of 15 m/s. (a) When does the stone again touch the ground? (b) What is the maximum height it reaches?

   Answer: (a) 8.1 s
   (b) 30.4 m

   (Compare these answers with those obtained in Exercise 21. With the lower gravity of Mars, the stone goes higher and remains airborne longer.)

24. A movie scene in which a train is supposed to fall from a cliff 40 m high is filmed with a model train that falls from a cliff 1 m high. By how much must the film of the faked catastrophe be slowed so that it mimics the real catastrophe?

   Answer: 6.4 times

*25. A stone is thrown vertically upward with a speed of 45 ft/s from a railway overpass. On the way down, the stone just misses the bridge and hits the ground 4.0 s after having been thrown. Find the height of the bridge above the ground.

   Answer: 76.2 ft

26. A balloonist floating at an altitude of 300 ft drops a bag of sand used as ballast overboard and starts to ascend at a speed of 5 ft/s. How high will the balloon be when the ballast hits the ground? (Neglect air friction.)

   Answer: 321.7 ft

*27. A fireworks mortar fires a star shell vertically upward at the same time a flare is dropped from a helicopter directly overhead hovering at an altitude of 250 ft. At what height will the two meet if the star shell's muzzle velocity was 100 ft/s? (Neglect air friction.)

   Answer: 150 ft

28. In a physics lab experiment, weights are attached to a long strip of paper that passes under a stylus. The weight are allowed to fall, causing the paper strip to accelerate from rest. The stylus puts a mark on the strip at regular time intervals. If the first mark is at a distance of 20 cm from the start of the strip, how much paper has passed under the stylus when the third mark is made?

   Answer: 1.8 m

*29. A rocket is launched upward with a constant acceleration of 150 m/s^2. Six seconds later, the fuel is exhausted. Find the following: (a) the highest velocity the rocket attains, (b) the highest altitude, (c) the total time of flight, and (d) the velocity with which the rocket strikes the ground. (Neglect air friction.)

   Answer: (a) 900 m/s
   (b) 44 km
   (c) 3.2 min
   (d) 929 m/s

*30. In a movie stunt, a package is dropped from a hovering helicopter and reaches a speed of v_f just before it hits the ground. An automobile traveling at a constant speed v_f reaches a point directly under the hovering helicopter just as the package hits the ground. Show that at the instant the package was released, the car had to be a distance away from the point where it catches the package that was twice the altitude of the hovering helicopter.

The wind is a vector, having both magnitude (strength) and direction. Since sailboats are powered by the wind, sailors must take the vector nature of the wind into account, especially if they want to go in a direction against the wind. Sailors accomplish this by tacking into the wind.

Vectors

At the end of this chapter, the reader should be able to:

1. Remember that a vector quantity has both *magnitude* and *direction.*

2. Solve vector-based problems both *analytically* and *graphically.*

3. Separate the horizontal and vertical components from any given **vector.**

4. Use the four equations of motion described in Chapter 1 to simultaneously solve motion in both the horizontal and the vertical directions.

2.1   Graphical Representation of Vectors

In Chapter 1, we developed a mathematical description for motion in one dimension. However, motion has another component besides its magnitude, We all know that if we're moving, we're going somewhere. Thus, there is *direction* as well as *magnitude* to motion. In Chapter 1, we did touch on direction in a limited sense. If you're traveling in a straight line, you can travel both forward and backward. This sense of direction crept into our equations of motion as a positive (forward) or negative (backward) sign. If we're traveling over the surface of a plane (such as the surface of the earth over relatively short distances), we can move off on any bearing from 0° to 360°. Our next step in furthering our mathematical description of motion is to incorporate this notion of direction into our theory.

In Chapter 1, we generally gave speed (or velocity) the symbol v in our equations. In this chapter we'll give velocity the symbol **v.** Notice the **bold** type; this bold type tells you that we're talking about a ***vector*** quantity as opposed to a scalar quantity. *A **scalar** quantity has only a magnitude associated with it.* For instance, temperature is a scalar. It makes no sense to associate a direction with it. The mass of an object is also a scalar. An object that has a mass of 2 kg has that mass whether it's on the surface of the earth or the surface of the moon, orbiting Jupiter or on an outward bound trajectory to the galaxy Andromeda. The mass of an object is an intrinsic property of that object. Weight, however, is a vector quantity. It's the earth's attraction for the apple that makes it fall to the ground. Weight is a force, and force has a direction associated with it. The direction of the force of gravity is toward the center of the planet that attracts the mass. "Down" on the earth is toward the planet's center. What we feel as "weight" is actually this force of attraction.

We can further symbolize a vector by an arrow. The direction of the arrow is the vector's direction, and its length is drawn proportional to its magnitude. In this way we can treat a vector graphically. For instance, the vector **v** in Figure 2.1 might be interpreted as a velocity of 30 m/s in a northeast direction, with 1 cm being a speed of 10 m/s. (The arrow is 3.0 cm long.)

Scalar: A quantity that has only magnitude.
Vector: A quantity that has both magnitude *and* direction.

Figure 2.1 An Arrow Representing a Vector

We can represent the trip someone takes in a car as a series of vectors laid head to tail, as in Figure 2.2. A coordinate system is created in which to place the trip, the vertical and horizontal axes being the principal compass directions north, south, east, and west. Beginning at the origin (0), the car first drives east for 15 mi, then north for 10 mi, and then west for 8 mi, arriving at its destination. The odometer in the car would have registered a trip distance of 33 mi, but according to the plot of the trip, the car is much closer to its starting point than that, because the trip wasn't in a straight line. If the plot in Figure 2.2 had been accurately scaled, all that would be necessary to find the resultant displacement of the car from its starting point would be to measure the length of the resultant vector to get the magnitude of the displacement, and to use a protractor to measure the angle between the $+x$ axis and the resultant to find the resultant's direction.

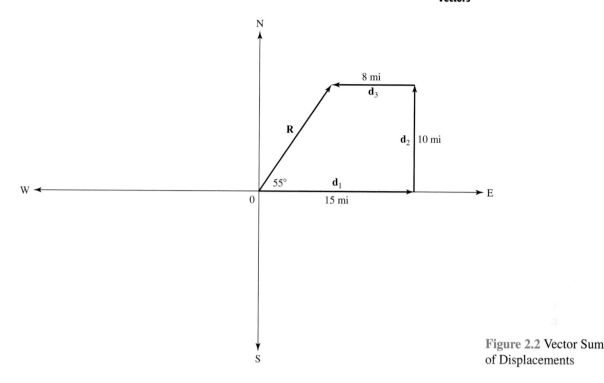

Figure 2.2 Vector Sum of Displacements

2.2   Vectors and Angles

Mathematically, the trip in Figure 2.2 is summed up in the resultant vector **R** and can be written $\mathbf{R} = \mathbf{d}_1 + \mathbf{d}_2 + \mathbf{d}_3$. Notice that vector $\mathbf{d}_3$ is in the opposite direction from vector $\mathbf{d}_1$. If we assume that a positive displacement is in the $+x$ (eastern) direction, then it is also natural to assume that a displacement to the west $(-x)$ is given a negative sign. When a group of vectors is added together head to tail, subtraction is actually the addition of a negative displacement. In Figure 2.2, we have only one vertical vector $(\mathbf{d}_2)$, and to that is assigned a positive displacement to the north $(+y)$. It follows that a negative displacement is assigned to the $-y$ (southerly) direction. Analytically these three vectors $\mathbf{d}_1$, $\mathbf{d}_2$, and $\mathbf{d}_3$ can be resolved into x and y (horizontal and vertical) components. We can rewrite our vector equation for **R** to give

$$\mathbf{R} = (d_{1x} + d_{1y}) + (d_{2x} + d_{2y}) + (d_{3x} + d_{3y})$$

where

$$\mathbf{d}_1 = d_{1x} + d_{1y} \qquad \mathbf{d}_2 = d_{2x} + d_{2y} \quad \text{and} \quad \mathbf{d}_3 = \mathbf{d}_{3x} + \mathbf{d}_{3y}$$

The subscripts tell you along which axis the component points. The x components can be added directly to the x components, and the y components can be added directly to the y components. From Figure 2.2, we can resolve vectors $\mathbf{d}_1$, $\mathbf{d}_2$, and $\mathbf{d}_3$.

$$\mathbf{d}_1 = 15_x + 0_y$$
$$\mathbf{d}_2 = 0_x + 10_y$$
$$\mathbf{d}_3 = -8_x + 0_y$$

Since vectors $\mathbf{d}_1$, $\mathbf{d}_2$, and $\mathbf{d}_3$ are either purely horizontal or purely vertical, we don't need any trigonometry to resolve them. We can now write our next equation for the resultant.

$$\mathbf{R} = (15_x + 0_y) + (0_x + 10_y) + (-8_x + 0_y)$$

Rearranging, we have

$$\mathbf{R} = (15_x + 0_x - 8_x) + (0_y + 10_y + 0_y)$$
$$\mathbf{R} = 7_x + 10_y$$

The resultant is made up of an x component 7 units to the right and a y component 10 units up. It forms a right triangle, as shown in Figure 2.3. Thus, we can use the Pythagorean theorem to obtain the magnitude of the resultant, which is the hypotenuse.

$$R = \sqrt{(7 \text{ mi})^2 + (10 \text{ mi})^2}$$

or, $R = 12.2$ mi. Since the resultant can be found through the use of a right triangle, trigonometry can be used to find the bearing of the resultant vector.

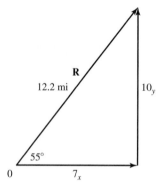

Figure 2.3 Displacements in the x and y Directions Summed to Give the Resultant

Earlier above it was stated that the resultant was pointed in roughly a northeast direction.

$$\tan \theta = \frac{10 \text{ mi}}{7 \text{ mi}}$$

$$\tan \theta = 1.43$$

$$\theta = \arctan(1.43)$$

$$\theta = 55°$$

The angle θ is 55° counterclockwise to the $+x$ axis. If northeast is at 45°, then our true bearing is more north-northeasterly (NNE). In obtaining both the bearing and the magnitude of the resultant from directly measuring the graph, we may find that some small errors can occur in both magnitude and direction. These errors come from the inaccuracies of our measuring tools (protractor and ruler). The analytical approach will usually give a more accurate answer. All of the units in the problem were in miles. The units must be consistent within the same problem. Then when we put the variables in a problem into a mathematical equation, we're sure to have a consistent answer. Notice that this problem can be worked in two different ways: graphically and analytically. The two methods are complementary, and there is often more than one way to work a problem.

Let's take another look at a single vector and use trigonometry to resolve it into its x and y components. In Figure 2.4 we have a single vector **v**. We have placed this vector into an x-y Cartesian coordinate system, with the tail of the vector at the origin. Note that **v** has projections onto both the x and the y axes. A projection is much like a shadow in that the v_x projection would be the shadow of vector **v** if **v** were illuminated from directly above. Similarly, v_y is the projection of **v** on the y axis if **v** were illuminated directly from the side. The lengths of these shadows are $v_x = 12$ units and $v_y = 6$ units if vector **v** is 13.4 units long with an angle of 26.6° counterclockwise from the $+x$ axis. Using trigonometric relationships for a right triangle, we can derive equations for resolving any vector into its x and y components.

$$\sin \theta = \frac{v_y}{v}$$

$$v_y = v \sin \theta \qquad \text{(Equation 2.1)}$$

$$\cos \theta = \frac{v_x}{v}$$

$$v_x = v \cos \theta \qquad \text{(Equation 2.2)}$$

$$\tan \theta = \frac{v_y}{v_x} \qquad \text{(Equation 2.3)}$$

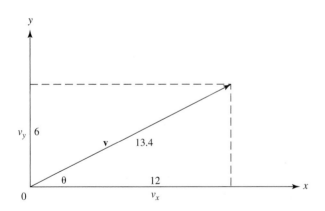

Figure 2.4 Resolving a Two-Dimensional Vector into Its x and y Components

Shorthand Vector Notation

Knowing that a vector has both magnitude and direction, and that an angle is associated with the direction, we can use an elegant shorthand type of notation to specify any vector. Vector **v** becomes $\mathbf{v} = A\angle\theta$, where A is the amplitude of the vector, and θ is the angle it makes with the $+x$ axis in a counterclockwise direction. Take, for example, the vector $7.2\angle213.7°$, as shown in Figure 2.5. Notice that the x-y plane has been divided up into four quadrants, as indicated by the Roman numerals I, II, III, and IV. If we stick to our

Figure 2.5 Four Quadrants with Angles and Component Signs

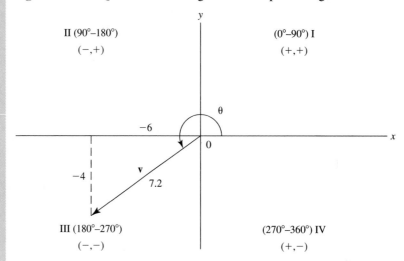

convention that the angle is counterclockwise from the $+x$ axis, then quadrant I has the angles $0°$ to $90°$, quadrant II the angles $90°$ to $180°$, quadrant III the angles $180°$ to $270°$, and quadrant IV the angles $270°$ to $360°$. When we resolve a vector into its x and y components, the quadrants also give us the sign of the x and y units. Our example vector is $7.2\angle 213.7°$, and it therefore falls into the third quadrant. In this quadrant, both x and y are negative. The angle θ is $213.7°$, but this is actually an angle of $33.7°$ underneath the $-x$ axis. With a magnitude of 7.2, the projection onto the $-x$ axis is 6 units, and that onto the $-y$ axis 4 units.

$$(7.2)(\sin 33.7°) = 4 \qquad (-y \text{ axis projection})$$
$$(7.2)(\cos 33.7°) = 6 \qquad (-x \text{ axis projection})$$

The angle $213.7°$ is $(33.7° + 180°)$. If we turn $180°$, we're going in the opposite direction.

<small>EXAMPLE 2.1</small>

Sum the two vectors $5.39\angle 21.8°$ and $5\angle 233.1°$.

<small>SOLUTION</small>

This problem can be written as an equation: $\mathbf{R} = \mathbf{v}_1 + \mathbf{v}_2$. The process is exactly the same as in Figure 2.2, where we put the vectors together head to tail. In this case, we put the tail of $\mathbf{v}_2$ at the head of $\mathbf{v}_1$, as shown in Figure 2.6. Note that it makes no difference in which order $\mathbf{v}_1$ and $\mathbf{v}_2$ are added, because the resultant will be the same.

First, we must resolve each vector into its x and y components.

$$v_{1x} = (5.39)(\cos 21.8°) \qquad v_{1x} = 5$$
$$v_{1y} = (5.39)(\sin 21.8°) \qquad v_{1y} = 2$$

$$\mathbf{v}_1 = 5_x + 2_y$$

Vector $\mathbf{v}_1$ is plotted in Figure 2.6, with its tail at the origin and its head at point (5,2). To find the angle of a vector that doesn't have its tail at the origin, place a tempo-

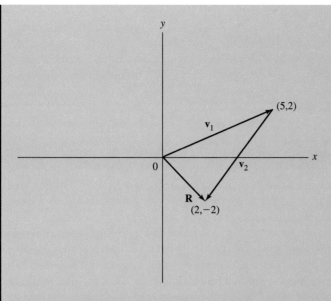

Figure 2.6 Adding Two
Vectors: $\vec{v}_1 + \vec{v}_2 = \vec{R}$

rary origin at the tail of the vector in question with its axes aligned with the coordinate system's x and y axes, as shown in Figure 2.6. Note that $\mathbf{v}_2$ has an angle of 233.1° associated with it, with both the x and the y signs negative. Most of the electronic calculators that we use for problem solving will give us both the sign and the magnitude of the x and y components directly.

$$v_{2x} = (5)(\cos 233.1°) \qquad v_{2x} = -3$$
$$v_{2y} = (5)(\sin 233.1°) \qquad v_{2y} = -4$$

$$v_2 = -3_x - 4_y$$

In plotting $\mathbf{v}_2$, we start at the head of $\mathbf{v}_1$ and move 3 units to the left and 4 units down. This brings us to the position of the head of $\mathbf{v}_2$, which is at the point $(2,-2)$. Analytically what we've done is shown below.

$$\mathbf{v}_1 + \mathbf{v}_2 = (5_x + 2_y) + (-3_x - 4_y)$$
$$\mathbf{v}_1 + \mathbf{v}_2 = (5_x - 3_x) + (2_y - 4_y)$$
$$\mathbf{v}_1 + \mathbf{v}_2 = 2_x - 2_y$$
$$\mathbf{R} = 2_x - 2_y$$

If the tail of the resultant vector $\mathbf{R}$ is at the origin, then the position of its head will be 2 units to the right and 2 units down. This puts it in the fourth quadrant, where $(x,y) = (+,-)$, where θ is between 270° and 360°.

$$\tan \theta = -2/2 \qquad \tan \theta = -1$$
$$\theta = 315°$$

$$\mathbf{R} = 2.83\angle315°$$

The magnitude of **R** is $[2^2 + (-2)^2]^{1/2}$, or 2.83, from the Pythagorean theorem. (The superscript 1/2 means that you take the square root of the quantity within the brackets.)

Trying to sum vectors analytically without plotting them can lead to undue confusion. You will find that the graphical plot is a great help in keeping track of what is going on.

2.3   Motion in a Vertical Plane

When a ball is thrown, the ball actually undergoes two different types of motion simultaneously: It is accelerating vertically under the influence of gravity, and in the absence of air friction it is also moving horizontally, with the magnitude of its horizontal motion remaining constant. Therefore, for projectile motion, we can apply the four equations for uniformly accelerated motion investigated in Chapter 1 in the vertical direction, and equations for uniform velocity in the horizontal direction. If, for instance, we have two balls on the surface of a flat table and they both drop off the edge at the same time vertically with no horizontal velocity (not rolling), they will hit the ground simultaneously. However, what happens if they both drop off the edge simultaneously, but one is rolling while the other has no horizontal component of velocity? Will they both hit the ground at the same time? Yes! The horizontal and vertical components of motion are essentially independent of one another, and both balls are falling under the influence of the same gravitational accelerations of 32 ft/s^2 (9.8 m/s^2). They'll both hit the ground at the same time, even though their individual trajectories will appear quite different.

Look at Figure 2.7. The times t_1, t_2, and t_3 are evenly spaced intervals in time. Notice that both balls A and B fall vertically the same distance in the same times. The only difference between balls A and B is that ball B has a horizontal velocity superimposed onto its fall, but this horizontal velocity does not at all influence the time that it takes ball B to fall. This fact allows us to dissect projectile problems into two parts: the horizontal velocity and the vertical free fall.

Figure 2.7 Two Balls Simultaneously Rolling off Table Top with Different Speeds

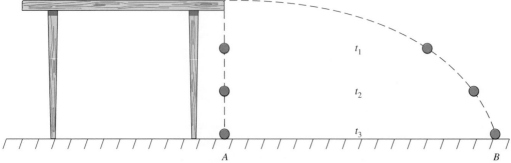

EXAMPLE 2.2

Ball B in Figure 2.7 is rolling at 4 ft/s when it rolls over the edge of the table, and the table is 3 ft high. (a) How far away from the table does B strike the ground? (b) What is the *velocity* of ball B 0.2 s after it rolls off the table top?

SOLUTION

(a) In this case we can use the fact that ball B falls 3 ft to find the time that B is in the air. We choose one of the four equations for uniformly accelerated motion.

$$h = v_0 t + \frac{1}{2}gt^2 \qquad v_{0y} = 0 \qquad h = \frac{1}{2}gt^2$$

$$t^2 = \frac{2h}{g} \qquad\qquad t = \sqrt{\frac{2h}{g}}$$

$$t = \sqrt{\frac{2(3 \text{ ft})}{32 \text{ ft/s}^2}} \qquad t = 0.433 \text{ s}$$

So ball B takes 0.433 s to fall, just as does ball A.

Knowing how long the ball is in the air, we can now use the horizontal velocity to calculate how far B translates to the right before striking the ground.

$$x = v_x t \qquad x = (4 \text{ ft/s})(0.433 \text{ s}) \qquad x = 1.73 \text{ ft}$$

Ball B strikes the ground 1.73 ft to the right of the table.

(b) We know what the x component of the velocity is: It's the speed with which the ball rolled off the table, or $v_x = 4$ ft/s. The y component can be calculated from our equations for uniformly accelerated motion.

$$g = \frac{\Delta v}{\Delta t} \qquad g = \frac{v_{fy} - v_{0y}}{t} \qquad g = \frac{v_{fy}}{t}$$

$$v_{fy} = gt$$
$$v_{fy} = (32 \text{ ft/s}^2)(0.2 \text{ s})$$
$$v_{fy} = 6.4 \text{ ft/s}$$

Previously we resolved our vector into its x and y components. Now we reverse this procedure to combine our x and y components into the vector **v.** Refer to Figure 2.8.

$$v = \sqrt{(4 \text{ ft/s})^2 + (6.4 \text{ ft/s})^2} \qquad v = 7.55 \text{ ft/s}$$

$$\tan \theta = \frac{6.4 \text{ ft/s}}{4 \text{ ft/s}} \qquad \theta = 58°$$

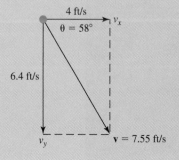

Figure 2.8 Ball in Free Fall with an Unchanging x Component and a Changing y Component of Velocity

The velocity of ball B 0.2 s after it has fallen off the table is 7.55 ft/s, with the direction of the velocity vector given as 58°.

From Figure 2.7 we can see that the path the ball follows is parabolic. A parabola is a particular type of mathematical curve, a portion of which is shown by the path of ball B in Figure 2.7. The reason the path is curved can be understood from Figure 2.8. Note that v_x is unchanging in time, but v_y gets larger as time progresses, and the resultant points more and more downward. Therefore, as time goes on, the direction of the velocity of ball B will point more vertically down. The ball will follow the curved parabolic trajectory shown in Figure 2.7.

2.4   Ballistics

Any ballistic projectile follows a parabolic path. A bullet is a classic ballistic projectile. It's ejected from the barrel of a gun with a certain muzzle velocity. The barrel is usually tilted at some angle θ from the horizontal, and the bullet follows a parabolic trajectory until it strikes the ground downrange of the gun some distance R at the same height from which it was launched. Figure 2.7 shows the parabolic path followed by a ball that rolls off a table top. However, that's only the latter half of the path followed by an object such as like a bullet. Figure 2.9 shows the ballistic trajectory of a bullet from when it's fired until it strikes the ground.

We want to come up with an equation for the range R of the projectile. We will make two simplifying assumptions in this derivation. The first is that there is no air resistance, and the second is that we'll ignore the height of the muzzle above the ground. Only when the muzzle is horizontal does its height become critical. In most cases, the height reached by the projectile is far greater than the height of the muzzle above the ground; muzzle height has little effect on the range and can safely be ignored.

From elementary geometry and the vector diagram in Figure 2.9, we see that the velocity can be broken up into its horizontal (v_x) and vertical (v_y) components.

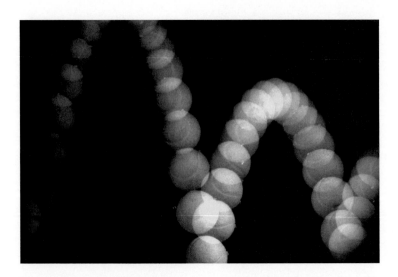

A tennis ball follows a parabolic trajectory between bounces.

Figure 2.9 x and y Velocity Components of a Ballistic Projectile with a Parabolic Trajectory

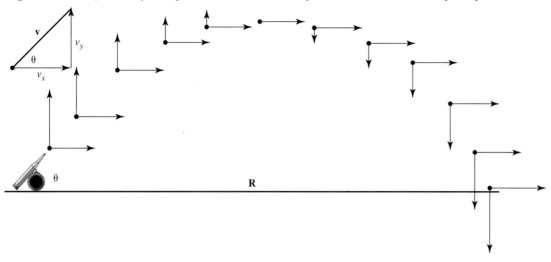

$v_x = v_0 \cos\theta$ (Constant because of no air resistance, and unaccelerated)

$v_{yi} = v_0 \sin\theta$ (The initial vertical component of velocity. It will change with time because of the acceleration of gravity.)

where v_0 is the speed at which the bullet initially emerges from the barrel of the gun. We can now treat the vertical component of the velocity (v_y) as a free-fall problem, just as in Chapter 1.

We're interested in the time of flight of the projectile, which we'll call T. We know from the previous chapter that at the peak of the trajectory, the vertical velocity is zero and the projectile has completed half of its trip. Using the definition of acceleration, we get

$$a = \frac{v_f - v_i}{t}$$

$$v_f = 0 \qquad v_i = v_{yi} = v_0 \sin\theta \qquad a = -g$$

$$-g = \frac{0 - v_0 \sin\theta}{t}$$

$$-gt = -v_0 \sin\theta \qquad gt = v_0 \sin\theta$$

$$t = \frac{v_0 \sin\theta}{g}$$

Remember that the time derived above is one-half the time of flight of the projectile.

$$\frac{T}{2} = \frac{v_0 \sin\theta}{g} \qquad T = \frac{2v_0 \sin\theta}{g}$$

The range is simply the time of flight times the x component of the velocity.

$$R = v_x T$$

or $\qquad R = (v_0 \cos \theta)\left(\dfrac{2v_0 \sin \theta}{g}\right)$ or $\quad R = \dfrac{2v_0^2 \sin \theta \cos \theta}{g}$

From trigonometry we can usefully employ the following identity:

$$\sin \theta \cos \theta = \frac{1}{2} \sin 2\theta$$

(Try this identity equation on your calculator. For a given angle, multiply the sine of the angle by the cosine, and store the result in memory. Then multiply the angle by 2, find the sine of the doubled angle, and divide this by 2. The result should agree with what you previously stored in the memory.) Thus, R becomes

$$R = \frac{2v_0^2(\frac{1}{2}\sin 2\theta)}{g}$$

(Equation 2.4)

$$R = \frac{v_0^2 \sin 2\theta}{g}$$

Equation (2.4) is the range equation for a ballistic projectile that lands *at the same elevation from which it was launched,* in the absence of air friction. Notice that since a sine function has a magnitude between -1 and $+1$, we get our greatest range when $\sin 2\theta = 1$. Then

$$R = \frac{v_0^2}{g}$$

$$\sin 90° = 1 \qquad 2\theta = 90° \qquad \theta = 45°$$

The greatest range for the projectile occurs when the barrel is elevated at an angle of $45°$ from the horizontal.

EXAMPLE 2.3

A mortar fires a round with a muzzle velocity of 100 m/s. (a) What is its greatest range? (b) What two angles of elevation would allow the mortar to hit a target 500 m away?

SOLUTION

(a) The greatest range occurs when $\theta = 45°$. Since $R = v_0^2/g$,

$$R = \frac{(100 \text{ m/s})^2}{9.8 \text{ m/s}^2} \qquad R = 1020 \text{ m}$$

(b) $\qquad R = \dfrac{v_0^2}{g} \sin 2\theta \qquad \dfrac{Rg}{v_0^2} = \sin 2\theta$

See Calculator Tip 2.1 in the chapter Summary.

$$\sin 2\theta = \frac{(500 \text{ m})(9.8 \text{ m/s}^2)}{(100 \text{ m/s})^2} \qquad \sin 2\theta = 0.49$$

$$2\theta = 29.3° \qquad \qquad \theta = 14.7°$$

The second angle would be $90° - 14.7°$, or $\theta_2 = 75.3°$. To check it, we substitute it into the range equation.

$$R = \frac{(100 \text{ m/s})^2}{9.8 \text{ m/s}^2} \sin[2(75.3°)]$$

$$R = 500 \text{ m}$$

The first angle of elevation is relatively shallow (see Figure 2.10), which means that the mortar round has a relatively high v_x or a high horizontal velocity to compensate for the fact that its time of flight is going to be short. The second angle of elevation is much greater, and thus the round's time of flight will be greater than in the first case. However, the horizontal component of velocity will be much reduced over the first case. Both angles will give the same range, however. If the object is to lob the round inside of something such as a fort, then the higher elevation will be chosen, since the round will enter the fort almost vertically. With the shallower angle of elevation, the round will hit the fort's walls instead of going inside.

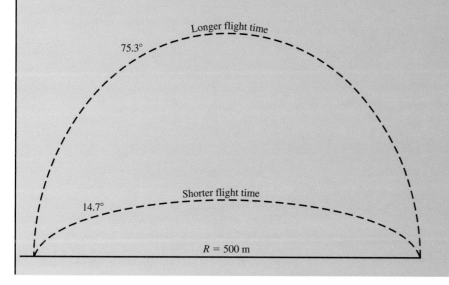

Figure 2.10 Two Angles Giving the Same Range for a Ballistic Projectile

Summary

1. All vectors (**x**, **v**, **a**, and so on) have both magnitude *and* direction. A vector can be presented graphically by using an arrow pointing in the appropriate direction, with its length symbolizing the magnitude. Or a vector can be represented analytically in the following form:

$$\mathbf{B} = B_x \mathbf{x} + B_y \mathbf{y}$$

where vector **B** has an x component of B_x and a y component of B_y. Also, you can work in reverse, finding the magnitude and direction of a vector from the analytical expressions

$$B^2 = B_x^2 + B_y^2 \quad \text{and} \quad \tan \theta = \frac{B_y}{B_x}$$

Calculator Tip 2.1: When finding the resultant direction of a vector, you must often use the arctangent ($\tan^{-1}$) function on a calculator. When finding an angle, divide the y component of the magnitude by the x component; then press the $\boxed{\tan^{-1}}$ button on your calculator. The resulting number will be the value of the angle as measured off the x axis.

2. Motion in two dimensions can be treated independently by using the same four equations of motion found in Chapter 1, with the exception that the horizontal motion will use the x components of the vectors, and the vertical motion will use the y components of the vectors.

$$v_x = v_{0x} \pm a_x t$$

$$x = x_0 + v_{0x}t + \left(\frac{1}{2}\right)a_x t^2$$

$$2a_x(x - x_0) = v_x^2 - v_{0x}^2$$

$$x = (v_{0x} + v_x)\frac{t}{2}$$

and

$$v_y = v_{0y} \pm a_y t$$

$$y = y_0 + v_{0y}t + \left(\frac{1}{2}\right)a_y t^2$$

$$2a_y(y - y_0) = v_y^2 - v_{0y}^2$$

$$y = (v_{0y} + v_y)t/2$$

Problem Solving Tip 2.1: Often, but not always, a_y is equal to the acceleration due to gravity (g), and a_x is equal to zero.

Key Concepts

SCALAR: A quantity that has only magnitude.

VECTOR: A quantity that has both magnitude *and* direction.

Important Equations

(2.1) $v_y = v \sin \theta$ where θ = angle between horizontal axis and vector

(2.2) $v_x = v \cos \theta$ where θ = angle between horizontal axis and vector

(2.3) $\tan \theta = \dfrac{v_y}{v_x}$

(2.4) $R = \dfrac{v_0^2 \sin 2\theta}{g}$

Conceptual Problems

1. Which of the following units could be associated with a vector quantity?
   a. m/s^2 b. kg/s
   c. h   d. mph

2. The magnitude of the resultant of two vectors is a minimum when the angle between them is:
   a. 0°   b. 45°
   c. 90°   d. 180°

3. An airplane has an air speed of 250 km/h and is flying in a wind of 40 km/h. Its speed has a range between a maximum and minimum ground speed of:
   a. 40 and 250 km/h   b. 250 and 290 km/h
   c. 210 and 290 km/h   d. 0 and 40 km/h

4. What is the resultant displacement of a trip that takes you 4 km north and then 3 km west?
   a. 1 km   b. 5 km
   c. 7 km   d. 12 km

5. Ball *A* is launched horizontally with some horizontal speed, and ball *B* is dropped straight down with no initial horizontal speed at the same time from the same height as ball *A*. Which of the following statements is true?
   a. Ball *A* reaches the ground first.
   b. Ball *B* reaches the ground first.

   c. Ball *A* hits the ground with the greater speed.
   d. Ball *B* hits the ground with the greater speed.

6. Two boats with identical speeds start at the same time to cross a river with a current. Boat *A* points its bow directly at the other bank and goes until it hits the bank of the other shore. Boat *B* points its bow upstream so that when it arrives at the other bank, it is directly opposite its starting point.
   a. Which boat arrives at the opposite bank first?
   b. Which boat travels the shortest distance relative to the water?
   c. Which boat travels the longest distance relative to the earth?

7. Three balls are thrown with the same speed simultaneously from a roof. Ball *A* is thrown upward; ball *B* is thrown horizontally; and ball *C* is thrown straight down.
   a. What is the order in which the balls reach the ground?
   b. List the speeds of the three balls from the fastest to the slowest as they hit the ground.

8. A hunter aims his rifle directly at a squirrel 100 yd away on the limb of a tree. The squirrel sees the muzzle flash of the rifle as it fires. The squirrel naturally wants to avoid the bullet speeding toward him. Should the squirrel remain on the branch or drop from the branch to avoid the bullet. Why?

9. Does the speed of a ballistic projectile fired at some angle above the horizontal less than 90° vary in its path? If so, where is the speed the least, and where is it the greatest?

10. You are riding a bicycle on a windy day. When you ride directly into the wind, the force of the wind feels much stronger than when you are standing still. Why? Is there a way that you could ride the bike so that you feel no wind? Give answers using reasoning based on vectors.

11. For the range equation, ballistic projectiles fired at 20° and 70° with the same speed have the same range. How do the times of flight for these two trajectories and the horizontal velocities contrast?

12. You are in a train and look out the window after a nap. You see another train on an adjacent track apparently moving in the opposite direction. If your compartment is well isolated so that you can't tell through vibration whether or not your train is moving, give three possible reasons that would explain the apparent motion that you see out the window. Is there any way to tell which motion is actually occurring if you can see only the other train?

Exercises

Graphical Representation of Vectors

1. A car drives 4 mi east and 7 mi south at a speed of 30 mph. How much time would the driver have saved if he had driven directly to his destination?

Answer: 5.9 min

2. Two cars leave a road crossing at the same time. One travels northwest at 100 km/h, and the other northeast at 60 km/h. How far apart are they in (a) 1 h and (b) 3 h?

Answer: (a) $d_1 = 116.6$ km
(b) $d_3 = 350$ km

3. A car traveled 10 mi east and then 20 mi south. The trip took $\frac{1}{2}$ hr. (a) What was the average speed for the trip? (b) What was the magnitude of the velocity for the trip?

Answer: (a) 60 mph
(b) 44.8 mph

Vectors and Angles

4. A 50-lb sack of potatoes is hung vertically from a rope. A man pushes the sack of potatoes to the side until the rope makes an angle of 20° from the vertical. (a) With what force is the man pushing on the sack of potatoes? (b) What is the tension in the rope?

Answer: (a) 18.2 lb
(b) 53.2 lb

5. A horizontal force and a vertical force combine to give a resultant force of 40 N that acts at an angle of 35° above the horizontal axis. Find the magnitudes of the horizontal and vertical forces.

Answer: $F_x = 32.8$ N; $F_y = 22.9$ N

6. A 20-lb wagon is pulled by a rope that has 10 lb of tension on it. (Tension is a force.) The rope makes an angle of 15° with the ground. (a) How much of this tension goes into the horizontal component of the force pulling the wagon? (b) What force does the wagon exert on the ground?

Answer: (a) 9.66 lb
(b) 17.4 lb

7. On a rainy, windy day, raindrops falling on a windowpane are found to make an angle of 30° with the vertical. If the wind is blowing horizontally and steadily at 15 mph, what is the terminal velocity of a raindrop? (The terminal velocity is the maximum vertical downward velocity that the raindrop will attain because of air resistance. Air resistance will limit the maximum velocity that an object will fall. If it didn't, getting caught in a rainstorm could be a fatal experience, as a drop falling from a great height would have the velocity of a high-speed bullet!)

Answer: 26 mph

8. An aircraft is climbing at a speed of 250 mph at an angle of 5° to the horizontal. (a) What is its rate of climb? (b) What is its ground speed?

Answer: (a) 32 ft/s
(b) 249 mph

*9. Two forces act on an object in a horizontal plane. The first is 4 N at a direction 35° counterclockwise from the $+x$ axis. The second is 9 N at a direction 65° clockwise from the $+x$ axis. A third force is applied to the object to keep it in equilibrium with the other two forces. What are its magnitude and its direction?

Answer: 9.2 N 40° CW from the $-x$ axis

*10. A boat with a speed of 2.5 m/s is used to ferry people and freight across a river that has a current of 1.1 m/s. (a) At what angle must the boat's direction be offset to compensate for the current if it is to end up at its dock directly across the river? (b) To the people on the dock on the opposite side of the river, the boat approaches head-on. At what apparent speed is the boat closing on the dock on the opposite side of the river?

Answer: (a) 26.1°
(b) 2.24 m/s

*11. A prop-driven airliner with a maximum air speed of 320 km/h takes off on a flight to destination due east at a distance of 1590 km. The airliner should arrive 5 h after take-off. If a cross wind starts blowing from the north at 70 km/h at take-off, will the airliner be able to keep to the schedule? Prove it with calculations.

Answer: No.

Motion in a Vertical Plane

12. A WWII bomber is flying at an altitude of 30,000 ft at a speed of 175 mph. How far ahead of the target must the bombardier drop his bombs to hit it?

Answer: 2.1 mi

13. A marksman shoots at a bull's eye on a target 100 yd away. The rifle is aimed directly at the bull's eye. If the muzzle velocity of the bullet is 800 ft/s, how far below the bull's eye will the bullet strike?

Answer: 2.25 ft

14. A ball rolls off the edge of a horizontal roof top and lands 4 ft away. The ball's speed just before it left the roof top was 3 ft/s. How high is the roof?

Answer: 28.3 ft

15. A paint ball is ejected horizontally from the roof of a 30-ft building. It is to mark the hood of a car 40 ft away in a parking lot. The paint ball ejector is powered by compressed air, and the speed at which the ball is ejected can be selected. At what speed should the paint ball be ejected to hit the hood of the car?

Answer: 29.2 ft/s

16. A light aircraft is flying at an altitude of 300 ft directly above a road at a speed of 50 mph. A car is moving on the road away from the aircraft at a speed of 30 mph. The pilot of the aircraft wishes to drop a paint ball onto the car's roof to mark it for future identification. At what distance from the car should the pilot drop the ball if it's to hit the car's roof?

Answer: 127 ft

17. A water balloon is thrown at 15 ft/s from the roof of a building 280 ft high at an angle of 30° below the horizontal. At what height above the ground will the balloon splat the side of another building 25 ft away from the first?

Answer: 59 ft

*18. A ball rolls off the top of a stairway with a horizontal velocity of 6 ft/s. The steps are 1 ft wide and 8 in. high. Which step will the ball hit first?

Answer: 2nd step

19. A rifle with a muzzle velocity of 1100 ft/s shoots a bullet at the bull's eye of a target 100 yd away. How high above the target must the gun be aimed so that the bullet will hit the target? Assume that the rifle and the target are at the same level.

Answer: 1.19 ft

*20. An arrow was launched horizontally from a small rise above the ground. It struck the ground 2.5 s after it was released at an angle of 55° from the horizontal. With what speed and from what height was the arrow launched?

Answer: $h = 30.625$ m; $v = 17.2$ m/s

*21. A cannon is aimed outward from the parapet of a fort at an angle θ to the horizontal. Find an equation for the time it takes a shell to hit the level ground below in terms of known variables: the height of the parapet h, the muzzle velocity v, the direction θ, and the acceleration of gravity g.

Answer: $T = \dfrac{2v \sin \theta}{g} + \sqrt{\dfrac{2h}{g}}$

Ballistics

22. Find the minimum initial speed of a Nerf dart that travels a horizontal distance of 15 ft.

Answer: 21.9 ft/s

23. A fireworks mortar can fire a bomb 250 m straight up. (a) What is its maximum horizontal range? (b) How long should the fuse burn if a fireworks display is to explode at the peak of its trajectory when fired so as to have maximum range?

Answer: (a) 500 m
(b) 5.05 s

24. A shotput can be hurled a maximum distance of R. If the shotput is launched upward with the same initial speed, how high will it go?

Answer: $R/2$

25. What percentage increase in initial speed is needed to increase the maximum range of a solid-fuel ballistic rocket by 15%?

Answer: 7%

*26. A baseball is struck by a batter and caught 250 ft away 6 s later. (a) What was the baseball's initial velocity? (b) At what angle to the horizontal was the ball hit? (c) What was its maximum height? (d) What was its speed 4 s after being struck? (e) What was its height at that time?

Answer: (a) 104.7 ft/s
(b) 66.5°
(c) 144 ft
(d) 52.6 ft/s
(e) 128 ft

*27. A shell is fired from a howitzer with a muzzle velocity of 250 m/s and at an elevation of 20°. (a) How far does the shell go? (b) What was its time of flight? (c) What was its maximum altitude? (d) At what other angle could the shell have been fired to have the same range? (e) What would the time of flight be in this case? (f) What would its peak altitude be in this case?

Answer: (a) 4.1 km
(b) 17.45 s
(c) 373 m
(d) 70°
(e) 47.9 s
(f) 2.8 km

*28. Show that the maximum height h that a golf ball rises when hit from the ground at an angle θ is $h = 0.25R \tan \theta$, where R is the range.

An aircraft carrier's catapult applies a tremendous force to a jet fighter-bomber, accelerating it to take-off speed before it leaves the carrier's deck.

Force and Motion

At the end of this chapter, the reader should be able to:

1. Describe the difference between **mass** and *weight.*

2. Separate forces, which are *vectors,* into *x* and *y* components.

3. Draw *free-body diagrams* and label the appropriate forces.

4. Use **Newton's second law** to solve free-body diagrams.

5. State the difference between *frictional force* and the *coefficient of friction.*

Up to this point we have been studying motion through its displacement, velocity, and acceleration, without regard as to what causes the motion. Our discussion was thus largely mathematical. In this chapter we shall discuss the causes of motion. If we are given a particle whose characteristics we know, and if we place the particle in a known environment, we should be able to predict its subsequent motion. This problem was largely solved by Sir Isaac Newton (1642–1727) when he formulated his three laws of motion. The three laws of motion were empirical laws in that they were formulated through observation. Galileo linked the fact that under acceleration, the distance covered goes as the square of the time. Newton took Galileo's work one step further and gave it the necessary mathematical underpinning that made it an exact science.

3.1   Newton's First Law

Up to the time of Galileo, people thought that the natural state of a body was at rest. If the cause of motion was removed, such as a hand pushing a block across a surface, the body (the block) soon ceased to move and came to rest. However, Galileo saw to the heart of the problem and argued that if the surface of the block were made smoother and smoother, and if the surface over which the block moved were lubricated with something such as oil, it took more and more time and a greater distance for the block to come to rest. He was able to show this principle empirically, through simple experiments. He extrapolated to say that a body's true state was in motion, and only a force could change that state of motion, in this case, friction. Galileo was able to successively reduce the force of friction.

Friction is a force that is omnipresent in our world, and until it was identified as a force, it was quite natural to assume that the natural motion of a body was at rest. Without friction, a body in motion would continue to move indefinitely in a straight line with the same speed. Newton adopted Galileo's concept and with it formulated *Newton's*

Newton's first law: Every body persists in its state of rest or of uniform motion in a straight line unless it is compelled to change that state by forces impressed upon it.

first law: Every body persists in its state of rest or of uniform motion in a straight line unless it is compelled to change that state by forces impressed upon it.

Friction is a force because it changes a body's state of motion: It decelerates the body until it eventually stops. Only when a body is at rest does the force of friction become zero. Only then does the body's state of motion continue uninterrupted: It remains at rest! This brings us to the concept of force as the cause of a change of motion. In Chapter 1, the laws of uniformly accelerated motion were formulated, and we placed acceleration at the top of the hierarchy. *Acceleration* was defined as the rate of change in velocity over time. We also saw that acceleration could be negative and that negative acceleration would cause a body to slow down. Friction causes a body to slow down; therefore, it is a force changing the state of motion of a body.

Forces cause accelerations or decelerations. This principle can be expressed mathematically as follows:

$$\mathbf{a} \propto \mathbf{F}$$

The above statement reads, "Acceleration is *proportional* to the applied force." The greater the applied force, the greater the acceleration or deceleration, provided that the mass remains constant. This mathematical statement is on the way to becoming an equation.

3.2   Newton's Second Law

All material objects possess mass. The usual way to measure a mass is to weigh it, but it's important to make the distinction that weight and mass are not the same thing. Weight is actually a force. It is the force exerted by the gravitational field of the earth on an object with mass. This force pulls the object toward the center of the earth. Force is a vector and thus has both magnitude and direction. The direction of the force of gravity, or weight, is always down toward the center of the earth. This force of gravity, or weight, will cause a body to accelerate. A dropped stone picks up speed as it falls to the earth. But since the acceleration of gravity is high enough to make it difficult to measure the acceleration of an object ($9.8\,\text{m/s}^2$ near the earth's surface), we need something that relies on gravity but is somewhat slower and therefore easier to observe.

Consider the apparatus shown in Figure 3.1. Assume that the table top is a frictionless surface. Upon that surface we place a box of mass M that can have various masses placed in it. We have attached this box to a smaller mass, m, hanging over the edge of the table, by a string that runs over a frictionless pulley. The force of gravity pulling down on m (mg) creates a tension in the string that tugs on the mass M sitting on the table. If the mass m is allowed to fall, the box will start to accelerate to the right. If we begin to add mass to the box and measure the acceleration, we will see that the acceleration will decrease as the mass in the box increases. The greater the mass in the box, the lower the acceleration of the system. We begin to get a feeling that ***mass** is the property of a body that determines its resistance to a change in its motion.* This resistance to acceleration is also called *inertia.*

Mass: The property of a body that opposes a change in its motion. Resistance to motion is also called *inertia.*

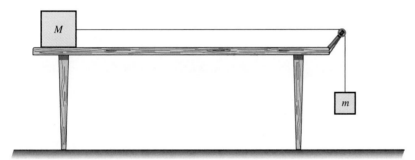

Figure 3.1 Apparatus for Accelerating Block M

We can make the mathematical statement that acceleration is inversely proportional to mass.

$$a \propto \frac{1}{m}$$

The above statement says that the greater the mass being accelerated, the smaller the acceleration will be. We have already seen that acceleration is directly proportional to the applied force. These two mathematical statements can be brought together into a single equation.

$$\text{If } \mathbf{a} \propto \mathbf{F}, \text{ and if } a \propto \frac{1}{m}, \text{ then } \mathbf{a} = \frac{\mathbf{F}}{m}.$$

This can be rearranged into the famous equation

$$\mathbf{F} = m\mathbf{a}$$

(Equation 3.1)

The variable **F**, standing for force, is in **bold** type because force is a vector. On the right side of the equation, **a** is also in bold type because acceleration is a vector. In the case of the box on the table (Figure 3.1), the direction of the acceleration is to the right. However, the symbols for mass (*M* and *m*) are in *italic* type because mass is a scalar with only magnitude but no direction. (The mass of a body is an intrinsic property of that body.) A scalar times a vector is a vector. Equation (3.1) is ***Newton's second law.***

Newton's second law: A net force applied to a massive object causes the object to accelerate (**F** = *m***a**).

In Figure 3.1, the distinction between mass and weight can be seen. The small mass *m*, hanging over the edge of the table, provides a constant force, *m***g** (acting down), to accelerate the large mass *M*. This vertical force of the weight acting downward is transformed into a horizontal pull on *M* by the tension in the string. If *m* isn't changed, then we have available a constant force with which to evaluate the acceleration of *M*.

$$\mathbf{w} = m\mathbf{g}$$

The above equation gives the relationship between weight and mass, where **g** is the acceleration of gravity. Near the earth's surface the acceleration of gravity is $9.8\,\text{m/s}^2$ or $32\,\text{ft/s}^2$ depending on which set of units you use. Mass is an intrinsic property of an object, but the weight of an object depends on where the object is. On the surface of the moon, where the gravity is $\frac{1}{6}$ that of earth's, an object will weigh $\frac{1}{6}$ as much as on the earth. Yet in both places the object will have the same mass. On a trajectory from the earth to Jupiter, the spacecraft *Magellan* is essentially weightless, but it possesses all of the mass that it had when launched at the earth's surface.

3.3   Newton's Third Law

Forces never act alone. They work in pairs. Consider a book lying on a table. That book has mass, and therefore there is a force acting on it: the force of gravity. Why doesn't the book move? The reason is that the force pulling it toward the center of the earth is exactly balanced by the table pushing up on the book. The sum of the two forces is zero, and there is no net force. With no net force, there is, consequently, no motion. Remember that force is a vector, and in this simple example two forces are acting in opposite directions: (1) the force of gravity acting down on the table and (2) the push from the table top acting upward on the book. These forces exactly cancel each other out. A dynamic equilibrium is in operation on the book lying inert on the table. The two forces that oppose each other act in pairs.

Another example of equal and opposite paired forces is water flowing out of a garden hose with a high velocity. As the water leaves the hose, the hose exerts a recoil on the hand of the person holding the hose. Another example is a rocket. The violently burning exhaust gases streaming out the rocket's nozzles at the rear provide the thrust that propels the rocket upward. The backward-streaming exhaust has a reaction force, which is the thrust propelling the rocket forward. A hose emitting a high-speed stream of water has the characteristics of a rocket nozzle, and you can feel the reaction force pushing on you as you hold the hose. Newton first stated this property of forces in ***Newton's third law: To every action there is always opposed an equal reaction; or, the mutual actions of two bodies upon each other are always equal, and directed to contrary parts.*** If you kick a door open, you exert a force on the door, and the door exerts a force on your toe. In other words, if the door is strong and heavy, you will very likely painfully stub your toe, that being the reaction force.

Newton's third law: For every action there is an equal and opposite reaction.

3.4   Units of Force

Up to now nothing much has been said about units. The units for velocity are either ft/s or m/s. But now that we have a force, we have a derived unit, or one based on other, more fundamental units. In the English system, the unit of force is the pound (lb). In the SI system the unit of force is the newton (N). Because $\mathbf{F} = m\mathbf{a}$, force has the fundamental SI units of

$$\text{newtons} = (\text{kg})(\text{m/s}^2)$$

or kg-m/s^2. The standard abbreviations are N for newtons and kg for kilograms.

In the SI system, when we "weigh" an object, that weight is translated directly into its mass. A kilogram is a unit of mass. If we have a balance calibrated in kilograms, we read the "weight" of the object as its mass in kilograms because at the surface of the earth, the acceleration of gravity ($\mathbf{g}$) is 9.8 m/s^2. We treat $\mathbf{g}$ as a constant factor. From $\mathbf{F} = m\mathbf{g}$, we can rearrange the equation to find the mass.

$$m = \frac{\mathbf{F}}{\mathbf{g}}$$

When we weigh an object, we are actually measuring the force of gravity on that object. But knowing what the acceleration of gravity is, we can divide by this constant factor and easily come up with the object's mass. In the English system, an object's weight is given directly as a force. The unit of a pound is a unit of force. In the English system, the unit of mass is the slug. To get the mass in slugs, we must divide the weight in pounds by the acceleration of gravity, 32 ft/s^2. Thus, the units for force in the English system are

$$\text{pounds} = (\text{slugs})(\text{ft/s}^2)$$

or lb = slug-ft/s^2. The abbreviation for pounds is usually lb. Remember that a pound is not a mass, but a force.

3.5   Applications of Newton's Laws

Throughout these applications, we'll be using heavily the facts that a force is a vector and that for every action there is an equal and opposite reaction. We'll construct force diagrams and from them be able to visualize and solve the problem. We'll also make much use of the concept of tension. We've already seen tension at work in Figure 3.1 where a weight acting down was translated into a horizontal pull by the tension in a string. We will also assume in this section that all surfaces, pulleys, and so on, are frictionless; we'll add friction in the next section. The first application deals with static equilibrium.

EXAMPLE 3.1

For this example, we want to find the tension in cables 1 and 2 in Figure 3.2. The mass m is in equilibrium, which means that all of the forces that act on m are balanced. There is no motion in a vertical or horizontal direction. We know, however, that the force of gravity is acting to pull mass m to the ground. If m is in equilibrium, then there is an equal and opposite force acting up. Mathematically, we can say

$$\mathbf{F}_{up} = m\mathbf{g}$$

That upward force comes from the y component of the tensions in cables 1 and 2.

$$F_{up} = T_1 \sin \theta_1 + T_2 \sin \theta_2$$

See Appendix A, Section A.2, "Simultaneous Equations."

We know the two angles, but we have only one equation and two unknowns, tensions T_1 and T_2. We can't solve for either T_1 or T_2 without another equation describing the system. To get the second equation, we use the fact that there is no sideways motion. That is, the x components of the tension in the two cables must exactly balance out, or

$$T_{1x} = T_{2x}$$

$$T_1 \cos \theta_1 = T_2 \cos \theta_2$$

We now have two equations and two unknowns and can solve for the tensions in the cables.

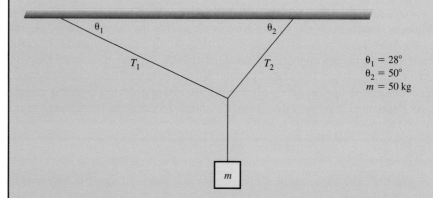

$\theta_1 = 28°$
$\theta_2 = 50°$
$m = 50 \text{ kg}$

Figure 3.2 Static Equilibrium of a Mass Suspended by Two Cables

First take a look at the force diagram in Figure 3.3. It's a picture of all of the forces acting on the mass m. All of the vertical forces are equal and opposite, as are the horizontal forces. Setting them equal, we have the following two equations:

$$m\mathbf{g} = T_1 \sin \theta_1 + T_2 \sin \theta_2$$

$$T_1 \cos \theta_1 = T_2 \cos \theta_2$$

We will solve the bottom equation first.

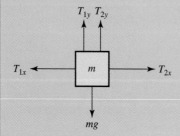

Figure 3.3 Free-Body Diagram of Forces Acting on Mass m

$$T_1 \cos 28° = T_2 \cos 50°$$

$$T_1(0.883) = T_2(0.643)$$

$$T_2 = T_1[(0.883)/(0.643)]$$

$$T_2 = 1.373T_1$$

We substitute this value for T_2 into the first equation and solve for T_1.

$$mg = T_1 \sin \theta_1 + T_2 \sin \theta_2$$

$$mg = T_1 \sin \theta_1 + (1.373T_1) \sin \theta_2$$

$$mg = T_1(\sin \theta_1 + 1.373 \sin \theta_2)$$

$$T_1 = \frac{mg}{\sin \theta_1 + 1.373 \sin \theta_2}$$

$$T_1 = \frac{(50 \text{ kg})(9.8 \text{ m/s}^2)}{(0.469) + (1.373)(0.766)}$$

$$T_1 = 322 \text{ N}$$

Knowing T_1, we can solve for T_2.

$$T_2 = 1.373T_1$$

$$T_2 = (1.373)(322 \text{ N})$$

$$T_2 = 422 \text{ N}$$

Thus, $T_1 = 322$ N, and $T_2 = 442$ N. If in doubt, substitute the numbers into the force diagram of Figure 3.3, and you'll quickly see that all of the forces do indeed balance.

EXAMPLE 3.2

An elevator whose mass is 1000 kg is suspended by a cable whose maximum tension is not to exceed 18,000 N (see Figure 3.4). What is the maximum upward acceleration that the elevator can have?

18,000 N
T

a

W
9800 N

Figure 3.4 Force Diagram

SOLUTION

$$T - \mathbf{W} = F_{net}$$

$$\mathbf{W} = (9.8 \text{ m/s}^2)(1000 \text{ kg})$$

$$\mathbf{W} = 9800 \text{ N}$$

$$F_{net} = 18,000 \text{ N} - 9800 \text{ N}$$

$$F_{net} = 8200 \text{ N}$$

$$F_{net} = m\mathbf{a}$$

$$\mathbf{a} = \frac{F_{net}}{m}$$

$$\mathbf{a} = \frac{8200 \text{ N}}{1000 \text{ kg}}$$

$$\mathbf{a} = 8.2 \text{ m/s}^2$$

Figure 3.4 is called a *force diagram* because it shows all of the forces acting on an object. All of the forces in this diagram are either up or down. The positive direction is that in which the elevator is accelerating, or going upward. The net force causes the actual acceleration of the object for which the force diagram was drawn. If the elevator accelerates downward, the tension in the cable must be less than the weight of the loaded elevator, and thus the tension will be less than 9800 N. If the elevator car is in free fall, or falling at 9.8 m/s², then the tension in the cable is zero. If the acceleration is upward, the tension in the cable will be greater than the weight of the elevator car. As a cable can only pull and not push, the maximum downward acceleration occurs when the car is in free fall, or at 9.8 m/s².

EXAMPLE 3.3

In Figure 3.5, two masses are sitting on a frictionless table top. They are both connected by a string, and the pair is further connected by a string to a weight hanging over the edge of the table top. The string goes over a frictionless pulley. What are (a) the acceleration of the system and (b) the tensions in the strings?

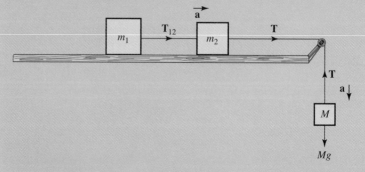

Figure 3.5 System of Masses Moving with Acceleration **a**

SOLUTION

We start with the mass M hanging over the edge of the table. It will accelerate downward with an acceleration **a**. Mathematically we can write

$$Mg - \mathbf{T} = M\mathbf{a}$$

This equation has two unknowns in it: the tension **T** and the acceleration **a**. (We assume that we know the masses of all of the blocks and the acceleration of gravity downward.) The acceleration is downward for the mass M hanging over the edge of the table, so down is the positive direction for M. The masses m_1 and m_2 both undergo the same acceleration **a** because they're tied together and to mass M, but this acceleration is to the right because of the tension in the string. The force providing this acceleration is the tension **T**. The tension in the string causes both masses tied to the string on top of the table (m_1 and m_2) to accelerate. The two masses on top of the table don't know that there is a weight hanging down from the end of the string, or whether there is someone pulling on the string, or whether there is a rocket firing at the string's end whose thrust pulls the string down. All they know is that there is a tension on the string that pulls them to the right. Mathematically,

$$\mathbf{T} = (m_1 + m_2)\mathbf{a}$$

We now have another equation with the same two unknowns as in the first equation: the tension **T** and the acceleration **a**. If we're first looking for the acceleration of the system, we can take the equation directly above and substitute it into the first equation for the tension. Substituting for the tension in the first equation, we have

$$Mg - (m_1 + m_2)\mathbf{a} = M\mathbf{a}$$

(a) We now solve for the acceleration of the system, **a**.

$$Mg - m_1\mathbf{a} - m_2\mathbf{a} = M\mathbf{a}$$

$$Mg = m_1\mathbf{a} + m_2\mathbf{a} + M\mathbf{a}$$

$$\mathbf{a}(m_1 + m_2 + M) = Mg$$

$$\mathbf{a} = \frac{Mg}{m_1 + m_2 + M}$$

The system's acceleration is directly proportional to the amount of mass hanging over the table's edge M and inversely proportional to the total mass of the system ($m_1 + m_2 + M$). The right side of the equation has units of force over mass, which is an acceleration. There is another way to write the equation for acceleration:

$$\mathbf{a} = \frac{M}{m_1 + m_2 + M}\mathbf{g}$$

Notice that the acceleration, **a**, for the system is always *less* than the acceleration of gravity, **g**. The denominator of the fraction is always greater than the numerator, and thus the fraction containing the masses is always less than 1. As a result, **a** is always less than **g**. As M gets increasingly large, the fraction gets closer and closer to 1, and thus the upper limit of the acceleration of the system is **g**, or that of gravity.

$$\frac{M}{m_1 + m_2 + M} < 1 \qquad \text{Therefore, } \mathbf{a} < \mathbf{g}$$

where the $<$ sign means "is less than." The tension $\mathbf{T}_{12}$ is merely the pull causing mass m_1 to accelerate. We already know what the acceleration of the system is, and m_1 has the same acceleration as the system.

$$\mathbf{T}_{12} = m_1\mathbf{a}$$

The tension in the string supporting the weight hanging over the edge of the table is

$$\mathbf{T} = (m_1 + m_2)\mathbf{a}$$

Notice that we've done the entire example without assigning any values to the masses. We will now do that. Assume that $m_1 = 1$ kg, $m_2 = 2$ kg, and $M = 0.5$ kg. We calculate an acceleration of

$$\mathbf{a} = \frac{(0.5 \text{ kg})(9.8 \text{ m/s}^2)}{1 \text{ kg} + 2 \text{ kg} + 0.5 \text{ kg}} \qquad \mathbf{a} = 1.4 \text{ m/s}^2$$

(b) We can now calculate the tensions in the two strings. For the string hanging over the edge of the table,

$$\mathbf{T} = (m_1 + m_2)\mathbf{a} \qquad \mathbf{T} = (1 \text{ kg} + 2 \text{ kg})(1.4 \text{ m/s}^2) \qquad \mathbf{T} = 4.2 \text{ N}$$

For the string between the two blocks sitting on the table,

$$\mathbf{T}_{12} = m_1\mathbf{a} \qquad \mathbf{T}_{12} = (1 \text{ kg})(1.4 \text{ m/s}^2) \qquad \mathbf{T}_{12} = 1.4 \text{ N}$$

EXAMPLE 3.4

In this example we talk about a device known as an *Atwood's machine,* in which two masses are slung over a single pulley (again assumed massless and frictionless), as in Figure 3.6. Mass m_2 is greater than m_1, so m_2 will accelerate downward, with m_1 moving up. The force diagrams for the two masses are shown next to them. We can write a mathematical expression for each one:

$$m_2\mathbf{g} - \mathbf{T} = m_2\mathbf{a} \quad \text{and} \quad \mathbf{T} - m_1\mathbf{g} = m_1\mathbf{a}$$

(Again it is assumed that the direction of the acceleration of the system is positive.) We can now eliminate $\mathbf{T}$ and solve for the acceleration of the system.

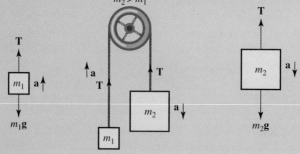

Figure 3.6 Atwood's Machine

$$\mathbf{T} = m_1\mathbf{a} + m_1\mathbf{g}$$

$$m_2\mathbf{g} - (m_1\mathbf{a} + m_1\mathbf{g}) = m_2\mathbf{a}$$

$$m_2\mathbf{g} - m_1\mathbf{a} - m_1\mathbf{g} = m_2\mathbf{a}$$

$$m_2\mathbf{a} + m_1\mathbf{a} = m_2\mathbf{g} - m_1\mathbf{g}$$

$$\mathbf{a}(m_2 + m_1) = (m_2 - m_1)\mathbf{g}$$

$$\mathbf{a} = \frac{m_2 - m_1}{m_2 + m_1}\mathbf{g}$$

The acceleration of the system is directly proportional to the *difference* of the two masses (as expected if you think about it) and again inversely proportional to the sum of the masses. As in Example 3.3, the acceleration will be less than that of gravity.

We can now give values for m_1 and m_2 and calculate the system's acceleration and tension in the string: $m_1 = 1$ kg, and $m_2 = 2$ kg.

$$\mathbf{a} = \frac{2 \text{ kg} - 1 \text{ kg}}{2 \text{ kg} + 1 \text{ kg}}(9.8 \text{ m/s}^2) \qquad \mathbf{a} = 3.27 \text{ m/s}^2$$

$$\mathbf{T} = m_1\mathbf{a} + m_1\mathbf{g}$$

$$\mathbf{T} = (1 \text{ kg})(3.27 \text{ m/s}^2) + (1 \text{ kg})(9.8 \text{ m/s}^2)$$

$$\mathbf{T} = 13.07 \text{ N}$$

EXAMPLE 3.5

In this example, a block is sliding down an inclined plane, as in Figure 3.7. As in our previous examples, the inclined plane is frictionless. The inclined plane makes an angle θ with the horizontal, as shown. For the block of mass m, its weight ($m\mathbf{g}$) is acting straight down. The smaller triangle shown in Figure 3.7, called a *force triangle,* is generated from the weight of the block m acting straight down. The weight acting straight down forms the hypotenuse of a right triangle, where the other sides are components of the weight pulling the block down the plane ($\mathbf{F}$) and pushing the block onto the plane ($\mathbf{F}_n$) where $\mathbf{F}_n$ is downward in a direction that is perpendicular to the surface of the top of the inclined plane and is called the *normal force.* The force $\mathbf{F}$ acts to accelerate the block down the plane and is parallel to the surface of the inclined plane. The smaller force triangle is at right angles to the larger triangle formed by the inclined plane; the two triangles are similar; and the angle inside the smaller force triangle is the same angle theta (θ) that the incline makes relative to the horizontal. The force down the plane and the normal force are always at right angles to each other. (The small boxes in the drawing mean that the two intersecting lines are at right angles to each other.) The force accelerating the block down the plane can be calculated using trigonometry.

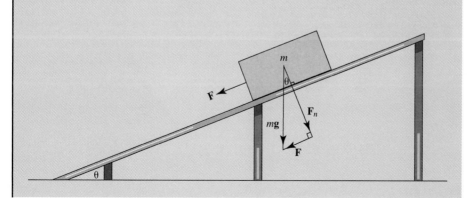

Figure 3.7 Block on a Frictionless Inclined Plane

$$\sin\theta = \frac{\mathbf{F}}{m\mathbf{g}}$$

(Equation 3.2)

or $\qquad\qquad \mathbf{F} = m\mathbf{g}\sin\theta$

There is no force acting down the plane if the inclined plane is flat ($\theta = 0°$), but the force down the plane continuously increases until $\theta = 90°$ where $\sin\theta = 1$, and the force acting down the plane is the full weight of the block.

In the next section we'll talk about friction, and the normal force will come into play. We'll also use some of the above examples again but incorporating friction to see how it changes things.

3.6   Frictional Forces

Up to now, we have been dealing with blocks on table tops and assuming a frictionless contact. That is, if we give a block on a table top an initial shove, it will continue in a straight line with constant velocity forever, or as long as the surface remains frictionless. This state of no friction is what gave rise to Newton's first law. Newton was able to see the implications of a frictionless state and turn it into a universal law. The solar system is a classical mechanical system: What allows it to run, or the planets to orbit, in an almost timeless manner is this lack of friction. Newton was able to see this in a flash of insight. However, in ordinary life, it is almost impossible to get rid of friction; it is considered mostly a nuisance, a force to be combated. For example, wouldn't it be wonderful to have frictionless motors and machines! The oil we change in our automobile's engine every 3000 miles or so is there primarily to combat friction. Without it, the engine would destroy itself in very short order!

We started studying physics by deriving the equations for uniformly accelerated motion, but we must realize by now that such equations will hold strictly true only on a planet or planetesimal without an atmosphere, such as our moon. Atmospheric friction quickly makes a mockery of these equations. For example, the first electronic digital computer was built toward the end of World War II in order to calculate range cards for artillery pieces. However, the calculations became too complex and time-consuming for humans to do, and the process had to be automated.

Automotive braking systems work on the principle of friction. The friction between the brake pads and disc slows and stops the vehicle. The friction dissipates as heat, as demonstrated in the photo, where the disc is glowing from the heat.

But friction also makes things we take for granted possible. For instance, without friction we couldn't walk. Driving a car wouldn't be possible without it; there must be friction between tire and road for traction in order for us to accelerate and brake. Without friction, we wouldn't be able to hold anything in our hands, use tools, or write or draw. Friction is both a bother and a necessity simultaneously. A frictionless surface is the anomaly. Only in specially contrived situations do you ever see an almost frictionless surface. An air track or an air table in a physics lab is one place where such a surface can be seen, but their surfaces are usually very restricted. Everyday life implies friction. In the overwhelming majority of cases, when a block slides over a table top, it slows and then stops. Frictional forces change its state of motion. They oppose the block's motion in every case. Moreover, the block doesn't necessarily have to be in motion for frictional forces to exist between the block and the table top.

In this section we'll investigate the force laws for sliding, not rolling, friction. Did you ever consider the case of a tire rolling over the road? This is a case of static friction. That part of a tire in contact with the road doesn't slip over the road. It is motionless with respect to the road. The auto's forward motion comes from the tire's turning about the axle of the car, and not from slipping against the road. Too much slippage, and the car goes nowhere. Try driving on glare ice! Only a little slippage causes very heavy tire wear. A dragster usually gets only a few runs on its slicks because the powerful motor in it causes the tires to slip and the rubber to heat, and the friction literally burns off the tires.

So what happens between two surfaces when they slide against one another? A couple of mechanisms can explain a wide variety of effects. If two rough, unlubricated surfaces are placed in contact, the roughness acts like a saw-toothed edge, and the teeth of the two surfaces catch on one another. If the surfaces are originally locked together, a relatively large force will be required to initially break the surfaces apart; then once they are sliding, a smaller force will be required to keep them sliding. If motion is maintained, the surfaces can't lock tightly together again. If, on the other hand, we have two highly polished and extremely clean pieces of steel and place them in contact, they will form microwelds between the two pieces. Again, it will take a relatively high force to initially break them apart, and a lesser force to keep the pieces sliding once they're in motion. Once moving, the welds continuously make and break, forming a weaker bond. These two mechanisms are very different, but notice that in both cases, it takes a higher force to initially get the surfaces sliding than it does to keep them moving. *The initial frictional force is called the force of* **static friction,** *and the lesser force required to keep the two pieces moving is called the force of* **dynamic** *or* **kinetic friction.**

These forces can be characterized by coefficients and the normal force that we talked about in Example 3.5, the inclined plane. Look at the block on the flat table top in Figure 3.8. The weight of the block acts directly downward, and in this case the weight *is* the normal force because it's perpendicular to the table top. This force is always there and is exactly balanced out by the force of the surface pushing up on the block. Frictional forces that oppose the motion of the block are proportional to the normal force. The proportionality constant for static friction is μ_s, and the equation for the force of static friction is

$$F_s = \mu_s m\mathbf{g}$$

Static friction: The force of contact between two bodies at rest opposing the sliding of one body over another.
Kinetic friction: The force of contact between two bodies in motion that opposes that motion.

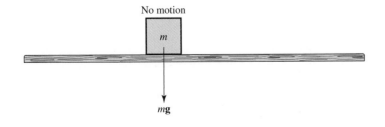

Figure 3.8 Block on a Flat Table Top

Table 3.1 Coefficients of Static and Kinetic Friction

Materials in Contact	Coefficient of Static Friction, μ_s	Coefficient of Kinetic Friction, μ_k
Wood on wood	0.5	0.3
Wood on stone	0.5	0.4
Steel on steel (smooth)	0.15	0.09
Metal on metal (lubricated)	0.03	0.03
Leather on wood	0.5	0.4
Rubber on dry concrete	1.0	0.7
Rubber on wet concrete	0.7	0.5
Glass on glass	0.94	0.4
Steel on Teflon	0.04	0.04

The proportionality constant for kinetic friction is μ_k, and the equation for the force of kinetic friction is

$$F_k = \mu_k m\mathbf{g}$$

These proportionality constants depend on the materials in contact, and some coefficients are listed in Table 3.1. There are no units for the coefficient of friction; it is just a pure number.

Next we exert a force from the right on the block as in Figure 3.9. If F is less than the force of static friction, no movement of the block will occur. When $F = \mu_s m\mathbf{g}$, then the block breaks loose and begins to move. This force is the minimum force necessary to get the block to move. The moment the block begins to move, we have the situation depicted in Figure 3.10.

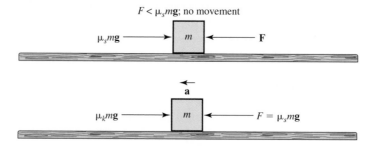

Figure 3.9 Force of Static Friction Acting on a Block

Figure 3.10 Force of Kinetic Friction Acting on a Block

Remember that it takes a greater force to overcome static friction than it does to keep the block moving against kinetic friction. Thus, the moment the block breaks loose, it begins to accelerate to the left. We can calculate what this acceleration is.

$$m\mathbf{a} = \mu_s m\mathbf{g} - \mu_k m\mathbf{g}$$

$$m\mathbf{a} = (\mu_s - \mu_k)m\mathbf{g}$$

$$\mathbf{a} = (\mu_s - \mu_k)\mathbf{g}$$

Of course, the force applied from the right can be higher than that used to just break the block loose. However, if we drop the force from the right to that of kinetic friction after the block has broken loose, the block slides along at a constant velocity, and we have the situation depicted in Figure 3.11. The force from the right is exactly balanced by the frictional force

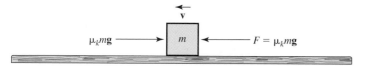

Figure 3.11 Dynamic Equilibrium of Forces (No Acceleration)

from the left, and there is no acceleration of the block. It moves at a constant velocity. Remember that just after the block broke loose, it started to accelerate. This implies that its velocity to the left increased. The longer it was allowed to accelerate, the greater was the velocity. At the time the force from the right dropped to just match kinetic friction, the block continued on at the velocity it had at that time.

The coefficients of static and kinetic friction depend, of course, on the materials the surfaces consist of, as the abbreviated list in Table 3.1 illustrates. In the next section, we will further explore friction through a number of examples.

3.7   Applications of Frictional Forces

EXAMPLE 3.6

First we will start with a block of wood on a wooden table top. The block of wood has a mass of 1 kg. (a) What minimum force will break the block loose? (b) What will the block's acceleration be the moment it breaks loose? (c) What is the applied force at which the block slides at a constant velocity? (d) If the time between when the applied force equals static friction and then is dropped to equal kinetic friction is 0.3 s, what is the constant velocity? (e) If the applied force suddenly drops to zero, how long does it take the block to come to rest? (f) How far does the block move in this time?

SOLUTION
For wood on wood, $\mu_s = 0.5$, and $\mu_k = 0.3$.

(a)
$$F = \mu_s m\mathbf{g}$$
$$F = (0.5)(1 \text{ kg})(9.8 \text{ m/s}^2)$$
$$F = 4.9 \text{ N}$$

(b)
$$\mathbf{a} = (\mu_s - \mu_k)\mathbf{g}$$
$$\mathbf{a} = (0.5 - 0.3)(9.8 \text{ m/s}^2)$$
$$\mathbf{a} = 1.96 \text{ m/s}^2$$

(c)
$$F = \mu_k m\mathbf{g}$$
$$F = (0.3)(1 \text{ kg})(9.8 \text{ m/s}^2)$$
$$F = 2.94 \text{ N}$$

(d)
$$v = \mathbf{a}t \qquad \mathbf{a} = 1.96 \text{ m/s}^2 \qquad t = 0.3 \text{ s}$$
$$v = (1.96 \text{ m/s}^2)(0.3 \text{ s})$$
$$v = 0.59 \text{ m/s}$$

(e) First we must find the deceleration.

$$F = \mu_k mg \qquad F = ma \qquad ma = \mu_k mg$$

$$\mathbf{a} = \mu_k \mathbf{g}$$

$$\mathbf{a} = (0.3)(9.8 \text{ m/s}^2)$$

$$\mathbf{a} = 2.94 \text{ m/s}^2$$

Now using the laws of uniformly accelerated motion, we can calculate the time the block takes to come to rest.

$$\mathbf{a} = \frac{\Delta v}{\Delta t} \qquad \Delta t = \frac{\Delta v}{\mathbf{a}}$$

$$\Delta t = \frac{v_f - v_i}{\mathbf{a}}$$

$$\Delta t = \frac{0 - 0.59 \text{ m/s}}{-2.94 \text{ m/s}^2}$$

$$\Delta t = 0.2 \text{ s}$$

(f)
$$x = v_0 t - \frac{1}{2} \mathbf{a} t^2$$

$$x = (0.59 \text{ m/s})(0.2 \text{ s}) - \frac{1}{2}(2.94 \text{ m/s}^2)(0.2 \text{ s})^2$$

$$x = 0.0592 \text{ m} \quad \text{or} \quad 5.9 \text{ cm}$$

EXAMPLE 3.7

In this example, we use the apparatus of Example 3.3 (Figure 3.5) again, but we analyze it with friction between the blocks and the table top. We assume that the pulley has so little friction as to be insignificant. The blocks are of steel, as is the table top, and $m_1 = 1$ kg, $m_2 = 2$ kg, and $M = 0.5$ kg. (a) Will the blocks on the table move from rest without being nudged? (b) What is the acceleration of the system? (c) What are the tensions in the strings?

SOLUTION
To solve the problem, we first draw the force diagram as shown in Figure 3.12. Our force diagram is like that of Figure 3.5, with the addition of the frictional forces $\mathbf{F}_{f1}$ and $\mathbf{F}_{f2}$. For a steel-on-steel surface, $\mu_s = 0.15$, and $\mu_k = 0.09$.

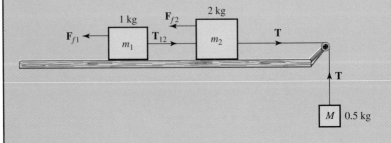

Figure 3.12 Blocks on Table Top with Friction Present

(a) We use the coefficient of static friction to find out what frictional forces are required to initially break the steel blocks loose.

$$\mathbf{F}_{f1} = \mu_s m_1 \mathbf{g} \qquad \mathbf{F}_{f1} = (0.15)(1 \text{ kg})(9.8 \text{ m/s}^2) \qquad \mathbf{F}_{f1} = 1.47 \text{ N}$$

$$\mathbf{F}_{f2} = \mu_s m_2 \mathbf{g} \qquad \mathbf{F}_{f2} = (0.15)(2 \text{ kg})(9.8 \text{ m/s}^2) \qquad \mathbf{F}_{f2} = 2.94 \text{ N}$$

$$\mathbf{F}_f = \mathbf{F}_{f1} + \mathbf{F}_{f2} \qquad \mathbf{F}_f = 1.47 \text{ N} + 2.9 \text{ N} \qquad \mathbf{F}_f = 4.41 \text{ N}$$

The tension **T** in the string attached to mass M must have enough pull to overcome the force of friction of 4.41 N. The tension **T** equals the weight of block M, if it is initially unmoving.

$$\mathbf{T} = (0.5 \text{ kg})(9.8 \text{ m/s}^2) \qquad \mathbf{T} = 4.9 \text{ N}$$

Since $\mathbf{T} > \mathbf{F}_f$, the blocks on the table break loose and begin to slide.
(b) We should expect the acceleration of the system to be less than that found in Example 3.3 because of friction. We can write a force equation for the mass M hanging over the edge of the table.

$$M\mathbf{g} - \mathbf{T} = M\mathbf{a}$$

We can also write a force equation for **T** and masses m_1 and m_2 on the table.

$$\mathbf{T} - (\mathbf{F}_{f1} + \mathbf{F}_{f2}) = (m_1 + m_2)\mathbf{a}$$

Since we're looking for the acceleration of the system, we combine the two equations above and eliminate **T**.

$$\mathbf{T} = M\mathbf{g} - M\mathbf{a}$$

$$(M\mathbf{g} - M\mathbf{a}) - (\mathbf{F}_{f1} + \mathbf{F}_{f2}) = (m_1 + m_2)\mathbf{a}$$

$$M\mathbf{g} - M\mathbf{a} - \mathbf{F}_{f1} - \mathbf{F}_{f2} = m_1\mathbf{a} + m_2\mathbf{a}$$

$$m_1\mathbf{a} + m_2\mathbf{a} + M\mathbf{a} = M\mathbf{g} - \mathbf{F}_{f1} - \mathbf{F}_{f2}$$

$$\mathbf{a}(m_1 + m_2 + M) = M\mathbf{g} - \mathbf{F}_{f1} - \mathbf{F}_{f2}$$

We now substitute the formulas for $\mathbf{F}_{f1}$ and $\mathbf{F}_{f2}$ using the coefficient of *kinetic* friction, 0.09.

$$\mathbf{a}(m_1 + m_2 + M) = M\mathbf{g} - \mu_k m_1 \mathbf{g} - \mu_k m_2 \mathbf{g}$$

$$\mathbf{a} = \frac{(M - \mu_k m_1 - \mu_k m_2)\mathbf{g}}{m_1 + m_2 + M}$$

$$\mathbf{a} = \frac{[M - \mu_k(m_1 + m_2)]\mathbf{g}}{m_1 + m_2 + M}$$

$$\mathbf{a} = \frac{[0.5 \text{ kg} - (0.09)(1 \text{ kg} + 2 \text{ kg})](9.8 \text{ m/s}^2)}{1 \text{ kg} + 2 \text{ kg} + 0.5 \text{ kg}}$$

$$\mathbf{a} = 0.9 \text{ m/s}^2$$

The acceleration of the system with friction is 0.9 m/s², which is indeed less than the acceleration predicted for the system without friction, or 1.4 m/s².
(c) We now solve for the tension, **T**, in the string attached to the weight hanging over the edge of the table.

$$Mg - T = Ma$$

$$T = Mg - Ma$$

$$T = M(g - a)$$

$$T = (0.5 \text{ kg})(9.8 \text{ m/s}^2 - 0.9 \text{ m/s}^2)$$

$$T = 4.45 \text{ N}$$

For T_{12},

$$m_1 a = T_{12} - F_{f2}$$

$$T_{12} = m_1 a + F_{f2}$$

$$T_{12} = m_1 a + \mu_k m_1 g$$

$$T_{12} = m_1 (a + \mu_k g)$$

$$T_{12} = (1 \text{ kg})[0.9 \text{ m/s}^2 + (0.09)(9.8 \text{ m/s}^2)]$$

$$T_{12} = 1.78 \text{ N}$$

We would expect that the tensions in the strings with friction would be greater than without friction. This is indeed the case. Without friction as in Example 3.3, we got $T = 4.2$ N, and $T_{12} = 1.4$ N.

EXAMPLE 3.8

In this example, we'll again examine the inclined plane in Example 3.5, but this time with friction added. Figure 3.13 is almost exactly like Figure 3.7, except for the force of friction pointing up the plane. In Example 3.5, we didn't assign any values to the mass or the angle; in Figure 3.13, we have done so. First we must calculate the force pulling the block down the plane. From Example 3.5, we know what this force must be.

$$F = mg \sin \theta$$

$$F = (1 \text{ kg})(9.8 \text{ m/s}^2)(\sin 16°)$$

$$F = 2.7 \text{ N}$$

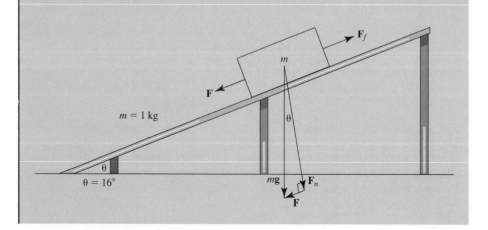

Figure 3.13 Block on an Inclined Plane with Friction between the Surfaces

From this, we can compute the acceleration that the block undergoes over the frictionless surface.

$$\mathbf{F} = m\mathbf{a}$$

$$\mathbf{a} = \frac{\mathbf{F}}{m}$$

$$\mathbf{a} = \frac{2.7 \text{ N}}{1 \text{ kg}}$$

$$\mathbf{a} = 2.7 \text{ m/s}^2$$

Knowing what the block does on a frictionless surface, we can now add friction. The block is steel, and the surface of the inclined plane is also steel ($\mu_s = 0.15$, and $\mu_k = 0.09$). We want to know the following: (a) Will the block break loose without being nudged? (b) What is the acceleration down the block after it begins sliding?

SOLUTION

(a) $$\mathbf{F}_f = \mu_s \mathbf{F}_n$$ (Equation 3.3)

The normal force $\mathbf{F}_n$ is the force the block exerts perpendicular to the surface of the inclined plane, or the force pushing the block onto the upper surface of the inclined plane. As the plane is tilted, it will be less than the weight of the block, until at 90° the normal force disappears.

$$\cos \theta = \frac{\mathbf{F}_n}{m\mathbf{g}}$$ (Equation 3.4)

$$\mathbf{F}_n = m\mathbf{g} \cos \theta$$

$$\mathbf{F}_n = (1 \text{ kg})(9.8 \text{ m/s}^2)(\cos 16°)$$

$$\mathbf{F}_n = 9.4 \text{ N}$$

$$\mathbf{F}_f = (0.15)(9.4 \text{ N})$$

$$\mathbf{F}_f = 1.4 \text{ N}$$

Since $\mathbf{F} = 2.7$ N pointing down the plane, and the force of friction is only 1.4 N, the block will start sliding without being nudged.

(b) As the block is now moving, the coefficient of friction drops from the static value to the kinetic value.

$$\mathbf{F}_f = \mu_k \mathbf{F}_n$$

$$\mathbf{F}_f = (0.09)(9.4 \text{ N})$$

$$\mathbf{F}_f = 0.85 \text{ N}$$

$$\mathbf{F} - \mathbf{F}_f = m\mathbf{a}$$

$$\mathbf{a} = \frac{\mathbf{F} - \mathbf{F}_f}{m}$$

$$\mathbf{a} = \frac{2.7 \text{ N} - 0.85 \text{ N}}{1 \text{ kg}}$$

$$\mathbf{a} = 1.85 \text{ m/s}^2$$

The acceleration on a frictionless surface is 2.7 m/s². The acceleration down the plane should be less with friction present, and it is at 1.85 m/s².

Summary

1. **Newton's first law:** *Every body persists in its state of rest, or of uniform motion in a straight line, unless it is compelled to change that state by forces impressed upon it.* Or, more simply, *An object in motion remains in motion, and an object at rest remains at rest, unless acted upon by an outside force.*

2. **Newton's second law:** *Force is proportional to the acceleration by a factor known as mass.* In other words, *the **mass** of a body is its resistance to acceleration.* This law can be expressed as

$$\mathbf{F} = m\mathbf{a}$$

Problem Solving Tip 3.1: When using Newton's second law, break up every force present, and the acceleration, into their respective *x* and *y* components. Then, for the forces, add all of the *x* components and *y* components separately. (Remember: Vectors that point in opposite directions have opposite signs!) Use the summed forces (*x* and *y*) as the **F** in **F** = *m***a**. In other words,

$$\text{Sum of } \mathbf{F}_x = m\mathbf{a}_x$$

$$\text{Sum of } \mathbf{F}_y = m\mathbf{a}_y$$

3. The *weight* of an object is its *mass* times the acceleration due to gravity.

$$\mathbf{W} = m\mathbf{g}$$

Note: This is simply a rewrite of Newton's second law! Remember, weight is a type of force.

4. **Newton's third law:** *To every action there is always opposed an equal reaction; or, the mutual actions of two bodies upon each other are always equal, and directed to contrary parts.*

5. In SI units, the unit of mass is the kilogram (kg), and the unit of force is the newton (N) [or, in English units, the pound (lb)].

$$\mathbf{F} = \frac{(m)(l)}{(s)^2}$$

Problem Solving Tip 3.2: Be very careful to make sure that all of the numbers you are using have been converted to the same type, for instance, that all units of time have been converted to seconds (s), all units of mass have been converted to kilograms (kg), and so on. It is often a good idea to work only in units of meters, kilograms, and seconds (the SI system) when solving problems, since most of the other units (such as the newton), depend on these.

6. **Static friction** is the force that resists the onset of motion, whereas **kinetic friction** is the force that resists the continuation of motion.

7. Frictional force always opposes the direction of motion of an object and is equal to the coefficient of friction (μ) multiplied by the normal force ($\mathbf{F}_n$) acting on that object.

$$\text{Static friction:} \quad \mathbf{F} = \mu_s \mathbf{F}_n$$

$$\text{Kinetic friction:} \quad \mathbf{F} = \mu_k \mathbf{F}_n$$

Key Concepts

KINETIC FRICTION: The force of contact between two bodies in motion that opposes that motion.

MASS: The property of a body that opposes a change in its motion. Resistance to motion is also called *inertia.*

NEWTON'S FIRST LAW: Every body persists in its state of rest or of uniform motion in a straight line unless it is compelled to change that state by forces impressed upon it.

NEWTON'S SECOND LAW: A net force applied to a massive object causes the object to accelerate (**F** = *m***a**).

NEWTON'S THIRD LAW: For every action there is an equal and opposite reaction.

STATIC FRICTION: The force of contact between two bodies at rest that opposes motion.

Important Equations

(3.1) $\mathbf{F} = m\mathbf{a}$ (Newton's second law)

(3.2) $\mathbf{F} = m\mathbf{g} \sin \theta$ (Force down the plane)

(3.3) $\mathbf{F}_f = \mu \mathbf{F}_n$ (The frictional force is proportional to the coefficient of friction times the normal force.)

(3.4) $\mathbf{F}_n = m\mathbf{g} \cos \theta$ (Normal force pushing object against plane)

Note that Equations (3.2) and (3.4) are derived from the force triangle in Figure 3.7.

Conceptual Problems

1. Why do packages slide off the seat of a car that is braked hard?

2. The acceleration of gravity on Mars is 3.7 m/s^2. Compared with her mass and weight on the earth, an astronaut on Mars has:

   a. less mass and less weight

   b. less mass and the same weight

   c. the same mass and less weight

   d. the same weight and less mass

3. An object on a surface exerts a normal force on that surface. The normal force:

   a. comes from the friction between the object and the surface

   b. is equal to the object's weight

   c. is parallel to the surface

   d. is perpendicular to the surface

4. When you "weigh" an object on a double pan balance by placing calibrated standard masses on the other pan, will this pan balance still give the correct answer on the moon or on Mars using the same set of calibrated standard masses? Why or why not?

5. When an object is "weighed" on a spring balance such as a bathroom scale, is it reading the object's mass or its weight?

6. A glass stands on a sheet of paper on a table top. If a force is suddenly applied to the paper, it can be removed from under the glass without disturbing the glass. If the force is applied slowly to the paper, both glass and paper will slide across the top of the table. Explain the difference between the two cases.

7. At either end of a table is a pulley. A line is passed over both pulleys, and a 5-lb weight is tied to each end of the line. What is the tension in the line, and why?

8. A "bottle rocket" has a long, thin tail that is inserted in an empty bottle that serves as the launcher. In which position does the rocket have the greater initial acceleration: when the bottle is vertical or when it is horizontal? Why?

9. You are standing on a spring scale inside an elevator. Does your "weight" increase or decrease when the elevator starts moving upward? Why?

10. A box rests on a rough floor. What happens to the force of friction when the box is pulled along the floor with a rope that makes a positive upward angle with the horizontal? Why?

11. A juggler standing on a stage juggles three balls. A trap door beneath him suddenly opens, and he goes into free fall. While in free fall, can he continue juggling? Why?

12. An hourglass with sand in it is placed on a scale with all of the sand in the lower portion of the glass. The weight is noted. The glass is then turned over and the weight recorded while the sand flows from the upper chamber into the lower chamber. Is the weight that the scale records more than, the same as, or less than in the previous case when the sand wasn't flowing?

Exercises

Newton's Laws

1. An empty car of mass 1000 kg has a maximum acceleration of 4 m/s^2. What acceleration does it have if it carries passengers and baggage with a load of 300 kg?

   Answer: 3.08 m/s^2

2. The driver of a sports car pushes down on the accelerator so that the engine pushes the car forward with a force of 2500 N, starting from rest. How fast will the car be going 20 s later?

   Answer: 180 km/h

3. A 50,000-lb aircraft is launched from the deck of an aircraft carrier by a steam catapult. The aircraft reaches a speed of 150 mph in 2.5 s. What force does the catapult exert on the aircraft during launch?

   Answer: 138,000 lb

4. A 250-lb motorcycle carrying a 150-lb rider comes to a stop in 150 ft from a speed of 60 mph. (a) Find the force exerted by the brakes on the road. (b) What force did the rider experience?

   Answer: (a) 322.5 lb
   (b) 121 lb

5. Two objects whose masses are 20 lb and 30 lb are suspended by string as shown in Figure 3.14. Find the tension in each string.

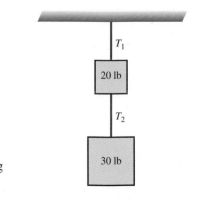

Figure 3.14 Masses Suspended by a String (Exercise 5)

Answer: T_2 = 30 lb; T_1 = 50 lb

6. Two masses of 7 kg and 5 kg are suspended from each side of a massless and frictionless pulley. What is the acceleration of the system?

Answer: 1.63 m/s²

7. (a) A 1500-kg elevator has a 1.5-m/s² acceleration downward. What is the tension in the cable? (b) This same elevator has a 1.5-m/s² acceleration upward. What is the tension in the cable now?

Answer: (a) 12,450 N
(b) 16,950 N

8. A hot-air balloon of mass *M* floats along at a constant altitude. The balloonist throws over some ballast of mass *m*, and the balloon starts ascending with a constant acceleration *a*. How much ballast has been thrown overboard? Neglect air friction.

$$Answer:\ m = \frac{Ma}{g + a}$$

9. A person is standing on a scale in an elevator. This person weighs 150 lb, but the scale reads 135 lb. (a) Is the elevator moving at constant velocity, or is it accelerating? (b) If the elevator is accelerating, what are the magnitude and the direction of the acceleration?

Answer: (a) Accelerating
(b) $a = 3.2$ ft/s², downward

10. A 5-kg block and a 15-kg block are both sliding down a frictionless inclined plane. The plane makes an angle of 15° with the horizontal. What is the acceleration of each block?

Answer: $a = 2.54$ m/s²

11. A load of bricks is hauled to the top of a wall by a rope over a pulley. If the load weighs 1000 lb, and if the rope can safely support no more than 1250 lb, what is the maximum acceleration that the load can undergo without breaking the rope?

Answer: 8 ft/s²

12. Three blocks are joined by string as in Figure 3.15. A force of 150 N pulls up on the top block. Find the tensions in the strings between the blocks.

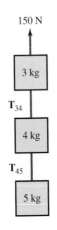

Figure 3.15 Blocks Being Accelerated Vertically (Exercise 12)

Answer: $T_{34} = 112.5$ N; $T_{45} = 62.5$ N

13. Find the acceleration of the blocks in Figure 3.16 and the tensions in the strings that join them. Assume that the surface is frictionless; that the pulley has negligible mass and friction; and that $m_1 = 1$ kg, $m_2 = 2$ kg, $m_3 = 3$ kg, and $M = 0.75$ kg.

Figure 3.16 Blocks Accelerating on a Frictionless Table Top (Exercise 13)

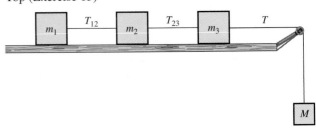

Answer: $a = 1.09$ m/s²; $T = 6.53$ N; $T_{23} = 3.27$ N; $T_{12} = 1.09$ N

14. At the launch of one of the moon missions, the *Saturn-Apollo* rocket has a mass of 5.4×10^5 kg. What is the initial acceleration of the rocket if the combined thrust of the five first-stage engines is 7.4×10^6 N?

Answer: 3.9 m/s²

*15. When an archer pulls on the bowstring of a certain bow to launch an arrow, the bowstring makes an angle of 60° with the arrow above and below the arrow. The pull that the archer must exert is 12 lb. What is the tension in the bowstring?

Answer: 12 lb

*16. A mass *m* is suspended as shown in Figure 3.17. The line is fastened securely at point *C* but is fastened at point *A* only with adhesive tape, and it will pull loose when the line tension exceeds 12 N. What is the greatest mass that can be supported by the line before the line pulls the tape loose?

Figure 3.17 Mass Suspended between Two Walls (Exercise 16)

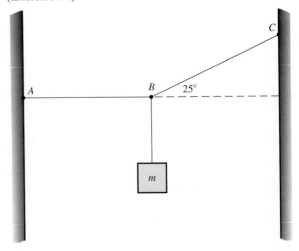

Answer: 0.57 kg

*17. A planter is hung from two wires anchored to the ceiling as shown in Figure 3.18. What is the tension in each wire if the planter has a weight of 10 lb? The weight of the wire is negligible.

Answer: $T_{35} = 7.16$ lb; $T_{45} = 8.33$ lb

*18. A 7-kg block is placed on a frictionless inclined plane and connected to a smaller 2-kg block over a frictionless pulley as shown in Figure 3.19. (a) What will the angle θ be if the two blocks remain motionless? (b) What is the acceleration of the system if the angle of the inclined plane with the horizontal is 35°?

Answer: (a) $\theta = 16.6°$
(b) $a = 2.2$ m/s^2

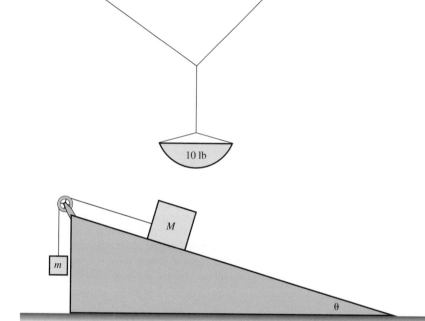

Figure 3.18 Planter Suspended from Ceiling (Exercise 17)

Figure 3.19 Inclined Plane with Adjustable Tilt (Exercise 18)

Friction

19. A man prevents a brick from falling by pressing it against a vertical wall. The brick has a mass of 1 kg. The coefficient of static friction between the brick and the wall is 0.6. (a) How much force must the man exert to prevent the brick from sliding down the wall? (b) If the surface between the brick and the wall is frictionless, will the man be able to prevent the brick from falling by applying a sideways force on it? Prove your answer.

Answer: (a) 16.3 N
(b) No

20. A driver driving on a concrete roadbed on a wet day sees an accident up ahead and brakes to a stop in 35 m. He brings the car to a safe halt without locking the brakes. What was the car's speed when the brakes were first applied?

Answer: 79.2 km/h

21. (a) What is the maximum acceleration that a front wheel–drive car can have on a level road, with 60% of the car's

weight on the front wheels? The coefficient of static friction is μ. (b) What is the maximum acceleration for a rear wheel–drive car with 40% of its weight on the rear wheels? (c) What is the maximum acceleration for a four wheel–drive car?

Answer: (a) $0.6 \, \mu g$
(b) $0.4 \, \mu g$
(c) μg

22. A 0.5-kg hard rubber hockey puck with an initial speed of 10 m/s slides along a smooth ice surface for 60 m before coming to a stop. What is the coefficient of kinetic friction between the puck and the ice?

Answer: 0.085

23. A truck carrying a 2000-lb steel girder is traveling along a highway at 60 mph. The girder is initially securely lashed to the truck bed. (a) If the roadbed is dry concrete, what is the minimum distance in which the truck can come to a stop? (b) If the girder isn't lashed down, and if the coefficient of static friction between the truck bed and girder is 0.4, what is the minimum distance in which the truck could stop without having the girder crash through the truck's cab?

Answer: (a) 121 ft

(b) 302.5 ft

24. A skier slides down a snow-covered slope at a constant speed. If the slope makes an angle of 15° with the horizontal, what is the coefficient of kinetic friction μ_k between the skier and the snow?

Answer: $\mu_k = 0.27$

25. A wooden crate of mass 200 kg slides down a wooden inclined ramp that makes an angle of 15° with the horizontal. If the crate is to slide down the ramp without accelerating, how much force should be applied, and where should it be applied?

Answer: 60.7 N, pushing down the plane

26. In Figure 3.20, what is the minimum acceleration that block M can have to the right if block m doesn't fall off? The surface along which block M slides is frictionless, and the coefficient of static friction between blocks M and m is μ_s.

Figure 3.20 Small Block Held in Place by Acceleration of Big Block (Exercise 26)

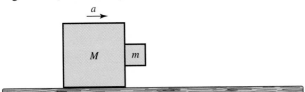

Answer: $a = g/\mu_s$

27. (a) In Figure 3.21, block M sits on a frictionless surface and is attached to block m by a string over a pulley with negligible mass and friction. What is the acceleration of block M? (b) The coefficient of kinetic friction between block M and the surface over which it slides becomes μ_k. What is block M's acceleration now?

Figure 3.21 Block Sliding over Table Top with Variable Friction (Exercise 27)

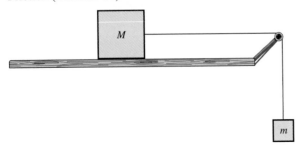

Answer: (a) $a = \dfrac{m}{M + m}g$

(b) $a = \dfrac{m - \mu_k M}{m + M}g$

*28. Two wooden blocks rest on a wooden surface as in Figure 3.22. What force is necessary to move the 4-kg block at constant speed? The pulley has negligible friction.

Figure 3.22 Large Block Moving at Constant Speed across Surface (Exercise 28)

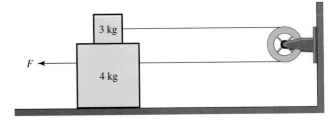

Answer: $F = 38.2$ N

*29. A boy pulls a sled at constant speed over packed, level snow. His sister is on the sled and has a mass of 50 kg. The coefficient of kinetic friction between sled and snow is 0.09. The tow rope makes an angle of 25° with the ground. What force does the boy exert in pulling the sled?

Answer: 46.7 N

*30. Two blocks joined by a string are sliding down an inclined plane as in Figure 3.23. The coefficient of dynamic friction between the 4.0-kg block and the plane is 0.4, and that between the 1.5-kg block and the plane is 0.1. (a) What is the acceleration of the system? (b) What is the tension in the string?

Figure 3.23 Blocks Tied Together Sliding Down Surface with Friction (Exercise 30)

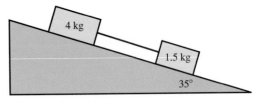

Answer: (a) 3.07 m/s²

(b) 2.64 N

31. The force of air resistance on a falling object of mass m is proportional to the square of its velocity, so that $F_r = kv^2$, where k is the proportionality constant that depends on the properties of the object. Find the terminal speed of the object.

Answer: $v = \sqrt{mg/k}$

*32. With the apparatus shown in Figure 3.24, find the mass of block M that will allow the 4-kg mass to remain motionless relative to the ceiling. Assume that the pulleys have negligible weight and friction.

Answer: M = 8 kg

Figure 3.24 Double Atwood's Machine (Exercise 32)

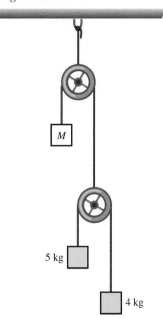

M

5 kg

4 kg

A roller coaster is designed around potential energy at the top of a rise being converted into kinetic energy of motion in the valleys. Successive rises are lower in height because of friction between the car and track.

Work and Energy

At the end of this chapter, the reader should be able to:

1. Find the **work** done on an object after knowing the force applied (in any direction) and the distance pushed (or pulled).

2. Calculate the gravitational **potential energy** of an object.

3. Calculate the **kinetic energy** of an object.

4. Recognize that a body can possess more than one type of energy simultaneously, and identify the types properly.

5. Use the concept of **conservation of energy** to solve physics problems.

6. Retroactively apply **Hooke's law** ($\mathbf{F} = k\mathbf{x}$) to free-body diagrams, as learned in the previous chapters.

7. Identify *nonconservative forces* in an exercise.

8. Use energy and time to calculate **power,** and work
   backward by using the power
   rating of a system to find the
   work done or the time
   required to do such work.

Energy doing work is a very common process. For example, the energy stored in gasoline is turned into motion as we drive. Coal, oil, and natural gas store energy so that when
these fuels are burned, the stored chemical energy is converted into heat that generates
the electricity needed to run our highly technological society. The energy stored in the
water at the top of a hydroelectric dam flows through turbine blades at the bottom of the
dam, causing the blades to spin and generate electricity. The energy in the wind can turn
the blades of a windmill that can in turn pump water or generate electricity. There is energy
in sunlight, and solar cells turn that energy directly into electricity. Electricity is itself a
very mobile form of energy. Its mobility is what makes it the most common energy source
used by most people in their everyday lives [besides the internal combustion engine (automobiles)].

But what is actually meant by the word *energy?* In the situations described in the paragraph above, energy is being converted into useful work. When a fuel is burned, the
stored energy in that fuel is converted into heat, and that heat can do work. This work done
is often in the form of motion—whether it be a piston moving down in a cylinder or the
spinning of turbine blades. Stored energy is often called *potential energy,* and energy of
motion is often called *kinetic energy.* A simple gravity-driven pendulum is constantly converting the potential energy of the bob's height into the kinetic energy of its speed and back
again. A spring is a very useful device because energy can be stored in it by compressing
or extending it. A compressed spring can do work, such as causing the gear wheels in a
watch to move. Potential energy, kinetic energy, and work are all interrelated. To better
define the relations among these three concepts, we will examine the concept of work first.

4.1   Work

Work: The work done by a force
F acting on an object undergoing
a displacement **x** is equal to the
magnitude of the force
component in the direction of the
displacement multiplied by the
magnitude of the displacement.

As explained in Chapter 3, a force acting on a body can change its state of motion. Work
is done on a body by changing its state of motion. Therefore, when a force is applied to
a body over some distance, work is done on this body. In Figure 4.1, a force is applied
to a body at the angle θ (theta) shown. Such a force might be the tension in a rope
pulling a box along a surface. If the force F isn't strong enough to lift the box off the
surface but is strong enough to overcome friction, then the box will move to the right.
The work done on the box will be the work done in moving the box to the right. *Work
is defined as the magnitude of the force component in the direction of the displacement
multiplied by the magnitude of the displacement.* In this case, the box moves horizontally, and the force component in the direction of the displacement is the horizontal
component of force F_x or $F \cos \theta$. Mathematically,

$$W = \mathbf{F} \cdot \mathbf{x}$$

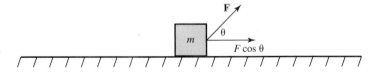

Figure 4.1 Force Applied to a
Body at an Angle

A waterfall illustrates gravitational potential energy converted into kinetic energy.

This product is called a *dot product* and is read, "*F* dot *x.*" In the case of Figure 4.1, the equation becomes

$$W = Fx \cos \theta$$

(Equation 4.1)

where F is the applied force, x is the displacement, and θ is the angle between the direction of the force and displacement. The vertical force component perpendicular to the displacement ($F \sin \theta$) doesn't contribute to the work done. The two variables on the right are vectors, force and displacement, but work (W) is a scalar with a magnitude but no direction. A dot product between two vectors becomes a scalar with no direction associated with it. *Both work and energy are scalars.*

The unit of energy is the newton-meter (abbreviated N-m). It is also commonly called the joule (abbreviated J), named after the English physicist James Joule (1818–89). Since energy and work are intimately related, the unit for energy is also the joule.

EXAMPLE 4.1

A tugboat pulls a barge along behind it as shown in Figure 4.2. The tension in the cable between the tug and the barge is 1000 N, and the angle that the cable makes with the horizontal is 15°. How much work does the tug do in pulling the barge 1 km?

$\theta = 15°$

Figure 4.2 Tug Pulling a Barge

SOLUTION

$$W = Fx \cos \theta$$

$$W = (1000 \text{ N})(1000 \text{ m})\cos 15°$$

$$W = 965{,}926 \text{ N-m (joules)}$$

The energy to do this work must come from somewhere, and in this case it comes from the heat generated in burning fuel in the tug's engine to rotate the tug's propeller to move the tug to pull the barge.

EXAMPLE 4.2

A man pulls a 50-lb wooden box across a wooden floor a distance of 70 ft. The man pulls the box with a rope that makes an angle of 30° with the horizontal (see Figure 4.3). The tension in the rope is 30 lb. The coefficient of kinetic friction between box and floor is 0.4. (a) How much work is done? (b) How much work was done overcoming friction? (c) How much work went into accelerating the box?

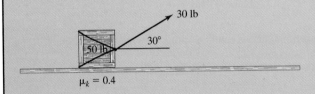

Figure 4.3 Rope Pulling Wooden Box across Wooden Floor

SOLUTION

(a)

$$W = Fx \cos \theta$$

$$W = (30 \text{ lb})(70 \text{ ft})\cos 30°$$

$$W = 1819 \text{ ft-lb}$$

(b) The force of friction (F_f) is the normal force (F_n) times the coefficient of kinetic friction. In this instance, we must take the vertical component of the force into account, because the normal force is less than mg because of it.

$$W_f = F_f x$$

$$F_f = \mu_k F_n$$

$$W_f = \mu_k F_n x$$

$$F_n = mg - F \sin \theta$$

$$W_f = \mu_k (mg - F \sin \theta)x$$

$$W_f = (0.4)[50 \text{ lb} - (30 \text{ lb})\sin 30°](70 \text{ ft})$$

$$W_f = 980 \text{ ft-lb}$$

The work done against friction is lost, being dissipated as heat.

(c)
$$W_a = W - W_f$$

$$W_a = 1819 \text{ ft-lb} - 980 \text{ ft-lb}$$

$$W_a = 839 \text{ ft-lb}$$

A little less than one-half of the total work done is used to increase the speed of the box across the floor.

4.2   Gravitational Potential Energy

Whenever an object is lifted, work is done against gravity. This work is stored in the lifted objects as gravitational potential energy. *Potential energy is energy stored in the position of an object.* In the case of lifting, the position is the object's height.

How much energy is stored in lifting a mass m to a height h?

$$W = Fx \cos \theta$$

The force F is the weight of the object, mg. The displacement, x, is the vertical height, h. When work is done against gravity, only the height h from the starting point is important. For example, if you climbed from one floor to another on a circular staircase, the distance you traveled would be greater than the height you climbed. However, only the vertical height would count in the work you did against gravity. (In moving an object parallel to the earth's surface, you do no work against gravity.) When an object is raised the angle between the weight and the displacement is 180°, with the weight pointing down and the displacement vertically up.

$$W = (mg)(h)\cos 180°$$

$$\cos 180° = -1$$

$$W = -mgh$$

The negative sign is usually ignored because it's understood that the higher an object is lifted, the greater the stored gravitational potential energy.

$$PE = mgh$$

(Equation 4.2)

Potential energy: Energy stored in the position of an object.

4.3   Kinetic Energy of Motion

An object in motion possesses energy, and the greater the speed, the greater the energy. A hammer does work on a nail by driving the nail into a piece of wood. A really big hammer, such as a pile driver, drives pilings into the ground to form the foundations for buildings, bridges, and piers. The energy in the hammer's moving head is converted into work. *Kinetic energy is the energy in the motion of an object.*

Kinetic energy: Energy in the motion of an object.

To find how much energy is stored in the speed of an object, we again start from the basic definition of work.

$$W = Fx \cos \theta$$

It's assumed that the force used to accelerate the object up to velocity v is pointed in the same direction as the object's velocity.

$$\cos 0° = 1$$

$$W = Fx$$

To find the total work done in accelerating the object to speed v, we break the work up into small packets ΔW, corresponding to a small change in the total displacement, Δx. Later, when all of the packets have been calculated for all of the displacement segments over the entire range of the displacement, we add all of the work packets together to find the total work done.

$$\Delta W = F \, \Delta x$$

From Newton's second law, $\mathbf{F} = m\mathbf{a}$,

$$\Delta W = m\mathbf{a} \, \Delta x$$

The magnitude of a linear acceleration is the rate of change of speed with time.

$$a = \frac{\Delta v}{\Delta t}$$

$$\Delta W = m\left(\frac{\Delta v}{\Delta t}\right)\Delta x$$

The above equation can be rearranged as follows:

$$\Delta W = m\left(\frac{\Delta x}{\Delta t}\right)\Delta v$$

However, $\Delta x/\Delta t = v$, so

$$\Delta W = mv \, \Delta v$$

Everything on the right side of the equation is now in terms of a constant (the mass) and the velocity. But the two terms v and Δv interact. In finding the total work, we add all of the little bits of work ΔW. In Chapter 1 in Section 1.3, "Displacement," an equation like the one above was discussed.

$$\Delta v = v_f - v_i$$

$$\Delta v = v - 0$$

$$\Delta v = v$$

When starting from rest, $\Delta v = v$. The equation is illustrated in Figure 4.4. If the force is constant, so is the acceleration, and the velocity will increase linearly with time ($v = at$). The change in velocity (Δv) also increases linearly with time, since it depends on the velocity. (If the velocity is doubled, so is the change in velocity.) The product $v \Delta v$ is the area under the diagonal line in Figure 4.4, $\frac{1}{2}v^2$. When we add up all of the little bits of work (ΔW), the total work becomes the area inside the right triangle multiplied by the mass.

$$W_T = \frac{1}{2}mv^2$$

The total work done in accelerating the object to speed v is the kinetic energy of that object.

$$KE = \frac{1}{2}mv^2 \qquad \text{(Equation 4.3)}$$

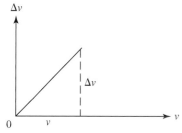

Area under curve $= \frac{1}{2}v \, \Delta v$

Area under curve $= \frac{1}{2}v^2$

Figure 4.4 Change in Velocity Increases Linearly with Velocity

Mass times velocity squared has the units of work. Breaking the newton-meter down into its fundamental units, we have

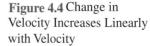

$$\text{N-m} = \frac{(\text{kg})(\text{m})}{\text{s}^2}(\text{m}) = (\text{kg})\frac{\text{m}^2}{\text{s}^2}$$

4.4 Conservation of Energy

Energy can be converted from one form into another. For example, if a bowling ball is lifted onto a shelf in a closet, work is done on the ball. This work is stored as the gravitational potential of the ball relative to the floor. If the ball falls off the shelf, the potential energy of the ball is converted into kinetic energy. As the ball falls, it loses potential energy and gains kinetic energy because it is speeding up. On the shelf the ball has an initial potential energy, and just before the ball hits the floor, it has a final kinetic energy. Without frictional losses, the initial potential energy equals the final kinetic energy. Energy has changed form, but the total energy stays the same, which is the basis of the principle of the conservation of energy.

Conservation of energy:
Energy conversion from one form to another without losses.

The simple gravitationally driven pendulum shown in Figure 4.5 is a good example of the conservation of energy. If the bob is lifted to position (1) and held there, the bob has potential energy equal to *mgh*. If it is then released, the bob falls and accelerates to reach its highest speed at the equilibrium position (2). [Position (2) is called the *equilibrium position* because it's the position where the pendulum bob would come to rest if the energy in the system dropped to zero.] The moving bob then begins to climb, and it slows down as its kinetic energy is converted back into potential energy. When the bob again rises to height *h* on the other side of position (2), it momentarily comes to a stop and then starts falling again, starting the cycle all over again. The bob will oscillate back and forth forever, converting PE to KE and back to PE if there are no losses to friction.

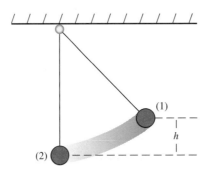

Figure 4.5 Simple Pendulum (Gravity-Driven)

At position (1), the energy is all potential. At position (2), the energy is all kinetic. At an intermediate position between positions (1) and (2), the energy is both kinetic and potential. However, regardless of where the bob is in its path, the energy of the system remains constant. If *E* is the total energy in the system,

$$E = mgh + \frac{1}{2}mv^2$$

The total energy in the system can be fixed by raising the bob to the initial height of h_0.

$$E = mgh_0$$

$$mgh_0 = mgh + \frac{1}{2}mv^2$$

$$gh_0 = gh + \frac{1}{2}v^2$$

$$2gh_0 = 2gh + v^2$$

The bob's speed at every point in its path can be determined if we know the height of that point from the lowest point of its trajectory.

$$v^2 = 2gh_0 - 2gh$$
$$v = \sqrt{2gh_0 - 2gh}$$
$$v = \sqrt{2g(h_0 - h)}$$

The trajectory of the pendulum bob is the arc of a circle. A similar relationship between the speed and the height of fall can be quickly derived for the bowling ball that rolled off the shelf.

$$mgh = \frac{1}{2}mv^2$$
$$2gh = v^2$$
$$v = \sqrt{2gh}$$

(Equation 4.4)

Compare Equation (4.4) with the one derived for conservation of energy in an oscillating pendulum. If $h = 0$, the bob's speed equation is exactly the same as Equation (4.4). It doesn't matter if the speed is vertical, as in the case of the bowling ball, or horizontal, as in the case of the simple pendulum. Energy is a scalar and has no direction associated with it. *The principle of **conservation of energy** states that the total energy of a system is constant despite the fact that the energy in the system is constantly changing back and forth between PE and KE.* If energy is conserved, the work done by the conservative forces (gravity in the case of the pendulum and the bowling ball) is ***path independent.*** The pendulum bob travels along the arc of a circle, and its velocity at the bottom of the arc is horizontal and has the same magnitude that it would have if it had been dropped straight down. The speed is independent of the path the object took.

Path independence: The work done by conservative forces is independent of the path taken in doing the work.

4.5   Hooke's Law and Springs

A spring is a well-known device used for storing energy. *A spring obeys **Hooke's law**, which states that the greater the compression or extension of the spring from its equilibrium position, the greater must be the force exerted on it.*

Hooke's law is stated mathematically as

$$\mathbf{F} = k\mathbf{x}$$

(Equation 4.5)

Hooke's law: The greater the extension or compression of a spring from its equilibrium position, the greater the force on it: $\mathbf{F} = k\mathbf{x}$.

where $\mathbf{F}$ is the force used to stretch or compress the spring, $\mathbf{x}$ is the spring's displacement from equilibrium, and k is the spring constant. The bigger k is, the stiffer the spring (k has units of newtons per meter, N/m).

Energy can be stored as potential energy in the compressed or extended spring. This stored energy can serve as the power source for watches and clocks. From

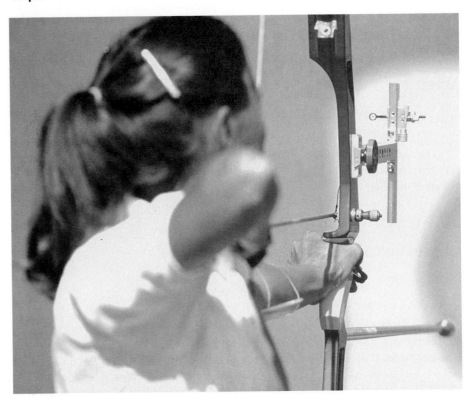

A bow converts the potential energy of its elastic material into the kinetic energy of the arrow after release.

Hooke's law and the definition of *work,* a formula can be derived for the amount of potential energy stored in a compressed spring. Because the force changes as the spring is compressed, the spring will be compressed a small amount. The work done in compressing it a small amount can be measured, and then the spring further compressed another small amount. Then we add all of the small amounts of work done in compressing to find the total amount of work done in compressing the spring. (We've seen this argument before in deriving the formula for kinetic energy.) The calculations are as follows (see Figure 4.6):

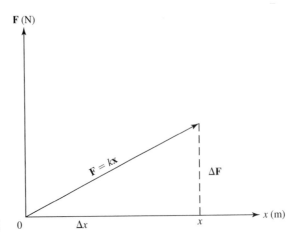

Figure 4.6 Change in Force on a Spring Increases Linearly with the Distance That the Spring Is Compressed

$$W = fx \qquad \Delta W = F \, \Delta x$$

$$F = kx \qquad \Delta W = kx \, \Delta x$$

where k, the spring constant, doesn't change. From Hooke's law, the force applied to the spring increases linearly with displacement. The total work done in compressing (or extending) the spring is the area inside the triangle of Figure 4.6.

$$W_T = \frac{1}{2}\Delta x \, \Delta F$$

$$\Delta x = x_f - x_i \qquad \Delta x = x - 0 \qquad \Delta x = x$$

$$\Delta F = kx_f - kx_i \qquad \Delta F = kx - 0 \qquad \Delta F = kx$$

$$W_T = \frac{1}{2}(x)(kx)$$

$$W_T = \frac{1}{2}kx^2$$

The potential energy stored in a compressed or stretched spring is

$$PE = \frac{1}{2}kx^2 \qquad \text{(Equation 4.6)}$$

where x is the amount of compression or extension.

4.6   Nonconservative Forces

A simple pendulum will run forever if no friction is present. It will continuously convert gravitational PE into KE and back again. The system is without losses. Energy that goes into overcoming the force of friction, however, cannot be recovered and reconverted. It is gone, dissipated as heat. Let us look again at Example 4.2, where the box slides across the floor.

EXAMPLE 4.3

In pulling the box in Example 4.2 across the floor, the person did 1819 ft-lb of total work. Of that work, 980 ft-lb was done against friction and was dissipated. The other 839 ft-lb went into accelerating the box, or into kinetic energy of motion. If the box started from rest, what is the final velocity of the box?

SOLUTION
This question can be answered in two ways. The first way is through an energy argument. (The mass of a 50-lb box is 1.56 slugs.)

$$W_a = 839 \text{ ft-lb}$$

$$W_a = \frac{1}{2}mv^2$$

$$mv^2 = 2W_a$$

$$v^2 = \frac{2W_a}{m} \qquad\qquad v = \sqrt{\frac{2W_a}{m}}$$

$$v = \sqrt{\frac{2(839 \text{ ft-lb})}{1.56 \text{ slugs}}} \qquad v = 32.8 \text{ ft/s}$$

The same speed can be found from another approach: force and uniformly accelerated motion. We start with the work done in accelerating the box.

$$W_a = 839 \text{ ft-lb}$$

$$W_a = F_a x$$

$$F_a = \frac{W_a}{x} \qquad F_a = \frac{839 \text{ ft-lb}}{70 \text{ ft}} \qquad F_a = 12 \text{ lb}$$

$$a = \frac{F_a}{m} \qquad a = \frac{12 \text{ lb}}{1.56 \text{ slugs}} \qquad a = 7.7 \text{ ft/s}^2$$

$$v_f^2 - v_0^2 = 2ax \qquad v_f^2 = 2ax \qquad v_f = \sqrt{2ax}$$

$$v_f = \sqrt{2(7.7 \text{ ft/s}^2)(70 \text{ ft})} \qquad v_f = 32.8 \text{ ft/s}$$

The two answers are the same, so both approaches work. One is not more "correct" than the other. Usually, the energy argument will yield an answer with fewer steps than force and uniformly accelerated motion. But regardless of the approach, if the physics is correct, the answer will be also!

For nonconservative forces such as friction, we can also write an equation for the total energy. For the box in Example 4.3, the equation is

$$E = \text{KE} + W_f$$

The work done in pushing the box across the floor equals the kinetic energy of the box's motion plus the work done against friction. The above equation tells you where the energy goes, but it is not an energy conservation equation because the work done against friction is dissipated as heat.

4.7   Power

Power: The rate at which work is done.

(Equation 4.7)

***Power* is the rate at which work is done.**

$$P = \frac{W}{t}$$

The units for power are the joule per second, or watt (W). As power is often measured in thousands of watts, the kilowatt (kW) is another often-used unit for power. One kilowatt is 1000 watts (1 kW = 1000 W). In the English system of units the common unit of power is horsepower (hp). Usually when we think of power in terms of horsepower, we think of the output of an internal combustion engine. When we think of the rate at which we use electrical energy, we usually think in terms of watts. But 1 hp equals 745.7 W. Both watts and horsepower are units of power. If we wished, we could calibrate the meters that measure the electricity we use in terms of horsepower-hours instead of kilo-watt-hours, and we could rate automobile engines in kilowatts instead of horsepower.

EXAMPLE 4.4

An electric motor powers an elevator for a 10-story building. A fully loaded elevator has a mass of 1500 kg. If the elevator rises at a speed of 2 m/s, what is the power required from the motor?

SOLUTION

$$P = \frac{W}{t}$$

The work done by the elevator in lifting its load is against gravity.

$$W = mgh$$

$$P = \frac{mgh}{t} \qquad P = mg\left(\frac{h}{t}\right)$$

The height the elevator rises over the time it took is the vertical velocity.

$$P = mgv$$

$$P = (1500 \text{ kg})(9.8 \text{ m/s}^2)(2 \text{ m/s})$$

$$P = 29,400 \text{ W} \quad \text{or} \quad 29.4 \text{ kW (39.4 hp)}$$

We can make further use of Example 4.4. In the example, we derived an equation for power involving velocity.

$$P = (mg)v$$

The product mg is simply the weight of the object being acted upon, and that is a force. So we can rewrite the above.

$$P = Fv$$

This is a general result. Let's start again with the definition for work.

$$W = Fx$$

Divide both sides by the time t.

$$\frac{W}{t} = \frac{Fx}{t}$$

$$P = F\left(\frac{x}{t}\right)$$

$$\left(\frac{x}{t}\right) = v$$

(Equation 4.8)

$$P = Fv$$

We now have an equation for power involving velocity.

In preceding sections of this chapter, we have seen that when nonconservative forces come into play, some of the work done is dissipated. This dissipation is usually caused by friction. It is common engineering practice to give devices such as motors an efficiency rating. This rating takes into account the amount of work or power lost in friction or heat. The efficiency rating usually gives the rating as a percent, and that is the part that goes into useful work. For instance, if we again look at Example 4.3, we see that of the total work done (1819 ft-lb), 980 ft-lb was against friction and was dissipated. The rest (839 ft-lb) went into kinetic energy and could be recovered later; it was stored. Thus,

$$\frac{839}{1819} = 46.1\%$$

The process of dragging the box across the floor against friction was 46.1% efficient. In general, the power delivered by any machine will be less than the power supplied to it because of the inevitable presence of friction, or for an electric motor resistive current losses. (A resistor dissipates power through heat.) In an electric motor, the current flowing through the windings in it will heat it up, and that heat energy is lost.

EXAMPLE 4.5

A hydraulic ram with a 2.5-hp motor can apply a force of 20,000 lb. If the ram is 90% efficient, how fast does it move (1 hp = 550 ft-lb/s)?

SOLUTION

$$P = Fv$$

$$v = \frac{P}{F}$$

$$(2.5 \text{ hp})\left(\frac{550 \text{ ft-lb/s}}{\text{hp}}\right) = 1375 \text{ ft-lb}$$

Only 90% of this power is available to do useful work.

$$P = (0.9)(1375 \text{ ft-lb})$$

$$P = 1237.5 \text{ ft-lb}$$

$$v = \frac{1237.5 \text{ ft-lb}}{20,000 \text{ lb}}$$

$$v = 0.0619 \text{ ft/s}$$

$$v = 0.742 \text{ in./s}$$

Power is a scalar, because its definition contains only scalars.

Summary

1. Units of *energy* in the SI system are joules (J), which can also be expressed as newton-meters (N-m) or kg-m/s^2.

2. Both work and energy use the same unit (J), and therefore the terms can often be interchanged to help you understand a problem.

3. An object can have more than one type of energy at the same time, such as kinetic energy, gravitational potential, and spring potential.

Problem Solving Tip 4.1: It is often useful to look at a "snapshot" of the exercise *just before* the problem "starts." Then look at a snapshot of the exercise *just after* the problem "ends." For example, let us look at a vertical pinball machine, where a steel ball is pushed down onto a spring, compressing it. Then the ball is released, allowing it to shoot upward. Before we even know what it is that we are looking for, we can set up the problem. Just before the action that we are interested in starts, we know the following fact: A spring is compressed, and therefore we have spring potential energy. (The ball is as far down as it will ever get in our problem, so there is no gravitational potential, and the ball is not moving *yet*, so there is no kinetic energy.) Just after the action that we are interested in, the spring has just fully extended, and the ball has been released to fly upward. The energies present are **kinetic energy** (as the ball is moving) and gravitational potential (as the ball is now higher than when it started). (There is no spring energy, because it is neither compressed nor stretched in this snapshot of time.) We can use **conservation of energy** (energies before = energies after), so

$$\text{Energies before} = \frac{1}{2}kx^2$$

$$\text{Energies after} = \frac{1}{2}mv^2 + mgh$$

where k is the spring stiffness, x is the initial compression of the spring, m is the mass of the ball, v is the velocity of the ball, and h is the height above the *initial position* of the ball. Therefore,

$$\frac{1}{2}kx^2 = \frac{1}{2}mv^2 + mgh$$

We can now solve for whatever the problem asked for, whether it be the velocity of the ball, the mass of the ball, the height the ball has traveled, the spring stiffness, or the initial compression of the spring!

Remember that the above formula we have created works only in this instance. Use the same procedure to find the correct formula for any problem you are interested in (no matter which chapter!), as long as you are not concerned with direction (θ) or time (t).

4. The unit of *power* in the MKS system is the watt (W), which can be expressed as joules per second (J/s) in the SI system. In English units, power is expressed in horsepower (1 hp = 745.7 W).

5. *Nonconservative forces* (such as friction) must be dealt with carefully. Remember that the work done in a system where friction is present is equal to the resultant energy imparted (E) (whether kinetic, gravitational, or spring) plus whatever work was necessary to overcome the friction.

$$W = E + W_f$$

Problem Solving Tip 4.2: Remember that work equals the dot product of force and distance ($W = \mathbf{F} \cdot \mathbf{x}$), so $W_f = F_f x \cos \theta$. Also remember that the force of friction equals the coefficient of kinetic friction (the object *is* moving!) times the normal force ($F_f = \mu F_n$).

Key Concepts

CONSERVATION OF ENERGY: Energy conversion from one form to another without losses.

HOOKE'S LAW: The greater the extension or compression of a spring from its equilibrium position, the greater the force on it: $\mathbf{F} = k\mathbf{x}$.

KINETIC ENERGY: Energy in the motion of an object.

PATH INDEPENDENCE: The work done by conservative forces is independent of the path taken in doing the work.

POTENTIAL ENERGY: Energy stored in the position of an object.

POWER: The rate at which work is done.

WORK: The work done by a force $\mathbf{F}$ acting on an object undergoing a displacement $\mathbf{x}$ is equal to the magnitude of the force component in the direction of the displacement multiplied by the magnitude of the displacement.

Important Equations

(4.1) $W = Fx \cos \theta$ (where θ is the angle between the direction of the force and displacement)

(4.2) $PE = mgh$ (Gravitational potential energy)

(4.3) $KE = \dfrac{1}{2}mv^2$ (Kinetic energy of a moving object)

(4.4) $v = \sqrt{2gh}$ (Speed of an object converting PE to KE by falling)

(4.5) $\mathbf{F} = k\mathbf{x}$ (Hooke's law for springs)

(4.6) $PE = \dfrac{1}{2}kx^2$ (Potential energy stored in a compressed or an extended spring)

(4.7) $P = \dfrac{W}{t}$ (Mathematical definition of *power*)

(4.8) $P = Fv$ (Power being expended on an object moving with a speed v and having a force of magnitude F on it)

Conceptual Problems

1. Against a nonconservative force, the work done in moving an object from A to B:

   a. is completely turned into heat

   b. doesn't depend on the path taken between A and B

   c. can't be completely recovered by moving the object from B to A

2. A pendulum bob and a dropped ball convert potential energy into kinetic energy. Gravity is an example of:

   a. a dissipative force

   b. a nonconservative force

   c. a conservative force

   d. any of the above depending on the zero-height reference level

3. Even if the net force acting on an object isn't zero, under what circumstances is no work done on that object even if the object is moving?

4. A golf ball and a ping-pong ball are both dropped at the same time down a glass cylinder with no air inside (a vacuum). Compare and contrast their speeds, potential energies, and kinetic energies as they fall.

5. Two climbers decide to climb the same peak. One of the climbers gets to the summit by climbing up a sheer rock wall, and the other climbs to the summit from the other side up a slope that has an angle with the horizontal of about 40°. Compare the amount of work that they do against gravity.

6. Where is the KE of a pendulum bob a maximum? Where is it a minimum? Where is the PE of a pendulum bob a minimum? Where is it a maximum?

7. Explain the basic design principle behind a roller coaster in terms of work and energy.

8. A vertical projectile has an initial speed of v_0 and is launched upward. When it returns to the point at which it was launched, what is the total work done on the projectile, and what has done this work? Assume for the first case no air friction; then assume a case in which there is air friction. Contrast the two cases.

9. Suppose that the total mechanical energy of an object is conserved. If the KE increases, what happens to the PE? If the PE remains the same, what happens to the KE?

10. The inclined plane is a simple machine that makes it easier for us to do work. Other examples of simple machines are the screw and the lever. How do these simple machines help us, and do they save us any work?

11. Spring A has a greater spring constant than spring B. If they are both stretched by the same amount, on which spring is the greater work done? If they are both stretched by the same force, on which spring is the greater work done?

12. A parachutist jumps out of an airplane, the parachute opens, and the chutist lands safely on the ground. Is energy conserved during the chutist's descent?

13. A stone is tied to a string and whirled in a vertical circle. Can you whirl this stone around the circle at a constant speed? Can you do so at a constant energy? Assume negligible air friction.

Exercises

Work, Potential Energy, and Kinetic Energy

1. Three kilojoules (3 kJ) of work is done lifting a 40-kg mass. If the mass is at rest before and after it is lifted, how high does it go?

Answer: 7.65 m

2. A box is pushed at a constant speed up a frictionless ramp that is 10 m long and at an angle of 20° to the horizontal (see Figure 4.7). The box weighs 3 kg. (a) How much work is done by the force in pushing the box up the ramp? (b) How much work is done in lifting the box straight up?

Figure 4.7 Box on a Frictionless Ramp (Exercise 2)

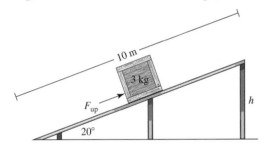

Answer: (a) 100 J
(b) 100 J

3. A 1.5-kg object initially at rest has a horizontal force of 7.0 N applied to it. (a) What is the speed of the object after it has gone 15 m? Use an energy argument. (b) Answer Part (a) using a force and uniformly accelerated motion argument.

Answer: (a) and (b) 11.8 m/s

4. A 5-kg stone is dropped from a height of h_0. (a) Derive an equation that gives the stone's velocity at any intermediate height between 90 m and the ground. (b) What is the stone's kinetic energy when it is 30 m above the ground? (c) What

is its potential energy at a height of 30 m? (d) What is its total energy at a height of 30 m?

Answer: (a) $v = [2g(h - h_0)]^{1/2}$
(b) 2941 J
(c) 1470 J
(d) 4411 J

5. A ball is dropped from a height of 75 m. (a) How high is the ball above the ground when $\frac{2}{3}$ of its total energy is kinetic energy? (b) How fast is it falling at this height?

Answer: (a) 25 m
(b) 31.3 m/s

6. A 3-lb hammer attains a speed of 15 ft/s when swung to hit a nail. If the nail is driven 1 in. into a horizontal wooden board when the hammer hits it, what is the average force on the nail?

Answer: 129.6 lb

7. A man lifts a 100-kg bale to a height of 10 m with a block and tackle. The object is at rest afterward. He exerts a force of 200 N on the rope and pulls 60 m of the rope through the block and tackle. (a) How much work has the man done? (b) What is the potential energy of the 100-kg bale at a height of 10 m? (c) If there is a difference between the answers to Parts (a) and (b), explain it.

Answer: (a) 12 kJ
(b) 9800 J
(c) Friction

*8. An object of mass m is uniformly accelerated horizontally from rest for a time t, during which time it covers a distance x. Find the work done on the object.

Answer: $W = \dfrac{2mx^2}{t^2}$

Conservation of Energy

9. A 90-lb girl on a swing is 7 ft above the ground at her highest point and 2 ft above the ground at the lowest point. What is her speed when she is closest to the ground?

Answer: 17.9 ft/s

10. A 0.25-kg ball is pushed across a frictionless surface by a 4-N force for a distance of 4 m. (A) What is its final kinetic energy if the ball starts from rest? (b) What is its speed?

(c) The same force is used to lift the ball to a height of 4 m. What is the ball's final kinetic energy if it starts from rest? (d) What is its vertical velocity?

Answer: (a) 16 J
(b) 11.3 m/s
(c) 6.2 J
(d) 7.04 m/s

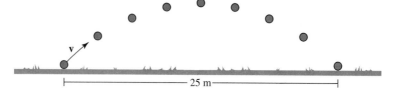

Figure 4.8 Trajectory of a
Shot Put Hurled 25 m
(Exercise 11)

25 m

11. A 20-kg shot put is thrown 25 m (see Figure 4.8). What is
its minimum initial kinetic energy?

Answer: 2450 J

*12. Figure 4.9 shows a simple pendulum of length L. (a) How
fast will the pendulum bob be going when it reaches the bottom of its arc? (b) The nail in the figure is located a distance
d below the point of suspension. How far below the point of
suspension in terms of L must the nail be located for the bob
to swing into a position vertically above the nail?

Answer: (a) $v = \sqrt{2gL}$
(b) $d = L/2$

Figure 4.9 Falling Pendulum Bob (Exercise 12)

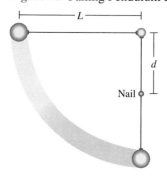

L

d

Nail

13. A 30-kg child sits in a swing suspended by a 2-m rope.
When the swing is started swinging, the child's father pulls
the swing back until the rope makes an angle of 15° with the
vertical. How fast is the child moving at the bottom of the
arc of the swing's path?

Answer: $v = 1.1$ m/s

Hooke's Law and Springs

14. A 55-kg load stretches a vertical spring 0.6 m from its equilibrium position when it is hung from it. (a) How much
energy is stored in the stretched spring? (b) The spring is cut
in half, and the two new springs are hung side by side. What
is the effective spring constant of the two side-by-side
springs acting together to support the load?

Answer: (a) PE = 161.6 J
(b) $k = 1797$ N/m

*15. When sitting on a spring, a mass depresses the spring with a
spring constant of 8 N/m a total of 3 cm. If the same mass is
dropped from a height of 10 cm onto the spring, how far is
the spring compressed before springing back?

Answer: Compression = 11.3 cm (0.113 m)

Nonconservative Forces

16. A 100-kg box is pushed across a level floor with a horizontal
force of 500 N. The coefficient of kinetic friction is 0.4.
(a) How much work is done in pushing the box 25 m? (b) How
much work went into overcoming friction? (c) How much
work went into accelerating the box? (d) If the box started from
rest, how fast would it be moving after being pushed 25 m?

Answer: (a) 12,500 J
(b) 9800 J
(c) 2700 J
(d) 7.3 m/s²

17. In Example 4.3, 839 ft-lb of work was used to accelerate a
box to a speed of 32.8 ft/s. It slid along a wooden floor,
where the coefficient of kinetic friction was 0.4. (a) If the
box were released after reaching the above speed, how
much further would it have slid? Use an energy argument.
(b) Answer Part (a), but use a force and uniformly accelerated motion argument.

Answer: (a) and (b) 42 ft

18. A fisherman hooks a 5-lb catfish, and it swims off at 8 ft/s.
By applying the line drag on his reel to the line, the fisherman tries to stop the catfish within 2 ft before it swims
under a log. If the line in his reel is only 5-lb test, would it
have parted? Prove your answer by finding the tension on
the line.

Answer: No

*19. A 1.5-kg block is pulled up a ramp by a constant force of
25 N parallel to the ramp. The ramp is 3 m long and at an
angle of 35° to the horizontal. The coefficient of kinetic friction between ramp and block is 0.4. What is the speed of the
block at the top of the ramp?

Answer: 6.86 m/s

20. A sled at the top of a hill is given a push and coasts down the
hill. At the top of the hill the sled's speed is 3 m/s, and at the
bottom 17 m/s. If the hill is 45 m high, how much work was
done against friction? The sled with its rider weighs 100 kg.

Answer: 34,750 J

21. A 25-kg block is pulled 15 m up a plane that is 20° above the horizontal at constant speed (see Figure 4.10). The applied force acts 23° above the plane. The coefficient of kinetic friction between the block and the plane is 0.4. (a) What is the magnitude of the force **F**? (b) How much work does the applied force do?

Figure 4.10 Block Being Pulled Up an Inclined Plane (Exercise 21)

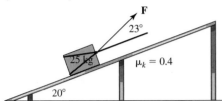

Answer: (a) 163.3 N
(b) 3560 J

*22. A wooden block with an initial speed of 5 m/s starts sliding up a wooden inclined plane that is at an angle of 20° to the horizontal. (a) If the coefficient of kinetic friction between block and plane is 0.3, how far up the plane does the block slide? (b) What is the block's speed at the bottom of the ramp when it slides back down?

Answer: (a) 2.04 m
(b) 1.55 m/s

Power

26. A 20-kW motor is used to hoist a 1000-kg steel beam to the fifteenth floor of a building under construction. The beam is lifted to a height of 68 m. How long does it take to lift the beam if there are no energy losses?

Answer: 33.3 s

27. An elevator motor must be able to lift a total load (elevator plus people) of 1500 kg at a speed of 1 m/s. What is the minimum rating of the motor?

Answer: 14.7 kW

28. A car needs 25 hp to move along a level road at 60 mph. What is the total force of air resistance and rolling friction at that speed?

Answer: 156 lb

29. A lift whose efficiency is 85% has a 10-kW motor. How long does it take the lift to raise a 1000-kg load to a height of 100 m?

Answer: 1.9 min

30. A man pulls a loaded wagon at a constant speed with a force of 20 lb for 150 ft. The rope attached to the wagon is at an angle of 25° above the horizontal. (a) How much work is done? (b) If the man moves the wagon 150 ft in 75 s, find his power output in horsepower.

Answer: (a) 2719 ft-lb
(b) 0.066 hp

31. An escalator carries people from the ground floor of a building to a gallery 20 ft above. The escalator has a capacity of

23. A block with a mass of 7 kg sits on a level floor, and the coefficient of friction between the block and the floor is 0.3. A spring is connected horizontally between the block and a nearby wall. The spring has a spring constant of 25 N/m. How much work does it take to move the block 15 cm to the right if the spring is initially in its equilibrium position?

Answer: W = 3.37 J

*24. The coefficient of kinetic friction between a hockey puck and the ice is μ. Show that if the puck has an initial speed of v, the puck will go a distance s before coming to stop, where

$$s = \frac{v^2}{2\mu g}$$

*25. A 1.5-kg mass sits on a level table top. The coefficient of friction between the mass and the table top is 0.25. A 4.5-kg mass hanging vertically is connected to the 1.5-kg mass on the table by a string passed over a frictionless pulley. The 4.5-kg mass is 1 m from the floor. What speed does the 4.5-kg mass have just before hitting the floor?

Answer: v = 3.67 m/s

150 people per minute. The average weight of a person is 150 lb, and the motor is 80% efficient. (a) What must the power output of the motor be in horsepower? (b) What is the power input in watts?

Answer: (a) 17.05 hp
(b) 12.7 kW

32. What is the power output of a 2400-lb race car that can go from 80 mph to 150 mph in 9 s? Neglect air resistance and rolling friction.

Answer: 262.4 hp

*33. A waterfall 70 ft high has 1.5×10^5 gal of water flowing over it every minute. (a) If a gallon of water weighs 8.3 lb, what is the output of the waterfall in horsepower? (b) If the waterfall's power is converted into electricity with an efficiency of 60%, how many watts can it generate?

Answer: (a) 2636 hp
(b) 1.18 MW

*34. A 1500-kg car moving at 100 km/h on a level road requires 40 hp. (a) How much resistance does the car meet at this speed? (b) How much power is needed if the car must climb a 10° grade at this speed?

Answer: (a) 1073 N
(b) 135 hp

35. If electricity costs 11¢ per kilowatt-hour, how much does it cost to light a 300-W halogen bulb for 6 h?

Answer: Cost = 20¢

Force is defined as the change in momentum over time, and it is this change that gives a reaction motor (rocket) its thrust.

Momentum

At the end of this chapter, the reader should be able to:

1. Associate (and solve for) force and **momentum** through the definition of an **impulse.**

2. Find horizontal and vertical components of a **momentum** vector.

3. Use the concept of conservation of momentum to sum all relevant components of momentum and solve for unknowns in both one and two dimensions.

4. State the definition of *mass flow,* and apply it to problems that use the concept of *thrust.*

5. Explain the difference between **inelastic** and **elastic collisions.**

6. Use information given to you (whether velocities or heights) to find the **coefficient of restitution** of an object.

The concepts we have developed so far (force, work, and energy) are often used and understood in ways that are far different from the definitions given in this book. We often find ourselves using these words in our daily speech, such as, "I have a lot of *energy* this morning and am wide awake"; or, "I'd like to take the day off but unfortunately must go to *work*"; or, "I'll *force* them to see things my way." These uses of the concepts of force, work, and energy are not as precise as the definitions found in a physics text, but they do show that we have a somewhat intuitive feel for the concepts. The terms *work* and *energy* are often used synonymously, and, in fact, both derive from how a force acts over distance. So, conversely, we can use the concept of force to help us define work and energy.

5.1   Linear Momentum

Let's take our original, simple definition of *work* and start from there to redefine a force in terms of work and energy.

$$W = Fx$$

In the presence of a constant force, if the force moves an object a small amount, Δx, we do a small amount of work on it, ΔW.

$$\Delta W = F \Delta x$$

We rearrange the equation to solve for force.

$$F = \frac{\Delta W}{\Delta x} = \frac{\Delta E}{\Delta x}$$

We see from the above that a force is actually the way the energy content of an object changes in space. If the energy of an object changes as a function of displacement, a force is acting on it. However, this chapter concerns momentum. Although the concepts of work and energy are somewhat intuitive, the concept of momentum is a bit more difficult! For example, if we're playing a game of cards, perhaps "Twenty-One," and we begin to win consistently, we might say, "I'm on a roll and must keep my momentum up!" This remark implies that something is moving and we want to keep it moving.

Momentum is linked to Newton's first law, which we can restate in these words: *A body at rest tends to remain at rest, and a body in motion tends to remain in motion.* In other words, inertia has something to do with this tendency for a body to remain in motion. However, if inertia is the resistance to a change in motion, a body can possess inertia without motion. Since momentum is linked to inertia *and* motion, then momentum includes both the mass *and* the velocity of a body. We know that we need a force to change a body's state of motion, so we can also define *force* in terms of momentum. *Force* is the rate of change of momentum over time. The normally accepted variable for momentum is *p,* so a force is redefined as

(Equation 5.1)

$$F = \frac{\Delta p}{\Delta t}$$

We now see that force has been redefined in terms of momentum. Up to now we have accepted the idea that force is more fundamental than work, energy, or momentum. Yet we find that a force is actually itself defined by these quantities! Thus, the quantities that define force must be more fundamental. The fundamental quantities that define the context of an object's motion are space and time. We built upon this notion in Chapter 1 in defining the concepts of velocity and acceleration and in deriving the equations of uniformly accelerated motion. Mass is another fundamental quantity. The greater a body's mass, the more reluctant it is to move. A body possesses mass. A body can also possess energy and momentum, but not force. A force acts upon a body, but a body does not possess it. It is *external* to the body.

*The mathematical definition of **momentum** is*

$$\mathbf{p} = m\mathbf{v}$$

Momentum: The product of the mass and velocity of a body. If the momentum of a body changes, a force has acted on that body.

(Equation 5.2)

A body can possess both mass and velocity and can therefore possess momentum. Momentum is a vector quantity. Whereas energy is a scalar and has magnitude only, momentum has both magnitude and direction. Therefore, we must treat every problem concerning momentum as a vector problem, as illustrated in the following example.

EXAMPLE 5.1

Two tennis balls are hurtling toward each other. Each ball has the same speed and mass. What is the total momentum of the two balls?

SOLUTION
The total momentum is the sum of the momentum of each ball.

$$\mathbf{p}_T = \mathbf{p}_1 + \mathbf{p}_2$$

Both balls are moving along the same line, but one is moving in the opposite direction from the other, so one of the velocities will have a negative sign.

$$\mathbf{p}_T = m_1 v_1 + m_2(-v_2) \qquad m_1 = m_2 = m \qquad v_1 = v_2$$

$$\mathbf{p}_T = m(-v_2) + mv_2$$
$$\mathbf{p}_T = m(v_2 - v_2)$$
$$\mathbf{p}_T = 0$$

The total momentum is zero despite the fact that each ball has a momentum of its own. This zero value arises because of the vector nature of momentum: Because the two balls are moving in opposite directions, they cancel each other out.

5.2   Impulse

If we apply a uniform force to an object for a short period of time, we will change the state of the object's motion, for example, a bat hitting a baseball, or a tennis racket hitting a tennis ball, or a golf club driving a golf ball downrange. In each case, the ball accelerates during the impulse.

The impulsive force of a racket on a tennis ball deforms the ball momentarily.

What we actually do by giving each ball an impulse is to change its momentum. Because the mass of each ball remains the same during the impulse, this change in momentum results in an acceleration. It's actually another way to state Newton's second law. We start with the definition of force as a change in momentum with time.

$$F = \frac{\Delta p}{\Delta t}$$

$$p = mv$$

$$F = \frac{\Delta(mv)}{\Delta t}$$

where $\Delta(mv)$ means that we can change either variable (mass or velocity) inside the brackets. Since the mass of the ball stays constant in each case, the force equation becomes

$$F = \frac{m\,\Delta v}{\Delta t}$$

A change in velocity over a change of time is an acceleration, so the above equation is actually a restatement of Newton's second law, $\mathbf{F} = m\mathbf{a}$. However, we plan to use the above equation to give us a mathematical definition of impulse. *An **impulse** is generally regarded as a large force that acts over a very short time,* such as a baseball bat hitting a baseball or a golf club driving a golf ball.

Impulse: A large force acting on a body over a very short period of time.

(Equation 5.3)

$$F\,\Delta t = m\,\Delta v$$

EXAMPLE 5.2

A golfer tees off and drives his golf ball downrange at a speed of 75 m/s. If the head of the club was in contact with the ball for $\frac{3}{4}$ ms, what was the average force exerted on the ball? The ball has a mass of 46 g.

SOLUTION

$$F \, \Delta t = m \, \Delta v$$

$$F = \frac{m \, \Delta v}{\Delta t}$$

$$F = \frac{(0.046 \text{ kg})(75 \text{ m/s} - 0)}{0.75 \times 10^{-3} \text{ s}}$$

$$F = 4600 \text{ N}$$

This is over 1000 lb! According to Newton's third law, for every action there must be an equal and opposite reaction. Therefore, the golf club feels a recoil force of over 1000 lb. The club does flex when it hits the ball, but it springs back to its original shape. If it weren't for the fact that the impulse acts for only a very short time, the club would be permanently bent.

5.3 Conservation of Momentum

In Chapter 4, we spoke of the conservation of energy. We know that energy can be dissipated by friction. We can always account for where the energy went, but kinetic energy is not always conserved. Momentum, on the other hand, is always conserved between objects free to move. Assume that we have two objects hurtling toward each other as shown in Figure 5.1. The system is isolated; that is, no external forces are acting on it.

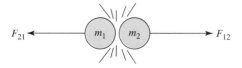

Figure 5.1 Two Masses in Collision

    A force can be defined as the rate of change in momentum over time. If F_{ext} is the external force acting on the system, then

$$F_{\text{ext}} = \frac{\Delta p}{\Delta t}$$

$$F_{\text{ext}} = 0$$

$$\frac{\Delta p}{\Delta t} = 0$$

$$p = \text{constant}$$

Something that is constant doesn't change over time. Therefore, its rate of change over time must be zero. If the external forces are zero and *force* is defined as the rate of change of momentum over time, then the rate of change of momentum over time must be zero. Thus, the change of momentum must be zero, or the momentum must be constant. The momentum before the collision must be equal to the momentum after the collision. But what happens during the collision, when forces are generated by the impulse of the bodies upon one another? The small time of contact is the same for each ball. Therefore, during the time of impulse, the forces generated are equal and opposite according to Newton's third law.

$$F_{21}\,\Delta t = -F_{12}\,\Delta t$$

The impulse is equal to the change in momentum, Δp.

$$\Delta p_{21} = -\Delta p_{12}$$

$$\Delta p_{21} + \Delta p_{12} = 0$$

The total change of momentum during the collision is also equal to zero, and the system's momentum is the same before, during, and after the collision. *Momentum is always conserved.* We don't have to know whether the balls stuck together or rebounded apart after the collision. Nor do we have to know how the force changed with time during the collision. The argument for the conservation of momentum is independent of the requirement for such detailed knowledge.

EXAMPLE 5.3

A 40-kg girl tosses a 6-kg pack at 4 m/s to a 55-kg boy, who catches it (see Figure 5.2). Both the boy and the girl are wearing ice skates and are standing on the frictionless surface of a frozen-over pond. Initially, neither one is moving. After the boy catches the pack, at what relative velocity do the boy and the girl move apart?

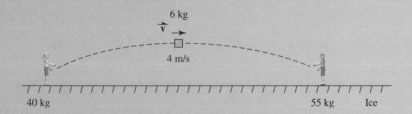

Figure 5.2 Boy and Girl Standing on Frozen Pond in Skates

6 kg
$\vec{v}$
4 m/s

40 kg   55 kg   Ice

SOLUTION
Before the girl tosses the pack, she is motionless, and thus her initial momentum is zero. When she tosses the pack, the total momentum of both the girl and the pack must still be zero if momentum is conserved.

$$\mathbf{p}_i = 0$$

Therefore,
$$\mathbf{p}_f = 0$$

$$\mathbf{p}_f = \mathbf{p}_i$$

$$\mathbf{p}_f = m_g\mathbf{v}_g + m_p\mathbf{v}_p$$

$$m_g\mathbf{v}_g + m_p\mathbf{v}_p = 0$$

$$m_g\mathbf{v}_g = -m_p\mathbf{v}_p$$

As the above equation illustrates, the girl and the pack are moving in opposite directions.

$$\mathbf{v}_g = -\frac{m_p}{m_g}\mathbf{v}_p$$

$$\mathbf{v}_g = -\frac{6\,\text{kg}}{40\,\text{kg}}(4\,\text{m/s})$$

$$\mathbf{v}_g = 0.6\,\text{m/s}\,(\text{to the left})$$

As shown in Figure 5.2, the pack is moving to the right. Therefore, the girl moves to the left at a velocity of 0.6 m/s as a reaction to her tossing the pack to the boy.

This problem has two acts. The second act is the boy catching the pack. For the second act, the initial momentum is the pack's momentum, since the boy is initially motionless.

$$\mathbf{p}_i = m_p\mathbf{v}_p$$

The final momentum is equal to the product of the boy's velocity times the sum of the mass of the boy and the pack.

$$\mathbf{p}_f = (m_p + m_b)\mathbf{v}_b$$

$$\mathbf{p}_f = \mathbf{p}_i$$

$$(m_p + m_b)\mathbf{v}_b = m_p\mathbf{v}_p$$

$$\mathbf{v}_b = \frac{m_p}{m_p + m_b}\mathbf{v}_p$$

Note from the above equation that the boy's velocity is in the same direction as the pack's.

$$\mathbf{v}_b = \frac{6\,\text{kg}}{6\,\text{kg} + 55\,\text{kg}}(4\,\text{m/s})$$

$$\mathbf{v}_b = 0.39\,\text{m/s}\,(\text{to the right})$$

The girl is moving to the left at 0.6 m/s as a result of her having tossed the pack, and the boy is moving to the right at 0.39 m/s as a result of his having caught the pack. Therefore, their relative velocity is (-0.6 m/s $-$ 0.39 m/s), or 0.99 m/s away from each other.

5.4   Thrust

In Example 5.3, we were actually talking about thrust, even though it was disguised as the reaction from throwing a pack. According to conservation of momentum, when the girl tossed the pack, if the pack was tossed to the right, the girl would move to the left. If she started to move, a force was acting on her, and that force came as a sudden impulse. Now assume that instead of throwing a single large object, she threw in succession many small objects. If the pack were loaded with many small stones of roughly the same size, and if she threw them one after the other at small, regularly spaced intervals, she would experience a relatively constant force during that time interval. If the timing sped up, and if the stones became infinitely small, say, as small as molecules, then we would have a mass flow, as when water flows out of a hose. Water appears to flow continuously from the hose because we can't see that water is actually made up of separate H_2O molecules spaced very closely together exiting the hose at some velocity. However, if the flow from the hose is fast enough, we feel a thrust from the water pushing back on the nozzle. So we can see that the impulse the girl receives from throwing the pack and the thrust from the nozzle of a garden hose are the same thing. The only real difference is that with impulse we have a discrete event, and with flow a continuous one.

The derivation of the equation for thrust emphasizes this principle. We again start with the definition of force in terms of a change of momentum in time.

$$F = \frac{\Delta p}{\Delta t}$$

The delta (Δ) symbol stands for a relatively large change, but something that appears continuous (such as the flow of water out of a hose) is made of many, very small changes blending together in such a way as to appear seamless. In this case, a d is used to show that each change is almost infinitesimally small.

$$F = \frac{dp}{dt}$$

If the momentum is changing with time, and if there are two variables (mass and velocity) in the equation for momentum, then it's possible for either variable to change.

$$F = \frac{d(mv)}{dt}$$

In the case of a rocket or a water hose, mass is constantly being ejected out the nozzle. We therefore have a mass flow, or a change of mass with time. The mass often exits with a constant exhaust velocity. Our force equation then becomes

(Equation 5.4)

$$F = v\frac{dm}{dt}$$

The above equation is the *thrust equation,* where v is the exhaust velocity, dm/dt is the mass flow rate, and F is the amount of thrust produced.

A rocket is probably the most common item that comes to mind when we think of a device based on the principle of thrust. A rocket can be used in the vacuum of space. It doesn't need anything to push against to generate its thrust. It comes solely from the conservation of momentum via the mass that it's ejecting out the nozzle. A common misconception is that the exhaust velocity is an upper limit to the maximum speed that the rocket can attain. The exhaust velocity is relative to the rocket, and this relationship doesn't change, regardless of the rocket's speed relative to the observer. The rocket's maximum speed is a function of the time that it accelerates ($v = at$), and its acceleration is a function of Newton's second law ($\mathbf{F} = m\mathbf{a}$). As long as the thrust continues, the rocket's speed continues to increase.

Thrust is directly proportional to the exhaust velocity and mass flow, and in most rockets today, that exhaust velocity is determined by a chemical reaction involving combustion. In general, the hotter the burn, the greater the thrust. However, a rocket does not necessarily have to be a chemical rocket. A water rocket is a simple toy in which water under pressure provides the thrust, just as in a hose nozzle. There is no chemical reaction providing the thrust. NASA is currently designing the next generation of rocket engines for interplanetary travel; these engines won't have to rely on a burn to provide thrust. Instead, an electric field accelerates an ionized substance to simultaneously provide both exhaust velocity and mass flow. An ion is an atom or a molecule with a positive electric charge. This net electric charge is the handle that the electric field uses to grab the ion and cause it to accelerate. Such an engine provides low thrust, but it can be run continuously. Using these engines in interplanetary space, travel could be accomplished through low accelerations (and decelerations) over very long time periods instead of through high accelerations over a short time as in conventional chemical rockets. These engines would be used in traveling from, say, an earth orbit to a Mars orbit. They wouldn't be used for traveling from the surface of the earth to orbit the earth because they wouldn't have enough thrust to counter the force of gravity. The currently more powerful chemical rockets would still be used for that job.

Another possible surface-to-orbit rocket engine is the nuclear rocket. A small, nuclear fission reactor would be the heat source, and the reactive material could be something as common as water converted to superheated steam. Much engineering has already been devoted to this idea, but the drawbacks have kept such an engine out of production. Most current rocket boosters are used once and then discarded. No one wants a highly radioactive fission reactor falling back to the earth in the form of space junk! Even with the reusable engines of the shuttle, the risk of an accident is too great.

EXAMPLE 5.4

A rocket well out in space performs a long burn. While the rocket's motor is running, it pumps 28 kg/s of fuel, with an exhaust velocity of 3200 m/s. If the initial mass of the rocket is 15,000 kg, and if the burn lasts 180 s, what are (a) the initial and (b) the final accelerations?

SOLUTION

(a)
$$F = v\frac{dm}{dt}$$

$$F = (3200 \text{ m/s})(28 \text{ kg/s})$$

$$F = 89{,}600 \text{ N}$$

$$a_i = \frac{F}{m_0}$$

$$a_i = \frac{89{,}600 \text{ N}}{15{,}000 \text{ kg}}$$

$$a_i = 5.97 \text{ m/s}^2$$

(b)

$$a_f = \frac{F}{m_f}$$

$$m_f = 15{,}000 \text{ kg} - (28 \text{ kg/s})(180 \text{ s})$$

$$m_f = 9960 \text{ kg}$$

$$a_f = \frac{89{,}600 \text{ N}}{9960 \text{ kg}}$$

$$a_f = 9 \text{ m/s}^2$$

Since the rocket is constantly losing mass because of the burning of fuel, the acceleration increases. If we wanted to find the final velocity of the rocket after the burn, we would normally use the equation $v_f - v_i + at$. However, since the acceleration isn't constant but is smoothly and continuously changing, we would have to use the average acceleration a_{avg} to calculate the final velocity.

5.5   Inelastic Collisions

In a collision, two or more objects meet in such a way that the energy and the momentum of each change. The objects exert a mutual force on each other for a very short period of time in the manner of an impulse. We've already seen that this momentary force can be very high. In the case of a car with a tree, it can be downright destructive! Examples of typical collisions are a bat with a ball, a basketball with a backboard, two molecules in a gas, subatomic particles in an accelerator, a meteorite with a planet, billiard balls on a pool table, or a pie in the face. In each case, the mutual interaction is over a small time period. Since we usually think of a collision as occurring quickly with a high impulsive force, it would be natural to assume that the event could be effectively dealt with using the concept of momentum, particularly conservation of momentum. This is indeed the case. In *all* collisions involving freely moving bodies, the law of conservation of momentum holds. The momentum of each object involved in a collision will change, but the overall momentum of the system will stay constant. Kinetic energy is involved too. Movement implies kinetic energy as well as momentum. However, kinetic energy is not always conserved. *In an **inelastic collision** (the type we're discussing in this section), kinetic energy isn't conserved.*

Inelastic collision: A collision in which kinetic energy is not conserved.

In Chapter 4, "Work and Energy," we discovered that we could always account for where all of the energy of a system went, but that energy wasn't always conserved. Energy could be dissipated, for instance, as heat generated from friction. This heat energy doesn't reappear as kinetic energy after the impulsive force of the collision. Indeed, in certain collisions, all of the kinetic energy of the system could disappear after the event!

EXAMPLE 5.5

Two balls of clay moving in opposite directions with identical velocities (v_1 and v_2) collide and stick together (see Figure 5.3). (a) What is the momentum of each ball before the collision? (b) What is the momentum of the system before the collision? (c) What is the velocity of the combined mass after the collision?

(d) What is the kinetic energy change for the system? (e) Where did that kinetic energy go?

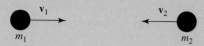

Figure 5.3

SOLUTION

(a)
$$\mathbf{v}_1 = -\mathbf{v}_2 \qquad m_1 = m_2$$

$$p_1 = m_1\mathbf{v}_1 \quad \text{(momentum of ball 1)}$$

$$p_2 = m_2\mathbf{v}_2 \qquad p_2 = m_1(-\mathbf{v}_1)$$

$$p_2 = -m_1\mathbf{v}_1 \quad \text{(momentum of ball 2)}$$

(b)
$$p_T = p_1 + p_2 \qquad p_T = m_1\mathbf{v}_1 + (-m_1\mathbf{v}_1)$$

$$p_T = 0$$

Each ball has a well-defined momentum, but the total momentum of the system is zero.

(c)
$$p_i = p_f \qquad p_i = 0 \qquad \text{Therefore, } p_f = 0.$$

$$p_f = m_f\mathbf{v}_f \qquad m_f = m_1 + m_2 \qquad p_f = (m_1 + m_2)\mathbf{v}_f$$

$$(m_1 + m_2)\mathbf{v}_f = 0 \qquad 2m_1\mathbf{v}_f = 0$$

$$2m_1 \neq 0 \qquad \text{Therefore, } \mathbf{v}_f = 0.$$

(d)
$$\text{KE}_i = \frac{1}{2}m_1\mathbf{v}_1^2 + \frac{1}{2}m_2\mathbf{v}_2^2 \qquad \text{KE}_i = \frac{1}{2}m_1\mathbf{v}_1^2 + m_1(-\mathbf{v}_1)^2$$

$$\text{KE}_i = \frac{1}{2}m_1\mathbf{v}_1^2 + \frac{1}{2}m_1\mathbf{v}_1^2 \qquad \text{KE}_i = m_1\mathbf{v}_1^2$$

$$\text{KE}_f = \frac{1}{2}(m_1 + m_2)\mathbf{v}_f^2 \qquad \text{KE}_f = \frac{1}{2}(m_1 + m_1)\mathbf{v}_f^2 \qquad \text{KE}_f = m_1\mathbf{v}_f^2$$

$$\mathbf{v}_f = 0 \qquad \text{KE}_f = m_1(0) \qquad \text{KE}_f = 0$$

$$\Delta\text{KE} = E_i - E_f \qquad \Delta\text{KE} = m_1\mathbf{v}_1^2 - 0$$

$$\Delta\text{KE} = m_1\mathbf{v}_1^2$$

(e) The lost kinetic energy went into heat. None of the initial kinetic energy of the system reappeared after the collision as kinetic energy. It had to go somewhere, however. The initial kinetic energy reappeared after the collision as heat in the amount of $m_1\mathbf{v}_1^2$. The clay balls warmed up as a result of the collision. Later on when we study thermodynamics, we'll find that on a macroscopic level, heat and temperature are actually the kinetic energy and velocity of molecules and atoms at a microscopic level.

5.6   Elastic Collisions

Elastic collision: A collision in which kinetic energy is conserved.

Elastic collisions are those in which both momentum and kinetic energy are conserved. As mentioned in Example 5.5, temperature is actually a collision process involving the atoms and molecules in a gas. When molecules and atoms collide, the collisions are elastic. The macroscopic temperature is the average kinetic energy of the molecules that make up the gases of the atmosphere. The molecules collide with each other, and since the collisions are elastic, no energy is lost. Thus, the process can go on almost indefinitely. If, for instance, billiard balls on a pool table made perfectly elastic collisions with each other and the bumpers on the side of the table, then when someone took a shot, the action on the table would never die down! As anyone who plays pool knows, eventually all motion comes to a stop. With each collision, a bit of kinetic energy is lost. Perfectly elastic collisions are relatively rare macroscopically in that most collisions, like the billiard balls, lose a little kinetic energy with each interaction. However, billiard balls approximate elastic collisions closely enough that they are often used as a model of elastic collisions. But a billiard ball rolls, and some of the kinetic energy after a collision goes into rotational kinetic energy.

A good model for elastic collisions is a popular physics toy that has six steel balls, all with the same mass, each suspended by two threads hung in a row, as shown in Figure 5.4. Collisions between the steel balls happen in a straight line, and all of the kinetic energy of the collisions is kinetic energy of motion ($\frac{1}{2}mv^2$). Collisions between steel balls are elastic, and that is what makes the apparatus in Figure 5.4 such an interesting toy. If, as in Case (A), one ball from the left is collided with the remaining balls as shown, only one ball on the right will fly off to the right. If, as in Case (B), two balls from the left are collided with the remaining balls, only two balls from the right will fly off together to the right. When the collisions are analyzed, it can be shown that the system's behavior can be explained only by the conservation of both momentum *and* kinetic energy. That makes the collisions, by definition, elastic.

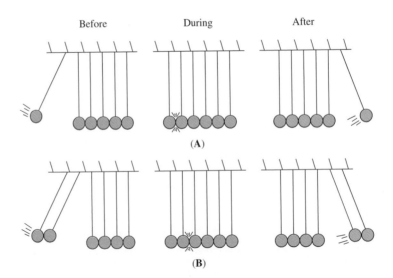

Figure 5.4 In-Line Pendulums Physics Toy

Assume, as in Case (A), that one ball from the left is allowed to collide with the five remaining balls. Momentum must be conserved, and that tells us that it would be okay if two balls from the right moved to the right with one-half of the left ball's velocity.

$$p_i = p_f$$

$$mv = m\left(\frac{1}{2}v\right) + m\left(\frac{1}{2}v\right)$$

$$mv = mv$$

Momentum is conserved with two balls moving off to the right, but what about the kinetic energy?

$$E_i = \frac{1}{2}mv^2$$

$$E_f = \frac{1}{2}m\left(\frac{v}{2}\right)^2 + \frac{1}{2}m\left(\frac{v}{2}\right)^2$$

$$E_f = \frac{1}{4}mv^2$$

Kinetic energy isn't conserved. However, the system didn't function with two balls moving off to the right, but instead with only one ball moving off to the right. In addition, when you observe the height to which the right ball rises, you will see that it rises to the same height to which the left ball was raised to initiate the collision process in the first place. Each suspended ball is a simple pendulum, so gravitational potential energy is converted into kinetic energy of motion when the collision occurs. After the collision, kinetic energy of motion is reconverted into gravitational potential energy (mgh). If both the left and the right balls reach the same height, they have the same energy. Energy is conserved! If both momentum and kinetic energy are conserved, the collision is elastic. With only one ball participating in the collision from the left and from the right, analysis yields the following results:

$$p_i = p_f$$

$$mv = mv$$

Conservation of momentum says that both the left and the right balls must have the same velocity, just as they have the same mass. Therefore,

$$E_i = \frac{1}{2}mv^2$$

$$E_f = \frac{1}{2}mv^2$$

Kinetic energy is also conserved with only one ball participating from the left and from the right. Prove it to yourself with a mathematical analysis similar to the one we just did.

5.7   Impedance

Impedance: An abrupt change from one medium to another involving energy transfer that can cause loss of energy through a rebound or a reflection.

In examining elastic collisions, we are also looking at energy transfer. Energy transfer occurs in many other types of physical processes, such as wave motion, ac and dc electric circuits, and optical systems. If we are interested in energy transfer, whether we wish to maximize it or minimize it, we are actually concerned with *impedance* and *impedance matching*. *Impedance is defined as energy transfer from one system to another.*

Most electric power is carried as alternating current (ac), and efficient energy transfer is very important. The most efficient way of transferring electric power over long distances is to use high voltage and low current. At an electrical power generating plant, voltage is usually stepped up to voltages as high as 200,000 V to be placed on high-voltage transmission lines. However, these voltages are much too high for use in the home and must be stepped down in a number of stages before reaching homes. At each stage, energy can be lost, and losses can show up as reflections down the line and excess heat that must be dissipated. Consider an optical system with multiple lenses. If we wish to maximize the amount of light entering the lenses, then we wish to eliminate reflection at the lens faces, for light would be lost with each reflection. This system is a wholly different one from collisions between steel balls and ac energy transfer, but it is also a process in which impedance matching is important. In such cases, we wish to maximize energy transfer and minimize losses. These new concepts about impedance and impedance matching with elastic collisions will be important in many other areas of physics radically different from colliding steel balls!

Consider a head-on elastic collision between two objects. Both momentum and energy are conserved, and both objects continue on the same straight line after the collision. What we want to know is the relationship between the relative approach velocity before the collision and the relative separation velocity after the collision. For momentum conservation, we can write

$$m_1 v_{1i} + m_2 v_{2i} = m_1 v_{1f} + m_2 v_{2f}$$

We can rewrite the above equation in the following way:

$$m_1 v_{1f} - m_1 v_{1i} = m_2 v_{2i} - m_2 v_{2f}$$
$$m_1(v_{1f} - v_{1i}) = m_2(v_{2i} - v_{2f})*$$

For conservation of energy, we can write

$$\frac{1}{2}m_1 v_{1i}^2 + \frac{1}{2}m_2 v_{2i}^2 = \frac{1}{2}m_1 v_{1f}^2 + \frac{1}{2}m_2 v_{2f}^2$$

It can also be rewritten as

$$m_1 v_{1i}^2 + m_2 v_{2i}^2 = m_1 v_{1f}^2 + m_2 v_{2f}^2$$
$$m_1 v_{1f}^2 - m_1 v_{1i}^2 = m_2 v_{2i}^2 - m_2 v_{2f}^2$$
$$m_1(v_{1f}^2 - v_{1i}^2) = m_2(v_{2i}^2 - v_{2f}^2)*$$

We can then divide the two equations with asterisks (*) by one another.

$$\frac{m_1(v_{1f}^2 - v_{1i}^2)}{m_1(v_{1f} - v_{1i})} = \frac{m_2(v_{2i}^2 - v_{2f}^2)}{m_2(v_{2i} - v_{2f})}$$

The masses cancel.

$$\frac{(v_{1f}^2 - v_{1i}^2)}{(v_{1f} - v_{1i})} = \frac{(v_{2i}^2 - v_{2f}^2)}{(v_{2i} - v_{2f})}$$

The numerator of each fraction on each side of the equal sign can be expanded.

$$(v_{1f}^2 - v_{1i}^2) = (v_{1f} - v_{1i})(v_{1f} + v_{1i})$$

$$(v_{2i}^2 - v_{2f}^2) = (v_{2i} - v_{2f})(v_{2i} + v_{2f})$$

This gives us the expanded equation

$$\frac{(v_{1f} - v_{1i})(v_{1f} + v_{1i})}{(v_{1f} - v_{1i})} = \frac{(v_{2i} - v_{2f})(v_{2i} + v_{2f})}{(v_{2i} - v_{2f})}$$

Canceling like terms on each side, we get

$$(v_{1f} + v_{1i}) = (v_{2i} + v_{2f})$$

Rearranging in terms of initial and final relative velocities yields

$$v_{2i} - v_{1i} = v_{1f} - v_{2f} \quad \text{or} \quad v_{2i} - v_{1i} = -(v_{2f} - v_{1f})$$

So what have we accomplished with all of this complex algebraic manipulation? *The above equation tells us that in an elastic one-dimensional collision, the relative velocity of approach before collision is equal to the relative velocity of separation after collision.* We can use this fact in trying to form an intuitive mental picture of energy transfer between participants in a collision. Assume that we again have an elastic one-dimensional collision between two objects, but that the second object is stationary relative to the first $(v_{2i} = 0)$. Momentum conservation gives us the following:

$$m_1 v_{1i} = m_1 v_{1f} + m_2 v_{2f}$$

The argument above about the equivalence of the relative velocity of approach versus separation simplifies to give us

$$v_{1i} = v_{2f} - v_{1f}$$

If we now solve the momentum equation for the final velocity of mass m_1, we get

$$m_1 v_{1f} = m_1 v_{1i} - m_2 v_{2f}$$

We can rearrange the relative velocity equation to solve for v_{2f}.

$$v_{2f} = v_{1i} + v_{1f}$$

We insert this into the momentum equation.

$$m_1 v_{1f} = m_1 v_{1i} - m_2(v_{1i} + v_{1f})$$

$$m_1 v_{1f} = m_1 v_{1i} - m_2 v_{1i} - m_2 v_{1f}$$

$$m_1 v_{1f} + m_2 v_{1f} = m_1 v_{1i} - m_2 v_{1i}$$

$$(m_1 + m_2)v_{1f} = (m_1 - m_2)v_{1i}$$

$$v_{1f} = \frac{m_1 - m_2}{m_1 + m_2} v_{1i}$$

The above equation tells us that if m_1 equals m_2, the final velocity of ball 1 is zero. Ball 1 stops dead, and none of the kinetic energy before the collision is carried off by the first ball. All of the kinetic energy must be carried off by the second ball. This is optimal energy transfer, with no rebound (reflection).

Now contemplate what the above equation tells us if masses m_1 and m_2 are unequal. Since v_{1f} is nonzero, the total kinetic energy after the collision is shared by both masses. If m_2 is greater than m_1, then v_{1f} is negative. Mass 1 rebounds in the opposite direction. If m_1 is greater than m_2, then both m_1 and m_2 move off in the same direction. Consider also the case in which m_1 is *very* much greater than m_2, or, in other words, that m_2 is *very* much smaller than m_1. In either case, the result is that $v_{1f} = v_{1i}$. That's no collision at all! We can theoretically reduce the mass of the second ball until it is

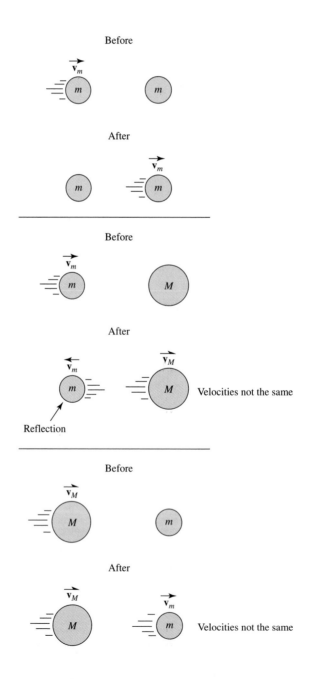

Figure 5.5 Impedance Matching (Reflection and Transmission) of Energy

infinitesimally small. In other words, the first ball doesn't strike a second ball but continues to the right with its original velocity. This fact may seem trivial to state, but it does show how expressive the mathematical modeling of a system can be if you only know how to read it. Such mathematical modeling works well at the extreme limits of the system (see Figure 5.5).

Now let's solve the momentum equation for m_2 and see what that implies for the collision.

$$m_1 v_{1i} = m_1 v_{1f} + m_2 v_{2f}$$

$$m_2 v_{2f} = m_1 v_{1i} - m_1 v_{1f}$$

We rearrange the relative velocity equation to solve for v_{1f}.

$$v_{1f} = v_{2f} - v_{1i}$$

We insert the above into the momentum equation.

$$m_2 v_{2f} = m_1 v_{1i} - m_1 (v_{2f} - v_{1i})$$

$$m_2 v_{2f} = m_1 v_{1i} - m_1 v_{2f} + m_1 v_{1i}$$

$$m_2 v_{2f} + m_1 v_{2f} = m_1 v_{1i} + m_1 v_{1i}$$

$$(m_1 + m_2) v_{2f} = 2 m_1 v_{1i}$$

$$v_{2f} = \frac{2 m_1}{m_1 + m_2} v_{1i}$$

The above equation tells us what happens if m_2 is much, much greater than m_1 ($m_2 \gg m_1$). As mass m_2 grows very large, v_{2f} approaches zero. This situation is similar to bouncing a golf ball against a concrete wall: When you throw the golf ball at the wall, the wall doesn't move. However, the golf ball rebounds back at you with all of the energy it had when it hit the wall. If we apply the same reasoning to the equation we derived for v_{1f} ($m_2 \gg m_1$), then we arrive at the result that $v_{1f} = -v_{1i}$. The golf ball rebounds from the wall in the opposite direction with the same speed. The wall has no energy transferred to it. If the wall doesn't move, it also has no momentum transferred to it. *Conservation of momentum doesn't apply if one of the objects involved is immobile.*

5.8 Coefficient of Restitution

In the equation for v_{2f} derived above, we arrived at the case of an object rebounding from a stationary object, such as a wall or a floor. The wall or floor doesn't have any energy transferred to it, and in a perfectly elastic collision, the colliding object rebounds in the opposite direction, carrying away all of the collision's kinetic energy. However, most macroscopic collisions are inelastic, and some kinetic energy is dissipated. If a ball is dropped from a height and reaches a velocity v_i just before it hits the floor, its velocity v_f as it rebounds from the floor is somewhat less, and the height to which it rebounds is a bit less than its original height. *How much less the velocity is after the collision than before the collision can be quantified in a term called the coefficient of restitution. The coefficient of restitution (e) is defined as the speed relative to the surface after the collision*

Coefficient of restitution (e): The ratio of the speed of an object just after a collision with an immobile surface to the speed just before the collision.

divided by the speed relative to the surface before the collision. If some of the kinetic energy is dissipated in the collision, the relative speed after the collision is less than the relative speed before the collision, and $e < 1$. We've already shown that if we have a perfectly elastic collision, the relative speed before the collision is equal to the relative speed after the collision, and $e = 1$. In certain cases of inelastic collisions, such as two soft clay balls colliding (as in Example 5.5), the relative velocity after the collision is zero, and therefore $e = 0$. The limits for the coefficient are $0 \le e \le 1$ (e is greater than or equal to 0 and less than or equal to 1).

EXAMPLE 5.6

A golf ball is dropped on the floor from a height of 1 m (see Figure 5.6). The ball rebounds back up to a height of 0.85 m. What is the coefficient of restitution?

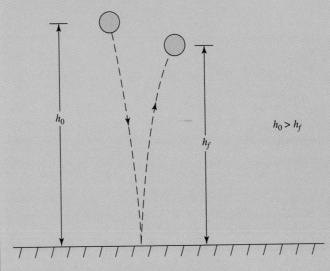

Figure 5.6 Loss of Energy during a Vertical Collision

SOLUTION

First we must find the relative speed of approach between ball and floor. From gravitational potential energy going into kinetic energy as the ball falls, we can write

$$v_i = \sqrt{2gh_0} \qquad v_f = \sqrt{2gh_f}$$

$$e = \frac{v_f}{v_i} \qquad e = \frac{\sqrt{2gh_f}}{\sqrt{2gh_0}}$$

$$e = \sqrt{\frac{2gh_f}{2gh_0}}$$

$$e = \sqrt{\frac{h_f}{h_0}}$$

(Equation 5.5)

$$e = \sqrt{\frac{0.85 \text{ m}}{1.0 \text{ m}}} \qquad e = 0.92$$

The coefficient of restitution can be defined in terms of speed, but from a practical standpoint, it's easier to measure the height to which a ball bounces than to measure its speed immediately after a collision.

As we have seen, when a collision occurs between a moving object and a stationary object (such as a wall or a floor), the immobile object has no kinetic energy transferred to it. Since the immobile object certainly has mass (it can be modeled as infinite), the complete lack of transferred energy implies that the object doesn't begin moving because of the collision, and no momentum is transferred to it. Thus, momentum isn't conserved in a collision between a moving object and a stationary one. There is a change in momentum. In fact, if $e = 1$, energy is conserved and momentum is not! This case comes about because energy is a scalar and momentum is a vector. If we throw an object against a wall, and if the coefficient of restitution between wall and object is 1, the rebound speed will be equal to the incident speed, or

$$v_i = -v_f$$

$$\text{KE} = \frac{1}{2}mv^2$$

$$\text{KE}_i = \frac{1}{2}mv_i^2$$

$$\text{KE}_f = \frac{1}{2}m(-v_f^2)$$

$$\text{KE}_f = \frac{1}{2}mv_f^2$$

Squaring a negative number gives a positive number.

$$\frac{1}{2}mv_i^2 = \frac{1}{2}mv_f^2$$

$$\text{KE}_i = \text{KE}_f$$

However, in the case of momentum, the negative sign for v_f doesn't go away. If momentum were conserved, the change in momentum would be zero ($\Delta p = 0$).

$$\Delta p = 0$$

$$\Delta p = p_i - p_f$$

$$p_i = mv_i$$

$$p_f = mv_f$$

$$\Delta p = mv_i - mv_f$$

$$\Delta p = m(v_i - v_f)$$

$$v_f = -v_i$$

$$\Delta p = m[v_i - (-v_i)]$$

$$\Delta p = m(2v_i)$$

$$\Delta p = 2mv_i$$

$$\Delta p = 2p_i$$

We see that the change in momentum isn't equal to zero; therefore, momentum isn't conserved in this case. We can see that the law of conservation of momentum doesn't apply in every case. (Is nothing sacred?) So where does the law apply, and where is it inapplicable? It applies in collisions where all of the involved bodies are free to move, with no external forces. It is inapplicable in cases that involve collisions with immobile objects, such as walls, floors, or ceilings, and in cases where there is an external force. An example of an external force is a golfer hitting a golf ball with a club. The golfer supplies a force to swing the club. A batter hitting a baseball also supplies force to swing the bat. From the equation for impulse, we get

$$F \, \Delta t = m \, \Delta v$$

$$F \, \Delta t = \Delta(mv)$$

$$F \, \Delta t = \Delta p$$

To be able to effectively use the concept of momentum, you must be aware of the context of the physical situation of the problem. The law of conservation of momentum is a powerful tool, but (like any other tool) it has its limits. We must work within those limits. The concept fails us when we try to apply it beyond its limits.

5.9   Conservation of Momentum in Two Dimensions

Momentum is a vetor and thus has a direction associated with it. Up to now in this chapter, all of the examples have been in one dimension, or along a straight-line path. The vector nature of momentum came in as a negative sign for velocity if a body was moving the other way. But what happens if the motion after a collision isn't along a straight line? Suppose that we have two pucks on an air table, which is like an air track except that we have an entire surface with an area that is essentially frictionless. What happens when these two pucks collide? If they collide head-on, the resultant motion of the two pucks will be along a straight line. But if the collision is off center, then the pucks will fly off at an angle, as in Figure 5.7. Both pucks 1 and 2 are free to move, so momentum is conserved.

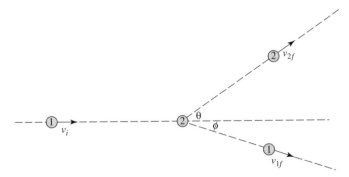

Figure 5.7 Conservation of Momentum in Two Dimensions

We can start describing the situation mathematically by writing equations for the conservation of momentum. But now, since there are two dimensions, we must write two equations for momentum conservation: one for the x axis and one for the y axis. Starting with the horizontal or x axis, we have

$$\mathbf{p}_{xi} = \mathbf{p}_{xf}$$

Before the collision, all of the momentum is concentrated in mass 1 moving along the x axis. After the collision, both masses 1 and 2 have x components of velocity.

$$m_1 v_i = m_1(v_{1f} \cos \phi) + m_2(v_{2f} \cos \theta)$$

$$m_1 v_i = m_1 v_{1f} \cos \phi + m_2 v_2 \cos \theta *$$

See Appendix A, Section A.2, "Simultaneous Equations."

We can now write a momentum conservation equation for the y axis. Initially, though, nothing is moving vertically, so the initial momentum along the y axis is zero.

$$\mathbf{p}_{yi} = \mathbf{p}_{yf}$$

$$\mathbf{p}_{yi} = 0$$

Therefore,

$$\mathbf{p}_{yf} = 0$$

Both masses 1 and 2 have y components of velocity.

$$m_1(v_{1f} \sin \phi) + m_2(v_{2f} \sin \theta) = 0$$

$$m_1 v_{1f} \sin \phi + m_2 v_{2f} \sin \theta = 0*$$

The equations followed by an asterisk (*) can now be solved together, because they both describe different aspects of the same situation. We have seven variables: two masses, two angles, and three velocities. If we have two equations, we can have only two unknowns if we expect to get an answer for the problem. We need to know five of the seven variables numerically, and we can solve for the other two. For instance, if we know the initial velocity, the masses, and the angles, we can solve for the final velocities.

$$m_1 v_i = m_1 v_{1f} \cos \phi + m_2 v_{2f} \cos \theta$$

$$m_1 v_{1f} \sin \phi + m_2 v_{2f} \sin \theta = 0$$

See Section A.2 in Appendix A if you need help solving simultaneous equations. But if the angles are both 0°, the equations above reduce to one equation which we have already solved for momentum conservation in one dimension. If you don't know whether or not mass 1 is going to rebound from mass 2, set up the equation as above. If you get a negative sign for v_{1f}, then you know that you chose wrong in thinking that mass 1 wouldn't rebound. The math is self-consistent, and the answer is correct regardless of whether you guessed wrong as to what went where. The negative sign means simply that mass 1 is going in the opposite direction from what you had guessed. The magnitude is correct.

Summary

1. The units of momentum (in the MKS system) are newton-seconds (N-s).

2. Momentum is a vector and therefore must be separated into x and y components.

3. **Momentum** within a system of freely moving particles is *always* conserved unless the system is acted upon by an outside force. In a collision with an immobile object, momentum is *not* conserved.

Problem Solving Tip 5.1: Remember the above statement! The conservation of momentum, along with the conservation of energy, can and will be a very useful tool in this chapter and future ones! When looking at a problem (in any chapter), ask yourself the following questions:

- Is there a collision or an explosion (a collision in reverse!) happening? If so, use conservation of momentum.

- Are energies changing form in the problem (for example, kinetic into gravitational potential, or spring potential into gravitational potential, and so on)? If so, try using conservation of energy.

- If the answer to both questions is yes, then use both conservation laws. It may look like a messy two-equation problem (an algebra problem with two unknowns), but it will probably be quicker than any other method.

  These first two questions can often save you more time (and heartache) than you can possibly believe!

Problem Solving Tip 5.2: Treat the conservation of momentum the same as you did with conservation of energy. Take a snapshot of momenta *just before* a collision, and set them equal to the snapshot of momenta *just after* the collision. Remember, momentum is a *vector,* and components (and sign convention) must be taken into account!

Key Concepts

COEFFICIENT OF RESTITUTION (*e*): The ratio of the speed of an object just after a collision with an immobile surface to the speed just before the collision.

ELASTIC COLLISION: A collision in which kinetic energy is conserved.

IMPEDANCE: An abrupt change from one medium to another involving energy transfer that can cause loss of energy through a rebound or a reflection.

IMPULSE: A large force acting on a body over a very short period of time.

INELASTIC COLLISION: A collision in which kinetic energy is not conserved.

MOMENTUM: The product of the mass and velocity of a body. If the momentum of a body changes, a force has acted on that body.

Important Equations

(5.1) $F = \dfrac{\Delta p}{\Delta t}$ (Redefinition of force as the rate of change of momentum)

(5.2) $\mathbf{p} = m\mathbf{v}$ [Mathematical definition of momentum (a vector quantity)]

(5.3) $F \Delta t = m \Delta v$ (Impulse)

(5.4) $F = v\dfrac{dm}{dt}$ (Thrust)

(5.5) $e = \sqrt{\dfrac{h_f}{h_0}}$ (Coefficient of restitution in terms of initial versus rebound height)

Conceptual Problems

1. An object at rest can have:
   a. velocity   b. momentum
   c. kinetic energy   d. potential energy

2. An object in motion doesn't necessarily have:
   a. velocity   b. momentum
   c. kinetic energy   d. potential energy

3. What doesn't change when two mobile objects collide with one another?
   a. The KE of each object
   b. The total KE of all of the objects
   c. The total momentum of all of the objects
   d. The momentum of each object

4. In an elastic collision:
   a. kinetic energy is conserved, but momentum isn't
   b. momentum is conserved, but kinetic energy isn't
   c. both kinetic energy and momentum are conserved
   d. neither kinetic energy nor momentum is conserved

5. In an inelastic collision:
   a. kinetic energy is conserved, but momentum isn't
   b. momentum is conserved, but kinetic energy isn't
   c. both kinetic energy and momentum are conserved
   d. neither kinetic energy nor momentum is conserved

6. In an elastic collision, the relative speed between two objects after the collision is:
   a. greater than before the collision
   b. less than before the collision
   c. the same as before the collision.

7. A ball strikes a wall elastically. If the ball's initial momentum is **p**, what is the change in the ball's momentum?
   a. **p**   b. 0   c. 2**p**   d. −2**p**

8. Object *A* collides with object *B* which is initially at rest. If maximum energy is transferred to *B*:
   a. mass *A* is greater than mass *B*
   b. mass *B* is greater than mass *A*
   c. both masses are the same

9. An open trailer is coasting along a frictionless road, and it starts raining. What happens to the trailer, and why?

10. What is the physical difference between firing a rifle loaded with a blank and firing a rifle loaded with a bullet?

11. If you were standing in the middle of a frozen pond on ice that has negligible friction and were holding a pair of stones, how would you get back to shore?

12. A ball of mass m moving to the right strikes a ball of mass M that is originally stationary. Mass M is greater than mass m. After the collision:

a. both balls move off to the right

b. both balls move off to the left

c. ball m moves off to the right, and ball M moves off to the left

d. ball m moves off to the left, and ball M moves off to the right

13. Air bags in autos help to prevent injuries. How do air bags work in terms of impulse?

14. In a collision between a bus and a car, why are the passengers in the bus less likely to be injured than the passengers in the car?

Exercises

Linear Momentum

---

*1. Water hits a cupped turbine blade as shown in Figure 5.8. The entrance and exit velocities of the water flow are the same, 20 ft/s. If the water is flowing at a rate of 200 gal/min, what force is exerted on the turbine blade? (One gallon weighs 8.3 lb.)

Answer: F = 34.6 lb

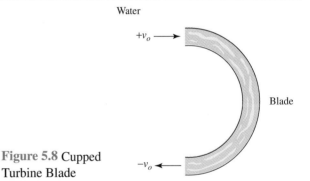

Water

$+v_o \longrightarrow$

Blade

$-v_o \longleftarrow$

Figure 5.8 Cupped Turbine Blade

Impulse

---

2. A 2000-lb car hits a tree at 30 mph and comes to rest in 0.3 s. Find (a) the initial momentum and (b) the average force on the car during the collision. (c) How many g's does the driver experience during the crash?

Answer: (a) 2750 slug-ft/s
(b) 9167 lb
(c) 4.6 g

3. A 40-ton airliner cruises at 590 mph. If its engines develop 76,000 lb of thrust, what is the minimum time it will take the jetliner to reach this speed using the full thrust of the engines? Solve using impulse, and ignore air friction.

Answer: 28.4 s

4. A 5-oz baseball is accidentally thrown into the windshield of a car moving in the opposite direction. The baseball has a speed of 45 ft/s, and the car has a speed of 25 mph. What is the average force on the windshield if the impact lasts 1 ms (1 ms = 0.001 s)?

Answer: 725 lb

*5. A 400-kg cannon on a wheeled carriage is at the foot of a 25° slope with respect to the horizontal. The gun fires a 25-kg shell at a muzzle velocity of 400 m/s in a direction away from the slope. (a) What is the cannon's initial speed up the slope? (b) How far along the slope does the cannon travel before it comes to rest and starts to roll back down? Disregard friction.

Answer: (a) 25 m/s
(b) 75 m

*6. A cannon has a barrel 8 m long and fires a 60-kg projectile with a muzzle velocity of 450 m/s. Use three different ways to find the average force (F_{avg}) on the projectile while it is in the barrel. (*Hint:* Use uniformly accelerated motion, work and energy, and momentum.)

Answer: $F_{avg} = 7.59 \times 10^5$ N

Conservation of Momentum

---

7. An astronaut weighing 125 kg is drifting toward a space shuttle at 0.2 m/s. A mission specialist in the shuttle tosses him a tool, which he catches. The tool weighs 3 kg and is tossed with a velocity of 10 m/s. After the astronaut has caught the tool, what is his velocity, and in what direction?

Answer: 0.04 m/s, away from the shuttle

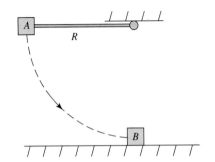

Figure 5.9 Swinging Magnet
Striking an Iron Cube

8. A space shuttle moving at 3.0 km/s, and almost in orbit, parts with its hydrogen/oxygen liquid fuel tank. The tank has 20% of the mass of the shuttle. Both tank and shuttle continue moving in the same direction. If the tank's speed is 1 km/s, what is the shuttle's speed?

Answer: 3.5 km/s

9. A skyrocket has a horizontal velocity of 300 m/s. It explodes and breaks into two parts. One part, which has a mass of 25% of the total weight, flies off in the opposite direction with a speed of 200 m/s. Find the speed and the direction of the other piece.

Answer: 467 m/s in the original direction

10. A 1-oz projectile strikes a 6-lb wooden block, becoming embedded in it. The block is suspended from a rope and rises in an arc to a height of 6 in. because of the collision. What was the speed of the projectile?

Answer: 543.6 ft/s

11. A 6-kg object moving at 5 m/s strikes a stationary 2-kg object and continues in the same direction at 3 m/s. What are the speed and the direction of the 2-kg object after the collision?

Answer: v_{2f} = 6 m/s in the original direction

*12. Magnet A is attached to a string of length R that is hung from a support as shown in Figure 5.9. The magnet is held in a position where the string is horizontal. The magnet is then released, and it swings down in an arc and strikes an iron cube of twice the magnet's mass. The cube sticks to the magnet, and both swing up. What is the maximum angle θ that the string will make with the vertical? Ignore friction.

Answer: θ = 27.3°

*13. A 30-g dart is thrown horizontally at a speed of 25 m/s at a 2-kg target supported from behind with a spring. The target recoils 2 cm. What is the spring constant?

Answer: k = 693 N/m

*14. A 600-g cart travels on an air track at a speed of 0.4 m/s. A 100-g projectile is carried on top of the cart and can be launched at a speed of 0.15 m/s. What is the final velocity of the cart if the projectile is launched in the same direction that the cart is moving?

Answer: v_f = 0.375 m/s

Thrust

15. Water flows out of a hose at a rate of 30 gal/min and at a speed of 20 ft/s, striking a car door that is being washed. What force does the water exert on the car door? One gallon of water weighs about 8.3 lb.

Answer: 2.6 lb

16. A 1.5-kg bucket sits atop a scale. Water is poured into the bucket from a height of 2.0 m at a rate of 200 g/s. (a) What does the scale read *after* water has been poured into the bucket for 1.5 min? (b) What does the scale read just *before* the pouring stops at 1.5 min?

Answer: (a) 19.5 kg
(b) 19.63 kg

17. A man in a rowboat accidentally drops his oars overboard. The bottom of the boat is filled with large stones used as ballast. To get the boat moving again toward the shore, he throws seven of these stones overboard toward the rear of the boat. The stones had an average weight of 4 oz, and the

man threw them with an average velocity of 60 ft/s. With what velocity did the boat reach the shore? The boat and all that it contained weighed 300 lb. Neglect water resistance and the loss of the mass from the stones thrown overboard.

Answer: 0.35 ft/s

18. A rocket motor is being static tested on the ground. It burns fuel at a rate of 5 kg/s with an exhaust velocity of 2500 m/s. The motor has a mass of 200 kg. If the motor is mounted so that its exhaust nozzle points down, what force does it exert on its restraints?

Answer: 10,540 N

19. A rocket (assumed weightless) launched from earth orbit ejects 20% of its mass as fuel with an exhaust velocity of 2500 m/s in the first 10 s of its flight. (a) What is the initial acceleration of the rocket? (b) What is its acceleration after 10 s?

Answer: (a) 50 m/s^2
(b) 62.5 m/s^2

20. What would the accelerations be for Exercise 19 if the rocket were launched from the earth's surface? Ignore variations in g with height.

Answer (a) 40.2 m/s^2
(b) 52.7 m/s^2

21. How high does the rocket in Exercise 20 climb in the first 10 s?

Answer: 2.3 km

Inelastic Collisions

23. A 40-kg boy running at 4 m/s jumps on a stationary 12-kg sled lying on the ice of a frozen-over lake. (a) How fast does the boy on the sled move initially? (b) How much kinetic energy was lost?

Answer: (a) 3.08 m/s
(b) 73.4 J

24. A 1500-kg car moving in the $+x$ direction at 100 km/h overtakes and collides with a truck of mass 2000 kg moving in the $-x$ direction at 60 km/h. The two vehicles stick together. (a) What are the direction and the initial speed of the wreckage? (b) How much kinetic energy was lost?

Answer: (a) 2.37 m/s; $+x$ direction
(b) 98.9%

22. At a coal-fired electric power station, coal falls onto a horizontal conveyer belt at a rate of 75 kg/s. The belt is moving at 3 m/s. What is the power needed to keep the belt moving at a constant speed?

Answer: $P = 337.5$ W

*25. A 0.5-kg piece of modeling clay is suspended 1 m above a floor. Another identical piece of clay is suspended 0.75 m above it. The upper piece is dropped, the falling piece hits the lower piece, and the two pieces stick together. The collision easily dislodges the lower piece, and the two pieces fall together to the floor. (a) How long does it take the two of them to reach the floor from the time of the collision? (b) Does the upper piece take more time or less time to reach the floor than if it had dropped without a collision? Prove it.

Answer: (a) $t = 0.297$ s
(b) More time

Elastic Collisions

*26. A neutron of mass 1.67×10^{-27} kg and speed 1.5×10^5 m/s collides head-on with a stationary deuteron of mass 3.34×10^{-27} kg. The deuteron, an atom of heavy hydrogen, has both a neutron and a proton in its nucleus. Since a proton and a neutron weigh almost the same, the deuteron weighs twice what the neutron weighs. The neutron has no electrical charge, and the proton has a positive charge. Collisions between atomic particles are essentially elastic. What are the final velocities of (a) the neutron and (b) the deuteron after the collision?

Answer: (a) $v_{nf} = -5 \times 10^4$ m/s
(b) $v_{df} = 1 \times 10^5$ m/s

*27. A 300-g air track cart moving at 0.5 m/s collides elastically with a 500-g cart initially at rest. The air track isn't friction-

less but has a coefficient of friction of 0.01 between the cart and the track. How far does the 500-g glider go up the track if the track slopes upward at an angle of 0.6°?

Answer: $s = 0.35$ m

*28. A ball of mass 2 m is launched straight up with a speed of v_0. Another ball of mass m is hung directly above the first ball from a light thread a height h above the floor. The lower ball collides with the upper ball in an elastic collision. Derive an expression for the height above the floor that the ball of mass m will rise as a function of v_0, h, and g.

Answer: $h_{\max} = \dfrac{8v_0^2 - 7gh}{9g}$

Coefficient of Restitution

29. A hard rubber ball is dropped on the floor from a height of 2 m and on its second bounce reaches a height of 0.75 m. What is the coefficient of restitution between the floor and the ball?

Answer: $e = 0.78$

*30. A ball is dropped onto a hard surface from a height of 1.5 m and rebounds to 70% of its original height. The average force that the ball exerted on the surface while it was in contact was 9 times the weight of the ball. For how long was the ball in contact with the surface?

Answer: $t = 11$ ms

Conservation of Momentum in Two Dimensions

31. A hockey puck moving at 5 m/s strikes another stationary puck and moves off at 4 m/s in a direction 35° from its original line of motion. What are (a) the final velocity of the originally stationary puck and (b) its direction? Both pucks have the same mass.

Answer: (a) $v_{2f} = 2.86$ m/s
(b) 53.1°

*32. A 1000-kg car traveling north at 100 km/h collides with a 2000-kg truck moving at 60 km/h traveling east. The vehicles stick together after the collision. What are the initial speed and the direction of the wreckage?

Answer: 52.15 km/h at 39.8° north of east

The linear motion of the pistons in a car's engine is converted into the circular motion of the crank-shaft that turns the wheels of the car. The circular motion of the wheels is converted back into linear motion as the car moves over streets, roads, and highways.

Circular Motion

At the end of this chapter, the reader should be able to:

1. Explain the difference between units of *degrees* and units of **radians,** and convert quickly between them.

2. Convert between linear concepts (such as **x**, **v**, **a**, and so on) and angular concepts (such as θ, ω, α, and so on).

3. Apply the four equations of uniformly accelerated motion (as learned in Chapter 1) to problems that deal with rotations.

4. Explain the relationship between *mass* and *inertia.*

5. State the difference between the *rotational kinetic energy* of a body and the *translational kinetic energy* of a body, and explain how a body can simultaneously have both forms of energy.

6. Calculate the *rotational kinetic energy* of an object.

7. Describe the difference between *torque* and *force*.

8. Use *Newton's second law for rotation* to solve problems dealing with torque and rotating bodies.

9. Use the concept of *conservation of angular momen-*

tum to solve exercises that concern rotating objects.

10. Utilize the concept of *power* (as learned in Chapter 4) to find the work done on (or by) a rotating object, or the time required to do such work.

6.1   Radian Measure

This chapter will be concerned with circular motion. Up to now we have considered only linear motion, and the units of displacement have been linear, such as meters and feet. With rotation, the units for displacement are angular, since angles are associated with a circle. If something turns 360°, it has turned in a full circle. But instead of using a degree as the basic unit for measuring an angular displacement, we will use the radian instead. *The **radian** (rad) is based on pi (π), which is the ratio of a circle's circumference to its diameter, where 2π rad = 360°.*

Radian: An angular measure for the circle based on pi (π), where 2π radians = 360°.

$$\pi = \frac{C}{D} \qquad D = 2r$$

$$\pi = \frac{C}{2r} \qquad C = 2\pi r$$

where r is the radius of the circle. This ratio leads directly to the formula for the circumference (C) of a circle, as shown above. The above equation says that there are 2π radii in the circumference of each circle. If we were to stand in the center of a circle and trace out a complete circumference, we would turn a full circle of 360°. We would also have traced out 2π radii. If we had turned less than 360°, we would have turned through a discrete angle θ, as shown in Figure 6.1. In turning through an angle θ, we would also trace out a small part of the circle's circumference, s. This pie-shaped wedge is bounded on two sides by the circle's radius, r. If the radian is used to measure the angle, there is a simple linear relationship between the angle (θ) and the subtended arc length (s).

(Equation 6.1)

$$s = r\theta$$

Equation (6.1) is true only if the angle θ is measured in radians and *not* in degrees. To divide a circle up into radians instead of degrees, we assign the radius a length of 1. This is the unit circle.

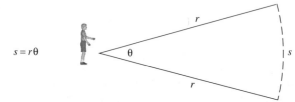

Figure 6.1 The Interior Angle in Radians Relates a Circle's Radius to an Arc Length, s.

$s = r\theta$

There is a direct relationship between degrees and radians. There are 360° for every 2π radians, or 2π rad. Every time we turn in a complete circle (360°), we trace out a complete circumference of length 2π.

$$C = 2\pi r \qquad r = 1 \qquad C = 2\pi$$

$$\frac{360°}{2\pi \text{ rad}} = \frac{180°}{\pi \text{ rad}} = 57.3°/\text{rad}$$

In turning a complete circle, we trace out a complete circumference of $C = 2\pi r$, and the arc length is

$$s = r\theta \qquad s = r(2\pi \text{ rad}) \qquad s = 2\pi r$$

6.2   Uniformly Accelerated Rotational Motion

Because the subtended angle relates directly to the distance traveled on the circle's circumference, radian measure is the natural unit for circular motion. Assume that we have a uniform circular disc rotating at a constant angular speed. All parts of the disc will go through the same angle in the same amount of time, but depending on the radius, some parts will trace out a greater arc length than others. The farther out on the disc radius, the greater the linear speed.

$$s = r\theta \qquad \Delta s = r\Delta\theta$$

$$\frac{\Delta s}{\Delta t} = r\frac{\Delta\theta}{\Delta t}$$

or

$$v = r\omega \tag{Equation 6.2}$$

where ω is the commonly used symbol for angular speed in radians per second (rad/s).

We can use the same reasoning as above to find the magnitude of the translational acceleration experienced by a point on the rim of a circle that is spinning faster and faster.

$$v = r\omega \qquad \Delta v = r\Delta\omega$$

$$\frac{\Delta v}{\Delta t} = r\frac{\Delta\omega}{\Delta t}$$

or

$$a = r\alpha \tag{Equation 6.3}$$

where α is the angular acceleration of the disc as a whole. If, for instance, a motor is spinning up a flywheel, the flywheel begins to rotate faster and faster. Its angular speed increases, and it undergoes an angular acceleration. Because of the direct link between rotational and translational (linear) motion, there is a direct link between uniform linearly accelerated motion and uniform angularly accelerated motion. Indeed, we can start with an equation for uniformly accelerated translational motion and directly derive its rotational equivalent.

$$v_f^2 - v_0^2 = 2ax$$

$$v = \omega r \qquad a = \alpha r \qquad x = r\theta$$

$$(\omega_f r)^2 - (\omega_0 r)^2 = 2(\alpha r)(r\theta)$$

$$\omega_f^2 r^2 - \omega_0^2 r^2 = 2r^2\alpha\theta$$

Cancel the r^2 terms out on both sides of the equation.

(Equation 6.4)

$$\omega_f^2 - \omega_0^2 = 2\alpha\theta$$

In exactly the same way, we can derive the rest of the equations for uniformly accelerated angular motion that are the direct counterparts to their translational equations. Table 6.1 lists the equations for uniformly accelerated translational motion and their rotational equivalents.

Table 6.1

Translational Motion	Rotational Motion	
$v_f^2 - v_0^2 = 2ax$	$\omega_f^2 - \omega_0^2 = 2\alpha\theta$	(Equation 6.4)
$x = v_0 t + \dfrac{1}{2}at^2$	$\theta = \omega_0 t + \dfrac{1}{2}\alpha t^2$	(Equation 6.5)
$x = \left(\dfrac{v_0 + v_f}{2}\right)t$	$\theta = \left(\dfrac{\omega_0 + \omega_f}{2}\right)t$	(Equation 6.6)
$v_f - v_0 = at$	$\omega_f - \omega_0 = \alpha t$	(Equation 6.7)

(Equation 6.4)

(Equation 6.5)

(Equation 6.6)

(Equation 6.7)

The uses of the equations for uniformly accelerated rotational motion are exactly like those for uniformly accelerated translational motion learned in Chapter 1. Why treat rotational and translational phenomena as two different things when they are the same mathematically? Just remember that angular parameters are measured in radians, and not in degrees!

EXAMPLE 6.1

A large flywheel is spun up to 1000 rpm from rest in 35 s. (a) What was its angular acceleration? (b) How many revolutions did the flywheel make in this time?

SOLUTION

(a) $$\alpha = \frac{\omega_f - \omega_0}{t} \qquad \alpha = \frac{\omega_f}{t}$$

$$\omega_f = \left(\frac{1000 \text{ rev}}{60 \text{ s}}\right)\left(\frac{2\pi \text{ rad}}{\text{rev}}\right) \qquad \omega_f = 104.7 \text{ rad/s}$$

$$\alpha = \frac{104.7 \text{ rad/s}}{35 \text{ s}} \qquad \alpha = 3 \text{ rad/s}^2$$

(b)

$$\omega_f^2 - \omega_0^2 = 2\alpha\theta \qquad \omega_f^2 = 2\alpha\theta$$

$$\theta = \frac{\omega_f^2}{2\alpha} \qquad \theta = \frac{(104.7 \text{ rad/s})^2}{2(3 \text{ rad/s}^2)}$$

$$\theta = (1827 \text{ rad})\left(\frac{1 \text{ rev}}{2\pi \text{ rad}}\right) \qquad \theta = 291 \text{ rev}$$

6.3 Rotational Energy and Moment of Inertia

Through direct conversion of the equations for uniformly accelerated translational motion, we arrive at the equations for uniformly accelerated rotational motion. Other concepts carry over directly from the translational to the rotational realm also. For example, energy can be stored in the rotation of a large cast-iron flywheel. That is the reason that internal combustion engines have a flywheel attached to the crankshaft. When the flywheel is put in motion by the crankshaft, it stores enough energy through rotation that the piston will be able to exhaust the just-burned gases, draw in a fresh charge through the carburetor, and compress the fresh charge before the spark plug fires it. The energy stored in a flywheel also acts to smooth out the running of an internal combustion engine. The flywheel acts like an energy buffer, and its inertia allows energy to be added to and taken from it smoothly.

How much energy does a flywheel store in terms of its angular speed? Let's simplify things somewhat, and instead of a flywheel say that we have a small iron ball at the end of a massless string swinging around in a horizontal circle. The ball is rotating at the end of the string with a constant speed v. The ball therefore has kinetic energy stored in it.

$$KE = \frac{1}{2}mv^2$$

$$v = \omega r$$

$$KE = \frac{1}{2}m(\omega r)^2$$

$$KE = \frac{1}{2}m\omega^2 r^2$$

$$KE = \frac{1}{2}mr^2\omega^2$$

$$KE = \frac{1}{2}(mr^2)\omega^2$$

$$I = mr^2$$

$$KE = \frac{1}{2}I\omega^2$$

The above equation is the kinetic energy of the spinning iron ball in terms of the angular speed. It is mathematically the same as the one for translational kinetic energy ($\frac{1}{2}mv^2$). We have, however, introduced a new concept called *moment of inertia* (I). In the case of a ball spinning around on the end of a string, the ball acts as a point mass of mass m spinning around at the end of a string of length r. Compare the two formulas for kinetic energy.

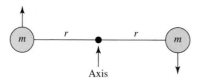

Figure 6.2 Moment of Inertia of a Dumbbell-Shaped Object

$$KE_t = \frac{1}{2}mv^2$$

(Equation 6.8)

$$KE_r = \frac{1}{2}I\omega^2$$

Moment of inertia: The resistance to the rotation of a body about a given axis.

Notice that m and I are analogous or have a one-to-one relationship. A mass, when moving, has inertia, and the term *moment of inertia* obviously implies inertia. *The only thing different about the **moment of inertia** is that it takes into account the spatial distribution of the spinning mass elements (the length of the radius). If there is more than one mass element, then the total moment of inertia is the summation of those spinning mass elements in space.* For a dumbbell-shaped object spinning about the center of the axle, we find the moment of inertia in the following way (assume that the mass of the axle is negligible; see Figure 6.2):

$$I = m_1 r_1^2 + m_2 r_2^2$$

Both masses are the same, and both are equidistant from the center.

$$m_1 = m_2 = m \qquad r_1 = r_2 = r$$
$$I = mr^2 + mr^2 \qquad I = 2mr^2$$

$$I = 2mr^2$$

If we have many mass elements, all spinning about a common center, the moment of inertia becomes

$$I = m_1 r_1^2 + m_2 r_2^2 + \cdots + m_n r_n^2$$

where n is the number of elements, and r_n is the distance from the center that mass element n is spinning. Of course, most spinning objects are continuous geometric shapes, such as a spinning hoops or discs, as opposed to a collection of discrete point masses. Nevertheless, we can treat a continuous object like a collection of discrete point masses and sum them all together to find the moment of inertia for a particular geometric shape.

$$I = \Delta m_1 r_1^2 + \Delta m_2 r_2^2 + \cdots + \Delta m_n r_n^2$$

We can use a summation sign, which is a shorthand way of stating the above equation.

$$I = \sum_1^n \Delta m_n r_n^2$$

To find the moment of inertia of a continuous, extended object, we assume that the mass elements become very small and of equal size.

Assume that we want to find the moment of inertia of a thin hoop rotated about its center, as in Figure 6.3. The hoop has a constant thickness, height, and density. We can state the general formula for finding the moment of inertia.

$$I = \sum_1^n \Delta m_n r_n^2$$

In the summation above, we sum only over what changes. Since the radius of the loop is constant, we can pull it out from behind the summation sign.

$$I = r^2 \sum_1^n \Delta m_n$$

Vertical axis is in and out of the page.
Height of hoop = h
Hoop width (w) is out of the page.

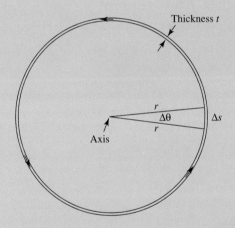

Figure 6.3 Thin Hoop
Rotated about Its Center

Now we must deal with the small mass elements, Δm_n. We can divide the hoop into very small divisions of arc length Δs and get our mass elements from them. Density is the mass per unit volume of a material.

$$\Delta m = \rho V \quad \text{(where } V = \text{volume)}$$

$$V = tw\,\Delta s$$

$$\Delta m = \rho tw\,\Delta s$$

where ρ is the density of the material of which the hoop is constructed, t is the thickness of the hoop, w is the width of the hoop, and Δs is the arc length.

$$I = r^2 \sum_1^n \rho tw\,\Delta s$$

We again sum only over those quantities that change, removing the constant quantities from behind the summation sign.

$$I = r^2 \rho tw \sum_1^n \Delta s$$

We can obtain an expression for the element of arc using radians.

$$\Delta s = r\,\Delta\theta$$

The summation now becomes

$$I = r^2\rho tw \sum_{1}^{n} r\,\Delta\theta$$

$$I = r^3\rho tw \sum_{1}^{n} \Delta\theta$$

So we end up summing over the angle θ, and for a complete circle the change in θ is 2π.

$$I = r^3\rho tw(2\pi)$$

Rearranging yields

$$I = (2\pi rtw\rho)r^2$$

The term inside the brackets is the mass of the hoop, $m = [(2\pi r)(w)(t)]\rho$.

$$I = mr^2$$

From Example 6.2, we can infer that a moment of inertia will have to be computed for each geometric configuration. However, if the same hoop were spun around a different axis, perhaps around an axis that lies along its diameter as in Figure 6.4, we would get a different moment of inertia for the same loop.

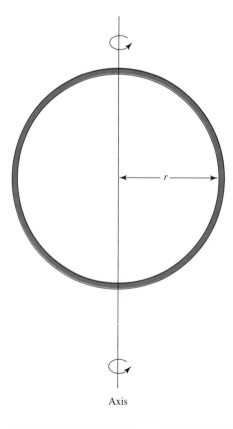

Figure 6.4 Moment of Inertia for a Thin Hoop Rotated about a Diameter

Axis

$$I = \frac{1}{2}mr^2$$

Table 6.2 gives the moment of inertia for some of the more common geometric shapes and axes of rotation.

Table 6.2 Moments of Inertia for Various Geometric Configurations and Axes of Rotation

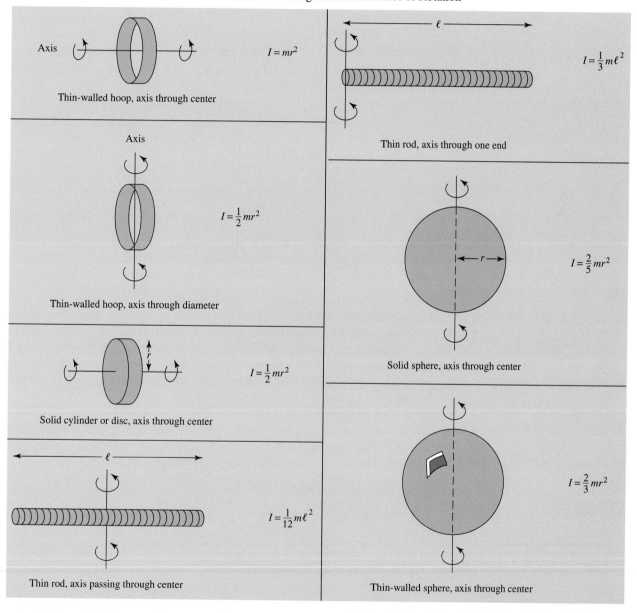

6.4 Combined Rotational and Translational Motion

When a car is traveling east at 60 mph, the car's speed as a whole is translational. Yet the car is moving forward because it is sitting on four rolling tires. The tire's angular speed is translated into a linear speed. When we began this chapter on circular motion, we started from the point of view of translational motion and developed the equations for uniformly accelerated rotational motion. The reverse argument is also true: We could have started from rotational motion and developed the equations for uniformly accelerated linear motion. The car is actually sitting on the axles that are attached to the four wheels. Each axle serves as the axis for that particular wheel. If the car is moving forward at 60 mph, then the axle of each wheel is also moving forward at 60 mph. If the car's linear speed is constant, then the tire's angular speed is also constant. From the equation that translates linear speed into rotational speed, we have

$$v = \omega r \quad \text{or} \quad \omega = \frac{v}{r}$$

If the tire has a diameter of 24 in., and if the car's speed is 60 mph, we can calculate the rotational speed of the tire.

$$60 \text{ mph} = 88 \text{ ft/s}$$

$$r = 12 \text{ in.} = 1 \text{ ft}$$

$$\omega = \frac{88 \text{ ft/s}}{1 \text{ ft}}$$

$$\omega = 88 \text{ rad/s}$$

$$\omega = \left(\frac{88 \text{ rad}}{\text{s}}\right)\left(\frac{1 \text{ rev}}{2\pi \text{ rad}}\right)\left(\frac{60 \text{ s}}{\text{min}}\right)$$

$$\omega = 840.34 \text{ rpm}$$

At 60 mph, a tire with a diameter of 24 in. rotates at about 840 rpm (or 88 rad/s).

Our perspective so far has focused on the central axis of rotation which transfers the tire's rotational motion to its translational motion relative to a stationary observer. Let us now shift our perspective to the point where the tire meets the road. At this point, the tire is stationary relative to the road. It is unmoving relative to an observer standing at the side of the road. The entire tire, including the car's axle, pivots about this point as shown in Figure 6.5. We can find the translational velocities for any point along the vertical diameter through the axle by using $v = \omega r$. At the pivot point, $v = 0$ because $r = 0$. At the axle,

$$v = (88 \text{ rad/s})(1 \text{ ft})$$

$$v = 88 \text{ ft/s} \quad \text{or} \quad v = 60 \text{ mph}$$

So the axle has a linear speed of 60 mph, as expected.

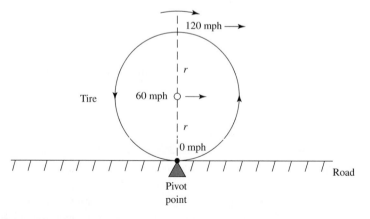

Figure 6.5 Automobile Tire Pivoting about Its Point of Contact with the Road

Now let's compute the speed at a point on the top of the tire.

$$v = (88 \text{ rad/s})(2 \text{ ft})$$

$$v = 176 \text{ ft/s} \quad \text{or} \quad v = 120 \text{ mph}$$

A point on the top of the tire has a linear speed of 120 mph! If we computed speeds for each point on the rim of the tire, we would see that the speeds vary smoothly from 0 mph at the point of contact between road and tire to 120 mph at a point on the top of the tire. But we know that the car as a whole has a linear velocity of 60 mph, which is the translational speed of the axle. The axle is the tire's axis of rotation. Our usual perspective on a rotating tire is from the point of view of the axle, with the tire rotating about the axle. If an observer were to move alongside a car moving at 60 mph at the same speed, this is exactly what he would see. The observer's relative speed with respect to the car would be zero. However, if the observer were again placed stationary at the side of the road and the car passed him at 60 mph, then the only sensible perspective to adopt concerning the tire would be one in which we accounted for the car's translational velocity. This would be the perspective of Figure 6.5. Both perspectives are valid, because both situations are simultaneously real. Being able to change one's perspective is a very valuable tool in problem solving, but also one of the hardest to acquire.

6.5   Rotational Kinetic Energy

Let's take another look at rolling and translating objects. If we roll a hoop along the ground, it has simultaneously both rotational and translational motion. We know from a previous chapter that kinetic energy is stored in the translational motion of the hoop. Since the hoop is also spinning, energy is also stored in the rotation of the hoop. How can the total kinetic energy of the hoop (both translational and rotational) be expressed? Since energy is a scalar, two completely different types of energy can be added directly together. If potential energy can be added to kinetic energy, so can translational and rotational kinetic energies.

$$KE_T + KE_r = KE_{\text{total}}$$

The translational kinetic energy is $\frac{1}{2}mv^2$, and the rotational kinetic energy is $\frac{1}{2}I\omega^2$. Therefore,

$$KE_{\text{total}} = \frac{1}{2}mv^2 + \frac{1}{2}I\omega^2$$

We can look at the spinning and translating hoop from either a rotational or a translational perspective. First we'll look at the hoop's kinetic energy from a translational point of view. For a hoop with the axis of rotation through its center perpendicular to the plane of the hoop,

$$I = mr^2$$

$$KE = \frac{1}{2}mv^2 + \frac{1}{2}(mr^2)\omega^2 \qquad KE = \frac{1}{2}mv^2 + \frac{1}{2}m(\omega r)^2$$

$$v = \omega r \qquad\qquad\qquad KE = \frac{1}{2}mv^2 + \frac{1}{2}mv^2$$

$$KE = mv^2$$

So all of the hoop's kinetic energy can be summarized in its translational motion only, although it has energy stored in translation and rotation.

Now let's examine the hoop's kinetic energy from a rotational point of view.

$$KE = \frac{1}{2}mv^2 + \frac{1}{2}I\omega^2 \qquad\qquad v = \omega r$$

$$KE = \frac{1}{2}m(\omega r)^2 + \frac{1}{2}I\omega^2 \qquad KE = \frac{1}{2}m(\omega^2 r^2) + \frac{1}{2}I\omega^2$$

$$KE = \frac{1}{2}mr^2\omega^2 + \frac{1}{2}I\omega^2 \qquad KE = \frac{1}{2}(mr^2)\omega^2 + \frac{1}{2}I\omega^2$$

For a hoop,

$$I = mr^2$$

$$KE = \frac{1}{2}I\omega^2 + \frac{1}{2}I\omega^2$$

$$KE = I\omega^2$$

Thus, we can also summarize the hoop's kinetic energy strictly in terms of its rotational motion.

EXAMPLE 6.3

At the top of an inclined plane stand a hoop and a disc, as shown in Figure 6.6. Both have the same diameter and total mass. If they are both released from rest from the same starting position, which one will reach the bottom first?

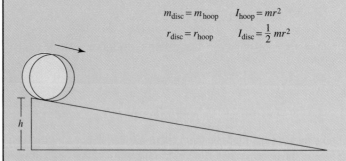

$$m_{disc} = m_{hoop} \qquad I_{hoop} = mr^2$$
$$r_{disc} = r_{hoop} \qquad I_{disc} = \frac{1}{2}mr^2$$

Figure 6.6 Disc and Hoop of Same Mass and Radius Racing Each Other to the Bottom of an Incline

SOLUTION
Since the hoop and the disc are racing each other to the bottom, our perspective should be the translational speed of each. This problem is a conservation of energy problem. The potential energy at the top is converted into kinetic energy at the bottom. Since both disc and hoop have the same initial potential energy, they will both have the same total kinetic energy at the bottom. However, since they have different moments of inertia, they will have different speeds at the bottom. The object

with the greater translational speed at the bottom will be the one to reach the bottom first.

We can summarize all of the hoop's kinetic energy by its translational speed.

$$KE = mv_{hoop}^2$$

We must also do this for the disc.

$$KE = \frac{1}{2}mv^2 + \frac{1}{2}I\omega^2$$

For a disc, $$I = \frac{1}{2}mr^2$$

$$KE = \frac{1}{2}mv^2 + \frac{1}{2}\left(\frac{1}{2}mr^2\right)\omega^2 \qquad KE = \frac{1}{2}mv^2 + \frac{1}{4}m(\omega r)^2$$

$$v = \omega r \qquad\qquad\qquad KE = \frac{1}{2}mv^2 + \frac{1}{4}mv^2$$

$$KE = \frac{3}{4}mv_{disc}^2$$

For the hoop, the potential energy at the top was converted into kinetic energy at the bottom.

$$PE_{hoop} = KE_{hoop}$$

$$mgh = mv_{hoop}^2$$

$$gh = v_{hoop}^2$$

$$v = \sqrt{gh} \qquad \text{(Hoop)}$$

We can now do the same analysis for the disc.

$$PE_{disc} = KE_{disc}$$

$$mgh = \frac{3}{4}mv_{disc}^2$$

$$gh = \frac{3}{4}v_{disc}^2$$

$$v = \sqrt{\tfrac{4}{3}gh} \qquad \text{(Disc)}$$

The disc has the greater linear speed at the bottom, so it wins the race! The disc also had the lower moment of inertia, so less of its total energy was in rotation, and more in translation. If the disc has a greater linear speed at the bottom than the hoop, it should also be rotating faster. If we had used angular speed as our point of view, we would have found that the disc was indeed rotating faster at the bottom of the inclined plane than the hoop. Either perspective, properly interpreted, would have shown that the disc arrived at the bottom of the plane before the hoop.

6.6 Torque

Torque is a force applied to a lever arm in an attempt to rotate something. A torque on an object such as a flywheel will cause it to spin faster and faster (a rotational accelera-tion). The linear equation for force is $F = ma$. Its rotational analog is $\tau = I\alpha$.

$$F = ma$$

(Equation 6.9)

$$\tau = I\alpha$$

Torque is a vector with a direction associated with it, just like force. (We'll deal with the direction of angular quantities in Section 6.7 on the conservation of angular momentum.)

Torque can also be defined in terms of leverage, which is the applied force multiplied by the leverage that force has on the pivot point.

Cross product: The vector product between two vector quantities and the sine of the angle between them.

$$\vec{\tau} = \vec{r} \times \mathbf{F}$$

*The above is called a **cross product**,* where r is the distance from the pivot to the point where the force is applied, and $\mathbf{F}$ is the applied force. It translates into the following equation:

(Equation 6.10)

$$\tau = rF \sin \theta$$

where the angle theta (θ) is between the direction of the lever arm and the direction of the applied force.

The example of opening a door will illustrate how a cross product works. If we look down on a door from above, it will appear as in Figure 6.7. The door's axis of rotation is the hinge line. We open the door (most of the time) using the doorknob. By using the knob, we apply maximum torque (turning force) to the door. The position of the knob at the outside edge of the door maximizes r, and therefore the torque. (If we try to open a door by pushing on it near the hinge line, we must push a lot harder than if we use the doorknob.) If you push or pull on the doorknob, the angle between the force on the knob and the door is 90°. Since sin 90° = 1, the torque is again maximized. If the angle is zero, regardless of the magnitude of the force or the length of the lever arm (r), the torque is zero (sin 0° = 0). This action would be equivalent to trying to open the door by pushing on the outside edge of the door toward the hinge line (see Figure 6.8).

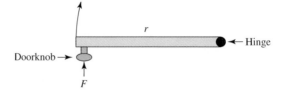

Figure 6.7 Torque Applied to a Door

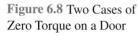

Figure 6.8 Two Cases of Zero Torque on a Door

The unit for torque is the newton-meter (N-m). These units are exactly the same as those for work! There is much confusion over this point. The distance unit in N-m for work is the distance that the force moves the object. The effective force component and the object's displacement are in the same direction. Force multiplied by displacement gives units of N-m. For torque we again have a force multiplied by a distance. However, this time the distance is the lever arm, and the effective component of the force is at 90° to the lever arm. So for work the N-m implies a force doing work over a distance, and for torque the same unit of the N-m implies a force applying leverage to rotate an object.

EXAMPLE 6.4

A thick, metal flywheel with an axis of rotation through its center is spun up to 750 rpm from rest in 35 s. The flywheel has a diameter of 0.5 m and a mass of 75 kg. (a) What torque is acting on the flywheel? (b) If the torque to the flywheel is supplied by a leather belt operating tangent to it as in Figure 6.9, what is the tension in the belt?

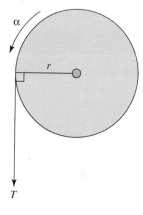

Figure 6.9 Torque on a Flywheel

SOLUTION

(a)
$$\tau = I\alpha$$

$$I = \frac{1}{2}mr^2 \qquad I = \frac{1}{2}(75 \text{ kg})(0.25 \text{ m})^2 \qquad I = 2.34 \text{ kg-m}^2$$

$$\alpha = \frac{\Delta\omega}{\Delta t} \qquad \alpha = \frac{\omega_f - 0}{t} \qquad \alpha = \frac{\omega_f}{t}$$

$$\omega_f = \left(\frac{750 \text{ rev}}{60 \text{ s}}\right)\left(\frac{2\pi \text{ rad}}{\text{rev}}\right) \qquad \omega_f = 78.54 \text{ rad/s}$$

$$\alpha = \frac{78.54 \text{ rad/s}}{35 \text{ s}} \qquad \alpha = 2.24 \text{ rad/s}^2$$

$$\tau = (2.34 \text{ kg-m}^2)(2.24 \text{ rad/s}^2) \qquad \tau = 5.24 \text{ N-m}$$

(b)
$$\tau = r \times F \qquad \tau = rF \sin\theta$$

$$\theta = 90° \qquad r = 0.25 \text{ m}$$

$$F = \frac{\tau}{r \sin\theta} \qquad \sin 90° = 1$$

$$F = \frac{\tau}{r} \qquad F = \frac{5.24 \text{ N-m}}{0.25 \text{ m}}$$

$$F = 20.96 \text{ N}$$

6.7   Angular Momentum and Conservation of Angular Momentum

Consider the gyroscope. It is both an interesting toy and a very useful device. As a toy, it can do amazing balancing acts on a string and apparently defies gravity. Its practical uses include navigation and stabilization. A good gyroscope is essentially a well-balanced, spinning mass mounted in bearings with negligible friction. If a gyroscope is spun up with its axis of rotation horizontal, it will precess (move in a horizontal circle) with the supported end at the origin of the circle. The unsupported end hangs in midair with apparently no support. What holds it up? It defies gravity!

A gyroscope "defies gravity" and can also be used as a compass because of conservation of angular momentum.

A gyroscope can also act as a compass. If the gyroscope is mounted in a ship in low-friction bearings and is spun up and pointed north, it will continue to point north regardless of the orientation of the ship or its direction. The gyroscope isn't sensing the direction of the earth's magnetic field as a normal compass does, so why does it continue to point north?

The secret of a gyroscope's performance lies in its angular momentum and the fact that angular momentum is conserved, just as linear momentum is. We can use what has been learned about linear momentum from preceding chapters to develop the concept of angular momentum. By direct analogy, one of the equations for angular momentum can be derived. The equation for linear momentum is

$$\mathbf{p} = m\mathbf{v}$$

and that for angular momentum

(Equation 6.11)

$$\mathbf{L} = I\boldsymbol{\omega}$$

The **bold** type indicates that both linear and angular momentum are vector quantities. When conservation of momentum problems were solved for linear momentum, the vector nature of momentum was always critical in setting up the problem. (In one-dimensional momentum problems, the vector nature of momentum showed up in a negative sign for a velocity in the opposite direction.) To understand why a gyroscope does what it does, you must also understand the vector nature of angular momentum. There is a second equation for angular momentum, and you can understand it by considering a small ball of mass m being whirled around in a horizontal circle. The speed of the ball times its mass is the linear momentum of the ball. But as with torque, there is a center of rotation, and the distance from the center of rotation must be taken into consideration. *Like the formula for torque, the second formula for **angular momentum** is a cross product.*

$$\mathbf{L} = \mathbf{r} \times \mathbf{p}$$

Equation (6.12) gives the magnitude of the angular momentum by multiplying the distance from the center of rotation (r) times the linear momentum of the spinning mass element (p) times the sine of the angle between them.

$$L = rp \sin \theta$$

Both torque and angular momentum are vector quantities, and therefore both have a direction as well as a magnitude. In Section 6.6 on torque, the directional nature was not emphasized, since it was not important in the examples given there. But to explain the seemingly magical operation of the gyroscope, we need first to explain the directional nature of angular momentum. *To find the direction of the torque or angular momentum, we use **right-hand rule 1** associated with cross products.*

Assume that we have a ball swinging around in a horizontal circle with a constant speed, as shown in Figure 6.10. This figure is a view from above the whirling ball. Also assume that the radius vector points to the ball at the end of the string. To find the direction of the angular momentum, point the thumb of your right hand in the direction of the radius vector, and the fingers of your right hand in the direction of the ball's linear velocity. The open palm of your right hand is then pointing in the direction of the angular momentum. For Figure 6.10, the direction of the angular momentum would be directly at the reader, out of the page, or "up."

Angular momentum: The product of an object's moment of inertia and its angular speed.

(Equation 6.12)

Right-hand rule 1: The right-hand rule used to find the direction of a torque or an angular momentum.

$\vec{\mathbf{L}}$ (angular momentum) is out of the page, in the direction of the open palm of the right hand.

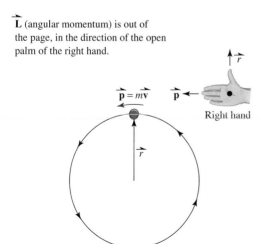

Figure 6.10 Top View of a Ball Moving in a Horizontal Circle at a Constant Speed

The direction of the angular momentum is at right angles to the plane formed by the radius vector and the linear momentum vector. The linear velocity (and therefore the momentum) of the ball is always changing. The speed of the ball is constant, but its direction is always changing. (At the top of the circle, the direction of the linear velocity is to the left, and at the bottom it is to the right.) If the angular momentum has a given direction (out of the page) under these circumstances, it must take this constantly changing linear velocity into account. With the right-hand rule, the direction of the angular momentum stays out of the page in Figure 6.10 regardless of the direction of the ball's velocity (and therefore momentum). If the angular momentum in Figure 6.10 is out of the page, and if $\mathbf{L} = I\boldsymbol{\omega}$, then the angular velocity has the same direction as the angular momentum.

In Section 5.1 on linear momentum, you learned that force is a change in momentum over time. Mathematically, an exact analogy between force and torque as changes in momentum over time can be made.

$$\mathbf{F} = \frac{\Delta \mathbf{p}}{\Delta t}$$

(Equation 6.13)

$$\boldsymbol{\tau} = \frac{\Delta \mathbf{L}}{\Delta t}$$

To change the angular momentum of a spinning object, an external torque (turning force) must be applied. Since angular momentum is a vector quantity, there are two ways to change it. Either the magnitude can be changed, or its *direction* can be changed. In a gyroscope, the spinning gyro is usually mounted in good, nearly frictionless bearings. To compensate for what friction still remains, power is fed to the gyro to keep it rotating at a constant angular velocity. The gyroscope mount is also fitted with gimbal bearings so that the gyroscope is effectively isolated from outside influences. Therefore, no external torque is acting on it. With no external torques acting on the gyroscope, its angular momentum remains constant, and it will continue to point in the direction in which it was first aligned. This is what makes a gyro compass work.

A bicycle wheel mounted on a short axle acts as a good gyroscope. An oft-repeated classroom demonstration goes thusly: One end of the axle of a bicycle wheel is tied to a rope that hangs from the ceiling, as shown in Figure 6.11. The wheel is then held by the other end so that the wheel's axle is horizontal. A torque is applied to the wheel's rubber

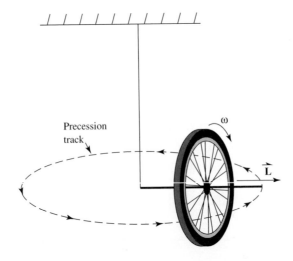

Figure 6.11 Bicycle Wheel Gyroscope

rim, and the wheel is spun up to a high rpm. The axle is then released, and the spinning bicycle wheel remains with its axle approximately horizontal and precesses (revolves) in a horizontal circle.

The unsupported end seemingly defies gravity. What holds it up? Seeing is believing, and the audience knows that some mysterious force is supporting the end of the axle and keeping it from falling. Figure 6.12 provides the answer.

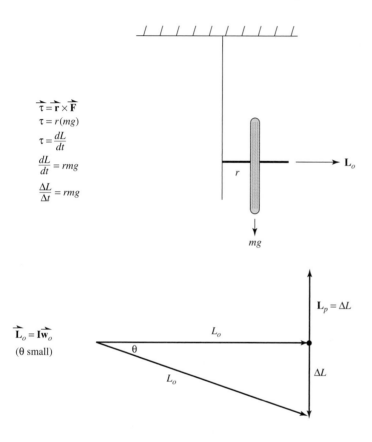

$$\vec{\tau} = \vec{r} \times \vec{F}$$

$$\tau = r(mg)$$

$$\tau = \frac{dL}{dt}$$

$$\frac{dL}{dt} = rmg$$

$$\frac{\Delta L}{\Delta t} = rmg$$

$$\vec{L}_o = I\vec{w}_o$$

$(\theta \text{ small})$

Figure 6.12 Conservation of Angular Momentum in the Bicycle Wheel Gyroscope

The spinning bicycle wheel is essentially free to rotate without friction; no external torques are acting on it. If it starts with its angular momentum horizontal before release, it tries to keep it horizontal after release. However, a torque is acting on the system. It is caused by the weight of the bicycle wheel, mg, acting down at the end of the lever arm of length r. This torque acts to try to bend the initial angular momentum of the system downward. (The right triangle in Figure 6.12 shows this situation.) If the spinning wheel hasn't lost any of its angular velocity, it still has its original angular momentum L_o, but it is now angled downward a little bit by the angle θ. There is now a vertically downward component of the angular momentum (ΔL), which is the change in angular momentum. Because of the principle of conservation of angular momentum, the initial angular momentum must equal the final angular momentum if no external torques are present. If initially there was no vertical component of the angular momentum (it was all horizontal), then finally there should be no vertical component. Something must provide an upward-pointing component of angular momentum to cancel the downward one from the torque. Notice that the wheel is precessing. It turns slowly in a counterclockwise horizontal circle if viewed from above. This precession provides the vertically upward component of angular momentum (L_p) to the system needed to cancel out the vertically downward one from the torque of the wheel's weight. The final total vertical component of the angular momentum is now zero, just as it was initially, and the wheel seems to float in the air!

In the demonstration with the suspended bicycle wheel, if the wheel is spun very fast, the downward drop of the axle isn't very noticeable. But as the wheel loses angular speed, the drop becomes more pronounced, and the precession velocity increases. The greater the change in angular momentum (the vertical component), the greater the precession angular velocity must be to compensate for it.

*Remember, the angular velocity has a direction as well, and that direction is found by using **right-hand rule 2.*** Curl your right hand into a loose fist at the center of rotation. The fingers of your right hand curl in the direction the object is rotating. Extend the thumb of your right hand from your fist, and point it in the direction of the angular velocity, ω. In this way, the angular velocity has a constant direction regardless of the fact that the linear velocity of the object changes constantly.

Right-hand rule 2: The right-hand rule is used to find the direction of an angular velocity.

EXAMPLE 6.5

When a figure skater begins to spin and pulls her arms in, she spins even faster. Why does her speed increase?

SOLUTION
Consider the skater to be an isolated system: She spins essentially without friction. With no outside influence (no outside torques), the skater's angular momentum stays the same. We can model the skater as two point masses spinning opposite each other in a horizontal circle (see Figure 6.13). Assume that the masses are connected by a rigid rod and that both are a distance r from the center of rotation and of equal mass.

$$L_o = I_o\omega_o$$

$$I_o = 2(mr^2)$$

$$L_o = (2mr^2)\omega_o$$

$$\omega_o = \frac{L_o}{2mr^2}$$

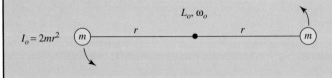

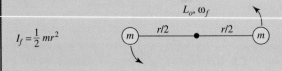

Figure 6.13 Conservation of Momentum for a Spinning, Dumbbell-Shaped Object

Now the masses are pulled in to one-half their original radius, just as the skater's arms are.

$$L_f = I_f \omega_f$$

$$I_f = 2m\left(\frac{r}{2}\right)^2$$

$$I_f = \frac{1}{2}mr^2$$

$$L_f = \left(\frac{1}{2}mr^2\right)\omega_f$$

$$L_f = L_o$$

$$L_o = \left(\frac{1}{2}mr^2\right)\omega_f$$

$$\omega_f = \frac{2L_o}{mr^2}$$

$$\frac{\omega_f}{\omega_o} = \frac{2L_o/mr^2}{L_o/2mr^2}$$

$$\frac{\omega_f}{\omega_o} = 4$$

$$\omega_f = 4\omega_o$$

By pulling her arms in, the skater decreases her moment of inertia. If her angular momentum remains the same, then she must spin faster to compensate for the changed moment of inertia.

6.8   Power

Thus far we have shown how the subjects of displacement, velocity, acceleration, force, energy, and momentum apply to both linear and circular motion. The only topic yet undiscussed is power. But before we talk about power, we must have an expression for work in terms of rotational motion. We can again use the idea of the direct link between rotational and linear quantities. If linear work is force times linear distance, then work done in rotating something is torque times angular displacement in radians.

$$W = \tau\theta \qquad \text{(Equation 6.14)}$$

The definition for power is the same whether for linear or for circular motion: It's the work done over time.

$$P = \frac{W}{t}$$

$$W = \tau\theta \qquad P = \frac{\tau\theta}{t}$$

$$\omega = \frac{\theta}{t}$$

(Equation 6.15)

$$P = \tau\omega$$

$$P = Fv$$

The final equation for power in angular terms is torque times angular velocity. The relationship between angular and linear terms is again shown with the angular power equation directly over the linear one. Power is power whether it is generated translationally or rotationally. As mentioned previously, the unit for power is the watt (W), which is 1 joule per second (J/s), or 1 newton-meter per second (N-m/s). The unit for torque is the newton-meter (N-m), and for angular velocity it is the reciprocal of the second. (A radian is a dimensionless number.) When these rotational units are combined in the power equation, the units are also watts, which is as it should be!

EXAMPLE 6.6

An electric motor turning at 450 rpm has a winding drum 1 m in diameter bolted to the output shaft. It lifts a mass weighing 50 kg. What power does the motor generate lifting the mass?

SOLUTION

$$P = \tau\omega$$

$$\tau = rF \qquad F = mg$$

$$\tau = rmg$$

$$P = rmg\omega$$

$$\omega = \left(\frac{450 \text{ rev}}{60 \text{ s}}\right)\left(\frac{2\pi \text{ rad}}{\text{rev}}\right) \qquad \omega = 47.1 \text{ rad/s}$$

$$P = (0.5 \text{ m})(50 \text{ kg})(9.8 \text{ m/s}^2)(47.1 \text{ rad/s})$$

$$P = 11{,}540 \text{ W} = 15.5 \text{ hp}$$

The same answer is obtained by multiplying the weight of the object being lifted by the speed with which it is being lifted, Fv.

$$v = \omega r \qquad v = (47.1 \text{ rad/s})(0.5 \text{ m}) \qquad v = 23.55 \text{ m/s}$$

$$P = (mg)v$$

$$P = (50 \text{ kg})(9.8 \text{ m/s}^2)(23.55 \text{ m/s})$$

$$P = 11{,}540 \text{ W}$$

Summary

1. The traditional unit of angular displacement (θ) is the *radian* (rad).

2. The four equations of uniformly accelerated motion can still be used in cases where objects are rotating instead of, or in addition to, translating.

Problem Solving Tip 6.1: When in doubt, remember that the physics in Chapters 1–5 is identical to the physics in this chapter. For example, the first equations of uniformly accelerated motion, Equation (1.1) and Equation (6.7), read, "Final velocity equals initial velocity plus acceleration multiplied by time." The terms *linear* velocity/acceleration and *angular* velocity/acceleration are there merely to remind you what units (whether linear or angular) we are using!

3. The unit for moment of inertia (I) is the *kilogram-meter squared* (kg-m^2).

Problem Solving Tip 6.2: People are often confused as to what inertia is, primarily because people are confused as to what *mass* is. If you begin to look at mass as the "resistance to linear acceleration," then the moment of inertia, being located in this chapter on circular motion, could be viewed as the "resistance to angular acceleration." Thus, other equations begin to make more sense. For example, $\mathbf{P} = m\mathbf{v}$ can be read, "Linear momentum equals linear mass multiplied by linear velocity"; and $\mathbf{L} = I\omega$ can be read, "Angular momentum equals angular mass multiplied by angular velocity." In both cases, momentum equals mass times velocity. Notice the strong similarity between this and Problem Solving Tip 6.1. This same analogy can be done with all of the equations found in this chapter! Try it for yourself!

4. An object can have both rotational kinetic energy and translational kinetic energy simultaneously.

Problem Solving Tip 6.3: Now you have an additional type of energy to use when applying Problem Solving Tip 4.1! When solving a problem, try the "snapshot" trick, and look for the types of energy that you know. [At this point, you know translational kinetic energy, rotational kinetic energy, work ($W = Fx$) due to friction (or another force), gravitational potential energy, and spring potential energy.]

5. Torque is the "force" applied in rotating an object and is measured, in the MKS system, in newton-meters (N-m). In the English system, torque is measured in foot-pounds (ft-lb).

Problem Solving Tip 6.4: One of the most confusing steps of solving a torque problem is starting it! See if the following steps help you solve torque problems more easily: If you are given a force applied to turn an object:

- Find where nature will cause the object to rotate. This is your "pivot point."

- Draw a line between this pivot point and the point on the object where the force is applied. We'll call this the *line of action*.

- Take the sine of the angle between the force vector and your line of action. (The angle is often 90°, so sin 90° = 1.)

- Multiply the magnitude of the force, the length of your line of action, and the sine of the angle as found above. This is the torque applied to your object.

  If you are not given any forces:

- Find the moment of inertia I of your object.

- Calculate the angular acceleration α of the object using the equations of uniformly accelerated motion.

- Multiply I and α together; this is the torque (and it is a direct application of Newton's second law!).

6. As in the linear case, if there are no outside torques, angular momentum is *always* conserved.

Problem Solving Tip 6.5: Refer to Problem Solving Tip 5.1, and use what you know from Problem Solving Tip 6.3.

Key Concepts

ANGULAR MOMENTUM: The product of an object's moment of inertia and its angular speed.

CROSS PRODUCT: The vector product between two vector quantities and the sine of the angle between them.

MOMENT OF INERTIA: The resistance to the rotation of a given body about a given axis.

RADIAN: An angular measure for the circle based on pi (π), where 2π radians = 360°.

RIGHT-HAND RULE 1: The right-hand rule used to find the direction of a torque or an angular momentum.

RIGHT-HAND RULE 2: The right-hand rule used to find the direction of an angular velocity.

Important Equations

(6.1) $s = r\theta$ (Arc length using radian measure for angle)

(6.2) $v = r\omega$ (Relationship between rotational and translational speed)

(6.3) $a = r\alpha$ (Relationship between rotational and translational acceleration)

(6.4) $\omega_f^2 - \omega_0^2 = 2\alpha\theta$ (Uniformly accelerated rotational motion)

(6.5) $\theta = \omega_0 t + \dfrac{1}{2}\alpha t^2$ (Uniformly accelerated rotational motion)

(6.6) $\theta = \left(\dfrac{\omega_0 + \omega_f}{2}\right)t$ (Uniformly accelerated rotational motion)

(6.7) $\omega_f - \omega_0 = \alpha t$ (Uniformly accelerated rotational motion)

(6.8) $\text{KE}_r = \dfrac{1}{2}I\omega^2$ (Rotational kinetic energy)

(6.9) $\tau = I\alpha$ (Newton's second law for rotational motion)

(6.10) $\tau = rF \sin \theta$ (Magnitude of torque in terms of leverage and applied force)

(6.11) $\mathbf{L} = I\boldsymbol{\omega}$ (Angular momentum)

(6.12) $\mathbf{L} = \mathbf{r} \times \mathbf{p}$ (Angular momentum in terms of a spinning mass element)

(6.13) $\tau = \dfrac{\Delta \mathbf{L}}{\Delta \mathbf{t}}$ (Torque is the rate of change of angular momentum.)

(6.14) $W = \tau\theta$ (Work done by an applied torque in rotating a body)

(6.15) $P = \tau\omega$ (Power absorbed by a body with an applied torque rotating at a given angular speed)

Conceptual Problems

1. An object spins around an axis of rotation a distance R from that axis. The particle:

   a. has a linear speed inversely proportional to R

   b. has an angular speed inversely proportional to R

   c. has a linear speed proportional to R

   d. has an angular speed proportional to R

2. What in rotational motion corresponds to force in linear motion?

   a. Weight   b. Angular momentum

   c. Torque   d. Moment of inertia

3. What is the rotational analog of mass?

   a. Angular speed   b. Angular momentum

   c. Torque   d. Moment of inertia

4. The moment of inertia of an object doesn't depend on:

   a. the location of the axis of rotation

   b. the angular speed

   c. the distribution of the mass about the axis

   d. the mass

5. A hoop and a disc with the same radius and mass roll down the same inclined plane. At the bottom, they each have the same:

   a. KE of rotation   b. total KE

   c. linear KE   d. angular speed

6. A brass hoop A rolls down an inclined plane, and an identical brass hoop B *slides* without rolling down a frictionless inclined plane of the same slope. Which of the following statements is true?

   a. Both hoops reach the bottom simultaneously.

   b. Hoop A reaches the bottom first.

   c. Hoop B reaches the bottom first.

   d. Both hoops have the same angular speed at the bottom.

7. In the designing of a flywheel, it's advantageous to put most of the mass around the rim. What is the advantage of this?

8. Which car will coast down a hill faster: one with light tires and wheels or one with heavy tires and wheels? Why?

9. A person riding a bicycle pushes down on the pedal with his foot on the forward half of the circle described by the pedal's rotation. Where is the torque on the pedal a maximum? Where is it a minimum?

10. Why does a diver tuck in her knees when performing somersaults during a dive?

11. Why is there a small propeller rotating in a vertical plane at the rear of helicopters that have only one large rotor rotating in a horizontal plane?

12. Are the equations for angular kinematics valid when the angle is expressed in degrees instead of radians? Why?

13. Why does a point on the rim of a tire have an acceleration even if the car is moving at a constant speed?

14. Sometimes it's difficult to break loose a nut that is tightly screwed down with a wrench. Putting a length of pipe over the wrench handle and then trying again often succeeds in breaking the nut loose. Why does this trick work?

15. How is it possible for a large force to produce a small or zero torque? How can a small force produce a large torque?

16. A contest is held between two people of the same weight to get to the bottom of a hill first using identical tires. One of the people curls up inside the tire and rolls to the bottom. The other builds a light saddle above the wheel attached to a light axle through the wheel and lets the tire carry him down the hill. Which of the two contestants gets to the bottom of the hill first?

17. Why does a gyroscope rotate around and around (precess) faster and faster as the angular speed of the gyro's rotor decreases because of friction?

Exercises

Radian Measure

1. How many radians does the hour hand of a clock turn through in 5700 s?

    Answer: 0.83 rad

2. The human eye can resolve an angular separation of 1 minute (1′), with 60′ = 1°. (a) How many radians is 1 minute of arc? (b) What is the smallest detail that can be resolved of an object 100 ft away?

    Answer: (a) 2.91×10^{-4} rad
    (b) 0.349 in.

3. A spy satellite in a low earth orbit of altitude 90 mi can easily resolve an object as small as a soft-drink can (about

5 inches high and $2\frac{1}{2}$ inches in diameter). How much better is the resolving power of the satellite's "eyes" than a pair of human eyes?

    Answer: 664 times

*4. Wedge-shaped pieces are cut from a pie. The length of the outer crust of each piece is equal to the pie's radius. (a) How many such pieces can be cut from a single pie? (b) What is the apex angle in radians of the remaining piece?

    Answer: (a) 6 pieces
    (b) 0.283 rad

Uniformly Accelerated Rotational Motion

5. What is the angular speed of the minute hand of a clock?

    Answer: 1.45×10^{-4} rad/s

6. The blade of a rotary lawn mower is 2.5 ft in diameter and rotates at 3600 rpm. What is the tip speed of the lawn mower's blade?

    Answer: $v = 321$ mph

7. A radio-control model race car has a speed of 35 mph with tires 2 inches in diameter. How fast are the tires spinning at this speed?

    Answer: 2933 rpm

8. A record player turntable revolving initially at $33\frac{1}{3}$ rpm comes to a stop in 15 s. (a) What was the angular deceleration of the turntable? (b) How many revolutions did the turntable make before coming to a stop?

    Answer: (a) 0.23 rad/s²
    (b) 4.2 rev

*9. A particle speed selector consists of two discs mounted on a common shaft, as shown in Figure 6.14. Particles such as neutrons with a wide range of velocities arrive at the left disc and can pass beyond it when the slot in the disc is vertical. If the particle's speed is just right, then the slot in the right disc is vertical when the particle arrives, and the particle is passed. If the angular separation between the slits is 1 min of arc (60′ = 1°), what speed allows a parti-

cle such as a neutron to pass completely through the selector?

    Answer: $v = 5.4 \times 10^5$ m/s

*10. A thin rod of length L is rotating about a vertical axis (see the dotted line in Figure 6.15) with an angular speed of 3 rad/s. The upper end of the rod moves through a circular arc whose length equals the length of the rod in $\frac{3}{4}$ s. What is the angle θ?

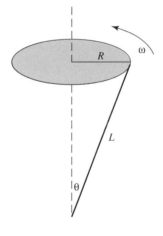

Figure 6.15 Rod Spinning about a Vertical Axis (Exercise 10)

    Answer: $\theta = 26.4°$

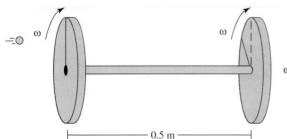

$\omega = 3000$ rpm

Figure 6.14 Particle Speed Selector (Exercise 9)

Rotational Energy and Moment of Inertia

11. The wheel shown in Figure 6.16 has a rim of mass 0.5 kg with a radius of 30.5 cm. Each of the four spokes has a mass of 0.1 kg. What is the moment of inertia of the wheel? ($I = \frac{1}{3}mL^2$ for a thin rod with the axis of rotation through one end. $I = \frac{1}{12}mL^2$ for a thin rod of length L with the axis through the center.)

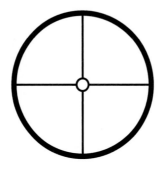

Figure 6.16 Four-Spoked Wheel (Exercise 11)

Answer: $I_T = 0.0589$ kg-m^2

12. What is the total kinetic energy for a bowling ball of mass 7 kg rolling at 6 m/s? ($I = \frac{2}{5}mr^2$ for a solid sphere rotating about its center.)

Answer: KE = 176.4 J

13. The wheel in Exercise 11 is modified so that the spokes are thin, rectangular strips set at an angle of 45°, just like the blades of a child's helicopter toy. (The modification of the spoke's shape from cylinder to rectangle doesn't affect the wheel's moment of inertia.) If the wheel is given an initial angular velocity of 20 rad/s such that the wheel rises into the air, how high will the wheel rise? Assume minimum friction and 100% blade efficiency.

Answer: $h = 1.3$ m

Combined Rotational and Translational Motion/Rotational Kinetic Energy

*14. A thin hoop and a hollow ball are released from rest at the top of an inclined plane (see Figure 6.17). Both have the same mass and diameter. (a) Which object will arrive at the bottom of the ramp first? (b) How much later does the slower object arrive at the ramp's bottom than the faster? ($I = mr^2$ for a thin hoop; $I = \frac{2}{3}mr^2$ for a hollow sphere with a thin wall.)

Figure 6.17 Race Down an Incline between a Hoop and a Hollow Ball (Exercise 14)

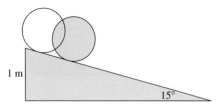

1 m

15°

Answer: (a) Ball
(b) 0.2 s

15. A ball is rolling with a translational velocity of 4 m/s. It encounters a rise of height 0.1 m and tries to roll up it. (a) Will the ball be able to roll to the top of the rise? (b) If so, what is the ball's linear speed at the top of the rise? (c) If not, how high up the rise does the ball go? Assume that the ball is a uniform sphere.

Answer: (a) Yes
(b) 1.4 m/s

*16. A yo-yo's axle rides on a narrow inclined plane, and when released from the top, the yo-yo slowly rolls down the plane and speeds up when it hits the floor. The diameter of the axle is $\frac{1}{5}$ the diameter of the outer discs. (a) What speed does the yo-yo have at the bottom of the incline before it hits the floor? (b) What is the linear speed of the yo-yo after it hits the floor? The height of the plane is h, and the mass of the yo-yo's axle is negligible.

Answer: (a) $v = \sqrt{\dfrac{4gh}{27}}$

(b) $v = 5\sqrt{\dfrac{4gh}{27}}$

Torque

17. A rope 1.5 m long is wound around the rim of the flywheel of a large gyroscope. The flywheel has a diameter of 20 cm and a moment of inertia of 0.025 kg-m^2. The rope is pulled with an average force of 3.5 N until it completely unwinds without slipping. (a) If the flywheel rotates with negligible friction, what is its final angular velocity? (b) What is its

kinetic energy? (c) What is its angular momentum? (d) How much work is done by the force unwinding the rope?

Answer: (a) 20.5 rad/s
(b) KE = 5.25 J
(c) $L = 0.51$ kg-m^2/s
(d) $W = 5.25$ J

18. A thin, hollow cylinder with its axis of rotation at its center has a string wrapped around it that is pulled with a force equal to the cylinder's weight. What is the string's acceleration?

Answer: a = g

*19. A thin rod has a pivot at one end (see Figure 6.18). The rod is held in a horizontal position and then released. What is the rod's maximum angular velocity? (*Hint:* To find the PE, use the rod's center of gravity which is located halfway along its length.)

Figure 6.18 Pivoted Rod Allowed to Fall (Exercise 19)

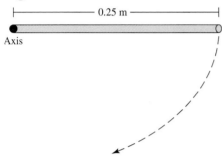

Answer: 10.8 rad/s

20. A weight *m* is tied to a string of length *h*, and the string is wound around the rim of a uniform, disc-shaped flywheel of mass *M* and radius *R* with its axis of rotation horizontal. How fast is the weight *m* falling after it unwraps the string? Assume that there is negligible friction and that the string completely unwraps without slipping.

Answer: $v = 2\sqrt{\dfrac{mgh}{2m + M}}$

*21. A shaft extends from a closed box. An electric motor is connected to the shaft and when started goes from 0 to 500 rpm in 3 s. The motor applies a constant torque to the shaft of 0.13 N-m. The shaft is supposed to be connected to a solid disc flywheel of mass 1.37 kg and radius of 7.5 cm inside the box. Is the object inside the box the supposed flywheel, or is it something else? If something else, what can it be? Prove it.

Answer: A hoop with the same mass and radius

22. A metal blade is pressed against the rim of a grinding wheel with a diameter of 20 cm and a mass of 1.6 kg. It is turned by an electric motor at a speed of 2400 rpm. The metal-to-wheel contact has a coefficient of friction of 0.8. When the motor is shut off, with how much force must you press the blade against the wheel to stop the wheel in 20 s? Assume negligible friction in the bearings.

Answer: F' = 1.26 N

23. Two masses of mass *m* and 2*m* are hung down from a frictionless, disc-shaped pulley of mass *m* to make an Atwood's machine. Show that the acceleration of the masses hung over the pulley is $\frac{2}{7}g$. (*Hint:* The string doesn't slip on the pulley, and thus the tension in each line is of a different magnitude.)

*24. A mass of 1 kg is attached to a shaft by string wrapped around it. The shaft has a radius of 7 cm and negligible mass. The shaft is horizontal. Attached to one end of the shaft is a dumbbell-shaped flywheel. The center of the dumbbell's rod is connected perpendicularly to the shaft's end. At each end of the dumbbell is a mass of 0.4 kg, and the rod connecting the two masses is of negligible mass. The mass is released from rest 0.75 m above the floor, and it has a speed of 3 m/s just before hitting the floor. How far apart are the two masses on the dumbbell?

Answer: D = 23 cm

Angular Momentum and the Conservation of Angular Momentum

25. A uniform disc of mass *m* and radius *r* is rotating about its center on bearings with negligible friction, with angular velocity ω_0. A thin hoop of the same mass and radius is dropped concentrically on the horizontally rotating disc so that their rims align. What is the final angular velocity of the system? ($I_{\text{hoop}} = mr^2$; $I_{\text{disc}} = \frac{1}{2}mr^2$.)

Answer: $\omega_f = \frac{1}{3}\omega_0$

26. A hockey puck of mass *m* is on a frictionless table. The puck is tied to a string that passes through a hole in the center of the table. The string is tied to a hook in the floor under the table. The puck is initially moving in a circle of radius *R* at angular speed ω. (a) If the string is pulled down so that the radius of the puck's circular path becomes *R*/2, what is the puck's new angular velocity? (b) How much work is done in pulling the string down?

Answer: (a) $\omega_f = 4\omega$

(b) $\omega = \frac{3}{2}mR^2\omega^2$

27. A merry-go-round in a playground has a 200-kg, disc-shaped platform with a radius of 1.8 m. It rotates with negligible friction. A 35-kg boy is at the center of the merry-go-round while a friend turns it up to a speed of 0.3 rev/s. If the boy in the center then walks to the rim of the merry-go-round, how fast will it be turning?

Answer: ω_f = 0.22 rev/s

28. Find an expression for the precessional velocity of the bicycle wheel of Figure 6.11 in terms of the wheel's mass, radius, and rim speed (assume a hoop shape for the wheel); the small radius *r* of suspension; and the angle θ of the axle's droop from the vertical.

Answer: $\omega_p = \dfrac{Rv\theta}{r^2}$

Power

29. A dynamometer is an instrument used to find the power that an engine is capable of delivering. If an internal combustion engine is put on a dyno and produces 25 N-m of torque at 3600 rpm, what is the output in horsepower?

    Answer: $P = 12.6$ hp

30. A freight elevator must be able to lift 7000 kg at a speed of 3 m/s. (a) If the winding drum is 2 m in diameter, what must the minimum output power of the motor be? (b) At what rpm is this power delivered?

    Answer: (a) $P = 276$ hp
    (b) 28.6 rpm

31. An electric motor develops 2.5 kW at 1800 rpm. (a) If a drum 0.25 m in diameter is bolted to the motor's output shaft, what is the heaviest mass that the motor can lift? (b) At what linear speed will the weight be lifted?

    Answer: (a) $m = 10.8$ kg
    (b) $v = 23.6$ m/s

A satellite orbits
because of the balance
between gravitational
and centripetal forces.

Centripetal Force, Centrifugal Force, and Gravitation

At the end of this chapter, the reader should be able to:

1. Explain the difference between *tangential acceleration* and *centripetal acceleration.*

2. Draw a free-body diagram of an object moving in a vertical circle, and use centripetal acceleration $\mathbf{a}_c$ to solve exercises that require Newton's second law.

3. Describe how *gravity* keeps objects in orbit around the earth.

4. Solve problems that involve freely orbiting bodies, and calculate such facts as the period of revolution and the velocity of an orbiting body and the radius of its orbit.

5. Solve for the minimum and maximum speeds that a vehicle would need to stay on any given banked turn, with or without friction present.

7.1   Centripetal Force

The natural trajectory for a moving object with no forces acting on it is a straight line. An object can also move in a straight line with a force acting on it if the force is in the direction of the object's motion. In studying uniformly accelerated motion in one dimension, we considered only a straight-line trajectory. Since acceleration is a vector, the possibility existed that the direction of the acceleration could be opposite that of the object's velocity. The object would then decelerate, or slow down, but the trajectory would still be rectilinear (a straight line). However, if the force is not in the direction of the object's velocity, then the object will not follow a straight line. If an object moves in a path that isn't straight, then there is a force acting on it.

Consider a ball that is thrown to someone some distance away. It is under the influence of gravity and follows a parabolic (curved) path. Such a path is shown in Figure 7.1. The force of gravity acts vertically downward. The initial velocity of the ball just as it is thrown $(\mathbf{v}_0)$ has a horizontal component (v_{0x}) and a vertical component (v_{0y}). In the absence of air friction, the horizontal component remains constant because no force acts on it horizontally. However, there is a force acting vertically: gravity. Gravity causes the ball's vertical component of velocity to decrease to zero as it climbs to its maximum height and then to increase in the opposite direction as it falls back down to the earth. The force of gravity "bends" the path of the ball out of the rectilinear into a parabola. The force of gravity is always acting at some angle to the ball's velocity and never directly in line with the velocity unless the ball is initially thrown straight up, as in Chapter 1.

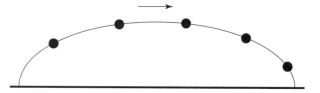

Figure 7.1 An Object Following a Curved Path Has a Net Force on It.

Let's look at simple circular motion. A string is tied to a ball and the ball whirled in a horizontal circle with a constant speed, as shown in Figure 7.2. If the string is cut, the ball will fly off at a tangent to the circle in a straight line, as shown in the figure. The uncut string keeps the ball following a circular path, and the tension in the string furnishes the force that bends the ball's path from the linear into the circular. The force ceases when the string is cut, and with no forces acting on the ball, its motion becomes a straight line. The ball flies off in a straight line tangent to the circle.

From this simple demonstration two facts can be deduced: (1) A force is acting on the ball (it doesn't follow a linear path). (2) That force is pointing toward the center of the circle (something is constantly pulling the ball inward, or it would fly off at a tangent to the circle in a straight line). *This inward-pointing force is called **centripetal force.*** In this particular case the centripetal force is supplied by the tension in the string. But what *is* centripetal force?

From Newton's laws we know that a force equals mass times acceleration.

$$\vec{\mathbf{F}} = m\vec{\mathbf{a}}$$

Centripetal force: The force generated from the change in direction of a velocity directed toward the center of the circle in which the object moves.

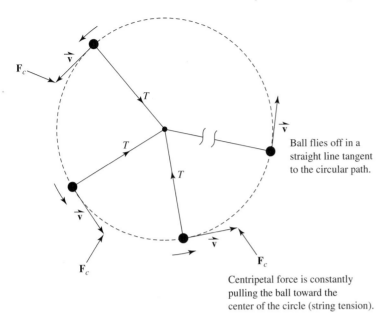

Ball flies off in a straight line tangent to the circular path.

Centripetal force is constantly pulling the ball toward the center of the circle (string tension).

Figure 7.2 Top View of a Ball Moving in a Circle at a Constant Speed

The acceleration is a changing velocity with respect to time. The equation for force can be rewritten as

$$\vec{\mathbf{F}} = m\frac{\overrightarrow{\Delta \mathbf{v}}}{\Delta t}$$

Velocity is a vector with both a magnitude *and* a direction. Velocity is constantly changing, thereby causing the centripetal force. If the speed of the ball whirling in the circle remained constant, then the magnitude of the velocity could not be changing. Instead, the velocity's *direction* was constantly changing, causing the acceleration that in turn caused the centripetal force. (This acceleration, by the way, is called *centripetal acceleration.*)

In the large circle of Figure 7.3A, the ball is moving clockwise at a constant speed. The instantaneous velocity **v** is drawn at point A tangent to the circle's circumference. The velocity's direction is tangent to the circle because the ball would follow that path at that point if the string were cut. At some time Δt later, the ball has traveled to point A' on the circle's circumference. The instantaneous velocity now has direction **v′**, as shown. In the time Δt, the radius vector sweeps out an angle $\Delta\theta$, and an arc length $v\,\Delta t$ is traced out as well. This situation is displayed in Figure 7.3B. In the same time Δt, the velocity changes direction from **v** (to the right and down) to **v′** (to the left and down). If the two vectors **v** and **v′** are placed tail to tail, the angle between the two velocity vectors is also $\Delta\theta$, as in Figure 7.3C. (The lengths of the sides of the isosceles triangle in Figure 7.3C are the same because the magnitudes of the velocities are the same.) The triangles in Figures 7.3B and 7.3C are similar isosceles triangles, and direct comparisons can be made between them. The ratio of a side to the base is the same for both triangles. (Other assumptions made here are that the angle $\Delta\theta$ is small and that the arc length $v\,\Delta t$ approximates the length of the dotted line of the base of the triangle in Figure 7.3B.)

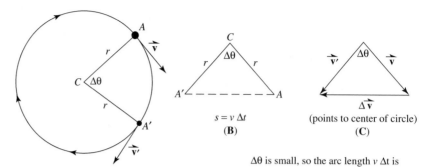

Figure 7.3 Vector Origin of
Centripetal Force

$\Delta\theta$ is small, so the arc length $v\,\Delta t$ is
approximately equal to length $A'A$.

$$\frac{\Delta v}{v} = \frac{v\,\Delta t}{r}$$

Algebraic manipulation of the above equation yields the acceleration.

$$\frac{\Delta v}{\Delta t} = \frac{v^2}{r}$$

$$a = \frac{\Delta v}{\Delta t}$$

(Equation 7.1)

$$a = \frac{v^2}{r}$$

Equation (7.1) is for the magnitude of the centripetal acceleration. A force is mass times acceleration. If the mass of the ball spinning in a circle is m, then the equation for centripetal force becomes

(Equation 7.2)

$$F_c = \frac{mv^2}{r}$$

Equation (7.2) is for the centripetal force. Look again at Figure 7.3C. The figure is drawn with **v** and **v′** to scale with the circle of Figure 7.3A. Notice that the change in velocity (Δ**v**) is pointed toward the center of the circle. The direction of the change in velocity gives the direction to the centripetal force. *This change in velocity is always toward the center of the circle. Therefore, the **centripetal force** is always directed toward the center of the circle.* It's the centripetal force that keeps the ball moving in the circle. In the example above, the tension in the string directed inward supplies this centripetal force, but later we'll see that tension isn't the only force that can keep things moving in a circle.

7.2 Centripetal and Angular Acceleration

Equation (7.1) for centripetal acceleration was derived purely from kinematics. It is worth repeating that this acceleration came *not* from a change in the magnitude of the velocity, but from the change in its direction. (The ball at the end of the string was spinning with a constant speed.) Anything that rotates will have a centripetal acceleration associated with it. If, as in Chapter 6, an angular acceleration is also associated with a spinning object (it rotates faster and faster), then a direction is also associated with the angular acceleration.

For example, if the ball rotating in the circle in Figure 7.4 possessed both centripetal *and* angular acceleration, the direction of the ball's instantaneous velocity would still be tangential to the circular path. But the angular acceleration causes the magnitude of the ball's velocity to increase. This increase is in the direction of the ball's instantaneous velocity (tangent to the circle), and the direction of the angular acceleration is also tangential to the circular path. However, as shown in Section 7.1, the direction of the centripetal acceleration is always toward the circle's center. Since both the angular and the centripetal acceleration act simultaneously, and since both are vectors, the resultant total acceleration is the vector sum of these two perpendicular accelerations.

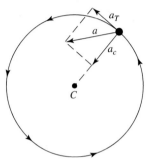

Figure 7.4 Centripetal and Angular Acceleration Acting Simultaneously on an Object

EXAMPLE 7.1

A ball at the end of a string is being rotated in a horizontal circle. The radius of the circle is $\frac{1}{2}$ m, and the angular acceleration of the system is 1 rad/s^2. What is the total acceleration of the ball at the instant it has a speed of 1.5 m/s?

SOLUTION

$$a_T = r\alpha \qquad a_T = (0.5 \text{ m})(1 \text{ rad/s}^2)$$

$$a_T = 0.5 \text{ m/s}^2$$

$$a_c = \frac{v^2}{r} \qquad a_c = \frac{(1.5 \text{ m/s})^2}{0.5 \text{ m}}$$

$$a_c = 4.5 \text{ m/s}^2$$

Since the angular and centripetal accelerations are perpendicular to each other, we can use the Pythagorean theorem to find the total acceleration.

$$a = \sqrt{a_T^2 + a_c^2}$$
$$a = \sqrt{(0.5 \text{ m/s}^2)^2 + (4.5 \text{ m/s}^2)^2}$$
$$a = 4.53 \text{ m/s}^2$$

In Example 7.1, the angular acceleration seems to contribute little to the total acceleration of the ball. However, with an angular acceleration, as time goes on, the speed of the ball continues to increase. The centripetal acceleration is a function of the ball's speed, and it also continues to increase. Since the centripetal acceleration goes as the square of the speed, the total acceleration bends more and more in the direction of the circle's center.

If there is an angular acceleration, there is *always* a centripetal acceleration associated with it. However, it doesn't always follow that if there is a centripetal acceleration, there is an angular acceleration as well. Although a centripetal and an angular acceleration may act simultaneously, in many problems they can be treated singly.

7.3   Centrifugal Force

Centrifugal force: The reaction (from Newton's third law) to the centripetal force. We experience it as the normal force from the object in which we're riding when the object moves in a curved path.

It is worth taking a second look at the ball tied to a string and spun in a horizontal circle. In this example, however, we replace the string with a long, thin rubber band. The rubber band is elastic and will stretch with increasing tension. If the ball is again spun in a horizontal circle, the radius of the circle will increase as the ball is spun faster and faster. Thus, the tension in the rubber band is also increasing. If the ball is spun fast enough, the rubber band will break, and the ball will fly off tangent to the circle as in Figure 7.2. We already know that the centripetal force acting along the string pointing toward the center of the circle is what keeps the ball rotating in the circle. But the tension in the string seems to act in the opposite direction. As the velocity increases, the tension tries to pull the string apart. We know that the string is under tension because it wouldn't remain taut under compression.

We've already seen that tension can act in both directions in the force diagrams drawn to find the acceleration of a system. From the whirling ball's point of view, the string's tension acts as a force constantly pulling it toward the center of the circle. But the centrifugal force acts in the opposite direction from the centripetal force, and it has the same magnitude. The centrifugal force tries to pull the string apart. ***Centrifugal force*** *is a reaction force; it's a reaction to the centripetal force.*

Centripetal force is the primary force and causes (from Newton's third law) a reaction to it. For example, when we drive around a fast turn in a car, we experience centrifugal force as a force pushing us to the outside of the turn. This force is used in theme-park rides. You can feel the *g* forces when going around a high-speed turn on a roller coaster, for instance. In another ride, you are locked into a cage, and the cage is initially spun in a horizontal circle. As the speed comes up to its maximum, the plane of the circle is tilted until it is vertical. The locking bar is then removed, and at the top you seem to defy gravity and don't fall out of the cage. What you experience as the centrifugal force (you feel pushed against the cage's outer surface) is actually the normal force of the cage pushing you toward the center of the circle.

The military pilots of highly maneuverable, fast jet fighters can be endangered by the centrifugal forces caused by a high-speed turn or loop. The forces can cause the pilot to lose consciousness and black out. These forces are typically called *g* forces because they are measured in terms of multiples of a standard gravity, 9.8 m/s^2. A pilot experiencing 3*g*'s weighs three times his normal weight. Since the plane can usually stand many more *g*'s than the pilot can, it is important to keep the radius of a turn above a certain minimum at a given speed, or the speed below a certain maximum for a given radius. Thus, it is very important to know the *g* forces that a pilot can stand before blacking out. In order to find out what types of limits should be imposed on the pilots of fighter aircraft, a device called a *centrifuge* is used to expose test subjects to high *g* forces. The centrifuge is simply a cage at the end of a long, horizontal, metal boom, and the subject is strapped into the cage. The cage is then spun around at different speeds to generate different *g* forces.

There is a way in which this centripetal force can seemingly be experienced; it is in some ways akin to bungee jumping. Assume that you have a belt around your waist, and a strong rope is tied to the belt. If you were then spun in a horizontal circle, you would appear as in Figure 7.5. Centrifugal force would keep the rope taut and would throw your arms and legs toward the outside of the circle. But what you would experience would be the rope pulling at your waist toward the center of the circle. The faster you spun, the harder the rope would pull at your midsection. You would feel as if the tension in the rope were pulling you toward the center of the circle. If your eyes were closed and there was no wind, you couldn't tell that you were spinning in a circle, but only that there was a force on the rope accelerating you in the direction of the rope's tension.

Velocity is out of the page, toward the viewer.

Figure 7.5 Side View of a Person Being Whirled around in a Horizontal Circle

EXAMPLE 7.2

Fifteen-lb test fishing line will break when a tension of 15 lb is applied to it. If a large steel ball bearing weighing 6 lb is tied to such fishing line and swung around in a horizontal circle of 2-m radius, at what rpm would the line break?

SOLUTION

$$T = \frac{mv^2}{r}$$

$$v^2 = \frac{rT}{m} \qquad v = \sqrt{\frac{rT}{m}}$$

$$m = \frac{6 \text{ lb}}{32 \text{ ft/s}^2} \qquad m = 0.1875 \text{ slug}$$

$$v = \sqrt{\frac{(2 \text{ m})(15 \text{ lb})}{0.1875 \text{ slug}}} \qquad v = 12.65 \text{ ft/s}$$

$$v = \omega r$$

$$\omega = \frac{v}{r}$$

$$\omega = \frac{12.65 \text{ ft/s}}{2 \text{ m}}$$

$$\omega = 6.33 \text{ rad/s}$$

$$\left(\frac{6.33 \text{ rad}}{\text{s}}\right)\left(\frac{1 \text{ rev}}{2\pi \text{ rad}}\right)\left(\frac{60 \text{ s}}{\text{min}}\right) = 60.5 \text{ rpm}$$

7.4   Motion in a Vertical Circle

In our example system of a ball tied to a string, the ball has, up to now, been spun in a horizontal circle. The reason for this has been to minimize the effect of gravity on the ball. If the plane of the circular path is vertical, then the influence of gravity must be taken into account. In a horizontal circle, the speed of the ball could be constant if no angular acceleration were present. In a vertical circle, however, the speed of the ball is constantly changing. At the top of the circle, the speed is at a minimum, and at the bottom it is at a maximum. At the circle's top, gravitational potential energy is at a maximum, and at the bottom it is at a minimum. In moving from the top of the circle to the bottom, the ball converts gravitational PE to KE of motion, speeding up. As the ball climbs from the bottom of the circle, it slows down because the KE is being converted back to PE. In a vertical circle, there is always an angular acceleration (or deceleration) because of the constantly changing speed of the ball, as shown in Figure 7.6.

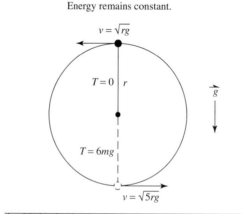

Figure 7.6 Object Rotating in a Vertical Circle

What are the minimum conditions for a vertical circular path? The ball must not fall out of the circle at the top, so at the top of the circle, the centrifugal force up must exactly balance the force of gravity (weight) down.

$$F_c = W \qquad F_c = \frac{mv^2}{r} \qquad W = mg$$

$$\frac{mv^2}{r} = mg \qquad \frac{v^2}{r} = g \qquad v = \sqrt{rg}$$

The ball must have a minimum speed of $(rg)^{1/2}$ at the top, or the string will go slack and the ball will fall out of the circle. Conservation of energy can be applied to this situation, and the total energy E_T at the top of the circle can be calculated.

$$E_T = KE + PE \qquad KE = \frac{1}{2}mv^2_{top}$$

$$PE = mgh \qquad h = 2r \qquad PE = 2mgr$$

$$E_T = \frac{1}{2}mv^2_{top} + 2mgr \qquad v_{top} = (rg)^{1/2}$$

$$E_T = \frac{1}{2}m(\sqrt{rg})^2 + 2mgr$$

$$E_T = \frac{1}{2}mrg + 2mrg$$

$$E_T = \frac{5}{2}mrg$$

If the ball has the total minimum energy above, what is its speed at the bottom of the circle, and what is the tension in the string? At the bottom of the circle, all of the energy is kinetic, and since energy is conserved, the total energy remains the same.

$$E_T = \frac{1}{2}mv^2_b \qquad E_T = \frac{5}{2}mrg$$

$$\frac{5}{2}mrg = \frac{1}{2}mv^2_b \qquad 5rg = v^2_b$$

$$v_b = \sqrt{5rg}$$

At the bottom of the circle, the tension in the string is a combination of the centrifugal force and the weight. (Both act down at the bottom.)

$$T = mg + \frac{mv^2_b}{r}$$

$$T = mg + \frac{m(\sqrt{5rg})^2}{r}$$

$$T = mg + \frac{5mgr}{r}$$

$$T = 6mg$$

At the bottom of the circle, the ball experiences a maximum force of 6g's. If we were to take the place of the ball in the above example, we would also experience 6g's at the bottom of the circle. However, at the top we would be essentially weightless, since the centrifugal force exactly counterbalances the weight. As the ball moves around the circle, its speed is constantly changing from its maximum at the bottom to its minimum at the top of the circle. A pilot executing a perfect loop at minimum conditions in an airplane (the plane doesn't fall out of the loop) experiences the same conditions described above. Doing such a loop in an aircraft that has an open cockpit can be quite disorienting the first few times, especially at the top of the loop. The plane spends more time at the top of the loop than it does at any other point on the circle. As the plane approaches the top, the g forces lessen until they are momentarily zero. The pilot is hanging upside down in the open a few thousand feet above the ground, seemingly about to fall out of the cockpit! The temptation to grab onto something to keep from falling out is overwhelming. However, even if the pilot weren't wearing the aerobatic harness, it would be impossible for him to fall out of the cockpit at that point. An object "falls" only when it possesses weight. At 0g, the pilot can't fall. If the plane were flying along in a straight line inverted (upside down), then the pilot could fall out, and the aerobatic harness would definitely be necessary. The pilot would feel his full weight being supported by the harness.

7.5   Gravitation

Gravity is the force that keeps us firmly fixed to the surface of this planet and simultaneously keeps the planets orbiting the sun. It regulates the universe at large and yet is the weakest of all of the known forces. The force of gravity has been used in this book quite frequently. It has appeared in the form of g, the downward-pointing acceleration of an object moving within the gravitational field of the earth. Since all objects close to the surface of the earth fall with the same acceleration, g was seldom referred to as the result of a force, but simply as an acceleration. But the acceleration of a falling object *is* a direct consequence of the attraction of the earth for that object. The force of gravity is also always attractive. A falling object falls straight down, so the line of force is along a straight line. To be more exact, a falling object falls radially inward toward the center of the earth. The form that the equation for the gravitational force takes is an inverse square law.

(Equation 7.3)

$$F = G\frac{m_1m_2}{r^2}$$

Thus, the force is inversely proportional to the square of the distance between the objects. The force is also directly proportional to the product of the two masses between which this force acts. Since force is a vector, it has both a magnitude and a direction. Its direction is along the line between the two masses. The objects exert equal and opposite forces on each other, and G is the proportionality constant that properly scales the force. Since there is a force between the two objects, it is natural to assume that if they were free to move, they would accelerate toward one another. Such is indeed the case. Whenever something is dropped, it acts in accordance with this law.

7.6   Weighing the Earth

The earth creates a strong gravitational field close to its surface. We use this nearly constant gravitational field near the surface in weighing things. We place an object upon a scale and read the weight. A scale relies upon the gravitational attraction between the object and the earth to compress or stretch a spring or to counterbalance another calibrated weight.

But this approach is hardly practical when trying to obtain the mass of something as large as a planet. Yet the mass of the earth is known. How was the earth "weighed"?

First, the form of the gravitational law had to be determined, and it was deduced from astronomical observations. Since the orbits of the planets around the sun are elliptical (a stretched circle), the force between the sun and the planet had to be an inverse square law, as given in Equation (7.3). Next, some data were needed as to the strength of the gravitational field in the vicinity of a massive body. These data are readily at hand in the acceleration of gravity, g, near the surface of the earth. Knowing the inverse square equation for gravity, and knowing that force is equal to a mass times an acceleration, we can arrive at an expression for the mass of the earth.

$$F_w = mg \qquad F = G\frac{mm_e}{r^2} \qquad F_w = F$$

$$mg = G\frac{mm_e}{r^2}$$

$$g = G\frac{m_e}{r^2}$$

$$m_e = \frac{r^2 g}{G}$$

Knowing g, we must also know the distance r separating the two bodies and the gravitational proportionality constant G. The distance r is the radius of the earth, r_e. Sir Isaac Newton used calculus to determine that a spherical massive body such as the earth acted as if all of its mass were concentrated at its center. The radius of the earth was already known, having been calculated with some accuracy as far back in time as ancient Greece. This left the determination of G, which was calculated for the first time by Lord Cavendish in 1798. The apparatus used consisted of a torsional pendulum and four massive lead spheres, as shown in Figure 7.7. Two identical spheres (m_1) were mounted on the torsional pendulum. The torsional pendulum is a device that hangs from a steel thread that acts like a spring. But instead of being compressed or stretched to store energy, it stores energy by being twisted. This apparatus can measure very small forces, on the order of 1×10^{-6} N. The two other masses (m_2) were mounted on a platform that could be rotated about a central pivot that was directly under the vertical torsional spring.

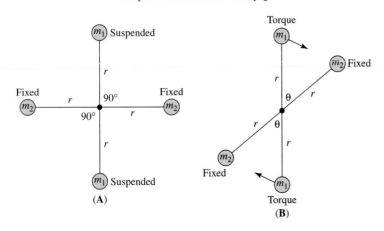

(A)

(B)

Figure 7.7 Torsional Pendulum Apparatus Used to Find the Gravitational Constant, G

Initially, the two arms (of length $2r$) had an angular separation of 90° (see Figure 7.7A). In this position, all of the gravitational forces between the masses canceled out, and there was no torque on the spring. Then the fixed platform was rotated to a small angle of separation θ, as shown in Figure 7.7B. This rotation caused a net torque upon the spring, and it twisted (rotated) a small amount in response to the gravitational attraction of the spheres. The rotation was measured by bouncing a light beam from a plane mirror attached to the thread and observing how far from its original position a spot on the wall moved. In this way the gravitational proportionality constant was measured.

In measuring G, the gravitational constant, we run into Hooke's law again, but this time it's for a spring that's in rotation. The law for the rotating spring has exactly the same mathematical form as Hooke's law for linear extension.

(Equation 7.4)

$$\tau = k\theta$$

where τ is the torque on the spring, θ is the angle through which the spring rotates when the torque is applied to the spring, and k is the spring constant in N-m/rad.

EXAMPLE 7.3

Knowing the torsional spring constant (k), the radial length of the arms of the torsional balance apparatus (r), the two masses m_1 and m_2, and the angle through which the spring rotates (θ), find the universal gravitational constant, G. Assume that θ is small and in radians.

SOLUTION

$$\tau = k\theta \qquad \tau = 2(r \times F) \qquad \tau = 2rF \qquad 2rF = k\theta$$

$$F = \frac{Gm_1m_2}{s^2} \qquad s = r\theta \qquad F = \frac{Gm_1m_2}{r^2\theta^2}$$

$$2r\left(\frac{Gm_1m_2}{r^2\theta^2}\right) = k\theta \qquad \frac{2Gm_1m_2}{r\theta^2} = k\theta$$

$$G = \frac{rk\theta^3}{2m_1m_2}$$

Now with G measured, we can calculate the mass of the earth.

$$m_e = \frac{r^2g}{G}$$

$$r_e = 6.38 \times 10^6 \text{ m} \qquad g = 9.8 \text{ m/s}^2$$

$$G = 6.67 \times 10^{-11} \text{ N-m}^2/\text{kg}^2$$

$$m_e = \frac{(6.38 \times 10^6 \text{ m})^2(9.8 \text{ m/s}^2)}{6.67 \times 10^{-11} \text{ N-m}^2/\text{kg}^2}$$

$$m_e = 5.95 \times 10^{24} \text{ kg}$$

7.7   Orbiting Bodies

Since the Renaissance, it has been known (but not universally accepted) that the earth is a spherical planet that orbits the sun. The earth is one of nine planets that orbit a rather ordinary star within the outer regions of our galaxy. The earth's orbit isn't really circular, but elliptical. (An ellipse is an elongated circle.) Thus, part of the earth's orbit is closer to the sun than the rest of the orbit. The same holds true for the planets and moons, and also for most of the asteroids in the asteroid belt between Mars and Jupiter. Comets have highly elliptical orbits, and their orbits can't be approximated as circular; but the earth's elliptical orbit is close enough to circular that to a first approximation it can be considered circular. In this book we'll assume that, to a first approximation, orbits are circular.

It is the attraction of the sun's gravitational field that causes the earth to orbit. The force that caused the legendary apple to fall upon Newton's head is the same one that causes the planets to orbit the sun, and our moon to orbit the earth. The gravitational force between two massive objects acts in a straight line between them. So what keeps the earth from falling into the sun, and the moon from falling into the earth? They *would* fall if they were stopped in their orbits, but the earth, moon, and planets are not stationary. The earth, for instance, is *constantly* falling inward toward the sun. But for every meter it falls inward, its orbital motion displaces it sideways so that it always maintains the same distance from the center of the sun.

The set of points that are all equidistant from a single point describe the surface of a sphere. Any slice through this sphere and through the central point will describe a circle centered on that point. Since this constant sideways motion is circular, a centripetal force is present. Remember, centripetal force requires a force that acts in a straight line between the object traveling in a circular path and the center of the circle. That force was the tension in a string in the example with the ball. The force had to point along the string in the direction of the center of the circle, since the string was always tugging the ball inward. In the case of our earth orbiting the sun, the force is the gravitational force. That gravitational force is always pulling the earth inward toward the sun. The centripetal force generated by motion in a circle is equal to the gravitational tug of the sun on the earth.

$$F_G = G\frac{m_e m_s}{r^2} \qquad F_c = \frac{m_e v^2}{r} \qquad F_G = F_c$$

$$G\frac{m_e m_s}{r^2} = \frac{m_e v^2}{r}$$

$$G\frac{m_s}{r} = v^2$$

where m_e is the mass of the earth, m_s is the mass of the sun, and r is the distance from the earth to the sun.

The equation above governs the earth's orbit about the sun. Notice that the mass of the earth cancels out. *Any* body orbiting at the distance from the sun that the earth is now, regardless of its mass, *must* orbit at the same speed. The above equation governs the orbits of all of the planets about the sun, from Mercury to Pluto. The mass of the sun alone dictates the orbital speed of the planet at its distance from the sun. Although the planet's mass cancels out for the above equation, that doesn't mean that the mass is gone! The planet's mass is the "handle" that allows the sun to interact with the earth gravitationally and that makes centripetal force possible.

If the sun is continually tugging at the earth, the earth is forever falling into the sun (see Figure 7.8). In the case of the ball being swung around in a circle at the end of a string, if the string were cut, the ball would fly off tangent to the circle. Similarly, if the gravitational attraction between the earth and the sun were cut off, the earth would fly off at a tangent to its orbit, as shown in Figure 7.8. However, because the sun is continuously pulling at the earth, the earth stays in its circular orbit. For every meter that the earth's tangential velocity would move the earth "above" its circular orbit, the earth falls 1 m toward the sun. So the earth is continuously falling toward the sun! As a matter of fact, if the earth's orbital velocity were suddenly reduced to zero, it *would* fall directly into the sun. The same is true of the moon. If it were stopped in its tracks, it would fall straight into the earth. In fact, if the orbital velocity of the earth were reduced slowly, the earth would spiral into the sun. There is no friction present in space (space is a nearly perfect vacuum), so any reduction in orbital speed is a rather remote possibility.

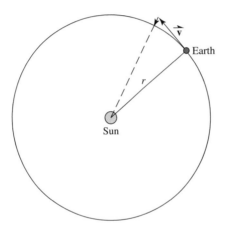

Figure 7.8 The Earth Orbiting the Sun

The orbital equation derived above says that as the radius of the orbit increases, the necessary orbital velocity drops. For example, the orbital period for Mercury, the planet closest to the sun, is 88 days. Mercury has a high orbital speed and completes an orbit quickly. However, the orbital period of Pluto (the planet farthest from the sun) is 248 *years*. Remember that in our derivation of the orbital equation above, the centripetal force and the gravitational force were equal. Since the centripetal force goes as velocity squared, reducing the orbital speed reduces the centripetal force. The two are no longer balanced, and the gravitational tug is stronger. The slow spiral into the sun would then begin! The retrorockets of early spacecraft such as the *Mercury* and *Gemini* capsules were used in exactly this way to deorbit the capsule. The rockets were fired in the direction of the orbit, reducing the orbital speed, and the capsule spiraled down to a landing on the earth.

How much energy would it take to leave the earth entirely? This energy is usually expressed in terms of escape velocity, and any body possessing this velocity in a direction radially outward from the earth will escape its gravitational pull forever. The reason the energy required is expressed as a velocity will be apparent in the following derivation. Since the earth's gravitational field is a force field, it requires work to overcome it. The work done against a force is the product of the force times the distance that an object moves against this force.

$$W = Fr$$

In calculating the work done in raising a body to a height h near the earth's surface, we found that the work done was simply $(mg)h$. The height that the body was raised was insignificant compared to the earth's radius, and the gravitational field (as expressed in terms of g) remained essentially constant. However, as an object escapes the earth, the gravitational force will not remain constant. It goes as 1 over r squared $(1/r^2)$. So to find the energy required to leave the earth, we have to add up many small increments of work to arrive at the total work. Each increment is small enough so that within the increment, the force is approximately constant.

$$\Delta W = F \, \Delta r$$

To find the total work done, we add up the work increments for each small displacement.

$$W_T = \Sigma \, F \, \Delta r$$

where Σ is the symbol for a summation and is referred to as a *summation sign*. On the left of the equal sign, if we sum up all of the infinitesimally small bits of work, we obtain the total work done.

$$F = G\frac{m_s m_e}{r^2}$$

$$W_T = \Sigma \, G\frac{m_s m_e}{r^2} \, \Delta r$$

$$W_T = Gm_s m_e \, \Sigma \, \frac{\Delta r}{r^2}$$

The constants are taken out from behind the right side of the summation sign, because we add over only what changes. We deal with what remains to the right of the summation sign as discussed in the accompanying Optional Topic.

O P T I O N A L   T O P I C

Derivation of Gravitational Potential Energy

To solve this summation, we will not use a graphical method. Instead, we will return to a solution by first looking at the definition of the slope of a line. The slope is the change in the y coordinate with respect to the change of the x coordinate over the same interval. Since $y = mx + b$ is the equation of a straight line where m is the slope, then the slope is $\Delta y/\Delta x$. The slope is constant for a straight line because the tilt is the same everywhere along it. We can generalize this idea of the slope to any function. If y is a function of x, or $y = f(x)$, then we can rewrite the slope formula as slope $= \Delta f(x)/\Delta x$. The slope is how a function is changing over an interval, and the change has a direction. If we were to look at the slope at any point on the rim of a circle, we would find that the slope is always tangent to the circle. It would be the direction in which the instantaneous velocity would be pointing for the ball in our example of the horizontal circle. (The speed of the ball is $\Delta s/\Delta t$ where s is the distance traveled along the circumference.)

With this new definition of the slope in mind, let's choose as our function of x the function $f(x) = 1/x$. We now apply the new definition of slope to $f(x)$.

$$\frac{\Delta f(x)}{\Delta x} = \frac{f(x + \Delta x) - f(x)}{\Delta x} \qquad f(x) = \frac{1}{x}$$

$$\frac{\Delta f(x)}{\Delta x} = \frac{\dfrac{1}{(x + \Delta x)} - \dfrac{1}{x}}{\Delta x}$$

$$\frac{\Delta f(x)}{\Delta x} = \frac{\dfrac{x}{x(x + \Delta x)} - \dfrac{x + \Delta x}{x(x + \Delta x)}}{\Delta x}$$

$$\frac{\Delta f(x)}{\Delta x} = \frac{-\Delta x}{\Delta x(x)(x + \Delta x)}$$

$$\frac{\Delta f(x)}{\Delta x} = \frac{-1}{x(x + \Delta x)}$$

We now let Δx go to zero ($\Delta x \rightarrow 0$).

$$\frac{\Delta f(x)}{\Delta x} = -\frac{1}{x^2}$$

If the general variable x becomes the displacement r, then

$$\frac{\Delta f(r)}{\Delta r} = -\frac{1}{r^2} \qquad \Delta f(r) = -\frac{\Delta r}{r^2} \qquad -\Delta f(r) = \frac{\Delta r}{r^2}$$

$$f(r) = \frac{1}{r} \qquad -\Delta\left(\frac{1}{r}\right) = \frac{\Delta r}{r^2}$$

$$W_T = Gm_s m_e \Sigma\left[-\Delta\left(\frac{1}{r}\right)\right]$$

$$W_T = -Gm_s m_e \Sigma \Delta\left(\frac{1}{r}\right)$$

In finding the total work, we're summing over small intervals of the function $1/r$.

(Equation 7.5)

$$W = -\frac{Gm_s m_e}{r}$$

Equation (7.5) gives us what we want: the expression for the amount of work done in overcoming a gravitational field. Yet it still remains to be evaluated. We must evaluate it over the limits of the radial distance r.

Since the spacecraft will be starting out from the surface of the earth, one of the limits will be r_e, the radius of the earth. Because we wish to escape entirely from the gravitational pull of the earth, the other limit for r will be infinity (∞). We can use infinity as a limit because the work at an infinite distance is zero. So the total work done against gravity in escaping from the earth is

$$W = -\frac{Gm_s m_e}{r} \bigg|_{r_e}^{\infty}$$

$$W = \frac{-Gm_s m_e}{\infty} - \frac{-Gm_s m_e}{r_e}$$

$$W = 0 + \frac{Gm_s m_e}{r_e}$$

$$W = \frac{Gm_s m_e}{r_e}$$

The spacecraft is given kinetic energy equal to this amount of work to enable it to escape from the earth.

$$\frac{1}{2}m_s v^2 = \frac{Gm_s m_e}{r_e}$$

Solving this equation for v, we get

$$v = \sqrt{2\frac{Gm_e}{r_e}}$$

Using the values obtained earlier for G, m_e, and r_e, we can calculate a value for the escape velocity of 11,154 m/s, which is roughly 25,000 mph. If the spacecraft has anything less than this speed, it will eventually return to the earth.

Since the escape velocity is a radial velocity, a spacecraft with less than the escape velocity will rise until its velocity drops to zero; then it will fall back to the earth. However, if the craft has a tangential velocity (parallel to the earth's surface) as well as a radial velocity (directly away from the earth), it has the possibility of orbiting the earth. Using the work done above for escape velocity, we can calculate the velocity necessary for the craft to rise to a given altitude.

$$\frac{1}{2}m_s v^2 = -G\frac{m_s m_e}{r} \bigg|_{r_e}^{h}$$

$$\frac{1}{2}m_s v^2 = -Gm_s m_e\left(\frac{1}{h} - \frac{1}{r_e}\right)$$

$$v = \sqrt{2Gm_e\left(\frac{1}{r_e} - \frac{1}{h}\right)}$$

The height h is the height to which we wish the spacecraft to rise relative to the earth's center. For example, assume that we wish to orbit a pay load in a low earth orbit, say, 100-mi altitude. We can calculate the radial velocity needed from the above equation. (Remember that the altitude is calculated from the center of the earth, not from the earth's surface. The gravitational field of the earth acts as if all of the earth's mass is concentrated at its center.) Therefore, the value that we use for h is 6.54×10^6 m, and the radial velocity needed to attain this altitude is 1745 m/s, or 3926 mph. However, if we want the spacecraft to orbit at this height, we also need the tangential speed dictated by the centripetal force equation earlier in the chapter.

$$v_T = \sqrt{\frac{Gm_e}{h}}$$

Again using $h = 6.54 \times 10^6$ m for our height, we obtain a value of 7790 m/s (17,527 mph) for the required tangential velocity. The tangential and radial velocities necessary for an orbit of 100-mi altitude add vectorially, as shown in the following equations and in Figure 7.9.

$$v = \sqrt{(17{,}527 \text{ mph})^2 + (3926 \text{ mph})^2}$$

$$v = 17{,}961 \text{ mph}$$

$$\tan\theta = \frac{3926 \text{ mph}}{17{,}527 \text{ mph}} \qquad \theta = 12.6°$$

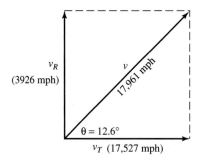

Figure 7.9 The Speed and Angle of Launch Needed to Orbit a Pay Load at 100 Miles

To orbit at an altitude of 100 mi, the spacecraft would have to have a velocity of 17,961 mph directed at 12.6° above the horizon. As the craft rose, its vertical speed would decrease until it was zero at an altitude of 100 mi. It would still be left with its tangential speed, however, which would be correct for orbit at that altitude. The values given are approximately correct for a space shuttle in a low earth orbit. They are an approximation because we didn't take into consideration the velocity of the earth's rotation, the plane of the orbit, the fact that the orbit would be elliptical instead of circular, and other such nuances. An orbital flight is actually quite an orchestration of many factors, all operating simultaneously, and it is no wonder that computers are necessary in the space program.

7.8   Centripetal Force on a Banked Track

A race car moving around a circular track will experience a centripetal force acting on it. With a flat track, the maximum speed around a corner is limited by the coefficient of friction between the tire and the track, because centrifugal force tries to throw the car off the track. Banking the track allows a higher top speed around the corners, and the amount of banking is quite noticeable on high-speed race tracks. For example, Daytona and Michigan International have very high banks, and the 500 track at Indianapolis has a moderate amount of banking. Centrifugal force also throws the car down onto the track and uses up much of the travel in its suspension.

The centripetal force on a car traveling around a banked turn allows the car to take the turn at a higher speed.

Figure 7.10 shows a car on a banked track and the centrifugal force acting on it. The centrifugal force always acts radially and is directed radially outward from the axis of the turn. (The arrow F_c for the centrifugal force is horizontal in the drawing.) The magnitude of the centrifugal force equals that of the centripetal force because it's a reaction to the centrifugal force. The horizontal centrifugal force can be resolved into a component up the banking (F_{cu}) and a component that is normal to and down into the track's surface (F_{cN}) as shown.

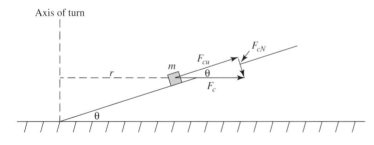

Velocity is out of the page.

$$F_c = \frac{mv^2}{r}$$
$$F_{cu} = F_c \cos \theta$$
$$F_{cN} = F_c \sin \theta$$

F_c = centrifugal force
F_{cu} = centrifugal force up the plane
F_{cN} = centrifugal force normal to the plane

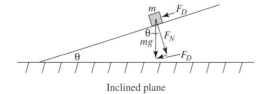

Inclined plane

Figure 7.10 Forces on a Race Car at Speed on a Banked Track

Remember that a banked track is also an inclined plane and that the normal force of the car on the track ($mg \cos \theta$) must be added to F_{cN}. The total normal force is increased, and therefore so is the force of friction opposing any movement up the track by the centrifugal force. This same increase in the normal force is also what pushes the car down onto its suspension so strongly. There is also a force down the plane from the inclined plane, and it is opposed by the centrifugal force component up the plane, F_{cu}. The force of friction will, as always, be in the direction opposite the motion of the car, either up or down the track. When solving a problem dealing with a banked track, draw the triangle for the inclined plane, and find the components parallel to the track and normal to the track. Then draw the centrifugal force triangle, and find the components parallel to and normal to the track. Only then combine the inclined and centrifugal parallel components, and the inclined and centrifugal normal components.

EXAMPLE 7.4

What is the maximum speed that a 1000-kg race car can have on a 15° banked race course going around a turn with a radius of 100 m? The coefficient of static friction between tire and track is 0.6.

SOLUTION
To keep the car from flying off the course, the forces up the track must be equal to the forces down the track.

$$F_{cu} = F_D + F_f$$

where F_f = the force of friction.

$$F_{cu} = F_c \cos \theta \qquad F_{cu} = \frac{mv^2}{r} \cos \theta$$

$$F_D = mg \sin \theta$$

$$F_f = \mu F_{\text{normal}} \qquad F_{\text{normal}} = F_N + F_c \sin \theta$$

$$F_{\text{normal}} = mg \cos \theta + \frac{mv^2}{r} \sin \theta$$

$$F_f = \mu \left(mg \cos \theta + \frac{mv^2}{r} \sin \theta \right)$$

$$\frac{mv^2}{r} \cos \theta = mg \sin \theta + \mu \left(mg \cos \theta + \frac{mv^2}{r} \sin \theta \right)$$

$$\frac{v^2}{r} \cos \theta = g \sin \theta + \mu \left(g \cos \theta + \frac{v^2}{r} \sin \theta \right)$$

$$\frac{v^2}{r} \cos \theta - \mu \frac{v^2}{r} \sin \theta = g \sin \theta + \mu g \cos \theta$$

$$v^2 \cos \theta - \mu v^2 \sin \theta = rg(\sin \theta + \mu \cos \theta)$$

$$v^2(\cos\theta - \mu\sin\theta) = rg(\sin\theta + \mu\cos\theta) \qquad v = \sqrt{\frac{\sin\theta + \mu\cos\theta}{\cos\theta - \mu\sin\theta}}\sqrt{rg}$$

$$v = \sqrt{\frac{\sin 15° + (0.6)\cos 15°}{\cos 15° - (0.6)\sin 15°}}\sqrt{(100\text{ m})(9.8\text{ m/s}^2)}$$

$$v = 31.84\text{ m/s} = 71.6\text{ mph}$$

Summary

1. **Centripetal force** is a force (whether a normal force, gravity, tension, and so on) that keeps a body moving in a circular path.

2. When an object is far away from the surface of a planet, the approximation of mgh can no longer be used. The true form for gravitational potential energy is

$$PE = \frac{-GMm}{r}$$

Key Concepts

CENTRIFUGAL FORCE: The reaction (from Newton's third law) to the centripetal force. We experience it as the normal force from the object in which we're riding when the object moves in a curved path.

CENTRIPETAL FORCE: The force generated from the change in direction of a velocity directed toward the center of the circle in which the object moves.

Important Equations

(7.1) $a = \dfrac{v^2}{r}$ (Centripetal acceleration)

(7.2) $F_c = \dfrac{mv^2}{r}$ (Centripetal force)

(7.3) $F = G\dfrac{m_1 m_2}{r^2}$ (Universal force of gravitation between two objects)

(7.4) $\tau = k\theta$ (Hooke's law for a spring in torsion)

(7.5) $PE = \dfrac{Gm_1 m_2}{r}$ (Gravitational PE between two objects separated by a distance r)

Also see the force triangle for centripetal force on a banked track in Figure 7.10.

Conceptual Problems

1. If an object undergoes uniform circular motion at a constant speed, its acceleration is constant in:

   a. neither magnitude nor direction   b. direction only

   c. both magnitude and direction   d. magnitude only

2. The centripetal force on a car rounding a curve on a level road is provided by:

   a. its brakes   b. friction between the tires and the road

   c. gravity   d. the torque on the steering wheel

3. The centripetal force needed to keep the earth in orbit around the sun is provided by:

   a. the gravitational pull of the moon

   b. the earth's rotation on its axis

   c. the gravitational pull of the sun

   d. the inertia of the earth

4. The acceleration of gravity of an object at the earth's surface:

    a. is directly proportional to the earth's radius

    b. does not depend on the earth's mass

    c. is a universal constant

    d. does not depend on the object's mass

5. A hole is drilled through the earth along its diameter. A stone is dropped into the hole, and the stone reaches the earth's center. At the center, the stone:

    a. has unchanged weight and zero mass

    b. has both mass and weight equal to zero

    c. has unchanged mass and zero weight

    d. has both unchanged mass and unchanged weight

6. The speed needed to put a satellite in orbit does not depend on:

    a. the mass of the satellite   b. the value of g at the orbit

    c. the radius of the orbit   d. the value of G

7. An astronaut is "weightless" in a spacecraft:

    a. in all orbits

    b. only if escape velocity is exceeded

    c. only on ballistic flights

    d. only when the moon's gravity pulls in the opposite direction

8. An iron ball is whirled in a vertical circle. At what point in the circle is the line supporting the ball most likely to break? Why?

9. Where is the centripetal acceleration due to the earth's rotation the greatest? Where is it least? Why?

10. If a track team went to the moon, could it easily set new records in the high jump? Could it easily set new records for the 100-yard dash? Why?

11. Explain how the spin cycle works in removing water from the clothes in a washing machine.

12. What kinds of problems would you and a friend have if you were playing catch on a rotating merry-go-round?

13. You are in a train compartment with no windows, and the suspension is so good that you can't feel any vibration from the tracks. Using only a pendulum bob and string, how can you tell whether the train is accelerating or whether it is rounding a corner?

14. How could artificial gravity be created in a space station?

15. An interstate highway has a speed limit of 65 mph, but the cloverleaf exit from the highway has a speed limit of only 30 mph. Why are the speed limits different?

16. If you were in the passenger seat of a car making a left-hand turn, you would be pushed into the door. This force is often called the *centrifugal force*, but physicists consider it a false force because it really doesn't exist if we take centrifugal to mean "center fleeing." Yet we experience it all the same. What actually is centrifugal force? Explain in terms that would please a physicist.

Exercises

Centripetal and Centrifugal Force

1. A pilot will black out if he experiences more than $5g$'s for an extended period. (a) What is the smallest radius of a horizontal circle in which a pilot flying an aircraft at 550 mph can turn? (b) For how long a time will the pilot experience the $5g$'s if he turns through 180°?

    Answer: (a) 4070 ft; about $\frac{3}{4}$ mi

    (b) 15.8 s; a long time at $5g$'s!

2. How much less does a person weigh at the equator than at the north or the south pole? (Assume that the person has a mass of 70 kg.)

    Answer: 2.4 N or about $\frac{1}{2}$ lb

3. A turn on an asphalt road has a radius of 500 ft. If the coefficient of static friction between rubber and a wet asphalt surface is 0.8, what is the fastest speed at which a car can take the curve if the curve isn't banked?

    Answer: 77 mph

4. A wooden box is resting on the wooden surface of a flatbed trailer, where the coefficient of static friction between box and flooring is 0.5. If the truck is traveling at 60 mph, what is the radius of the tightest turn that the truck can make without having the box slip off the truck?

    Answer: r = 484 ft

5. A coin is placed 18 cm from the center of a rotating turntable. If the coefficient of static friction between coin and turntable is 0.6, at what rpm will the coin fly off?

    Answer: ω = 54.4 rpm

*6. Control-line flying of model aircraft involves powered aircraft that is tethered by a wire cable and is flying around in a horizontal circle. The aircraft has two cables attached to it which are connected to the aircraft's elevator. The pilot on the ground holds a handle connected to the other end of the wires and flies the aircraft from the center of a circle. If the model has a mass of 0.75 kg and each lap takes 4 s, what is the net tension on the cables? Assume that the pilot's arm makes an angle of 20° with the horizontal and that the line is 10 m long.

    Answer: T = 16.1 N

*7. A boy on a merry-go-round is holding a string with a ball on its end. The boy is sitting 6 m from the merry-go-round's center, and the string makes an angle of 4° from the vertical. How long does it take the merry-go-round to make one revolution?

Answer: t = 18.6 s

8. An electron with a mass of 9.11×10^{-31} kg moves with a speed of 2×10^6 m/s in a circle of 5.7-cm diameter in a magnetic field. In the same magnetic field, a proton with a mass of 1.67×10^{-27} kg moves in the same plane with the same speed. It experiences the same centripetal force. What is the radius of the proton's orbit?

Answer: r = 52.2 m

*9. Passengers in an airplane ride at an amusement park are spun around a vertical tower. The airplane is hung from cables, and the cables make an angle of 50° with the vertical when the plane is at top speed. The plane is 45 m from the radial tower. What is the speed of the plane?

Answer: v = 22.93 m/s

*10. A pendulum bob of mass m with a line length of 0.75 m is swung around in a horizontal circle at constant speed, and the line makes an angle of 15° with the vertical. What is the speed of the ball?

Answer: v = 0.714 m/s

Motion in a Vertical Circle

11. A cord 0.75 m long is used to whirl a stone with a mass of 0.7 kg in a vertical circle. At the top of the circle, the stone's speed is 4 m/s. What is the tension in the cord (a) at the top of the circle and (b) at the bottom of the circle?

Answer: (a) T = 8.1 N
(b) T = 49.3 N

12. You're in an aircraft that is flying straight and level at 150 mph. The aircraft then executes a vertical circular loop that leaves you upside down and weightless at the top. Assume that the pilot throttles back the engine during the loop. Neglect air friction. (a) What is the diameter of the loop? (b) If at the bottom of the first loop the pilot decides to do a consecutive second loop just like the first, how many g's do you experience at the bottom?

Answer: (a) D = 605 ft
(b) 6g's

13. A road has a hump 15 m in radius. What is the fastest speed at which a car can travel over the hump without becoming air-borne?

Answer: v = 43.6 km/h

14. A simple pendulum has a cord 0.6 m long. If the tension in the string is 1.5 times the weight of the pendulum bob at the bottom of the swing, what is the bob's velocity?

Answer: v = 1.7 m/s

15. An aircraft with a speed of 550 km/h pulls out of a dive, and the pilot experiences 8g's. What is the radius of the arc?

Answer: r = 340.3 m

16. A bucket of water is swung *at constant speed* in a vertical circle of 1.5-m diameter by a force applied at the proper time in the swing. At what minimum rpm does the bucket move if the water stays in the bucket at the top of the arc?

Answer: ω = 34.5 rpm

*17. A string that is hung vertically can support a weight of *Mg* before breaking. The string is doubled, and a mass *M* is hung from the end of it. If the mass is swung in a vertical circle of 1.5-m diameter at constant speed, what is the maximum rpm?

Answer: ω = 34.5 rpm

*18. The drum in a dryer rotates around a horizontal axis at a constant speed. The drum has a diameter of 0.7 m, and the clothes fall away from the drum's edge when they reach a position of 60° as measured vertically from the top of the drum with the drum's axis. How fast is the drum spinning?

Answer: 71.5 rpm

Gravitation

19. The average orbital radius of the earth is 1.5×10^{11} m, and its average orbital speed is 3×10^4 m/s. What is the sun's mass?

Answer: $m_s = 2 \times 10^{30}$ kg

20. How fast does the moon fall toward the earth? The radius of the moon's orbit is 3.84×10^8 m.

Answer: $a = 2.7 \times 10^{-3}$ m/s^2

21. On the *Apollo* missions to the moon, there was a point in the trip where the gravitational pull of the earth on the capsule

was just equal to the moon's gravitational pull. Relative to the earth's surface, what was the altitude of the capsule at this point? The moon's mass is 7.3×10^{22} kg, and its orbital radius is 3.8×10^8 m.

Answer: $h = 3.36 \times 10^8$ m

22. An apple of mass 0.2 kg falls to the earth. Calculate the apple's acceleration.

Answer: g = 9.8 m/s^2

23. What is the acceleration of gravity 1000 km above the earth's surface? If this value is so close to the value at sea level, why are objects weightless in orbit?

Answer: $g = 7.3$ m/s^2

*24. An *Apollo* command module is halfway along a line between the earth and the moon. What acceleration of gravity do the astronauts feel, and in what direction? The mass

of the moon is 7.35×10^{22} kg, and the radius of its orbit is 3.84×10^8 m.

Answer: $g = 0.011$ m/s^2 in the direction of the earth

25. If the moon suddenly stopped in its orbit and fell straight into the earth, at what speed would it collide with the earth?

Answer: 1.44 km/s

Orbiting Bodies

26. A geosynchronous orbit is one in which the orbital period is 24 h. A satellite placed in geosynchronous orbit will remain over the same spot on the earth because it rotates synchro- nously with the earth. What is the altitude of a geosynchro- nous orbit?

Answer: Altitude $= 3.59 \times 10^7$ m

Centripetal Force on a Banked Track

27. A curve on an interstate highway has a radius of 350 m. At what angle should the road be banked for traffic traveling at a speed of 100 km/h? Assume worst conditions for driving, or a coefficient of friction of 0 between road and tire (glare ice).

Answer: $\theta = 12.7°$

*28. The banking at an oval race course is at an angle of 7°, with the radius of the turns being $\frac{1}{4}$ mi. (a) If the coefficient of static friction between track and tire is 1.0, what is the maxi- mum theoretical speed with which a car can take the turn? (b) What angle of banking would place no limit on the high- est speed at which a car could take the turn?

Answer: (a) $v = 158.5$ mph
(b) $\theta = 45°$

29. A highway curve with a radius of 400 m is banked so that if there were no friction between tire and road, a car could travel around it at 120 km/h. If a 1000-kg Triumph TR7 rounds the curve at 200 km/h, how much must the fric- tional force be to keep the Triumph from sliding into the guard rail?

Answer: $F_f = 4741$ N

*30. A coin is placed on a phonograph record of 30-cm diameter. It begins to slide when the record is tilted at an angle of 40° from the horizontal. Will the coin slide off the record if it is rotated at $33\frac{1}{3}$ rpm? Prove it.

Answer: The coin doesn't slide off.

String instruments produce music because of standing sound waves on strings, and wind instruments produce music from standing sound waves in a cavity. In a guitar the standing waves on the strings are amplified by standing sound waves in the cavity of the guitar body.

Wave Motion

At the end of this chapter, the reader should be able to:

1. Explain and discuss the concept of **superposition.**

2. Calculate the *fundamental frequency* of a vibrating string.

3. Solve for the *velocity, frequency,* and/or *wavelength* of a pulse traveling down a string under tension.

4. Use the concept of the **Doppler effect** to calculate the shift in frequency of a moving sound source.

5. Calculate the **intensity** of a received signal a known distance from a given power source.

6. State the difference between **transverse** and **longitudinal waves.**

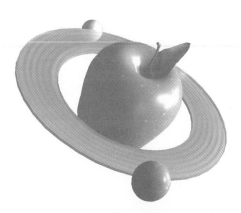

The last two chapters dealt with rotational motion. From the study of rotation and centripetal force arises the concept of periodicity. Rotating bodies have a tendency to repeat a certain motion over and over again. This periodicity is a characteristic of wave motion, and any periodic motion can be described with the mathematical description of waves. An orbiting planet has a characteristic orbital period. Our own is very well known: 365 days, or a year. The note sounded on a guitar is from a standing wave on a string, and its frequency is the inverse of the wave's characteristic period. The sound heard from a saxophone is generated from a standing sound wave in a tube open at one end, and its frequency is also the inverse of the wave's characteristic period. Light is an electromagnetic wave, and the unique wavelength of each color determines how we perceive the color. For example, we see red light as red because it has a characteristic wavelength of about 700 nanometers (700 nm $= 700 \times 10^{-9}$ m). Blue light has a characteristic wavelength of around 400 nm, and the eye will sense any electromagnetic wave with that wavelength as blue light.

All that we see and hear can be sensed only through the interaction of waves on our eyes and ears. Since so much of the information about our physical world comes to us as information carried by waves, it is important in trying to understand the physical world to understand wave phenomena.

8.1   Pulses on a String

If a long cord is tied to a wall and then stretched horizontally, it becomes a path for an energy pulse on which to travel back and forth. If we pluck the end held in our hand with a finger, we will see a pulse move down the cord to the wall, be reflected at the wall, and travel back up the cord to our hand again. By plucking the taut cord, we put an energy pulse on it. This traveling energy pulse is a wavelet. If we played with the cord a little bit, we would see that the more tightly the cord is pulled, the faster the pulse moves back and forth. We would also see that the thinner (lighter) the cord, the more quickly the pulse moves. The two cord variables that we could control were the tension on it and its linear density. (The linear density is the mass per unit length.) We can describe the speed of the pulse on the cord in terms of these two parameters.

If we were to take a snapshot of a pulse traveling down the cord, we would see Figure 8.1A. The pulse would bulge the string upward (or downward) as shown. The pulse is traveling to the right in the picture. Although the pulse in the figure has a circular profile, such is not necessarily the case. If we were to take a look at a pie-shaped wedge at the top of the arc of the circle, we would have Figure 8.1B. The wedge is small, so the angle θ is small. The pulse is energy traveling along the cord and is localized only in the arc of the circle, traveling to the right with a speed v. Whereas the cord itself does not move to the right (only up and down), the energy does. However, each element of the cord experiences a force throwing it outward as the energy of the pulse moves through it. This force is just balanced by the radial component of the tension trying to pull the cord element back to its original position. Tension is a force and therefore a vector. It operates in the cord but has components that are parallel to the cord and perpendicular to the cord. If we look at the top of the pie-shaped wedge drawing of Figure 8.1B, we can see that the tension on the topmost cord element works to both the right and the left tangent to the surface of the cord. Since the cord doesn't move sideways, the parallel tension components on the element at the top must be equal and opposite and thus must cancel. However, each element contributes a tension element downward,

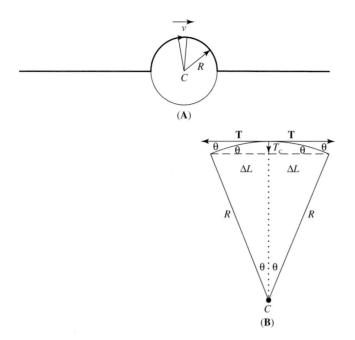

Figure 8.1 Wave Pulse
Traveling along a String

T_d. But the angle θ is small, and from what we know about small angles in radians, we
can write

$$T\theta = T_d$$

Since there are two cord elements, there are therefore two downward-directed tension components.

$$T_D = 2T\theta$$

The energy of the pulse traveling along the cord causes the mass elements in the circular arc to be thrown outward, but the mass elements can't move along the arc. Yet each mass element has a velocity component tangent to the arc because of the energy passing through it, and this energy passes through the mass elements with the propagation velocity v. A mass element with an instantaneous velocity tangent to the circular arc, and with an inward-pointing force due to the tension pointing toward the center of curvature, is experiencing a centripetal force. The centripetal force equals the tension components pointing inward toward the center of the circle.

$$\frac{\Delta m v^2}{R} = 2T\theta$$

The radius R is the distance from the cord element to the center of curvature (C) of the circular cord segment. The velocity v is the velocity of the pulse traveling to the right in the cord element. The mass Δm is the mass of the two cord elements at the top of the cone, each of length ΔL. The linear mass density of the cord is known.

$$\Delta m = (2 \, \Delta L)\mu$$

where μ is the symbol for the linear density.

Our emerging mathematical description of what is happening in the circular arc of the cord through which the energy passes now becomes

$$\frac{2\,\Delta L\mu v^2}{R} = 2\mathbf{T}\theta$$

We now have an equation not only in terms of what we can measure (linear density and tension), but also in terms of quantities whose magnitudes can't easily be measured (R and θ). However, from the pie-shaped cone, the angles at the bottom of the cone are equal to θ because of similar triangles. Therefore, we can find the angle in terms of L and R.

$$R\theta = \Delta L \qquad \theta = \frac{\Delta L}{R}$$

$$\frac{2\,\Delta L\mu v^2}{R} = 2\mathbf{T}\frac{\Delta L}{R}$$

$$\mu v^2 = \mathbf{T}$$

$$v = \sqrt{\frac{\mathbf{T}}{\mu}}$$

(Equation 8.1)

Equation (8.1) describes the speed of a wavelet traveling down a taut cord in terms of the properties of the medium, namely, its density and the tension in it.

As an example of Equation (8.1), consider the bull-whip, an item that not many people have a chance to handle. However, these whips are often seen in the movies. They are generally very thick near the handle and taper to a fine point at the end, and they are usually quite long. If handled properly, when whipped, they produce a loud crack much like a gunshot. The loud crack is actually the sonic boom of the whip's end passing a segment of the whip near the end. From Equation (8.1), you can see that the speed of the energy pulse on the whip is an inverse function of the linear density. If the whip is tapered, the density decreases toward the whip's end, causing the pulse to accelerate as it moves down the whip toward the end. If the whip is long enough, the speed of the pulse will exceed that of sound, and a loud crack will be heard.

8.2   Reflections at a Boundary

Now that we have the pulse traveling down the cord, we might ask, What happens when the pulse comes up against a boundary? The cord is, after all, tied to the wall. In Chapter 5, we studied collisions and looked a little at the phenomenon of impedance. We noticed that when a ball collided with something immovable, such as a wall, if kinetic energy was conserved, the ball bounced back (was reflected) in a direction opposite its initial velocity. At the wall we looked at the changes in kinetic energy and momentum. Since a wavelet traveling on a string possesses both energy and momentum, we can look at the collision process in much the same way (see Figure 8.2). When the wavelet collides with the wall, if no energy or momentum is transferred to the wall, the wavelet should bounce back from the wall with its original energy. It does bounce, but its shape is modified. If the cord is firmly tied to the wall, when the pulse reaches the wall, the energy in the pulse exerts an upward force on the wall, just as it exerts an upward force on the cord, giving the wavelet its characteristic hump. By the law of action and reaction (Newton's third law), the wall then exerts a downward force on the cord. If the arc of the pulse was upward before the collision with the wall, then after the collision it will be pointing downward. The down-

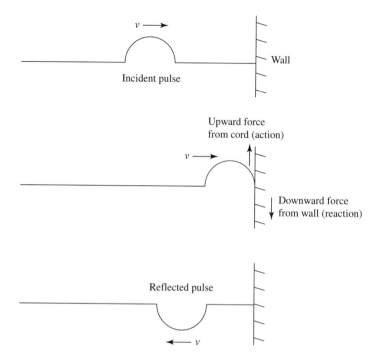

Figure 8.2 Wave Pulse
Reflected from a Wall

ward reaction force of the wall on the cord gives the cord a downward shake, and the cord
returns with a downward-pointing arc.

8.3   Superposition Principle

We have seen what happens when a single wavelet travels down a cord to the wall, is
reflected, and then travels back the way it came as an inverted pulse. Now we are going
to complicate things a bit further by starting another noninverted pulse down the cord
toward the wall while the reflected, inverted pulse is traveling back toward its origin. The
pulses will eventually meet somewhere in between. What happens when the two wavelets
meet? If you watch the process, you'll see the two wavelets come together, momentarily cancel one another out, and then continue on after passing *through* each other as if nothing had ever happened.

This phenomenon, illustrated in Figure 8.3, is called **superposition,** *which is the inter-*
ference of one wave with another, whether destructive or constructive. One wavelet is
superimposed upon another, and what appears (or doesn't appear in this case) is their simple sum. If the noninverted pulse can be represented as a $+1$ and the inverted pulse as a
-1, at the moment they're superimposed, the total wave function is $A = (+1) + (-1)$,
or $A = 0$, where A is the wave's amplitude. Both the $+1$ (noninverted pulse) and the -1
(inverted pulse) still exist, but they cancel. Notice that to momentarily cancel, the pulses
must run through one another. They occupy the same space at the same time, unlike matter, which can't occupy the same space at the same time. If there were a collision between
two billiard balls, at all times before, during, and after the collision the billiard balls would
make contact only on the balls' outer surfaces, and the balls wouldn't merge into one
another. But the wavelets exist unchanged before and after the collision. During the event,
they momentarily seem to wink out of existence. During the collision, they interfere
destructively, and the total amplitude decreases momentarily to zero. Then, as the pulses
begin to move past each other, the original forms and directions of the two pulses

Superposition: Constructive or
destructive interference of one
wave with another. The waves
occupy the same space at the
same time as they move through
one another.

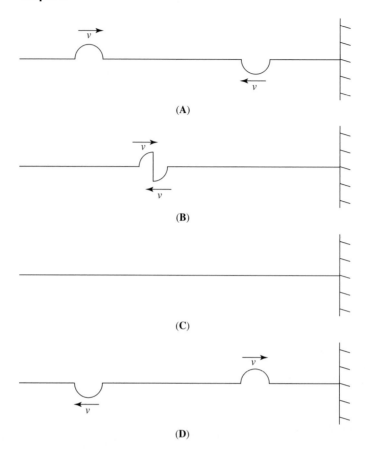

(A)

(B)

(C)

Figure 8.3 Collision of Two
Wave Pulses on a String

(D)

reemerge. They don't really wink out of existence, though. That is just an illusion, because they both appear after the collision as if nothing had ever happened. During the collision, the energy of the two pulses doesn't disappear.

The pulse traveling up and down the cord is a wavelet, and a wavelet doesn't have any mass associated with it. A massive particle can be associated with a particular point in space, but the same can't be said about a wave. At first glance the pulse (or wavelet) *seems* to be localized, but when you take a good look at the wavelet's arc, you realize that the energy in the pulse is spread over the entire arc. We can talk about our theoretical point particle, but classically we can't talk about a "point" wavelet. A wave's energy is spread out in space; it is not localized.

8.4   Wave Equation

Up to now we have been dealing with single pulses. We will next create a pulse train on our stretched string wave path. Assume that we have an oscillator (for example, your finger) that will pluck the string with short, sharp pulses that are separated by a relatively long, constant spacing in time. As we successively pluck the string, we'll create a series of pulses that look somewhat like those in Figure 8.4. The pulses as illustrated have both an inverted and a noninverted arc. When you pluck the string with your finger, the string

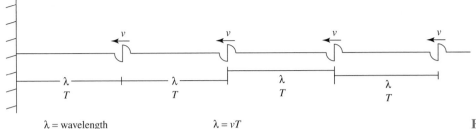

λ = wavelength   λ = vT

T = period

Figure 8.4 Pulse Train on a String

will first follow your finger up (if you're plucking the string from the bottom). But since the taut string has a springiness or resilience because of the tension, it will also descend below its equilibrium position after you have plucked it. (Actually these positive and negative oscillations will continue with an ever-decreasing amplitude until they are finally damped out because of the inevitable friction in the string and air resistance.) However, if you pluck the string at constant time intervals, the spacings between the pulses on the string will also be constant. We call the uniform spacing the *wavelength,* for reasons that will become clear in a bit. The usual symbol for wavelength is λ, the Greek letter lambda. The period between the pulses has the symbol T, not to be confused with the tension in the string. (Only a limited number of symbols are traditionally used to name variables, and sometimes the same symbol will be used in two different roles. When you choose a formula to use in solving a problem, be careful that you understand what the symbols actually mean. The symbol T can stand for tension in some cases, and for the period in others.) The relationship between the period, wavelength, and speed of the pulse on the string is

$$\lambda = vT$$

The velocity can be isolated by dividing both sides by the period.

$$v = \frac{\lambda}{T}$$

The inverse of the period is just the frequency of the vibrations in cycles per second, or hertz (Hz).

$$f = \frac{1}{T}$$

$$v = \lambda f \qquad \text{(Equation 8.2)}$$

Equation (8.2) is the classical wave equation that describes the relationship between the wavelength and the frequency for a wave traveling in a medium with a characteristic velocity v. This equation holds true for any periodic phenomenon.

8.5   Standing Waves on a String

We'll now take our system one step further in its evolution. We've gone from isolated pulses to a wave train. At this stage, we'll connect an oscillator to our stretched string, and the oscillator will have the ability to run continuously, with its frequency controllable from zero to infinity. With a continuously running oscillator, we can continuously pump energy into the stretched string at various frequencies and observe the results. If we start the oscillator running at a very low frequency, we'll see the isolated pulses running up and down the string. If the frequency is increased a little more, the single pulses will start to interfere, adding here and canceling there. If the frequency is increased still further, things are happening too fast for us to be able to notice individual pulses. The string quivers and quakes, and its behavior is, in general, quite chaotic. If the frequency is then increased still further, something rather startling happens: The chaos suddenly turns into order. The system's first (and lowest) resonant frequency is reached, and a pattern much like that in Figure 8.5A pops into view. If the frequency is increased a bit more, the first resonance disappears, and the chaotic behavior of the string reappears. A continuous increase in frequency will eventually find the system's second resonant point, which will look like Figure 8.5B. Cranking the frequency knob still higher, we will find the third resonance point, which is displayed in Figure 8.5C. In fact, we can increase the frequency to infinity, passing through an infinite number of resonances on the same string, and we will quickly notice a pattern. The number of "bumps" that each resonance pattern contains increases with frequency, and as the frequency gets higher and higher, the frequency knob must be increased less and less for the system to reach the next resonance. Obviously something systematic is happening here. What is it?

To explain what is shown in Figure 8.5, we must first digress a bit. Keep in mind that with standing waves, the entire space of the string is filled with waves. There isn't any space between pulses as in Figure 8.4. A full wave has both a noninverted and an inverted arc, as shown in Figure 8.6. The wave shown is a sine wave. It has the same characteristics as the sine function that we've been using in trigonometric functions. If the sine wave's maximum amplitude is 1, then $\sin(0°) = 0$, $\sin(90°) = 1$, $\sin(180°) = 0$, $\sin(270°) = -1$, and so on. Both angular and radian representations can be used. Try finding the values of the above points on the sine-wave graph from your calculator in radians. Almost all calculators can give values of the trigonometric functions in both angular and radian measure. Note that the actual wave seen on the string, particularly Figure 8.5B, has the form of Figure 8.6. If we were to use a strobe light to freeze the string in time, we would see Figure 8.6 exactly. As it is, at the maxima positions, the string has both a positive *and* a negative maximum at that position. It is blurred to our eyes because of its rapid oscillations up and down between the positive and negative maxima.

We see the patterns of Figure 8.5 only when our oscillator hits on a particular resonance. What are the conditions for resonance? Each end of the string is firmly tied to the wall, and the amplitude there must be zero. These points on the string of zero amplitude are called *nodes*. Conversely, the points that have maximum amplitude (either positive or negative) are called *antinodes*. At resonance, an integral number of half-wavelengths can fit neatly onto the string, because each half-wavelength is terminated at each end by nodes. Figure 8.5 shows the first three resonances. Because the entire space on the string must be filled with the waves, we can get an equation for each of Figures 8.5A–C.

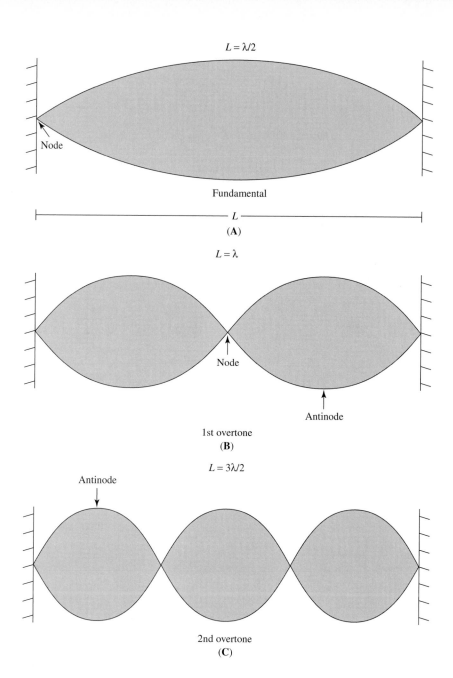

$L = \lambda/2$

Node

Fundamental

L

(A)

$L = \lambda$

Node

Antinode

1st overtone
(B)

$L = 3\lambda/2$

Antinode

2nd overtone
(C)

Figure 8.5 Standing Waves on a String

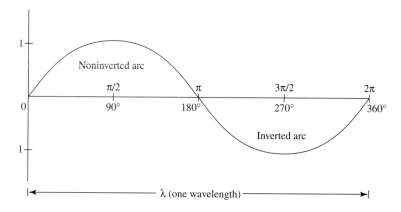

Noninverted arc

$\pi/2$ π $3\pi/2$ 2π
90°   180°   270°   360°

0

Inverted arc

λ (one wavelength)

Figure 8.6 Sine Wave

$$(\text{Figure 8.5A}) \qquad 1\left(\frac{\lambda}{2}\right) = L$$

$$(\text{Figure 8.5B}) \qquad 2\left(\frac{\lambda}{2}\right) = L$$

$$(\text{Figure 8.5C}) \qquad 3\left(\frac{\lambda}{2}\right) = L$$

There is theoretically no upper limit on the number of half-wavelengths that can fit onto the string of length L. The wavelength just gets shorter and shorter as more half-wavelengths are packed into the limited space. A general equation for the resonances available for a string of length L is

$$L = n\left(\frac{\lambda}{2}\right) \qquad \text{where } n = 1, 2, 3, \ldots$$

(Equation 8.3)

or

$$\lambda = \frac{2L}{n}$$

For Equation (8.3), as n gets larger, the wavelength (λ) must become shorter since the length of the string is fixed.

The standing wave is like a snapshot in that it is frozen in time. If we are using the sine wave to represent the spatial amplitude of the wave at any distance along the string and for any arbitrary amplitude, we can start with

$$A = A_0 \sin \theta$$

where A_0 is the maximum amplitude of the wave. Since θ is an angle, we must somehow represent it in terms of the horizontal distance L.

$$\theta = kL$$

So

$$A = A_0 \sin(kL)$$

where k is a constant that we must determine and the units would be radians per meter (rad/m). Whenever $L = n(\lambda/2)$, kL (or θ) must equal $n\pi$. (If a full wavelength is 2π rad, then when a resonance pattern fits on the stretched string, there must be an integral number of half-wavelengths on it.) Bringing these conditions together, we have

$$k\left(\frac{n\lambda}{2}\right) = n\pi$$

Rearranging and canceling, we get

$$k = \frac{2\pi}{\lambda}$$

Our wave equation now becomes

$$A = A_0 \sin\left(\frac{2\pi}{\lambda}L\right)$$

Whenever L is equal to an integral number of half-wavelengths, a resonance appears on the string, just as in Figure 8.5. The above expression can be generalized even further by allowing L to become any x, or horizontal distance. Our wave equation now becomes

$$A = A_0 \sin\left(\frac{2\pi}{\lambda}x\right)$$

(Equation 8.4)

Equation (8.4) describes the amplitude at any point on a string when standing waves are present.

In Equation (8.4), if $x = n\lambda$, then the amplitude is zero. These are the nodes of the standing wave. At these points on the string, the string has zero amplitude at all times. However, if we instead shift our attention to a spot on the string where the amplitude is always changing, such as the points of maximum amplitude in Figure 8.5 (antinodes), we can see that the amplitude does indeed change in time. The blurring of the space between the top and bottom of the envelope in Figure 8.5 at an antinode is caused by the rapid up-and-down fluctuation of the string at that point, and our eyes can't follow the string quickly enough.

The oscillator driving the stretched string has a frequency associated with it. The oscillator attached to the string could be as pictured in Figure 8.7. The spinning disc is driven by an electric motor whose speed we can select simply by turning a knob. This speed is its angular velocity. Suppose that there is a short post with a hole in it projecting from the face of the disc, and the stretched string is passed through this post. When the disc revolves at an angular velocity ω, the string will be forced to follow with a time-varying, up-and-down amplitude. The amplitude-versus-time plot of this motion is the sine wave of Figure 8.6.

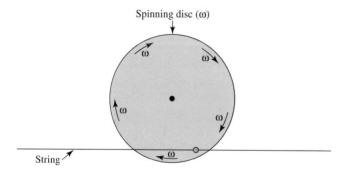

Figure 8.7 Oscillator Pumping Energy into a Stretched String

If we were to plot a graph of the amplitude versus time at an antinode, we would get the same picture as that of Figure 8.6, with the horizontal axis time instead of an angle θ. Thus, we should be able to convert time into an angle at the antinodes, just as we converted the spatial dimension x into an angle at the nodes of our standing wave pattern. We have already done this, though, when studying circular motion in Chapter 6.

$$\theta = \omega t$$

We now have a sinusoidal equation that describes how the amplitude of the stretched string at the antinodes varies with time.

(Equation 8.5)

$$A = A_0 \sin(\omega t)$$

Both Equations (8.4) and (8.5) describe a standing wave: (8.4) in terms of space, and (8.5) in terms of time.

Now suppose that we have a sinusoidal wave traveling to the right. In a time t, the wave as a whole would have traveled to the right a distance of vt. We can modify Equation (8.4) so that it is an equation for a traveling wave.

$$A = A_0 \sin \frac{2\pi}{\lambda}(x + vt)$$

Remember that a standing wave is formed by incident and reflected waves interfering constructively and destructively. The resultant pattern on the string is the sum of a wave traveling to the right plus a similar wave traveling to the left.

$$A = A_0 \sin \frac{2\pi}{\lambda}(x + vt) + A_0 \sin \frac{2\pi}{\lambda}(x - vt)$$

See Appendix A, Section A.3, "Trigonometric Identities."

Through the use of a trigonometric identity, we can reduce the equation above to a much more manageable form.

$$A_0 \sin \frac{2\pi}{\lambda}(x + vt) = A_0 \sin \frac{2\pi}{\lambda}x \cos \frac{2\pi v}{\lambda}t + A_0 \cos \frac{2\pi}{\lambda}x \sin \frac{2\pi v}{\lambda}t$$

$$A_0 \sin \frac{2\pi}{\lambda}(x - vt) = A_0 \sin \frac{2\pi}{\lambda}x \cos \frac{2\pi v}{\lambda}t - A_0 \cos \frac{2\pi}{\lambda}x \sin \frac{2\pi}{\lambda}t$$

Adding the right sides of two equations together, we have

$$A = 2A_0 \sin \frac{2\pi}{\lambda}x \cos \frac{2\pi v}{\lambda}t$$

$$\frac{2\pi}{\lambda} = k \qquad (\text{where } k = \text{wave number})$$

$$\omega T = 2\pi \qquad (\text{where } T = \text{period of the wavelength})$$

$$A = 2A_0 \sin kx \cos \frac{\omega t}{\lambda}vT$$

But $vT = \lambda$, so

$$A = 2A_0 \sin kx \cos \frac{\omega \lambda}{\lambda}t$$

(Equation 8.6)

$$A = 2A_0 \sin kx \cos \omega t$$

At the point $x = \lambda/2$ in Figure 8.5B, the amplitude is zero regardless of the time (a node). However, if $x = \lambda/4$ in Figure 8.5B (an antinode), then the amplitude of the string will fluctuate up and down between the positive and negative maximum values of A_0 according to the function $\sin(\omega t)$, blurring the string to our eyes. Of course, the function disappears whenever $\omega t = n\pi/2$, where $n = 1, 3, 5, \ldots$. If we took a strobe light and synchronized it properly with the string of Figure 8.5B for one the above times, we would see a perfectly flat string. The strobe light is able to freeze the entire length of the string in time, much like the standing wave pattern freezes the string in space at the nodes.

Let's return to Figure 8.7 and take another look at the sine-wave generator. Note that it's spinning with an angular speed of ω. This term is also represented in the time-varying part of Equation (8.6), since it must vary with the frequency of the oscillator. The maximum amplitude is now twice the initial amplitude for either the incident or the reflected wave. For constructive interference where the two waves add their amplitudes, this result would be expected.

8.6　Sound Waves

Standing waves on a string produce sounds, as anyone who has heard the music from a guitar well knows. The vibrating string interacts with the air and produces a sound at the same frequency as the vibrating string. Both the sound and the vibrating string can be described by the same mathematical entity, the sine wave, but the characters of their waves differ greatly. The standing waves on a string are transverse waves. Look again at Figure 8.6. The amplitude of the sine wave is in the plane of the page (the $+y$ and $-y$ directions), and the velocity of propagation is in the $+x$ direction. *In a **transverse wave**, the amplitude is at right angles (90°) to the direction of propagation. The amplitude variation of the sound wave, however, is in the direction of its propagation. For this reason, it is called a **longitudinal wave**.*

In a note from a guitar, the longitudinal sound waves are produced by the transverse standing waves, as in Figure 8.8. The vibrating string is moving from the left to the right and back again in the figure. This back-and-forth motion alternately compresses and rarefies the air. The pressure differential is propagated through the air at the speed of sound and falls upon the eardrum. The eardrum vibrates in unison with these pressure fluctuations, and we "hear" the fundamental frequency from the guitar. The maxima and minima of the sound wave, which are the pressure fluctuations, are in the same direction as the propagation. When someone speaks into a microphone, the microphone acts like the eardrum and translates the pressure variations in the air into a back-and-forth motion of the microphone's diaphragm. This motion is translated into an electrical signal which eventually will cause another diaphragm perhaps far removed from the first to vibrate in air and reproduce virtually exactly the original sound.

Transverse wave: A wave in which the direction of propagation is 90° to the amplitude.

Longitudinal wave: A wave in which the amplitude variation is in the direction of the propagation.

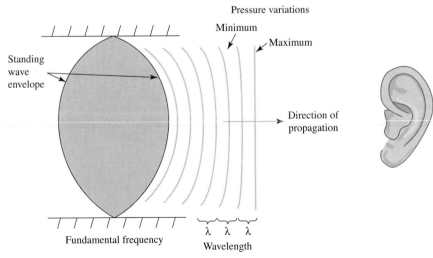

Figure 8.8 Production of a Sound Wave

Sound is a series of pressure variations of the air propagated through the air at the speed of sound. Air is a medium much like the string discussed at the beginning of this chapter, with a characteristic wave speed that can be derived from the properties of the medium. The speed of sound in air is 343 m/s, or 1125 ft/s, at standard temperature and pressure (1 atm and 20°C). This speed will vary with temperature and pressure, much as the wave's speed on the string will vary with tension and linear density. Since the pressure variations vary like a sine wave, and since the wave propagates through the air according to the wave equation $(v = f\lambda)$, the conditions for creating standing waves in air are almost exactly like those for standing waves on a string. A wind instrument, such as an organ or a trombone, produces music from standing waves in air. The standing waves are usually produced in a tube that is open at one end and closed at the other, as in Figure 8.9. At the closed end, the pressure is constrained to be at a minimum. At the open end, the pressure will be at a maximum at resonance. The length of the tube determines what resonances are possible.

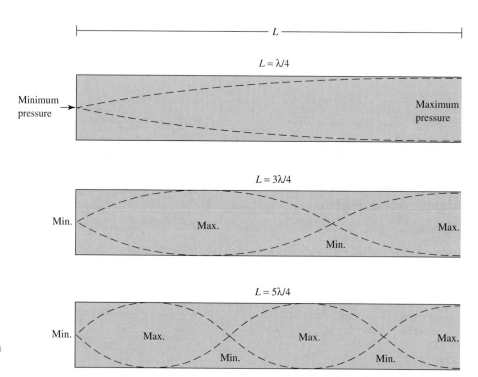

Figure 8.9 Standing Waves in Air

In Figure 8.9, the fundamental resonance occurs when the length of the tube is one-quarter the wavelength of the sound wave. Succeeding figures will show that a resonance occurs whenever an odd multiple of $\lambda/4$ equals the length of the tube.

$$L = (2n + 1)\left(\frac{\lambda}{4}\right)$$

where $(2n + 1)$ is always odd, and $n = 0, 1, 2, 3, \ldots$. In an organ each note has an individual pipe, with the fundamental frequency being the prominent note of the pipe. The other overtone frequencies also exist simultaneously in the pipe, but with amplitudes less than the fundamental, or strongest, resonance. If you have ever looked at a collection of organ pipes, you probably have noticed that the longer the pipe, the lower the frequency, a fact that can be confirmed with the wave equation.

$$\lambda f = v$$

If the speed of sound in air is a constant, and if the wavelength grows larger, the frequency must decrease.

8.7　The Doppler Effect

When a source of sound moves toward or away from you, the frequency of the sound is shifted. If the source is moving toward you, the sound is shifted to a higher frequency; if it is moving away, then the sound shifts to a lower frequency. *This shift is called the Doppler effect.* You have probably noticed that whenever an ambulance or a police car blasts past you with the siren wailing, the pitch is relatively high when it is approaching. Then, as the vehicle passes you, the siren immediately decreases in pitch. The effect is the same if the source is stationary and the listener is moving, since the motion between source and listener is relative.

Doppler effect: An apparent frequency shift of a wave due to the relative movement of the source toward or away from you.

　　Figure 8.10 shows what happens to a sound wave whenever there is relative motion between the source and the listener. With no relative motion between source and listener, the sound waves move out from the source in circular wavefronts, as in Figure 8.10A. The circles are concentric and evenly spaced, with the spacing being equal to a wavelength. The speed of sound in air is constant, so if the frequency of the sound is constant, the spacing between wavefronts will also be evenly spaced. In Figure 8.10A, listeners 1 and 2 will both hear the sound at its original frequency f_0. However, if the source is moving, both listeners 1 and 2 will hear different frequencies. Listener 2 will hear a higher-pitched frequency than f_0 because the wavelength is shortened, and listener 1 will hear a frequency lower than f_0 because the wavelength is lengthened. What causes this difference in perceived frequencies from the same source?

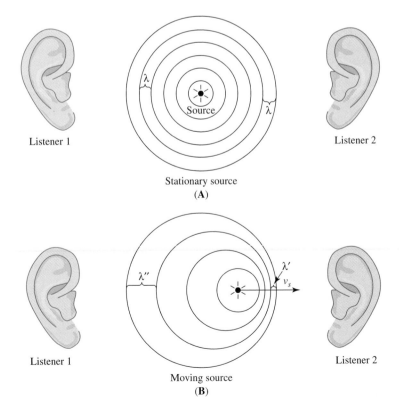

Listener 1　　　　　Source　　　　　Listener 2

Stationary source
(A)

Listener 1　　　　　v_s　　　　　Listener 2

Moving source
(B)

Figure 8.10 Doppler Shift

In Figure 8.10B, the spacing between one pressure maximum and another is equal to one wavelength, or λ_0. The time between maxima from the point of view of the source is T_0. However, according to listener 2, the spacing between adjacent pressure maxima is shorter by a factor of $v_s T_0$ because the source moved closer to listener 2 by that distance in time T_0. If the spacing between adjacent maxima is still equal to one wavelength from the point of view of listener 2, then the wavelength will have been shortened according to her perspective. The new spacing between adjacent maxima is

$$D' = \lambda_0 - v_s T_0$$

which is just equal to the shifted wavelength λ' as heard by listener 2.

$$\lambda' = \lambda_0 - v_s T_0$$

Of course, you don't hear the wavelength. You hear the frequency of the sound, or the time variations in the pressure maxima.

The frequency heard by listener 2 is

$$\lambda' f' = v \qquad f = \frac{v}{\lambda'}$$

(Equation 8.7)

$$f' = \frac{v}{\lambda_0 - v_s T_0}$$

where v is the speed of sound in the air and doesn't change regardless of the motion of the source. (It's constant.) But since the wavelength is shorter, the frequency is higher. For listener 1, the argument is the same, except that the perceived wavelength is now longer by the amount $v_s T_0$.

$$\lambda'' = \lambda_0 + v_s T_0$$

and

$$f'' = \frac{v}{\lambda_0 + v_s T_0}$$

Since the wavelength is longer for listener 2, the perceived frequency is lower.

8.8   Sonic Boom

From Figure 8.10, we saw what happens when the source is moving in the medium. Because the speed of sound is finite (767 mph at standard temperature and pressure), it is possible for the source to move faster than the speed of sound in air. What happens when the source moves faster than the speed of sound? An aircraft moving through the air disturbs the air, and the resulting pressure waves move out at the speed of sound in spherical shells surrounding the aircraft. If the aircraft is moving at the speed of sound, these pressure ridges pile up in front of the craft and create a wall of high pressure known as the "sound barrier." If the craft weren't properly designed, this wall of pressure could destroy the aircraft as it tried to push through the barrier. Much attention was given to this problem in the early stages of jet aircraft design. It became an issue near the close of World War II, when jets were becoming operational, and the speeds that could be attained were in excess of the speed of sound.

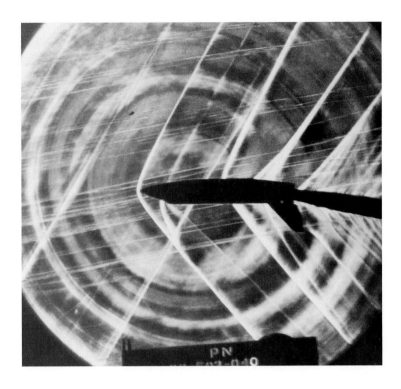

Shock waves caused by a model in a wind tunnel where the air is moving at a Mach number of 12.

Once through this barrier, though, an aircraft outdistances the sound wave that it produces. It is at the apex of a cone such as that pictured in Figure 8.11. If we were in the bow of a stationary rowboat and the boat were rocking up and down, the bow would disturb the water and create circular wavefronts moving out from it, as in Figure 8.10A. As the boat began to move forward, the symmetric, circular, concentric wavefronts of Figure 8.10A would be distorted by the boat's motion, as in Figures 8.10B and 8.11A. The wavefronts to the front would begin to bunch up, and those behind to spread out, leading to the Doppler shift. But the Doppler shift is *heard* as an audible frequency when the frequency is shifted by motion. The advantage of this bow wave example is that the effect can be *seen.* When the boat was being rowed at the speed of a wave in the water, the observer at the bow would see Figure 8.11B. The wavefronts would cease to race out ahead of the boat as circular ridges, since the boat has caught up with the successively emitted wavefronts. The boat's bow is now always where the emitted wavefront in the forward direction would be. It is at this speed that the observer would begin to see the familiar conical bow wave generated by a boat moving through the water. If the boat were being rowed faster than the speed of a wave on the surface of the water, a well-developed bow wave such as that in Figure 8.11C would be observed.

The conical bow wave actually consists of the summation of many successive waves along a common front, as in Figure 8.11C. The boat's bow is the center of the next circular wavefront before it begins to expand. The faster the boat moves through the water, the narrower is the cone. The angle between the axis of motion and the conical wavefront is therefore a function of the source's speed and the velocity of a wave in that medium.

$$\sin \theta = \frac{v_s t}{v t}$$

$$\sin \theta = \frac{v_s}{v}$$

(Equation 8.8)

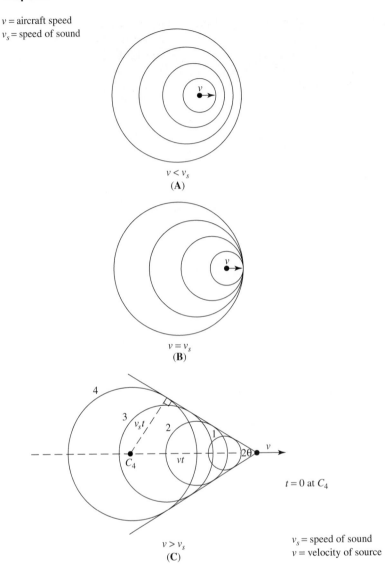

v = aircraft speed
v_s = speed of sound

$v < v_s$
(A)

$v = v_s$
(B)

$v_s t$

vt

2θ

v

$t = 0$ at C_4

$v > v_s$
(C)

v_s = speed of sound
v = velocity of source

Figure 8.11 Sonic Boom

The distance $v_s t$ is the distance that the conical wavefront moves in a time t. The wavefront is always tangent to the surface of successive, expanding, spherical surfaces formed when a single disturbance expands outward, as when a pebble is dropped into still water. The distance vt is the distance that the boat moves along its axis of motion in the same time t. The ratio of v_s to v is the *inverse* of what is known as the *Mach number.* A conical bow wave doesn't form until the center of the disturbance (the bow) is moving faster than or equal to the speed of the wave in that particular medium. Up until the speed of the source equals the speed of the wave in that medium, we have a Doppler effect. There is a frequency shift. When the speed of the source equals or exceeds the speed of the wave in that medium, then we have a Mach effect and the conical bow wave. If the source of the disturbance is a jet aircraft, then the sound of its engines becomes shriller and shriller as its speed approaches that of sound. When the aircraft equals or exceeds the speed of sound, what we hear is the "sonic boom" so often cited in the newspapers and literature. What we hear is no longer the frequency associated with successive wave crests hitting our eardrum, but the single, sharp shock wave of the conical sonic bow wave.

If we heard the drone of a jet aircraft's turbines coming from the left, we would naturally look upward and to the left to see the aircraft. If the aircraft were traveling at the speed of sound, when we heard the high-pitched drone of its turbines, we would have to look directly overhead to see it. If we were to hear the sudden, sharp sonic boom of a supersonic aircraft in our left ear, then we would have to look upward and to the right to see the aircraft, since the aircraft would be outpacing its own sound waves!

8.9   Light Waves

As mentioned at the beginning of the chapter, the light that hits our eyes is also a wave. It is a transverse wave, just like a pulse traveling down a string and waves on the surface of the water. Since a light wave is a transverse wave, its amplitude is at 90° to its direction of motion. A light wave is an electromagnetic wave and has an amplitude associated with its electric field and one associated with its magnetic field. Both amplitude components are at 90° to each other and at 90° to the direction of propagation of the wave. Both the side view and the frontal view are given in Figures 8.12A and B.

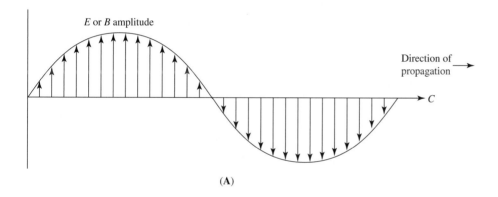

E = electric field amplitude
B = magnetic field amplitude
C = speed of light (186,000 mi/s)

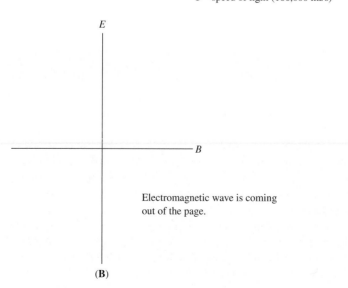

Figure 8.12 Transverse Electromagnetic Wave Propagating in Space

Since light is a wave, it has both a frequency and a wavelength. It obeys the wave equation, $\lambda f = v$, where v in this case is the speed of light. The speed of light is very high, 186,000 mi/s (3×10^8 m/s). The speed of light is the fastest known quantity in the universe. Nothing can go faster than the speed of light, and therefore there is no possibility for a sonic boom when an object moves faster than the speed of light. However, matter, time, and space undergo weird transformations as an object's speed approaches that of light. The theory of special relativity is built on the foundation that the speed of light is an absolute upper limit to the relative velocity between any two objects. (More on this topic will follow much later in this book.) Moreover, light is not only a wave; it is also a particle. Light is emitted both as a wave and as a particle. This particle is called a *photon.* It acts as a particle in that it carries both energy and momentum, but unlike all of the particles we have studied up to now (point masses), it has no mass and moves only at the speed of light. Light therefore has a dual nature. But this chapter deals with waves, so we'll concentrate on the wave nature of light here.

A light wave is essentially an energy ripple traveling through space. The frequencies and wavelengths with which we are most familiar are concentrated in the visible spectrum, or those wavelengths that the eye can readily detect. These wavelengths are centered at about 590 nm (590×10^{-9} m), which is yellow light. As the wavelengths become longer, the light reddens. The longest visible wavelength corresponds to about 700 nm. As the wavelength shortens, the light is shifted toward the blue end of the spectrum. The shortest visible wavelength is about 400 nm, or in the nearly ultraviolet region. Since light obeys the wave equation, these limits of the visible spectrum (400 to 700 nm) also correspond to a frequency spectrum (7.5×10^{14} to 4.3×10^{14} Hz). Of course, the possible wavelengths of electromagnetic radiation aren't restricted to the visible spectrum. At wavelengths much longer than those of the visible spectrum, we have long- and short-wave radio, microwaves, television, and radar. The electromagnetic waves carry information through space in most cases, and energy in others. A microwave oven will cook food, and the microwave radiation therefore deposits its energy in whatever is being cooked. Infrared radiation has wavelengths just a little longer than what we can see. If you plug in an electric resistance heater, the elements will glow a red-orange. This color is at the long-wavelength end of the visible spectrum. However, the heat you feel coming from the heater isn't visible, but it obviously can be felt. At wavelengths shorter than visible light, which are more energetic, we have X rays and gamma rays, X rays are energetic, penetrating rays used in medical imagery and in examining luggage at airports. Gamma rays come from intergalactic space and are of extremely high energy.

8.10   Intensity

A small flashlight bulb placed at the center of a large room acts very much as a point source of light. Up close, the bulb is very intense, and we can't look at it directly because our eyes become overloaded. We must squint to look directly at the bulb. At some distance from the bulb, we can easily read and do close work such as sewing and writing. Further away, reading is impossible, but we can still see objects and discern outlines. Much further away, the bulb is a glowing point that illuminates nothing, similar to starlight. The light dims with distance. The reason that it does so is the way a light wave, or any wave in general, propagates.

Figure 8.13 illustrates this principle. The bulb is in the center of the picture. It radiates light *isotropically;* that is, it emits light evenly in all directions. Figure 8.13 actually shows only a slice or section of such radiation. Remember that the light is being emitted not only in the plane of the page, but also out of the page and into the page. However,

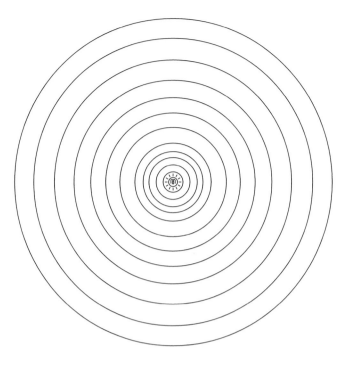

Figure 8.13 Wave Propagation from a Point Source

isotropically means that any section that is taken through the source will look like Figure 8.13. *Isotropic* also means symmetric. If the bulb is a flashlight bulb, its rated power might be of the order of 5 W. It's constant. However, notice in the drawing that as the light radiates into space, each wavefront must be stretched to cover a larger circumference in the picture. In three dimensions, it must cover the increasingly larger surface of a sphere. *The definition of **intensity** is the power per unit area, or watts per square meter in the metric system.*

Intensity: The power received per unit area from a distant power source.

$$I = \frac{P}{A}$$

$$A = 4\pi r^2$$

$$I = \frac{P}{4\pi r^2}$$

(Equation 8.9)

The surface area of a sphere is $4\pi r^2$ and is in the denominator of the fraction of Equation (8.9). So the light intensity drops off quite rapidly. This equation holds for such dissimilar emissions as light and sound.

8.11 Radar and Sonar

Light waves and sound waves are two very dissimilar waves. One is a transverse wave, and the other is a longitudinal wave. One carries its electric and magnetic field amplitudes along with it; the other propagates through pressure variations in the medium (air or water). The propagation velocity of light in a vacuum is nearly instantaneous as far as our everyday experience is concerned (186,000 mi/s), and that for sound in air is a paltry 0.213 mi/s. The difference in speeds becomes obvious at a fireworks display. Whenever a star shell explodes, we see the spectacular display well before we actually hear the boom of the exploding shell. If we are 1 mile away from an exploding shell traveling at the speed of light, it takes the light only 5.38×10^{-6} s ($5.38\ \mu$s) to reach us. This is very nearly instantaneous as far as our everyday experience is concerned. On the other hand, it takes 4.7 s for the boom to reach us. But despite their dissimilarities, both are waves.

Sonar and radar are both used for determining distance to a distant object and for determining the motion, both rotational and linear, of the object. (A weather Doppler radar uses the Doppler effect to "see" into a storm.) For determining distance, a short, sharp pulse is emitted, and the time it takes the signal to return to the transmitter determines the distance, as shown in Figure 8.14. The pulse takes a time t to travel to the object being detected and then back again. Since the pulse travels twice the distance D, the equation describing the distance is

$$2D = vt$$

or
$$D = \frac{vt}{2}$$

Whether the pulse is sonar or radar, just substitute the proper propagation speed into the equation to find the distance. The speed of sound in water is 5022 ft/s.

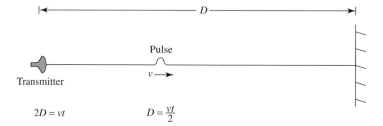

Figure 8.14 Radar and Sonar Ranging

Two methods can be used to measure a linear velocity toward or away from a transmitter. The first relies on obtaining two ranges to an object at a time Δt apart.

$$D_0 = \frac{vt_0}{2} \qquad D_1 = \frac{vt_1}{2}$$

$$D_0 - D_1 = \frac{vt_0}{2} - \frac{vt_1}{2}$$

$$D_0 - D_1 = \frac{v}{2}(t_0 - t_1)$$

$$\Delta S = D_0 - D_1$$

$$\Delta S = \frac{v}{2}(t_0 - t_1)$$

The distance that the object moved between the times t_0 and t_1 is readily calculated, and the time Δt is also easy to find. So the velocity of the object toward or away from the transmitter can be calculated.

$$v_{\text{object}} = \frac{\Delta S}{\Delta t} \qquad \left[\text{where } \Delta t = (t_0 - t_1)\right]$$

Of course, the object can vary its speed or change its heading in the time Δt that it takes to find the two distances. A better way to determine the distance is from the Doppler shift. Both light and sound exhibit a Doppler shift. Whenever a pulse hits a moving object, the wavelength of the reflected pulse changes relative to its original wavelength. If the object is moving toward the transmitter, the wavelength shortens, and if it is moving away from the transmitter, the wavelength lengthens. The greater the wavelength (or frequency) shift, the greater the relative velocity toward or away from the transmitter. We start with our equation for the Doppler shift.

$$f' = \frac{v}{\lambda_0 \pm v_s T_0}$$

where f' is the shifted wavelength, λ_0 the original emitted wavelength, T_0 the period of the emitted wavelength, v the speed of the wave in the medium, and v_s the speed of the object in the medium. In the denominator of the fraction, use the plus sign if the object is moving away from the transmitter, and the minus sign if it is moving toward the transmitter. In the time T_0, the original wavelength is lengthened or shortened by the amount $v_s T_0$. But

$$T_0 = \frac{1}{f_0}$$

$$f' = \frac{v}{\lambda_0 + v_s \frac{1}{f_0}}$$

$$f' = \frac{v f_0}{\lambda_0 f_0 \pm v_s}$$

But $\lambda_0 f_0 = v$, so

$$f' = \frac{v f_0}{v \pm v_s}$$

$$\frac{1}{f'} = \frac{v \pm v_s}{v f_0}$$

$$\frac{v f_0}{f'} = v \pm v_s$$

$$\pm v_s = \frac{v f_0}{f'} - v$$

$$\pm v_s = v \left(\frac{f_0}{f'} - 1 \right)$$

$$\pm v_s = v \left(\frac{f_0}{f'} - \frac{f'}{f'} \right)$$

$$\pm v_s = v \left(\frac{f_0 - f'}{f'} \right)$$

$$\Delta f = f_0 - f'$$

$$\pm v_s = \frac{v}{f'} \Delta f$$

For radar, v is the speed of light (c). For sonar, v is the speed of sound in air for a bat, and the speed of sound in water for a submarine. The greater the difference in frequencies, the greater the relative speed. If f' is higher than f_0, then the object is moving toward the transmitter. If f' is lower than f_0, then the object is moving away from the transmitter.

The Doppler shift can be very small, so sophisticated electronics is necessary to implement this technique in radar systems. A widespread use of Doppler radar is in the radar systems used to track storms and weather systems. Not only can a Doppler system give relative linear velocities, but it can also tell if an object is rotating, and in what direction. If an object is rotating, one side will be moving toward and the other away from you. If the signals returning from an object are shifted at the edges relative to the signal from the center, then the object's rate of rotation can be determined. This information can tell much about the internal dynamics of intense storms and can spot tornadoes before they fully form. A bat uses sonar and the Doppler shift similarly to Doppler radar. Since bats are nocturnal creatures, they can't rely on visible light to catch their prey. The bat emits a chirp of high pitch, and in that short chirp is a frequency spectrum. When the bat hears the returning echo, it knows not only where its prey (usually an insect) is, but how the prey is moving relative to the bat. The bat can readily hear and interpret the Doppler shift in the original chirp and can translate that information into the prey's position and relative speed. The bat doesn't have to see its prey to catch it; it must only hear its echo.

8.12   Beats

In Section 8.3, the superposition principle was discussed. *Superposition* simply means that waves can be added together, as can standing waves on a string. Because a wave is a moving pulse of energy, two waves can occupy the same space at the same time. They can add in such a way that they exhibit constructive and destructive interference, as illustrated in Figure 8.15. Notice that for both constructive and destructive interference, the waves are *added* in both cases. The wave A_2 in Figure 8.15B is shifted 180° to the right or left from A_2 in Figure 8.15A. Subtraction is really the addition of a negative number; 180° is half a wavelength, and the two waves A_1 and A_2 are out of phase by 180°. If a vertical line is dropped at each time t, the resultant wave will be the simple sum of A_1 and A_2 at each t.

Beats occur when two waves of slightly different frequencies interfere. If we take two tuning forks of 440 Hz, set both into vibration, and listen, we will hear a single note. However, if we take a small weight and attach it to a tine of one of the forks and set both again into vibration, we will hear a slow rising and falling of the sound intensity. This slow-frequency beat is caused by two waves of slightly different frequencies constructively and destructively interfering. When one of the tuning forks has the weight attached, its frequency of vibration is slightly altered from the original 440 Hz. Figure 8.16 illustrates what happens. There are two waves with frequencies f_1 and f_2, where f_1 is larger than f_2. When the maxima of both waves occur together, there will be an intensity maximum (loud; times t_1, t_3, and t_5). Whenever both waves are 180° out of phase, there will be an intensity minimum (times t_2 and t_4). A beat occurs whenever one of the waves has undergone one cycle more than the other (between times t_3 and t_5). Between times t_3 and t_5, the upper sine wave has undergone four cycles, and the middle sine wave with the higher frequency has undergone five cycles. This brings them both back to the beginning, and if they started in phase, they are again in phase.

$$T_b = n_1 T_1 \qquad T_b = n_2 T_2$$

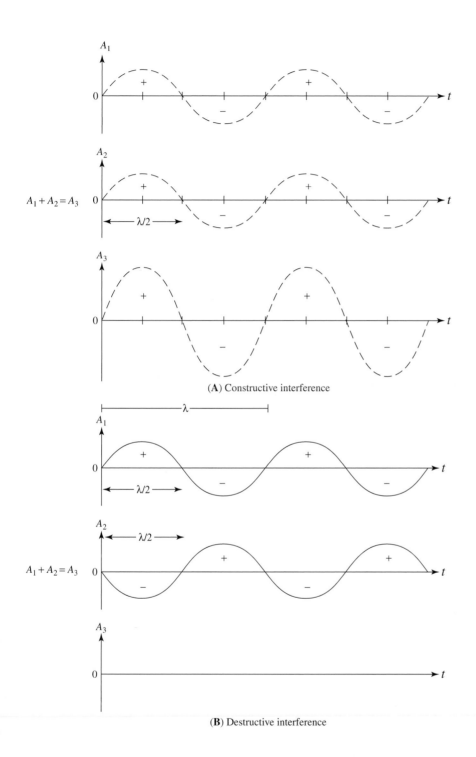

(A) Constructive interference

(B) Destructive interference

Figure 8.15 Constructive and Destructive Interference

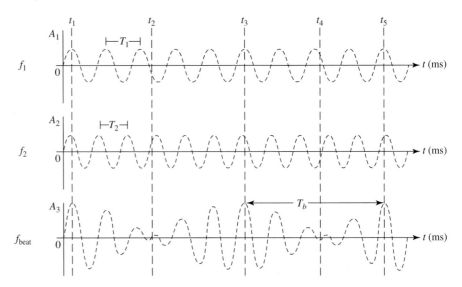

Figure 8.16 Temporal
Interference or Beats

$$n_1 - n_2 = 1$$

where T is the period of the associated wave, T_b is the period of the beat, and n is an integer number of wavelengths f r the given wave.

$$\frac{T_b}{T_1} - \frac{T_b}{T_2} = 1$$

$$T_b\left(\frac{1}{T_1} - \frac{1}{T_2}\right) = 1$$

$$\frac{1}{T_1} - \frac{1}{T_2} = \frac{1}{T_b}$$

(Equation 8.10) $$f_1 - f_2 = f_b$$

where f_b is the beat frequency.

Summary

1. Energy often propagates through a medium in the form of a *wave*. When this propagation causes the material in the medium to oscillate up and down, the wave is called a **transverse wave.** When the propagation causes the material to oscillate in the direction of propagation, the wave is called a **longitudinal wave.**

2. The *frequency* and the *wavelength* of a standing wave can be related through use of the wave equation, Equation (8.2).

Problem Solving Tip 8.1: Equations (8.1) and (8.2) can be used together to simplify otherwise difficult problems. If $v = \lambda f$, and $v = (T/\mu)^{1/2}$, then $\lambda f = (T/\mu)^{1/2}$. These equations can be very useful, considering that we often do not care about the velocity of the pulse along the string, but we do care about these other terms!

3. Sound travels through air, at typical classroom conditions, with a velocity of 343 m/s.

Key Concepts

DOPPLER EFFECT: An apparent frequency shift of a wave due to the relative movement of the source toward or away from you.

INTENSITY: The power received per unit area from a distant power source.

LONGITUDINAL WAVE: A wave in which the amplitude variation is in the direction of the propagation.

SUPERPOSITION: Constructive or destructive interference of one wave with another. The waves occupy the same space at the same time as they move through one another.

TRANSVERSE WAVE: A wave in which the direction of propagation is 90° to the amplitude.

Important Equations

(8.1) $v = \sqrt{\dfrac{T}{\mu}}$ (Speed of a pulse on a stretched string where T = tension and μ = linear density)

(8.2) $v = \lambda f$ (Wave equation where λ = wavelength and f = frequency)

(8.3) $\lambda = \dfrac{2L}{n}$ (Wavelength of a standing wave on a string of length L where n = number of bumps)

(8.4) $A = A_0 \sin kx, \ k = \dfrac{2\pi}{\lambda}$ (Sine wave described in terms of displacement where A = amplitude)

(8.5) $A = A_0 \sin \omega t$ (Sine wave described in terms of time where ω = angular frequency)

(8.6) $A = 2A_0 \sin kx \cos \omega t$ (Equation of a standing wave)

(8.7) $f' = \dfrac{v}{\lambda_0 + v_s T_0}$ (Doppler shift where v = speed of the wave in the medium, v_s = speed of the source, λ_0 = unshifted wavelength, and T_0 = unshifted period)

(8.8) $\sin \theta = \dfrac{v_s}{v}$ (Sonic cone where v_s = speed of sound in air and v = speed of source)

(8.9) $I = \dfrac{P}{4\pi r^2}$ (Intensity where P = power)

(8.10) $f_1 - f_2 = f_b$ (Beat frequency)

Conceptual Problems

1. Which one of the following properties of a wave is independent of all of the others?
   a. Frequency
   b. Amplitude
   c. Wavelength
   d. Speed

2. In a transverse wave, the particles of the medium:
   a. move parallel to the direction of travel
   b. move in circles
   c. move perpendicular to the direction of travel
   d. move in ellipses

3. Which of the choices below is a longitudinal wave?
   a. A wave in a stretched string
   b. An electromagnetic wave
   c. A water wave
   d. A sound wave

4. Which of the following will lower the pitch of a guitar string?
   a. Decreasing the tension in the string
   b. Using a lighter string

   c. Increasing the tension in the string
   d. Lengthening the string

5. The amplitude of a sound wave determines its:
   a. resonance
   b. pitch
   c. overtones
   d. loudness

6. A pure tone of the correct frequency causes a tuning fork to vibrate. This is an example of:
   a. interference
   b. an overtone
   c. resonance
   d. harmonics

7. When a wave leaves one medium and enters another, the quantity that remains unchanged is:
   a. wavelength
   b. speed
   c. frequency
   d. amplitude

8. A pulse sent down a stretched string dies away and disappears. What happens to the energy in the pulse?

9. What property of a sound wave governs its loudness? What governs its pitch?

10. Stars that rotate around a common center of gravity are called "double stars." They are so close together that a telescope can't resolve such a system into two points of light. Yet astronomers know that there are two stars instead of one from information gleaned through the telescope. What could this information be?

11. The speed of sound in helium is faster than in air. Why, then, if you fill your lungs with helium and begin to talk, do you sound like Donald Duck?

12. When waves meet and interfere with each other, the resultant waveform is determined by:

    a. diffraction   b. reflection

    c. superposition   d. refraction

13. Is energy passing through the nodes of a standing wave? Why or why not?

14. When a standing wave forms on a stretched string and the second harmonic is present, how many wavelengths are present on the length of string?

15. Interference plays a major role in:

    a. the Doppler effect   b. beats

    c. diffraction   d. reflection

16. A wave pulse moves along a stretched string with a constant velocity. Does this mean that the particles of a string have zero acceleration? Justify your answer.

17. How long would it take a pulse to travel from one end to another of a taut, "massless" rope?

18. When a car is at rest, its horn emits a frequency of 600 Hz. If a blind person hears the horn at a frequency of 580 Hz, should he jump out of the way? Why?

Exercises

Pulses on a String

1. A heavy rope hanging down from the ceiling is given a short, sharp transverse shake at its lower end. Show that the speed with which the pulse travels up the rope increases from bottom to top.

*2. A block of mass 2.1 kg sits on a horizontal, frictionless table. A string is attached to the block, goes over a frictionless pulley, and is attached to a smaller block weighing 0.25 kg and hanging over the table's edge. If the linear density of the string is 9×10^{-4} kg/m, what is the velocity of a pulse on the string?

    Answer: v = 49.2 m/s

*3. A control-line aircraft flies in a horizontal circle with a radius of 10 m. The aircraft is attached to a handle held by

the pilot at the circle's center by two steel cables of linear density 8×10^{-4} kg/m. If the aircraft completes a lap in 2.1 s and has a mass of 1.7 kg, how long would a pulse take to travel from the pilot to the plane and back again?

    Answer: t = 64.9 ms

4. A long, steel cable is used as a guy wire in bracing a tall antenna with a height of 35 m at the top. The guy wire is given a sharp rap, and the returning pulse can be heard 3.5 s later. The type of cable used for the guy wire has a linear density of 0.08 kg/m, and the angle between the cable and the ground is 35°. What is the tension in the cable?

    Answer: T = 97.4 N

Wave Equation

5. Small, transverse bumps are laid across a highway around a gradual curve. They are there so that a sleepy driver will hear and feel them when she drives over them and hopefully will be alerted in time to avoid driving off the curve. If the bumps are spaced 10 ft apart, and if the curve is taken at 60 mph, at what frequency will the vibrations and sound be generated?

    Answer: 8.8 Hz

6. A ripple tank has a length of 1.0 m with an oscillator at one end. It takes a wave 0.6 s to move the length of the tank, and the wave crests are 70 mm apart. What is the frequency of the oscillator?

    Answer: 23.9 Hz

*7. A disc rotating about its center has 12 equally spaced holes drilled into its perimeter. If a jet of air is blown through the holes while the disc rotates, a siren is produced. The disc has a radius of 0.5 m. A 25-g steel ball is tied to a string 0.25 m long and is attached to the disc's perimeter. If the string would break at a tension of 9 N, what will the siren's frequency be just before the string breaks?

    Answer: f = 83.6 Hz

Standing Waves on a String

8. A guitar string is 35 cm long and has a linear density of 4.7 g/m. If the fundamental frequency is to be 440 Hz (the musical note A), to what tension should the string be adjusted?

Answer: $T = 446$ N

9. If the guitar string in Exercise 8 is to vibrate at a frequency of 659 Hz (the musical note E), at what distance from one end should the string be fingered?

Answer: $D = 11.7$ cm

*10. A standing wave exists on a stretched string 1.2 m long. The standing wave has a maximum amplitude of 10 cm and is the second overtone. (a) Where are the nodes located? (b) If the oscillator frequency is 120 Hz, what is the amplitude of the midpoint 3 ms after the oscillator is switched on ($3 \text{ ms} = 3 \times 10^{-3}$ s)? (c) How much time does it take the midpoint to go from maximum amplitude to zero amplitude?

Answer: (a) 0, 0.4, 0.8, and 1.2 m
(b) $A = 6.4$ cm
(c) $t = 2.08$ ms

*11. A string with a linear density of 7 g/m is tied to the ceiling of an elevator. A 4-kg mass is hung from the string 0.75 m from the ceiling. If the elevator accelerates upward at 0.2 m/s^2, what is the fundamental frequency of the string?

Answer: $f = 50.4$ Hz

*12. A taut string has four loops when driven at a frequency of 120 Hz. The string is tied to the wall at one end and is passed over a pulley with a weight hanger tied to the other end. If there is a mass of 1.5 kg on the weight hanger for the four loops, how much mass must be on the hanger for only two loops if the frequency remains 120 Hz?

Answer: $M = 6$ kg

*13. A guy wire supports the mast of a sailing schooner from the rear and is under a large tension. The wire has a length of 9 m and a linear density of 0.25 kg/m. A sailor can push on the wire sideways at its midpoint with a force of 145 N, thereby deflecting the wire by 6 cm. If the sailor plucks the wire like a guitar string, what will the fundamental frequency of its vibration be?

Answer: $f = 11.6$ Hz

Sound Waves

*14. A small explosive charge detonates at the sea's surface. An oceanographic survey ship that is some distance away and is equipped with passive sonar records a difference of 7 s between the time that it heard the sound in the water and the time that it heard the sound in the air. If the speed of sound in air is 343 m/s, and that in seawater is 1531 m/s, how far is the ship from the detonation?

Answer: 3092 m

*15. You can determine the speed of sound in a gas by finding the fundamental resonance of a column of the gas in a water-filled glass tube. A 440-Hz tuning fork is held over a tall glass tube filled with the gas, and water is added until the fundamental resonance is detected. If the gas column is 24 cm high, what is the speed of sound in the gas?

Answer: $v = 422.4$ m/s

16. A square room has a length and a width of 25 ft. If someone inside were playing a boom box loudly, at what fundamental frequency would the room resonate and induce the risk of having the doors and windows blown out?

Answer: $f_0 = 22.5$ Hz

17. A tube open at both ends has a length of 70 cm. What is the fundamental frequency at which it will oscillate?

Answer: $f = 245$ Hz

18. In frustration a student heaves his physics text out a window 15 m above the ground. He sees someone standing below the window on the ground. Feeling guilty, he decides to warn the person below the window. How long can he wait before the warning is too late?

Answer: 1.706 s

19. Low-frequency standing waves can be generated in tall, cylindrical structures, such as unused smokestacks. At what frequency would a 50-m-tall stack resonate? Assume that the stack is closed at the bottom and the fundamental frequency is the strongest one heard.

Answer: $f = 1.715$ Hz

20. The rule for finding the distance of a lightning flash from you is to count the seconds from when you see the flash until you hear the roll of the thunder. Divide the seconds by 5, and the result should give you the distance in miles. What is the percent error under normal conditions?

Answer: About 6.5%

Doppler Effect

21. You are driving on a divided highway at 60 mph, and an ambulance is approaching you at 60 mph on the other side. The ambulance's siren has a frequency of 440 Hz. (a) How much higher is the pitch that you hear on approach? (b) After the ambulance has passed, how much lower than f_0 (the ambulance's original frequency) does the pitch that you hear seem?

Answer: (a) $f_0 - f' = 81$ Hz
(b) $f'' - f_0 = 60$ Hz

*22. Jockey Two-Pockets is late for a dinner engagement at the Restaurant at the End of the Universe and runs a red light in the vicinity of Uranus. When stopped by a traffic cop in a prowling cruiser, Jockey insists that the light was green. The cop takes Jockey at his word and then promptly fines him a zirconium for every km/s that the stoplight was Doppler-shifted from the red. What was Jockey's fine? (Red light has a wavelength of 6.7×10^{-7} m, and green light 5.4×10^{-7} m.)

Answer: 57,600 zirconia!

*23. Derive an expression for the frequency as a function of time that you would hear if you dropped a source of frequency f_0 down a very deep mineshaft.

Answer: $f = \dfrac{v}{v + gt} f_0$

Sonic Boom

24. An aircraft has a speed of 850 km/h at an altitude of 3.5 km. A person on the ground hears a sonic boom coming from directly overhead. How far away is the point on the ground that is actually directly under the aircraft?

Answer: $D = 3.7$ km

25. A model rocket in a wind tunnel develops a shock cone with an angle of 30° when the gas is flowing at 425 m/s. What is the speed of sound in the gas?

Answer: $v_s = 213$ m/s

Intensity

26. A 100-W light bulb is in the middle of a square room. A photocell 0.5 m on a side is placed 1.5 m from the bulb. If the conversion efficiency of the photocell is 15%, what is the maximum power that the cell can deliver?

Answer: $P = 0.133$ W

Radar and Sonar

27. An active sonar detects an incoming torpedo. The sonar operates with a short chirp of 100 kHz. The returning pulse is detected 1 s later and is Doppler-shifted to 102 kHz. (a) How far away is the torpedo when first detected? (Use first approximation, without taking into account the torpedo's motion.) (b) What is the torpedo's velocity relative to the sonar? (c) If the torpedo is on a collision course, how long will it be to impact after it was first detected if no evasive action is taken?

Answer: (a) 1531 m
(b) $v_s = 30$ m/s
(c) $t_i = 48.5$ s

Beats

28. In the cabin of a turboprop commuter airliner, a beat frequency of 5 Hz is heard from the props. (a) If the rpm of one prop is 2500, what is the rpm of the other prop? (b) If the optimum rpm of an engine is 2500, by how much must the rpm of the other engine be either increased or reduced? (c) Why was the pilot applying right rudder to keep the liner on a constant compass heading?

Answer: (a) $f = 2200$ or 2802 rpm
(b) 300 rpm
(c) The right engine was producing the excess of rpm. With no rudder input, the plane would fly in a circle to the left because of the difference in torques between the two engines. To compensate, the pilot *had* to give right rudder.

*29. Two identical guitar strings are plucked simultaneously and have a beat frequency of 3 Hz. The "tuned" string has a fundamental frequency of 440 Hz. If both strings have the same length and are made of the same materials, how much higher or lower is the tension in the untuned string?

Answer: ± 1.4 %

*30. Two identical strings are each 0.75 m long and have a mass of 0.2 g apiece. One string is fixed at both ends by clamps, and the other is attached to the ceiling of an elevator with a 5-kg mass hanging from it. With the elevator at rest, both strings vibrate with the same fundamental frequency. When the elevator accelerates downward, a beat frequency of 2 Hz is heard. What is the acceleration of the elevator?

Answer: $a = 0.137$ m/s^2

The fixed period of the swinging pendulum of an antique clock allows the clock to keep time.

Simple Harmonic Motion

At the end of this chapter, the reader should be able to:

1. Discuss the relationship between sine and cosine functions.

2. Explain how circular motion relates to **simple harmonic motion (SHM).**

3. Solve for the frequency and/or the period of oscillation for both spring and mass pendulums.

4. Apply the concept of conservation of energy when solving problems that deal with oscillating systems.

5. State the relationship between the displacement, velocity, and acceleration of an object exhibiting simple harmonic motion.

**Simple Harmonic Motion
(SHM):** A periodic phenomenon
in which a system naturally
oscillates sinusoidally.

In Chapter 8, we studied how energy can propagate through a medium as a wave. Wave motion implies periodic motion, and we saw how a sine function (as well as a cosine function) could satisfy the condition of periodicity. This chapter again deals with a periodic phenomenon: *simple harmonic motion (SHM).*

A pendulum of any type has the common characteristic of periodic motion: It oscillates back and forth with some fixed, constant period. A simple pendulum is driven by gravity. A mass on the end of a spring is driven by the energy stored in a stretched (or compressed) spring. An electronic tank circuit with interconnected capacitors and inductors is driven by the energy stored in electric and magnetic fields and oscillates at a fixed frequency. All of these systems are fundamentally different as to how the oscillations are produced, but the underlying mathematics of all of the systems is the same. This mathematical similarity will be stressed in this chapter.

9.1   The Traveling Wave

Let's take another look at our sine wave. This time we're interested in deriving a complete wave function that describes how a traveling wave changes in both space and time.

$$A = A_0 \sin \theta$$

In this equation, A is the amplitude of the wave at any time or place; A_0 is the maximum amplitude and is an initial condition and as such is arbitrarily chosen. (It can have any initial value that we choose to assign to it.) The sine function gives us our necessary periodicity.

Figure 9.1 is a graph that shows two complete cycles of a sine function. The vertical axis of the graph has been specified as amplitude, A, while the horizontal axis is the angle theta (θ). Two complete wavelengths are shown. The first goes from radian angles 0 to 2π, and the second from 2π to 4π. The horizontal axis can be seen through two different perspectives. One perspective is the wavelength of the sine wave, which is labeled above the graph. In this case, the horizontal axis would have units of distance. Note that every time one whole wavelength is completed, the angle θ goes through 2π. If the spatial coordinate is one wavelength, $\theta = 2\pi$; if the spatial coordinate is two wavelengths, $\theta = 4\pi$; and so on. Knowing this, we can redefine θ in terms of the spatial coordinate, x.

$$\theta = 2\pi\left(\frac{x}{\lambda}\right)$$

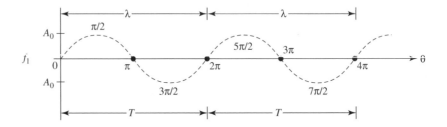

Figure 9.1 Sine Wave

Every time the distance x is one complete wavelength, or $n\lambda$, the function $\sin \theta$ will be zero and will give us a node on the graph. The ratio x/λ assures us that $\theta = 2n\pi$ every time $x = n\lambda$ ($n = 0, 1, 2, 3, \ldots$). Redefining θ in terms of x now gives a repetitive function for all x. This function is just like the standing wave function that was studied in Chapter 8 on waves. We now have

$$A = A_0 \sin\left(2\pi \frac{x}{\lambda}\right)$$

Our emerging traveling wave equation above is now defined in terms of the wave's wavelength.

Since the wave has simultaneously a wavelength *and* a frequency, the next step is to derive a periodic sine function in terms of time. In Chapter 6 on circular motion, the symbol for angular velocity was ω. In one complete revolution an object turned through 2π rad, and the angle through which the object turned was defined in terms of the angular velocity and time.

$$\theta = \omega t \qquad \omega = 2\pi f$$

$$\theta = (2\pi f)t$$

$$f = \frac{1}{T}$$

$$\theta = 2\pi\left(\frac{1}{T}\right)t$$

where t is time and T is the period. Whenever the time goes through one complete period, the angle goes through 2π. The periodic sine function is now redefined in terms of the time t.

$$A = A_0 \sin\left(2\pi \frac{t}{T}\right)$$

The sine function is now defined in terms of both space and time. How do we bring them together? If a wave is traveling to the right on the x axis with a speed v, at a time t later, it will have traveled a distance d given by

$$d = x + vt$$

Both sides of this equation can be multiplied by 2π.

$$2\pi d = 2\pi(x + vt)$$

In the time of one period, the wave will have traveled one complete wavelength.

$$vT = \lambda$$

We can define v by the above equation.

$$v = \frac{\lambda}{T}$$

$$2\pi d = 2\pi\left(x + \frac{\lambda}{T}t\right)$$

Now divide both sides by lambda (λ).

$$\frac{2\pi d}{\lambda} = 2\pi\frac{x}{\lambda} + 2\pi\frac{t}{T}$$

The term on the left has now become the angle θ through which the traveling wave has moved.

$$\theta = 2\pi\frac{x}{\lambda} + 2\pi\frac{t}{T}$$

$$\theta = 2\pi\left(\frac{x}{\lambda} + \frac{t}{T}\right)$$

We now have the complete expression for a traveling wave.

$$A = A_0 \sin 2\pi\left(\frac{x}{\lambda} + \frac{t}{T}\right)$$

If the wave had been traveling to the left, the plus sign would be replaced by a minus sign, giving

$$A = A_0 \sin 2\pi\left(\frac{x}{\lambda} - \frac{t}{T}\right)$$

By common usage, the terms are rearranged somewhat and are given other symbols.

$$\frac{2\pi}{\lambda} = k \qquad \frac{2\pi}{T} = \omega$$

The final form for the traveling wave equation is now

(Equation 9.1)

$$A = A_0 \sin(kx \pm \omega t)$$

which is a general wave equation. For a special case, such as a standing wave on a string frozen in time, $t = 0$. For an object revolving about a stationary axis, $x = 0$.

9.2 Sine-Cosine Relationships

Figure 9.2 shows three sine waves successively shifted by 90° to the left. The upper curve is a sine wave. The middle curve is a cosine wave. Just from the name *cosine,* you can guess that a cosine curve has a close relationship with a sine curve. If you look at the two graphs, the relationship becomes apparent. A cosine curve *is* a sine curve but is shifted by 90°. A cosine curve is simply a sine curve with the coordinate system origin shifted $\pi/2$ rad to the right. (You can prove this to yourself on a calculator. Find the cosine of any angle that you choose. Then add 90° to the original angle, and find the sine of the result. The two results should agree.) The two different formulas for the cosine curve are given to the right of the middle graph. But if the curve is shifted to the left, why is the $\pi/2$ phase shift added to, and not subtracted from, the original angle? If the sine wave were moving to the right with a velocity v, the observer would have to move to the right also at the same velocity for the sine wave to appear motionless. The origin moves with the observer.

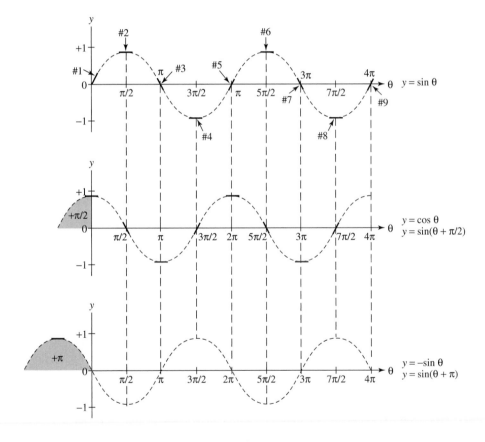

Figure 9.2 Sine-Cosine Relationships

However, there is a deeper connection between a sine curve and a cosine curve that gives rise to the 90° phase shift. To show this relationship, we must focus on the slope of the line at given points on the upper sine-wave graph. First let's review what the slope of a line is. The generic equation for a straight line is $y = mx + b$. If the line goes through the origin, $b = 0$, and we're left with $y = mx$. If x is plotted on the horizontal axis and y is plotted on the vertical axis (the standard layout), then m is the slope of the line. The slope is merely how much the line is tilted. The larger the slope, the more vertical the line. The slope is the rise over the run. An infinite slope is a vertical line. A slope of zero is a horizontal line. If the slope is positive, the line is tilted up from left to right. If the slope is negative, the line is tilted down from left to right. Mathematically, the slope is $\Delta y/\Delta x$, or the change in the vertical coordinate over the change in the horizontal coordinate. Since all points on a straight line have the same slant or slope, we can pick *any* pair of points, no matter how far apart, to find the slope. However, if our line isn't straight but curves as does the sine wave, then we must use a pair of points that are very close together to find the slope of a particular section of the plot. This approach is necessary because on a sine wave, the slope is always changing, and we want the slope only at a specific spot. Finding the slope of two points that are very close together will give us the tangent to the line at that point.

> *Review the relation between tangent and slope of line in Appendix A, Section A.3, "Trigonometric Identities."*

We are now prepared to derive the cosine curve from the sine curve by plotting the slope of the sine curve directly below. A good sine curve can be plotted using values from a calculator and a large sheet of graph paper. The more points that are plotted, the better the graph. If an infinite number of points are plotted, then the dotted line becomes a filled-in line, and the tangent to the line at any point can be drawn. On the upper sine curve in Figure 9.2, nine tangent lines are drawn. The slope for each tangent line will be plotted directly underneath as a point. Look at slope #1; this is where the sine curve goes through the origin. The slope is positive because it rises from left to right. If the sine curve is carefully drawn, you can see that it makes an angle of 45° with the x axis at the origin. The tangent of 45° is $+1$, so the value of the slope at that point is $+1$. (The tangent of the angle between the horizontal axis and the line is the slope of the line.) It is plotted on the graph directly below.

Now look at slope #2. It is taken at the positive peak of the sine curve, so the tangent line is horizontal. Its slope is 0. It is plotted directly below. Now look at slope #3. The slope of this line segment at this point also makes an angle with the x axis of 45°, but it slants down from left to right. The slope has a negative sign, so the slope of line segment #3 is -1. It is plotted directly below. It should now be easy to see that the slope of line segment #4 is 0 because the tangent is again a horizontal line. It is plotted directly below. The slope of line segment #5 is exactly like that of #1, so its value will be $+1$. It is plotted directly below. One complete cycle of the sine wave has now been plotted. You should plot the slopes of segments #6 through #9 to become familiar with how the process works.

With slopes #1 through #9 plotted, we now have a graph of how the sine curve changes for two complete cycles. Remember that a slope is the rise over the run, or how a curve changes at a given point. Figure 9.2 shows only the slopes at maximum and minimum values of the sine curve. If more points were desired, between slopes #1 and #2, for instance, the angle of the tangent line between the origin and the first positive maximum would be plotted. Inspection would show that the slope will decrease from $+1$ to 0 along this quarter-cycle. In the next quarter-cycle the slope will continue to decrease from 0 to -1 and will then begin to increase again. So the slope of a sine curve is constantly changing. After plotting a number of points on the middle axis, we would fill in the dots, and the resulting curve would be a cosine curve.

If we were now to apply the same procedure to the new cosine curve and plot the change directly below on the lower axis, we would get our sine curve back again, but inverted. The coordinate axis origin would have been shifted to the right by a full 180°. You should again make the effort to plot the negative sine curve from the cosine curve to become familiar with the reasoning behind the relationship between sine and cosine curves. The inverted sine curve occurred the second time we plotted the tangent of the curve directly above. We plotted the change in the original sine curve with respect to the angle twice, and we obtained the original sine curve, but inverted. We've seen a mathematically identical concept in an acceleration. An acceleration is the second change in displacement with respect to time. [The first change is the change in displacement with respect to time. The second change is the change in the velocity (based on the displacement) with respect to time.] This fact is a very valuable mathematical relationship that finds use in just about any periodic relationship. Since simple harmonic motion is periodic, an understanding of the relationships among the three plots in Figure 9.2 is important.

9.3   The Connection between Circular Motion and Simple Harmonic Motion (SHM)

Now our task is to connect a physical system swinging back and forth like a simple pendulum to the sine wave discussed above. If the amplitude of a pendulum were plotted against time, a sinusoidal wave would describe the motion.

$$A = A_0 \cos \omega t$$

At time $t = 0$, when the pendulum is released, it has its maximum amplitude. The Greek letter ω (omega) implies motion with a constant angular speed, but the pendulum's motion seemingly has little to do with a constant angular speed. It is always accelerating and decelerating. Figure 9.3 shows the connection between a swinging pendulum and a disc revolving with constant angular speed. In each of the three drawings, a gravity-driven, simple pendulum is oscillating back and forth above a disc with a peg inserted on its rim revolving at a constant angular speed of ω. If the angular speed is adjusted properly, then the pendulum bob will always stay directly above the peg as it moves around the circumference of the circle. The peg and the bob will always stay vertically aligned. Uniform circular motion and simple harmonic motion (SHM) are directly related. The key is that the angle is a linear function of time, but the function of the angle is not!

But what angle are we talking about? The angle that the suspension string makes with the vertical is ϕ (phi). The peg on the rim of the disc goes 2π rad for every revolution. This angle is described by the equation

$$\theta = \omega t$$

The angles θ and ϕ are *not* the same angle. The angle θ will go through a full 2π for each revolution. The angle ϕ is usually small, where $\sin \phi = \phi$ in radians. (Try this on your calculator. Take a small angle in radians, and find the sine of that angle in radians. The results will be very close together. You should go through a few small angles to prove it to yourself.)

It's worth discussing the relative time that the bob spends at certain points in its trajectory. In the middle drawing of Figure 9.3, the bob is at the bottom of its trajectory and is moving at its fastest speed. At that time, the peg on the rim of the disc has a purely horizontal motion, so the horizontal component of the peg's motion is at a maximum. In the lower drawing, the bob is at its maximum amplitude and has its lowest horizontal component of velocity (zero). At this point the horizontal component of the peg's velocity reverses direction and must therefore, for an instantaneous period of time, have a horizontal speed of zero. At the same time, the peg's vertical speed is at a maximum. The peg is still moving with constant angular speed, but its translational motion is strictly vertical. The bob seems to linger at the two endpoints and to hurry through the lowest point in its trajectory where its translational speed is greatest. The bob spends most of its time

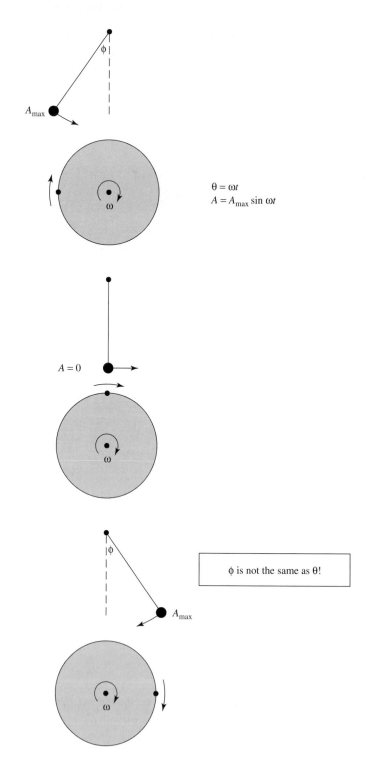

$$\theta = \omega t$$
$$A = A_{max} \sin \omega t$$

$A = 0$

ϕ is not the same as θ!

Figure 9.3 Relationship between SHM and Angular Speed

near the amplitude maxima, and the least amount of time where its amplitude is zero. The point where the amplitude is zero is the equilibrium position. This position is where the bob would hang if it were not oscillating.

9.4   The Simple Pendulum

We are now ready to analyze the simple, gravity-driven pendulum. From observation of a simple pendulum, we can show that the period (the time that the pendulum takes to go through one complete oscillation) is constant for small angles, regardless of the amplitude of the swing. This property has made the pendulum useful in constructing timepieces. For example, a grandfather clock uses a long pendulum to regulate its clock mechanism. A smaller cuckoo clock uses a pendulum for the same reason. A mechanical wristwatch has a torsional pendulum driven by a spring inside. What is the physical reason for the remarkable stability of a pendulum's period? To find out, we'll derive the period of the pendulum's motion from the basic forces acting on the pendulum bob.

Figure 9.3 shows the relation between simple harmonic motion and uniform circular motion. Figure 9.4 shows the same situation, but with a bit more geometry grafted onto the figure. The peg is moving around the rim of the disc with constant speed v. The time that it takes the peg to move around the disc once is the circumference of the disc over the speed.

$$T = \frac{2\pi R}{v}$$

or, rearranging somewhat,

$$T = 2\pi \frac{R}{v}$$

To get a swing to rise to high amplitude, the riders must pump energy into the swing in resonance with the pendulum's period.

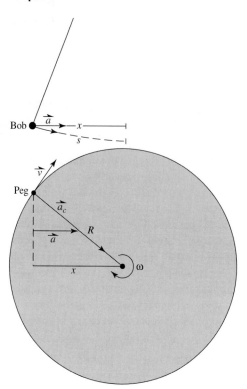

Figure 9.4 Derivation of the Period of a Harmonic Oscillator

   In Figure 9.4, two similar right triangles are nested one inside the other. The bigger outer triangle has as the hypotenuse the radius of the disc, R. The horizontal leg consists of the distance x of the peg from the center of the circle, which is also the distance of the pendulum bob from its equilibrium position. The inner smaller triangle is derived from the two accelerations that the peg sees. Remember that we were interested in the horizontal component of the peg's motion, and that as the peg's horizontal velocity was changing, a horizontal acceleration was associated with it. The horizontal leg of the smaller inner triangle is the horizontal acceleration that the peg sees toward the disc's center. It is also the horizontal acceleration that the pendulum bob experiences toward its equilibrium position. The inward-pointing radial acceleration is the centripetal acceleration that anything moving in a circular path undergoes. By similar triangles,

$$\frac{a_c}{a} = \frac{R}{x}$$

$$a_c = \frac{v^2}{R} \qquad \text{(Centripetal acceleration points radially inward.)}$$

$$\frac{v^2/R}{a} = \frac{R}{x} \qquad \frac{v^2}{a} = \frac{R^2}{x}$$

$$\frac{R^2}{v^2} = \frac{x}{a} \qquad \frac{R}{v} = \sqrt{\frac{x}{a}}$$

Therefore,

$$T = 2\pi\sqrt{\frac{x}{a}}$$ (Equation 9.2)

The period of the peg about the rim of the revolving disc is now in terms of motion that is common to both disc and pendulum. Equation (9.2) is general in that it holds true for *any* system undergoing simple harmonic motion.

Now we must interpret Equation (9.2) in terms of the forces that act explicitly on the simple, gravity-driven pendulum. Figure 9.5 illustrates those forces, and again we can use the geometric property of similar triangles to derive the period of the pendulum. The upper larger triangle is constructed from the physical dimensions of the system, and the lower smaller triangle is derived from the forces at work on the pendulum. The forces at work are gravity pointing straight down, the tension in the suspension string pointing along the string, and the restoring force that is always pointing in the direction of the equilibrium position. The apex angle ϕ is the same for both triangles. From the upper and lower similar triangles

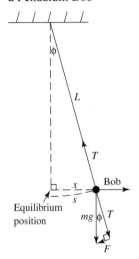

Figure 9.5 Forces Acting on a Pendulum Bob

$$\frac{L}{x} = \frac{mg}{F}$$

$$F = ma$$

where a is the linear acceleration toward the equilibrium position of Figure 9.4. The path that the bob actually travels is s, which is the arc of a circle; x is approximately equal to s *only* if the angle ϕ is small.

$$\frac{L}{x} = \frac{mg}{ma} \qquad \frac{L}{x} = \frac{g}{a}$$

$$\frac{x}{a} = \frac{L}{g}$$

Therefore, $$T = 2\pi\sqrt{\frac{L}{g}}$$ (Equation 9.3)

Equation (9.3) is for the period of a simple pendulum. The period is independent of the weight of the pendulum bob.

O P T I O N A L   T O P I C

The Equation of Motion for a Simple, Gravity-Driven Pendulum

We will derive the period of the simple pendulum *again*, but without using anything external to the system (such as the rotating disc). In Figure 9.6A, the pendulum bob is moving away from the equilibrium position, and in Figure 9.6B it is moving toward the equilibrium position. The restoring force F is always pointing toward the equilibrium position. At the equilibrium position, the restoring force is zero. When the bob is at its farthest from the equilibrium position, the restoring force is at its greatest. A force triangle is drawn

Figure 9.6 Restoring Force Acting on a Pendulum Bob

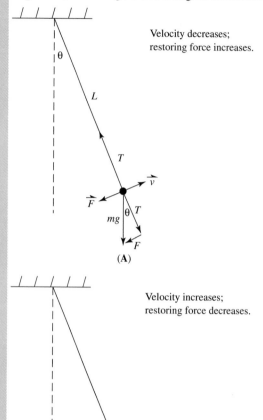

(A)

Velocity decreases;
restoring force increases.

Velocity increases;
restoring force decreases.

(B)

under the bob of Figure 9.6A. From it, we can see that the restoring force is proportional to sin θ.

$$F = mg \sin \theta$$

The restoring force and velocity of the bob are always out of phase with one another. As the velocity increases, the restoring force decreases. This negative feedback keeps the bob from launching itself into orbit. Since the velocity is changing along the trajectory of the bob, there is a force along this path.

$$F = m\frac{\Delta v}{\Delta t} \qquad F = mg \sin \theta$$

$$mg \sin \theta = -m\frac{\Delta v}{\Delta t}$$

Remember that the restoring force and velocity are always out of phase; therefore, the negative sign is necessary.

$$g \sin \theta = -\frac{\Delta v}{\Delta t}$$

$$\frac{\Delta v}{\Delta t} = \frac{\Delta}{\Delta t}\left(\frac{\Delta s}{\Delta t}\right) \qquad \frac{\Delta v}{\Delta t} = \frac{\Delta^2 s}{\Delta t^2}$$

The acceleration is the change with respect to time of the velocity, or the second change with respect to time of the displacement along the arc.

$$g \sin \theta = -\frac{\Delta^2 s}{\Delta t^2}$$

Mathematically, the above derivation is very similar to what we did in Figure 9.2 with the sine curve, where we took the second change in the original sine function with respect to the angle θ of the sine curve and got an inverted sine curve back again. Now the displacement along the trajectory (s) must be changed into a function of the angle.

$$s = L\theta$$

where s is the arc length, L is the length of the pendulum's suspension, and θ is the angle in radians. If the angle θ undergoes a small change $\Delta\theta$, then the arc length described is Δs.

$$\Delta s = L \, \Delta \theta$$

which gives us

$$g \sin \theta = -L\frac{\Delta^2 \theta}{\Delta t^2}$$

But the angle θ (in radians) is small, and therefore $\sin \theta = \theta$. So we now have the equation

$$g\theta = -L\frac{\Delta^2 \theta}{\Delta t^2}$$

The above equation can be rearranged into the following:

$$\frac{\Delta^2 \theta}{\Delta t^2} + \frac{g}{L}\theta = 0$$

We already solved this equation in principle when we discussed the sine-cosine relationships. We know that the second change of a sine function with respect to the angle θ returns an inverted sine curve. The amplitude of the pendulum can be expressed in terms of the angle θ, so we can use the following equation for the amplitude:

$$\theta = \theta_0 \sin \omega t$$

(Remember that the angle ωt is the *phase* angle of the pendulum's swing, and not the same as the angle θ used above. Theta is limited to small angles, whereas ωt goes through a full 2π.) The independent variable in the above equation is the time t, with the angular speed ω a constant.

The second change of the sine function with respect to time will give us an inverted sine function. However, since ωt is the angle ϕ, and ω is a constant, ω doesn't change with time.

$$\phi = \omega t \qquad \Delta\phi = \omega\,\Delta t \qquad \frac{\Delta\phi}{\Delta t} = \omega$$

$$\frac{\Delta x}{\Delta t} = \frac{\Delta\phi}{\Delta t}\frac{\Delta x}{\Delta\phi} \qquad \frac{\Delta x}{\Delta t} = \omega\frac{\Delta x}{\Delta\phi}$$

where $\Delta x/\Delta\phi$ is the change in the function (sine or cosine) with respect to the phase angle, as in Figure 9.2.

$$\frac{\Delta\theta}{\Delta t} = \omega\theta_0 \cos \omega t$$

$$\frac{\Delta^2\theta}{\Delta t^2} = -\omega^2\theta_0 \sin \omega t$$

Next we substitute the expression above into

$$\frac{\Delta^2\theta}{\Delta t^2} + \frac{g}{L}\theta = 0$$

and obtain

$$-\omega^2\theta_0 \sin \omega t + \frac{g}{L}\theta_0 \sin \omega t = 0 \qquad -\omega^2 + \frac{g}{L} = 0$$

$$\omega^2 = \frac{g}{L} \qquad \omega = \sqrt{\frac{g}{L}}$$

From this calculation we can find the period of the pendulum's motion.

$$2\pi f = \sqrt{\frac{g}{L}} \qquad f = \frac{1}{2\pi}\sqrt{\frac{g}{L}}$$

$$T = \frac{1}{f}$$

(Equation 9.3)

$$T = 2\pi\sqrt{\frac{L}{g}}$$

Equation (9.3) is the same as the one for the period of the pendulum's swing found earlier in this section. We took two entirely different paths to the answer, but we arrived at the same place in spite of the different paths. In physics, an answer can often be found through one or more methods. Both of the above methods are equally valid, and one is not better than the other. Different people see a given situation through different perspectives. If the physics is valid, the path will lead to the same answer, regardless of the starting point. There is a third way of finding the pendulum's period. However, it is left as an exercise at the end of this chapter.

The above derivation of the period of a simple pendulum might seem complicated compared to the derivation done with the revolving disc. However, the advantage of the approach directly above is that it applies to *any* simple harmonic oscillator, regardless of its principle of operation. A simple pendulum is mechanically quite simple with a weight at the end of a string being gravity driven. Why it oscillates isn't too difficult to answer using the conservation of energy, but arriving at an expression for the period of the oscillation is much more difficult. Even simple systems have a subtlety and depth that require us to take much more than a superficial look. But once the secret of the system's dynamics has been unlocked, the key will fit many more locks than the one it was originally designed for. We can now examine an entire class of seemingly unrelated phenomena, and a surprising underlying unity will emerge.

9.5   Mass and Spring Pendulum

A spring with a mass on the end of it is a very different system mechanically from the simple, gravity-driven pendulum, but both systems exhibit the same kind of behavior: simple harmonic motion (SHM). Figure 9.7 illustrates the motion of a spring-driven pendulum. A mass *m* lying on a frictionless horizontal surface oscillates back and forth with a period *T*. If its amplitude were plotted as a function of time, it would trace out a sine wave. It therefore exhibits simple harmonic motion. It will yield the oscillator equation that we derived for the gravity-driven pendulum.

First, let's derive the spring pendulum's period using the rotating disc method. The general formula for the period of any system exhibiting SHM is

$$T = 2\pi\sqrt{\frac{x}{a}}$$

$$F = kx \qquad x = \frac{F}{k}$$

$$F = ma \qquad a = \frac{F}{m}$$

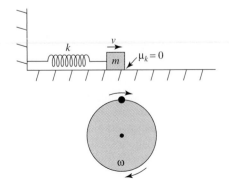

Figure 9.7 Spring-Driven Harmonic Oscillator

$$\frac{x}{a} = \frac{m}{k}$$

(Equation 9.4)

$$T = 2\pi\sqrt{\frac{m}{k}}$$

(Try working out the units to prove that you find the period in units of seconds.)

Next, we'll derive the oscillator equation from the forces acting on the system.

$$F = -kx \qquad F = m\frac{\Delta^2 x}{\Delta t^2}$$

$$-kx = m\frac{\Delta^2 x}{\Delta t^2}$$

$$\frac{\Delta^2 x}{\Delta t^2} + \frac{k}{m}x = 0$$

As mentioned above, this equation has the general form of the *oscillator equation.* Whenever an equation is derived for a system from the point of view of forces and has this general mathematical form, that system will exhibit simple harmonic motion.

Again, it is worth pointing out that acceleration is the second change in the displacement with respect to time. The first change in the displacement with respect to time is velocity, and the change in velocity with respect to time is the acceleration. The above equation has the same mathematical form as the one derived for the gravity-driven pendulum. It is solved in the same way mathematically as is the equation for the gravity-driven pendulum, with the angular speed being the square root of the constant coefficient of the x term.

$$\omega = \sqrt{\frac{k}{m}}$$

or

$$T = 2\pi\sqrt{\frac{m}{k}}$$

This equation for the period is the same as the one obtained with the rotating disc method, as it should be. Physics is consistent. We took two different paths, but nevertheless we arrived at the same answer.

9.6   The Torsional Pendulum

The torsional pendulum is another spring-driven pendulum, but instead of the masses moving back and forth as in the gravity-driven and spring-driven pendulums in the previous examples, the masses move around in a circle. The motion is illustrated in Figure 9.8. We will set up the oscillator equation from the torques acting on the system shown in the figure. The pendulum bob in this example undergoes circular motion instead of linear motion. It oscillates because of a torque on the thin suspension thread from which the bob

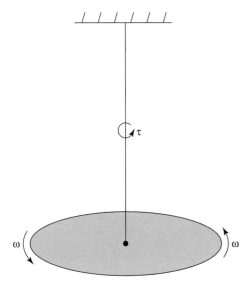

Figure 9.8 Torsional Pendulum

is hanging. The suspension string acts like a spring in that the torque is directly proportional to the amount that the string is twisted from its equilibrium position.

$$\tau = -k\theta$$

Energy can be stored in the spring by twisting it, and this stored energy becomes potential energy.

$$PE = \frac{1}{2}k\theta^2$$

where k has the units of torque per degree.

To find the period of the torsional pendulum's oscillations, we set up the equation of motion for the pendulum.

$$\tau = -k\theta \qquad \tau = I\alpha$$

$$\alpha = \frac{\Delta^2\theta}{\Delta t^2} \qquad \tau = I\frac{\Delta^2\theta}{\Delta t^2}$$

$$I\frac{\Delta^2\theta}{\Delta t^2} = -k\theta$$

$$\frac{\Delta^2\theta}{\Delta t^2} + \frac{k}{I}\theta = 0 \qquad \text{(Equation 9.5)}$$

We've seen Equation (9.5) three times now for three types of mechanically different harmonic oscillators. The above equation is the generic harmonic oscillator equation. Like linear acceleration, angular acceleration (α) is the second change of the angular displacement (θ) with respect to time.

The angular frequency of oscillation can now be readily determined. The angular frequency is always the square root of the coefficient of the linear displacement variable, whether it is for a rotational or a translational displacement.

$$\omega = \sqrt{\frac{k}{I}}$$

(Equation 9.6)

$$f = \frac{1}{2\pi}\sqrt{\frac{k}{I}}$$

The moment of inertia for a torsional pendulum with a flat disc as the bob is $\frac{1}{2}mr^2$, and the frequency becomes

$$f = \frac{1}{2\pi}\sqrt{\frac{k}{\frac{1}{2}mr^2}} \qquad f = \frac{1}{2\pi}\sqrt{\frac{2k}{mr^2}}$$

We could have used the rotating disc method for finding the period of the pendulum, but it is easier to get the equation of motion from the torques in the system and recognize the equation of motion as the harmonic oscillator equation. *All* harmonic oscillators will be described by the oscillator equation. Learn to recognize it, and be able to read the angular velocity from it.

9.7   Energy Conservation in SHM Systems

The simple harmonic motion systems that we have studied up to now have all conserved energy. The oscillations of the system are an interplay between potential and kinetic energy. Without friction, a simple pendulum would continue to oscillate to and fro forever. However, friction is rather difficult to eliminate entirely. Both a cuckoo clock and a grandfather clock use simple pendulums to regulate the movement of the clock's hands. However, without weights, the pendulum's oscillations would decrease because of friction, and the pendulum would eventually come to rest in the equilibrium position. The weights hanging from the clocks serve to replace the energy lost to friction. As the weights slowly fall, their gravitational potential energy is converted into the energy that the clock's mechanism inevitably loses to friction. If there were no friction, then the weights wouldn't be needed.

In a frictionless simple pendulum, the gravitational potential energy at the top of the pendulum's swing is converted into kinetic energy at the equilibrium position; then the kinetic energy is converted back to potential energy again on the return swing. At the top of the swing, all of the energy is potential (the speed is zero), and at the bottom of the swing, all of the energy is kinetic (it's at its lowest point). At intermediate positions, the energy is a combination of both potential and kinetic. Mathematically, the relationship is given by the following equation:

$$E = mgh + \frac{1}{2}mv^2$$

where the total energy E is constant, not changing with time. The above equation could be written

$$mgh_0 = mgh + \frac{1}{2}mv^2$$

where h_0 is the height of the pendulum bob above its equilibrium position at its maximum amplitude. (By pulling the bob up to the arbitrary height h_0, we can give the system any initial energy E that we please.) Now, knowing what the total energy is, we can calculate the bob's speed at any point along its trajectory. We can also find how high the bob is above its equilibrium position if we can find its instantaneous speed. Let's solve the above equation for the speed of the bob at any point in its trajectory.

$$mgh_0 = mgh + \frac{1}{2}mv^2$$

$$\frac{1}{2}v^2 = g(h_0 - h)$$

$$v = \sqrt{2g(h_0 - h)}$$

We now have the bob's velocity at any point in its trajectory from energy considerations.
    We can come up with a similar equation for the spring-driven pendulum.

$$E = \frac{1}{2}kx^2 + \frac{1}{2}mv^2$$

$$E = \frac{1}{2}kx_0^2$$

$$\frac{1}{2}kx_0^2 = \frac{1}{2}kx^2 + \frac{1}{2}mv^2$$

$$mv^2 = k(x_0^2 - x^2)$$

$$v = \sqrt{\frac{k}{m}(x_0^2 - x^2)}$$

The above equation also gives the speed of the bob for any position in its trajectory. The same calculation can be done for the torsional pendulum, but it is left as an exercise at the end of the chapter.
    The equations that we have built have inherent limitations mathematically as well as physically. Notice that for the gravity-driven pendulum, h_0 is always greater than or equal to h. It can't be less than h, or the quantity under the square root sign would be negative, and you can't take the square root of a negative number. (At least you can't until you begin the study of complex numbers. As far as we're concerned in this book, the square root of -1 is forbidden!) When h equals h_0, the speed is zero, and all of the energy is potential energy. When h is zero, the speed is at its maximum at the equilibrium position, and all of the energy is kinetic. (When we say that $h = 0$, we mean that the bob is the lowest it's going to get. It could still be 10 ft above the floor at the lowest point in its trajectory, but in the reference frame of the pendulum, the potential energy is zero.) Often, looking at an equation and properly interpreting it will give us some of the limitations imposed upon the system. If we start out with a maximum height h_0 for the gravity-driven pendulum, it can't go higher because its total energy is fixed. The mathematics tells us this by limiting square roots to positive numbers. When you derive an equation to describe a system, apply this technique to it, and see if the equation mirrors the behavior of the system.

9.8   Velocity and Acceleration in SHM Systems

In the preceding section the velocity was found for a bob at any point in its trajectory. When velocities are found from energy considerations, time plays no part, and as a result the calculations are often simplified. However, the velocity of the bob can also be found from the equation for the amplitude of the bob at any time t.

$$A = A_0 \cos \omega t$$

Refer to Figure 9.4. The amplitude of the pendulum's swing would be the extremes of the bob's horizontal motion. In Figure 9.4, either extreme would be a distance R horizontally from its equilibrium position and to each side of it. As a result, the above equation would be modified to give

$$A = R \cos \omega t$$

If the amplitude is really the horizontal displacement from the equilibrium position, then

$$A = x$$

(Equation 9.7)   and $$x = R \cos \omega t$$

But the velocity is just the change in displacement over time. If Equation (9.7) is treated as the graph of the amplitude (on the vertical axis) versus the time (on the horizontal axis), then the velocity is just the slope at any point (recall Figure 9.2).

$$\frac{\Delta x}{\Delta t} = -\omega R \sin \omega t$$

(Remember that ωt is the phase angle and that whenever we find the change of a function with respect to this phase angle, it's the time t that changes, not the phase velocity ω. Therefore, we must pull out the coefficient ω every time we look at how the function changes with time.)

$$\phi = \omega t \qquad \Delta\phi = \omega \, \Delta t \qquad \frac{\Delta\phi}{\Delta t} = \omega$$

$$\frac{\Delta x}{\Delta t} = \frac{\Delta\phi}{\Delta t}\frac{\Delta x}{\Delta\phi} \qquad \frac{\Delta x}{\Delta t} = \omega\frac{\Delta x}{\Delta\phi}$$

where $\Delta x/\Delta\phi$ is the change in the function (sine or cosine) with respect to the phase angle, as in Figure 9.2.

(Equation 9.8) $$v = -\omega R \sin \omega t$$

Equation (9.8) gives the velocity of the bob at any time t. There now exists an equation for the velocity at any position in the trajectory, and for any time t. Of course, the bob is also accelerating and decelerating, and an acceleration is the change in velocity over the change in time. It is therefore easy to find the acceleration from Equation (9.8) by looking at how it changes with time once again.

$$\frac{\Delta v}{\Delta t} = -\omega^2 R \cos \omega t$$

$$a = -\omega^2 R \cos \omega t \qquad \text{(Equation 9.9)}$$

The negative sign means that when the bob is approaching the maximum amplitude ($\omega t = 0, \pi, n\pi$), the acceleration is in the opposite direction, toward the equilibrium position. When the speed is zero, the acceleration is *not* zero. This concept makes sense only if you realize that at the maximum amplitude, the bob changes direction. For an infinitesimal moment, its velocity is zero. The restoring force is, however, at a maximum. The velocity is zero, but the acceleration is at a maximum.

9.9   The Damped Harmonic Oscillator

Any material object that possesses length has a resonant frequency. For example, a bridge of length L can vibrate up and down (transversely) as in Figure 8.5. It has a fundamental frequency of vibration, a first overtone, a second overtone, and so on, with the wavelength of the different modes being determined by the length of the span. If one end of the span were unsupported, as in a diving board, the modes of vibration would be as in Figure 8.9. In both cases, a characteristic frequency is associated with each resonant mode. If energy is pumped into the bridge at one of the resonant frequencies, the bridge will begin to oscillate at that frequency. Energy over time is power, and a small amount of power at the right frequency will eventually cause a very large amplitude of oscillation. If this amplitude is large enough, it might destroy the bridge span. Such a bridge is an undamped harmonic oscillator. Damping implies friction, and with damping, the oscillations of an oscillating system will slowly die away. An undamped system will continue to oscillate long after the original stimulus has stopped.

A swing in a playground is also a harmonic oscillator. It is actually a gravity-driven pendulum of length L, with you, the rider, as the pendulum bob. If the height of the frame of the swing is 10 ft, and if there is a 2-ft clearance under the swing's seat, then the length of the pendulum is roughly 8 ft. The period of the swing's oscillations will be roughly

The Tacoma Narrows Bridge failed because of an undamped resonance between the wind and the suspended span.

3 s. To get the swing to carry you high into the air, you must pump your legs at the right time in every cycle to get it swinging and to increase the amplitude of the oscillation. The right time to extend your legs is as you reach the maximum height when facing rearward, and the right time to pull them in again is as you pass through the equilibrium position nearest to the ground. By slowly pumping energy into the system in this way, you can easily find yourself 8 ft in the air at each amplitude maximum. A swing is a damped system because if you stop pumping your legs, the oscillations will noticeably decrease after several oscillations. There is friction in the system.

A car's suspension is also an oscillatory system. The car's springs each act as a spring-driven pendulum, as in Figure 9.7. The mass on the end of the spring is the tire and axle of that particular wheel. Since the function of the suspension system is to aid handling and improve the ride, undamped oscillations are definitely unwanted. Any evenly spaced series of bumps in the road, if driven over at the correct speed, will cause the springs in the suspension to oscillate at their resonant frequency and will cause the car to become unmanageable. If these oscillations are undamped, they will eventually tear the suspension right out of the car. Therefore, a damper of these unwanted resonant oscillations is necessary; this damper is the shock absorber. The shock absorbers allow the springs to work, but they will damp out the amplitude of the oscillations to zero in less than one cycle. The shock absorber is filled with oil which is allowed to flow through a small orifice whenever the spring is compressed or extended. The damping is controlled by the viscosity of the oil or the diameter of the orifice. A car's suspension system is a critically damped system. We want the oscillations to die away very quickly, hopefully within one period of the oscillation. A shock absorber is designed to do just that.

Summary

1. The *period* (T) of a system is measured in seconds (s) and is the time required to complete one *entire* oscillation.

Problem Solving Tip 9.1: Be very careful from this point onward with regard to symbols. It is at this point in your study that the repetition of symbols often becomes problematical. For example, T stands not only for the *period of oscillation* but also for *tension*! Also, do not confuse capital T with lower-case t, which stands for *time*. A little carelessness with symbols can lead to complete disaster, whether you're working on homework or on an exam!

As we continue on in the text, we will see more and more symbols being replicated to mean different things. Such repetition cannot be helped, as there are only 26 letters in the Roman alphabet (which we use), and we've already used some of the Greek alphabet as well!

2. The velocity of an object exhibiting **SHM** is at a minimum ($v = 0$) when its displacement from equilibrium is at its maximum, and the velocity is at its maximum when its displacement is at its minimum ($x = 0$).

Problem Solving Tip 9.2: You can see this concept from conservation of energy arguments. Refer to Problem Solving Tip 4.1, and consider the following. An object exhibiting SHM has two forms of energy: kinetic and potential. A block oscillating back and forth on a frictionless surface by means of a spring has kinetic energy and spring potential energy. Using our "snapshot" method, we can look at the block at *any point in its motion* and call it our *initial state*. Its energies would then be

$$\text{Energy "before": } \quad \frac{1}{2}mv_i^2 + \frac{1}{2}kx_i^2$$

At any later time, it also has kinetic and spring potential energy.

$$\text{Energy "after": } \quad \frac{1}{2}mv_f^2 + \frac{1}{2}kx_f^2$$

(Remember, x in this case is the displacement from the object's equilibrium position.)

Thus, as long as we know any information concerning the system, we have a good chance of solving this problem for either velocity or displacement (which is very common) in one step! Remember, in some cases the displacement and/or velocity will be equal to zero!

Key Concept

SIMPLE HARMONIC MOTION (SHM): A periodic phenomenon in which a system naturally oscillates sinusoidally.

Important Equations

(9.1) $A = A_0 \sin(kx \pm \omega t)$ (Traveling wave equation)

(9.2) $T = 2\pi\sqrt{\dfrac{x}{a}}$ (Period of SHM in terms of the horizontal displacement from the equilibrium position and acceleration toward the equilibrium position)

(9.3) $T = 2\pi\sqrt{\dfrac{L}{g}}$ (Period for a simple pendulum where L is the length and g is the acceleration of gravity)

(9.4) $T = 2\pi\sqrt{\dfrac{m}{k}}$ (Period for a mass spring harmonic oscillator of mass m and spring constant k)

(9.5) $\dfrac{\Delta^2\theta}{\Delta t^2} + \dfrac{k}{I}\theta = 0$ (Generic harmonic oscillator equation)

(9.6) $f = \dfrac{1}{2\pi}\sqrt{\dfrac{k}{I}}$ (Frequency of a torsional pendulum where k is the torsional spring constant and I is the moment of inertia)

(9.7) $x = R\cos\omega t$ (Horizontal displacement of a harmonic oscillator from its equilibrium position)

(9.8) $v = -\omega R\sin\omega t$ (Velocity of a harmonic oscillator)

(9.9) $a = -\omega^2 R\cos\omega t$ (Acceleration of a harmonic oscillator)

[For Equations (9.7) through (9.9), the motion and displacement variables can be either rotational or translational.]

Conceptual Problems

1. The product of the period and the frequency is equal to:
   a. the speed of the pulse   b. 1
   c. 2π d. the wavelength

2. The amplitude of a simple harmonic oscillator is independent of:
   a. maximum KE   b. maximum speed
   c. maximum acceleration   d. frequency

3. The maximum speed of a simple harmonic oscillator occurs when:
   a. it's at the equilibrium position
   b. it's at a maximum amplitude
   c. time is zero
   d. the phase angle is 0° or 180°

4. The maximum acceleration of a simple harmonic oscillator occurs when:
   a. it's in the equilibrium position
   b. it's at a maximum amplitude
   c. the phase angle is 90°
   d. its velocity is highest

5. The KE of a harmonic oscillator:
   a. is greatest at a maximum amplitude
   b. is greatest when the PE is greatest
   c. is greatest at the equilibrium position
   d. is greatest when the acceleration is at a maximum

6. The PE of a harmonic oscillator:
   a. is greatest at a maximum amplitude
   b. is greatest when the KE is greatest
   c. is greatest at the equilibrium position
   d. is greatest when the acceleration is zero

7. The sum of the KE and the PE for a simple harmonic oscillator without friction:
   a. varies with the amplitude
   b. varies with the acceleration
   c. is constant regardless of the instantaneous amplitude or acceleration
   d. is zero when the acceleration is zero

8. The period of a simple harmonic oscillator is the same no matter what the initial amplitude. A larger amplitude means that a bob must go farther. How can this relationship be reconciled with the fact that the constant period is independent of the amplitude?

9. Two children are sitting on playground swings with equal chain lengths. One child is much heavier than the other. One is pulled 5° from the vertical, and the other is pulled 10° from the vertical. If they are both released simultaneously, will they both come back to their starting points at the same time? Why or why not?

10. The pendulum of London's Big Ben is noted particularly for its accuracy. The pendulum is 3.98 m long and is adjusted by pennies placed on top of the bob. How can this adjustment affect the period of the pendulum?

11. A spring of force constant k has a mass m suspended from it and oscillates at a frequency of f. The spring is cut in half, and the same mass is hung from it and then started into oscillation. How is the frequency of the shorter spring affected relative to the longer one?

12. The two halves of the spring in Problem 11 are suspended side by side and are connected at the bottom to the mass m used above. What is the relation to the frequency of the original oscillator?

13. How can a simple pendulum be used to trace a sinusoidal curve?

Exercises

The Simple Pendulum

1. A simple, gravity-driven pendulum 0.5 m long oscillates with a frequency of 1.0 Hz. (a) What is the value of the acceleration of gravity in the location where the reading was made? (b) If the place is an elevator car on the earth, what are the magnitude and the direction of the car's acceleration?

   Answer: (a) $g' = 19.7 \text{ m/s}^2$
   (b) $a = 9.9 \text{ m/s}^2$ up

2. A hole is bored through the earth along a diameter. A stone dropped into this hole executes simple harmonic motion. What is the frequency of this motion? Assume that the earth's density is constant.

   Answer: $f = \dfrac{1}{2\pi}\sqrt{\dfrac{4}{3}\pi\rho G}$

*3. Derive the harmonic oscillator equation for a simple, gravity-driven pendulum using a torque argument.

   Answer: $\dfrac{d^2\theta}{dt^2} + \dfrac{g}{L}\theta = 0$

*4. Solve the simple harmonic motion equation derived in Exercise 3 with a cosine function, and find the period of oscillation.

   Answer: $T = 2\pi\sqrt{\dfrac{L}{g}}$

5. A pendulum bob is suspended by a string of length L. Directly below the suspension point, a smooth peg is fastened to the wall a distance $L/2$. Write an expression for the period of this odd oscillator for small oscillations.

   Answer: $T = 1.707\pi\sqrt{\dfrac{L}{g}}$

6. A clock pendulum is adjusted so that it has a period of 2 s, so that each swing is 1 s long. What is this pendulum's period on the moon?

   Answer: $T = 2.46$ s

Mass and Spring Pendulum

7. The weight needed to stretch a spring by 4 cm when hung vertically is 65 g. (a) What is the force constant of the spring? (b) How much work will gravity do on the spring? (c) If the spring were set into oscillation, what would the frequency of oscillation be? (Neglect the force of gravity.)

   Answer: (a) $k = 15.9 \text{ N/m}$
   (b) $W = 0.0127 \text{ J}$
   (c) $T = 0.4$ s

8. An object weighing 2.5 kg is dropped from a height of 3 m onto a spring. The maximum compression of the spring is 2 cm. What is the force constant of the spring?

   Answer: $k = 369{,}950 \text{ N/m}$

9. A spring with a spring constant of 330 N/m is compressed 70 mm, and a 5-g rubber ball is placed onto the cocked spring. When the spring is released, what is the speed of the ball?

   Answer: $v = 18 \text{ m/s}$

10. (a) A set of springs k_1 and k_2 are connected side by side to a block of mass m. The mass is then pulled a small distance x horizontally, stretching both springs. What is the potential energy of the system? (b) The springs are now reconnected to the block end to end. The block is again pulled horizontally a small distance x, stretching the springs. What is the potential energy of the system now?

   Answer: (a) $PE = \dfrac{1}{2}(k_1 + k_2)x^2$
   (b) $PE = \dfrac{1}{2}\left(\dfrac{k_1 k_2}{k_1 + k_2}\right)x^2$

11. In orbit or in space, the condition of weightlessness doesn't allow the use of a gravitational field to weigh an object, as with a conventional scale. The oscillations of a spring-driven pendulum can, however, "weigh" an object. An object is attached to a spring with a spring constant of 7 N/m, and the system oscillates with a period of 2.3 s. What is the mass of the object?

   Answer: $m = 0.938 \text{ kg}$

*12. A mass on a frictionless surface has two springs attached to it. Both springs can store energy in compression and tension.

One of the springs with a constant force of 15 N/m is attached on the left, and the other spring with a constant force of 8 N/m is attached on the right. If the mass is 1.5 kg, what is the period of oscillation of the spring-driven pendulum?

Answer: t = 1.6 s

13. An object of unknown mass is suspended from a spring, and the system is set into oscillation with a frequency of 3.5 Hz. If the object were to hang from the spring without oscillating, by how much would it stretch the spring?

Answer: Δy = 2 cm

*14. A mass of 0.25 kg is connected to a spring with a force constant of 12 N/m. The mass is placed upon a frictionless horizontal surface, and the spring is stretched a distance of 30 cm and then released. (a) What is the amplitude? (b) What is the period of oscillation? (c) What is the acceleration at maximum amplitude? (d) What is the maximum velocity? (e) What is the total energy of the system? (f) At what times is the force on the mass zero?

Answer: (a) A_0 = 30 cm
(b) *T* = 0.907 s
(c) *a* = 14.4 m/s^2
(d) v_{max} = 2.08 m/s
(e) E_T = 0.54 J
(f) *t* = 0.227 s, 0.68 s, 1.13 s, 1.59 s, …

*15. When a 2.5-kg mass is hung from a spring, it stretches the spring by 6 cm. The same mass is placed on a frictionless horizontal surface and is connected to the spring which is also horizontal and connected to a fixed support. The block is pulled 4 cm from the equilibrium position and released. What is the initial acceleration of the block?

Answer: a = 6.5 m/s^2

*16. An engineer wants to construct a ball launcher that when horizontal will give a 25-g steel ball an acceleration of

15 m/s^2 when released. The spring can be compressed only a maximum of 12 cm because of size constraints. (a) What spring constant must the spring have? (b) What will the speed of the ball be just after release?

Answer: (a) k = 3.125 N/m
(b) *v* = 1.34 m/s

*17. A 3400-lb Chevy Impala is supported by four identical worn shock absorbers whose springs have a force constant of 5200 lb/ft apiece. (a) What is the natural frequency of oscillation for the car? (b) Raised tar strips are placed 54 ft apart along the highway to alert a tired driver of a turn. How fast must the car be going when the frequency of the bumps is the same as the resonant frequency?

Answer: (a) f = 2.22 Hz
(b) *v* = 81.8 mph

*18. (a) Determine the equation of motion for a mass spring pendulum oriented vertically. Put the equation of motion in a form similar to the harmonic oscillator equation. (b) The period of the vertical mass spring oscillator is the same as a horizontally oriented one, despite the extra term. Why?

Answer: (a) $\dfrac{d^2y}{dt^2} + \dfrac{k}{m}y = g$

(b) When the oscillator is moving down, gravity assists its motion; when it is moving up, gravity opposes it. The influence of gravity averages out to zero.

19. A spring with a constant of 90 N/m is placed vertically on the floor. A 0.7-kg block is placed on the top of the spring and is pushed down to start the system oscillating. At what amplitude will the block leave the spring?

Answer: A = 7.6 cm

The Torsional Pendulum

20. The bob of a clock based on a torsional pendulum looks like Figure 9.9 when viewed from above. If the torsional spring constant *k* is 1.5 N-m/rad, *m* = 50 g, and *R* = 8 cm, what is the period of the pendulum?

Answer: T = 0.058 s

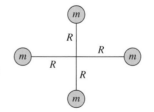

Figure 9.9 Top View of Torsional Pendulum Bob

Energy Conservation in SHM Systems

21. The bob of a gravity-driven Foucalt pendulum is moved from its equilibrium position to a position 5 cm higher and is then released. The bob has a mass of 0.5 kg. (a) What is the total energy of the system? (b) What is the bob's velocity at the equilibrium position? (c) What is the bob's velocity when the bob is at a height of 3 cm above the equilibrium position? Now suppose that the support string makes an angle of 8° from the vertical when the bob is at its maximum displacement, and the bob has been moved along a cir-

cular arc of length 71.8 cm. (d) What is the length of the pendulum? (e) What is its frequency? (f) If this length were cut by 2/3, what would the new frequency of oscillation be?

Answer: (a) E_T = 0.245 J
(b) *v* = 0.99 m/s
(c) *v* = 0.626 m/s
(d) *L* = 5.13 m
(e) *f* = 1.38 Hz
(f) *f* = 0.22 Hz

*22. The bob of a simple, gravity-driven pendulum of length L is pulled to one side through a small angle θ and released. (a) What is the speed of the bob at the equilibrium position? (b) What is the maximum amplitude of the pendulum's motion?

$$\textit{Answer: (a)} \ \ v = \sqrt{2gL(1 - \cos\theta)}$$
$$\textit{(b)} \ \ A_0 = L\sin\theta$$

23. A 1000-kg wrecking ball hangs from the end of a crane by a 30-m cable. (a) If the crane operator quickly moves the crane sideways 4 m, how much time lag is there before the ball passes directly under the crane? (b) How much momentum does the ball have when it is directly under the

crane again? (c) If a standing wall stops the ball in 5 ms, with how much average force does the ball hit the wall?

$$\textit{Answer: (a)} \ \ t = 2.75 \ \text{s}$$
$$\textit{(b)} \ \ p = 2280 \ \text{kg-m/s}$$
$$\textit{(c)} \ \ F = 102{,}376 \ \text{lb}$$

24. Show that for a horizontal mass spring oscillator, the displacement x and the velocity v have the following relation, where v_0 is the highest velocity at the equilibrium position and x_0 is the maximum displacement:

$$x = x_0\sqrt{1 - \left(\frac{v}{v_0}\right)^2}$$

Velocity and Acceleration in SHM Systems

25. A simple, gravity-driven pendulum has a length of 5.13 m and a maximum amplitude of 71.8 cm as measured horizontally from its equilibrium position. The equation for its amplitude is given by $R = R_0 \cos \omega t$. (a) What is the bob's acceleration at its maximum displacement? (b) What is the bob's acceleration at the equilibrium position? (c) What is the bob's velocity at the equilibrium position? (d) What is the maximum height above the equilibrium position that the bob can attain? Use conservation of energy. (e) Using the amplitude, obtain the maximum height. The support cord makes a maximum angle of 8° with the vertical. Remember that the angle is small.

$$\textit{Answer: (a)} \ \ a = -1.37 \ \text{m/s}^2$$
$$\textit{(b)} \ \ a = 0$$
$$\textit{(c)} \ \ v = -0.99 \ \text{m/s}$$
$$\textit{(d)} \ \ h = 5 \ \text{cm}$$
$$\textit{(e)} \ \ h = 5 \ \text{cm}$$

26. Electronic gear is tested to see if it can withstand acceleration. A test rack undergoes SHM with an amplitude of 25 mm. What frequency will give the test gear in the rack a maximum acceleration of $20g$'s?

$$\textit{Answer:} \ 14.08 \ \text{Hz}$$

27. A piston weighing $\frac{1}{4}$ lb is able to withstand a maximum force of 450 lb. If the stroke of the engine is 3 in., what is the maximum rpm that the engine can be allowed to turn without overstressing the piston?

$$\textit{Answer:} \ 4588 \ \text{rpm}$$

*28. The system shown in Figure 9.10 is made to oscillate. What is the maximum amplitude that the oscillations can have before the top block slips on the bottom block?

Figure 9.10 SHM Accelerated System

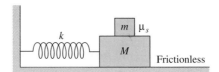

$$\textit{Answer:} \ A_0 = \frac{\mu_s g(m + M)}{k}$$

*29. By substituting the time-dependent expressions for v and x into the equation for the total energy of a mass spring horizontal oscillator, show that the total energy is independent of time. Also show that energy is conserved in the absence of friction.

$$\textit{Answer:} \ E_T = \frac{1}{2}kA^2$$

*30. Show that the maximum tension in the suspension string of a simple pendulum is $mg(3 - 2\cos\theta_m)$, where θ_m is the maximum angular amplitude. Assume a small angle.

Damped Harmonic Oscillator

31. A bob with a mass of 0.35 kg lies on a horizontal surface, and the coefficient of friction between the bob and the surface is 0.15. The bob is attached to a spring with a constant of 15 N/m, and it is stretched 0.5 m and then allowed to

oscillate. How far does the bob travel before the bob damps down and comes to rest??

$$\textit{Answer:} \ x = 3.64 \ \text{m}$$

The gearing in an automotive transmission amplifies the torque produced by a vehicle's engine before the torque is applied to the wheels.

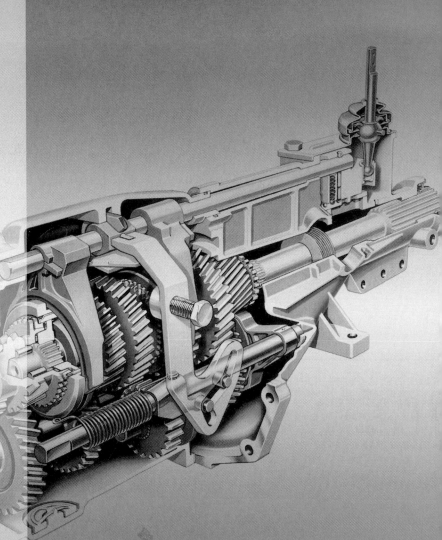

Equilibrium and Simple Machines

At the end of this chapter, the reader should be able to:

1. Use Newton's second law to solve problems that involve objects in *equilibrium.*

2. Explain the concept of a **center of mass.**

3. Calculate the exact center of mass for a two-dimensional body.

4. Calculate the moment of inertia for each of the bodies shown in Table 6.2 rotating about an axis *other* than their center of mass.

5. State the difference between a *simple pendulum* (defined in Chapter 9) and a *physical pendulum.*

6. Give several examples of tools that use the concept of the *inclined plane* and the *lever.*

7. Solve for the mechanical advantage that a given *simple machine* imparts to its user.

10.1 Translational Equilibrium

Translational equilibrium:
The condition that occurs when the sum of the forces acting on a body equals zero.

The word *equilibrium* implies that something is at rest, or ummoving. *Unmoving* means that no forces act on the body at rest, or the sum of the forces acting on the body at rest add up to zero. The first case is trivial; *it is the second case,* **translational equilibrium**, *that interests us in this section.*

$$\mathbf{F}_1 + \mathbf{F}_2 + \mathbf{F}_3 + \mathbf{F}_4 + \cdots + \mathbf{F}_n = 0$$

or

(Equation 10.1)

$$\Sigma \, \mathbf{F}_n = 0$$

Force is a vector, and a vector is described using a coordinate system. We'll use the standard *x-y-z* rectangular coordinate system as the backdrop for our force vectors. To further simplify our discussion, the examples and problems in this chapter will be confined to two dimensions (*x, y*). Once you understand calculations in two dimensions, extending from two dimensions into three dimensions if necessary is a relatively simple matter mathematically. The problem will take longer to solve, but the mathematics is at the same level of difficulty.

To demonstrate translational equilibrium in two dimensions, we'll start with an example.

EXAMPLE 10.1

In Figure 10.1 an object of mass *m* is hanging from the ceiling, supported by two ropes *A* and *B*. What is the tension in each rope?

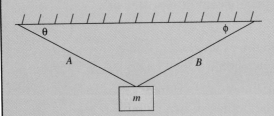

Figure 10.1 Object Suspended by Two Ropes

SOLUTION

To answer the question, we must first construct a vector diagram of all of the forces acting on the object. The real forces act through the tensions in the two ropes, with the directions shown in Figure 10.2. The tensions T_A and T_B can be resolved into horizontal (*x*-direction) components and vertical (*y*-direction) components. For equilibrium, the sum of the forces in the *x* direction must be zero. (The object doesn't move horizontally.) The sum of the forces in the *y* direction must also equal zero. (The force of gravity is counterbalanced by a force acting up.)

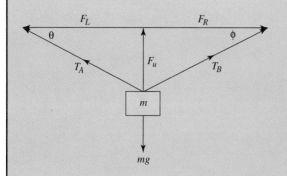

Figure 10.2 Force Components from Tension in Ropes

In an actual physical situation, we could easily weigh the object to find the force of gravity, and we could also easily measure the angles θ and ϕ. What is not always so easy is measuring the tension in the two ropes. In a laboratory situation we could measure the tensions by first attaching the ropes to spring scales and then measuring the tensions directly. But in the real world it's usually easier and more useful to compute the tensions in the ropes. This computation is often necessary from an engineering point of view to ensure that the maximum tension doesn't exceed the breaking strength of the rope.

Mathematically, the weight acting down must be exactly counterbalanced by an upward-acting force.

$$mg = F_u$$

This upward-acting force is the sum of the vertical components of the tension in each rope.

$$F_u = T_A \sin \theta + T_B \sin \phi$$
$$mg = T_A \sin \theta + T_B \sin \phi$$

We're solving for the tension in each rope, and now we have one equation but two unknowns. We need another equation to be able to solve for both tensions. The second equation comes from the fact that the object doesn't have any horizontal motion, which means that the horizontal components of the tensions in the ropes are balanced.

$$F_L = F_R$$
$$T_A \cos \theta = T_B \cos \phi$$

We now have a system of two equations and two unknowns and should be able to solve for T_A and T_B in terms of what we know.

$$mg = T_A \sin \theta + T_B \sin \phi$$
$$T_A \cos \theta = T_B \cos \phi$$

See Appendix A, Section A.2, "Simultaneous Equations."

Since we know the two angles, we know the sines and cosines of the angles as well. We'll solve for T_A first.

$$T_B = T_A \frac{\cos \theta}{\cos \phi}$$

$$mg = T_A \sin \theta + \left(T_A \frac{\cos \theta}{\cos \phi} \right) \sin \phi$$

$$mg = T_A \sin \theta + T_A \cos \theta \tan \phi$$

$$T_A(\sin \theta + \cos \theta \tan \phi) = mg$$

$$T_A = \frac{mg}{\sin \theta + \cos \theta \tan \phi}$$

We can now solve for T_B.

$$T_B = T_A \frac{\cos \theta}{\cos \phi}$$

$$T_B = \left(\frac{mg}{\sin \theta + \cos \theta \tan \phi} \right) \frac{\cos \theta}{\cos \phi}$$

$$T_B = \frac{mg \cos \theta}{\sin \theta \cos \phi + \cos \theta \sin \phi}$$

$$T_B = \frac{mg}{\sin \phi + \tan \theta \cos \phi}$$

We now have an equation for the tension in each rope. To check whether the equation for each is correct, think of what happens if one of the ropes breaks. If rope B breaks, θ becomes 90°, and, as the equation for T_A tells us, the tension in rope A is now mg! The equation for the tension in rope B tells us a similar story if rope A breaks.

EXAMPLE 10.2

An object of mass m is suspended from a rigid boom of negligible mass projecting horizontally from a wall, and the boom is held in place by a cable attached to the outer end of the boom, as in Figure 10.3. (a) What is the tension in the cable? (b) What is the force of compression on the boom?

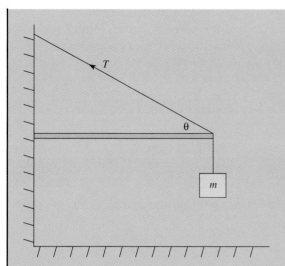

Figure 10.3 Object Suspended from a Boom and a Guy Wire

SOLUTION

(a) We know that the weight of the box acting down must be counterbalanced by an upward force if the system is in equilibrium. Any tensive or compressive forces acting in the strut are horizontal. The strut can have nothing to do in contributing a force to counterbalance the weight of the object acting down. The upward force is therefore the vertical component of the tension in the cable.

$$F_u = mg$$

$$F_u = T \sin \theta$$

$$mg = T \sin \theta$$

$$T = \frac{mg}{\sin \theta}$$

Both the math and the equation are simple, yet they tell a dynamic instead of merely a static story. If the strut were removed, the angle θ would go to 90°, and the tension would be mg. The tension in the cable is holding the strut and the object in place. The tension in the cable would increase as the angle θ decreased. As a matter of fact, the tension in the cable would have to be infinite to hold the strut horizontal if the cable were laid out along the top of the strut! There has to be a vertical component of the tension to counter the weight. Without it ($\theta = 0$), the weight can't be countered. Notice also that the tension in the cable will always be greater than the weight of the object that it must support.

(b) Knowing the tension in the cable, we can calculate the compressive force in the strut. It is the horizontal component of the cable's tension.

$$F_C = T \cos \theta$$

$$F_C = \left(\frac{mg}{\sin \theta} \right) \cos \theta$$

$$F_C = \frac{mg}{\tan \theta}$$

Again, what does the equation tell us about the system? The compressive forces on the strut become enormous as the angle θ becomes small. So as θ decreases, the tension in the cable increases, and the compressive forces in the strut also increase. If θ goes to zero, then either the strut will snap or the cable will break, regardless of the weight hanging from the system!

10.2   Rotational Equilibrium

Rotational equilibrium: The condition that occurs when the sum of the torques acting on a body equals zero.

Rotational equilibrium is achieved when the sum of the torques acting on a body equals zero.

$$\tau_1 + \tau_2 + \tau_3 + \tau_4 + \cdots + \tau_n = 0$$

or

(Equation 10.2)

$$\Sigma \tau_n = 0$$

EXAMPLE 10.3

Figure 10.4 shows a massless beam of length L. At each end of the beam hangs a mass. Since the beam can rotate about the pivot, the mass on the left causes a counterclockwise torque, and the one on the right a clockwise torque. If the system is in equilibrium (that is, if it doesn't rotate), the clockwise torque must equal the counterclockwise torque.

$$\tau_{cw} = \tau_{ccw}$$

$$\tau_{cw} = (L - r)mg \qquad \tau_{ccw} = rMg$$

$$(L - r)mg = rMg$$

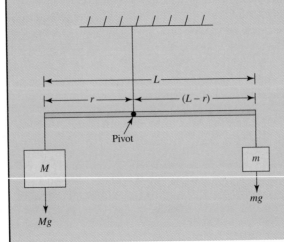

Figure 10.4 Rotational Equilibrium of Two Masses Suspended from a Pivoted Beam

$$m(L - r) = rM \qquad mL - mr = Mr$$

$$mr + Mr = mL \qquad r(m + M) = mL$$

$$r = \left(\frac{m}{m + M}\right)L$$

Common sense says that if the weights at each end are equal, the balance point (or pivot) should be midway between the two ends. If $m = M$ in the above equation, then $r = \frac{1}{2}L$, as logic predicts.

In Example 10.3, the pivot point was specified, and the condition for rotational equilibrium about the pivot was calculated. What if there were no pivot, and we had to calculate the point about which the system balances? The problem is the same but viewed from a different perspective; the approach is the same.

We've looked at simple examples of translational and rotational equilibrium, but now let's take Example 10.2 and look at it through the perspective of *rotational* equilibrium. Remember that this problem was first solved for translational equilibrium. We can consider the point where the strut contacts the wall to be the potential pivot. If the cable were to break, the system would pivot about that point. Once the pivot is established, we can analyze the clockwise and counterclockwise torques about it.

$$\tau_{cw} = \tau_{ccw}$$

$$\tau_{ccw} = LT \sin \theta \qquad \tau_{cw} = mgL$$

$$LT \sin \theta = mgL \qquad mg = T \sin \theta$$

This result is the same as that obtained analyzing the system from the perspective of translational equilibrium, which is as it should be. Physics is consistent!

A teeter-totter is a playground device based on equal and opposite torques.

Now let's analyze Example 10.3 (the beam with the weights hanging from the ends) from the point of view of *translational* equilibrium. The sum of the vertical forces must equal zero, and the sum of the horizontal forces must also be zero. Since no horizontal forces are acting on the system, the forces must necessarily be zero. However, the weight acting down must be balanced by the tension acting up.

$$T = Mg + mg$$
$$T = (m + M)g$$

We could find the tension in the rope using rotational equilibrium, but it would be much harder conceptually. If a system is in equilibrium, we can arbitrarily assign our pivot point anywhere we like. In our derivation of the condition for rotational equilibrium, the choice of the point where the rope was attached to the beam was perfectly natural. The system would naturally rotate about that point. However, the torque that the rope exerts about that point is zero, since the lever arm is zero, and the tension doesn't enter into the equation. But if we're interested primarily in finding the tension, we can't use that point as our pivot, because the tension drops out! Instead, we'll use the ends of the beam as our possible pivot points. If the left end is used, we have

$$\tau_{cw} = \tau_{ccw}$$

$$mgL = rT$$

The tension can't be in terms of length, so we need another equation to remove the lengths. We'll now use the right end of the beam as our pivot.

$$\tau_{cw} = \tau_{ccw}$$

$$(L - r)T = LMg$$

$$r = L - \frac{Mg}{T}L$$

$$\left(L - \frac{mgL}{T}\right)T = LMg$$

$$LT - mgL = LMg \qquad T - mg = Mg$$

$$T = mg + Mg$$
$$T = (m + M)g$$

The above equation is exactly what we obtained in the first instance. This method wasn't the most elegant way to find the tension in the rope, but it again points to the fact that physics is consistent. However, in solving for the tension in the above way, we had to shift our perspective on the problem twice, and such shifting of perspective is one of the hardest things to do in problem solving.

10.3 Center of Mass (CM)

We've seen the center of mass (often referred to as the *center of gravity*) before when we studied rotating bodies, but we haven't looked at it directly, but only peripherally. *The center of mass (CM) is that point at which all of the mass of a body seems to be concentrated.* When we calculated a rotating body's moment of inertia, the body was rotating about its center of mass in all cases. In the case of a ring or a disc, or a solid or hollow sphere, the center of mass is at the center, whether or not there is mass at the center! The center of mass for a uniformly rigid object usually occupies the spot of maximum symmetry. A body rotates most easily about its center of mass. We can also find the moment of inertia for an object that rotates at a point not on its center of mass, but that's easiest to calculate by first finding the center of mass.

Center of mass (CM): The point at which all of a body's mass seems to be concentrated.

We've also studied torques. A torque causes a body to rotate. If we have a free body such as the disc in Figure 10.5A, and if the direction of the force associated with the torque is through the center of mass (CM) ($\sin \theta = 0$), the torque is zero. A force whose direction passes *through* the center of mass will cause translational motion only. In Figure 10.5B, the force acts only tangentially at the rim. Since no component of this force is pointed directly at the CM, the effect will be a pure rotation. A force whose direction doesn't go through the center of mass can cause a body to both translate *and* rotate. Figure 10.5C is a case in point. If a force F acts on the rim of the disc in the direction indicated, there will be both a component of the force pointing directly at the CM and a component tangential to the rim. The component pointing directly at the CM will cause the disc to start a translational acceleration, but the component tangential to the rim will also cause the disc to rotate simultaneously.

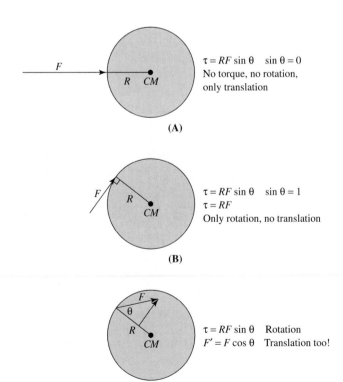

$\tau = RF \sin \theta \quad \sin \theta = 0$
No torque, no rotation, only translation

(A)

$\tau = RF \sin \theta \quad \sin \theta = 1$
$\tau = RF$
Only rotation, no translation

(B)

$\tau = RF \sin \theta \quad$ Rotation
$F' = F \cos \theta \quad$ Translation too!

(C)

Figure 10.5 Torque Acting on a Disc

Equilibrium is described in terms of the center of mass of a rigid object, and the balance point for a rigid body is at its center of mass. Equilibrium can be both stable and unstable. For example, consider a solid rectangle of uniform density. Its center of mass would be at the center of the rectangle, the place where all of the rectangle's mass acts as if it is concentrated. If the rectangle lies on one of its long faces, it has a wide base, as in the upper drawing of Figure 10.6, and its center of gravity is low (a good prescription for stability). As a result, the rectangle will resist a large range of perturbations (disturbances) before toppling over. In the upper left drawing, the block lies flat, and as we start tilting the block about its lower right corner, we'll use that as our pivot. The upper left drawing is in stable equilibrium, so the sum of the counterclockwise torques must equal the sum of the clockwise torques.

$$\tau_{ccw} = \tau_{cw}$$

$$\tau_{ccw} = R(mg)\sin\theta$$

The force associated with the clockwise torque is the normal force and operates at 90° to a lever arm ($R\sin\theta$) from the pivot.

$$\tau_{cw} = (R\sin\theta)mg$$

$$R(mg)\sin\theta = (R\sin\theta)mg$$

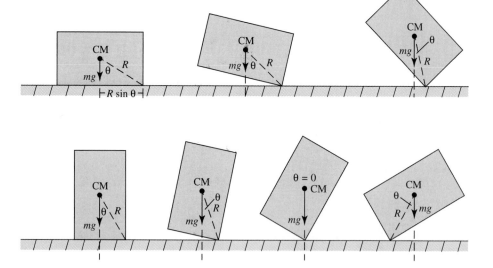

Figure 10.6 Stability of a Rectangular Block Relative to Its CM

In the upper middle drawing of Figure 10.6, the block has been tilted up a bit about the lower right corner, and now the counterclockwise torque isn't balanced by a clockwise torque. Thus, the block will fall back to its original equilibrium position. The normal force operates *only* when a face of the block is resting on the surface. If the block is tipped up on one of its corners, then the surface can no longer push up on the block. In this case there is no balancing clockwise torque. There is only the counterclockwise torque which will cause the block to rotate back to a position where the torque provided by the normal force can again balance the torques, and the block will come to rest in its original equilibrium position. In the upper right drawing, the block is tilted a bit more about the lower right corner, but again when released it will return to its equilibrium position.

The reason is that as long as the line of action of the weight at the center of mass stays within the original baseline, the torque produced by tilting the block about an edge will always tend to return it to its position of stable equilibrium.

Now look at the sequence of the lower drawings of Figure 10.6. The rectangular block is the same, but the baseline is narrower and the center of gravity higher, a bad prescription for stability. The lower configuration is easier to knock over than the upper one. However, as long as the line of action remains within the original baseline, the torque will return the block to its stable equilibrium position. Notice, though, that as the block is tilted over farther and farther, the angle θ gets smaller and smaller. Eventually we reach a position where $\theta = 0$. Since there is now no restoring torque, this position must be in equilibrium. It is an *unstable* equilibrium, though, because any slight perturbation could cause it to topple to either side. A further tilt of the block to the right will cause the line of action to pass through the long face, and the block will finally fall over.

Up to now most of the objects whose motion we've discussed have been treated as point particles. However, most moving objects can't be treated as point particles. For example, the inner wheels of a car negotiating a turn are rotating more slowly than those on the outside of the turn. A spacecraft in orbit might be tumbling and rolling in a complicated way as it circles the earth. After being launched into the air, an exploding fireworks starshell has fragments moving every which way. Certainly these objects can't be treated as point particles if their respective motions are to be treated in detail. However, if the center of mass of the object can be found, then the motion of the object can often be reduced to the motion of the center of mass. The exploding starshell is a particularly good case in point. After the explosion, the collection of fragments will follow the original trajectory of the unexploded shell, regardless of the movement of an individual fragment. A detailed analysis of the motion of a tumbling body can be greatly simplified if you know its center of mass. Its subsequent motion can then be thought of as the motion of its center of mass (translational) and the motion of the individual mass elements that make up the object about the object's center of mass (rotation).

10.4   Finding the Center of Mass

Since the center of mass is so useful, it follows that it is worth knowing how to calculate it. Assume that we have 12 point masses, each of mass m distributed as shown in Figure 10.7. The center of mass for the system of particles is the sum of each mass times its position, divided by the total mass of the system.

$$x_{cm} = \frac{\Sigma\, x\, \Delta m}{\Sigma\, \Delta m}$$

(Equation 10.3)

Observe that the 12 particles are symmetrically laid out about the origin. An educated guess as to where the center of mass for the system of particles should be is the origin. Using the above formula for the CM, we shall prove that the center of mass is indeed the origin. Since there are two coordinates (x and y), we'll have to calculate an x coordinate (x_{cm}) and a y coordinate (y_{cm}) for the CM. (If this were a three-dimensional problem, we would also have to calculate z_{cm}.)

$$x_{cm} = \frac{3m(-L) + 2m(-L/2) + 2m(0) + 2m(L/2) + 3m(L)}{12m}$$

$$x_{cm} = 0$$

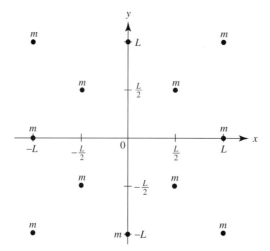

Figure 10.7 Finding the CM for a System of Point Particles

By symmetry we'll get the same result for the y axis: $y_{cm} = 0$. The CM will be at the center of the square at the origin, which is what we assumed using an educated guess.

A solid object is in reality a collection of a huge number of very small particles (atoms and molecules), each with a very, very small mass. Thus, in principle we can use the mathematical method demonstrated above to calculate the CM for any solid object. The only trouble is that the numerator of the equation for the x or y component of the center of mass will be an almost infinite sum. However, if density (mass per unit volume) and thickness are constant over a suitably chosen small slab of some material, we'll obtain an equation that we can handle. Let's calculate the CM for a flat, solid, square object of side length L and thickness t. This time we'll displace the plate so that the origin is at the square plate's lower left corner, as in Figure 10.8. Since the plate is square, you would again guess that the CM is at the square's center. Let's find out by calculating the x component of the center of mass. We'll arbitrarily divide the square up into six vertical slabs, each of mass Δm. Each vertical slab rests on the x axis, and each slab has its own center of mass at the center of the slab. Knowing this, we can set up the equation for the x component of the center of mass.

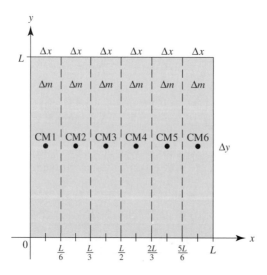

Figure 10.8 Finding the CM for a System of Point Particles

$$x_{cm} = \frac{\Delta m(L/12) + \Delta m(3L/12) + \Delta m(5L/12) + \Delta m(7L/12) + \Delta m(9L/12) + \Delta m(11L/12)}{6 \Delta m}$$

$$x_{cm} = \frac{L}{2}$$

Therefore, the calculated CM again turns out to be where we instinctively thought that it should. If we calculated for the y component, it would also come out at $y_{cm} = L/2$. It wouldn't make any difference how many slabs we cut the square up into.

Let's start with a generalization of the CM equation for the x coordinate and again calculate x_{cm}.

$$x_{cm} = \frac{\Sigma \, x \, \Delta m}{\Sigma \, \Delta m} \qquad\qquad \text{(Equation 10.3)}$$

where the Σ sign means that the individual mass elements along the x axis are added together. Density (ρ) is the mass per unit volume, or in this example the mass per unit area since the thickness of the slab is constant. Using density, we can convert the mass element into the density times an area.

$$\Delta m = \rho \, \Delta x \, \Delta y$$

where ρ is the density (mass per unit area), and $\Delta x \, \Delta y$ is a small area element.

$$x_{cm} = \frac{\Sigma \, x\rho \, \Delta x \, \Delta y}{\Sigma \, \rho \, \Delta x \, \Delta y}$$

where the constants come out from behind the summation sign. Since we're looking for x_{cm}, Δy isn't summed over either.

$$x_{cm} = \frac{\rho \, \Delta y \, \Sigma \, x \, \Delta x}{\rho \, \Delta y \, \Sigma \, \Delta x}$$

Quantities to the left of the summation sign can be canceled.

$$x_{cm} = \frac{\Sigma \, x \, \Delta x}{\Sigma \, \Delta x}$$

The sum in the denominator adds up to x. We've seen the sum in the numerator before, but as $v \, \Delta v$ as in Section 4.3. It becomes

$$\Sigma \, x \, \Delta x = \frac{1}{2} x^2$$

$$x_{cm} = \frac{\frac{1}{2} x^2}{x}$$

$$x_{cm} = \frac{1}{2} x$$

If x is the length of one of the sides (L in our example), then $x_{cm} = \frac{1}{2}L$, which is again as expected.

Figure 10.9 Find the CM Using a Plumb Bob

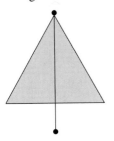

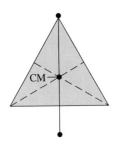

Chapter 10

There is another way in which the center of mass can be found for any two-dimensional object without resorting to any calculations. The object can be suspended from any point around its rim, and a plumb bob hung from the suspension point. (A plumb bob is a weight at the end of a string. When it is suspended from the string, it hangs down and gives a truly vertical line, as shown in Figure 10.9.)

If we wanted to find the center of mass for an equilateral triangle, we would hang a plumb bob from one of its apexes (although any point on its rim would do). With the plumb bob hanging from the suspension point, we would draw a line using the string for a guide. The center of mass would be somewhere along that line. (If the center of mass is where all of the mass of the object acts as if it's concentrated, then it too would hang straight down and be under the string at some point.) We would next hang the object from another apex and repeat the procedure with the plumb bob all over again. The center of mass would be where the lines intersect. For an intersection, a minimum of two lines is needed. In Figure 10.9, three lines are used. This method can be used to find the center of mass of any two-dimensional object, regardless of how irregular its outline.

10.5　The Parallel-Axis Theorem and the Physical Pendulum

In Section 6.3, the moment of inertia of a rotating body was introduced. In Table 6.1, which gives the moments for various rotating bodies, the axis of rotation was through the center of mass. Example 6.2 showed how to calculate the moment of inertia using the center of mass as the point around which the moment of inertia was calculated. However, what if the axis of rotation doesn't go through the center of mass? A simple relationship exists between the moment of inertia of the center of mass and the moment of inertia of the same body when the axis of rotation is not through the center of mass. This relationship, known as the *parallel-axis theorem,* is

(Equation 10.4)

$$I = I_{cm} + Md^2$$

where I_{cm} is the moment of inertia as calculated from the center of mass, M is the total mass of the rotating body, and d is the straight-line distance from the center of mass to the axis of rotation. Knowing the moment of inertia of the center of mass and the location of the center of mass, we can now calculate the moment of inertia of the body for any rotational axis through it. The center of mass can be either calculated or found using a plumb bob as discussed in the previous section.

When an object is set spinning about its center of mass, it simply rotates as does a wheel about its axle. But when an object is given a push to get it rotating about a pivot that is not at the center of mass, it oscillates with simple harmonic motion as a pendulum does. As a matter of fact, such an object is called a *physical pendulum.* How would we calculate the frequency of oscillation for a physical pendulum?

EXAMPLE 10.4

In Figure 10.10, a solid disc is pivoted about a point on its rim. What is the frequency of its oscillations?

Figure 10.10 Physical Pendulum

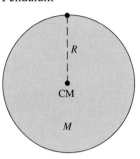

SOLUTION

To find the frequency, we'll use a torque argument. All of the mass in the disc *acts* in some cases as if it is concentrated at the center of mass, but it is actually distributed *around* it. The moment of inertia *must* take this into account. A common mistake in trying to find the frequency of oscillation for a physical pendulum is taking the center of mass to be where all of the object's mass is physically concentrated, when in reality it obviously is not! In the derivation below, the angle θ is between a vertical line through the pivot, and the line through the pivot and the center of mass.

$$\tau = R \times F \qquad \tau = R(Mg)\sin\theta \qquad \tau = RMg\sin\theta$$

$$\tau = I\alpha \qquad \alpha = \frac{\Delta^2\theta}{\Delta t^2} \qquad \tau = I\frac{\Delta^2\theta}{\Delta t^2}$$

$$I\frac{\Delta^2\theta}{\Delta t^2} = -RMg\sin\theta \qquad \sin\theta = \theta \qquad \text{(Small oscillations)}$$

$$I\frac{\Delta^2\theta}{\Delta t^2} = -RMg\theta$$

Rearranging, we obtain

$$\frac{\Delta^2\theta}{\Delta t^2} + \frac{RMg}{I}\theta = 0$$

The above equation is the by now (hopefully) well-known harmonic oscillator equation, and the physical pendulum's frequency can be read directly from the equation.

$$\omega = \sqrt{\frac{RMg}{I}}$$

$$f = \frac{1}{2\pi}\sqrt{\frac{RMg}{I}}$$

(Equation 10.5)

If the distance R were zero, the disc would be suspended from its center of mass. The oscillation frequency would be zero. The disc will rotate only if it's suspended from its center of mass. However, if it is suspended from the rim, we still have to find the moment of inertia for the disc with the pivot at the rim. The moment of inertia from Table 6.1 for a solid disc rotating around the center of mass is $\frac{1}{2}MR^2$. From the parallel-axis theorem,

$$I = \frac{1}{2}MR^2 + MR^2 \qquad I = \frac{3}{2}MR^2$$

We substitute this into the equation for frequency.

$$f = \frac{1}{2\pi}\sqrt{\frac{RMg}{3/2MR^2}} \qquad f = \frac{1}{\pi}\sqrt{\frac{g}{6R}}$$

If we had assumed that all of the mass was concentrated at the center of mass, we would have obtained for the frequency the square root of g over R. The above solution can be verified experimentally.

The equation for a pendulum's frequency,

$$f = \frac{1}{2\pi}\sqrt{\frac{RMg}{I}}$$

also gives us a way to find the moment of inertia of a body experimentally. It has already been shown that a body's center of mass can be found experimentally with a plumb bob. Knowing the object's center of mass, we can suspend it from the desired axis of rotation, set it oscillating, and measure its frequency. It's then a straightforward matter to calculate the body's moment of inertia from the above equation.

10.6 Mechanical Advantage

Assume that we wish to lift a box weighing 50 lb straight up a distance of 10 ft. The work we would have to do would be 500 ft-lb. If there were another person to help us, we could probably both manage to lift the box (with some effort) to the desired height, with each person shouldering a load of 25 lb. However, if we were alone, the task might be beyond our strength. What to do? A block and tackle allows a single person to lift much heavier loads than by simple brute force. The block and tackle operates on the principle of mechanical advantage, which is much like leverage in a torque problem. It lessens the necessary force, but we must still do the same amount of work in the end.

Let's assume that we can easily lift a 25-lb load (see Figure 10.11). From the equilibrium examples earlier in this chapter, we know that the two ropes holding up the box must have a tension of 25 lb apiece. However, if the two ropes are in reality a single rope curled twice around the blocks, then the rope has a tension in it of only 25 lb. This block-and-tackle arrangement is a force amplifier with an amplification of 2. If there were four ropes holding up the box, the necessary tension to lift the box would drop to only $12\frac{1}{2}$ lb. However, we don't get something for nothing! It still takes 500 ft-lb to lift the box 10 ft straight up. For the block and tackle with two ropes, for each 2 ft of rope pulled through your hands, the box would move up only 1 ft. So to lift the box 10 ft, you would have to

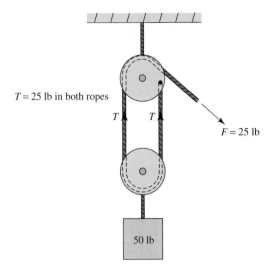

T = 25 lb in both ropes

T   *T*

F = 25 lb

50 lb

Figure 10.11 Block and Tackle

pull 20 ft of rope through the block and tackle. The work is still 500 ft-lb (*W* = 20 ft × 25 lb, or 500 ft-lb).

EXAMPLE 10.5

(a) What is the mechanical advantage of the block-and-tackle arrangement shown in Figure 10.12? (b) If the load weighs 160 N and must be raised 5 m, how much work must be done? (c) How much rope must be pulled through the top pulley to raise the load the required 5 m?

SOLUTION
(a) In Figure 10.11, we had a single block and tackle with a mechanical advantage of 2. In Figure 10.12, we have three interconnected blocks and tackles, each with a mechanical advantage (*MA*) of 2.

$$MA = (2)(2)(2) \qquad MA = 8$$

If, however, Figure 10.12 were only a single block and tackle with six ropes attached to two pulleys, the mechanical advantage would only be 6.

(b) $$W = (160 \text{ N})(5 \text{ m}) \qquad W = 800 \text{ N-m}$$

(c) $$\frac{160 \text{ N}}{8} = 20 \text{ N} \qquad (20 \text{ N})(L) = 800 \text{ N-m}$$

$$L = 40 \text{ m}$$

Figure 10.12 Compound Block and Tackle

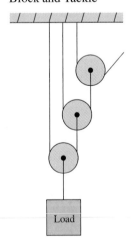

Load

10.7   The Lever

The lever is another simple machine that can be classified as a force amplifier. However, this force amplifier is constructed around the concept of torque.

A lever is illustrated in Figure 10.13. The solid rod or plank of wood rotates about the pivot, or fulcrum. The counterclockwise torque created by a hand pushing down on the left side of the lever (F_{in}) causes an equal torque on the right side.

$$\tau_{ccwL} = \tau_{ccwR}$$

$$\tau_{ccwL} = F_{in}L_{in} \qquad \tau_{ccwR} = F_{out}L_{out}$$

$$F_{in}L_{in} = F_{out}L_{out}$$

We're interested in the ratio of the output force to the input force, so we rearrange the above equation to get

$$\frac{F_{out}}{F_{in}} = \frac{L_{in}}{L_{out}}$$

Since we want the ratio of the output force over the input force to be as large as possible, the input lever arm should be as long as possible, and the output lever arm should be as short as practicable. Common devices built around the concept of the lever are a crowbar, a wheelbarrow, a pair of pliers, and a pair of scissors. A definition of *mechanical advantage* is the ratio of the output force to the input force, and a lever gives us a mechanical advantage. It's also worth remembering that you don't get something for nothing. The work done by the lever is equal to the work you do in pushing on the lever.

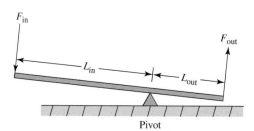

Figure 10.13 The Lever

Assume that we're using a lever to lift a weight (mg) to a height h_{out} from the ground.

$$W_{in} = W_{out}$$

$$W_{in} = F_{in}h_{in} \qquad W_{out} = (mg)h_{out}$$

$$F_{in}h_{in} = (mg)h_{out}$$

If the lever turns through a small angle (in radians) in lifting the weight, then

$$h_{in} = L_{in}\theta \quad \text{and} \quad h_{out} = L_{out}\theta$$

(Both the input side and the output side will move through the same angle.) This gives us

$$F_{in}L_{in}\theta = (mg)L_{out}\theta$$

The angle cancels on both sides, and we have

$$F_{in}L_{in} = (mg)L_{out}$$

If we assume that the weight we're lifting with the lever is equal to the output force, we can rearrange the above to get

$$\frac{(mg)}{F_{in}} = \frac{L_{in}}{L_{out}}$$

This expression is almost exactly the same as the one we obtained earlier from the perspective of torque. Physics is consistent!

10.8 Gearing and Torque

Figure 10.14 shows a torque amplifier. It usually consists of two pulleys connected by a belt or chain, but it could also be two toothed gears of different diameters meshing together. In the case of the block and tackle and the lever, we discovered that although we had mechanical advantage, we didn't get something for nothing. The same is true here. The work that we put into the simple machine is equal to the work at the output.

$$W_{in} = W_{out}$$

$$W_{in} = \tau_{in}\theta_{in} \qquad W_{out} = \tau_{out}\theta_{out}$$

$$\tau_{in}\theta_{in} = \tau_{out}\theta_{out}$$

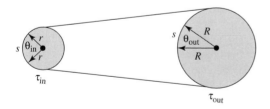

Figure 10.14 Torque Amplifier (Gearing)

If the belt of chain doesn't slip on the pulley, then the distance the belt travels (s) is the same for both pulleys. From the definition of an angle in radians, we can find the angle through which the input pulley travels and the angle through which the output pulley travels in terms of s and the input and output radii.

$$s = r\theta_{in} \qquad \theta_{in} = \frac{s}{r}$$

$$s = R\theta_{out} \qquad \theta_{out} = \frac{s}{R}$$

$$\tau_{in}\left(\frac{s}{r}\right) = \tau_{out}\left(\frac{s}{R}\right)$$

$$\frac{\tau_{in}}{r} = \frac{\tau_{out}}{R}$$

$$\frac{\tau_{out}}{\tau_{in}} = \frac{R}{r}$$

Since R is larger than r, the output torque is greater than the input torque.

In Figure 10.14, the smaller pulley will have a larger angular speed than the larger pulley. If the input pulley were larger than the output pulley, then the output pulley would have a faster rpm than the output pulley, but a reduced torque. The transmission of an automobile usually has three or four gear ratios that it can change as the car speeds up. As the car accelerates from rest, a large torque is needed at the drive wheels to get the car moving. The engine usually has a cruising rpm of around 3000 to 4000 rpm and will shift gears at the upper end of its cruising rpm. In first gear, at 4000 rpm the car might have a speed of only 20 mph despite its high acceleration to that speed. If the car is to go any faster, then either the engine must increase its rpm or the driven gear must have a smaller radius so that its rpm will increase, thus increasing the speed of the auto. So the transmission shifts gears to the smaller radius output gear, or you manually shift if the car is equipped with a stick shift. If the car were to accelerate from rest with an output gear that had too small a radius (for example, pulling away from a stoplight in third gear), then there would be insufficient torque at the drive wheels for the car to pull away quickly. Indeed, such abuse causes the engine to labor too hard and can cause as much damage in the long run as over-revving the engine for long periods. Trying to accelerate away from a stoplight in third gear would probably stall the engine. (If you did manage to get moving, the clutch would pay a heavy price!)

10.9 The Inclined Plane and Screw

The inclined plane shown in Figure 10.15 is another simple machine that allows us to raise an object without having to use a force equal to that object's weight. If we wanted to raise the mass m to a height h, then we would have to do an amount of work equal to mgh. If we raised the object straight up, then we would have to use a force at least equal to the object's weight, mg. If the object in Figure 10.15 were a 5-ton limestone block to be used in the construction of Khufu's pyramid on the Ghisa Plateau in Egypt, then raising the block straight up vertically would have been beyond the then-current technology. The ramp, or inclined plane, would be about the only viable alternative.

In Figure 10.15, the force triangle has been drawn for the 5-ton block. In the case of an inclined plane, friction is usually present. Let's assume that we want to push the block

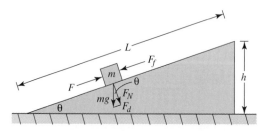

Figure 10.15 The Inclined Plane

up the plane with a constant velocity, and let's compare that force with the weight of the block.

$$F = F_d + F_f$$

where F_d is the weight down the ramp caused by the block's weight, F_f is the force of friction opposing the block's motion up the ramp, and F is the force pushing the block up the ramp at constant speed.

$$F_d = mg \sin \theta \qquad F_f = \mu_k mg \cos \theta$$

$$F = mg \sin \theta + \mu_k mg \cos \theta$$
$$F = (\sin \theta + \mu_k \cos \theta)mg$$

The term in parentheses $(\sin \theta + \mu_k \cos \theta)$ is the fraction of the block's weight with which the team of men must push or pull the block (F) to move it up the ramp. Note that if friction were entirely eliminated, the force would be only $mg \sin \theta$, and $\sin \theta$ is small for a small angle. However, since friction is present, $\cos \theta$ would be almost 1 for a small angle, and you can see how crucial a part the coefficient of kinetic friction plays in determining F. For instance, if the angle θ of the inclined plane were only $10°$, and if μ_k were only 0.1, the fraction of the block's weight needed to push the block up the plane would be 27.2% or 2720 lb. To lower that force, we could attempt a further reduction for μ_k, and the plane's angle could be reduced further. However, a reduction in the plane's angle means that the block must travel farther to rise to the same height h. Juggling such factors in an attempt to optimize a solution with the resources on hand is what engineers do!

Without friction, the amount of work done on the block must be equal to mgh.

$$W = F_d L \qquad\qquad W = (mg \sin \theta)L$$
$$W = mg(L \sin \theta) \qquad L \sin \theta = h$$

$$W = mgh$$

However, with friction thrown into the bargain, the work that the team of men must do will exceed mgh.

$$W = F_d L + (\mu_k mg \cos \theta)L$$
$$W = mgh + (\mu_k mg \cos \theta)L$$

where the term $(\mu_k mg \cos \theta)L$ is the extra price in work that must be paid because of friction.

The screw is an inclined plane wrapped around a cylinder. Suppose that we take a piece of thin wire and place it on top of the inclined plane of Figure 10.15. We then place a cylinder with a small diameter standing vertically at the bottom of the inclined plane. Next we rotate the cylinder against the wall of the plane to the right, simultaneously wrapping the wire around the cylinder as we go. When we reach the end of the plane, we will have turned the cylinder with the wire wrapped around it into a screw. The screw has helical threads engraved around it, and it turns rotational motion into linear motion. Rotating a screw with a screwdriver causes it to move linearly into or out of a bolt.

A screw is also a force amplifier. We can take a nail and drive it vertically into a piece of wood, but we need a hammer that produces a large impulse to do so. It takes a lot of force to drive the nail directly into the wood. However, if we screw a screw into the wood, it takes a lot less force to bury it in the wood. We must, however, turn the screw through quite a few revolutions to drive it into the wood. In the end, whether we use a hammer or a screwdriver, we do the same amount of work.

The screw is used not only to fasten things together but also to propel ships and aircraft through fluid media such as water and air. Rotational motion is turned into translational motion. The pitch of a screw is inversely related to the angle with the horizontal of an inclined plane. The greater the pitch, the smaller the angle, and the greater the number of turns needed to advance the screw a given distance. Less torque is needed, but it must be applied through more rotations. For an aircraft propeller, changing the blade's pitch is like changing the gears in a car. The finer the pitch, the greater the thrust for a given rpm, but at the expense of air speed. Aircraft propellers can automatically change pitch so that maximum thrust is available both with a fine pitch for take-offs and climbing and with a coarse pitch for high-speed cruising.

Summary

1. **Equilibrium** is the state where all forces (and torques) on a body sum to zero.

Problem Solving Tip 10.1: Remember, Newton's second law states that the sum of all forces on an object equals mass times acceleration ($\mathbf{F} = m\mathbf{a}$). Thus, the concept of equilibrium implies that the acceleration of a body in equilibrium is equal to zero. But also remember that a body can be moving and possess no acceleration (for example, a hockey puck gliding across a frozen pond with a constant velocity). Thus, as long as a body has a constant velocity (linear or angular), we can use the methods shown in this chapter!

2. A freely rotating body will tend to revolve around its **center of mass (CM)**. Also, an object's center of mass can be used as the point that is pulled by gravity. Thus, the center of mass follows the same parabolic trajectory predicted by the equations of uniformly accelerated motion given in Chapter 1.

Problem Solving Tip 10.2: The concept of a center of mass is extraordinarily useful when solving many torque problems. By using the center of mass as the "pivot point" of the problem, we can usually solve the problem in the least number of steps. See Problem Solving Tip 6.4.

3. The parallel-axis theorem [Equation (10.4)] calculates the moment of inertia for an object not rotating about its center of mass.

4. Whereas a simple pendulum (Chapter 9) consists of a point mass swinging at the end of a long, massless rod, the much more realistic physical pendulum can be of any shape and size.

Problem Solving Tip 10.3: When using Equation (10.5), remember that R is the distance between the object's center of mass and the pivot point and I is the moment of inertia about the pivot point [see Equation (10.4)].

Key Concepts

CENTER OF MASS (CM): The point at which all of a body's mass seems to be concentrated.

ROTATIONAL EQUILIBRIUM: The condition that occurs when the sum of the torques acting on a body equals zero.

TRANSLATIONAL EQUILIBRIUM: The condition that occurs when the sum of the forces acting on a body equals zero.

Important Equations

(10.1) $\Sigma \mathbf{F}_n = 0$ (Condition for translational equilibrium)

(10.2) $\Sigma \tau_n = 0$ (Condition for rotational equilibrium)

(10.3) $x_{cm} = \dfrac{\Sigma x \, \Delta m}{\Sigma \Delta m}$ (Equation for finding the x coordinate for the center of mass. The same equation is used for the y and z axes. Simply substitute y or z into the equation.)

(10.4) $I = I_{cm} + Md^2$ (Equation for the moment of inertia of an object rotated at a point not on the center of mass. I_{cm} is the moment of inertia at the object's center of mass, M is the object's mass, and d is the distance the pivot is from the center of mass.)

(10.5) $f = \dfrac{1}{2\pi}\sqrt{\dfrac{RMg}{I}}$ (Oscillating frequency of a physical pendulum)

Conceptual Problems

1. An object is in equilibrium in two dimensions. The number of equations possible to solve for equilibrium conditions is:
   a. 1
   b. 2
   c. 3
   d. 4

2. An object in equilibrium cannot have:
   a. any torques acting on it
   b. velocity
   c. an acceleration
   d. any forces acting on it

3. A picture hangs from two wires. The tensions in the wires must:
   a. each equal the picture's weight
   b. each be one-half the picture's weight
   c. each have a tension greater than the picture's weight
   d. have a vector sum equal to the picture's weight

4. A weight is hung from the midpoint of a rope whose endpoints are at the same level. For the rope to be perfectly horizontal, a force must pull on each end of the rope which is:
   a. equal to the weight
   b. one-half of the weight
   c. infinite
   d. zero if the rope is tied to a wall

5. An object is in rotational equilibrium. If the sum of the torques at a given point is zero, then:
   a. the sum is zero about no other point
   b. the sum is zero about all other points
   c. the object is also in translational equilibrium
   d. the sum varies according to the point chosen

6. In an equilibrium problem, the point about which the torques are computed:
   a. must be at the object's center of mass
   b. can be located anywhere

   c. must be at the end of or on the rim of the object
   d. cannot be the pivot point

7. The center of mass of an object can be at a point where there is no mass. Explain how this can be.

8. A hammer is thrown through the air. As it flies through the air, it both rotates and follows a ballistic trajectory. If the path that the hammer followed had to be plotted, would a point at the end of the hammer's handle be a good choice? Why or why not?

9. You don't get something for nothing. What price do you have to pay to get a mechanical advantage?

10. Explain how leverage is used twice in a claw hammer used for driving and extracting nails.

11. The simple crank translates rotational motion into linear motion. Explain how you could increase the input torque to the crank and how the output force of a rope pulling a liquid-filled bucket could be increased.

12. What happens to the mechanical advantage that a motor has on the wheels as you shift up while moving onto a highway?

13. Stability and the height of the center of mass are related. How and why?

14. A wheelbarrow uses torque and leverage in its design. Explain the advantages and disadvantages of having the mass carried in the bin directly over the wheel.

Exercises

Translational Equilibrium

1. A stiff, horizontal beam of negligible mass projects perpendicularly to the wall of a building. A guy wire attached to the end of the beam makes an angle of $35°$ to the horizontal with the beam. A weight of 250 lb hangs from the end of the beam. (a) What is the tension in the guy wire? (b) With what force does the beam press against the wall?

   Answer: (a) 435.9 lb
           (b) 357 lb

2. An object weighing 75 kg is suspended from two ropes of equal length that make an angle of $60°$. What is the tension in each rope?

   Answer: $T = 424.3$ N

3. What is the tension in each rope in Figure 10.16?

Figure 10.16 (Exercise 3)

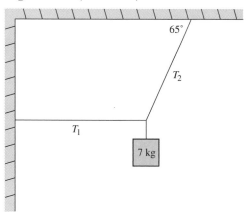

Answer: T_1 = 75.7 N; T_2 = 32 N

4. What is the mass M in Figure 10.17?

Figure 10.17 (Exercise 4)

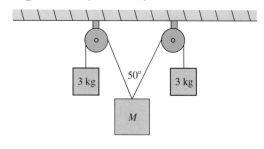

Answer: M = 5.44 kg

5. A 75-kg load is suspended by two ropes. One makes an angle of 35° with the vertical, and the other makes an angle of 15° with the vertical. What is the tension in each rope?

Answer: T_{15} = 550.6 N; T_{35} = 248.3 N

6. A load is suspended from two ropes. One of the ropes has a tension of 55 N and makes an angle of 60° with the ceiling. The other rope makes an angle of 40° with the ceiling.

(a) What is the tension in this rope? (b) What is the mass of the load?

Answer: (a) T = 35.9 N
(b) m = 7.2 kg

7. If the beam in Figure 10.18 can support a compressive force of only 6700 lb before buckling, what is the maximum load that the crane can lift?

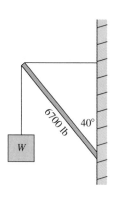

Figure 10.18 (Exercise 7)

Answer: W = 5132.5 lb

8. A 220-kg load is supported as shown in Figure 10.19. (a) What force does each leg exert on the ground? (b) What is the minimum coefficient of static friction to keep the legs from sliding out from under the load? (c) What is the compressive load in each leg?

Figure 10.19 Lever Arm (Exercise 8)

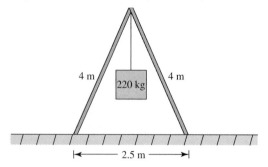

Answer: (a) 1078 N
(b) 0.329
(c) 1135 N

Rotational Equilibrium

9. Two children weighing 20 kg and 30 kg sit on opposite ends of a teeter-totter 3 m long. The teeter-totter is pivoted at its midpoint. If the 20-kg child sits at one extreme end, where must the 30-kg child sit to balance the teeter-totter?

Answer: L = 1 m

10. A man pulls down on the lower end of the lever in Figure 10.20 with a force of 150 N. What is the tension in the cord?

Figure 10.20 (Exercise 10)

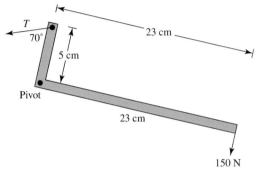

Answer: T = 734.3 N

11. A plank 9 ft long and weighing 11 lb has one end resting on the floor and its other end resting on the edge of a table 3 ft high. (a) What force does the plank exert on the table? (b) If a short length of rope were tied to the end of the plank at 90° to support it, and if the table were slid out from under the board, what would the tension in the rope be?

Answer: (a) F = 5.5 lb
(b) T = 5.2 lb

*12. An aluminum ladder 5 m long and weighing 20 kg rests against a frictionless wall at its upper end at an angle of 40° from the vertical. A 65-kg person stands on the ladder 1.5 m from the top. If the ladder is not to slip, what horizontal force must be applied to the bottom end of the ladder?

Answer: F_h = 185.8 N

*13. A rod with a mass of 10 kg has its lower end hinged to a wall as in Figure 10.21. A horizontal rope 0.7 m long joins the middle of the rod to the wall, and a 15-kg mass is suspended from the upper end of the rod. What is the tension in the rope?

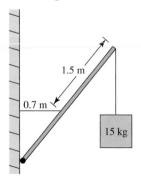

Figure 10.21 (Exercise 13)

Answer: T = 206.7 N

*14. Two sawhorses support 3 ft from each end a wooden plank 18 ft long with a weight of 65 lb. How near to either end can a 150-lb painter stand without causing the plank to tip?

Answer: 4.8 in.

*15. A box 6 ft wide falls over when tilted at an angle of 25° from the horizontal. How high above the ground is the center of gravity when the box is sitting level on the floor?

Answer: 6.4 ft

*16. A homogeneous rod with a mass of 12 kg is supported by two vertical lines as shown in Figure 10.22. A steel block with a mass of 35 kg is placed as in the figure. What is the tension in each line?

Figure 10.22 (Exercise 16)

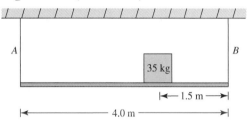

Answer: T_A = 187.4 N; T_B = 273.2 N

*17. The front and rear axles of a 12,000-lb truck are 15 ft apart. The center of gravity of the truck is located 6 ft behind the front axle of the truck. How much weight does each axle support?

Answer: W_F = 7200 lb; W_R = 4800 lb

18. When a car is rolled onto an appropriate scale, its front wheels weigh 800 lb, and the rear wheels register 1200 lb. If the axles of the car are 7 ft apart, where is the center of gravity of the car relative to the front axle?

Answer: 4.2 ft

19. A 2700-kg load is supported by a boom hinged at the base of a vertical mast as shown in Figure 10.23. What is the tension in the cable from the end of the boom to the top of the mast? The weight of the boom is negligible.

Figure 10.23 (Exercise 19)

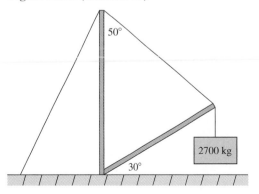

Answer: T = 24,500 N

20. A pair of forces equal in magnitude and opposite in direction act on a rigid rod. The forces both act perpendicular to the rod and act on the rod at locations a distance D apart. The rod is pivoted at the left end. The torque produced depends not on the location of the pivot, but only on the magnitude of the forces (F) and the distance between them (D). Find an expression for the torque on the rod only in terms of F and D.

Answer: $\tau_{\text{net}} = FD$

21. A uniform meter stick balances at the 50-cm mark. When two nickels, weighing 5 g apiece, are stacked one on top of the other at the 12-cm mark, the meter stick balances at the 45.5-cm mark. What is the mass of the meter stick?

Answer: M = 74.5 g

Center of Mass/Finding the Center of Mass

22. What is the location of the center of mass of the flat object shown in Figure 10.24?

Figure 10.24 (Exercise 22)

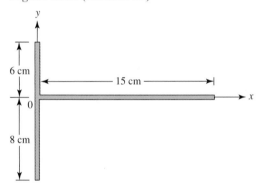

Answer: $CM_x = +3.9$ cm; $CM_y = -0.48$ cm

*23. The flat plate in Figure 10.25 has a rectangular hole punched in it as shown. Where is the center of mass relative to the origin? (*Hint:* Treat the rectangular hole as "negative mass.")

Answer: $(CM_x, CM_y) = (0.33$ m, 0.15 m)

*24. Two skaters of mass 80 kg and 55 kg stand 9 m apart, each holding the end of a rope stretched between them. If each pulls himself along the rope until they both meet, how far does each skater travel? Assume that friction is negligible.

Answer: 3.67 m and 5.33 m

*25. The dumbbell shown in Figure 10.26 is balanced on a frictionless surface. When the dumbbell falls over, the heavier upper ball falls to the right. When the dumbbell comes to rest, where is the smaller ball relative to the initially vertical rod connecting the two balls? The rod is thin and of negligible mass.

Figure 10.26 Dumbbell (Exercise 25)

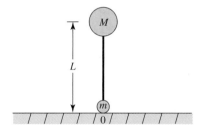

Answer: $\dfrac{M}{m + M}L$, to the right of the origin

Figure 10.25 (Exercise 23)

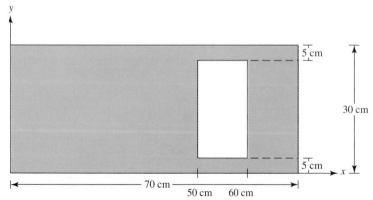

Mechanical Advantage

26. A block-and-tackle system has two pulleys, one of which is suspended from the ceiling. An object with a mass of 25 kg is suspended from the lower pulley. To keep the object suspended, a tension of 81.7 N is needed on the rope on which you haul. (a) How many lines are there between the two pulleys? (b) If you winch the object up 2 m, how much rope passes through your hands?

Answer: (a) Three lines
(b) 6 m

27. A block-and-tackle system has five lines between two pulleys. A tension of 35 N is required on the hauling rope to keep an object of 15 kg suspended. What is the mass of the lower pulley?

Answer: 2.86 kg

Gearing and Torque

28. A winch has a drive gear with 9 teeth and a driven gear with 54 teeth. (a) If the drive gear runs at 9000 rpm, how fast does the driven gear run? (b) If the motor driving the drive gear is rated at 8.2 kW at 9000 rpm, what is the output torque of the winch?

Answer: (a) 1500 rpm
(b) 4.64×10^7 N-m

29. What force does the rope in Figure 10.27 exert on whatever it's lifting if a force of 127 N is applied to the crank handle?

Figure 10.27 (Exercise 29)

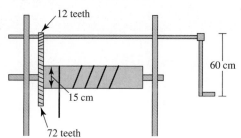

Answer: F = 6096 N

Inclined Plane and Screw

30. A man pushing on a load can exert, at most, a force of 250 N. A loaded cart with a mass of 60 kg is to be pushed up a ramp to a height of 10 m. (a) If the wheels on the cart lower the coefficient of friction to 0.25, what is the steepest angle that the ramp can have? (b) What is the length of the ramp? (c) How much of the total work done is lost to friction?

Answer: (a) 10.3°
(b) 55.9 m
(c) 57.9%

A rolling mill produces sheet steel by forcing the steel between rollers, thereby causing it to undergo plastic deformation.

Some Properties of Solids

At the end of this chapter, the reader should be able to:

1. Use the concept of **density** to solve for the mass of an object if given its volume, and solve for the volume of an object if given its mass.

2. State the difference between **tension, compression,** and **shear.**

3. Explain the difference between **stress** and **strain.**

4. Solve for the *breaking strength* of a material, if given its dimensions and composition.

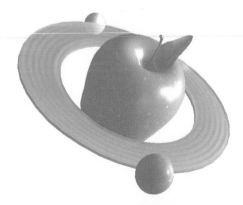

To be able to specify which solid materials should be used for construction of a structure, an engineer must know whether the material in question can stand up to the stresses and strains that will be imposed on the structure. Therefore, researchers have compiled a body of knowledge of the properties of construction materials by subjecting the materials to the forces of stress, strain, and shear. Researchers have learned under what conditions such materials can safely be used and, sometimes more importantly, under which conditions they would fail. This chapter will explore some of the properties of matter and some of the different types of forces that can be imposed upon different materials.

11.1   Density

Up to now, we've been treating solid matter as something that has mass. Initially we used the concept of a point mass, in which the mass of the object was concentrated in a point with no extent. When we talked about the center of mass in Chapter 10, we had to take into consideration that a massive object has a certain extent in space. We found that with the center of mass concept, we could indeed treat a massive object with spatial extent as a point particle for translational motion, since all of the mass of an object acts as if it is concentrated at the center of mass. The moment of inertia of an object (the rotational analogue of mass) takes into account the spatial distribution of the mass of that object.

Density: The ratio of a mass to the space occupied by that mass. The space is usually a volume, but it can be an area or a linear distance.

*Since all massive objects have an extent in space (even something as tiny as an atomic nucleus), massive objects have a property known as **density** (ρ).* In other words, the mass occupies a volume. For example, a sample of pure copper has a characteristic ratio of mass to volume. A pure sample of aluminum also has its characteristic ratio of mass to volume, as does pure lead. In short, every element and identifiable substance has a characteristic density. The mathematical definition of density is

(Equation 11.1)

$$\rho = \frac{m}{V}$$

where ρ is the symbol for density, m stands for mass, and V here stands for volume, not velocity.

There is more than one type of density. We already discussed density in a limited sense when we examined the speed of a pulse traveling on a stretched string. The string was under tension and was characterized by a linear density. *Linear density* was defined as the mass per unit length. A string is usually thought of as a length without width or breadth. But the string exists in all three dimensions, its length overwhelming the other two dimensions. If we look at the other two dimensions, we find that most pieces of string are circular in cross section. A string's cross-sectional area is πr^2, and the cross-sectional area times the length gives the volume. The string *could,* therefore, be given as a volumetric density, but since the string is thought of in terms of its length in the overwhelming number of cases, it's more common to think of it in terms of linear density instead of volumetric density.

Another type of density is the mass per unit area. If we deal with a thin plate of uniform thickness of a given material, we think of the plate as having area and not volume. Again, however, the plate occupies volume: Its area times its thickness equals its volume. But if the plate is very thin, then its thickness is often ignored. Similarly, we are usually concerned with the area of a piece of paper and ignore its thickness. Liquids and gases also have a density associated with them. But the densities of liquids and gases are most often thought of in terms of volumetric density, or the mass per unit volume, because liquids and gases will occupy the volume of whatever container holds them and can't be sliced up into sheets or lengths easily.

Table 11.1 gives the volumetric densities of some common substances. If you examine the table, you'll see that the density of ice is less than the density of liquid water. Thus, ice cubes float in water. Different woods have different densities which can vary quite a bit, but since the density of all woods is less than that of water, wood floats. One of the heaviest substances in the table is a liquid (mercury)! The density of a gas is orders of magnitude less than that of a solid or a liquid because the individual atoms and molecules that make up a gas aren't bound to each other and are in constant motion, therefore occupying much more space for a given mass.

Table 11.1 Densities of Various Substances at 20°C and 1 atm

Substance	Density (kg/m^3)	Phase
Air	1.3	Gas
Aluminum	2.7×10^3	Solid
Brass	8.4×10^3	Solid
Concrete	2.3×10^3	Solid
Copper	8.9×10^3	Solid
Gasoline	6.8×10^2	Liquid
Gold	1.9×10^4	Solid
Helium	0.18	Gas
Hydrogen	0.09	Gas
Ice	9.2×10^2	Solid
Iron and steel	7.8×10^3	Solid
Lead	1.1×10^4	Solid
Mercury	1.4×10^4	Liquid
Silver	1.05×10^4	Solid
Water	1.00×10^3	Liquid
Wood (balsa)	1.3×10^2	Solid
Wood (oak)	7.2×10^2	Solid
Wood (pine)	3.7×10^2	Solid

EXAMPLE 11.1

A 250-g figurine is supposedly pure gold. It might, however, be gold-plated lead. How can we tell without a chemical assay?

SOLUTION

We can determine the density of the figure and compare it to that of gold. If it's less than that of gold, the figurine isn't pure gold, but alloyed with something else. If the density is close to that of lead, then the figurine is probably gold-plated lead.

To find the volume of the piece, we can submerge it in a cylindrical glass of water. By measuring how much the water rises when the figurine is submerged, we know how much water the figurine displaced, and therefore the volume of the figurine. When we do this, we find that the figurine displaces 2.27×10^{-5} m^3.

$$\rho = \frac{m}{V} \qquad \rho = \frac{0.25 \text{ kg}}{2.27 \times 10^{-5} \text{ m}^3}$$

$$\rho = 1.1 \times 10^4 \text{ kg/m}^3$$

The density is the same as that of lead, and the figurine is actually lead with a very thin plating of gold!

11.2   The Structure of Solids

All matter is made up of atoms. For example, a piece of copper wire is made up of billions of individual copper atoms chemically bonded together. A lump of steel consists of countless iron atoms linked together with a little carbon thrown in to make it hard steel instead of relatively soft iron. A salt crystal is actually made up of two elements, sodium and chlorine, in equal measure. The sodium and chlorine atoms are bonded together in a regular array. Pure water is made up of molecules of H_2O, two hydrogen atoms chemically bonded to an oxygen atom. A molecule consists of a combination of different atoms. Since water is a liquid, these H_2O molecules are loosely bound to one another and are free to move about. This property allows liquids (and gases as well) to assume the shape of any container into which they are placed. The loose binding of liquids and gases allows us humans to pass through them with relative ease.

A solid, on the other hand, has a rigidity that comes from tight binding of each molecule or atom to other atoms or molecules close by. These bonds, which are broken only with difficulty, are electrical in nature. Think of the individual bonds as small springs between atoms or molecules, as in Figure 11.1. This figure shows a cube with an atom or a molecule at each apex. The more rigid the cube, the stiffer the springs between the atoms or molecules.

Many solids are crystalline, with their atoms and molecules arranged in regular patterns. If the cube in Figure 11.1 were extended in three dimensions, with each apex of the original cube becoming the apex of another cube, it would be an example of a crystalline solid. If you look at a salt crystal under a magnifying glass, you'll see that it has

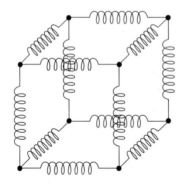

Figure 11.1 Springlike Bonding between Atoms or Molecules in a Solid

a cubic structure. If you break that crystal with the tine of a fork and look at the residue with a microscope, you'll again see a cube. The silicon used in most integrated circuits is a crystal. Some solids consist not of a single pure crystal, but rather of many tiny crystals with many different orientations. If you look at a piece of iron under a microscope, you'll see many individual crystals. Each will be different in area from its neighbors, and each will have a different orientation. Some solids, however, have no regular structure. These solids, such as wax and glass, are called *amorphous*.

11.3   Stress

Stresses are forces that tend to deform a body. If we apply a force to a solid, the little springs that connect the atoms and molecules (the electrical forces between the electrons in the outer shells of an atom or a molecule) are stretched or compressed. If the springs aren't stressed beyond their elastic limits, they spring back to their equilibrium position, and the solid snaps back to its original configuration. If the springs in a solid are stiff, it takes a lot of force to deform the solid in such a way that it remains deformed. Remember, the spring analogy is just that—an analogy. It succeeds up to a point in explaining intermolecular and interatomic bonds. Such bonds will be discussed in more detail in Chapter 27 on the elements. However, the spring analogy is borne out when we look at a tensile stress applied to a solid. *A tensile stress acts to cause a solid to lengthen.*

Tensile stress: An equal and opposite force pair applied to a body, causing it to lengthen.

A stone arch supports heavy loads by exploiting the fact that stone is much stronger in compression than in tension.

Figure 11.2 shows bodies acting under tension. In Figure 11.2A, two forces act on the solid from each side. The forces are equal and tend to try to pull the solid apart. The amount by which the solid is deformed is proportional to the applied tension, provided that it is not above the elastic limit of the solid, in which case it is permanently deformed. If not beyond the elastic limit, the tension acts on the body just as hanging a weight from a spring does. It follows Hooke's law.

$$F = kx$$

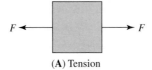

(A) Tension

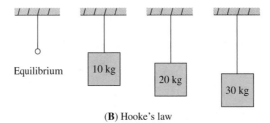

(B) Hooke's law

Figure 11.2 (A) Body Undergoing Tensile Stress; (B) Wire Undergoing Tensile Stress

The elongation of a spring is easy to see, but if the body is a thin steel wire, it might be hard to observe. If 10 kg causes the wire to stretch by 1 mm, then doubling the mass hanging from the wire will double the elongation to 2 mm, and so on. When the weight is removed, the wire will snap back to its equilibrium position, as in Figure 11.2B. However, if too much weight is hung from the wire (stretching it beyond its elastic limit), then the wire will be stretched so much that it will have a new equilibrium position that is longer than the original one. The result is *called **plastic deformation**, because in a metal, the material deforms instead of breaking. However, if more weight is again hung from the wire, it will break when it reaches its **ultimate strength**.*

A graph of this behavior is plotted in Figure 11.3. If the wire in Figure 11.2 were copper, as more and more weight were hung from it, the length versus tension would follow a graph much like that in Figure 11.3. Initially, the wire would obey Hooke's law and would snap back to equilibrium if the weight were removed. Eventually, though, if enough weight were added until the wire reached its elastic limit, it would start to deform permanently. Therefore, with any further stretching, the wire would keep its new length permanently. Notice that the graph bends at the point of the elastic limit into a dog-leg. After that point, it takes much less added weight to plastically deform the wire. As more weight were added, the wire would eventually reach its ultimate strength, when it would break.

Plastic deformation: The response of some materials such that when they are placed under enough tension, they will lengthen and not return to their original equilibrium position.

Ultimate strength: The tensile stress that causes a material to break.

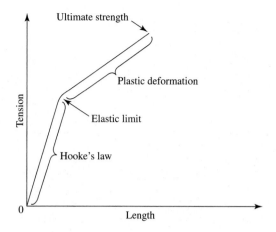

Figure 11.3 Plot of
Increasing Tension on a Wire

Plastic deformation is used in industry to shape a metal into a desired configuration. *A material, for example, a metal such as copper, that has a large plastic range in tension is called a **ductile material.*** Copper wire is a ductile material that is often manufactured by drawing it through a die, as in Figure 11.4.

Compression is a stress that is the opposite of tension. Compression tries to shorten an object. Some metals, such as iron, some steel alloys, aluminum, and so on, have a large plastic range in compression. These materials are called ***malleable materials.*** Industry uses the malleability of such materials to shape them into desired forms. Figure 11.5 illustrates a few processes that shape malleable materials: rolling, forging, and extrusion. Steel sheet is most often manufactured in a rolling mill. Steel panels and fenders with compound curves for automobile bodies are often formed by forging.

Ductile material: A material that can be deformed plastically under tension into a new shape.

Compression: A force pair applied to a material that causes the material to shorten.

Malleable material: A material that can be deformed into a new shape under compression.

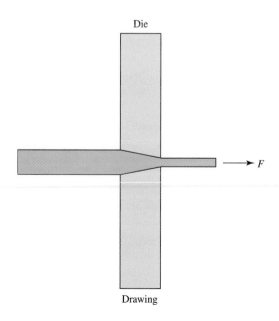

Figure 11.4 Plastic Deformation of a Metal Using a Die

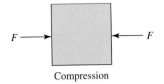
Compression

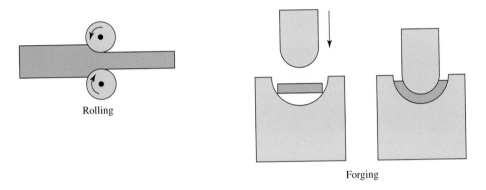
Rolling

Forging

Extrusion

Figure 11.5 The Forming of
a Metal through Compression

11.4 Young's Modulus

The concepts of compression and tension have been defined qualitatively in the previous section. Now what we need is a quantitative way to measure these stresses for a given body of a given composition. ***Stress is the application of equal and opposite forces to a body.*** In the cases that we've so far examined, stress causes either tension or compression. If the body is cylindrical and the forces are applied to each end, then these forces act down the length of the body and across the cross-sectional area.

Stress: The application of equal and opposite forces to a body.

$$\text{Stress} = \frac{\text{force}}{\text{area}}$$

$$\text{Stress} = \frac{F}{A}$$ (Equation 11.2)

If the area is circular, then the cross-sectional area is πr^2. If it is square, then $A = L^2$, where L is the length of one side, and so on. If the stress is compressive, then it causes a *pressure* across the cross-sectional area. We will discuss pressure more fully in Chapter 12 on fluids. The unit for stress is the N/m^2 (newtons per square meter). This unit is commonly called a *pascal* (abbreviated Pa) after the French scientist Blaise Pascal who pioneered research into pressure. In English units, the unit for stress is the $lb/in.^2$ (pound per square inch), often written psi. The conversion between the two systems is

$$1 \text{ psi} = 6.89 \times 10^3 \text{ N/m}^2$$

A pascal is a rather small unit. If you press down on a table hard with your thumb, that produces a pressure of about 1 million pascals!

Strain is the change in length of an object that undergoes stress (forces that cause either tension or compression).

Strain: The change in length of an object undergoing stress.

$$\text{Strain} = \frac{\text{change in length}}{\text{original length}}$$

$$\text{Strain} = \frac{\Delta L}{L_o}$$ (Equation 11.3)

Since strain is a ratio of lengths, it is unitless.

Below the elastic limit, strain is directly proportional to stress.

$$\text{Stress} = Y \times (\text{strain})$$

$$\frac{\text{Stress}}{\text{Strain}} = Y$$

where the proportionality constant Y is known as *Young's modulus.*

$$Y = \frac{F/A}{\Delta L/L_o}$$

Rearranging, we get $$Y = \frac{FL_o}{A \, \Delta L}$$ (Equation 11.4)

The units for Young's modulus are the same as for stress because strain is a pure number. Table 11.2 gives Young's modulus for some common materials.

Table 11.2 Young's Modulus for Some Common Materials

Material	($\times 10^{10}$ Pa)
Aluminum	7.0
Brass	9.1
Concrete	2.0
Copper	11.0
Glass	5.5
Iron	19.0
Lead	1.6
Steel	20.0

EXAMPLE 11.2

A steel wire is originally 1.5 m long and 2.0 mm in diameter. It must support a load of 20 kg. (a) How much is the wire stretched when loaded? (b) How much stress is applied to the stretched wire?

SOLUTION

(a)
$$Y = \frac{FL_o}{A\,\Delta L} \qquad \Delta L = \frac{FL_o}{AY}$$

$$\Delta L = \frac{(20\text{ kg})(9.8\text{ m/s}^2)(1.5\text{ m})}{\pi(0.001\text{ m})^2(20 \times 10^{10}\text{ N/m}^2)}$$

$$\Delta L = 4.68 \times 10^{-4}\text{ m} \quad \text{or} \quad \text{about } \tfrac{1}{2}\text{ mm}$$

(b)
$$\text{Stress} = \frac{F}{A}$$

$$\text{Stress} = \frac{(20\text{ kg})(9.8\text{ m/s}^2)}{\pi(0.001\text{ m})^2}$$

$$\text{Stress} = 6.24 \times 10^7\text{ N/m}^2$$

Stress and tension are *not* the same; the tension is only 196 N.

11.5   Shear and Shear Modulus

Shear: Two opposite and equal forces acting on a body, but *not* in the same line.

*The third type of stress that can be applied to a body is **shear**, in which the two forces that act on the body aren't acting on the same line,* as shown in Figure 11.6. If the object in the figure is in shear, one of the two forces might act parallel to the upper horizontal face, and the other might act parallel to the lower horizontal face of the object. The result is a skewing of the object. For example, if you place a heavy telephone book on a table and then push the upper end to the side with your finger, the book will be deformed much like the object in Figure 11.6.

Figure 11.6 Body Undergoing Shear

The shear angle is defined as the angle θ in Figure 11.6. If the material is a block of steel instead of a telephone book, then the angle θ will be very small.

$$\tan \theta = \frac{s}{L} \quad (\theta \text{ small})$$

$$\theta = \frac{s}{L} \quad (\theta \text{ in radians}) \qquad \text{(Equation 11.5)}$$

Strain is a change in length divided by the original length for compression and tension. For shear, the strain is measured by θ, which is also a dimensionless number. Shear stress is also proportional to shear strain for most materials, so a shear modulus can be defined.

$$\text{Shear modulus} = \frac{\text{shear stress}}{\text{shear strain}}$$

$$\text{Shear stress} = \frac{F}{A} \quad (\text{where } A = \text{cross-sectional area})$$

$$\text{Shear strain} = \theta$$

$$\text{Shear modulus} = \frac{F/A}{\theta} = \frac{F/A}{s/L}$$

$$\text{Shear modulus} = \frac{FL}{sA} \qquad \text{(Equation 11.6)}$$

Shear modulus also has units of pascals. Table 11.3 gives the shear modulus for some typical materials.

If you compare Table 11.2 for Young's modulus to Table 11.3 for shear modulus, you will quickly see that the shear modulus is lower than Young's modulus. Thus, most materials are deformed more easily by shearing than by using tension or compression. A common instrument that operates by using shear is a pair of scissors. (Scissors are often called *shears!*) Concrete doesn't have a shear modulus. If enough shear stress is applied to concrete, it breaks instead of deforming. The shear strength of a material is the amount of shear stress that must be applied to break it.

Table 11.3 Shear Modulus for Some Common Materials

Material	($\times 10^{10}$ Pa)
Aluminum	2.4
Brass	3.6
Concrete	None
Copper	4.2
Glass	2.3
Iron	7.0
Lead	0.56
Steel	8.4

EXAMPLE 11.3

A punch press operates by shearing to punch a hole through a thickness of metal. If mild steel ruptures when a shear stress of about 3.5×10^8 Pa is applied, how much force must be applied to the punch in order for it to punch a square hole 10 mm on a side through a sheet of 4-mm-thick steel?

SOLUTION

The shear stress operates parallel to the *walls* of the hole, not against the upper surface.

$$\text{Shear stress} = \frac{F}{A} \qquad F = A \times (\text{stress})$$

$$F = [4 \times (0.004 \text{ m})(0.010 \text{ m})](3.5 \times 10^8 \text{ N/m}^2)$$

$$F = 56{,}000 \text{ N} \quad \text{or} \quad \text{about 6.3 tons}$$

11.6　Stresses in Construction

Stone and wood were the first large-scale building materials used by man. One of the earliest techniques for using these materials was the post-and-beam construction shown in Figure 11.7. Some early examples of this technique still exist, for example, Stonehenge in England and an early Egyptian temple that stands near the sphinx (the Temple of Karnak). Both structures are remarkably similar in appearance and have withstood the test of time. The very survival of such monuments attests to the validity of the post-and-beam technique.

Figure 11.7 Post-and-Beam Construction

However, the middle, unsupported section of a post-and-beam structure sags because of the pull of gravity, putting the horizontal stone or wood beam under tensile stress. The greater the distance between posts, the greater the sagging caused by the load that the beam must support. The Young and shear moduli of wood are lower than for any other material, and the beam must be thicker and thicker to support a heavier load or to allow a greater distance between posts. Since beams were usually carved out of a single piece of wood, trees with very large diameters were needed for ambitious building projects. These trees were hard to come by in countries such as Egypt, where wood used as a building material had to be imported. (The ancient Egyptians imported cedars from Lebanon.) The large trees were hard to come by even in areas that had forests available for harvesting. Stone has higher Young and shear moduli than wood, but it is much weaker in tension than in compression. As a result, the unsupported length of a beam was limited, and buildings constructed out of wood or stone using the post-and-beam technique had narrow spans and narrow doorways and windows with many interior supports. The pyramids of Egypt are among the most ancient structures still standing. They were built primarily out of limestone blocks. But the pyramids are essentially a pile of these blocks, and the loading on the stone is primarily compressive. Stone handles compressive loads very well.

The arch is illustrated in Figure 11.8. An arch built from stone allows the stone blocks to work primarily in compression. Since stone handles compression much better than tension, a stone arch could support a larger load than its equivalent (with the same post spacing) post-and-beam structure, or the posts could be farther apart, giving a structure more free interior space. The arch was used extensively by the Romans. The Romans were also very good hydraulic engineers. Many of their ancient aqueducts are still standing in Europe and the Near East. These aqueducts were built primarily on the principle of the arch. The Egyptians also used the arch in a limited way. The interior rooms in some of the pyramids used what is called a *corbeled vault,* which is a peaked arch. The interior rooms had to support the weight of the thousands of tons of limestone blocks overhead, and an arch was necessary.

Both the post-and-beam and the arch construction techniques are supported at both ends. If one end were unsupported, the structure would be a cantilever structure; a diving board is such a structure. The ancients rarely built using the cantilever technique because of the relative tensile weakness of the commonly available building materials. Modern materials, such as steel and reinforced concrete, have good tensile strengths, and cantilever structures are now relatively common. A modern skyscraper is a good example of a cantilever construction. A modern skyscraper has vertical sides. The wind loads on the sides of a tall skyscraper can be considerable, and in consequence the sides must be braced against the wind. Instead of a load-bearing wall, many of these skyscrapers have steel skeletons made up of many interlocking trusses. Truss members are triangular in shape, and their individual members are stressed primarily in tension and compression, as opposed to shear. Thus, a truss is much more rigid than a solid beam, and often lighter (see Figure 11.9).

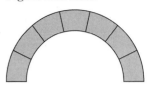

Figure 11.8 Arch

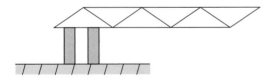

Figure 11.9 Cantilever Construction

The bow and arrow is an ancient weapon. A bow works in tension. As the string is pulled back, the bow acts like a big spring storing potential energy. When the arrow is released, the bow gives the stored potential energy to the arrow as kinetic energy. The famous English longbow was constructed of a solid piece of yew wood and was reputed to throw an arrow as far as a mile. But to throw that far, the bow had to be "long." Therefore, the English longbows were often as tall as a man. Remember that wood has a relatively low Young's modulus, so a bow constructed of wood that stored a lot of energy had to be relatively thick and long. The bow had to be shot from a standing position; it was ineffective from horseback. The Mongols fought from horseback, and their bows had to be short. A Mongol bow was a compound bow. Instead of being made from a single piece of wood, it was usually made from two different materials sandwiched together front to back. As the bowstring was stretched, the material at the front of the bow was put under a tensile stress, and the material at the back of the bow was put under a compressive stress. A compound bow usually had wood at its front, because wood is springy and can store energy efficiently in tension (much like a spring in tension). The rear part of the bow was constructed of a material such as animal horn which can store energy like a spring in compression. The combination of the two materials made for a superior bow over one constructed from one material only.

Summary

1. Forces can be applied to materials to *deform* them. The greater the *modulus* of the material, the harder it is to deform its shape.
2. Deformations can take many forms. Among them are **tensile stress,** which causes a material to lengthen; **compression,** which causes a material to shorten; and **shear,** which causes a material to "bend."
3. **Stress** is the force per unit area that an object experiences; it is measured in pascals (Pa), where 1 Pa = 1 N/m^2.
4. **Strain** is the amount of deformation that an object experiences. Being a ratio of change over original, it is a unitless quantity.

Key Concepts

COMPRESSION: A force pair applied to a material that causes the material to shorten.

DENSITY: The ratio of a mass to the space occupied by that mass. The space is usually a volume, but it can be an area or a linear distance.

DUCTILE MATERIAL: A material that can be deformed plastically under tension into a new shape.

MALLEABLE MATERIAL: A material that can be deformed into a new shape under compression.

PLASTIC DEFORMATION: The response of some materials such that when they are placed under enough tension, they will lengthen and not return to their original equilibrium position.

SHEAR: Two opposite and equal forces acting on a body, but *not* in the same line.

STRAIN: The change in length of an object undergoing stress.

STRESS: The application of equal and opposite forces to a body.

TENSILE STRESS: An equal and opposite force pair applied to a body, causing it to lengthen.

ULTIMATE STRENGTH: The tensile stress that causes a material to break.

Important Equations

(11.1) $\rho = \dfrac{m}{V}$ (Density, or mass per unit volume)

(11.2) Stress $= \dfrac{F}{A}$ (where F = force and A = cross-sectional area)

(11.3) Strain $= \dfrac{\Delta L}{L_o}$ (Change in length divided by the original length)

(11.4) $Y = \dfrac{FL_o}{A\,\Delta L}$ (Young's modulus, or the relation between stress and strain for a given material)

(11.5) $\tan\theta = \dfrac{s}{L}$ (Shear angle where L = thickness of material)

(11.6) Shear modulus $= \dfrac{FL}{sA}$ (Relationship between stress and shear)

Conceptual Problems

1. Equal and opposite forces are exerted on an object along different lines of action. Which of the following statements is true?
   a. The object is under compression.
   b. The object is under tension.
   c. The object is under shear.
   d. The object is being accelerated.

2. If a material is ductile, it:
   a. can break under tension
   b. can be permanently deformed
   c. shrinks under compression
   d. can be temporarily deformed

3. The stress on an object is equal to:
   a. Young's modulus
   b. the relative change in its dimensions
   c. the elastic limit
   d. the force per cross-sectional area

4. A mass is suspended from a wire. The stress does not depend on:
   a. the acceleration of gravity
   b. the wire's length
   c. the suspended mass
   d. the wire's cross-sectional area

5. Two students want to break a piece of strong cord. To break the cord, should they pull at it from each end as in a tug-of-war, or should they tie it to a tree and both pull from the opposite end? Why?

6. A three-legged stool has three identical legs: one of copper, one of aluminum, and one of steel. A load is placed at the center of the stool. Which leg is under the least stress? Explain.

7. A steel cable is replaced by another one that has a diameter three times that of the original one. How much does the load-carrying ability of the new cable increase over that of the old one?

Exercises

Density

1. A steel cylinder with a circular cross section has a diameter of 5 cm and a mass of 24 kg. What is its length?

Answer: L = 1.57 m

2. Humans have about the same density as fresh water. What is the volume of a person with a mass of 65 kg?

Answer: V = 0.065 m³

*3. A concrete slab with dimensions 2 m × 2 m × 10 cm has a mass of 1300 kg. (a) Does the concrete contain steel reinforcing rods, and if so what is their mass? (b) If the circular cross section of a rod can be seen at one face of the slab, and if its diameter is 1 cm, what length of rod is contained in the slab?

Answer: (a) Yes; 539 kg
(b) *L* = 879.8 m

*4. Water and oil are poured into a glass U tube with a constant cross-sectional area. When equilibrium is reached, the tube is as shown in Figure 11.10. What is the density of the oil?

Figure 11.10 (Exercise 4)

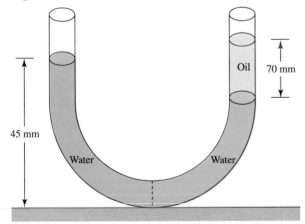

Oil   70 mm

45 mm

Water   Water

Answer: ρ = 643 kg/m³

*5. Gold can be beaten into very thin foil. (a) If 2 g of gold is beaten into a thin sheet with an area of 1.5 m², and if the density of gold is 1.9 × 10⁴ kg/m³, how thick is the sheet? (b) If the mass of a gold atom is 3.27 × 10⁻²⁵ kg, what is the radius of a gold atom? (Assume that atoms are spherical.) (c) How many atoms thick is the sheet?

Answer: (a) *t* = 7 × 10⁻⁸ m
(b) *r_a* = 1.6 × 10⁻¹⁰ m
(c) 218 atoms

*6. A 1.5-kW pump must raise water from a shallow 6-m well. What is the flow rate in liters per minute if the pump is 75% efficient?

Answer: 1140 liters/min

*7. The gravitational attraction of a spherical mass distribution is the same as that of a point mass located at the center of the sphere. What would be the radius of a lead sphere that would attract you with a gravitational force $\frac{1}{1000}$ your own weight if you were standing right next to it? The density of lead is 1.13 × 10⁴ kg/m³.

Answer: R = 3.1 km

*8. A newly found planet has no atmosphere, and a small moon orbiting very near its surface has an orbital speed of 3.5 km/s. The planet has an average density of 4 × 10³ kg/m³. What is the mass of the planet? Assume that the planet is spherical.

Answer: M = 6.08 × 10²³ kg

Stress

9. A spring is hung from a ceiling, and a number of weights are hung from the spring. The lengths of the stretched spring are given below:

Weight (N)	0	5.0	10.0	15.0	20.0	25.0	30.0
Length (cm)	15	16.2	17.6	18.8	20.0	27.0	36.0

(a) What is the force constant of the spring? (b) What is the approximate elastic limit of the spring?

Answer: (a) 394.7 N/m
(b) 21 N

*10. An adobe brick has a density of 2100 kg/m^3 and an ultimate strength in compression of 8.0 × 10^6 Pa. (a) How high can a wall of such bricks be built before the lowest bricks in the wall are crushed to powder? (b) How many bricks are above the lowest brick if each brick is 8 cm high?

Answer: (a) $h = 388.7$ m
(b) 4858 bricks

Young's Modulus

11. A nylon rope 12 mm in diameter breaks when a force of 28,000 N is applied to it. What force would cause a rope made of the same type of nylon to break if the rope had a diameter of 15 mm?

Answer: F = 43,750 N

12. A 1.5-m-long steel wire has a circular cross section with a diameter of 0.80 mm. If a mass of 9 kg is hung from the wire, how much will the wire stretch?

Answer: $\Delta L = 1.32$ mm

13. The maximum acceleration of a fast freight elevator is 3 m/s^2.
* The maximum load of the elevator is 7000 kg, and the elevator has a mass of 500 kg. If a steel cable is to be used to suspend the elevator, what should its minimum diameter be? The elastic limit of steel is 3.6 × 10^8 Pa.

Answer: D = 18.4 mm

*14. A metal wire 0.8 m long and with a diameter of 1.00 mm is supported at both ends and suspended horizontally. A 15-kg mass is then suspended from the wire's midpoint, and the wire sags vertically by 30 mm. What is the Young's modulus of the wire?

Answer: $Y = 5 \times 10^{11}$ Pa

15. The breaking stress of a certain steel is 11 × 10^8 N/m^2. What is the minimum diameter of a steel wire of circular cross section that will safely support a load of 90 kg?

Answer: D = 1 mm

16. A metal wire 1.5 mm in diameter and 2.5 m long has a 6-kg mass hanging from it. If the wire stretches 0.8 mm from the tension, what is the value of Young's modulus for the metal?

Answer: $Y = 1.04 \times 10^{11}$ N/m^2

Shear and Shear Modulus

17. Two aluminum sheets are riveted together with eight rivets along one edge. The maximum shear stress that a rivet can stand is 2.4 × 10^7 Pa, and the rivets each have a diameter of 4 mm. How much force applied parallel to the plates can the rivets stand before they are sheared off?

Answer: F = 2400 N

18. A punch press can exert a maximum force of 18,000 N. It punches seven square holes simultaneously in a piece of aluminum. If the holes are 0.5 cm on a side, what is the maximum thickness of aluminum that can be used? The shear strength of aluminum is 7 × 10^7 Pa.

Answer: t = 1.84 mm

An airplane flies because of the differing velocities of a fluid (air) above and below a wing's surface.

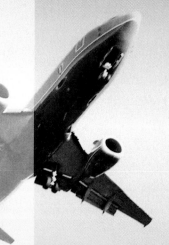

CHAPTER 12

F luids

At the end of this chapter, the reader should be able to:

1. Explain the differences among *atmospheric pressure,* **gauge pressure,** and **absolute pressure.**

2. Calculate the pressure at a given depth of a fluid.

3. Solve for the force necessary to lift a known weight using a hydraulic press, given radii of the columns of the press.

4. Calculate the **buoyant force** on a floating object and apply this force to Newton's second law.

5. Describe why the velocity of a fluid in a pipe increases as the radius of the pipe decreases.

6. Using *Bernoulli's equation,* solve for the pressure difference, the elevation difference, or the velocity difference of opposite ends of a pipe.

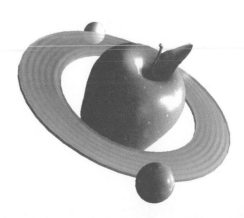

12.1   Pressure

We live at the bottom of a sea of air. Air is a gas composed of approximately 20% oxygen and 80% nitrogen. Gases have a low density, but if you live at the bottom of a column that is 20 to 30 mi high (depending on how you define the top of the atmosphere), even a low-density substance can add up to a lot of weight. The weight of the atmosphere above us exerts a pressure at the surface of the earth of about 14.7 lb/in.^{2} (1.013×10^5 Pa). We don't notice this pressure because the internal pressure in our bodies is the same as the external pressure. If for some reason our body's internal pressure were less than the external pressure, then the force pressing in on us would crush us. If the inverse were true, then we would explode! If you've ever had a bad cold and been in a jet airliner at 30,000 ft, you probably had an earache, which was caused by atmospheric pressure behind the eardrum and slightly less pressure in the ear canal. (The pressurized cabin of an airliner is actually less than 1 atm when the aircraft is at its cruising altitude.) This pressure difference causes an earache. In addition, breathing is possible only because of a pressure difference. The muscles in our rib cage collapse our lungs, creating an overpressure and forcing the air out. When these muscles relax and the lungs reexpand, the pressure inside the lungs is less than atmospheric, and the lungs refill with air because atmospheric pressure pushes air into the now low-pressure area until the pressures equalize.

Pressure can be measured with a gauge. One of the most familiar gauges is one that reads the air pressure in the tires of a car. Recommended pressure for many tires is 32 lb/in.^{2}; this pressure is what is called *gauge pressure*. The gauge itself is referenced to atmospheric pressure, and the 32 lb/in.^{2} is actually 32 lb/in.^{2} above atmospheric. Thus, **gauge pressure** *is actually a differential pressure (the difference between an inside pressure and an outside pressure). The **absolute pressure** in the tire is the gauge pressure plus atmospheric.*

Gauge pressure: The differential pressure between the inside and the outside of a volume, or the difference between the absolute pressure and the atmospheric pressure.

Absolute pressure: The total pressure on an object, or the gauge pressure plus the atmospheric pressure.

(Equation 12.1)

$$P = 32 \text{ lb/in.}^2 + 14.7 \text{ lb/in.}^2$$

$$P = 46.7 \text{ lb/in.}^2$$

Pressure has the same units as stress, or pascals (N/m^2), because pressure is the force divided by the area over which the force is exerted.

$$P = \frac{F}{A}$$

A body with an interior pressure greater than the exterior pressure will tend to expand, as, for example, a balloon that is being inflated. A body with an interior pressure less than the exterior pressure will tend to collapse inwardly or be crushed by the exterior pressure.

EXAMPLE 12.1

A barometer is an instrument that measures the pressure of the atmosphere. The atmospheric pressure varies from a high pressure somewhat above normal atmospheric on cold, clear days, to a low slightly below normal atmospheric on stormy days. Knowing that a severe storm is coming, the people inside a house close it up tightly so that inside the house the pressure remains at 1 atm (1.013 bar). At the height of the storm, the outside pressure falls to 980 millibars (0.98 bar). The roof of the house is flat, measures 30 ft by 26 ft, and weighs 18,000 lb. What is the net force on the roof, and in what direction (1 bar = 14.5 lb/in.^{2})?

SOLUTION

$$P = \frac{F}{A} \qquad F = PA \qquad F = \Delta PA$$

$$\Delta P = 1.013 \text{ bars} - 0.98 \text{ bar}$$

$$\Delta P = (0.033 \text{ bar})\left(\frac{14.5 \text{ lb/in.}^2}{\text{bar}}\right)$$

$$\Delta P = 0.4785 \text{ lb/in.}^2$$

$$A = (30 \text{ ft})(26 \text{ ft}) \qquad A = (780 \text{ ft}^2)\left(\frac{144 \text{ in.}^2}{1 \text{ ft}^2}\right) \qquad A = 112{,}320 \text{ in.}^2$$

$$F = (0.4785 \text{ lb/in.}^2)(112{,}320 \text{ in.}^2) \qquad F = 53{,}745 \text{ lb (up)}$$

$$F_{net} = F_{up} - W$$

$$F_{net} = 53{,}745 \text{ lb} - 18{,}000 \text{ lb}$$

$$F_{net} = 35{,}745 \text{ lb}$$

The excess pressure inside the house threatens to blow the roof right off the house! The upward force on the roof caused by the pressure difference between the inside and the outside of the house is almost three times the weight of the roof! When a storm removes the roof of a house, the cause in many cases isn't the high winds, but the pressure difference from inside to outside.

12.2   Depth and Pressure

In a body of water, the deeper something is, the more pressure it has to withstand. This fact imposes a severe design constraint on the engineers who must design deep-sea submersibles and submarines. It also imposes a constraint on those who do free diving (scuba divers, for instance). If a diver descends one hundred feet or so with scuba gear and then ascends too quickly, she can possibly get the dreaded and painful cramps called the "bends." A diver with the bends must be put into a high-compression chamber, and the pressure inside brought slowly down to atmospheric. The pressure at depth is caused by the weight of water above a diver's head or above a submarine or deep-sea submersible.

Assume that we have a column of water such as that in Figure 12.1. The column has the shape of a tall water glass of constant cross-sectional area A. What is the pressure at

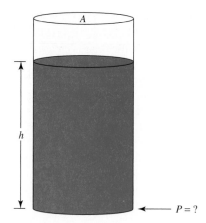

Figure 12.1 Column of Water

the bottom of the column of depth h? To solve this problem, we start with the definition of pressure.

$$P = \frac{F}{A} \qquad F = mg \qquad P = \frac{mg}{A}$$

$$m = \rho V \qquad \text{(where } \rho = \text{the density of water)}$$

$$V = Ah \qquad m = \rho Ah$$

$$P = \frac{\rho Ahg}{A}$$

(Equation 12.2)
$$P = \rho gh$$

The pressure increases linearly with depth, and the pressure is directly proportional to the density. The density of water is over 700 times the density of air. Suppose that we live under a column of air that is roughly 30 mi high and the pressure is 14.7 lb/in.^{2}. If the air were replaced by water, the pressure would be 11,307 lb/in.^{2} (roughly $5\frac{1}{2}$ tons)!

12.3   Properties of Pressure in a Fluid and Fluid Flow

This chapter deals with fluids, and many people think of a fluid as a liquid, with air being in another class, that of a gas. However, both a liquid and a gas can be considered a fluid since both exhibit fluid flow. For example, natural gas is piped from place to place much as oil and water are. Both liquids and gases are considered fluids because they have so many properties in common and can be described by the same laws, such as the one for pressure given in Equation (12.2). The only mathematical difference between the pressure at the bottom of a column of water and the pressure of a column of air is the value for the density. A big difference between a liquid and a gas is that a liquid is almost incompressible compared to a gas. But in many processes a gas can also be treated as incompressible. In this chapter we'll treat gases as incompressible fluids.

The pressure in a liquid is exerted equally in all directions. The pressure at the depth h in Figure 12.1 is caused by the weight of the water above, but the pressure P acts equally in all directions at that depth. As a result, the pressure on the sides of the tall glass of water is the same as the pressure acting down or up. The walls of a large dam are built with a triangular profile because of the pressure on the dam's wall. A large hydroelectric dam is constructed to be thicker at its base than at its apex because the pressure is greater at its base than at the top.

For a small volume of fluid, the pressure is the same throughout the volume. Take the case of a submarine at depth. If the submarine has a height of 15 ft from keel to upper deck, and if it is at a depth of 1000 ft, then the pressure doesn't vary much from the upper deck down to the keel. The variation is negligible. The fact that the pressure in hydraulic fluid is constant throughout the volume is the principle behind the operation.

Fluid flow is also a consequence of pressure. Assume that we have a small volume of fluid in which the difference in pressure from top to bottom is negligible. If pressure were exerted on this small volume of fluid from the outside, and if there were a pressure difference from one end of the fluid to another as in Figure 12.2, then the fluid would flow as a consequence of Newton's second law. Assume that Figure 12.2 is a small section of water hose with a constant cross-sectional area, and that the pressure on the left side is greater than that on the right.

$$P_1 > P_2 \qquad \Delta P = P_1 - P_2$$

$$\Delta P = \frac{F_1}{A_1} - \frac{F_2}{A_2}$$

$$A_1 = A_2$$

$$\Delta P = \frac{1}{A}(F_1 - F_2)$$

$P_1 \longrightarrow$ 　　　　　　　　　　　　　　$\longleftarrow P_2$

Figure 12.2 Fluid Flow in a Pipe Caused by a Pressure Differential

From the above, you can see that a pressure differential is the same as an unbalanced force, and the fluid will flow from a region of high pressure to that of a lower pressure. In Section 12.1 on pressure and depth, a pressure differential existed between the top and bottom of the fluid column, but because of Newton's third law (action and reaction), the fluid doesn't flow. The situation is the same as a book lying on a table. The book presses down on the table, and the table presses up on the book. The net force is zero, and the book stays put.

As we learned earlier, when no net force exists, a body won't accelerate. An unaccelerated body can have a constant velocity, or a velocity of zero. We can push a wooden crate across a rough floor at constant velocity and do a considerable amount of work in the meantime. But if the box is motionless at the end of all that pushing, then the energy spent by our effort has been dissipated by friction. The same can also be true in fluid flow. A fluid has a property known as *viscosity*. If you've ever bought motor oil, you probably know that it is often classified by its viscosity; for example, 30-weight oil is thinner than

50-weight. (An oil classified as 10W/30 means that at room temperature, it has a viscosity of 30, and at the operating temperature of the engine, it has the thinner viscosity of 10.) Viscosity is a measure of the internal friction that a fluid has to overcome to flow. A thick pancake syrup will flow less readily than vinegar because of its internal friction. In this book we often idealize motion by talking of movement without friction. A fluid without viscosity is analogous. Oddly enough, such a fluid exists. Liquid hydrogen cooled down very close to absolute zero has no viscosity! However, all other fluids have viscosity, and energy is lost overcoming it. A flowing fluid also has friction between itself and the walls of whatever container through which it is flowing. Fluid flow without friction, called *laminar flow,* is an idealized situation very rarely encountered in reality. The water in a garden hose moves through the pipe at a constant velocity despite the pressure difference because of the viscosity of the water and the friction between the water and the pipe.

EXAMPLE 12.2

A deep-sea submersible is at a depth of 1000 m. The interior of the submersible is at 1 atm. There is a quartz viewing port in the side of the submersible with a diameter of $\frac{1}{2}$ m. What force does the window have to withstand?

SOLUTION

$$P = \rho g h$$

$$P = (1000 \text{ kg/m}^3)(9.8 \text{ m/s}^2)(1000 \text{ m})$$

$$P = 9.8 \times 10^6 \text{ Pa}$$

The above pressure is the gauge pressure. To find the absolute pressure, we must add atmospheric pressure to the gauge pressure.

$$P_{abs} = P_g + P_{atm}$$

The absolute pressure is the pressure that is trying to crush the hull of the deep-sea submersible. However, the interior pressure is 1 atm, and the port must withstand the differential pressure across it.

$$\Delta P = P_{abs} - P_{atm}$$

$$\Delta P = P_g + P_{atm} - P_{atm}$$

$$\Delta P = P_g$$

$$\Delta P = 9.8 \times 10^6 \text{ Pa}$$

$$P = \frac{F}{A} \qquad F = PA$$

$$F = (9.8 \times 10^6 \text{ N/m}^2)(\pi)(0.25 \text{ m})^2$$

$$F = 1.92 \times 10^6 \text{ N} \quad \text{or} \quad 4.3 \times 10^5 \text{ lb!}$$

That's almost half a million pounds or 215 tons! At those pressures, *any* leak would quickly destroy the vessel.

12.4 Hydraulic Mechanical Advantage

In Chapter 10, simple machines and mechanical advantage were discussed. We learned that mechanical advantage implies an amplifier of either force or torque. A hydraulic press is another relatively simple machine that is a force amplifier. Most automobiles have one of two types of braking systems: disk brakes or drum brakes. With a disk system two calipers pinch a steel disk to slow the car down, and with a drum system two brake shoes make contact with the inside of a circular steel brake drum to slow and stop the car. Regardless of the type of brake, though, the overwhelming majority of braking systems are activated by hydraulic pistons. An input piston is connected to the brake pedal, and there are usually output pistons at each wheel. Copper brake lines filled with hydraulic fluid connect the input and output pistons. Since pressure is distributed equally in all parts of an incompressible fluid, the pressure created by pushing your foot down on the brake pedal is distributed equally to all four wheels. A 2500-lb car moving at 60 mph carries with it a lot of energy, and an equal amount of energy must be expended to slow and stop a car from this speed. Sticking your foot out the door and dragging it on the road to stop the car doesn't appear to be a very sensible design solution to a car's brakes (except maybe for Fred Flinstone). But every time the car is braked to a halt, it's your foot that halts it. The extra leverage needed comes from the mechanical advantage given by a hydraulic press, such as that shown in Figure 12.3.

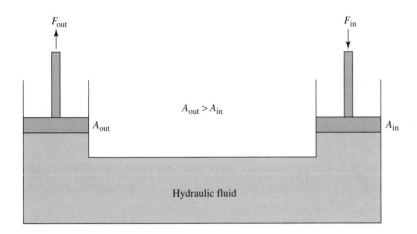

Figure 12.3 Hydraulic Amplifier

When your foot pushes down on the brake pedal, a certain amount of force is applied to the input brake piston of cross-sectional area A_{in}, causing an input pressure to the hydraulic fluid, P_{in}.

$$P_{in} = \frac{F_{in}}{A_{in}} \qquad P_{out} = \frac{F_{out}}{A_{out}}$$

$$P_{in} = P_{out}$$

$$\frac{F_{in}}{A_{in}} = \frac{F_{out}}{A_{out}}$$

$$F_{out} = \left(\frac{A_{out}}{A_{in}}\right)F_{in}$$

(Equation 12.3)

Since the output piston has a greater area than the input piston, the fraction A_{out}/A_{in} is greater than 1, and thus the output force is greater than the input force. The hydraulic press exhibits mechanical advantage and is a force amplifier.

Since the hydraulic fluid is incompressible, the relationship between the input stroke and the output stroke is easily derived. The volume of fluid displaced by the input piston is the same as the volume displaced by the output piston because of the hydraulic fluid pushing on it.

$$V_{in} = V_{out}$$

$$V_{in} = A_{in}h_{in} \qquad V_{out} = A_{out}h_{out}$$

$$A_{in}h_{in} = A_{out}h_{out}$$

$$\frac{h_{out}}{h_{in}} = \frac{A_{in}}{A_{out}}$$

The above equation means that the output stroke will be shorter than the input stroke, because the ratio A_{in}/A_{out} is less than 1. In Chapter 10 on simple machines, it was shown that you don't get something for nothing. The work done on the input will now be compared to the work done by the output.

$$W_{in} = F_{in}h_{in} \qquad W_{out} = F_{out}h_{out}$$

$$W_{out} = \left[\left(\frac{A_{out}}{A_{in}}\right)F_{in} \right]\left[h_{in}\left(\frac{A_{in}}{A_{out}}\right) \right]$$

$$W_{out} = F_{in}h_{in}$$

$$W_{out} = W_{in}$$

The work done on the input is the same as the work done by the output! By now this relationship should come as no surprise.

12.5   Buoyancy

Some things float; others don't. A block of steel certainly doesn't float on a pool of water, but a block of wood does. Why? The first thing that might come to mind is density. The steel is denser than the water and won't float on water. The wood is less dense than the water and will float. We might be tempted to conclude that the reason that objects either float or don't float is density. But there is more to it than that. Density is certainly a part of it, but not the whole story.

If a body floats in water, then the water is applying an upward force on it, much as a table exerts an upward force on a book lying on it. This upward force is in balance with the weight of the object, or it wouldn't float. Moreover, this upward force works on the object immersed in the liquid whether or not it floats! This force, called the *buoyant force,* acts just like any other force if we draw a free-body diagram. For example, if we drop a block of aluminum in water, it sinks. But a strange thing happens if we weigh the aluminum block in air and then weigh it again immersed in water. The block in the water weighs less than the block in air! The mass is certainly the same in the water as in the air, since the block doesn't shrink or become hollow. So what's going on? When the block is placed in the water, it displaces an amount of water equal in volume to the block. The buoyant force up equals the weight of the water displaced. *In*

Although steel has a higher density than water, steel ships float because the average density of the ship over the entire ship's volume is lower than that of water. (A ship contains a lot of empty space.)

general, the **buoyant force** equals the weight of the fluid displaced, regardless of whether the fluid is a liquid or a gas. A submarine floating in the sea is operating by the same principle as a blimp or a hot air balloon floating in the air. Both air and water are fluids, and objects can float in both of them.

   This definition of *buoyant force* can be converted into a mathematical formula.

$$F_b = W_f \qquad W_f = \rho_f V_f g$$

$$F_b = \rho_f V_f g$$

(Equation 12.4)

Buoyant force: The force acting up on an object immersed in a fluid that is equal to the weight of the displaced fluid.

where ρ_f is the density of the fluid, V_f is the volume of the fluid displaced, and W_f is the weight of the fluid displaced. Buoyant force was first discovered by the Greek Archimedes and thus is known as *Archimedes' principle.* An object immersed in a fluid has a higher pressure exerted on its bottom than its top because the bottom is at a greater depth than the top. The difference in pressure points up, and a net force up acts on the body. The pressures on the object's sides all point inward. The pressure on the left side points right, and that on the right side points left. They cancel out because they have equal and opposite magnitudes.

EXAMPLE 12.3

What you don't see can hurt you. The *Titanic* sank because it hit an iceberg. The ship hit the iceberg with a glancing blow, and the iceberg tore a long gash in the ship's side under the water line. The *Titanic* struck the submerged part of the iceberg. Most of an iceberg *is* submerged, with only a small part of its volume above water. How much of an iceberg is submerged?

   The density of ice is less than that of water. What keeps an iceberg afloat is its buoyancy. The buoyant force is equal to the weight of the water displaced.

$$F_b = V_w \rho_w g$$

The buoyant force must equal the iceberg's weight if the iceberg is to float.

$$F_b = V_i \rho_i g$$

$$V_w \rho_w g = V_i \rho_i g$$

$$V_w \rho_w = V_i \rho_i$$

$$\frac{V_w}{V_i} = \frac{\rho_i}{\rho_w}$$

The volume of the water displaced is equal to the submerged part of the iceberg.

$$V_w = V_{sub}$$

$$\frac{V_{sub}}{V_i} = \frac{\rho_i}{\rho_w}$$

The ratio of the density of ice to the density of water gives the fraction of the iceberg that is under water. The density of water is 1000 kg/m^3, and the density of ice is 920 kg/m^3.

$$\frac{\rho_i}{\rho_w} = \frac{920 \text{ kg/m}^3}{1000 \text{ kg/m}^3}$$

$$\frac{\rho_i}{\rho_w} = 92\%$$

Sea water contains salt, and its density is somewhat greater than that of fresh water. Thus, an iceberg floats a little higher in the water. Nevertheless, approximately nine-tenths of an iceberg is under water.

12.6   Bernoulli's Equation and Fluid Flow

Why does an airplane fly? An aircraft is immersed in a fluid, and that fluid is the air. As the plane moves through the air, its wings generate lift. That lift is a result of the fluid flow around the wing's profile.

Figure 12.4 shows the air flowing around a wing's profile. If the air mass is originally calm, the leading edge of the wing separates the air into air flowing above the wing and air flowing below the wing. Behind the trailing edge of the wing, the air recombines. Since the air flows around the wing, the air above the wing must travel a longer distance than the air below the wing. If two air molecules were adjacent to one another before the wing passed through the air mass, and if the molecules were again adjacent after the wing had passed through, then the molecule that flowed over the top of the wing would have to have traveled faster than the one below to arrive at the same place, because it had to go a bit farther as a result of the curved upper surface of the wing. Because the air above the wing flows faster than the air below the wing, the pressure is less above the wing than

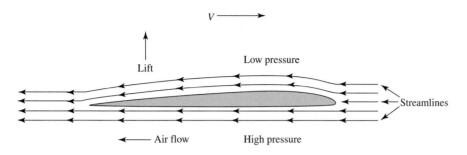

Figure 12.4 Air Flowing
around a Wing

below it. This difference in pressure creates lift, and the plane flies. This is the stock answer
to the question, "Why does an airplane fly?"

But why does the speed difference create lift? The answer to that question comes in
deriving Bernoulli's equation, named in honor of Daniel Bernoulli (1700–82), who first
derived it. Bernoulli was trying to find an equation that would describe fluid flow. Fluids
can flow in complex ways. Turbulent flow is chaotic and practically impossible to predict. Fluids also have viscosity, or internal friction, as mentioned earlier. Turbulent, vis-
cous fluids are far from ideal. All real fluids have viscosity and can flow turbulently.
However, we can get useful results by assuming an ideal fluid that flows without turbulence. An ideal fluid has no viscosity and is incompressible. *Flow without turbulence is
called* **laminar flow.**

In laminar flow, every particle of liquid passing a particular point follows the same
path. The paths are called *streamlines*. It is also assumed that the direction in which the
individual particle moves is always the same as the direction of the fluid as a whole. (When
individual particles begin bucking the current, turbulence sets in.) So laminar flow is a
smooth flow, and now we can define flow: ***Flow** is the rate at which mass passes a given
point.* Mathematically, we can define flow as

$$\text{Flow} = \frac{\Delta m}{\Delta t}$$

We can think of flow in terms of traffic. Traffic is made up of many individual cars,
trucks, buses, and motorcycles. A traffic flow rate could be the number of vehicles per
hour, for example. In this chapter the flow rate will be defined as a mass flow, and it's
easy to see that each vehicle has mass. However, for traffic, vehicles per hour makes more
sense than kilograms per second. But what if we had water flowing through a hose? Water
is made up of many individual water molecules of H_2O (two atoms of hydrogen bonded
to an atom of oxygen). As we observe water flowing out of the hose, the flow appears to
be a continuous process. We don't perceive individual molecules tumbling out of the nozzle, but rather a smooth, unbroken flow. Water doesn't act like traffic, but flows continuously. A liquid flow is often given as volume per unit time, but a volume of liquid also
has a mass. So the basic definition for *flow* will be mass per unit time.

Starting from this definition, we can define flow by following a small volume as it
moves through a pipe laminarly. If the pipe has a uniform cross-sectional area, then

$$\frac{\Delta m}{\Delta t} = \frac{\rho \, \Delta V}{\Delta t} \qquad \Delta V = A \, \Delta s$$

where Δs is the distance traveled through the pipe in time Δt.

Laminar flow: Fluid flow
without turbulence.

Flow: The rate at which mass
passes a given point, or mass
per unit time.

(Equation 12.5)

$$\frac{\Delta m}{\Delta t} = \frac{\rho A \ \Delta s}{\Delta t} = \rho A \frac{\Delta s}{\Delta t}$$

But $\Delta s/\Delta t$ is the speed at which the fluid flows through the pipe, and thus

$$\frac{\Delta s}{\Delta t} = v$$

$$\frac{\Delta m}{\Delta t} = \rho A v \qquad \text{Flow} = \rho A v$$

Having defined our ideal fluid and having defined flow, we can now proceed to derive Bernoulli's equation.

We will follow a small volume of liquid as it flows through a non-uniform pipe, as shown in Figure 12.5. If the fluid is incompressible, then the flow rate at all points in the pipe is the same. The volume element of fluid that we follow through the pipe will also have the same volume at all points in the pipe. When a fluid flows through a pipe, its speed changes with the diameter of the pipe if the flow rate is constant. Its speed is higher through a narrow section of the pipe, and lower through a wide section of pipe. The speed of the fluid also varies as the fluid rises and falls. As water falls, it speeds up, and as it rises, it slows down. We can verify all of these facts by looking at the flow from a work/energy perspective. Since we are looking at an ideal fluid flowing laminarly, conservation of energy holds. The work done on the fluid is equal to the change in potential energy plus the change in kinetic energy.

$$W = \Delta \text{PE} + \Delta \text{KE}$$

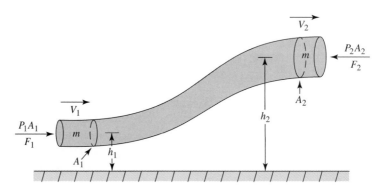

Figure 12.5 Bernoulli's Equation Derived from Fluid Flow

We will look at our volume element at two areas: Area 1 is where the pipe is low and narrow, and area 2 is where the pipe is high and wide. The work done in making the liquid rise is

$$W = F_1 s_1 - F_2 s_2$$

where s_1 and s_2 are the distances that the volume element travels in the narrow and wide areas in a time t, and $s = vt$.

$$P = \frac{F}{A} \qquad F = PA$$

$$W = P_1 A_1 s_1 - P_2 A_2 s_2$$

But $A_1 s_1$ = volume, and since both volume elements are the same, $A_1 s_1 = A_2 s_2 = V$.

$$W = P_1 V - P_2 V$$

The volume is just the mass over the density of the fluid.

$$V = \frac{m}{\rho} \qquad W = P_1 \frac{m}{\rho} - P_2 \frac{m}{\rho}$$

We now have the work done on the fluid in terms of characteristics of the fluid: its pressure, mass, and density. Positive work is done on the volume element. Remember from the example of the hose that fluid flows from a region of higher pressure to a region of lower pressure. Thus, area 2 must have a lower net pressure than area 1.

Next let's do the same thing for the difference in potential energy.

$$\Delta PE = mgh_2 - mgh_1$$

(The above was written because we followed the same volume, and therefore the same mass, from area 1 to area 2.) The change in kinetic energy is

$$\Delta KE = \frac{1}{2} m v_2^2 - \frac{1}{2} m v_1^2$$

We now put everything together.

$$W = \Delta PE + \Delta KE$$

$$\left(P_1 \frac{m}{\rho} - P_2 \frac{m}{\rho} \right) = (mgh_2 - mgh_1) + \left(\frac{1}{2} m v_2^2 - \frac{1}{2} m v_1^2 \right)$$

Next we rearrange the above equation and put everything concerning area 1 on the left and everything concerning area 2 on the right.

$$\frac{P_1}{\rho} - \frac{P_2}{\rho} = gh_2 - gh_1 + \frac{1}{2} v_2^2 - \frac{1}{2} v_1^2$$

$$P_1 - P_2 = \rho gh_2 - \rho gh_1 + \frac{1}{2} \rho v_2^2 - \frac{1}{2} \rho v_1^2$$

$$P_1 + \rho gh_1 + \frac{1}{2} \rho v_1^2 = P_2 + \rho gh_2 + \frac{1}{2} \rho v_2^2 \qquad \text{(Equation 12.6)}$$

Equation (12.6) is Bernoulli's equation. It describes fluid flow and leads to some rather surprising conclusions. But before discussing these conclusions, let's first return to Figure 12.4 and continue our discussion of why an aircraft is able to fly.

To use Bernoulli's equation in the context of Figure 12.4, we must realize that the two air molecules that traveled around the wing (one over and one under) end up back together again after the wing passes. Thus, $h_1 = h_2$, and the potential energy terms drop out of the equation. (In deriving Bernoulli's equation, we looked at beginning and end states in which the fluid rose. In this case the fluid is at the same height at both beginning and end.)

$$P_1 + \frac{1}{2}\rho v_1^2 = P_2 + \frac{1}{2}\rho v_2^2$$

We now distinguish between upper and lower flows around the wing.

$$P_u + \frac{1}{2}\rho v_u^2 = P_L + \frac{1}{2}\rho v_L^2$$

$$\frac{1}{2}\rho v_u^2 - \frac{1}{2}\rho v_L^2 = P_L - P_u$$

$$\frac{1}{2}\rho(v_u^2 - v_L^2) = P_L - P_u$$

The velocity above the wing is faster than the air's velocity below the wing, so the term on the left of the equation is positive. The term on the right of the equation must also be positive. The only way for this equation to be true is for the pressure below the wing to be greater than the pressure above the wing, and as a result, the wing develops upward lift!

The same argument can also be used for the venturi found in a carburetor. If you look into the throat of a carburetor, you'll see the throat constricted where the fuel is fed into the air stream. As the air moves from the wide throat of the carburetor into the venturi, the throat narrows and the air speeds up. Thus, in the venturi the air is below atmospheric pressure. The fuel coming from the tank is at atmospheric pressure, and the pressure difference between the tank and the venturi forces the fuel to flow from the tank into the carburetor's venturi. The carburetor doesn't suck fuel into itself; atmospheric pressure pushes it in!

12.7 Surface Tension and Capillary Action

The surface of a liquid acts like a membrane under tension. A membrane under tension is like a sheet of rubber stretched smooth and tight. In a volume of water, the water molecules are polarized and thus attract one another with a small electrical force. A water molecule immersed in the volume sees roughly the same force exerted on it from all sides since it is completely surrounded by other molecules. However, if the molecule were on the surface, then it would experience a downward force because there are no molecules above it. This inward force causes a drop of water on a slick surface to bead and a suspended drop to form a sphere. A soap film has a large surface tension. When a good bubble is blown from a bubble pipe, it will sometimes bounce when it falls slowly upon a slick floor. The bubble acts as if it is made from a very thin rubber membrane. A bubble that floats in undisturbed air forms a sphere because of the surface tension. A sphere has the least surface area for its volume, and surface tension minimizes surface area. That is why you'll see no solitary square bubbles (or triangular bubbles, for that matter)!

A steel needle can "float" on water because of the high surface tension of water, whereas a solid block of steel will sink in a pool of water. Steel has a much greater density than water and therefore can't float in water. But water's surface tension will supply enough force to counteract the weight of a small steel needle. To float a steel needle, all we will need is a large-mouthed glass filled with water, the needle, and some toilet paper. We first place a small piece of the toilet paper gently upon the water's surface. The paper's surface should be flat. Then we place the needle upon the paper; the needle is supported by the paper. Next we gently push the paper down into the water and away from the needle. When the toilet paper is soaked, it will sink, leaving the needle suspended on the surface of the water! To prove that the needle is really steel, you can use a magnet to move the needle about on the water's surface. But the membrane formed by the water's surface tension is fragile, and any rough handling of the needle will send it immediately to the bottom when the membrane is broken.

But is the needle floating? No. Flotation is caused by buoyancy, and the needle's density is about eight times that of water. The needle's suspension on the water's surface is caused by a phenomenon completely different from buoyancy: surface tension. Modern ships are made from steel, and they float because of buoyancy, not surface tension. In order to float, a modern ship with a steel hull must have an average density less than that of water. A steel ship encloses a lot of empty space. Since density is the mass per unit volume, when all of a ship's empty space is taken into account, the average density is lower. You can alter the density of a submarine by filling its ballast tanks with water or by blowing that water out. With no water in the ballast tanks (empty space), its density is less than that of water, and the submarine floats. With water in the tanks, the submarine's average density is greater than that of water, and it sinks.

We encounter tension every time we try to clean something with soap and water. Water is "wet," but what is meant by "wet"? When a liquid wets something, it penetrates it. If we place a small bit of one end of a dry sponge into water, the water will soak the end of the sponge. Moreover, it will rise through the sponge until the sponge is thoroughly wet, even though the sponge is not lowered into the water. If a drop of water is placed upon a paper towel, it will soak into the towel and make it soggy. Whenever a material draws a liquid into it, this action is called *capillary action.* However, if a drop of water is placed upon a slick, glassy surface, it will bead up (depending on the material of the surface) and not penetrate (wet) the surface. When a bead forms, the cohesive forces that bind the drop together are stronger than the forces that would make the water penetrate the solid surface. The drop will form a bead because of the surface tension mentioned above. If we have an object of clothing that we wish to wash, we want the water to soak into the cloth and remove whatever stain is present. Water alone can wash clothes, but water with a detergent is *much* better. A detergent acts on water to make the surface tension much less. Therefore, it much more effectively penetrates the cloth from which the article of clothing is made. In other words, a detergent makes water wetter! Substances such as alcohol and gasoline have low surface tensions, and they wet cloth much better than water. To prove to yourself that a detergent makes water wetter, try the experiment with the needle again, but with water with a little detergent mixed in. You'll find that you can't get surface tension to suspend the needle this time because of the drastically lowered surface tension. Moreover, if you drop some detergent in the water while it supports the needle with surface tension, the needle will immediately sink because of the reduced surface tension.

12.8   Power and Fluid Flow

Power in the context of fluid flow must take pressure into account. In Chapter 4 on work and energy, we started from the definition of *power* (work per unit time) and were able to derive another formula for power.

$$P = Fv$$

Now we'll start from the above equation and define *force* in terms of pressure. (In this derivation a lowercase p will be the symbol for pressure, and a capital P will be the symbol for power.)

$$p = \frac{F}{A} \qquad F = pA$$

$$P = pAv$$

$$v = \frac{x}{t}$$

(Equation 12.7)

$$P = p\frac{Ax}{t}$$

$$Ax = \text{volume}$$

So Ax/t is the volume flow, and therefore power equals the pressure times the volume flow. This definition is in contrast to the way flow is defined earlier in the book. Earlier it was defined as the mass per unit time. You must be careful in using equations in that you must know the context into which the derivation of a certain equation is embedded. For this reason, the derivations are given along with the equations. To know how to use them, you must know the situations they were derived to model.

Summary

1. Like the moduli learned in Chapter 11, pressure has units of pascals (Pa), which can also be described as force per unit area (N/m²).

2. Hydraulic mechanical advantage allows a force to be multiplied by increasing the output area of a hydraulic press, as shown in Equation (12.3).

3. The buoyant force [Equation (12.4)] on an object is caused by the weight of fluid that the object displaces.

Problem Solving Tip 12.1: You can use buoyant force as just another force when implementing Newton's second law. Simply add it in when drawing your free-body diagram. Thus, an anchor resting on the bottom of a lake would have its weight drawn downward, and the tension of the anchor/chain and the buoyant force would be drawn upward! Notice that when $\mathbf{F} = m\mathbf{a}$ is used on the above example, the tension in the chain will be less than the anchor's true weight. For this reason, objects feel lighter in water than they do in air! They don't lose mass, but instead the water that they displace (by being submerged) "wants" to help you lift the object and thus "lightens your load."

4. Strongly related to the conservation of energy, Bernoulli's equation states that pressure is also conserved as a fluid flows through a pipe. Equation (12.6) relates the outside pressure, the pressure caused by elevation, and the pressure caused by velocity from one point in a pipe to another.

Problem Solving Tip 12.2: When you're solving fluid flow problems using Bernoulli's equation, an extra unknown is often involved. It can almost always be found by remembering that flow remains constant throughout a pipe or by using the continuity equation: $\rho_1 A_1 v_1 = \rho_2 A_2 v_2$. [Note that the densities (ρ) are almost always equal and thus can be canceled from the equation.] Therefore, as the area decreases, the velocity increases. If you use this relationship with Bernoulli's equation, you will almost always have enough information to solve anything that is asked of you.

Key Concepts

ABSOLUTE PRESSURE: The total pressure on an object, or the gauge pressure plus the atmospheric pressure.

BUOYANT FORCE: The force acting up on an object immersed in a fluid that is equal to the weight of the displaced fluid.

FLOW: The rate at which mass passes a given point, or mass per unit time.

GAUGE PRESSURE: The differential pressure between the inside and the outside of a volume.

LAMINAR FLOW: Fluid flow without turbulence.

Important Equations

(12.1) $P = \dfrac{F}{A}$ (Pressure is the force divided by the area over which it is exerted.)

(12.2) $P = \rho g h$ (Pressure at depth is proportional to the depth.)

(12.3) $F_{out} = \left(\dfrac{A_{out}}{A_{in}}\right)F_{in}$ (Hydraulic mechanical advantage)

(12.4) $F_b = \rho_f V_f g$ (Buoyant force)

(12.5) $\text{Flow} = \dfrac{\Delta m}{\Delta t}$ (Mass flow)

(12.6) $P_1 + \rho g h_1 + \dfrac{1}{2}\rho v_1^2 = P_2 + \rho g h_2 + \dfrac{1}{2}\rho v_2^2$ (Bernoulli's equation)

(12.7) $P = p\left(\dfrac{Ax}{t}\right)$ (Power as related to pressure and volume flow)

Conceptual Problems

1. Why is a hydraulic press able to produce a mechanical advantage?
   a. At a given depth in the fluid, the pressure is the same in all directions.
   b. The force that a fluid exerts on a piston is always perpendicular to its face.
   c. The external pressure exerted on the fluid is the same throughout the fluid.
   d. The pressure gets greater as the output piston gets smaller.

2. In a hydraulic jack, the output piston equals the input piston in:
   a. speed
   b. force
   c. work
   d. displacement

3. The pressure at the bottom of a liquid-filled container depends only on:
   a. the shape of the container
   b. the height of the container
   c. the height of the liquid
   d. the area of the liquid surface

4. Buoyant force:
   a. is caused by an increase in pressure with depth
   b. is absent for objects that sink
   c. causes an object to sink if the object has a density smaller than that of the liquid in which it is immersed
   d. equals the volume of the object

5. A bar of soap placed in a water-filled bathtub sinks to the bottom. The buoyant force of the soap is:
   a. less than the weight of the bar
   b. more than the weight of the bar
   c. the same as the weight of the bar
   d. zero

6. Bernoulli's equation is based on:
   a. the conservation of momentum
   b. Newton's third law
   c. Newton's second law
   d. the conservation of energy

7. Adding detergent to water:
   a. increases the surface tension
   b. decreases the surface tension
   c. attracts the dirt into the water and out of whatever is being washed
   d. decreases the buoyancy of the dirt

8. What is the maximum pressure differential that can be achieved by sucking through a straw, and why?

9. What are the gauge pressure and the absolute pressure inside a flat tire?

10. A thumbtack is pushed into the center of a playing card, and the tip of the thumbtack is positioned into the bottom of the hollow axis of an empty spool used to carry thread. If you blow hard through the spool's axis and release the playing card, what will happen, and why?

11. A narrow column of air is blown upward from the exhaust of a vacuum cleaner, and a ping-pong ball is placed into the column of air. The ball rides the column of air with stability and moves back into it if pushed from its center. Why?

12. An ice cube floats in a glass of water that is full to the brim. What happens when the ice melts, and why?

13. A bridge in Sweden allows ships to sail over a highway. When a ship is in the canal over the highway, does the load that the bridge carries increase, decrease, or remain the same? Why?

14. If steel has roughly eight times the density of water, why does a steel ship float?

15. Why is ice on the wings of an airplane more dangerous than ice on its fuselage?

16. The surface of a baseball is rough and carries some air around with it when it spins. Use this fact to explain how a curve ball can be thrown.

Exercises

Pressure

1. A 5-kg brick has dimensions 10 cm × 20 cm × 40 cm. What pressure does the brick exert on a table when it is lying on each face successively?

> *Answer:* 2450 N/m²; 1225 N/m²; 612.5 N/m²

2. A large inner tube is lying on its side on a flat surface. A wooden platform weighing 600 lb is placed upon the inner tube. The pressure inside the inner tube is measured by a gauge as 32 lb/in.² (normal pressure for an automobile tire). (a) How much of the tube is in contact with the surface? (b) If the radius of the tube is 2 ft from the tube's center to the middle of the tube's width, how wide is the contact patch all the way around the tube?

> *Answer:* (a) 18.75 in.²
> (b) 0.124 in.

3. A jet weighing 100,000 lb has a wing with an area of 1000 ft². If the plane is flying level, what must the difference in pressure be between the upper and lower surfaces of the wing to keep the plane in the air?

> *Answer:* 0.694 lb/in.²

4. A plunger is a large, thick rubber cup with a wooden handle used for clearing clogged drains and toilets. (a) If a plunger cup has a diameter of 4.5 in., what is the maximum suction pressure that it can apply to a clogged drain? (b) With how much force must you pull on the handle to get the maximum suction?

> *Answer:* (a) 14.7 lb/in.²
> (b) 233.8 lb

*5. A person is drinking a soft drink from a 64-oz, Magna Slurp cup with a straw. If a person can reduce the pressure inside the straw above the liquid to 80% of atmospheric by vigorous sucking, what is the maximum height that the liquid can rise in the straw?

> *Answer:* $h = 2.07$ m

*6. In the apparatus shown in Figure 12.6, water and oil rise to the indicated heights when the suction pump is turned on. What is the density of the oil?

> *Answer:* $\rho = 833$ kg/m³

*7. A U-shaped tube of constant cross-sectional area is filled with equal volumes of water and mercury, as shown in Figure 12.7. Both the mercury and the water fill a 20-cm length of the tube, and the tube is open to the atmosphere at the top. What is the difference in the height of the upper surfaces?

> *Answer:* $h = 18.6$ cm

Figure 12.6 (Exercise 6)

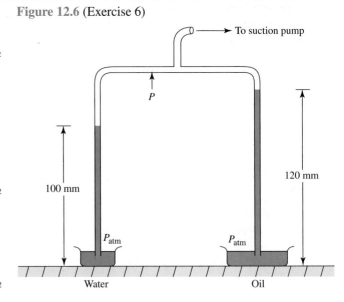

Figure 12.7 (Exercise 7)

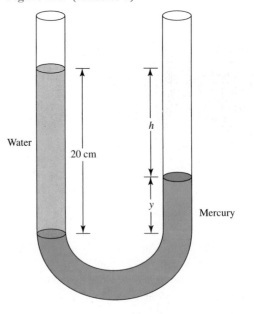

*8. A glass disc is held on the bottom of a cylindrical tube, and the combination is submerged in water until the disc is at a depth of 30 cm. When submerged, the glass disc is released. The tube and the disc have the same diameters, and the walls of the tube are negligibly thin. The tube is open to the atmosphere at the top. What is the maximum thickness that the glass disc can have before it falls away from the tube? The density of glass is 2.5×10^3 kg/m^3.

Answer: t = 20.0 cm

Depth and Pressure

9. A deep-sea submersible dives to a depth of 17,500 ft. What pressure must a round hatch with a diameter of 2.5 ft withstand?

Answer: 5,332,539 lb!

Properties of Pressure in a Fluid and Fluid Flow

10. Water is flowing at a speed of 1.5 m/s through tubing with an inside diameter of 15 mm. What should the inside diameter be if the speed must be reduced to 1 m/s? The rate of flow should remain constant.

Answer: d = 18.4 mm

11. A large pipe with an inside diameter of 40 mm contains a liquid that flows at a speed of 1.5 m/s. The large pipe feeds three smaller pipes with an inside diameter of 15 mm each. What is the flow speed in the smaller pipes?

Answer: v = 3.56 m/s

12. A water tower has a hole 9 mm in diameter 3.5 m below the top of the tower. What is the flow rate through the hole in liters per second?

Answer: Flow = 0.53 liter/s

*13. Gasoline is being siphoned out of a tank as shown in Figure 12.8. The density of gasoline is 680 kg/m^3, and the inside diameter of the hose is 1.5 cm. (a) At what speed does water emerge from the bottom of the hose? (b) What is the flow rate in liters per second?

Figure 12.8 (Exercise 13)

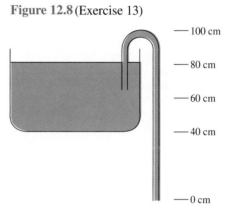

— 100 cm

— 80 cm

— 60 cm

— 40 cm

— 0 cm

Answer: (a) *v* = 3.96 m/s
(b) Flow = 0.7 liter/s

Hydraulic Mechanical Advantage

14. The input piston of a hydraulic jack is $\frac{1}{2}$ inch in diameter, and the output piston is 2 inches in diameter. A jack handle is attached to the input piston. The jack must lift the axle of a car, supporting 1500 lb a height of 9 in. (a) How much work must be done? (b) What must the force on the input piston be for the jack to lift the axle? (c) The input piston moves 1 in. per stroke of the jack handle. How far does the output piston travel? (d) How many strokes of the jack han-

dle are necessary to raise the axle the required amount? (e) How much work is done by the person pumping the lever?

Answer: (a) 1125 ft-lb
(b) 93.6 lb
(c) 0.062 in.
(d) 145 strokes
(e) 1125 ft-lb

Buoyancy

15. A wooden board 9 cm thick floats in still water with its upper surface parallel to the water's surface. If the upper 3.5 cm of the board's thickness is above the water, what is the density of the board?

Answer: ρ = 611 kg/m^3

16. A long, narrow raft 20 m long and 10 m wide has a mass of 1.15×10^5 kg. What minimum depth of water is needed to float the raft?

Answer: h = 57.5 cm

*17. A man is standing on a square raft 2.5 m on a side. If the man has a mass of 75 kg, how much does the raft rise out of the water when he dives off?

Answer: $h_0 - h_A$ = 1.2 cm

*18. A dock 2 m wide and 10 m long for small boats is made from 3-cm-thick pine planks whose density is $\rho = 370$ kg/m^3. Six empty 180-liter drums are lashed to the dock for extra flotation. The drums weigh 18 kg apiece. How much load can the dock carry before the dock's deck is awash?

Answer: $m = 1350$ kg

*19. A 25-kg balloon is filled with 120 m³ of hydrogen gas. What is the maximum load that the balloon will lift? The density of air is 1.3 kg/m³, and the density of hydrogen gas is 0.09 kg/m³.

Answer: Load $= 1178$ N

*20. A submarine with a mass of 1200 tons fully laden (with empty ballast tanks) is designed so that when its ballast tanks are full, it has a buoyancy equal to that of water. When the ballast tanks are empty, the submarine has a buoyancy of only 0.92 times that of water, with only 10% of its volume out of the water. What is the capacity of the ballast tanks?

Answer: $V_w = 4274$ ft³

*21. A thin-walled cylindrical tank with a mass of 3.5 kg and a diameter of 20 cm floats in water with its bottom flat side parallel to the water's surface. Gasoline, whose density is 680 kg/m³, is poured into the tank until the tank is half-submerged in the water. How far from the top edge is the gasoline? The tank has a height of 0.5 m.

Answer: $y = 0.296$ m

*22. A widget has an apparent mass of 271 g when immersed in an oil of density 700 kg/m³, and 250 g when immersed in water. What are (a) the mass and (b) the volume of the widget?

Answer: (a) $m = 637$ g
(b) $V = 3.87 \times 10^{-4}$ m³

*23. A hydrogen-filled weather balloon has a 5-m diameter and a total weight of 30 kg, including its instrumentation, but excluding the hydrogen. Before release, the balloon is tethered to the ground by a line, and the wind blows it to the side so the line makes an angle of 30° with the vertical. What is the tension in the line? The density of air is 1.3 kg/m³, and that of hydrogen 0.009 kg/m³.

Answer: $T = 557$ N

*24. A figurine is alleged to be solid gold, but it could be lead plated with a thin layer of gold. The figurine has a mass of 0.475 kg when weighed in the air, and the scale reads 0.437 kg when the figurine is immersed in water. (a) Is the figurine solid gold? (b) If not, what percentage by mass is lead?

Answer: (a) No
(b) 81.25%

*25. A block of pine with a density of 550 kg/m³ floats on a still pool. The block has dimensions 20 cm × 40 cm × 10 cm. (a) How much of the block is submerged? (b) If a block of iron is fastened *underneath* the pine, how much is needed to give the block a neutral buoyancy?

Answer: (a) $d = 5.5$ cm
(b) $m = 4.13$ kg

*26. An object of mass m and density ρ is submerged in a liquid with a smaller density ρ_o. Show that the effective weight of the submerged object is

$$W_{eff} = mg\left(1 - \frac{\rho_o}{\rho}\right)$$

Bernoulli's Equation

27. A pipe tilted at an angle of 60° with the horizontal is 3 m long and has an inside diameter of 5 cm at the low end. The pipe's diameter increases gradually until its diameter at the high end is 8 cm. If water is pumped into the low end at a speed of 7 m/s and a pressure of 2.5 bars, what are (a) the speed and (b) the pressure of the water at the upper end of the pipe?

Answer: (a) $v = 2.73$ m/s
(b) $P = 2.45$ bars

*28. A U-shaped tube contains water and is open at each end to the atmosphere. Air is blown over the top of one end with a velocity v. Develop a relationship between the air speed and the difference in height of the water column.

Answer: $v = \sqrt{\dfrac{2\rho_w g \, \Delta h}{\rho_a}}$

Power and Fluid Flow

29. Water emerges from the nozzle of a fire hose with a speed of 15 m/s and a flow rate of 45 liters/s. (a) What force does it take to hold the nozzle steady? (b) What is the required power of the pump, assuming a 60% efficiency?

Answer: (a) $F = 675$ N
(b) $P = 16.9$ kW

30. A 750-hp pump throws a column of water 90 m into the air in the fountain of Zurichsee in Zurich, Switzerland. (a) What is the speed of the water as it leaves the mouth of the fountain? (b) How many kilograms of water are thrown into the air per second, assuming a pump efficiency of 65%?

Answer: (a) $v = 42$ m/s
(b) Flow $= 206$ kg/s

Thermodynamics

The next three chapters (Chapters 13–15) deal with *thermodynamics,* or the science of heat and heat flow. Heat is the macroscopic manifestation of the kinetic energy of a collection of microscopic particles that make up all matter (atoms and molecules). To understand heat, you must understand kinetic energy, studied in Chapter 5 in Part I, "Mechanics." Thermodynamics is *not* a separate topic in physics, but a continuation of the mechanics already studied. The reason thermodynamics is often thought of as a separate topic, though, is that the mechanics that is fundamental to the study of heat and temperature is often subordinated to the background by the jargon associated with thermodynamics, such as *heat* and *temperature.* Both heat and temperature are manifestations of the kinetic energy of a group of molecules and atoms, and an intuitive feeling for the concepts of thermodynamics *must* be based on an intuitive feeling for the concept of work and energy. Without the latter, comprehension of the former is impossible.

P A R T   I I

Hot air balloons are buoyant and float because the density of the hot air inside is less than that of the cooler air surrounding the balloon. The expansion of the heated air lowers its density.

H Heat

At the end of this chapter, the reader should be able to:

1. Calculate the change in length (or volume), given the material and the change in temperature.

2. Convert quickly between **Fahrenheit** and **Celsius temperature scales.**

3. Solve for the average velocity of the molecules in a known gas if given the ambient temperature of the gas.

4. Define *absolute zero.*

5. Calculate the new temperature of a gas if both the volume and the pressure change by known amounts.

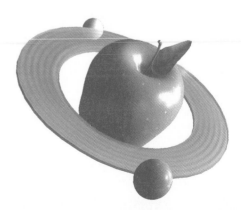

13.1   Hot and Cold

Celsius or Centigrade temperature scale: The temperature scale tied to the phase changes of water.

Thermometer: An instrument used to measure temperature.

Hot and cold are sensations that we can experience directly. We usually experience hot and cold in terms of whether or not we are comfortable. If it's too hot, we're uncomfortable. (Turn on the air conditioner!) If it's too cold, we're also uncomfortable. (Turn on the heat!) If we're comfortable, we usually don't notice the temperature. This comfortable temperature is loosely defined as *room temperature,* around 70 degrees Fahrenheit (70°F). If the temperature is too far above or below room temperature, it is dangerous, because we could either die of heat prostration or freeze to death.

Heat and temperature are linked, but what is hot and what is cold? *Hot* and *cold* are relative terms. During the winter, a day in the 50s (degrees Fahrenheit) is considered balmy and warm, but during the summer, the same 50° temperature is considered quite cool. Hot and cold are very subjective terms, and a temperature scale is used to quantify them. Water is a necessary and commonplace substance with which every living thing (on this planet anyway!) is familiar. Most of our body weight is water. Water can exist as a solid, liquid, or gas. Solid water is called *ice* and is usually considered cold. Gaseous water is called *steam* and is usually considered quite hot. Liquid water can be hot or cold depending on its temperature relative to what's considered comfortable (70°F or room temperature). By starting with an ice cube and adding heat to it, we can make it go from a solid to a liquid to a gas. We can also define a temperature scale by the phase changes of water. *The **Celsius (or Centigrade) temperature scale** is tied to the phase changes of water: 0°C is defined as the temperature of a melting ice cube, and 100°C is defined as the temperature of boiling water.* So somehow temperature is tied to the heat content of a substance, be it solid, liquid, or gas. The higher the temperature of something, the more heat it contains.

*The instrument used to measure temperature is a **thermometer**.* A common thermometer is a glass tube filled with either alcohol or mercury. As the temperature increases, the liquid expands. Since the glass tube expands much less than the liquid, the liquid moves up the tube with a rising temperature and down the tube with a falling temperature. Because both the expansion and the contraction of the liquid are linear, the temperature scale is also linear. Metals also expand and contract with heat. Another type of thermometer uses a coil spring that expands and contracts with the temperature. A pointer attached to this spring moves with the expansion and contraction of the spring. This movement is also linear with the temperature.

Both of the above-mentioned thermometers have thermal expansion of some substance in common. A thermometer can also be based on a device called a *thermocouple.* If two dissimilar metals are placed in contact, a small voltage exists across the junction. If the junction is heated, this small voltage difference increases with the temperature, and the increase can be calibrated in terms of temperature.

The color of an object can also be used as an indication of temperature. For example, if a piece of metal is heated until it just begins to glow, it will be glowing red. As it is heated further, it will go from red to orange to yellow to white to blue. Of course, this "thermometer" would be used for temperatures well above that of boiling water! Two pieces of iron that have been heated red-hot can be welded on a forge by hammering them together. Astronomers can use the color of a star to predict its temperature. A blue star is much hotter than a red one. Our sun is a relatively cool, and therefore long-lived, red star. Because of our sun's longevity, a planet orbiting this star had a chance to have life evolve upon its surface. A blue star has a much higher temperature than a red star and must burn its fuel at a much faster rate to generate the high temperature. As a result, it has a shorter life span in general, and the chances are smaller that a planet in orbit around such a star will have life.

13.2   Thermal Expansion

By far the most common thermometer found in the home is the one whose operating principle is thermal expansion. Most materials react to a change in temperature by a change in their volume. Most materials undergo an increase in volume with a temperature increase; a few actually decrease in volume when the temperature increases. Also, in most materials the volume increase or decrease is linear: If the temperature increase (or decrease) is doubled, the volume increase (or decrease) is doubled. This section will concentrate upon this common characteristic of most materials, although you should be aware that not all materials act in this way. Uranium, for example, when fabricated into fuel rods for a reactor, will expand with an increase in temperature, but it will not always return to its original volume when the temperature is lowered. In the next temperature-increase cycle, it gets bigger still. As a result, early fission reactors had many problems, including cracking of the cladding, because this characteristic of uranium was unknown at the time.

When a material is heated, its entire volume will undergo expansion. Thermal expansion is subdivided into two different categories: linear expansion and volume expansion. Often we are interested only in how a material changes in one dimension when heated, such as its length. This *linear expansion* is used for solids. For example, if a long iron rod is heated, it will increase in length. However, its diameter will also increase, although we may not be interested in the diameter's increase. Engineers must take these increases and decreases into account when designing a structure such as a long suspension bridge that uses steel cables. The difference in length caused by summer and winter temperatures could be as much as a meter or more. If not taken into account, the constant changing in length caused by temperature fluctuations throughout the year could fatigue and weaken the structure.

The linear expansion of a material is proportional both to the temperature and to the initial length of the material. The longer the material is initially, the greater its linear expansion when heated, assuming that the entire object is uniformly heated. If only a small segment of a rod is heated, for example, then only that small segment will expand thermally. Different materials will expand different amounts when exposed to the same temperature change. Each material has a characteristic coefficient of linear expansion, and its expansion when heated is also proportional to it. For instance, if an internal combustion engine has an aluminum piston in a steel bore, both piston and bore are subjected to the same temperature rise from the burning fuel. The piston will expand more than the steel bore, and if the clearance between the piston and the bore is too tight, the piston will become too large for the bore and become stuck in the bore. This is obviously an unwanted condition!

Pulling all of these proportionalities together, we have the equation for linear expansion.

$$\Delta L = aL_o \, \Delta T$$

(Equation 13.1)

where ΔL is the change in length, a is the coefficient of linear expansion, L_o is the original length before the material was heated (or cooled), and ΔT is the temperature difference between initial and final states. Some coefficients of linear expansion for common materials are given in Table 13.1. You can glean certain trends by examining the table. For example, metals expand most. Aluminum expands twice as much as steel, and both iron and steel have the same coefficient of expansion. (They should, since steel is mostly iron with just a little carbon and other materials alloyed with it to make it steel.)

Table 13.1 Coefficients of Linear Expansion

Material	Coefficient (°C)	Coefficient (°F)
Aluminum	$2.4 \times 10^{-5}/°C$	$1.3 \times 10^{-5}/°F$
Brass	$1.8 \times 10^{-5}/°C$	$1.0 \times 10^{-5}/°F$
Concrete	0.7 to 1.2 $(\times 10^{-5}/°C)$	0.4 to 0.7 $(\times 10^{-5}/°F)$
Copper	$1.7 \times 10^{-5}/°C$	$0.94 \times 10^{-5}/°F$
Iron	$1.2 \times 10^{-5}/°C$	$0.67 \times 10^{-5}/°F$
Lead	$3.0 \times 10^{-5}/°C$	$1.7 \times 10^{-5}/°F$
Quartz	$0.05 \times 10^{-5}/°C$	$0.028 \times 10^{-5}/°F$
Silver	$2.0 \times 10^{-5}/°C$	$1.1 \times 10^{-5}/°F$
Steel	$1.2 \times 10^{-5}/°C$	$0.67 \times 10^{-5}/°F$

Quartz and other glasses expand very little relative to the metals and concrete. Concrete can have steel reinforcing rods within it. With the rods, the concrete will (or must) expand as much as the steel embedded within. Without the reinforcing rods, the concrete won't expand as much.

EXAMPLE 13.1

(a) By how much does a steel girder 9 m long used in the construction of skyscrapers contract between its normal summer temperature of 25°C and its average winter temperature of 5°C? (b) What forces does this contraction impose on the surrounding structure?

SOLUTION
(a)
$$-\Delta L = aL_o \, \Delta T$$

The negative sign is there because the material will contract when it is cooled down.

$$\Delta T = T_f - T_i$$

$$-\Delta L = aL_o(T_f - T_i)$$

$$T_f = 5°C \qquad T_i = 25°C$$

$$-\Delta L = (1.2 \times 10^{-5}/°C)(9 \text{ m})(5°C - 25°C)$$

$$\Delta L = 2.16 \text{ mm}$$

This amount is large enough to be easily read on a ruler.
(b) To find the force that this contraction causes on the surrounding structure, we must use Young's modulus. The cross section of many girders is in the shape of a capital letter I, and thus they're called *I beams*. The cross-sectional area of this beam is 0.03 m^2.

$$Y = \frac{\text{stress}}{\text{strain}} \qquad Y = \frac{F/A}{\Delta L/L_o} \qquad F = YA\frac{\Delta L}{L_o}$$

$$F = (20 \times 10^{10}\ \text{N/m}^2)(0.03\ \text{m}^2)\left(\frac{2.16\ \text{mm}}{9\ \text{m}}\right)$$

$$F = 1{,}400{,}000\ \text{N}$$

or over 160 tons!

A liquid can flow and will take the shape of any container that holds it. Thus, the entire volume of the liquid must be taken into account when it expands because of heating. The equation for this *volume expansion* is mathematically the same as for linear expansion.

$$\Delta V = bV_o\,\Delta T \qquad\qquad \text{(Equation 13.2)}$$

where b is the coefficient of volume expansion, V_o is the initial volume, and ΔT acts the same as in linear expansion. Table 13.2 gives the coefficient of volume expansion for some common materials. Note that the fluids most commonly used in thermometers (ethyl alcohol and mercury) expand much more than the material that holds them (usually glass). Water isn't used in a thermometer because it freezes at too high a temperature. Ethyl alcohol is commonly used in thermometers because its freezing point is much lower than that of water ($-173°$F as opposed to $32°$F), and it isn't as toxic as mercury. (Mercury as a liquid is relatively innocuous, but mercury vapor is a *very* dangerous poison.) Alcohol is mixed with a red dye so that it is easily seen in the glass tube of the thermometer.

Compared to the other materials listed in Table 13.2, water behaves in a weird fashion around its freezing point. From 0°C to 4°C, the volume of water will *decrease* as it is heated. Only then will its volume begin to behave as expected in this section and increase with temperature. Equally unusual is the fact that water increases in volume when it freezes, which is opposite to how most materials act when frozen. This is why ice floats and water pipes can burst when frozen. Because ice floats, a body of water freezes from the top down. Ice is a good insulator, and once the surface of a pond is frozen, it is much harder to freeze the water underneath the ice. For this reason, many bodies of water are still liquid during the winter below a relatively thin skin of ice. If water behaved as most substances do, ice would sink, and a body of water would freeze from the bottom up. It would be relatively easy to completely freeze a body of water solid, and life on this planet would be radically different from what it is today.

Table 13.2 Coefficients of Volume Expansion

Material	Coefficient (°C)	Coefficient (°F)
Ethyl alcohol	$7.5 \times 10^{-4}/°$C	$4.2 \times 10^{-4}/°$F
Glass	$0.2 \times 10^{-4}/°$C	$0.1 \times 10^{-4}/°$F
Ice	$0.5 \times 10^{-4}/°$C	$0.3 \times 10^{-4}/°$F
Mercury	$1.8 \times 10^{-4}/°$C	$1.0 \times 10^{-4}/°$F
Water	$2.1 \times 10^{-4}/°$C	$1.2 \times 10^{-4}/°$F

EXAMPLE 13.2

A thermometer's bulb contains a volume of 5×10^{-8} m^3 of ethyl alcohol at 72°F. If a temperature increase of 60°F is to be represented by a 12-cm rise in the height of the column of alcohol, what must the column's diameter be?

SOLUTION

$$\Delta V = bV_o \Delta T$$

$$\pi \left(\frac{D}{2}\right)^2 \Delta L = bV_o \Delta T$$

$$D = \sqrt{\frac{4bV_o \Delta T}{\pi \Delta L}}$$

$$D = \sqrt{\frac{4(4.2 \times 10^{-4}/°F)(5 \times 10^{-8} \text{ m}^3)(60°F)}{\pi(0.1 \text{ m})}}$$

$$D = 0.127 \text{ mm}$$

When a material expands because of heating, the same mass of material is still present after the temperature rise as existed before the heating started. In other words, mass is conserved. Thus, the density of the material falls as the material expands. This effect, though, is so small in an example such as the one above that it can be ignored. The glass also expands as the temperature rises, but its expansion is so much less than that of the alcohol that it can be ignored in the first approximation.

13.3   Linear Temperature Scales

Most thermometers used to measure air temperature display temperature in degrees Fahrenheit and/or degrees Celsius. The two scales are often displayed together, so that when you view the top of the red (or silver) column, the right side might be calibrated in degrees Fahrenheit, and the left in degrees Celsius. A linear formula will convert degrees Celsius into degrees Fahrenheit, and vice versa.

The word *linear* has been used often in this chapter so far, but what does it actually mean? A linear amplifier is one in which if the input is doubled, the output is doubled; if the input is quadrupled, then the output is quadrupled; and so on. In a nonlinear amplifier, if the input is doubled, then the output might be squared or increased by four times. If the concept of linearity were depicted graphically, then for a linear amplifier, the plot of the input versus the output would be a straight line. A nonlinear amplifier's input-versus-output plot would be a curved line. A liquid thermal-expansion type of thermometer is linear because if the temperature doubles, then the height of the liquid in the column will rise to twice its original height. Thus, both the Celsius and the Fahrenheit temperature scales are also linear. To find the conversion formula, we need to know how to use the equation of a straight line.

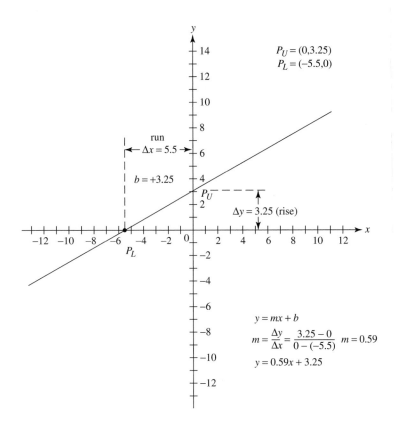

$P_U = (0, 3.25)$
$P_L = (-5.5, 0)$

$y = mx + b$

$m = \dfrac{\Delta y}{\Delta x} = \dfrac{3.25 - 0}{0 - (-5.5)}$ $m = 0.59$

$y = 0.59x + 3.25$

Figure 13.1 Equation of a Straight Line

A straight line is shown in Figure 13.1, and its formula is

$$y = mx + b$$

Normally the horizontal axis is the x coordinate, the vertical axis is the y coordinate, and b is the y intercept. (When $x = 0$, then $y = b$.) If the line goes through the origin, then the y intercept is 0. The slope of the line m tells how "tilted" the line is. The slope is the ratio of the rise over the run. If a line goes 2 units up when it travels 2 units to the right, then the slope is $+1$. This is also the tangent of 45°, so a line with a slope of 1 makes an angle of 45° with the horizontal. Mathematically, the slope is

$$m = \frac{\Delta y}{\Delta x}$$

If the slope is positive, then the line rises from left to right. If the slope is negative, then the line falls from left to right. *Anything* linear will have the equation of a straight line.

To find the linear conversion formula to convert degrees Fahrenheit to degrees Celsius, we can directly apply what we've learned about a straight line in the paragraph

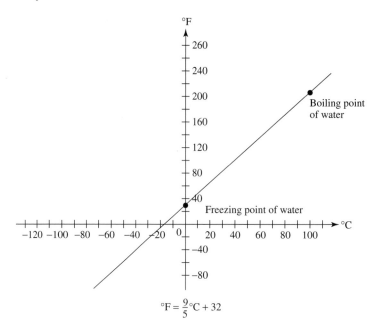

Figure 13.2 Conversion between °*C* and °*F*

$$°F = \frac{9}{5}°C + 32$$

above. First the axes must be labeled. In this instance the vertical axis will be degrees Fahrenheit, and the horizontal axis will be degrees Celsius, as shown in Figure 13.2. The equation of a straight line is to be the template for our conversion formula.

$$y = mx + b$$

Degrees Fahrenheit is substituted for y because the vertical axis displays degrees Fahrenheit. Degrees Celsius is substituted for x because the horizontal axis displays degrees Celsius.

$$°F = m°C + b$$

The freezing point of water can be used to determine the y intercept. At 0°C (the freezing point of water), the temperature in degrees Fahrenheit is 32°F. Therefore,

$$b = 32°F$$

$$°F = m°C + 32°F$$

The slope of the line must now be calculated.

$$m = \frac{\Delta y}{\Delta x}$$

The boiling and freezing points of water are used to calculate the slope.

$$\frac{\Delta y}{\Delta x} = \frac{212°F - 32°F}{100°C - 0°C}$$

$$\frac{\Delta y}{\Delta x} = \frac{9}{5}$$

We now substitute 9/5 for m, and we have a formula that converts from degrees Celsius into degrees Fahrenheit.

$$°F = \frac{9}{5}°C + 32$$ (Equation 13.3)

To convert from degrees Fahrenheit into degrees Celsius, we solve the above formula for °C on the left.

$$°C = \frac{5}{9}°F - 17.8$$

We could also have derived this formula directly by initially labeling the vertical axis as degrees Celsius and the horizontal axis as degrees Fahrenheit.

13.4   Atoms, Molecules, and Heat in Solids and Liquids

A solid surface, such as a wall, might be built out of many subunits, such as bricks. Each brick is firmly anchored in place relative to the other bricks in the wall. They can't move freely from one position to another. However, if your hand is on the wall and a heavy truck rolls by on the road outside, then you might feel vibrations in the wall from the truck. A solid block of copper is very much like the brick wall, except the "bricks" are much smaller. A block of copper is built up of individual atoms of copper. Copper is an element and can be found in the periodic table. (The periodic table lists the elements, each of which is an atom, that make up the material world around us.) In a block of copper, each copper atom is fixed in a lattice and can't move relative to its neighbors, except in the sense of the vibrations that the entire lattice undergoes. (A lattice is a framework. A brick wall has a rectangular lattice because most bricks are rectangular.) The bricks in a wall are cemented together with mortar, and that gives the wall its rigidity. The atoms of copper are "cemented" together by electrical forces. These forces act just like tiny springs. These springs give the copper block its rigidity.

The periodic table of the elements is given in Appendix E in the back of the book.

Figure 13.3 is an illustration of small spheres, one at each of the eight corners of a cube, and bound together with springs. This structure is a good analogy of the copper atoms in a block of copper. Of course, this is just a small portion of the block. For the entire block,

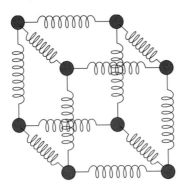

Figure 13.3 Atoms "Cemented" Together by Electrical Forces in a Solid

the cube is extended in all three dimensions as far as needed. The original cube is surrounded by cube clones in all directions.

What does this have to do with heat? The spring cube fixes each sphere so that it can't move very far out of position relative to its neighbors. However, since the spheres are connected by springs, the whole cube can jiggle and vibrate quite a bit, a little like Jello. The vibrations in the cube *are* the heat. If the cube is at absolute zero temperature, then there are no vibrations present in the lattice. As the cube warms up, the vibrations become stronger and stronger. If the temperature is high enough, then the solid melts. If the vibrations are strong enough, then the spheres have enough energy to leave the lattice and roam about freely. The springs are no longer strong enough to maintain the lattice structure. When we studied simple harmonic motion in Chapter 9, a mass on the end of a spring oscillated back and forth at a characteristic frequency. As more energy was fed into the system, the amplitude of the oscillations became bigger and bigger. The same is true of the copper atoms. They oscillate about an equilibrium position, and as the block heats up, the oscillations become bigger and bigger. If the block doesn't melt, then the copper atoms maintain their positions relative to one another, but they are in constant motion. If the oscillations become bigger as the block is heated, then the entire block expands slightly. This process is what is responsible for thermal expansion.

The next step up from a solid is a liquid. If we take solid water (ice) and add enough heat, then the ice melts and becomes a liquid. A solid is one phase of matter; liquid is another. In a liquid the elemental building blocks still exist, but instead of being bricks in a wall, they become like kernels of rice in a bag. The kernels of rice are all touching one another, but unlike in a solid where each building block is firmly cemented to one another creating a lattice, the rice kernels can easily change positions relative to one another. Water is the most common liquid on this planet. The "kernels" are molecules. Each molecule of water contains two atoms of hydrogen and one of oxygen. An atom has a positively charged nucleus and a negatively charged cloud of electrons about it. Since a molecule is composed of atoms, it too has a cloud of electrons around it. Like charges repel one another, and each molecule's electron cloud repels that of every other molecule around it. Thus, the molecules in a liquid are constantly bouncing off one another and milling around aimlessly.

A liquid can be thought of as many small spheres (the molecules) with springs attached closely packed together and bouncing off one another, as shown in Figure 13.4. The more heat that is added to a liquid, the faster the molecules move relative to one another, and the more vigorously they collide; also, the average distance from one molecule to another increases. This is the reason behind thermal expansion of liquids. Since a liquid is more loosely bound than a solid, its coefficient of thermal expansion should be greater. If you compare Table 13.1 (Coefficients of Linear Expansion) with

Table 13.2 (Coefficients of Volume Expansion), you'll notice that the coefficient for volume expansion for liquids is on average roughly 10 times that of linear expansion for solids. If too much heat is added to a liquid, the molecules in it acquire too much energy and bounce right out of the liquid. In other words, the liquid boils. The liquid then enters the third phase of matter, a gas.

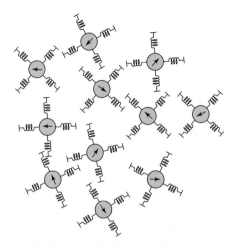

Figure 13.4 Molecules or Atoms in a Liquid

13.5   The Kinetic Theory of Gases and Absolute Zero

We live at the bottom of a sea of gases. This sea of gas, commonly called "the air that we breathe," is a combination of two gases: oxygen and nitrogen. Air is approximately 20% oxygen and 80% nitrogen by volume. Our body needs the oxygen in the air to sustain the slow combustion that is our metabolism. (The nitrogen is relatively inert, and although we breathe it in, it takes no part in the body's metabolism. A trace of other molecules are also present in air, but they are negligible compared to the two principal components.) Both oxygen and nitrogen are elements, but in air both are found in molecular form. Oxygen is found as O_2 (two atoms of oxygen bound together), and nitrogen is found as N_2 (two atoms of nitrogen bound together). You can easily move about in air as if it weren't there; but if you move fast enough, you feel a breeze across your skin. The breeze is made up of the individual molecules in the air hitting your skin.

The model for a gas is that of a collection of very tiny billiard balls making perfectly elastic collisions with each other and the walls of their container. Remember that a perfectly elastic collision conserves energy, and that none of the energy is lost to friction. In other words, a gas is like a perpetual motion machine. There is no friction to slow it down. In earlier chapters, we learned that the work lost to friction went into heat. This heat warmed something up, and we now know that this heat causes molecules and atoms to jiggle faster, or mill around more violently, or fly around more swiftly. At the molecular and atomic levels, heat is the system's internal energy, and the system is without friction. Internal energy can be added to or subtracted from the system, but there is no "friction" to cause it to wind down and stop. (The planets orbiting the sun are also like a perpetual motion machine. Since there is no air friction, and since the gravitational force of the sun does no work on the planets, the planets will theoretically revolve around the sun forever.) The more energy in the system, the faster the billiard balls move, and the more violently they collide. *A gas is a system macroscopically described by its temperature, but the microscopic cause of the **temperature** is the average kinetic energy of the molecules of the gas, which exchange energy through perfectly elastic collisions.*

Temperature: The average kinetic energy of the molecules of a gas.

Living at the bottom of this sea of air, we feel an even pressure over the entire surface of our body. The reason is that our body is huge relative to the size of the individual molecules striking it from all sides. On average, there are as many molecules striking our left side with the same force as there are striking our right side, and we feel no net force. However, the elastic, billiard-ball nature of the gases in our atmosphere can be seen through a phenomenon known as ***Brownian motion.*** Smoke is made up of many microscopic particles that are readily suspended in the air. If there is no breeze, the smoke from a fire will hang above the fire almost indefinitely. If a little smoke from a match is blown into a small chamber under the objective of a microscope, the small smoke particles can be seen individually. But the behavior of the smoke particles is strange. They jerk around every which way in a very erratic manner instead of just lazily floating in the air. The reason is that the smoke particles are so small that the differences between the molecules hitting each side don't average out to zero, and there is a net force on one side or another. One side always has a greater force imparted to it through the impulse of the colliding molecules than the other, causing the particles to jerk about and act in a jerky manner.

Brownian motion and the pressure on the top of a piston that forces the piston down in an automobile engine involve the same motion. After the gas-air mixture in the cylinder ignites, the gas in the cylinder is at a very high temperature. The molecules of this hot gas are moving at a very high velocity. When they strike the top of the piston, each one imparts an impulsive force to the piston's top. Since the piston can move in the cylinder, this impulsive force drives the piston down the cylinder bore. The pressure on the other side of the piston is near 1 atm, but this value is so small relative to the pressure on the hot gas side of the piston that it is negligible.

The situation in the piston is similar to the one shown in Figure 13.5. The rather violent Brownian motion does useful work by turning the wheels of the automobile. (Brownian motion was discovered by a fellow named Brown. However, it was explained initially by a fellow named Albert Einstein, who also explained the photoelectric effect, which someone else first noticed. Einstein is best known for his theories of special and general relativity, but the Nobel Prize that he received was for his discovery of the causes of Brownian motion and the photoelectric effect. A Nobel Prize is awarded for an actual discovery, not a theory. Einstein's theory of relativity therefore was ineligible for a Nobel

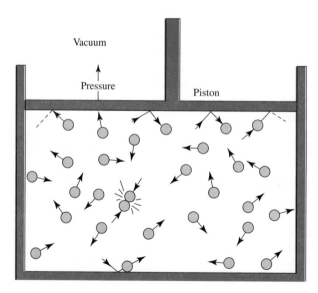

Figure 13.5 Pressure from Molecules in a Gas on a Piston

Prize. As another aside, Swedish chemist Alfred Nobel, founder of the Nobel Prize, was the inventor of dynamite, which also causes the violent expansion of a gas through the heat of an explosion!)

Qualitatively, the temperature of a gas is related to the average kinetic energy of the molecules that make up the gas. This qualitative statement can be stated quantitatively as a formula.

$$\frac{3}{2}kT = \text{KE}_{\text{avg}}$$

$$\frac{3}{2}kT = \left(\frac{1}{2}mv^2\right)_{\text{avg}}$$

$$\frac{3}{2}kT = \frac{1}{2}m(v^2)_{\text{avg}} \qquad \text{(Equation 13.4)}$$

where k is known as *Boltzmann's constant* and has a value of 1.38×10^{-23} J/K; T is the temperature in kelvins (K); and 0 K is absolute zero. (You can't get any colder!) The Kelvin temperature scale is an absolute scale because it is linked to the average kinetic energy of the molecules that make up the gas. In Equation (13.4), when $(v^2)_{\text{avg}} = 0$, then the temperature T must also be zero. We can calculate K from °C simply by adding 273.

$$\text{K} = {}^\circ\text{C} + 273 \qquad \text{(Equation 13.5)}$$

Thus, absolute zero is −273°C, or −459.4°F.

Not all of the molecules in a gas move at the same speed or in the same direction. They move as in Figure 13.5. Some of the molecules are moving faster than average, and some more slowly. The direction is also different for each molecule. Some are moving up, some down, some sideways, and some into or out of the page. Because of the random character of the motion of each molecule, this motion must be averaged statistically, because temperature is the average effect of the molecules' kinetic energy. (Again, our bodies are so much larger than an individual molecule that what we sense as the air temperature is actually the average kinetic energy of the collection of molecules in the gas. We can't feel the effect of a single molecule hitting our body.) If the direction of the motion of the molecules were averaged out, it would be as if one-third of the total number were moving left and right in Figure 13.5, one-third were moving up and down, and one-third into and out of the page. *Such division of molecular motion into three spatial directions is called **equipartition of energy*** and is responsible for the 3 in the $\frac{3}{2}kT$ in Equation (13.4).

Space has three dimensions, each of which adds equally to the average energy as a whole. Figure 13.6 shows the distribution of velocities in a sample of oxygen gas at 0°C (273 K). Some of the molecules are almost motionless, and some are traveling much faster than the root-mean-squared (rms) speed. The rms speed is not the same as the simple arithmetic average. With an arithmetic average, if a particle is moving up at 5 m/s, and if another particle is moving down at 5 m/s, then the average velocity is zero. However, we know that the average energy of the system of two particles is *not* zero. Since temperature is a function of kinetic energy, the simple arithmetic average speed can't be used to calculate the temperature. If Equation (13.4) relating absolute temperature to kinetic energy is used to calculate the speed of a molecule, the calculated speed is the rms speed. (More will be said about rms averages in Chapter 21 on sinusoidal currents and voltages.)

Equipartition of energy: The principle stating that the motion of molecules and atoms in each of the three spatial dimensions add equally to give the internal energy of a sample of gas.

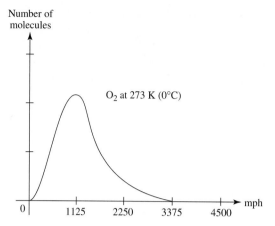

Figure 13.6 Velocity Distribution of Molecules in a Gas

In Figure 13.6, the greatest percentage of molecules are traveling at roughly 1125 mph. The speeds taper off on either side of the peak. To the left they taper off to zero, and to the right they taper off in terms of the percentage of molecules with the higher speeds to 3375 mph. For a given sample of gas, the area under the curve would total 100% of the gas.

If water is left long enough in an open pan at 70°F, all of the water will dry up, or evaporate. The water is too cool to boil, but a small fraction of the water molecules have enough speed and therefore enough energy to leave the water and become water vapor, a gas. As water molecules leave the liquid, its volume becomes smaller. Since there are always some molecules in the liquid that statistically have enough energy to leave, they will do so until none of the original water in the pan is left, even though the water never boils. This principle is the one behind clothes drying on the clothesline; it is a statistical phenomenon. The distribution of molecular speeds in a liquid is what allows water to evaporate.

Our sun is a great ball of gas, mostly hydrogen. The great pressure at the center causes a temperature of millions of degrees and an extremely high density. This high temperature causes the hydrogen atoms to come together so violently that they fuse into helium atoms, giving off energy in the process. This atomic fusion of hydrogen to helium is the source of the sun's energy. Oddly enough, however, the ignition temperature for fusion is so high that the temperature at the center of the sun is still too low to cause fusion. The speed distribution of the atoms at the sun's center are curved as in Figure 13.6, and only those that have speeds to the far right of the peak participate in the fusion of hydrogen to helium.

EXAMPLE 13.3 MOLECULAR SPEEDS

What are the rms speeds of the oxygen and nitrogen molecules in a sample of air at room temperature (20°C)? A molecule of O_2 has a mass of 5.31×10^{-26} kg, and a molecule of N_2 has a mass of 4.65×10^{-26} kg.

SOLUTION
First Equation (13.4) relating the absolute temperature of a gas to its average kinetic energy must be solved for the rms velocity.

$$\frac{3}{2}kT = \frac{1}{2}m(v^2)_{avg} \quad \text{(where the subscript "avg" is the same as "rms")}$$

$$v_{rms} = \sqrt{\frac{3kT}{m}}$$

The molecules of N_2 have the same average kinetic energy as the molecules of O_2, but because the oxygen molecules are heavier, they must have a lower rms speed.

$$T = 273 \text{ K} + 20°C \qquad T = 293 \text{ K}$$

$$v_{rms} = \sqrt{\frac{3(1.38 \times 10^{-23} \text{ J/K})(293 \text{ K})}{5.31 \times 10^{-26} \text{ kg}}}$$

$$v_{rms} = 478 \text{ m/s}$$

The above rms speed is for a molecule of oxygen. The speed for molecules of nitrogen is higher.

$$v_{rms} = \sqrt{\frac{3(1.38 \times 10^{-23} \text{ J/K})(293 \text{ K})}{4.65 \times 10^{-26} \text{ kg}}}$$

$$v_{rms} = 511 \text{ m/s}$$

13.6 The Ideal Gas Law

Early in this chapter, thermal expansion was explained for two of the three phases of matter only: solids and liquids. Nothing was said about gases, which also expand when heated. We said nothing about gases earlier because gases exhibit a peculiar property: A given number of molecules or atoms of a gas will fill the same volume at a given temperature and pressure *regardless of the composition of the gas*. At STP (standard temperature and pressure), 6.02×10^{23} atoms of neon gas (an inert element) will fill the same volume as the same number of air molecules, which is made up of two gases, nitrogen and oxygen, in molecular form. The number 6.02×10^{23} is the number of molecules or atoms in a mole of gas. A *mole,* a reference volume that chemists use, is 22.4 liters at STP. By definition, a mole (abbreviated mol) of gas contains 6.02×10^{23} molecules or atoms of a gas. The number 6.02×10^{23} is called *Avogadro's number.* Because of this unusual property, each gas does not have a separate coefficient of expansion as do solids and liquids. Rather, gases are governed by the ideal gas law, which has the mathematical form

$$PV = nRT \qquad\qquad \text{(Equation 13.6)}$$

where P is pressure, V is volume, n is the number of moles of the gas, R is the gas constant (0.0821 liter-atm/mol-deg), and T is the temperature. The temperature scale used for the ideal gas law is the absolute scale, the Kelvin scale. The three things that usually change (variables) are the pressure, volume, and temperature. The constant terms are the number of moles (n) and the gas constant. When the gas law is used, we'll be looking at how one of the variables changes between an initial condition and a final condition. A usually unstated condition is that the amount of gas between the initial and final conditions remains the same. This condition allows the ideal gas law to be rewritten as

$$\frac{PV}{T} = nR$$

If both initial and final conditions are equal to nR, then the ideal gas law can be further restated.

$$\frac{P_iV_i}{T_i} = nR = \frac{P_fV_f}{T_f}$$

(Equation 13.7) or

$$\frac{P_iV_i}{T_i} = \frac{P_fV_f}{T_f}$$

The above restatement of the ideal gas law enables us to use the law without having to deal with n and R.

Before the molecular nature of gases was known, two gas laws were formulated: Boyle's law and Charles's law. Boyle's law was the ideal gas law when the initial and final conditions were at a constant temperature. It reduced the ideal gas law to

(Equation 13.8)

$$P_iV_i = P_fV_f \quad \text{(Boyle's law)}$$

Charles's law was the ideal gas law at constant volume. It reduced the ideal gas law to

(Equation (13.9))

$$\frac{V_i}{T_i} = \frac{V_f}{T_f} \quad \text{(Charles's law)}$$

They were separate laws because their underlying interrelatedness was unknown. The kinetic theory of gases revealed the underlying molecular nature of gases and related microscopic quantities to macroscopic ones. The ideal gas law does something similar, but the microscopic quantities (n and Avogadro's number) drop out. We are left only with those quantities that are relatively easy to measure: pressure, temperature, and volume.

EXAMPLE 13.4

A gas stored in a tank at 20 atm with a capacity of 1.0 m^3 is allowed to expand quickly into a balloon in air at 1 atm. The tank is outside at 27°C. Immediately after filling, the balloon has a radius of 1.56 m. (a) What is the temperature of the gas in the balloon immediately after filling? (b) The gas in the balloon is allowed to warm to the ambient air temperature. What is the final volume of the balloon?

SOLUTION

(a) $$\frac{P_iV_i}{T_i} = \frac{P_fV_f}{T_f} \qquad T_f = \frac{P_fV_f}{P_iV_i}T_i$$

$$T_i = 273 \text{ K} + 27°\text{C} \qquad T_i = 300 \text{ K}$$

$$V_f = \frac{4}{3}\pi(1.56 \text{ m})^3 \qquad V_f = 16 \text{ m}^3$$

$$T_f = \frac{(1 \text{ atm})(16 \text{ m}^3)}{(20 \text{ atm})(1 \text{ m}^3)}(300 \text{ K}) \qquad T_f = 240 \text{ K}$$

$$T_f = 240 \text{ K} - 273 \text{ K} \qquad T_f = -33°\text{C}$$

The gas is quite a bit colder than when it came out of the tank! This cooling of a rapid expansion of a gas under pressure is the basis for the operation of air conditioners and refrigeration.

(b)
$$\frac{P_i V_i}{T_i} = \frac{P_f V_f}{T_f} \qquad \frac{V_i}{T_i} = \frac{V_f}{T_f} \qquad V_f = V_i \frac{T_f}{T_i}$$

$$V_f = (16 \text{ m}^3)\frac{(300 \text{ K})}{(240 \text{ K})}$$

$$V_f = 20 \text{ m}^3$$

13.7 The Determination of Absolute Zero

Both the Celsius and the Fahrenheit temperature scales have no upper or lower limits. When the scales were created, they were intended only to give a quantitative measurement of the very qualitative notion of hot or cold. Both temperature scales were designed to measure the temperatures that the human body normally experiences, and no one knew if there were an upper or lower limit to how hot or cold it could get. However, knowing something of the kinetic theory of gases, and knowing that temperature is linked to the average kinetic energy of the constituent molecules of a gas, we can establish a lower limit to temperature.

When the average kinetic energy of the molecules is zero, then the temperature is zero (absolute degrees on the Kelvin scale). If the kinetic energy is zero, then all of the little billiard balls have come to rest and are lying on a pile on top of one another. This pile comprises a much smaller volume than the billiard balls would occupy whizzing around at thousands of miles per hour, colliding violently with each other at room temperature. Effectively, at absolute zero the volume of the gas is zero, as predicted in the initial form of the ideal gas law. Knowing theoretically that there is an absolute lower limit to temperature, we can use the ideal gas law to obtain that temperature.

Figure 13.7 shows two plots of temperature on the horizontal scale and volume on the vertical scale. (The pressure is held constant at 1 atm.) The upper plot is for nitrogen, and the lower plot is for oxygen. The solid lines are the actual behavior of the gases as the temperature drops. Nitrogen becomes liquid at 77 K ($-196°$C), and oxygen becomes liquid at 90 K ($-183°$C). The dotted lines extrapolate the gas all the way back to absolute zero. Notice that both lines have the same slope, and both extrapolate back to $-273°$C for absolute zero. Absolute zero has never been obtained in the laboratory, although it is

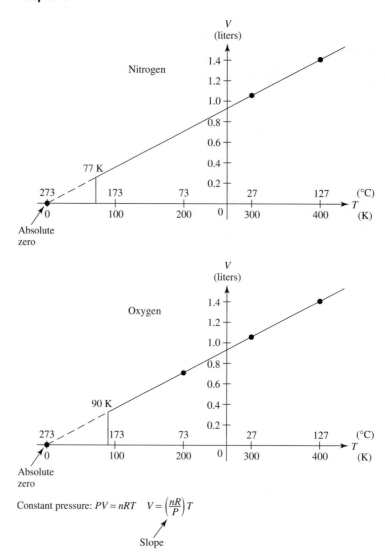

Figure 13.7 Extrapolating to Absolute Zero

now possible to get arbitrarily close. But you don't have to actually go somewhere to know where something is if somebody or something points it out to you. That's what the two plots in Figure 13.7 do: They both point the finger at absolute zero.

Summary

1. As a material is heated, the molecules that make up that material gain energy and begin to oscillate. This oscillation causes the material to expand slightly. The reverse is also true: As a material is chilled, it will shrink. This relationship is shown by Equations (13.1) and (13.2).

2. Both the English and the metric systems use the *degree* (°) as their unit of temperature, but they use different scales. The English temperature scale is the Fahrenheit scale, where water freezes at 32°F and boils at 212°F. The metric scale is the Celsius scale, where water freezes at 0° and boils at 100°. Note that there is a sizable difference between a degree Fahrenheit (°F) and a degree Celsius (°C). The rela-

tionship between these two temperature scales is shown by Equation (13.3).

3. The temperature of a material is a function of the molecular velocities within that material. In a gas, that relationship is given by Equation (13.4).

4. Boyle's law [Equation (13.8)] and Charles's law [Equation (13.9)] are both special cases of the ideal gas law [Equation (13.7)]. The ideal gas law relates the pressure, temperature, volume, and number of moles of a gas at one point in time to the same quantities at another point in time.

Problem Solving Tip 13.1: When confused about which law to use, try starting with the ideal gas law [Equation (13.7)]. See if the following steps help you:

- Find what is constant in your problem (R is always constant, but P, V, T, and n may also be constant as well). For example, we'll assume that only T and V remain constant.

- Isolate the constants in the equation by solving for them on the right side of the equation. In our example, that would leave us with

$$\frac{V}{T} = \frac{nR}{P}$$

- If we have arranged properly, our relationship must hold true at both the initial and the final stages of our problem.

$$\frac{V_1}{T_1} = \frac{nR}{P}$$

$$\frac{V_2}{T_2} = \frac{nR}{P}$$

- If the right sides are equal, then the left sides must also be equal.

$$\frac{V_1}{T_1} = \frac{V_2}{T_2}$$

Note that we have just derived Charles's law!

- What you have just found is the correct formula for you to use in your problem. Substitute the initial and final values that you know, and the remaining term should be what you have been asked to solve!

Key Concepts

BROWNIAN MOTION: The erratic motion of smoke particles caused by the particles being unevenly hit on the sides by the molecules of the air in which they are suspended.

CELSIUS OR CENTIGRADE TEMPERATURE SCALE: The temperature scale tied to the phase changes of water.

EQUIPARTITION OF ENERGY: The principle stating that the motion of molecules and atoms in each of the three spatial dimensions add equally to give the internal energy of a sample of gas.

TEMPERATURE: The average kinetic energy of the molecules of a gas.

THERMOMETER: An instrument used to measure temperature.

Important Equations

(13.1) $\Delta L = a L_o \, \Delta T$ (Thermal expansion)

(13.2) $\Delta V = b V_o \, \Delta T$ (Volume expansion)

(13.3) $°F = \dfrac{9}{5}°C + 32$ (Temperature conversion from Celsius to Fahrenheit)

(13.4) $\dfrac{3}{2}kT = \dfrac{1}{2}m(v^2)_{avg}$ (Relationship between temperature and the average KE of a collection of molecules)

(13.5) $K = °C + 273$ (Conversion from Celsius to absolute temperature scale)

(13.6) $PV = nRT$ (Ideal gas law)

(13.7) $\dfrac{P_i V_i}{T_i} = \dfrac{P_f V_f}{T_f}$ (Ideal gas law restated in terms of initial and final conditions)

(13.8) $P_i V_i = P_f V_f$ (Boyle's law at constant temperature)

(13.9) $\dfrac{V_i}{T_i} = \dfrac{V_f}{T_f}$ (Charles's law at constant pressure)

Conceptual Problems

1. The volume of a sample of gas is inversely proportional to:
   a. the Fahrenheit temperature
   b. the absolute temperature
   c. the Celsius temperature
   d. the pressure

2. Absolute zero is the temperature at which:
   a. hydrogen becomes a liquid
   b. water freezes
   c. molecules cease moving
   d. atoms and molecules disappear

3. The kinetic theory of gases says that at a given temperature:
   a. light gas molecules have a higher average energy than heavier gas molecules
   b. heavy gas molecules have a higher average speed than lighter gas molecules
   c. all molecules have the same average kinetic energy
   d. heavy gas molecules have a higher average energy than lighter gas molecules

4. The temperature of a volume of gas is raised while the volume is held constant. The interior pressure increases because:
   a. the molecules are in contact with the wall for a shorter time
   b. the masses of the molecules increase
   c. each molecule loses more KE when it strikes the wall of the container
   d. the molecules have higher average speeds and strike the walls of the container more often

5. The volume of a sample of gas is reduced at a constant temperature. The pressure increases because:
   a. the molecules strike the walls of the container with higher speeds
   b. the molecules strike the walls of the container with more force

   c. the molecules have more energy
   d. the molecules strike the walls of the container more often

6. Evaporation cools a liquid because:
   a. there is less volume
   b. the fastest molecules have a high probability of escaping
   c. the surface tension decreases
   d. the pressure over the liquid increases

7. Molecules actually attract one another slightly. Does this fact cause an increase or a decrease in the pressure calculated from the ideal gas law? Why?

8. At absolute zero, an ideal gas would occupy zero volume. A real gas doesn't. Why?

9. At absolute zero, molecular motion virtually ceases. What happens to the volume of a solid at absolute zero?

10. In the kinetic theory of gases, all of the collisions between molecules are elastic. What would happen to the air in a room if the collisions were inelastic instead?

11. The temperature of a volume of gas is increased, and the volume is held constant. What happens to the density of the gas, and why?

12. The temperature of a volume of gas increases while the pressure remains constant. What happens to the density of the gas, and why?

13. In the kinetic theory of gases, the particles:
    a. have negligible masses
    b. attract one another
    c. interact only through collisions
    d. lose energy through inelastic collisions

14. A jar is half-filled with boiling water, and the lid is quickly screwed on. Later, after the water has cooled to room temperature, the lid is extremely hard to remove. Why?

Exercises

Thermal Expansion

1. A steel bridge that crosses a river is measured to be 200 m long when the temperature is 30°C. Six months later, when the temperature falls to −10°C, what would the new length of the bridge be?

   Answer: 199.904 m

2. How large a gap should be left between steel rails that are 10 m long when laid at 60°F if they are to just barely touch one another (and therefore not buckle) at 100°F?

   Answer: 5.36 mm

3. In the winter (−20°C), a building is determined to be 100 m wide when measured with a steel tape measure. How much

   wider would the same building appear to be in the summer (30°C) when measured with an identical steel tape measure?

   Answer: 6 cm

4. A completely filled glass mug of water has a capacity of 0.5 liter at 15°C. Later, the temperature rises to 25°C. How much water overflows the mug?

   Answer: 1.05 milliliters

5. A machinist wishes to place an aluminum washer with an inside diameter of 3.99 cm over a 4.01-cm-diameter steel bolt. If she is to carry out her task, to what temperature must

she raise both items, assuming that they both began at 20°C?

Answer: 421°C

*6. A rivet is a metal pin used to fasten two metal plates together. It has a head and is heated before being inserted into a hole through the two metal plates. The pointed end is then hammered while it is still hot into a second head that won't pass through the hole in the plates. When the rivet cools, it shrinks and pulls the metal plates tightly together. What force does a steel rivet 10 mm in diameter exert on the two metal plates that it holds together when it cools down from 500°C (glowing red) to 20°C (room temperature)?

Answer: 9.05×10^4 N!

Linear Temperature Scales

9. The average internal body temperature for a human is 98.6°F. What is this temperature on the Celsius scale?

Answer: 37°C

10. There is a temperature at which both a Fahrenheit thermometer and a Celsius thermometer would give the same reading. What is this temperature?

Answer: −40°

Kinetic Theory of Gases and Absolute Zero

12. To what temperature must a gas sample initially at 0°C be raised in order for the average KE of its molecules to double?

Answer: 273°C

13. The surface of the sun consists primarily of very hot hydrogen atoms ($m = 1.67 \times 10^{-27}$ kg) heated to temperatures of 6000 K. What is the rms speed of these atoms?

Answer: 12.2 km/s

14. What is the average KE of a molecule in a gas (a) at 100°C and (b) at 0°C?

Answer: (a) 7.72×10^{-21} J
(b) 5.65×10^{-21} J

Ideal Gas Law

16. If the pressure on 2 m³ of air is decreased from 100 kPa to 50 kPa, and if the temperature remains constant, what would the new volume be?

Answer: 4 m³

17. At constant temperature, a piston compresses a sample of nitrogen (originally at a pressure of 100 kPa) from a volume of 0.1 m³ down to 0.01 m³. What is the new pressure?

Answer: 1000 kPa

18. A 10-liter fire extinguisher hangs on the wall of a burning building. If the initial pressure inside the extinguisher was 10 atm, what would the internal pressure be if the fire raised the temperature of the room from 20°C to 1300°C?

Answer: 53.7 atm

19. An air tank used for scuba diving has a safety valve set to open at a pressure of 280 bars. A full tank commonly has a

*7. A horizontal steel bar is 1.5 m long at 20°C and has its ends solidly fixed in place. The bar is then heated to 250°C. Assuming that the figure formed by the hot bar is triangular, how much higher is the center of the bar from its original position?

Answer: 11 cm

*8. A clock with a brass pendulum keeps good time at 20°C. The thermal coefficient of expansion for brass is 1.85×10^{-5}/°C. If the clock is run at 0°C for 24 h, (a) by how many seconds is it in error, and (b) is the error fast or slow?

Answer: (a) 173 s or about 3 min
(b) Fast

11. In the hottest part of the summer, the central desert of Australia can reach temperatures in excess of 45°C. What would a Fahrenheit thermometer read in such conditions?

Answer: 113°F

15. Far above a bonfire (where the rising air from the fire has very little effect), the smallest particles of suspended ash can reach average speeds of 3.7×10^{-3} m/s where the temperature has cooled to 290 K. Find the average mass of each particle.

Answer: 8.77×10^{-16} kg

pressure of 205 bars at a temperature of 22°C. To what temperature must such a tank be heated in order for the safety valve to open?

Answer: 131°C

20. It is a common practice to "inject" a gas into a container to raise its internal pressure and/or temperature. If a rigid container already had 30 moles of gas at STP, how many more moles would have to be added if the new temperature and pressure were to be 30°C and 6 atm?

Answer: 147 moles

*21. A 1-liter container contains air at a pressure of 0.6 MPa. Another container contains 2 liters of air at 0.3 MPa. If the two containers are connected together by piping and a valve, and if the valve is opened, what is the final pressure in the two containers?

Answer: 0.4 MPa

*22. A diesel engine sucks air into a cylinder, compresses it to raise the air's temperature, and then has diesel oil injected into it that ignites and burns. A particular engine has a compression ratio of 20 : 1. If the air is at 1 atm and 20°C before it is injected, and at 55 atm just before the oil is injected, what is the air's temperature?

Answer: 546°C

*23. A bubble of marsh gas generated at the bottom of a warm, deep pool doubles in volume as it rises to the surface. If the pond's temperature is uniform, how deep is the pond?

Answer: 31 m

*24. A glass tube of uniform cross-sectional area is closed at one end and open at the other. Inside the glass tube is a column of mercury riding on a cushion of air. When the tube is held vertically with the closed end down, a 100-mm mercury column rides on a 100-mm column of air. When the tube is held vertically with the open end down, the 100-mm

mercury column has a 131-mm air column above it. What is the atmospheric pressure?

Answer: 1.022×10^5 Pa, or slightly over 1 atm

*25. An air bubble 1.5 cm in diameter is released at the bottom of a 30-m-deep lake, where the water temperature is 5°C. What is the diameter of the bubble just before it reaches the surface where the water temperature is 17°C?

Answer: 2.4 cm

*26. An upright cylinder is 1 m tall and is closed at its lower end. It is fitted with a light piston that's free to slide and has a good seal. Initially the piston is in the center of the cylinder with air above and below. The upper part of the cylinder above the piston is then filled with water until the water is about to overflow the cylinder. What is the height of the piston above the bottom of the cylindrical chamber?

Answer: 0.45 m

Heat flows into the bottom of the pot by conduction. The heat causes the water in the pot to mix by convection. When the water temperature is high enough, the water changes state from liquid to gas (steam). As long as boiling water remains in the pot, the water temperature remains constant at the boiling point (100°C at STP).

Heat Capacity and Heat Conduction

At the end of this chapter, the reader should be able to:

1. Discuss the difference between temperature and heat.

2. Explain what is meant by *heat flow*.

3. Calculate the heat needed to raise the temperature of a given substance by a known amount.

4. Calculate the heat necessary to melt (or boil off) a known amount of a given material.

5. Solve for the heat flow through a wall of given construction if the interior and exterior surfaces experience different (but known) temperatures.

6. Describe the three methods of heat transfer: **radiation, convection,** and **conduction.**

7. Calculate the amount of heat radiated by an object of known material and temperature in a given amount of time.

14.1   Heat Flow

Heat flows from one body to another because of a temperature difference between the bodies. Heat always flows from a warmer body to a colder body. Temperature is linked to the average kinetic energy of the individual molecules of a gas, and these molecules were modeled as tiny billiard balls making perfectly elastic collisions with one another and with the walls of their container.

Suppose that we have a box with two chambers with a divider between the chambers. One chamber is filled with colliding tiny billiard balls with a given average kinetic energy. This average energy corresponds to a temperature T_o (K). The other chamber is also filled with a large number of these tiny billiard balls, but they are lying motionless in a pile at the bottom of the chamber. This latter situation corresponds theoretically to zero kinetic energy and a temperature of $0\,$K. If the divider is removed, the balls that are lying motionless in the pile will be bombarded by the balls originally whizzing about in the chamber at a temperature of T_o. As these balls bombard the motionless pile, the motionless balls begin to gain kinetic energy through collisions with the moving balls, and they start flying and bouncing about also. Because the original group of balls are imparting energy to the originally motionless pile, they will slow down somewhat and have less kinetic energy. Eventually all of the balls in the pile will have acquired some equilibrium kinetic energy and will be moving about at the new average kinetic energy for the enlarged system. (The fast balls will have cooled down, and the originally motionless balls will have warmed up.) All of the balls in the enlarged mixed system are now flying about with an average kinetic energy less than the original, but greater than zero. The originally moving balls will have lost some of their kinetic energy in stirring up the motionless balls.

Macroscopically, when a hot gas is mixed with a cool gas, both will reach an equilibrium temperature intermediate between the hot and cold temperatures. The "heat" flows from the hot gas to the cold gas. In the example above, the "cold" gas had no internal energy and therefore no heat to contribute to the mixture. The average kinetic energy of the mixed system was therefore somewhere between $\frac{3}{2}kT_o$ and 0. However, from this example we can extrapolate and see that when any two gases at different temperatures are mixed, the hotter gas will raise the average kinetic energy of the molecules of the colder gas, and heat will always flow from the warmer to the colder gas. This follows from the conservation of energy. Conservation of energy can be used for gas molecules because there is no internal friction to slow down and bring the system to a halt. The total energy of this isolated system would therefore remain constant. This principle also holds true for liquids and solids for essentially the same reasons.

What happens when a tray of water is placed in the freezer of a refrigerator to make ice cubes? The water cools down and freezes. The heat in the water flows out into the freezer chamber, warming it somewhat. When the heat flows out of the water, the water cools. If it cools enough, it forms ice. If the freezer is very large compared to the ice tray, then the amount by which the freezer compartment warms up is negligible compared to the amount by which the water cooled down. In the days before refrigeration, there were iceboxes, and the iceman had to come around periodically and place a large block of ice into the icebox. This large block of ice cooled everything in the icebox. But the ice eventually melted and therefore had to be replaced often. A modern refrigerator also has a cool material, a gas, passing through the S-shaped tubes in the walls of the refrigerator. This gas is in a closed system that constantly recools the gas. How it does this is a subject for the next chapter. Even now, refrigerators are sometimes called *iceboxes,* although they aren't the same thing.

14.2   Heat Capacity

Copper tubing used for plumbing is often soldered together, a process that requires a lot of heat, usually from a butane torch. If you grasp one end of a copper pipe in your hand and heat the other end, the heat eventually reaches your hand at the far end. The end grasped by your hand becomes hotter and hotter, and eventually you must drop the pipe to avoid a burn. This demonstration provides a good illustration of *heat flow* and *heat capacity*. The heat flows from the hot end to the cool end, and as heat is continually added to the pipe, its temperature rises. The pipe will stay hot for awhile, until it cools by radiation and convective heat flow into surrounding materials and the air.

A material can store heat energy. *The **heat capacity** of a material relates how much heat energy is stored in the material to the temperature rise or fall and to the mass of the material.* The equation for heat capacity is

Heat capacity: The ability of a material to store heat.

$$Q = mc\,\Delta T$$

(Equation 14.1)

where Q is the amount of heat energy, m is the mass of the material, c is the heat capacity of the material, and ΔT is the amount by which the temperature changes. Table 14.1 lists the heat capacities for various materials.

We can find the units for heat capacity by solving the heat capacity equation for c. Heat (Q) is in terms of joules (1000 J = 1 kJ) because heat is energy. The temperature scale is degrees Celsius, and mass is in terms of kilograms. There are other units for heat. One common unit found often on food packaging is the calorie. One calorie is equal to 4.185 J. However, on food packaging the calorie is actually equal to a kilocalorie, or 4185 J. Another unit for heat is the British thermal unit, which is equal to 1054 J. Confusing? You bet! If you consistently use joules as the unit for heat, you'll always know that heat is energy. If you need to use the other units for heat, look them up in the table of conversion factors

Table 14.1 Heat Capacities of Various Materials

Material	Specific Heat Capacity
Air (1 atm)	0.70 kJ/kg-°C
Alcohol (ethyl)	2.43 kJ/kg-°C
Aluminum	0.92 kJ/kg-°C
Concrete	2.90 kJ/kg-°C
Copper	0.39 kJ/kg-°C
Glass	0.84 kJ/kg-°C
Gold	0.13 kJ/kg-°C
Granite	0.80 kJ/kg-°C
Human body	3.47 kJ/kg-°C
Ice	2.09 kJ/kg-°C
Iron	0.46 kJ/kg-°C
Lead	0.13 kJ/kg-°C
Mercury	0.14 kJ/kg-°C
Silver	0.23 kJ/kg-°C
Steam	2.01 kJ/kg-°C
Water	4.19 kJ/kg-°C
Wood (pine)	1.76 kJ/kg-°C

in Appendix D at the back of the book. The unit of a calorie doesn't remind you that heat is energy, but using units of joules always will.

Some interesting facts can be gleaned from Table 14.1. Ice (2.09 kJ/kg-°C), water (4.19 kJ/kg-°C), and steam (2.01 kJ/kg-°C) have different heat capacities, regardless of the fact that all are made up of molecules of H_2O. Solid, liquid, and gas are the three phases of matter, and each may have a different heat capacity from the others. Metals have a low heat capacity but are excellent conductors of heat, just as they are good conductors of electric current.

EXAMPLE 14.1

Water flows through an instantaneous water heater at the rate of 2 liters/min. If the initial temperature of the water is roughly room temperature (20°C), and if the heating coils consume 4 kW of electricity, what is the temperature of the hot water, assuming that all of the heat input heats the water?

SOLUTION

$$Q = mc\,\Delta T \qquad \frac{\Delta Q}{\Delta t} = c\,\Delta T\frac{\Delta m}{\Delta t} \qquad \Delta T = \frac{\Delta Q/\Delta t}{c\,\Delta m/\Delta t}$$

$$T_f - T_i = \frac{\Delta Q/\Delta t}{c\,\Delta m/\Delta t} \qquad T_f = T_i + \frac{\Delta Q/\Delta t}{c\,\Delta m/\Delta t}$$

$$\frac{\Delta m}{\Delta t} = \left(\frac{2\ \text{liters}}{\text{min}}\right)\left(\frac{1\ \text{min}}{60\ \text{s}}\right)\left(\frac{1\times10^3\ \text{kg}}{\text{m}^3}\right)\left(\frac{1\times10^{-3}\ \text{m}^3}{1\ \text{liter}}\right)$$

$$\frac{\Delta m}{\Delta t} = 0.033\ \text{kg/s}$$

$$T_f = 20°C + \frac{4000\ \text{J/s}}{(4.19\ \text{kJ/kg-°C})(0.033\ \text{kg/s})}$$

$$T_f = 48.9°C$$

EXAMPLE 14.2

A man pours 0.15 kg of hot coffee at 85°C into a cup made of glass with a mass of 0.4 kg. The cup was originally at room temperature. If no heat leaves the coffee-cup system, what is the final temperature of the coffee?

SOLUTION
Both coffee and cup will reach an intermediate equilibrium temperature that is the same for both. The heat from the coffee warms the cup. The heat lost from the coffee equals the heat gained by the cup (conservation of energy).

$$Q_{lost} = Q_{gained}$$

$$m_c c_c\,\Delta T_c = m_{cup}c_{cup}\,\Delta T_{cup}$$

$$m_c c_c(T_{ic} - T_e) = m_{cup}c_{cup}(T_e - T_{icup})$$

Be careful of the signs of ΔT here. Both Q_{lost} and Q_{gained} are positive numbers. The equilibrium temperature should be less than 85°C and greater than 20°C.

$$m_c c_c T_{ic} - m_c c_c T_e = m_{cup} c_{cup} T_e - m_{cup} c_{cup} T_{icup}$$

$$m_c c_c T_e + m_{cup} c_{cup} T_e = m_c c_c T_{ic} + m_{cup} c_{cup} T_{icup}$$

$$T_e(m_c c_c + m_{cup} c_{cup}) = m_c c_c T_{ic} + m_{cup} c_{cup} T_{icup}$$

$$T_e = \frac{m_c c_c T_{ic} + m_{cup} c_{cup} T_{icup}}{m_c c_c + m_{cup} c_{cup}}$$

$$T_e = \frac{(0.15 \text{ kg})(4.19 \text{ kJ/kg-°C})(85°C) + (0.4 \text{ kg})(0.84 \text{ kJ/kg-°C})(20°C)}{(0.15 \text{ kg})(4.19 \text{ kJ/kg-°C}) + (0.4 \text{ kg})(0.84 \text{ kJ/kg-°C})}$$

$$T_e = 62.4°C$$

14.3 Changes of State

When a material undergoes a change of phase (ice to water or water to steam, for instance), the phase change happens at a constant temperature. If a thermometer is placed into a pot of water at room temperature and the pot heated on the stove, the temperature rises from room temperature to the boiling point of water (212°F or 100°C). But when the water begins to boil, the temperature remains constant until all of the water is boiled away, regardless of how fast heat is added. The amount of heat that is added per unit mass for a phase change to occur in a given material is a constant. This amount of heat is called the *heat of fusion* for the phase change from solid to liquid, and *heat of vaporization* for the change from liquid to gas. The heat equations for phase changes are

$$Q = mL_v \qquad \text{(For vaporization)} \qquad \text{(Equation 14.2)}$$

$$Q = mL_f \qquad \text{(For fusion)} \qquad \text{(Equation 14.3)}$$

Heat capacity is used when a material stays in the same phase and is either warmed up or cooled down. If the phase doesn't change, then the temperature of the material can change as heat is added or removed. The heat of vaporization is used only when a phase change takes place.

Figure 14.1 shows what happens when heat is added to a 1-kg ice cube initially at −50°C. There are three phases: solid, liquid, and gas. The rising parts of the graph show the three distinct phases where the heat capacity equation applies. Note that the temperature rises as heat is added. The slope of the line in this case is the inverse of the heat capacity for that phase because the mass of the water is 1 kg. The smaller the heat capacity, the steeper the line. The larger the heat capacity, the flatter the line. (If a material has a very large heat capacity, it will take a lot of heat to warm it up by just a few degrees.) But when the temperature of a phase change is reached, heat capacity no longer applies, and the equation for heat of vaporization or fusion does. (The fact that the line is flat during a phase change doesn't mean that its heat capacity is infinite! A fundamentally different process takes over. Moreover, the heat capacity is different before

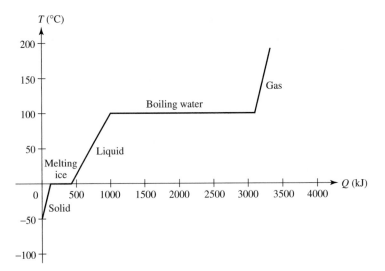

Figure 14.1 Phase Changes
of Water

and after the phase change.) So when asked in a problem to calculate the amount of heat required to take an ice cube at a temperature less than 0°C from solid to liquid, remember that there will be three distinct parts to the problem. Two will be using heat capacity, and one heat of fusion. Find the heat required to bring the ice cube to the melting point, then the heat required to melt the ice cube, and finally the heat required to bring the water up to the required temperature. Add the three amounts of heat together for the total. The total heat could be the amount of electricity consumed or the amount of a given fuel that must be burned.

Table 14.2 gives the melting point, heat of fusion, boiling point, and heat of vaporization for a number of common substances.

Popcorn pops because of a steam explosion. The volume occupied by a given mass of steam is much greater than that occupied by the same mass of liquid.

Table 14.2 Melting Point (MP), Heat of Fusion (L_f), Boiling Point (BP), and Heat of Vaporization (L_v) at 1 atm

Substance	MP	L_f	BP	L_v
Alcohol (ethyl)	$-114°C$	105 kJ/kg	78°C	854 kJ/kg
Copper	1083°C	134 kJ/kg	1187°C	5069 kJ/kg
Lead	330°C	25 kJ/kg	1170°C	732 kJ/kg
Mercury	$-39°C$	12 kJ/kg	358°C	297 kJ/kg
Nitrogen	$-210°C$	26 kJ/kg	$-196°C$	201 kJ/kg
Oxygen	$-219°C$	14 kJ/kg	$-183°C$	213 kJ/kg
Silver	961°C	88 kJ/kg	2193°C	2335 kJ/kg
Tungsten	3410°C	184 kJ/kg	5900°C	4813 kJ/kg
H_2O (water)	0°C	335 kJ/kg	100°C	2260 kJ/kg

EXAMPLE 14.3

An ice cube has an original mass of 0.5 kg. (a) How much electric power must be consumed to warm the block of ice from $-20°C$ to 70°C? Assume that all of the heat generated goes into warming the water. (b) If the heater is rated at 3 kW, how long will it take to heat the block of ice to 70°C? (c) If electricity costs 13 cents per kilowatt-hour, how much did it cost to get the hot water?

SOLUTION

(a) First the ice must be warmed to its melting point: $c_i = 2.09$ kJ/kg-°C.

$$Q = c_i m \, \Delta T \qquad Q = c_i m(T_f - T_i)$$

$$Q = (2.09 \text{ kJ/kg-°C})(0.5 \text{ kg})[0 - (-20°C)]$$

$$Q = 20.9 \text{ kJ}$$

Next the amount of heat necessary to melt the ice must be calculated: $L_f = 335$ kJ/kg.

$$Q = (335 \text{ kJ/kg})(0.5 \text{ kg}) \qquad Q = 167.5 \text{ kJ}$$

Now that the ice is water, the heat to warm it to 70°C must be calculated.

$$Q = (4.19 \text{ kJ/kg-°C})(0.5 \text{ kg})(70°C - 0°C) \qquad Q = 146.7 \text{ kJ}$$

$$Q_T = 20.9 \text{ kJ} + 167.5 \text{ kJ} + 146.7 \text{ kJ} \qquad Q_T = 335.1 \text{ kJ}$$

(b) $\qquad t = \dfrac{Q}{P} \qquad t = \dfrac{335 \text{ kJ}}{3 \text{ kJ/s}} \qquad t = 111.7 \text{ s} \qquad (\approx 2 \text{ min})$

(c) $\qquad 1 \text{ kW-h} = \left(\dfrac{1000 \text{ J}}{\text{s}} \right)(3600 \text{ s}) = 3.6 \times 10^6 \text{ J}$

$$\text{Cost} = \frac{335.1 \text{ kJ}}{3.6 \times 10^6 \text{ J/13¢}} = 1.2¢$$

14.4   Heat Conduction

Heat can flow through a material much as a fluid flows through a pipe. For example, if a piece of copper pipe is heated with a torch commonly used to solder plumbing joints together, the heat discolors the metal. A hot piece of copper tubing will have a bluish tinge to it. As the pipe is being heated by the torch, the blue area of the pipe will move from the hot end to the cold end as the pipe warms. Heat flow can be readily seen in this case. When heat flows through the pipe in the example, though, no material moves from one place to another as in fluid flow. What does flow is the heat energy. If energy is flowing down the pipe at a certain rate, then power is moving through the pipe. Heat flow is power. The molecules of the hot end of the pipe are jiggling violently, and these violently jiggling molecules in turn cause their neighbors to jiggle. This jiggling progresses down the pipe like a chain reaction or a row of falling dominoes.

What parameters influence this heat flow? First of all, heat continues to flow until both ends of the pipe are the same temperature. When this condition is reached, heat flow stops. For heat to flow, there must be a temperature difference between the ends of the pipe. The thickness of the material plays a part. The thicker the material, the longer it takes heat to flow from one side of it to the other; thus, heat flow in a material is inversely proportional to the thickness of the material. Surface area is also an important factor in heat flow. Greater surface area provides a greater area for heat to flow into, so the material "soaks up" the heat faster. Finally, each material is different and has its own characteristic ***thermal conductivity.***

Thermal conductivity: The ability of heat to flow through a material.

The equation for heat flow puts all of these parameters together.

(Equation 14.4)

$$\frac{Q}{t} = \frac{kA\,\Delta T}{d}$$

where k is the thermal conductivity, A is the surface area, and d is the material's thickness.

Some materials conduct heat better than others. Metals conduct heat much better than about any other material, whereas a material such as rubber or plastic conducts heat very poorly. The vast difference in heat conduction from metals to insulators is due to free electrons in the metal. A metal wire (usually copper) conducts electricity very well because electrons are free to move inside it. (More will be said about this subject in the sections dealing with electromagnetism.) These same electrons can also help to carry heat.

Table 14.3 gives the thermal conductivity for a number of materials. Note that body fat is a very good insulator, as are ice and snow. Whales have a very thick layer of body fat (blubber) over their entire bodies to protect them from the cold of extreme depths of arctic oceans. This blubber is one of the reasons whales were almost hunted to extinction: The blubber could be rendered into a fine oil used to light lamps and to lubricate fine machinery. The discovery of petroleum had a big impact on whaling, since kerosene was a good lamp oil, and good lubricating oils could be made from petroleum. Thus, petroleum became a substitute for whale oil. Oddly enough, the discovery of petroleum was a large factor in saving the whales from extinction because it killed the market for whale oil. Because ice and snow are a good insulating material, the igloo, constructed from blocks of snow, is good protection from the arctic cold. The body heat of the occupants coupled with a small heat source such as a 100-W light bulb can raise the temperature of the interior of an igloo above freezing when the temperature outside is $-30°F$ to $-40°F$.

Table 14.3 Thermal Conductivity of Some Common Materials

Material	Thermal Conductivity (k)
Metals:	
Aluminum	205 W/m-°C
Brass	109 W/m-°C
Copper	385 W/m-°C
Iron and steel	50 W/m-°C
Silver	406 W/m-°C
Insulators:	
Cork	0.04 W/m-°C
Glass wool	0.04 W/m-°C
Down	0.02 W/m-°C
Kapok	0.03 W/m-°C
Gases:	
Hydrogen	0.13 W/m-°C
Air	0.024 W/m-°C
"Vacuum"	0.0
Various Materials:	
Body fat	0.17 W/m-°C
Brick	0.60 W/m-°C
Concrete	0.80 W/m-°C
Glass	0.80 W/m-°C
Ice	1.60 W/m-°C
Water	0.60 W/m-°C
Wood (pine)	0.13 W/m-°C
Snow	0.18 W/m-°C

We can derive the units for thermal conductivity by solving the heat flow equation for k. In the metals category, silver is a better heat conductor than iron or steel by about a factor of 8. But iron or steel is still a much better conductor of heat than anything else in the table. Conversely, it is often desirable to insulate something to prevent heat from leaking out or in. For example, a refrigerator is surrounded by insulating material to keep external heat from leaking in. In such a case, the thermal conductivity should be as low as possible. The "material" with the lowest thermal conductivity, and therefore with the best insulating properties, is a vacuum. Although a vacuum isn't really a material at all (it is the complete absence of any mass in a given volume), it is used as a material in a thermos bottle. For heat conduction to take place by transfer of energy from molecule to molecule or from atom to atom, something must be able to carry the heat from inside to outside, or vice versa. With nothing between the inside or outside, heat can't be transferred! The only place where heat can be transferred from inside to outside in a thermos bottle is around the top, where the inner (usually glass) bottle is fused to the outer bottle. The glass there is thick, with little surface area, and heat conduction through this "bottleneck" is very slow.

EXAMPLE 14.4

A cylindrical hot water tank is insulated on all sides by a thickness of glass wool. The tank has a diameter of 0.75 m and a height of 2 m. The thickness of the glass wool is 3 cm. The water temperature is 74°C, and room temperature is 20°C. At what rate does the tank lose heat to the room?

SOLUTION

$$\frac{Q}{t} = \frac{kA \, \Delta T}{d}$$

The surface area of the tank must be calculated. (Assume that the thickness of the pink glass wool is negligible compared to the dimensions of the tank.)

$$A = 2(\pi r^2) + 2\pi rh$$

$$A = 2\pi(0.375 \text{ m})^2 + 2\pi(0.375 \text{ m})(2 \text{ m})$$

$$A = 5.6 \text{ m}^2$$

$$\frac{Q}{t} = \frac{(0.04 \text{ W/m-°C})(5.6 \text{ m}^2)(75°C - 20°C)}{0.03 \text{ m}}$$

$$\frac{Q}{t} = 410.7 \text{ W}$$

If the water is kept at a constant temperature, then the power supplied to the heater coils in the tank must be 410.7 W of electricity.

14.5   Thermal Resistance

Thermal resistance: The ability of a material to block heat flow, or to act as an insulator.

*The ability of a material to act as an insulator is often given in terms of its thermal resistance. Simply put, the definition of **thermal resistance** is the thickness of a material over its thermal conductivity.*

(Equation 14.5)

$$R = \frac{d}{k}$$

where R is the symbol for thermal resistance. The greater the thermal conductivity, the smaller the resistance of a material to the flow of heat. The thicker the material, the greater its resistance to the flow of heat. The equation for heat flow can be put in terms of the thermal resistance.

$$\frac{Q}{t} = \frac{kA \, \Delta T}{d} \qquad R = \frac{d}{k} \qquad \frac{Q}{t} = \frac{A \, \Delta T}{R}$$

The greater the thermal resistance, the smaller the heat flow, and the better the material acts as an insulator.

EXAMPLE 14.5

Compare the insulation of two types of window construction: The first is a single pane of glass 3 mm thick, and the second is two panes of 3-mm glass with a 6-mm air space in between.

SOLUTION

$$R_1 = \frac{d}{k} \quad R_1 = \frac{0.003 \text{ m}}{0.8 \text{ W/m-}°\text{C}} \quad R_1 = 0.00375°\text{C-m}^2/\text{W}$$

To calculate the thermal resistance of the double-paned window, we must add the thermal resistance of the air sandwich to the resistances of the two panes of glass. The thermal resistances add because the heat must pass through all three materials.

$$R = 2R_1 + R_2$$

$$R_2 = \frac{0.006 \text{ m}}{0.024 \text{ W/m-}°\text{C}}$$

$$R = 2(0.00375°\text{C-m}^2/\text{W}) + 0.25°\text{C-m}^2/\text{W}$$

$$R = 0.2575°\text{C-m}^2/\text{W}$$

The insulating property of the double-paned window is over 68 times better than that of the single-paned window! Most of the improvement comes from the sandwich of air between the two panes.

The units of thermal resistance are somewhat confusing and are often omitted when a calculation is made. An industry standard for thermal resistance is often used that has no units. The conversion is

$$R\text{-}1 = 0.085°\text{C-m}^2/\text{W}$$

where the term on the left is read, "R dash 1." In Example 14.5, the total thermal resistance for the double-paned window in terms of the industry standard is R-3.03.

14.6 Convection and Radiation

Heat can flow from one place to another in two ways other than conduction: convection and radiation. *In convection the fluid being heated mixes because heat causes density gradients in the fluid.* For example, when a pot of water is placed on the stove to boil, the water becomes hot primarily by convection. Cold water has a higher density than hot water and thus tends to sink. When the pot is placed on the burner, the bottom of the pot heats the low layers of water through conduction. When the lower layers of water become warmer than the upper layers, the upper, denser layers sink to the bottom, displacing the warm water. The cooler water then heats until its density is lower than that of the upper, cooler layers, and it is displaced by the sinking, cooler water. This process continues until the water in the pot is at the boiling temperature ($212°$F), when the temperature of the water can no longer be raised.

Convection: Heat flow in a fluid caused by mixing through density gradients.

Convection is also responsible for the breezes coming from offshore near a body of water on a sunny day. The temperature of a land surface rises more quickly than the temperature of water, and a convection cell forms (see Figure 14.2). The air over the warmer land heats up more rapidly than the air over the water. The cooler air with the greater density over the water sinks and moves in from offshore to displace the lighter, less dense, warm air. As the warm air rises, it expands and cools. It then sinks over the body of water, and the process begins all over again. Both conduction and convection are mechanical processes in which atoms and molecules cause each other to jiggle or in which relatively large volumes of a material move to cause heat flow.

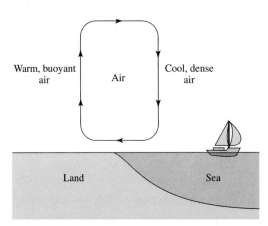

Figure 14.2 Offshore Breeze Caused by Convection

Radiation: Heat flow caused by the emission of photons.

Radiation is heat flow caused by the emission of photons, which are massless particles traveling at the speed of light. A photon moves at the speed of light because it *is* light. The photon carries energy, and we feel this energy as thermal radiation. For example, the warm glow that we feel near a radiator is caused by radiation, not conduction or convection. Photons will quite happily move in a vacuum. Examine an incandescent, glowing, 100-W light bulb. Its intense glow is due to a very hot filament at the bulb's center. The filament is heated by the electric current moving through it. Although you can't touch the filament directly, you know that it's hot because the glass bulb around the filament is too hot to touch. How does the heat get from the filament to the bulb's glass envelope? Inside the glass there is a pretty good vacuum. If the vacuum weren't there, the filament would quickly burn up. We've already learned that a vacuum has a thermal conductivity of zero and that convection can occur only when volumes of material move because of density gradients. Some other process must therefore be responsible for the heat so obviously flowing from the filament to the glass envelope. That process is radiation.

Figure 14.3 Correlation between Temperature and Color of a Glowing Object

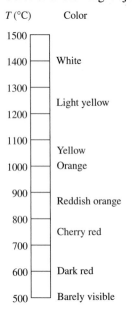

When an object is heated, it begins to glow. The filament of a light bulb glows white-hot. If the voltage to a 100-W light bulb could be continuously varied, we would see the filament in the bulb first glow red, then orange, then yellow, and finally white. If the voltage were further increased, the filament would glow blue, but the filament would melt before it got to that temperature. The color of the glow is independent of the material and is dependent only on the material's temperature. We can actually guess the temperature of a glowing object by observing the object's color. Figure 14.3 gives a correlation between an object's color and its temperature.

Our eyes use emitted photons to see. A photon has a wavelength and a frequency associated with it, and visible light is only a small part of the electromagnetic spectrum. Figure 14.4 shows the electromagnetic spectrum, which is a continuum that contains radio,

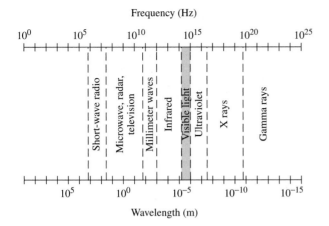

Figure 14.4 The
Electromagnetic Spectrum

television, infrared radiation (heat), visible light, ultraviolet (black light), ultraviolet, and gamma rays. At the top of the diagram is a frequency scale, and at the bottom is a wavelength scale. As the frequency increases, the wavelength decreases. The higher the energy, the higher the frequency, and the shorter the wavelength. If a photon has a wavelength associated with it, it should obey the wave equation. It does. The wave equation says that the frequency times the wavelength equals the propagation speed of the wave in the medium.

$$\lambda f = c$$

where c is the speed of light, 3×10^8 m/s. However, a photon is a particle with a discrete amount of energy. How can a particle exhibit a wavelike property such as a wavelength and a frequency? One of the great paradoxes of physics, this topic will be covered in the chapters on electromagnetics.

A hot body changes color as the temperature rises, as shown in Figure 14.3. We can see color: It is the visible part of the spectrum. However, most of the heat is in the infrared part of the spectrum and can't be seen. A radiator filled with hot water doesn't glow, but its heat can be felt. The implication is that when a body radiates heat, it does so over a wider range of the electromagnetic spectrum than that centered on the visible glow of a body such as a filament. (A 100-W light bulb is a much more efficient generator of heat than of light.) How much energy a hot body radiates away is given by a simple law called the *Stefan-Boltzmann law*.

$$r = e\sigma T^4$$

(Equation 14.6)

where r is the amount of energy that the object radiates in W/m^2 of surface area, e is the emissivity, σ (sigma) is a proportionality constant equal to 5.67×10^{-8} W/m^2-K^4, and T is the absolute temperature in kelvins (K). Emissivity is a number that varies from 0 to 1. For example, an object with an emissivity of 0 is a perfect reflector, and an object in which $e = 1$ is a perfect absorber; e might have a value of 0.07 for polished chrome steel and 0.97 for matte black paint. A perfect absorber looks black and is called a *blackbody*. The application of flat black paint will turn a surface into an almost perfect absorber of radiation. Conversely, a blackbody is also the best possible radiator.

Any object above absolute zero radiates electromagnetic radiation. Our eyes are sensitive to only a small part of the electromagnetic spectrum, but with the help of instrumentation we can "see" further into the electromagnetic spectrum than our eyes alone can. For example, infrared goggles allow our eyes to directly detect heat in the infrared and are especially useful at night. A radio antenna "sees" long-wavelength radiation that is impossible to detect with our eyes. Very low energy radiation comes in the form of very long wavelength radio waves. One of the proofs of the birth of our universe through the Big Bang came through the discovery of very long wavelength radiation at a temperature of roughly 3 K. Initially our universe was very hot and compact. As it expanded, it cooled, and at the present stage of the universe the background radiation (which should be constant over the entire sky) was predicted to be at its present 3 K. This fact was accidentally discovered by a pair of Bell Telephone engineers who were trying to calibrate a sensitive antenna. They couldn't account for or get rid of the long-wavelength signal which turned out to be the remnant of the birth pangs of our universe.

EXAMPLE 14.6

In the firing of clay or ceramics in an oven, regardless of the color of the pot or plate that is eventually produced, the color of the objects in the oven is the same and varies with temperature, just as in Figure 14.3. The door of the oven acts like a blackbody. A small hole in the wall of a cavity in an object acts like a blackbody because any radiation that falls on it is trapped by reflection from inside until absorbed by the cavity walls. At what rate does radiation escape from a small door in the wall of an oven 5 cm on a side whose interior temperature is 850°C?

SOLUTION

$$P = rA$$

The rate at which something radiates is power, and from the Stefan-Boltzmann law, the power produced equals r times the surface area of the blackbody.

$$P = e\sigma A T^4$$

where T is absolute.

$$T = 850°C + 273 \text{ K} = 1123 \text{ K}$$

$$P = (1)(5.67 \times 10^{-8} \text{ W/m}^2\text{-K}^4)(0.05 \text{ m})^2(1123 \text{ K})^4$$

$$P = 225.4 \text{ W}$$

14.7   The Triple Point of Water

Pressure affects the temperature at which water boils, as most cooks know. If you live in Denver (the Mile-High City), water boils at 96°C because of the lesser pressure at that altitude. Greater pressure raises the boiling point of water. A pressure cooker is a sealed pot in which food can be cooked faster than in an open pot. Water at a pressure of 2 atm will boil at 120°C instead of 100°C.

The vaporization curve of water is given in Figure 14.5. Everything to the left and above the curve is a liquid. Everything to the right and below the curve is a vapor. The upper limits for the vaporization curve of water are 374°C and 218 atm. Water cannot exist above this temperature as a liquid.

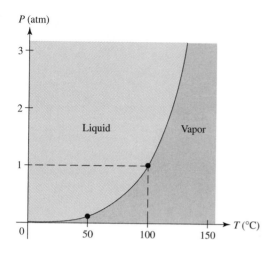

Figure 14.5 The Vaporization Curve of Water

Our sun is a great sphere of mostly hydrogen gas. As the center of the star is approached, the temperature and the density of the hydrogen increase until the temperature is in the millions of degrees and the density is such that it has a mass of thousands of kilograms per square centimeter. Yet at the center of the sun, the hydrogen is still a gas regardless of the pressure. The chemical bonding necessary to form the stable lattice of a solid can't form at those temperatures regardless of the enormous pressure. The necessary mixing and fusing of the hydrogen nuclei must necessarily take place in a gas, and the temperatures at the sun's center ensure that the hydrogen stays gaseous.

Figure 14.6 shows the fusion curve for water. Everything to the left of the line is a solid, and everything to the right is a liquid. For a given temperature, say, $-0.05°C$, if the pressure is increased enough, the ice can be melted without the addition of any heat. This fact allows ice skaters to skate. The weight of an ice skater is supported by the very small surface area of the skate's blade. For a given weight, the narrower the blade, the greater the pressure. Directly under the blade, the ice melts, and the skater glides along on a thin film of water. As soon as the skater passes a given point, the pressure is released, and the water quickly refreezes.

The fusion and vaporization curves of water intersect at a point of $0.01°C$ and 4.6 torr (4.6 torr is well below atmospheric pressure, 760 torr being roughly 1 atm). The triple-point

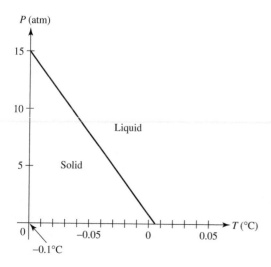

Figure 14.6 Fusion Curve for Water

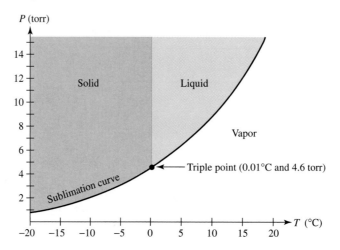

Figure 14.7 The Triple Point of Water

diagram for water is given in Figure 14.7. At the point of intersection, H_2O exists simultaneously in all three phases: solid, liquid, and vapor. Along the bottom edge of the fusion curve, sublimation can take place. At this point, water can go directly from the solid phase into the vapor phase. This fact is the basis for the freeze-drying process used to preserve foods and blood plasma. The material to be preserved is quickly cooled below the triple point of water in a gas-tight chamber. The chamber is then pumped down to a very low pressure with a vacuum pump to remove the water vapor. Freeze-drying doesn't affect the structure of biological materials as much as other methods of dehydration.

14.8   The Mechanical Equivalent of Heat

By now you should know that the heat content of a gas is the average kinetic energy of the collection of atoms and/or molecules that make up the gas. You should also know that the work done by the force of friction is dissipated as heat. If you have ever had to take a rough-cut piece of wood and sand it to its final shape with a piece of sandpaper, you've noticed that the sandpaper where you're sanding the wood gets hot. The physical act of sanding generates heat. A bullet speeding through the air has a lot of kinetic energy. You could even think of it as an ultraturbo molecule and associate a temperature with it. But when the bullet strikes a target and comes to rest inside the target without causing it to move, all of the kinetic energy of the bullet immediately goes into heat. The heat might even be enough to melt the bullet.

About 1910, a meteor struck in a wilderness area of Siberia in the region known as Tunguska. The strike was so violent that it was noticed as far away as Denmark and Sweden and lit up the eastern horizon. (It struck at night.) When a Soviet scientific expedition was sent to investigate the site in the early 1920s, the scientists found no impact crater, but they did discover an area of flattened forest directly below where the meteor burst in the air that was many miles in diameter. No remnants of the meteor were found. Geologists at the time were stumped by the evidence. Something tremendous had happened, but nobody knew exactly what. It wasn't until the late 1950s and early 1960s that geologists began to see that when a meteor traveling at many miles per second collides with the earth, it does so explosively because of the great kinetic energy that it carries. This kinetic energy is almost instantaneously converted into heat with an explosion equal to megatons of TNT. Such an explosion can be much greater than that of a thermonuclear device. The mechanical equivalent of heat is responsible for such explosions. The asteroid that wiped out the dinosaurs in the Jurassic period 65 million years ago was just such an explosion.

Summary

1. *Heat* is a form of energy and thus has the units of joules (J). Other units of heat are in common usage. Among them are the *calorie* (cal), which is the amount of heat needed to heat 1 kg of water 1°C (4.185 J = 1 cal); the *Calorie* (note the capital "C"), typically found on food packages (1 Cal = 1000 kcal = 4185 J); and the English unit for heat, the *British thermal unit* (1054 J = 1 Btu).

2. The **heat capacity** of a substance determines how much heat is necessary to raise it by a given amount of temperature. The mathematical expression for heat capacity is given by Equation (14.1).

3. The *latent heat of fusion* determines how much heat is required to completely melt a substance. The *latent heat of vaporization* determines how much heat is required to completely boil a substance. When working in reverse (that is, condensing or freezing a substance), the latent heat determines how much energy is liberated in the process. These concepts are given by Equations (14.2) and (14.3).

Problem Solving Tip 14.1: When solving problems that deal with the heat transfer from one object to another, consider the following. Look at what is *losing* heat and what is *gaining* heat. Place everything that is losing heat on one side of the equation, and everything that is gaining heat on the other. Don't forget both the $Q = mc\,\Delta T$ and $Q = mL$ terms!

4. As one part of an object becomes heated, its molecules begin to vibrate more quickly (as learned in Chapter 13). As these "hotter" molecules collide with neighboring, "cooler," molecules, they share some of their energy with them. It is in this manner that heat "flows" from one part of an object to another in a process known as *conduction*. The rate at which conduction happens depends on many factors, among them being material, cross-sectional area, length, and temperature difference [see Equation (14.4)].

5. The second way that heat can be transferred is **radiation.** An object will radiate heat away from itself at a rate given by Equation (14.6).

6. The third way that heat can be transferred is **convection.** As air molecules contact a hot object, they gain energy in much the same way as with conduction. These molecules, by vibrating more quickly, take up more space and thus have a lower effective density than their "cooler" neighbors. As we learned in Chapter 12, this difference causes a form of buoyancy, and the warmer air rises to be replaced by cooler air which is then heated by the "hot" object. As this cycle continues, heat is transferred away from the object and into the surrounding medium.

Key Concepts

CONVECTION: Heat flow in a fluid caused by mixing through density gradients.

HEAT CAPACITY: The ability of a material to store heat.

RADIATION: Heat flow caused by the emission of photons.

THERMAL CONDUCTIVITY: The ability of heat to flow through a material.

THERMAL RESISTANCE: The ability of a material to block heat flow, or to act as an insulator.

Important Equations

(14.1) $Q = mc\,\Delta T$ (Heat capacity)

(14.2) $Q = mL_v$ (Heat of vaporization)

(14.3) $Q = mL_f$ (Heat of fusion)

(14.4) $\dfrac{Q}{t} = \dfrac{kA\,\Delta T}{d}$ (Heat flow)

(14.5) $R = \dfrac{d}{k}$ (Thermal resistance where d is the material's thickness and k is the coefficient of thermal conductivity)

(14.6) $r = e\sigma T^4$ (Stefan-Boltzmann law for heat radiation, where r is the power per unit area, T is in K, e is the emissivity between 0 and 1, and σ is the proportionality constant 5.67×10^{-8} W/m^2-K^4)

Conceptual Problems

1. Heat flows in the direction of:
   - a. the higher temperature
   - b. the lowest pressure
   - c. the lowest temperature
   - d. solid to liquid

2. When steam condenses into water:
   - a. its temperature drops
   - b. it absorbs heat
   - c. it gives off heat
   - d. its temperature rises

3. A pressure cooker cooks food more rapidly because:
   - a. the higher pressure raises the boiling point of water
   - b. the pressure forces heat into the food
   - c. the pressure raises the heat capacity of the food
   - d. the high pressure lowers the boiling point of water

4. Heat transfer through conduction occurs:
   - a. only in solids
   - b. only in liquids
   - c. never in gases
   - d. in solids, liquids, and gases

5. Heat transfer through convection occurs:
   - a. in gases, liquids, and solids
   - b. in gases and liquids
   - c. only in gases
   - d. only in liquids

6. The materials with the highest heat conductivities are:
   - a. liquids
   - b. metals
   - c. solids
   - d. materials at high temperatures

7. In convection, a heat column of fluid moves because:
   - a. molecular collisions occur within it
   - b. molecular motions line up
   - c. its density is different from that of the surrounding fluid
   - d. the pressure is less in a heated fluid

8. Four pieces of tungsten are heated by an electric current. Which of the following statements is true?
   - a. The filament glowing white has the lowest temperature.
   - b. The filament glowing yellow has the highest temperature.
   - c. The filament glowing white has an intermediate temperature.
   - d. The filament glowing red has the lowest temperature.

9. To be able to radiate photons, an object must:
   - a. be hot enough to glow
   - b. be at a temperature greater than 0°C
   - c. have an electric current through it
   - d. be a blackbody

10. Why will a car radiator filled with antifreeze instead of water be more likely to overheat in the summer, disregarding the fact that the summer is hotter?

11. How does sweating help to cool the body?

12. If you think it's taking too long to hard-boil eggs in a pot of boiling water, would it make sense to turn up the heat?

13. In the winter, why does the metal handle of a snow shovel seem so much colder than the wooden shaft?

14. Why is the heating element of a water heater always installed at the bottom of the tank?

15. A thermos bottle has an inner and an outer glass container. The space between the containers is a vacuum. Why are the sides of the glass facing the vacuum chamber silvered?

Exercises

Heat Capacity

1. How much heat is needed to raise 4 m³ of water from 4°C to 14°C? (The heat capacity of water is 4186 J/kg-°C.)

   Answer: 167 MJ

2. (a) If 20 J of energy is pumped into a flask containing 100 g of water, by how much will the temperature rise? (b) If 20 J were removed from the same flask, by how much would the temperature fall?

   Answer: (a) 0.05°C
   (b) 0.05°C

3. A student, wanting a cool drink, drops an ice cube into a glass of soda. The ice cube initially has a mass of 20 g at −20°C, and the glass contains 200 g of soda at 20°C. What will the temperature of the soda be when the ice reaches 0°C?

   Answer: 19°C

4. After 10 kJ of heat was added to a substance weighing 1 kg, the temperature was found to have risen by 4.8°C. (a) What substance is this? (b) What would the substance be if the temperature had risen (b) by 2.4°C or (c) by 77°C?

   Answer: (a) Ice
   (b) Water
   (c) Gold

5. A piece of aluminum (c = 920 J/kg-°C) at 120°C is dropped into 10 kg of water at room temperature (22°C). The equilibrium temperature is found to be 30°C. What is the mass of the aluminum?

   Answer: 4 kg

6. How many liters of water at 4°C would you have to add to 20 kg of 54°C ethyl alcohol if you wanted the equilibrium temperature of the final solution to equal 24°C?

   Answer: 17.4 liters

*7. A 100-W heating element is placed into a large, insulating styrofoam cup filled with ice water. In the cup are 250 g of water at 20°C and 250 g of ice at −5°C. How long does it take the water to boil?

   Answer: 45.8 min

Changes of State

8. What heat is required to completely boil off 1 kg of water at (a) 100°C and (b) 50°C?

Answer: (a) 2.26 MJ
(b) 2.47 MJ

9. How many kilograms of lead (already at the melting point of 330°C) will 2.5×10^4 J liquefy?

Answer: 1 kg

10. A 10-kg block of ice at $-10°C$ is dropped into a bucket containing 100 kg of ethyl alcohol at 100°C. (a) What will the final temperature of the ethyl alcohol be at the moment the ice completely melts? (b) How much energy will be transferred from the ice to the alcohol?

Answer: (a) 85.4°C
(b) 3.56 MJ

*11. A blacksmith who wishes to cool down a newly forged, red-hot copper candlestick ($T = 800°C$) decides to pour large amounts of water at 20°C onto his creation until its temperature reaches 140°C. If the candlestick's mass is 2 kg, how much water will be required to properly cool the copper?

Assume that all steam created upon contact with the copper rapidly rises out of the influence of the hot metal. ($c_{copper} = 387$ liter/kg-°C.)

Answer: 0.2 kg

12. Steam at 125°C surrounds a 250-g block of ice at $-20°C$. How much of the steam is needed to melt the ice and bring it to 20°C?

Answer: $M = 43.5$ g

*13. A 5-kg iron ingot is taken from a forge at 1000°C and plunged into a bucket holding 5 kg of water at 20°C to cool it. After the ingot cools, how much water remains in the bucket?

Answer: 4.83 kg

14. A 60-g piece of iron and a 40-g piece of copper both at 90°C are dropped into an insulating styrofoam cup containing 300 g of water and 100 g of ice at 0°C. What is the equilibrium temperature?

Answer: $T_{eq} = 0°C$

Heat Conduction

15. A pane in a glass window measures 1.0 m by 0.75 m and is 5 mm thick. If the outside and inside temperatures are 10°C and 22°C, respectively, how much heat is lost through this window (a) in 1 s and (b) in 1 day?

Answer: (a) 1440 J
(b) 124 MJ

16. A basement wall has a cross section of 1 cm of drywall ($k = 0.04$ W/m-°C) nailed to 10 cm of concrete. The inside air (the drywall side) is 25°C, and the surrounding earth (the concrete side) is 54°C. (a) What is the temperature at the drywall-concrete interface? (b) What is the rate of energy loss per square meter through the basement wall?

Answer: (a) 44.3°C
(b) 77.2 W

17. The heat flow through the 10-cm brick wall of a kiln is found to be 1000 J/m² per second. If the temperature just outside the wall is measured to be 45°C, what is the temperature inside the kiln?

Answer: 211.7°C

*18. A swimming pool has a thin film of ice on it. The water directly under the ice is at 0°C. If the air temperature above the ice is $-15°C$, what is the rate of ice formation if the ice is 10 mm thick?

Answer: 2.6 cm/h

Thermal Resistance

19. How many centimeters of (a) brick, (b) iron, and (c) snow would yield the same thermal resistivity as 5 cm of glass wool?

Answer: (a) 75 cm
(b) 62.5 m
(c) 1 m

*20. Two layers of insulation are placed side by side in a wall. The insulation value of the first material is R-1, and the insulation value for the second material is R-2. Show that the equivalent insulation value for the wall is R-1 + R-2.

Convection and Radiation

21. A particular brick of material (10 cm × 20 cm × 5 cm) has an emissivity of 0.8 and is heated to 300°C. At what rate does it radiate energy?

Answer: 4.9 W

*22. Our sun (which can be thought of as a perfect blackbody) has a surface temperature of approximately 5500°C, and the solar energy emitted from the sun arrives at the earth at the rate of 1.4 kW/m². At what rate would solar energy arrive if the sun were just 5% cooler?

Answer: 1.14 kW/m²

*23. A certain blackbody is usually at a temperature of 1000 K. To what value should its temperature be changed in order for it to radiate energy at (a) twice the normal rate and (b) half the normal rate?

Answer: (a) 1190 K
(b) 840 K

Mechanical Equivalent of Heat

24. A 30-g bullet is traveling at 400 m/s. If we treat the bullet as a particle in a gas, what is its equivalent temperature?

Answer: $T = 1.16 \times 10^{26}$ K

25. A lead bullet emerges from the muzzle of the rifle that fired it at 100°C. The bullet strikes a piece of steel armor with enough energy to melt the bullet. What was its minimum speed?

Answer: $v = 234$ m/s

26. One kilogram of water at 0°C turns into ice at 0°C. If the energy that the water lost in turning from water into ice became kinetic energy, how fast would the 1-kg block of ice be moving?

Answer: $v = 818.5$ m/s

*27. A block of ice at 0°C with a mass of 55 kg is pushed along the floor of a large freezer also at 0°C for 25 m. Because of the friction, 30 g of ice melts. What is the coefficient of friction between block and floor?

Answer: $\mu = 0.746$

*28. A waterfall 5 m tall has water flowing over the top at 5 m/s. At the base of the fall, the water flows away at only 3 m/s. What is the maximum amount by which the temperature at the bottom can exceed the temperature at the top?

Answer: $\Delta T = 0.0136$°C

The technical infrastructure of our society depends on heat engines (steam turbines) that convert heat into work, which in turn generates electricity.

Thermo—dynamics

At the end of this chapter, the reader should be able to:

1. Describe what is meant by **isobaric, isothermal,** and **adiabatic** processes.

2. Calculate the work done by (or on) a gas in an isobaric process if given pressures, temperatures, and volumes.

3. Solve for the work done by (or on) a gas by using nothing other than a *P-V* (pressure-versus-volume) graph.

4. Solve for the *efficiency* of a given heat engine.

5. Calculate the change in *internal energy* of a gas if the work done and the heat generated by the gas are known.

6. Explain the **second law of thermodynamics.**

7. Briefly discuss how an *internal combustion engine* operates.

8. Briefly discuss how a *refrigerator* operates.

9. Explain what is meant by *entropy.*

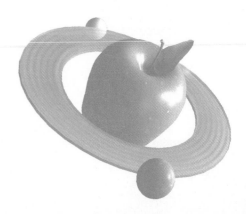

15.1   Work Done by a Gas

In an automobile internal combustion engine, a mixture of gasoline and air under pressure is ignited by a spark. This mixture then burns and becomes an expanding hot volume of mostly water vapor and carbon dioxide gas. The expanding gas at a high temperature pushes on a piston that turns a crankshaft. The expanding gas does work which turns the wheels of the car. The piston in the engine moves up and down in a cylinder of cross-sectional area A (usually circular). On the power stroke, the piston moves down a distance Δs from top dead center to bottom dead center and applies a pressure on the head of the piston. The force on the top of the piston is the pressure times the circular area of the head of the piston.

$$F = PA$$

The work that piston does is the force on the piston times the distance it travels down the bore.

$$W = F \, \Delta s$$

Combining the two, we have

$$W = (PA)\Delta s$$

$$A \, \Delta s = \text{change in volume} = \Delta V$$

(Equation 15.1)

$$W = P \, \Delta V$$

The work done by the expanding gas is equal to the pressure times the change in volume. *An expanding gas does work.* Conversely, work must be done on a gas to compress it.

In Chapter 14, the ideal gas law was discussed. Equation (15.1) for the work done on or by a gas fits neatly into the ideal gas law.

$$PV = nRT$$

$$P \, \Delta V = nR \, \Delta T$$

The volume on the left side of the equation is allowed to change, so something on the right side of the equation must also change to maintain the equality. Only temperature can change if the system is closed. If work is done on the gas in compressing it, the temperature of the gas rises. This fact is used in diesel motors to cause combustion. However, if work is done on the gas in getting it to expand, the temperature of the gas falls. This fact is used in air conditioning and refrigeration.

Isobaric process: A process in which there is constant pressure.

A process that occurs at a constant pressure is called an **isobaric process.** If the work done by the expanding gas in the cylinder of an internal combustion engine were done at a constant pressure, it would map out the rectangle shown in Figure 15.1. The work done by the expanding gas is the area under the *P-V* curve. Of course, the pressure in the cylinder of the internal combustion engine is not constant but might follow the curve shown in Figure 15.2. As the volume increases, the pressure will drop at constant temperature according to the ideal gas law. After ignition of the gas-air mixture in the cylinder, things happen so fast that a good approximation is that the expansion is happening at a constant temperature. *A process that happens at a constant temperature is called an* **isothermal process.**

Isothermal process: A process in which there is constant temperature.

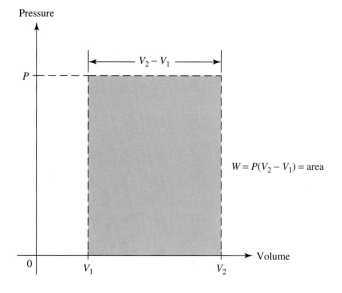

Figure 15.1 Work Done at Constant Pressure (Isobaric Process)

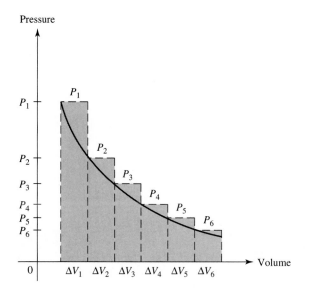

Figure 15.2 Work Done at Constant Temperature (Isothermal Process)

The work done by the expanding gas is still the area under the P-V curve in Figure 15.2. However, the area is broken up into a series of small expansions at a constant pressure. The work done is the sum of the areas of the six rectangles.

$$W = P_1 \, \Delta V_1 + P_2 \, \Delta V_2 + P_3 \, \Delta V_3 + \cdots$$

If the width of each rectangle is relatively large, then the area of the rectangles will be somewhat larger than the area under the curve. But as the width of each rectangle becomes smaller and the number of rectangles becomes larger, the sum of the area of the rectangles approximates more and more closely the actual area under the curve.

EXAMPLE 15.1

(a) How much work must be done on a kilogram of water (at the boiling point) to turn it into steam at STP? (b) How much of this heat does work in expanding the steam at 1 atm? (c) Where does the rest of the heat generated from boiling the water (2091 kJ) go?

SOLUTION

(a) The heat of vaporization of water is 2260 kJ/kg.

$$Q = L_v m$$

$$Q = (2260 \text{ J/kg})(1 \text{ kg})$$

$$Q = 2260 \text{ kJ}$$

(b) The density of steam at 1 atm is 0.6 kg/m³.

$$W = P \, \Delta V \qquad W = P(V_s - V_w)$$

$$V_s = \frac{1 \text{ kg}}{0.6 \text{ kg/m}^3} \qquad V_s = 1.67 \text{ m}^3$$

The volume of the water is only 0.001 m³. It can be ignored because the volume of the steam is so much greater.

$$W = (1.013 \times 10^5 \text{ Pa})(1.67 \text{ m}^3)$$

$$W = 169 \text{ kJ}$$

(c) The heat becomes the internal energy of the steam. Remember that the molecules in a gas act as millions of particles colliding in a perfectly elastic manner. There is no internal friction in this system to rob it of its internal kinetic energy. The remaining heat generated (2091 kJ) is the average kinetic energy of all of the H_2O molecules in the steam.

15.2   The First Law of Thermodynamics

First law of thermodynamics: Law stating that energy can be neither created nor destroyed but can change forms.

The internal combustion engine is a heat engine and operates according to the first law of thermodynamics. *In essence, the **first law of thermodynamics** says that energy cannot be created or destroyed but can be converted from one form to another.* In a heat engine, part of a flow of heat is converted into mechanical work. In all heat engines, three characteristic processes take place:

1. Heat is absorbed from a high-temperature source.
2. Mechanical work is done.
3. Heat is given off at a lower temperature.

In an internal combustion engine such as the four-cycle piston engine in most automobiles, process (1) takes place when the fuel-air mixture is ignited. The fuel gives off heat as it is burned. Process (2) takes place when this heat expands the gas in the cylinder against the piston, doing mechanical work. This heat also raises the internal energy of the gas. Process (3) takes place when the burned gas in the cylinder is exhausted to the atmosphere, giving off its heat.

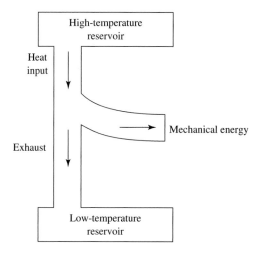

Figure 15.3 Heat Flow in a Heat Engine

A diagram of the heat flow in a heat engine is given in Figure 15.3. The heat flow is from the burning gases in the cylinder to the low-temperature reservoir that is the external atmosphere (usually at STP). Only part of this heat is tapped off directly as mechanical energy. This relationship (the first law of thermodynamics) can be quantified as

$$Q = \Delta U + W$$

(Equation 15.2)

Equation (15.2) states that the net heat input equals the change in internal energy plus the net work output. For an internal combustion engine, Q is positive because heat is added by the burning gas; W is positive because work is done on the surroundings by the system; and ΔU is also positive because the internal energy of the gas was raised. Example 15.1 demonstrated the work done by a gas on its surroundings. But it can also be used to illustrate the first law of thermodynamics as given in Equation (15.2). Of the total amount of heat added to the system (2260 kJ), most (2091 kJ) went into the internal energy of the system, and only a small fraction (169 kJ) went into work done on the surroundings.

15.3 The Efficiency of a Heat Engine

In Equation (15.2) for the first law of thermodynamics, whenever heat is liberated to do work, the internal energy of the gas doing the work is always raised as a by-product of the liberation of the heat. This heat is considered to be waste heat because it does no useful work. In the case of the internal combustion engine, this waste heat is exhausted to the atmosphere as a hot exhaust gas. The resulting implication is that an internal combustion engine has an efficiency of less than 100%. Its actual efficiency is only around 30%. In fact, most of the heat generated by an internal combustion engine is wasted (on the order of 70%)! An equation for efficiency can be derived from the first-law equation.

$$Q = \Delta U + W$$

$$\text{eff} = \frac{W}{Q}$$

$$\text{eff} = \frac{Q - \Delta U}{Q}$$

$$\text{eff} = 1 - \frac{\Delta U}{Q}$$

(Equation 15.3)

As long as heat is exhausted from the engine (ΔU), then the engine's efficiency will be less than 100%. For 100% efficiency, the low-temperature reservoir in Figure 15.3 would have to be at absolute zero. But any heat that is added to a reservoir at absolute zero will immediately raise the temperature of the reservoir above absolute zero. You can't win!

Example 15.1 has already had a great deal of utility so far in this chapter. We will use it again in this section to illustrate the efficiency of a heat engine. An internal combustion engine goes through repetitive cycles as it generates power. Example 15.1 can be considered as one cycle of a heat engine, and its efficiency can be calculated.

$$\text{eff} = 1 - \frac{\Delta U}{Q} \qquad \text{eff} = 1 - \frac{2091 \text{ kJ}}{2260 \text{ kJ}} \qquad \text{eff} = 7.5\%$$

The early Newcomen steam engines, invented in 1712, were called *atmospheric engines* at the beginning of the Industrial Revolution because the highest pressure in them was 1 atm. A partial vacuum was created in the cylinder by condensing the steam, and that caused atmospheric pressure on the other side of the piston to push the piston up or down the cylinder. In Example 15.1, steam was expanded at 1 atm. Because the early Newcomen steam engines expanded against 1 atm, they had low efficiencies similar to those of Example 15.1.

In Chapter 14, *temperature* was defined as the average kinetic energy of a collection of atoms or molecules in a gas.

$$\frac{3}{2}kT = \frac{1}{2}mv_{\text{avg}}^2$$

With this in mind, we can cast the efficiency of a heat engine in terms of the temperatures of the burning gases and the temperature of the gas when it is exhausted to the outer atmosphere.

$$\text{eff} = 1 - \frac{\Delta U}{Q}$$

$$Q = \frac{3}{2}kT_{\text{in}} \qquad \Delta U = \frac{3}{2}kT_{\text{exh}}$$

$$\text{eff} = 1 - \frac{3/2kT_{\text{exh}}}{3/2kT_{\text{in}}}$$

$$\text{eff} = 1 - \frac{T_{\text{exh}}}{T_{\text{in}}}$$

(Equation 15.4)

EXAMPLE 15.2

What is the theoretical efficiency of an internal combustion engine that burns its gas-air mixture at about 1200 K and exhausts it at about 500 K?

SOLUTION

$$\text{eff} = 1 - \frac{T_{\text{exh}}}{T_{\text{in}}} \qquad \text{eff} = 1 - \frac{500 \text{ K}}{1200 \text{ K}} \qquad \text{eff} = 58.3\%$$

The actual efficiency can be much lower because heat leaks through the engine's metal block, warming the entire engine. An actual efficiency of about 35% to 40% is normal. An automobile's radiator and cooling system are needed to rid the engine of this waste heat. At 35% efficiency the engine is better by a factor of 4 than the first crude steam heat engines, but it still wastes more heat than it converts to useful work.

15.4 The Second Law of Thermodynamics

Nature spontaneously tends toward disorder. Physical systems also tend to seek the lowest-energy state. These two tendencies are often in conflict. When they are in conflict, disorder usually wins. The second law of thermodynamics makes highly improbable certain processes that are energetically feasible. Take, for instance, the process in which an ice cube is dropped from a sufficient height. Without air friction, as the ice cube falls, it will increase its kinetic energy. When it hits the ground, an inelastic collision takes place, and all of its kinetic energy is transformed into heat. If the velocity was high enough, this heat melts the ice cube, and all that is left is a puddle on the ground. From an energy point of view, it's quite possible for this puddle of water to spontaneously jump into the air, becoming an ice cube as it does so. The puddle of water has a lot of internal energy. If some of it is converted into kinetic energy, the water loses internal energy, and the water cools. If the water loses enough internal energy, it turns back into an ice cube. However, the second law of thermodynamics makes this event all but impossible. The reason is that the puddle of water is a more disordered system than the ice cube, even though it has more energy. *One way of stating the **second law of thermodynamics** is to say that natural processes tend to move in the direction of greater disorder.*

Energetically it would be equally feasible to power a heat engine by extracting the internal energy from the air at STP, exhausting liquid air. The liquid air has much less internal energy than gaseous air, and the balance would be turned into work by the heat engine. But a heat engine must operate between a hot reservoir and a cold reservoir. Heat flows between the two, and energy in the form of work is extracted. The cold reservoir for most heat engines is the atmosphere at STP, and work is required to cool something from room ambient to a lower temperature. Liquid air is much cooler than 20°C. Also, a liquid is more ordered than a gas, and the second law of thermodynamics therefore forbids this process. The internal energy present in a volume of air at STP originates from the random motion of molecules and atoms. This random motion is disorder in the system. If a heat engine were constructed that takes heat from a source and converts it into an equivalent amount of work, the "gas" in the heat engine's exhaust would have to be at absolute zero. No cold reservoir is that cold. The exhaust gases from a heat engine will at best be exhausted at the ambient temperature of the air and will have the ambient temperature's equivalent internal energy. Another way of stating the second law is to say that it is impossible to transform a given amount of heat completely into useful work. That is, no heat engine can be 100% efficient, and some of its heat input must be rejected.

Second law of thermodynamics: Law stating that natural processes tend to move in the direction of greater disorder (entropy).

EXAMPLE 15.3

A 30-g ice cube at $-10°C$ strikes a solid surface with enough energy to completely melt the cube. What is its minimum speed when striking the surface?

SOLUTION

$$Q = mc\,\Delta T \qquad Q = (0.03\ \text{kg})(4.19\ \text{kJ/kg-°C})(10°C) \qquad Q = 1257\ \text{J}$$

Thus, 1257 joules is necessary to warm the ice cube to the melting point.

$$Q = mL_f \qquad Q = (0.03\ \text{kg})(335\ \text{kJ/kg}) \qquad Q = 10{,}050\ \text{J}$$

Therefore, 10,050 joules is needed to completely melt the ice.

$$Q_T = 1257\ \text{J} + 10{,}050\ \text{J} \qquad Q_T = 11{,}307\ \text{J}$$

$$Q = \frac{1}{2}mv^2$$

$$v = \sqrt{\frac{2Q}{m}}$$

$$v = \sqrt{\frac{2(11{,}307\ \text{J})}{0.03\ \text{kg}}}$$

$$v = 868\ \text{m/s}$$

That speed is over 1900 mph or 2864 ft/s, which is much faster than a bullet from a high-powered, flat-trajectory rifle! The internal energy stored in the puddle of water is surprisingly large.

15.5   The Carnot Engine

The efficiency equation for heat engines calculates the maximum efficiency that can be had when an engine operates between the temperature of the heat input and exhaust. This efficiency describes the operation of an ideal heat engine called a *Carnot engine,* after the French engineer Nicolas Sadi Carnot who first described it in the early 1800s. A Carnot engine is ideal in the sense that there are no losses to friction or losses of stored heat by conduction or radiation. Another requirement of an ideal heat engine is that every process in it must be reversible without losing any energy to outside the system. Heat flow at a constant temperature is reversible without losing energy. From the equation for the first law of thermodynamics, if a gas expands at a constant temperature, all of the heat added to the gas becomes work done by the gas because its internal energy doesn't change. Therefore, its temperature doesn't rise. As mentioned earlier in the chapter, a process that takes place at a constant temperature is an *isothermal process.*

    *Another theoretically reversible process, the **adiabatic process,** takes place with no heat transferred from the system to the outside.* Heat transfer takes time, and many processes in a heat engine occur fast enough that they are essentially adiabatic. The ideal gas law has three variables: pressure, volume, and temperature. Other heat-related

Adiabatic process: A process in which there is no heat transfer from inside a system to outside the system.

processes take place with the variables volume and pressure held constant. As mentioned previously, a constant pressure process is called *isobaric*.

An ideal Carnot heat engine uses only isothermal and adiabatic processes. The Carnot engine consists of a cylinder filled with an ideal gas and fitted with a piston. Initially, an amount of heat Q_o is added to the gas, and it expands isothermally according to the ideal gas law. If the expansion happens slowly, the pressure inside the cylinder remains constant. The piston moves up or down the cylinder to equalize the inside and outside pressures as the gas takes up heat from the source, so $P \Delta V = nRT$. Since $P \Delta V$ is the work done, the heat added exactly equals the work done. The heat source is then removed, and the expansion continues adiabatically. This further expansion is at the expense of the internal energy of the gas, and the temperature of the gas drops. In both the isothermal and the adiabatic processes, the gas expands, pushing the piston up or down the cylinder. The piston does work on whatever it is attached to.

The engine must now be returned to its original state to do further work. The gas in the cylinder is compressed first isothermally and second adiabatically to return it to its original condition. The Carnot cycle is shown in Figure 15.4. Isothermal and isobaric expansion takes place from points *a* to *b*. Adiabatic expansion takes place from *b* to *c*. The net work done by the expanding gas is the area under the curve *abc*. Now the gas must be returned to its original condition. Isothermal compression takes place from *c* to *d*, and adiabatic compression takes place from *d* to *a*. The work done on the gas is under curve *cda*. The net work done by the engine is the area under the curve *abc* minus the area under the curve *cda*, or the area inside the closed curve *abcd*. Because work must be done on the engine to return it to its original condition, even a Carnot engine cannot be 100% efficient.

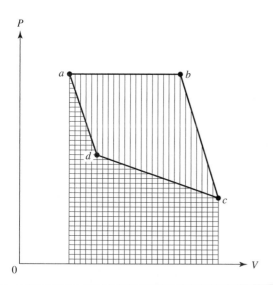

Figure 15.4 The Carnot Cycle

15.6   Internal and External Combustion Engines

The most common types of heat engines are internal combustion engines, which include jets and piston engines (both gas and diesel), and external combustion engines, which are steam engines and steam turbines. An engine with a piston in a cylinder is cyclical, going through cycle after identical cycle to produce power. The jet engine and the steam turbine operate essentially continuously.

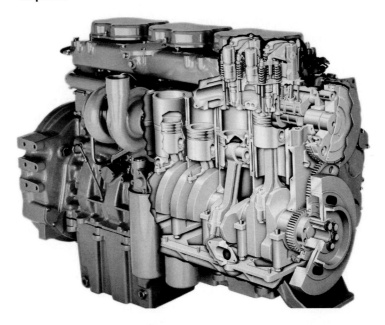

Cutaway drawing of an internal combustion engine.

The first type examined in some detail is the piston engine found in cars and trucks. These engines are essentially one of two types: gasoline or diesel. Both types have a cylinder with a piston in it, and at the top of the cylinder is a pair of valves that open and close depending on the part of the cycle the engine is in. Figure 15.5 shows the *intake stroke,* the first part of the cycle when the fuel-air mixture is sucked into the combustion chamber. The exhaust valve is closed, and the intake valve is open. When the piston moves down the bore, a partial vacuum is created in the bore above the piston, and the higher atmospheric pressure shoves a charge of fuel-air mixture into the cylinder.

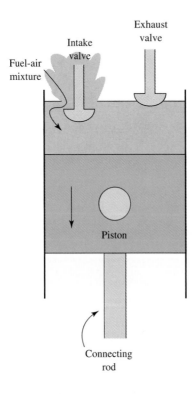

Figure 15.5 Intake Stroke of a Four-Cycle Engine

Figure 15.6 is the *compression stroke*. Both the intake and the exhaust valves are closed. As the piston moves up the bore, the gases above the piston are compressed. The connecting rod attached to the piston is connected to the crankshaft and converts the up-and-down vertical motion of the piston into the rotary motion of the crankshaft. The piston is at the top of the stroke in position to turn the crankshaft during the next part of the cycle, the power stroke. The compressed gases can now do work on the piston, pushing it down as they burn and expand.

Figure 15.7 is the *power stroke*. Both the intake and the exhaust valves are still closed. The gases are ignited by the spark from a spark plug and begin to burn. The burning gases expand and push the piston down the cylinder. The piston does work on whatever the crankshaft is attached to. Above the piston there is an overpressure, and the pressure differential between the top of the piston and the bottom of the piston pushes the piston down the cylinder.

$$F = PA$$

$$W = Fh \qquad \text{(where h is the distance moved down the bore)}$$

$$W = Pa(h) \qquad Ah = \Delta V$$

$$W = P \, \Delta V$$

Figure 15.8 is the *exhaust stroke*. The exhaust valve is open, and the intake valve is closed. As the piston moves back up the bore, it pushes the burned gases up the cylinder and out the opened exhaust valve. When the piston reaches the top of its stroke, the exhaust valve closes, and the intake valve opens. The piston and cylinder are again exactly as in Figure 15.5, and the whole cycle starts all over again.

A gasoline engine with valves is often called a *four-stroke engine* because each cycle can be broken up into four distinct stages. Those stages are shown in Figures 15.5 through 15.8. The crankshaft often has a heavy flywheel attached to it so that the crankshaft will keep turning after the power stroke. The flywheel stores kinetic energy of rotation, allowing the cylinder to go through the exhaust, intake, and compression stages to get the engine ready to fire again. The flywheel also smooths the running of the engine from power stroke to power stroke. The coordination of the valve openings and closings is achieved by a camshaft geared to the crankshaft; it ensures that the valves open and close at the proper times. A spark plug fires the gas-air mixture in a gasoline engine. The moment when the plug fires is important and must be coordinated to ignite the fuel-air mixture when the piston is near the top of the compression stroke. If the plug fires at the wrong time (at the beginning of the compression stroke or near the end of the power stroke, for example), the engine loses power and could be severely damaged.

A diesel engine doesn't have a spark plug and its associated spark timing gear. The strokes are the same as in Figures 15.5 through 15.8, but during the intake stroke, only air enters the cylinder. This air is compressed and heats up during compression. The fuel is then injected into the cylinder at the end of the compression stroke when the air is at maximum compression, and the heat of the compressed air is enough to ignite the fuel-air mixture. Compression ratios for diesel engines are larger than for a gasoline engine because the air must be compressed more to get it to the temperature for ignition of the fuel-air mixture. Conversely, the compression ratios of a gas engine must be kept relatively low so that the gas-air mixture doesn't ignite too early.

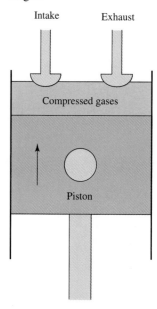

Figure 15.6 Compression Stroke of a Four-Cycle Engine

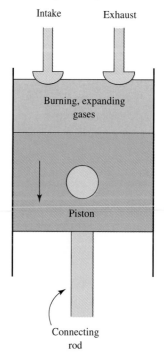

Figure 15.7 Power Stroke of a Four-Cycle Engine

Figure 15.8 Exhaust Stroke of a Four-Cycle Engine

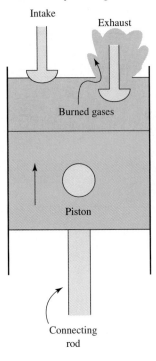

The spark times the ignition in a gasoline engine, and the fuel injection times the ignition in a diesel engine.

The indicator diagram for a four-stroke gasoline engine is given in Figure 15.9. The line from points a to b is the intake stroke. It takes place at constant pressure (atmospheric). The curve from b to c is the compression stroke. The volume is decreasing, and the pressure is increasing. Work is being done on the gas inside the cylinder. Ignition takes place from c to d. It takes place at the top of the piston's stroke, and the pressure rises rapidly before the piston gets a chance to move far from top dead center. The power stroke is represented by the curve from d to e. The pressure inside the cylinder is dropping, and the volume is increasing. The expanding gas inside the cylinder is doing work. The vertical line from e to b is when the exhaust valve opens. It opens when the piston is near the bottom of its stroke, and the overpressure inside the cylinder pushes most of the exhaust gases out before the piston moves very far from bottom dead center. The exhaust stroke, represented by the line from b to a, occurs at constant pressure (atmospheric) because the pressure inside the cylinder has already fallen near atmospheric in curve e to b.

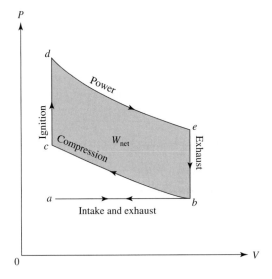

Figure 15.9 Indicator Diagram for a Four-Stroke Engine

Figure 15.9 is very similar to the indicator diagram for a Carnot engine given in Figure 15.4, which holds true for any heat engine. The only significant difference is the horizontal line from a to b (the intake and exhaust strokes). Both strokes occur at atmospheric pressure, and the charge of fuel and air is replenished in the cylinder. This replenishment is important in a real engine because it ensures that the working gas in the cylinder doesn't gradually heat up during running. If this were to occur, the engine's efficiency would drop as the working gas heated up. There is an existing engine called a *Stirling engine* which works by applying a heat source to one end of a double piston. It's simpler mechanically than either a gasoline or a diesel engine, but since the gas in the cylinder isn't exhausted and replenished, it heats up as the engine runs, and efficiency drops.

Two other types of heat engines are in common use: the steam turbine and the gas turbine. The steam turbine is a steam engine. In electric power generated by coal- or oil-burning power plants, as well as power generated by nuclear installations, steam turbines are used to convert the heat generated by burning fossil fuels or from nuclear fission into

Turbine blades of a large steam turbine used to generate electricity in a coal-, oil-, or gas-fired power plant.

mechanical work. The generated heat boils water and heats the resulting steam so that it becomes high-pressure steam. The high-pressure, high-temperature steam is fed into the first stage of a steam turbine such as the one diagramed in Figure 15.10. The steam expands through many stages in series (only two expansion stages are shown in the drawing), and the expanding steam does work by turning the turbine's axle. The steam pushes on the rotating blades attached to the turbine's shaft or axle, causing it to rotate just as the crankshaft does in a gasoline or diesel four-stroke engine. This rotating shaft is connected to a generator that converts the rotary motion into an alternating electrical signal that travels over the power lines to the outlets in the walls of our homes and offices. (The generator will be explained in succeeding chapters dealing with electromagnetism.) The expansion of the steam is essentially a continuous process from intake to exhaust, in contrast to the four distinct phases in the operation of a four-stroke engine. The final result is the same, though, with power being fed to a rotating shaft.

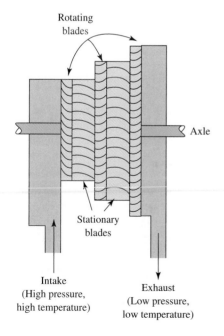

Figure 15.10 Steam Turbine

Since the steam turbine is a heat engine, it will have an indicator diagram similar to the one for a gasoline or diesel four-stroke engine. Such a diagram is graphed in Figure 15.11. Because the expansion of the steam is essentially continuous, and because heat is added externally instead of internally, the indicator diagram is somewhat simpler than for a four-stroke engine. At the intake, the turbine takes high-pressure, high-temperature steam and allows it to expand through multiple expansion stages inside the turbine. As the steam expands, its volume increases, and its pressure and temperature fall. At the exhaust, if the temperature is less than 100°C at 1 atm, it condenses back to liquid water. At this point it is again heated externally to reconvert it into high-pressure, high-temperature steam, and the process repeats itself.

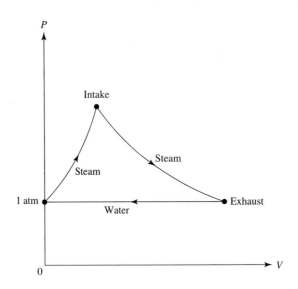

Figure 15.11 Indicator
Diagram for a Steam Turbine

A gas turbine is an internal combustion engine. The gases that expand are burned inside the engine as in a four-stroke engine, in contrast to the external heating of the water and steam in the steam turbine. But as in the steam turbine, the gases expand in a continuous fashion and turn the turbine's shaft. The work done by the expanding gases is used in two very different ways depending on whether the rotating shaft must do most of the work, such as turning a propeller or the rotors of a helicopter, or whether the turbine is to be used as a jet engine and the expanding gases are to provide thrust to push the aircraft forward. A turboprop aircraft such as the C-130 cargo aircraft uses gas turbines to turn the propellers of its four engines. A jet aircraft, however, turns the expanding gases directly into thrust. But regardless of whether the turbine turns a prop or produces thrust, there must be at least two stages of turbine blades. The first stage is used to compress and heat (through rapid compression) the incoming air before the fuel is injected into it. After the fuel-air mixture burns, the expanding gases pass through the second stage of turbine blades. In a jet engine this second stage is only large enough so that sufficient power is added to the shaft to run the compressor. After passing through the second stage, the gases continue to expand but provide thrust which pushes the aircraft forward. A jet engine is an internal combustion, gas turbine reaction engine! If the gas turbine's design is to turn a propeller, the second stage of turbine blades will be long (as in the steam turbine) because the expanding gases are expected to produce more shaft horsepower than thrust.

A chemical rocket engine is also a heat engine. Its thrust is provided by fast-moving exhaust gases expanding out its rear nozzles. The very high heat of combustion causes the exhaust gases produced to rapidly expand and produce thrust just as the jet engine does. Both the jet and the rocket are reaction motors. The jet can operate only in the atmosphere because it must use the oxygen in the atmosphere to burn the fuel (usually kerosene). A rocket carries its oxidizer with it (usually in the form of liquid oxygen) to add to its fuel (kerosene, alcohol, or liquid hydrogen). The oxidizer is carried as a liquid in a separate tank, and for this reason, the rocket engine can operate in the vacuum of space. The fact that a rocket engine can produce thrust in the vacuum of space proves that the engine doesn't need something to "push against" to produce thrust. The thrust is produced by conservation of momentum. The total momentum of the expanding exhaust gases leaving the rocket from behind must equal the momentum of the spacecraft moving in the opposite direction. That concept was covered in Chapter 5 dealing with momentum.

15.7   Refrigerators and Their Thermodynamics

If a gas is compressed rapidly, the temperature of the gas rises. Conversely, if a gas expands rapidly, its temperature drops. Most refrigeration equipment is based upon this principle.

The working substance of a refrigeration unit, called a *refrigerant,* operates in both the liquid and the vapor phases. The refrigerant is a gas that is easily liquefied, and for this reason most refrigerants are fluorocarbons. Freon is such a fluorocarbon. A fluorocarbon can be liquefied at moderate temperatures and pressures. Freon at room temperature can be liquefied at a little less than six times atmospheric.

A refrigeration system has four main components: a compressor, a condenser, an expansion valve, and an evaporator (see Figure 15.12.) The refrigeration cycle starts with the refrigerant (freon in this system) at roughly STP. It enters the compressor as a gas. The compressor has a cylinder-and-piston assembly much like that of an internal combustion engine. When the piston moves down, the intake valve opens, and freon is sucked into the cylinder. When the piston reaches the bottom of its stroke, the intake valve closes, and the piston moves back up the bore, compressing the freon. At a predetermined pressure (around 8 bars), the exhaust valve opens and discharges the compressed freon gas into the condenser. Because of the rapid compression, the freon gas is warmer than the original 20°C. In the condenser it is cooled back down to a bit above room temperature, but because of the relatively high pressure, it changes phase from gas to liquid. (The condenser is either air-cooled or water-cooled.) This high-pressure liquid freon now enters the expansion valve where it expands rapidly. The pressure drops from about 8 bars to a little above atmospheric. As the freon expands, its temperature drops well below the freezing point of water (around −10°C). From the expansion valve, the gaseous freon runs through the evaporator. In the evaporator the cold freon picks up heat from the various compartments of the refrigerator and warms

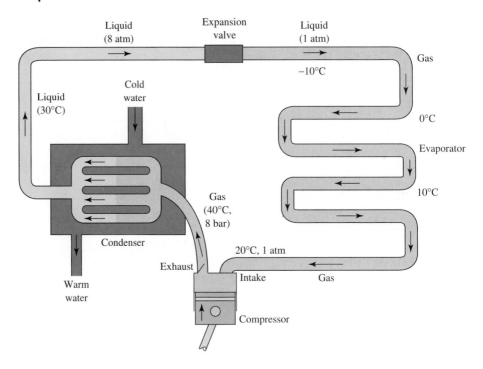

Figure 15.12 Refrigeration
Unit

up again. At the end of its tour through the evaporator, the freon is again at about room temperature and pressure and ready to be fed back into the compressor to start the cycle all over again. Remember that heat flows from a warm temperature to a cooler temperature, never the other way around. The heat in whatever item you put in a refrigerator to cool or freeze flows into the freon refrigerant that is constantly running through the evaporator coils.

The above description also applies to air conditioners. Both the refrigerator and the air conditioner are machines that extract heat at a low temperature and exhaust it at a high temperature. It is the Carnot cycle for heat engines in reverse. The first law of thermodynamics still applies, but now Q is the heat exhausted at high temperature, ΔU is the heat absorbed at low temperature, and W is the work that the refrigerator or air conditioner compressor has to do to absorb the heat.

Other compounds are now replacing freon as a refrigerant because of the damage that freon does to the ozone layer at the top of our atmosphere. Freon itself is relatively inert, but as a gas, when it gets into the atmosphere, it makes its way to the top where the fluorine in the freon destroys the ozone there. Ozone, a compound made up of three oxygen molecules, is formed by the intense ultraviolet radiation found in the upper reaches of our atmosphere. It quickly breaks down, but the fluorine from the fluorocarbons breaks ozone down even faster, and an area depleted of ozone (an ozone hole) appears in our atmosphere, allowing ultraviolet rays to reach the surface of the earth. If the ultraviolet radiation were at the same level at the surface of our planet as it is at the top of our atmosphere, it would quickly kill off life on this planet. The ozone in our atmosphere acts as a shield, and compounds such as freon are being taken off the market to help maintain the integrity of our atmospheric shield.

15.8   Entropy

The concept of entropy deals with the amount of disorder or order in a system. It's entropy that governs whether or not a process is irreversible. Remember Example 15.3 in which an ice cube was dropped from a great enough height so that when it hit the ground, it immediately melted because of the mechanical equivalent of heat? From an energy point of view, it would be okay if the puddle of water on the ground suddenly froze and the energy lost from the water by freezing turned into kinetic energy, hurling the ice cube back into the air. But have you ever seen such a thing really happen, except in a film that has been run backward? It's a highly improbable process. Things run down thermodynamically like a clock. If two materials at different temperatures are placed together, heat flows down the temperature gradient, energy is moved, and work is done. But if the two objects are left in contact long enough, they come to an equilibrium temperature, and no more work can be done. Energy is neither lost nor destroyed, but now no energy is available to do useful work. The energy is randomized throughout the system and is at a maximum level of disorder. Chaos proceeds spontaneously. Have you ever noticed how we're always picking up after ourselves and cleaning things? Dirty things stay dirty unless washed. A washed item is an item that has increased order. Entropy is related to disorder, and disorder is always increasing. Thus, the entropy of the universe increases with time.

The universe started as a very small, highly energetic point and started to expand. It's been expanding ever since for roughly the last 15 billion years. As the universe expanded, it cooled, until the background radiation left over from the formation of the universe (the Big Bang) has dropped to around 3 K. This radiation was detected only in the middle 1960s. Stars form and burn out, and the universe will eventually run down to a "heat death." The energy that was originally there will still be present, but it will be degraded until everything is at one temperature and at a maximum of disorder. The universe will have run down!

The direction of time flow at our level of organization always flows from the past into the future. Oddly enough, the reason for this is thermodynamic. The arrow points in the direction of greatest disorder and is governed by irreversible processes. The past is irretrievable for the same reasons that the puddle on the floor won't spontaneously become an ice cube and go shooting up into the air.

Summary

1. A process in which the pressure remains constant is called an **isobaric** (equal-pressure) **process** ($\Delta P = 0$). The work done by such a process is given by Equation (15.1).

2. A process in which the temperature remains constant is called an **isothermal** (equal-temperature) **process** ($\Delta T = 0$).

3. An **adiabatic process** is one in which there is no heat exchange with the outside environment ($Q = 0$).

4. As a heat engine operates, some of the energy available is inevitably wasted as heat. The ratio of useful work that can be done to the amount of available energy in a heat engine is called the *efficiency* of that engine and can be calculated by using Equation (15.3).

5. The Carnot efficiency [Equation (15.4)] is the highest efficiency possible for a given heat engine.

6. The **first law of thermodynamics** states that energy can be neither created nor destroyed in a process, but may change form. Thus, the internal energy of a gas may be converted into useful work, or it may be partially given off as waste heat. The relationship between these forms of energy are given in Equation (15.2).

7. The **second law of thermodynamics** states that naturally occurring process will move in the direction that ends with the greatest *entropy*, or disorder, in a system.

Key Concepts

ADIABATIC PROCESS: A process in which there is no heat transfer from inside a system to outside the system.

FIRST LAW OF THERMODYNAMICS: Law stating that energy can be neither created nor destroyed but can change forms.

ISOBARIC PROCESS: A process in which there is constant pressure.

ISOTHERMAL PROCESS: A process in which there is constant temperature.

SECOND LAW OF THERMODYNAMICS: Law stating that natural processes tend to move in the direction of greater disorder (entropy).

Important Equations

(15.1) $W = P \, \Delta V$ (Work done on or by a changing volume of gas)

(15.2) $Q = \Delta U + W$ (First law of thermodynamics, where Q is the heat input, ΔU is the rise in internal energy, and W is the work done)

(15.3) $\text{eff} = 1 - \dfrac{\Delta U}{Q}$ (Efficiency of a heat engine)

(15.4) $\text{eff} = 1 - \dfrac{T_{\text{exh}}}{T_{\text{in}}}$ (Efficiency of a heat engine in terms of combustion and exhaust temperatures in kelvins)

Conceptual Problems

1. A heat engine operates by taking in heat at a high temperature and:
   a. converting some into work and exhausting the rest at the same temperature
   b. converting it all into work
   c. converting some of it into work and exhausting the rest at a lower temperature
   d. changing the internal energy of the gas

2. Heat flows from a high temperature to a low temperature. The law responsible for this fact is:
   a. conservation of energy
   b. the first law of thermodynamics
   c. the second law of thermodynamics
   d. conservation of momentum

3. In an adiabatic process:
   a. the system's temperature remains constant
   b. no work is done on or by the system
   c. the system's pressure remains constant
   d. no heat enters or leaves the system

4. The maximum amount of heat that can be converted to useful work:
   a. depends on the intake and exhaust temperatures
   b. depends only on the intake temperature
   c. depends only on the exhaust temperature
   d. depends on the conservation of momentum between molecules

5. The work necessary to extract a given amount of heat:
   a. depends on the intake and exhaust temperatures
   b. depends only on the intake temperature
   c. depends only on the exhaust temperature
   d. depends on the conservation of momentum between molecules

6. In a P-V diagram, the area enclosed by one cycle of a heat engine:
   a. equals the work done on the engine per cycle
   b. equals the heat output per cycle
   c. equals the heat input per cycle
   d. equals the work done per cycle

7. A heat engine with no friction can be 100% efficient:
   a. if its exhaust equals its intake temperature
   b. if its exhaust temperature is 0 K
   c. if the change in internal energy is maximized
   d. if the heat input equals the intake temperature

8. A Carnot engine that operates between absolute temperature T_u and T_L:
   a. has the maximum theoretical efficiency
   b. has the same efficiency as a real engine operating between the same temperatures
   c. has input heat equal to the change in internal energy
   d. is 100% efficient

9. Rank these engines according to their efficiencies from least to greatest: steam turbine used in electrical generation plants; diesel engine; automobile gasoline engine; and Carnot engine.

10. Diesel fuel in a diesel engine is ignited by:
    a. rapid compression
    b. the hot, compressed gas into which it is injected
    c. a spark plug
    d. the hot exhaust gases left over from the previous cycle

11. The physical concepts behind the operation of a refrigerator are almost the same as those of:
    a. a heat engine   b. the freezing of water
    c. heat capacity   d. entropy

12. A refrigerator:

   a. causes heat to flow from a cold temperature to a warm temperature

   b. produces cold as opposed to heat in a heat engine

   c. removes heat from a volume and moves it elsewhere

   d. reverses the entropy process

13. A refrigerator exhausts:

   a. heat at a higher temperature than the temperature at which it is absorbed

   b. the same amount of heat that it absorbs from the interior

   c. heat at a lower temperature than the temperature at which it is absorbed

   d. less heat than it absorbs from its contents

14. The refrigerant used in most refrigerators and air conditioners is:

   a. a gas that won't liquefy

   b. a gas that will easily liquefy

   c. a liquid that won't solidify

   d. a solid that easily liquefies

15. Heat is absorbed by the refrigerant in a refrigerator when:

   a. the refrigerant melts

   b. the refrigerant vaporizes

   c. the refrigerant condenses

   d. the refrigerant is compressed

16. Why can't you cool a room by leaving the refrigerator door open?

17. List at least three energy sources that don't employ heat engines directly.

Exercises

Work Done by a Gas

1. In the indicator diagram shown in Figure 15.13, what is the total work done on (or by) the system?

Figure 15.13 (Exercise 1)

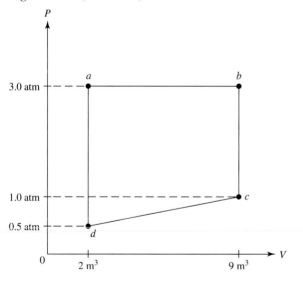

Answer: 160 kJ, by the system

2. The indicator diagram of a particular engine is shown in Figure 15.14. Find the work done in each part of the cycle: (a) from *a* to *b*; (b) from *b* to *c*; (c) from *c* to *d*; and (d) from *d* to *a*. (e) Find the total work done per cycle of this engine.

Figure 15.14 (Exercise 2)

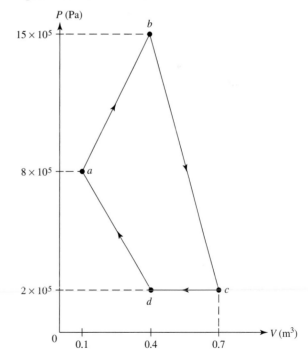

Answer: (a) W_{ab} = 350 kJ
(b) W_{bc} = 260 kJ
(c) W_{cd} = −60 kJ
(d) W_{da} = −150 kJ
(e) W_{tot} = 400 kJ

3. A four-cylinder, four-stroke motor develops 120 hp at 4000 rpm. The pistons are 4.25 inches in diameter, and the stroke is 5 in. What is the average overpressure on a piston during the power stroke?

Answer: 41.7 lb/in.^{2}

4. Three moles of a gas in a cylinder are expanded adiabatically by an airtight piston in a cylinder by doing 7 kJ of work in pulling the piston up the cylinder. What is the temperature change of the gas?

Answer: -280.6°C

First Law of Thermodynamics

5. An amount of heat equal to 1000 J flows into a system (containing 30 moles of an ideal gas) that does 200 J of work on its surroundings. (a) What is the change in the internal energy of the gas? (b) What is the corresponding change in temperature?

Answer: (a) 800 J
(b) 2.14°C

6. A Carnot engine has a combustion temperature of 600°C and an exhaust temperature of 150°C with a power output of 120 hp. (a) What is the rate at which the engine generates heat? (b) What is the rate at which the engine exhausts heat?

Answer: (a) 196.5 kW
(b) 95.2 kW

7. The ratio of the heat absorbed to the work done for a window air conditioner rated at 2500 Btu/h is 2 : 1. (a) How much power must be supplied to the air conditioner when it runs? (b) If electricity costs 10¢ per kilowatt-hour, how much does it cost to run the air conditioner for 12 h? A British thermal unit (Btu) equals 1054 J.

Answer: (a) 349 W
(b) 42¢

Efficiency of a Heat Engine

8. A given engine is 20% efficient, and it draws from a hot reservoir at a temperature of 1200 K. What is the exhaust temperature?

Answer: 960 K

9. It has been proposed that OTECs (ocean thermal energy converters) could be constructed in the near future as an alternative energy supply. These devices would act as heat engines, operating between 26°C (a temperature typical at the surface of a tropical ocean) and 5°C (the temperature that is commonly found at depths greater than 1 km). What would the maximum efficiency of such a system be?

Answer: 7%

10. A turboprop aircraft engine burns 250 g/h/hp delivered to the prop. What is the efficiency of the engine? Kerosene delivers 46 MJ/kg of heat when burned.

Answer: 23.4%

11. A power station burns 2000 tons of coal per day. The station's efficiency is 39%. How much power does the station produce? Coal releases 32.6 MJ of heat per kilogram when burned.

Answer: 267.5 MW

12. A 350-MW power plant operates at an efficiency of 41%. Water from a river flowing at 45 m^3/s is used to absorb the waste heat. What is the temperature rise of the river because of the power plant?

Answer: 2.67°C

13. The air resistance of a car is 275 lb when it is moving at 60 mph in a straight line on a level road. At that speed the car can go 26 mpg. What is the efficiency of the engine and drive train? The heat of combustion for gasoline is 47.3 MJ/kg.

Answer: 42%

Carnot Engine

14. A cylinder initially contains 10 liters of gas at 3 atm. A piston is then used to compress the gas to 5 liters *at the same pressure (an isobaric process).* Soon afterward, *while the volume remains constant,* the gas is heated until the internal pressure reaches 9 atm. The piston is then withdrawn slowly, and the gas expands *isobarically* back to 10 liters. Finally, the gas is cooled so that the internal pressure returns at constant volume to the original value of 3 atm. How much work was done on the system? (*Hint:* You should draw an indicator diagram.)

Answer: $W = 3000$ J

15. A Carnot engine absorbs 10 MJ of heat from a reservoir at 500 K and exhausts it into a reservoir at 300 K. How much work was done *on* the engine in contrast to the useful work that the engine does?

Answer: 5.45 MJ

16. A Carnot steam engine takes in 300 kJ of heat at 500 K and exhausts it at 250 kJ. What is the exhaust temperature?

Answer: 416.7 K

17. (a) A certain engine operating between 400°C and 120°C is found to have an efficiency of 20%. What would its efficiency be if it were a Carnot engine? Does this claim seem reasonable? (b) Another engine operating between 500°C and 300°C is claimed to have an efficiency of 40%. What would its efficiency be if it were a Carnot engine? Does this claim seem reasonable?

Answer: (a) 41%; yes

(b) 26%; no. No heat engine can have an efficiency higher than a Carnot engine.

*18. In a particular coal-burning power plant, the steam turbines work in pairs in series. The exhaust heat from the first turbine becomes the input heat for the second. The input steam temperature for the first turbine is 650°C, and the exhaust temperature is 450°C. The input steam temperature for the second turbine is 440°C, and the exhaust temperature is 250°C. (a) At what rate must coal be burned to generate 100 MW of power? The heat of combustion of coal is 28 MJ/kg, and the turbines operate at 65% of their Carnot efficiency. (b) At what rate must water pass through the plant if the waste heat can warm the water up by only 6°C?

Answer: (a) 1064 metric tons/day

(b) 3506 m³/h

19. A heat engine has an exhaust at 325°C and a Carnot efficiency of 38%. How much lower must the exhaust temperature go to achieve a Carnot efficiency of 45%?

Answer: 67.5°C

Refrigerators and Their Thermodynamics

20. A refrigerator with a Carnot efficiency exhausts heat at 25°C and extracts it from the freezer compartment at −5°C. How much work does the refrigerator have to do per joule of heat extracted?

Answer: $W = 0.11$ J

Electromagnetism

The study of electromagnetism begins with this part (Chapters 16–21). It might seem to be an abrupt break with the material that has come before, but remember that the study of heat and thermodynamics is based on a mechanics concept from the first part of this text: work and energy. The study of electromagnetism is also based on concepts learned in the mechanics section of this text: forces, fields, and flow. You should now understand how a force interacts with an object. The two new specific forces that are introduced are the electrical force acting on a charged particle or two mutually interacting magnetic fields. The concept of the volt is based on work, electric current on flow, and alternating current circuits on concepts studied when wave phenomena were investigated. Thus, everything that is "new" requires that you understand concepts already investigated in the first half of this text. Keep this principle in mind when studying electromagnetics, and you'll find that a deep understanding of electromagnetic concepts is based on the principles of mechanics!

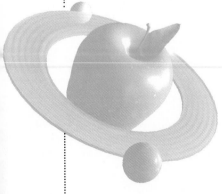

PART III

A static-electricity generator and three children with their hair standing on end demonstrate the repulsion of like charges.

Electric Charge

At the end of this chapter, the reader should be able to:

1. Explain the concept of an *electric charge.*

2. Discuss the nature of an *electric field.*

3. Solve for the electric field surrounding a given charge at a known radius.

4. Calculate the force between two known charges that have been separated a given distance.

5. Explain what makes a material a good *conductor* of electricity.

6. Describe the basic structure of an *atom.*

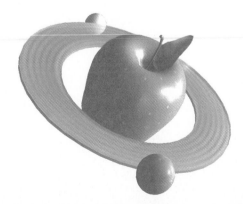

16.1   Electricity and Modern Society

Common use of electricity is a relatively recent phenomenon. At the turn of the century (1900), electricity was a novelty; today it is a necessity. It is so ubiquitous that it is taken for granted. A region without electricity and the conveniences that it makes possible is considered abnormal and primitive in this country. Indoor plumbing is made possible by electricity (electric pumps). Electric lighting turns night into day, and electrically run appliances make cleaning much easier than doing things "the old-fashioned way" (by hand). Power tools make mass production and automation possible. Heating and cooling are most often done electrically. Modern communications would be impossible without electronics. There would be no radio or television, no telephone networks, no computers or computer networks. In short, without electricity, modern civilization would cease to exist. We would be returned to a technological era equivalent to about the 1860s (without the telegraph, of course).

Electricity is for the most part a source of power. It is a power source to run our machines, heat and cool our homes, and light our living spaces. One of the great advantages of this power source is that it can be silently pumped into our homes just by hooking up a couple of wires. This power can be instantly switched on and off and generates no exhaust (at least at the point of use). The power delivered by the electricity is a function of the voltage between the two wires (usually 120 V_{ac} at an indoor outlet here in the United States) and the current running through the two wires.

Besides providing power, electricity is the basis for the construction of other useful devices using electricity. An amplifier can turn a weak force into a strong one. An electronic amplifier can turn a weak signal into a strong one. Electronic amplifiers and radio waves (electromagnetic energy) made long-distance wireless communications possible. Radio and TV are possible only because of the existence of electronic amplifiers. These amplifiers, originally built from vacuum tubes, were glass and metal cylinders that could amplify a stream of electrons (current). Later, semiconductors allowed the replacement of vacuum tubes with cheaper, smaller, and more reliable transistors. These transistors then made possible integrated circuits and the revolution that has become the present information age. Logic devices could also be built using electricity using a saturated (fully on) transistor to signal a logic 1 (on) and a logic 0 (off). The miniaturization of such logic elements led to mass use of the digital computer and the evolution of intelligent instrumentation, machinery, and robotics. In summary, there is almost no part of our daily lives that isn't touched by electricity. Thus, it wouldn't hurt to know something about it.

16.2   Electric Charge and the Structure of the Atom

When a light or a television or any other electrical appliance is plugged into the wall, the electricity from the socket delivers power to the appliance. That power is the product of the voltage times the current. The electric current is a flow of very small, negatively charged particles called *electrons*. The electron was discovered rather recently by British physicist J. J. Thomson in the early 1900s, but its effects were noted much earlier by the ancient Greeks. They discovered that when a piece of amber (petrified tree sap) was rubbed on a cloth, it acquired the property of being able to attract small, light fibers. The charged amber would also cause a small blue spark when touched to a metal implement. In fact, the word *electricity* originally meant "amber fire" in Greek.

When amber is rubbed on a piece of cloth, electrons are added to or removed from the surface. This leads to a net charge which creates an electric field to which other charged particles respond. The electron is one of the three fundamental particles that make up the *atom*, the basic building block of all matter. The atom has a small, very dense, positively charged nucleus with practically all of the mass of the atom. *The nucleus is surrounded by a swarm of negatively charged, orbiting electrons. The electron has a mass about 1800 times smaller than that of one of the two particles that comprise the nucleus. In the nucleus are positively charged particles called **protons,** and neutral particles called (naturally enough) **neutrons.*** They are about the same size and mass. A force exists between charged particles, and dissimilar particles attract whereas similar particles repel one another. A positively charged particle repels another positively charged particle, and a positively charged particle attracts a negatively charged particle, Thus, protons attract electrons, and protons repel other protons.

The atom shown in Figure 16.1 is a helium atom. It has the properties of helium because of the two protons in the nucleus. The neutrons keep the protons from getting too close to one another. The electron orbits are very far away from the nucleus.

Electron: A negatively charged particle that orbits the nucleus and that is much lighter than the proton or the neutron.

Proton: A positively charged particle in the nucleus of an atom. The number of protons in the nucleus defines the atom's elemental identity.

Neutron: A neutral particle in the nucleus of an atom that is used to separate the protons and that is about the same mass as the proton.

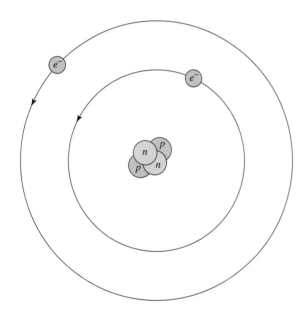

Figure 16.1 Helium Atom

Because the radius of an atom is defined by the radius of the outermost electron orbit, an atom is mostly empty space! The nucleus of an atom would fly apart because of the electrostatic forces among the protons if it weren't bound tightly by a much stronger force, the *nuclear force*. The nuclear force is a short-range force. Only when protons approach each other very closely does the nuclear force become stronger than the electromagnetic repulsive force and bind protons together. The number of protons in the nucleus determines the unique characteristics of an element, and all matter is made up of combinations of these elements. The nucleus is heavy and tightly bound together. It is very difficult to break a nucleus up into its protons and neutrons. However, the nucleus is surrounded by a cloud of orbiting electrons, and these electrons (at least the outer electrons) are relatively easy to remove from the atom. Most matter not part of a star is electrically neutral, so for each proton in the nucleus there must be a corresponding orbiting electron in the electron shell. The proton and the electron both have the same magnitude of charge, but they are opposite in sign. That is, $+q + (-q) = 0$, and the positive and negative charges cancel to make the atom electrically neutral despite the fact that it is made up of charged particles. The charge on an individual proton is $+1.6 \times 10^{-19}$ coulombs (abbreviated C), and that on an individual electron is -1.6×10^{-19} C, where a coulomb is the unit of electric charge.

When an electron is removed from an atom, an isolated negative charge exists. However, the atom from which the electron was removed is now no longer electrically neutral and now has a net positive charge. Such an atom is called an *ion*. Yet the mobility of the lighter, smaller electron is much greater than that of the ion. So when a piece of amber (or rubber, or glass) is rubbed with wool, silk, or some other fabric, it acquires a net charge because of the addition of electrons to or removal of electrons from the atoms of which the material is made. The atomic nucleus is bound tightly into the material being rubbed, and it is not free to leave, in contrast to the electrons. The site where the electron left will have a net positive charge. So separation of charge is primarily a function of losing and acquiring electrons. When amber is rubbed, an exchange of electrons takes place.

16.3 Coulomb's Law

Electric charges influence one another at a distance. This influence is a force, and for like charges it is a force of repulsion, for unlike charges a force of attraction. Since the force is between charges, there must be a minimum of two charges for a force to exist. It is therefore logical to assume that the force is proportional to the product of the two charges.

$$F \propto q_1 q_2$$

where the symbol $\propto$ means "is proportional to." The force weakens quickly with the distance of separation and goes as 1 over the square of the distance.

$$F \propto \frac{1}{r^2}$$

Putting these two facts together with a proportionality constant, we have an equation for the force between two charges.

$$F = k\frac{q_1 q_2}{r^2}$$

The proportionality constant is

$$k = \frac{1}{4\pi\varepsilon_0}$$

which gives a final form for the equation for the force between two charges.

(Equation 16.1)

$$F = \frac{q_1 q_2}{4\pi\varepsilon_0 r^2}$$

Equation (16.1) is known as *Coulomb's law.* The symbol ε_0 is the permittivity of free space of an electric force and is equal to 8.85×10^{-12} C^2/(N-m^2). This value for permittivity can change depending on the material that fills the space where the electric force exists.

Permittivity is a measure of the attenuation that the space surrounding an electrically charged object has to the force between charges. The greater the permittivity (or attenuation), the smaller the force. It's interesting to note that in the denominator of Coulomb's law is the term $4\pi r^2$, which is the formula for the surface of a sphere. Coulomb's law is spherically symmetrical. In other words, for two charges a distance r apart with one of the charges fixed in position, the surface that would be swept out by all of the positions where the force is equal to a given value in newtons would be the surface of a sphere. It's worth noting that if $r = 0$, the force between the two charges is infinite. (In mathematics, this is called a *singularity*.) But since two objects can't occupy the same space at the same time, for finite charge distributions there is no problem. However, for this reason, neutrons are needed in the nucleus: so that the repulsive force between protons can't grow so large that not even the nuclear force can keep the nucleus from flying apart!

EXAMPLE 16.1

Two identical charged particles each have a mass of 10 g and a charge of 1×10^{-3} C. (a) What force is required to hold the particles stationary at a distance 20 cm apart? (b) The charge on each particle is only 1 mC (one-thousandth coulomb). Therefore, a full coulomb of charge is a lot of charge, and that coulomb force is a very strong one! If the two charges are released, with what acceleration do they initially fly apart?

SOLUTION

(a)
$$F = \frac{1}{4\pi\varepsilon_0} \frac{q_1 q_2}{r^2}$$

$$F = \frac{(1 \times 10^{-3} \text{ C})^2}{4\pi(8.85 \times 10^{-12} \text{ C}^2/\text{N-m}^2)(0.2 \text{ m})^2}$$

$$F = 224{,}795 \text{ N} \quad \text{or} \quad \text{over 25 tons!!}$$

(b)
$$a = \frac{F}{m}$$

$$a = \frac{224{,}795 \text{ N}}{0.01 \text{ kg}}$$

$$a = 2.25 \times 10^7 \text{ m/s}^2$$

The acceleration found in Example 16.1 is much greater than that of gravity. It seems that electrical forces are much stronger than gravitational forces. The reason why this fact isn't obvious in everyday life is that most isolated charges are so small, usually on the order of a microcoulomb ($1\ \mu$C, about 10^{-6} C) or smaller. As has already been stated, the charges on the proton and the electron are much smaller than the microcoulomb by a factor of 10^{13}. Another and more important reason is the fact that charges can mask one another. Most atoms are electrically neutral because the positive nucleus is masked by the negative electron cloud surrounding it. The force of gravity, on the other hand, can't be masked, and all mass elements attract all other mass elements. For something as big as a planet, the cumulative effect of those infinitesimally small mass elements that comprise a planet add up to a significant gravitational field. The planet is made up of matter consisting of atoms in turn built up from charged particles. The reason there isn't a correspondingly large electric field is the masking effect of opposite charges.

16.4 The Electric Field

For an electric force to exist, there must be at least two charges. These two charges interact with one another at a distance through empty space, so the presence of the charges must alter that space in some way. But does it take both charges to alter that space, or only one charge? One charge is enough. A single charge will produce an electric field around it with spherical symmetry. If another charge enters the field, then a force between the charges is directed radially. But the field exists regardless of the presence of the second charge.

We exist within a significant gravitational field that we take for granted and notice only occasionally when we get clumsy and trip and begin to fall. The earth's gravitational field exists whether or not there are any external bodies (such as ourselves) around to acknowledge it. The earth and the other planets orbit the sun and travel in their orbits because of the gravitational field of the sun. But the sun's field is independent of the planets, asteroids, or comets that might orbit it. The field gives rise to the force, whether gravitational or electrical, not vice versa. The field is fundamental.

An *electric field* is defined as the force per unit charge and is given by

(Equation 16.2)

$$\mathbf{E} = \frac{\mathbf{F}}{q}$$

Since force $\mathbf{F}$ is a vector, so is the electric field $\mathbf{E}$: It has both magnitude and direction. Assume that we have two charges: a charge q that is responsible for the field and a test charge q_t with which we can test the field's strength. From Equation (16.2), we can write

$$\mathbf{E} = \frac{\mathbf{F}}{q_t} \qquad \mathbf{E} = \frac{\dfrac{1}{4\pi\varepsilon_0}\dfrac{qq_t}{r^2}\hat{\mathbf{r}}}{q_t}$$

(Equation 16.3)

$$\mathbf{E} = \frac{1}{4\pi\varepsilon_0}\frac{q}{r^2}\hat{\mathbf{r}}$$

where the $\mathbf{r}$ with a little hat on it ($\hat{\mathbf{r}}$) is a unit vector (length 1) in the direction along the line joining the two charges, and q is the charge causing the electric field. The units for the electric field are newtons per coulomb (N/C).

The gravitational field is also defined by the force per unit "charge." The gravitational "charge" is a test mass, m_t. Near the surface of the earth, the gravitational force on a test mass is just the weight of the test mass, $m_t g$. The formula for the gravitational field is then

$$E_G = \frac{m_t g}{m_t} \qquad E_G = g$$

which is the familiar acceleration of gravity ($g = 9.8$ m/s²). As an object moves farther and farther away from the earth, g becomes smaller and smaller, and just like the electric field it drops as 1 over r squared ($1/r^2$).

Electric field lines can actually be mapped. They would look like the illustrations in Figure 16.2. By definition, the field lines point away from a positive charge. The field lines are actually the path that a test charge would take in the presence of the charge causing

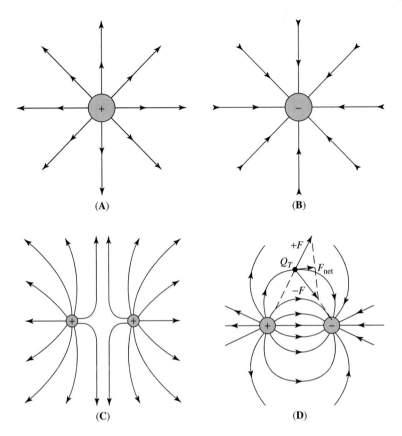

Figure 16.2 Electric Field Lines

the field. The test charge is positive and free to move, whereas the field charge is fixed and immobile.

In Figure 16.2A, the test charge would move radially outward in a straight line from the positive charge causing the field (repulsive force). In Figure 16.2B, the test charge would move radially inward toward the negative charge causing the field because of the attractive force. In Figure 16.2C, two positive charges are responsible for the field. A positive test charge would move along the paths indicated by the field lines. Figure 16.2D has two opposite charges responsible for the field, and a positive test charge would follow a path if placed in the field as indicated by the field lines.

In Figures 16.2A and B, the field lines are straight because a force acting on a test charge in the area would act along the straight line between the test charge and the source charge. However, in Figures 16.2C and D, the paths are curved because the force that the test charge experiences because of the field is a net force, and force is a vector. Our positive test charge is shown in Figure 16.2D at a position between and above the two fixed charges, with the forces acting on it from each charge shown as an arrow. The force from the positive charge is pushing the test charge up and to the right in the diagram, acting along a direct line between the positive source charge and the positive test charge. The negative source charge is simultaneously pulling the positive test charge down and to the right along a direct line between them. If the positive and negative source charges have equal magnitudes, then the lengths of the vectors (arrows) in Figure 16.2D will be the same. These two vectors add and can be broken up into horizontal and vertical components. The vertical components are opposite in direction and equal in magnitude, and therefore cancel. At the location shown in Figure 16.2D, there is no net vertical force. However, the two horizontal forces are in the same direction to the right and add directly. At the spot that q_t occupies, the net force on it would cause it to move horizontally to the right along the field line. As the test charge moves away from the location shown, the vertical forces will no longer cancel completely, and there will be a vertical as well as a horizontal component to the net force on the test charge. As a result, the field lines are curved.

16.5   Electrostatics

As was mentioned earlier, the first evidence of electricity came from rubbing a piece of amber and the resulting ability of the amber to attract small bits of light fabric. The piece of amber can be replaced with a hard rubber rod, and the bits of cloth with a light, hollow sphere of aluminum hung from a string (see Figure 16.3). Assume that both rod and sphere are initially uncharged. If the rod is rubbed by a piece of fur, the rod acquires a net negative charge and becomes the source of an electric field. If the charged rod is brought near the uncharged aluminum sphere, the sphere is attracted, as shown in Figure 16.3A. Why?

Aluminum is a conductor, and in a conductor the electrons are free to move about. When the charged rod is brought near the sphere, the rod's electric field causes the electrons in the sphere to try to move as far away from the rod as they can get. The side of the sphere away from the rod therefore acquires a net negative charge, and the side of the sphere near the rod acquires a net positive charge (see Figure 16.3B). There is an attraction between the positively charged side of the sphere and the rod. There is also a repulsion between the rod and the negatively charged part of the sphere. The attractive force is stronger than the repulsive force because the electrons causing the repulsive force are farther away on the sphere than the positive ions responsible for the attractive force. The result is that the sphere is attracted.

If the rod doesn't touch the sphere, the sphere has no net charge, regardless of the separation of the charge. The rubber rod has *induced* a separation of charge on the sphere. If all of the charges on the sphere were added up, the total charge would be zero. However, if the rod touches the sphere, then the sphere will acquire a net negative charge. Free electrons on the rod are attracted to the sites of net positive charge on the sphere and neutralize them, and in this way the sphere acquires a net negative charge. The rod and the sphere now have the same charge, and like charges repel. The entire dance between rod and sphere consists of three steps: (1) The charged rod is brought near the uncharged sphere, and the sphere is first attracted to the rod. (2) Rod and sphere touch, and then (3) the sphere is immediately repelled by the rod.

Assume now that we have a large, heavy steel ball bearing and that we put a net negative charge on the ball bearing. What is the electric field in the interior of the bearing? We can answer this question by knowing that the electrons are free to move within the volume of the metal and that being mutually repulsive they're going to try to get as far away from each other as possible. Thus, the excess electrons will move to the surface of the bearing and distribute themselves evenly on it. (The electrons are effectively trapped on the surface because it takes a much larger force than that of the mutual repulsion to remove an electron from the metal's surface.) When the excess electrons are evenly distributed over the surface, all electron migration ceases. Those electrons in the interior of the bearing don't move after the charge redistribution. No movement means no net force

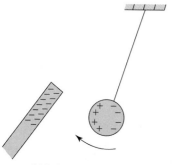

(**A**) Induced-charge separation; net attraction

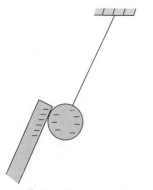

(**B**) Transfer of charge from rod to sphere

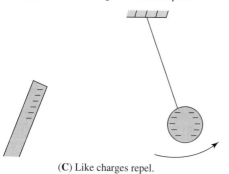

(**C**) Like charges repel.

Figure 16.3 Interaction between a Charged Rod and an Initially Uncharged Metal Sphere

on the electrons, and that means no electric field in the interior! Since the charges on the surface have reached an equilibrium and no longer move, there is also no electric field tangent to the surface of the bearing. The excess electrons on the surface will arrange themselves in such a way that they will be equidistant from each other. If one tries to move, a net force will be generated that will push it back to its equilibrium position.

Another interesting fact about the surface charge distribution discussed above is that to a test charge outside the sphere, it would appear as if all of the charge on the sphere were concentrated at the sphere's center. The earth's gravitational field acts in a similar manner. The entire mass of the earth acts as if it's concentrated at the earth's center, giving rise to the nearly constant value of the gravitational field at the earth's surface (9.8 m/s^2). All points on the earth's spherical surface are nearly the same distance away from the center.

EXAMPLE 16.2

Two charged, insulating spheres are stuck together by the force of electrical attraction, as shown in Figure 16.4. The charges aren't free to move from one sphere to another. One of the spheres has a negative surface charge density of -5×10^{-5} C/m^2 and a radius of 1 cm. The other sphere has a radius of 0.5 cm and a surface charge density of 15×10^{-5} C/m^2. What is the force holding the two spheres together?

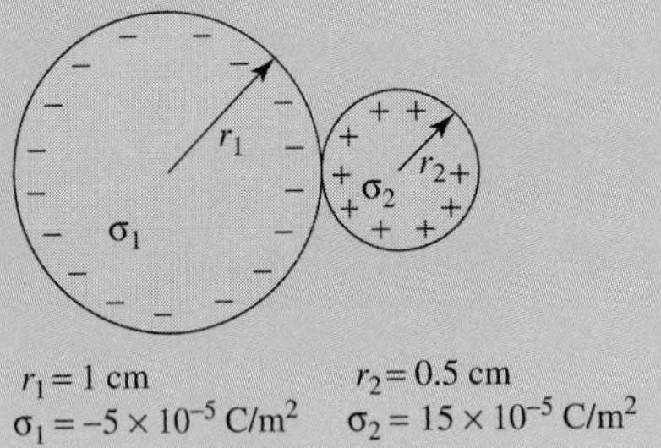

Figure 16.4 Two Charged Spheres in Contact

$r_1 = 1$ cm $r_2 = 0.5$ cm
$\sigma_1 = -5 \times 10^{-5}$ C/m^2 $\sigma_2 = 15 \times 10^{-5}$ C/m^2

SOLUTION

$$F = \frac{1}{4\pi\varepsilon_0} \frac{q_1 q_2}{r^2}$$

$$q_1 = (-5 \times 10^{-5} \text{ C/m}^2)4\pi(0.01 \text{ m})^2 \qquad q_1 = -6.28 \times 10^{-8} \text{ C}$$

$$q_2 = (15 \times 10^{-5} \text{ C/m}^2)4\pi(0.005 \text{ m})^2 \qquad q_2 = 4.71 \times 10^{-8} \text{ C}$$

$$r = 0.01 \text{ m} + 0.005 \text{ m} \qquad r = 0.015 \text{ m}$$

$$F = \frac{1}{4\pi(8.85 \times 10^{-12} \text{ C}^2/\text{N-m}^2)} \frac{(-6.28 \times 10^{-8} \text{ C})(4.71 \times 10^{-8} \text{ C})}{(0.015 \text{ m})^2}$$

$$F = -0.118 \text{ N}$$

Summary

1. The fundamental unit of charge is the coulomb (C).

Calculator Tip: Since charge is most often found in small amounts, we often encounter units such as a microcoulomb (μC; 10^{-6} C) and picocoulomb (pC; 10^{-9} C). It is a very common mistake, when plugging scientific notation into a calculator, to miss the correct answer by a factor of 10. This error probably occurs because of incorrect data entry into the calculator. For example, if we are trying to enter 10^{-6}, we must remember that the full expression is 1×10^{-6}! So we should hit $\boxed{1}$, then $\boxed{\text{EXP}}$ (or whatever key your calculator uses for exponents), then $\boxed{-6}$. (The common mistake is to

hit $\boxed{10}$, then $\boxed{\text{EXP}}$, then $\boxed{-6}$, which is the same as entering 10×10^{-6} or 1×10^{-5}!)

2. Electric fields can be expressed in units of newtons per coulomb (N/C).

3. A charge possesses an electric field that is proportional to the size of the charge and that decreases in strength with the square of the distance from the charge [see Equation (16.3)]. When another charge is placed within this field, it becomes subject to a force. The greater this added charge, the greater

the force between them. Both Equation (16.2) and Equation (16.1) describe this mathematically.

Problem Solving Tip 16.1: Another convenient way to write Equation (16.2) is $\mathbf{F} = q\mathbf{E}$, where q is the charge placed in the electric field, $\mathbf{E}$ is calculated using the original charge, and $\mathbf{F}$ is the force between them. Do not forget that electric force is still a vector, and we may need to break it up into appropriate components if the problem requires it.

4. An atom consists of a nucleus, which contains **protons** (positively charged particles) and **neutrons** (uncharged par-

ticles). Around this nucleus exists a cloud of **electrons** (negatively charged particles).

5. Electrons are free to move along the outer surface of a conductor.

6. As a charged body is brought close to an uncharged object, it may induce a charge on the second body by repelling like charges and attracting opposite charges. In this way, net attraction (or repulsion) may still occur between a charged object and a neutral body, as illustrated in Figure 16.3.

Key Concepts

ELECTRON: A negatively charged particle that orbits the nucleus and that is much lighter than the proton or the neutron.

NEUTRON: A neutral particle in the nucleus of an atom that is used to separate the protons and that is about the same mass as the proton.

PROTON: A positively charged particle in the nucleus of an atom. The number of protons in the nucleus defines the atom's elemental identity.

Important Equations

(16.1) $F = \dfrac{q_1 q_2}{4\pi\varepsilon_0 r^2}$ (Coulomb's law)

(16.2) $\mathbf{E} = \dfrac{\mathbf{F}}{q}$ (Definition of an electric field)

(16.3) $\mathbf{E} = \dfrac{1}{4\pi\varepsilon_0}\dfrac{q}{r^2}\hat{\mathbf{r}}$ (Electric field a distance r away from a charged object)

Conceptual Problems

1. Which of the following statements is false?
   a. The proton and the neutron have about the same mass.
   b. The electron and the proton have the same mass.
   c. Atomic nuclei contain only protons and neutrons.
   d. The electron and the proton have charges of the same magnitude but opposite sign.

2. The electrons in an atom:
   a. orbit the nucleus at some distance relatively far from the nucleus
   b. form an atom with protons and neutrons like the raisins in a plum pudding
   c. are always speeding through the nucleus
   d. are permanently attached to the atom

3. An object can have a positive electrical charge:
   a. because the nuclei of its atoms are electrically neutral
   b. if it has an excess of electrons
   c. if it has a deficiency of electrons
   d. if the electrons acquire a positive charge

4. Both a gravitational and an electrical force exist between two protons. The gravitational force:
   a. is stronger than the electrical force
   b. is weaker than the electrical force
   c. doesn't exist at the atomic level
   d. has a stronger attraction between protons than the electrical force of repulsion

5. A metal wire conductor:
   a. has too many electrons
   b. doesn't have enough electrons
   c. has a mobile "gas" of electrons inside
   d. locks the electron rigidly into place inside the crystal lattice

6. Electric field lines:
   a. point from negative to positive charges
   b. physically exist
   c. can be seen in a corona discharge
   d. are the trajectories that a charged particle would take in the presence of other charges

7. Twenty million free extra electrons are placed on an initially uncharged, solid metal sphere. Which of the following statements is true?

   a. The electrons distribute themselves uniformly throughout the sphere.

   b. The electrons are concentrated mostly near the bottom of the sphere.

   c. The electrons distribute themselves uniformly on the surface of the sphere.

   d. The electrons are concentrated at the center of the sphere.

8. Two charges of $+Q$ are placed a distance x apart. If one of the charges is replaced with a $-Q$ charge:

   a. the force goes from repulsive to attractive

   b. the magnitude of the force is smaller

   c. the magnitude remains the same

   d. the magnitude becomes zero

9. When two objects attract one another electrically, must both objects be charged? Why?

10. If two objects repel one another electrically, must both objects be charged? Why?

11. Name two differences between electrical and gravitational fields.

Exercises

Electric Charge and the Structure of the Atom

1. If a conducting sphere has a net charge of $-2\ \mu C$, how many more electrons than protons (or vice versa) are located on the sphere? One microcoulomb ($1\ \mu C$) is 1×10^{-6} C.

   Answer: 1.25×10^{13} more electrons

2. If the conductor in Exercise 1 had 1.3×10^{15} electrons removed from it, what would the net charge be on the sphere?

   Answer: $+208\ \mu C$

*3. Initially, four conducting rods (*A, B, C,* and *D*) are charged with $-1\ \mu C$, $2\ \mu C$, $-3\ \mu C$, and $4\ \mu C$, respectively. Then rod *A* is brought into contact with rod *B*. (a) What is the net charge on each rod after the rods are again separated? (b) Soon after, rod *B* is brought into contact with rod *C*. After the rods are separated, what is the net charge on each of these rods? (c) Rod *C* is then touched to rod *A* and to rod

D simultaneously. What is the net charge on each of the rods (*A, B, C,* and *D*) after this contact?

   Answer: (a) $A = +0.5\ \mu C; B = +0.5\ \mu C$
           (b) $B = -1.25\ \mu C; C = -1.25\ \mu C$
           (c) $A = +1.08\ \mu C; B = -1.25\ \mu C;$
               $C = +1.08\ \mu C; D = +1.08\ \mu C$

*4. A cube is placed on a very sensitive scale, and its weight is measured. It is then given a net charge of -7.5 pC (pico $= 10^{-12}$) and is weighed again. What is the difference between the two weights? An electron has a mass of 9.11×10^{-31} kg.

   Answer: 4.27×10^{-23} kg

5. A conducting metal sphere ($r = 10$ cm) is given $-65\ \mu C$ of net charge. What is the charge density on the surface of the sphere?

   Answer: 5.2×10^{-4} C/m^2

Coulomb's Law

6. What would the repulsive force be between a $+2$-μC point charge and a $+8$-μC point charge separated by (a) 4 cm, (b) 8 cm, and (c) 3 nm?

   Answer: (a) 90 N
           (b) 22.5 N
           (c) 1.6×10^{16} N

7. (a) How far away from a 1-C point charge would an identical charge have to be to produce a force of 1 N? (b) How far away from a 105-μC charge would an identical charge have to be to produce the same force?

   Answer: (a) 95 km
           (b) 10 m

8. Two small, neutral metal spheres are placed into a Cartesian coordinate system. Sphere *A* is located at the origin (0 m, 0 m), and sphere *B* is located 2 m away along the *x* axis (2 m, 0 m). If 10^{13} electrons are taken from sphere *A* and placed on sphere *B*, what would the magnitude and direction of the force be on a $+5$-μC charge placed halfway between *A* and *B*?

   Answer: 0.144 N to the right

9. Three charges are placed on a triangle as shown in Figure 16.5. Charge *A* has been given 4.5×10^{13} extra electrons. Charge *B* has been given 9×10^{13} extra electrons. Charge *C* has had 9×10^{13} electrons taken from it. What is the force

between (a) charges A and B, (b) charges B and C, and (c) charges A and C? Also tell whether the charges are repulsive or attractive.

Figure 16.5 (Exercise 9)

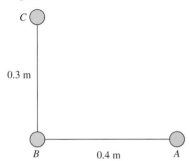

0.3 m

B 0.4 m A

Answer: (a) 5.8 N, repulsive
(b) 20.7 N, attractive
(c) 3.7 N, attractive

10. Two charges, each having $-4\ \mu C$ of charge and a mass of 40 g, are held 40 cm apart on a frictionless table. What would the initial acceleration of these charges be when the charges are released?

Answer: 22.5 m/s^2

*11. Two identical rubber balls hang from a common attachment point via two thin threads 50 cm long with negligible mass. Both balls have the same charge of $20\ \mu C$. If each string makes a 30° angle with the vertical, what is the mass of each ball?

Answer: 2.55 kg

12. One way to levitate an object is to use electrical forces. Figure 16.6 illustrates a charged ping-pong ball that is constrained to move only in a vertical line by a cardboard tube. Directly below this ball on the floor is a 2-μC charge. If the ping-pong ball has a mass of 10 g, what charge must the ball carry to levitate it 50 cm above the floor?

Figure 16.6 (Exercise 12)

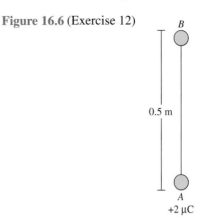

B

0.5 m

A
$+2\ \mu C$

Answer: $+1.36 \times 10^{-6}$ C

13. Assume that a proton is a very small sphere with a radius of approximately 1×10^{-15} m and that an electron is a dimensionless point charge. If we bring these particles into contact with one another, what normal force does the proton have to generate at its surface to keep the electron from falling through?

Answer: 230.4 N

*14. Two small, positively charged objects experience a mutually repulsive force of 1.52 N when they are 20 cm apart. The sum of the charges on the spheres is $6\ \mu C$. What is the charge on each sphere?

Answer: 4.5 μC and 1.5 μC

*15. The hydrogen atom is a single proton orbited by an electron. Assume that the electron has a circular orbital radius of 5.29×10^{-11} m. (a) What is the speed of the electron in its orbit? (b) What is the orbital frequency of the electron? The proton's mass is 1.67×10^{-27} kg, the electron's mass is 9.11×10^{-31} kg, and the magnitude of the charge on each is 1.6×10^{-19} C.

Answer: (a) 2.2×10^6 m/s
(b) 6.6×10^{15} Hz

Electric Field

16. Where on the line connecting charges A and B in Figure 16.7 is the resultant electric field zero? Give your answer relative to charge A given that (a) $A = +20\ \mu C$ and $B = +40\ \mu C$ and (b) $A = -20\ \mu C$ and $B = +40\ \mu C$.

Answer: (a) 0.21 m to the right of A
(b) 1.21 m to the left of A

Figure 16.7 (Exercise 16)

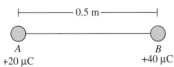

$\vdash$——— 0.5 m ———$\dashv$

A B
$+20\ \mu C$ $+40\ \mu C$

17. (a) Halfway between the two charges of Figure 16.7, what are the magnitude and the direction of the electric field?

(b) If a 5-μC charge were placed there, what resultant force would it experience?

Answer: (a) -2.88×10^6 N/C
(b) -14.4 N

*18. In Exercise 8, what would the magnitude and the direction of the electric field vector be at the point $(-1\ m, 1.5\ m)$? Draw a brief diagram showing the electric field line to which this vector belongs.

Answer: 3379.4 N/C at 67.1° north of west

*19. An electron is placed in a space of constant electric field of magnitude 5000 N/C. The electric field is directed across the space from left to right, and the distance from one side of the space to another is 4 cm. The electron is initially placed at the right side of the space. When the electron is released, what speed will it have when it hits the left side of the space?

Answer: 8.4×10^6 m/s

*20. Show that the kinetic energy stored in a particle of mass m and charge q released from rest in a uniform electric field E is $(Eqt)^2/2m$ where t is time.

*21. The corners of a square 2 cm on a side are occupied by point charges of magnitude $+1.5$ nC. What is the electric field (a) at the center and (b) at the midpoint of one edge?

Answer: (a) 0

(b) 4.83×10^4 N/C

*22. An electron is horizontally injected into a volume with a vertical electrical field E of constant magnitude and direction. Show that the motion of the electron is parabolic. (*Hint:* The equation for a parabola is $y = kx^2$, where k is a constant.)

Answer: $k = eE/mv_0^2$

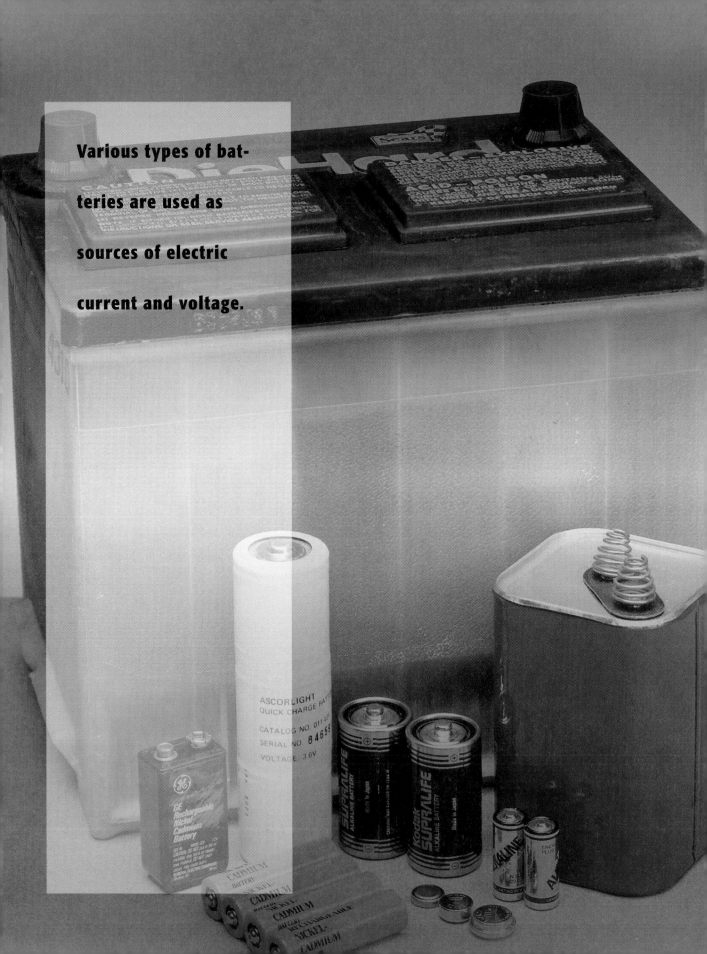

Various types of batteries are used as sources of electric current and voltage.

Electric Energy and Voltage

At the end of this chapter, the reader should be able to:

1. Explain what is meant by **voltage.**

2. Calculate the work done in separating two known charges.

3. Calculate the **capacitance** of a known electrical **capacitor.**

4. Solve for the energy stored in a given capacitor with and without a *dielectric* present.

5. Show examples of a *circuit* that is composed of capacitors in *series* and capacitors in *parallel.*

6. Solve for the *equivalent capacitance* of a group of capacitors in series and in parallel.

17.1 Charge in a Uniform Electric Field

Assume that a positive point charge of charge q, mass m, and initial velocity zero is placed into a region with a uniform electrical field directed downward, as in Figure 17.1. What happens to that charge?

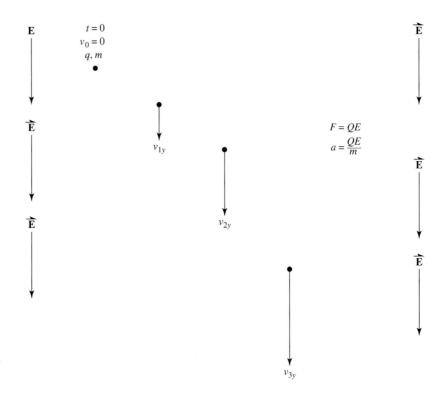

Figure 17.1 Positive Charge in a Uniform Electric Field

The charge has a force on it caused by the electric field ($\mathbf{F} = q\mathbf{E}$) and starts to accelerate vertically downward. Its acceleration is

$$a = \frac{\mathbf{F}}{m} \qquad a = \frac{Q\mathbf{E}}{m}$$

and is constant for a *uniform* electric field. Now that we know the charged particle's acceleration, the question posed above becomes a mechanics problem, exactly the same as those studied in the first chapters of this book. Knowing the acceleration, we can use the four equations of uniformly accelerated motion to find the particle's velocity and position at any future time t.

EXAMPLE 17.1

An electron with a horizontal velocity of 2×10^5 m/s is injected into an evacuated region (vacuum) with a uniform, vertical electrical field as shown in Figure 17.2. (a) How long is it before the electron strikes the bottom surface? (b) What is the

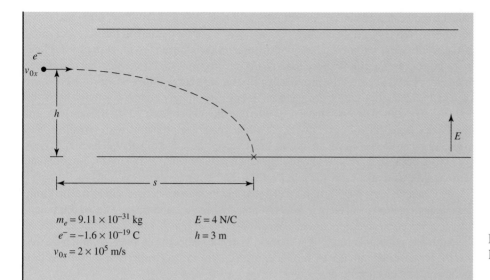

$m_e = 9.11 \times 10^{-31}$ kg $E = 4$ N/C
$e^- = -1.6 \times 10^{-19}$ C $h = 3$ m
$v_{0x} = 2 \times 10^5$ m/s

Figure 17.2 Ballistic Electron

electron's vertical velocity when it strikes the bottom surface? (c) Where does the electron strike the lower surface?

SOLUTION

(a)
$$h = v_{0y}t + \frac{1}{2}at^2 \qquad h = \frac{1}{2}at^2$$

$$a = \frac{F}{m} \qquad\qquad a = \frac{eE}{m}$$

$$a = \frac{(1.6 \times 10^{-19} \text{ C})(4 \text{ N/C})}{9.11 \times 10^{-31} \text{ kg}} \qquad a = 7.025 \times 10^{11} \text{ m/s}^2$$

$$t = \sqrt{\frac{2h}{a}} \qquad t = \sqrt{\frac{2(3 \text{ m})}{7.025 \times 10^{11} \text{ m/s}^2}}$$

$$t = 2.9 \times 10^{-6} \text{ s} = 2.9 \ \mu\text{s}$$

(b) $\qquad v_{fy}^2 - v_{0y}^2 = 2ah \qquad v_{fy}^2 = 2ah$

$$v_{fy} = \sqrt{2ah} \qquad v_{fy} = \sqrt{2(7.025 \times 10^{11} \text{ m/s}^2)(3 \text{ m})}$$

$$v_{fy} = 2.05 \times 10^6 \text{ m/s}$$

(c) $\qquad s = v_{0x}t \qquad s = (2 \times 10^5 \text{ m/s})(2.9 \ \mu\text{s})$

$$s = 0.58 \text{ m}$$

As illustrated in Example 17.1, electrons move quickly, and things happen fast. With such small masses and charges, it's a good idea to become proficient with the exponential function on your calculator, since you'll be using it often from now on. Note that an electron has both a mass and a charge, but no diameter or radius. The reason is that the electron can be considered a point particle, with no extension in space. A finite mass with no volume implies infinite density, but an electron isn't a particle in the classical sense of the word. More will be said about this topic in the chapters on modern physics later in the book (Part V). But for now, assume that an electron acts classically in the presence of an electrical field, with the proviso that it's a point charge. Since an electron can have such high accelerations, its speed can also be correspondingly high. It has mass, however, and particles with mass can't have a speed exceeding that of light (3×10^8 m/s) according to the special theory of relativity. So when you solve problems with speeding electrons and get an answer in which the electron exceeds the speed of light, go back and check your work because there's something wrong either with your work or with the problem. More will be said about this topic in Part V.

In Example 17.1, the electron accelerated and its kinetic energy increased. Therefore, a stationary electron in an electric field has potential energy, and when released this potential energy is converted into kinetic energy. Remember that an object near the surface of the earth has potential energy: $PE = mgh$. When an object is dropped from a height h, this PE is converted into KE. A similar expression for PE can be derived for the electron in the constant electric field.

$$F = eE \qquad W = F\,\Delta h$$

$$PE = eEh$$

From the above formula, we can quickly calculate the vertical speed of the electron just before it hits the floor in Example 17.1.

$$PE = eEh \qquad PE \rightarrow KE$$

$$\frac{1}{2}mv^2 = eEh$$

$$v = \sqrt{\frac{2eEh}{m}} \qquad v = \sqrt{\frac{2(1.6 \times 10^{-19}\text{ C})(4\text{ N/C})(3\text{ m})}{9.11 \times 10^{-31}\text{ kg}}}$$

$$v_y = 2.05 \times 10^6 \text{ m/s}$$

The velocity agrees with the one calculated in Example 17.1. Remember that for kinetics problems in which we're not interested in the time dependence of a particle, it's often much easier to solve from the perspective of energy than forces. In the study of electricity, energy is more often a much more advantageous viewpoint than forces.

17.2   Work Done in Separating Two Charges

If we want to store energy in an electrical system, we can do it by separating two charges. If we have a positive charge of $+q$ and a negative charge of $-q$, the charges attract one another, and it will take a positive amount of work to separate them. Once the charges have been separated, the work done in separating them is stored as potential energy in the system. If we want to separate the two charges by a distance r, how much work must we do? The work done is equal to the force times the change in distance. Thus,

$$\Delta W = F\,\Delta r \qquad F = \frac{1}{4\pi\varepsilon_0}\frac{QQ}{r^2}$$

$$\Delta W = \frac{Q^2}{4\pi\varepsilon_0} \frac{\Delta r}{r^2}$$

However, the force doesn't remain constant but changes with distance. How can we calculate the total work done in moving the positive test charge a distance r away? (We assume that the $-q$ charge remains fixed.) We move the test charge a small distance Δr_1 from the fixed charge, measure the small increment ΔW_1 of work done, and then move our test charge another small increment Δr_2. In this way the total work done will be the sum of the small increments of work.

$$W_T = \Delta W_1 + \Delta W_2 + \Delta W_3 + \cdots + \Delta W_n$$

where n is the total number of small increments moved. We can rewrite the above expression using the summation symbol.

$$W_T = \Sigma \Delta W_n$$

As Δr becomes very small or goes to zero, the number of increments goes to infinity. Our expression for the total work done now becomes a summation over an infinite number of very small increments.

$$W_T = \sum_{n=0}^{\infty} \Delta W_n$$

Substituting the expression derived from Coulomb's law for ΔW and moving all of the constant terms to the left of the summation sign, we're left with the following:

$$W_T = \Sigma \Delta W \qquad W_T = \Sigma \frac{Q^2}{4\pi\varepsilon_0} \frac{\Delta r}{r^2}$$

$$W_T = \frac{Q^2}{4\pi\varepsilon_0} \Sigma \frac{\Delta r}{r^2}$$

The only things remaining behind the summation sign are the variables being summed over. This summation is of the same type as we came up with in finding the gravitational PE in Section 7.7, where we used it to find the escape velocity. At this point it might be worth reviewing how we found the solution for the summation above. However, we will omit the review and give the answer below.

$$W = \frac{Q^2}{4\pi\varepsilon_0} \frac{1}{r}$$

The above equation gives the work done in separating two charges of $+Q$ and $-Q$. We don't need to do the infinitesimal summation every time two charges are separated. This equation holds true for any two charges separated by a distance r. Remember that the work done goes as $1/r$ and *not* $1/r^2$. Forgetting that fact is a very common mistake in problem solving. If the charges don't have the same magnitude, then

(Equation 17.1)

$$W = \frac{1}{4\pi\varepsilon_0} \frac{Q_1 Q_2}{r}$$

In separating our positive test charge from the stationary charge $-Q$, we find that the work is

$$W = \frac{1}{4\pi\varepsilon_0} \frac{Q_T Q}{r}$$

17.3   The Volt

The work done in separating two charges represents the potential energy of the system. This potential is defined as the work done per unit charge in separating a test charge from a charge Q.

$$\frac{W}{Q_T} = \frac{1}{4\pi\varepsilon_0} \frac{Q}{r}$$

Voltage: The work done per unit charge; a measure of the PE of a system.

*The work done per unit charge, called **voltage** (V), is a measure (in units of volts, V) of the potential energy of the system.*

    Batteries are rated according to their size and voltage. A typical AA battery will be rated at 1.5 V, and an automotive battery will be rated at 12 V. A battery is stored electrical (actually electrochemical) energy and is capable of doing work. The voltage is the amount of work that the battery is capable of doing per unit charge.

(Equation 17.2)

$$V = \frac{W}{Q} \qquad \text{(Definition of voltage)}$$

The units of the volt are joules per coulomb (J/C).

    Notice that a single charge can be the cause of a potential. The potential at a distance r from a charge q exists whether or not there is a second charge. This situation is similar to the one with the electric field, since a single charge will cause an electric field in the space surrounding it. The existence of a potential even when two charges aren't separated doesn't mean that you get something for nothing. The actual work done is equal to the voltage times the charge.

$$W = qV$$

No work is done until a charge moves across the potential difference (or charge separation takes place)!

    Most of us at one time have measured the voltage across a battery to judge whether it was discharged or still fresh. There are two poles on a battery: positive $(+)$ and negative $(-)$. Most voltmeters have color-coded leads. The red lead connects to the positive terminal, the black lead to the negative terminal. If a carbon battery is fresh, it will measure around 1.6 V, and if it is discharged, around 1.4 V. A lead-acid automotive battery (nominally 12 V) will read 13.2 V after being freshly charged and around 11 V if it is flat or in need of a recharge. Nickle-cadmium (nicad) batteries (nominally 1.2 V) read 1.4 V per cell after a fresh charge and about 0.9 V in a discharged state.

17.4   Energy Storage in an Electric Field

The energy stored in separating two charges is stored in the electric field. Look at the following two equations for the voltage and the electric field:

$$V = \frac{1}{4\pi\varepsilon_0}\frac{Q}{r} \qquad E = \frac{1}{4\pi\varepsilon_0}\frac{Q}{r^2}$$

They're almost identical. The potential (voltage) is derived from Coulomb's law, which is the base equation for the electric field. Thus, a mathematical link exists between the field and the potential. The field is the rate of change in space of the potential. We could open up a battery, place the negative terminal of a voltmeter at the negative terminal of the battery, and then place the positive terminal of the meter at the positive end of the battery and slide it down toward the negative terminal. We would see the reading go from 1.5 V to 1.2 V to 0.9 V to 0.6 V to 0.3 V and finally to 0 V where the two meter leads touch at the negative terminal. The rate at which the voltage (or electrical potential) changed in space is equal to the electric field inside the battery.

$$E = \frac{\Delta V}{\Delta r}$$

(Equation 17.3)

The potential (or voltage) is the sum of the field over a distance, as was shown immediately above in the derivation of the volt. In the text above, the force between charges was used, but the field is the force per unit test charge. The field is fundamental, and the potential is derived from the field. The electrical energy stored in the system is stored in the electric field in the form of a potential.

17.5   The Parallel-Plate Capacitor

*A **capacitor** is a device for separating charges and storing electrical energy.* The simplest capacitor is constructed from two parallel conducting plates of area A separated by a distance D with the space between filled with air. Such a capacitor is shown in Figure 17.3. Attached to the capacitor at the left is a battery connected to the plates by two wires. The polarity of the battery symbol is by convention positive for the long line and negative for the short line.

Capacitor: An electrical device that stores energy in an electric field produced through the separation of charges.

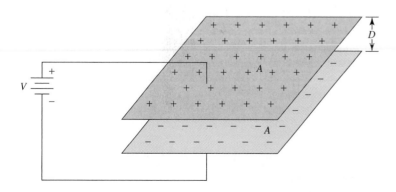

Figure 17.3 Parallel-Plate Capacitor

To store energy in the capacitor, we must first separate the charges. The external agent that we use to separate charge is the battery. Initially there is no voltage across the capacitor. But when the battery is connected to the conducting plates of the capacitor, the battery's electric field causes electrons in the upper plate to flow toward the positive terminal, causing an electron deficit in the upper plate and an overall positive charge density in the upper plate. These electrons end up on the bottom plate, because electrons attempt to flow away from the negative terminal of the battery, causing an overall negative charge density for the lower plate. Eventually, equilibrium is established when the voltage across the capacitor equals the voltage across the battery, and charge separation stops. If the capacitor doesn't leak, then the battery can be disconnected, and the capacitor will remain charged with an electric field between the plates and the electrical energy stored in the capacitor.

How much energy is stored in a capacitor when a battery with voltage V is attached to it? The work done in moving a positive test charge from the lower plate to the upper plate is

$$\Delta W = q_T \, \Delta V$$

Initially the voltage across the capacitor is zero, so separating the first charges requires no work. However, as the capacitor continues to charge, the voltage across the capacitor increases, making it harder and harder to separate further charges, and thus fewer and fewer charges are separated. Eventually, the voltage across the capacitor nears the voltage across the battery, and charge separation trickles down to zero. The total work done in separating a total charge Q is related to the average work done per unit charge in separating the charges, which is the average voltage across the capacitor as the capacitor charges.

$$W = Q\overline{V}$$

where $\overline{V}$ is the average voltage. Thus,

$$\overline{V} = \frac{V_i + V_f}{2} \qquad \overline{V} = \frac{0 + V}{2} \qquad \overline{V} = \frac{V}{2}$$

$$W = Q\frac{V}{2} \qquad W = \frac{1}{2}QV$$

$$U = \frac{1}{2}QV$$

(Equation 17.4)

In the above equations, Q is the total charge separated; V is the voltage across the battery and therefore the capacitor; and U is the total energy stored in the capacitor. The work done in separating the charges becomes the energy stored in the capacitor.

The electric field in the capacitor is the difference in voltage across the plates divided by the separation between the plates. If there is a potential difference of 6 V across the capacitor, then the voltage decreases as you move from the top plate to the bottom plate in a linear fashion.

$$E = \frac{V}{D}$$

Thus, the electric field can be expressed in the units of volts per meter (V/m) in addition to the units of newtons per coulomb (N/C). The volt per meter is the same as the newton per coulomb. Convince yourself of this fact by working out the units.

The electric field of a parallel-plate capacitor in the region between the plates looks like Figure 17.4. As shown in the drawing, the electric field is everywhere the same (uniform) in both magnitude and direction. The constant electric field used in Example 17.1 at the beginning of this chapter is actually found in a parallel-plate capacitor. The total charge on the upper plate is equal to the surface charge density times the area of the plate. Since the upper plate is a conductor, the surface charge density is constant over the entire plate. There is a relationship between the surface charge density and the electric field in the volume of the capacitor. That relationship is given below.

$$E = \frac{\sigma}{\varepsilon_0}$$

(Equation 17.5)

where σ (sigma) is the charge per unit area, and ε_0 is the permittivity constant for air or a vacuum. The permittivity constant is an index as to how the electric field is attenuated by the surrounding medium, in this case, air. (To *attenuate* something is to lessen it.) The electric field lessens as the permittivity constant gets larger. We saw this effect when we first looked at Coulomb's law and the electric field. Also, the greater the charge density, the greater the electric field, which is logical.

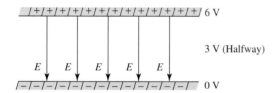

6 V

3 V (Halfway)

0 V

Figure 17.4 Electric Field in a Parallel-Plate Capacitor

Knowing the two relationships immediately above, we can now calculate the energy stored in a parallel-plate capacitor in terms of its electric field.

$$U = \frac{1}{2}QV \qquad V = ED$$

$$Q = \sigma A \qquad U = \frac{1}{2}(\sigma A)ED$$

$$\sigma = \varepsilon_0 E$$

$$U = \frac{1}{2}(\varepsilon_0 E)AED = \frac{1}{2}\varepsilon_0 E^2 AD$$

The term AD is the volume inside the parallel-plate capacitor, so

$$\frac{U}{V} = \frac{1}{2}\varepsilon_0 E^2$$

(Equation 17.6)

is the volume energy density of a capacitor. Since the term ε_0 is a fundamental constant of space itself, we have therefore shown that the electrical energy is stored in the electrical field inside the volume of a parallel-plate capacitor. The field is fundamental, and the energy of the system is stored in the electrical field.

17.6　Capacitance

Capacitance: The charge stored by a capacitor for a given voltage.

We've talked for pages about the parallel-plate capacitor without once defining *capacitance!* As defined earlier, a capacitor is an electrical energy storage device and stores electrical energy in an electrical field. *The definition of **capacitance** is*

(Equation 17.7)

$$C = \frac{Q}{V}$$

or the charge stored for a given voltage. The unit for capacitance is the farad (F), or coulombs per volt (C/V). As the voltage across the capacitor is increased, the amount of separated charge increases. Theoretically, there is no upper limit to the amount of charge that you can store in a capacitor. But practically there is a very real limit imposed by the dielectric strength of the materials used in fabricating the capacitor. For an air-gap parallel-plate capacitor, the limit is imposed by the maximum field strength in air that can be supported without arcing (sparking across the plates). This maximum field is about 3×10^6 V/m. Typical capacitors are found in these ranges: picofarads (pF; 1×10^{-12} F), nanofarads (nF; 1×10^{-9} F), and microfarads (μF; 1×10^{-6} F). Low-voltage capacitors in the farad range are now also being produced. These capacitors store enough energy that they can be substituted for batteries under certain conditions. The flash lamps on cameras use capacitors because the stored energy can be discharged quickly across the lamp to give an intense flash.

A given capacitor will have a certain capacitance whether it is charged or discharged. The volume of a box remains the same whether the box is full or empty. The box's volume is dependent solely on its geometry and is independent of whether or not it is full of something. The same concept also applies to the capacitor. Starting from the definition of *capacitance,* we can now derive the capacitance of a parallel-plate capacitor. The answer should be independent of whether or not there is a voltage across the capacitor, and dependent on geometric quantities such as the plate area and the spacing between the plates.

$$C = \frac{Q}{V} \qquad Q = \sigma A$$

$$V = ED \qquad C = \frac{\sigma A}{ED}$$

$$E = \frac{\sigma}{\varepsilon_0} \qquad C = \frac{\sigma A}{(\sigma/\varepsilon_0)D}$$

(Equation 17.8)

$$C = \frac{\varepsilon_0 A}{D}$$

Note that the final formula [Equation (17.8)] does indeed depend on the geometry of the capacitor and not on whether it's charged or discharged. The greater the plate area, the greater the capacitor's ability to store charge. The greater the separation between the plates, the smaller the capacitance. For a constant voltage across the plates, if the plate separation is increased, the electric field will drop. Anything that decreases the electric field will decrease the energy stored in the capacitor.

The energy stored in a parallel-plate capacitor is given by $U = \frac{1}{2}QV$. Now that we have a definition for *capacitance,* we can come up with two more formulas for the energy stored in a capacitor in terms of the capacitance.

$$U = \frac{1}{2}QV \qquad C = \frac{Q}{V}$$

$$Q = CV \qquad U = \frac{1}{2}(CV)V$$

$$U = \frac{1}{2}CV^2 \qquad \text{(Equation 17.9)}$$

$$U = \frac{1}{2}QV \qquad V = \frac{Q}{C}$$

$$U = \frac{1}{2}Q\left(\frac{Q}{C}\right)$$

$$U = \frac{Q^2}{2C} \qquad \text{(Equation 17.10)}$$

Equation (17.9) is in terms of capacitance and voltage, and Equation (17.10) in terms of capacitance and charge.

EXAMPLE 17.2

A parallel-plate capacitor has plates 2 cm on a side and a plate spacing of 0.05 mm. (a) If the space between the plates is filled with air, what is the capacitor's capacitance? (b) What is the maximum voltage across the plates that the capacitor can withstand without arcing? (c) If the capacitor is charged up to this voltage, how much charge has been separated? (d) How much energy is stored within the capacitor? (e) What is the charge density on the plates?

SOLUTION

(a) $\qquad C = \frac{\varepsilon_0 A}{D} \qquad C = \frac{(8.85 \times 10^{-12} \text{ C}^2/\text{N-m}^2)(0.02 \text{ m})(0.02 \text{ m})}{(0.05 \times 10^{-3} \text{ m})}$

$$C = 70.8 \times 10^{-12} \text{ F}$$

$$C = 70.8 \text{ pF}$$

(b) $\qquad E = \frac{V}{d}$

$$V = Ed \qquad V = (3 \times 10^6 \text{ V/m})(0.05 \times 10^{-3} \text{ m})$$

$$V = 150 \text{ V}$$

(c) $\qquad C = \frac{Q}{V}$

$$Q = CV \qquad Q = (70.8\ \text{pF})(150\ \text{V})$$

$$Q = 1.06 \times 10^{-8}\ \text{C}$$

(d) $\qquad U = \frac{1}{2}CV^2 \qquad U = \frac{1}{2}(70.8\ \text{pF})(150\ \text{V})^2$

$$U = 7.97 \times 10^{-7}\ \text{J}$$

(e) $\qquad\qquad E = \frac{\sigma}{E_0}$

$$\sigma = \varepsilon_0 E \qquad \sigma = (8.85 \times 10^{-12}\ \text{C}^2/\text{N-m}^2)(3 \times 10^6\ \text{N/C})$$

$$\sigma = 2.66 \times 10^{-5}\ \text{C/m}^2$$

17.7   The Dielectric Constant

The equation for capacitance has the electric permittivity constant for air in the numerator.

$$C = \frac{\varepsilon_0 A}{D}$$

Up to now, we've been dealing with capacitors that have either air or a vacuum between their plates. If the material between the plates is something other than air or vacuum, then the permittivity constant used in the equation for capacitance will be different from ε_0. Since ε_0 is the lowest permittivity constant, placing another material between the plates will increase the capacitance of the capacitor over that of just an air gap. The reason for this is as follows: Matter is made up of atoms and molecules, which are collections of charged particles. It is the behavior of atoms and molecules in an electric field that gives rise to the higher permittivity constant. Water has a permittivity constant that is 80 times higher than that of a vacuum. Water is a relatively simple molecule; its behavior in an electric field is shown in Figure 17.5.

A water molecule is made up of two atoms of hydrogen and one atom of oxygen. A cloud of electrons surrounds the entire molecule. The electrons are shared between the oxygen and hydrogen atoms. The cloud around the molecule is not symmetrical, however, and the molecule is slightly polarized. The end with the oxygen molecule has a slightly negative polarity (the electrons spend slightly more time there), and the end with the two hydro-

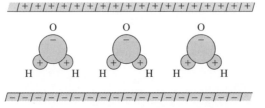

Figure 17.5 Water Molecules Aligned by an External Electric Field

gen molecules has a slightly positive polarity (the electrons in orbit around the molecule spend slightly less time there). In the presence of an electric field, a torque is exerted on the water molecule that tends to align it with the electric field. The negative end is attracted by the positive charge distribution on the upper plate, and the positive end is attracted by the negative charge distribution on the bottom plate. If water fills the space between the plates, then every water molecule will try to attain this orientation. Each water molecule still has a net charge of zero, but there is effectively a net negative charge density in the water near the interface between the water and the upper plate, and there is a net positive charge distribution in the water at the interface between the water and the lower plate. Because of the masking effect of positive and negative charges, the charge density of the upper and lower plates is reduced by the polarization of the water molecules. The reduction of the charge density reduces the electric field in the volume of the capacitor between the plates. From Equation (17.5) for the electric field in terms of the charge density, lowering the electric field is equivalent to lowering the charge density on the plates (which is what the polarization of the molecules does) or *increasing* the permittivity constant of the space inside the material (in this case, water). Mathematically it's the same thing. A greater permittivity constant means a greater capacitance. The greater permittivity constant means a smaller electrical field for a given voltage across the capacitor.

$$E = \frac{\sigma}{\varepsilon_0}$$

This smaller electrical field means that there is less stress between the plates, similar to increasing the volume of a box so that more can be stored inside. The capacitance increases, and ultimately more energy can be stored inside for a given voltage. The increased dielectric constant is given by a number K, which is a multiple of ε_0.

Table 17.1 gives the value of K for some common materials. The reason why a vacuum has such a low dielectric constant is that there is no matter to polarize. Water should be a marvelous dielectric material to use in a capacitor to increase its capacitance except for one major objection: Water can conduct a small current, and that would allow the charge on the plates to bleed off. The material used between the plates must be an insulator. Mica and waxed paper are used as dielectrics in capacitors because they're both insulators and they both have good structural stability.

Table 17.1 Dielectric Constants (K) for Some Common Materials

Substance	K
Air	1.0006
Ethyl alcohol	26
Glass	5–8
Ice	94
Mica	2.5–7
Neoprene	6.7
Sulfur	3.9
Teflon	2.1
Water	80
Waxed paper	2.2

EXAMPLE 17.3

The parallel-plate capacitor of Example 17.2 was left charged at 150 V across the plates. Now suppose that the battery is disconnected, and a neoprene dielectric is inserted between the plates. (a) What is the energy stored in the capacitor? (b) What is the voltage across the capacitor? (c) What is the electric field between the plates? (d) The energy in the capacitor is 6.7 times less than without the dielectric. Where did the extra energy go?

SOLUTION

(a) The charge remains the same.

$$C = \frac{Q_0}{V}$$

$$Q_0 = CV \qquad Q_0 = (70.8 \text{ pF})(150 \text{ V})$$

$$Q_0 = 1.06 \times 10^{-8} \text{ C}$$

$$U = \frac{Q_0^2}{2C'}$$

$$C' = \frac{K\varepsilon_0 A}{D} \qquad C' = \frac{(6.7)(8.85 \times 10^{-12} \text{ C}^2/\text{N-m}^2)(0.02 \text{ m})(0.02 \text{ m})}{0.05 \times 10^{-3} \text{ m}}$$

$$C' = 474 \text{ pF}$$

$$U = \frac{(1.06 \times 10^{-8} \text{ C})^2}{2(474 \text{ pF})} = 1.18 \times 10^{-7} \text{ J}$$

(b)
$$C' = \frac{Q_0}{V}$$

$$V = \frac{Q_0}{C'} \qquad V = \frac{1.06 \times 10^{-8} \text{ C}}{474 \text{ pF}}$$

$$V = 22.4 \text{ V}$$

(c)
$$E = \frac{V}{D} \qquad E = \frac{22.4 \text{ V}}{0.05 \times 10^{-3} \text{ m}}$$

$$E = 4.5 \times 10^5 \text{ N/C}$$

(d) The capacitor does work on the dielectric. A force of attraction exists between the dielectric and the space between the plates. The dielectric is pulled into the capacitor. To remove the dielectric would require that work be done on the dielectric by an external agent. This work done would go into extra energy stored in the capacitor.

17.8   Capacitors in Series and in Parallel

Capacitor circuits can be connected together in two basic configurations: *series* or *parallel* (or a combination of series and parallel). A series configuration is shown in Figure 17.6. If a charge wanted to travel from the positive terminal of the battery to the negative terminal, only one external path is available to it. That path is through capacitors C_1, C_2, and C_3, in that order. If there is only a single path through a number of components, then those components are connected in series. The battery, however, doesn't know that it is connected to three capacitors in series. All it knows is that it is able to separate a certain amount of charge. The battery sees a net total capacitance. What total capacitance does the battery see?

Stacking three capacitors in series as in Figure 17.6 is like increasing the plate separation for a single capacitor. Increased plate separation means lowered capacitance, so the three capacitors in series will have less total capacitance than the smallest of the three. Since there is only one charging path, the current is the same in all three capacitors. When the capacitors reach full charge, each has the same charge Q across it. Charging stops when the voltage across the three capacitors equals the battery voltage. Putting these two facts together, we can derive the total capacitance that the battery sees.

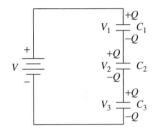

Figure 17.6 Capacitors in Series

$$V = V_1 + V_2 + V_3 \qquad C = \frac{Q}{V} \qquad V = \frac{Q}{C}$$

$$\frac{Q}{C_T} = \frac{Q}{C_1} + \frac{Q}{C_2} + \frac{Q}{C_3} \qquad Q\left(\frac{1}{C_T}\right) = Q\left(\frac{1}{C_1} + \frac{1}{C_2} + \frac{1}{C_3}\right)$$

$$\frac{1}{C_T} = \frac{1}{C_1} + \frac{1}{C_2} + \frac{1}{C_3} + \cdots + \frac{1}{C_n}$$

(Equation 17.11)

If there are more than three capacitors in series, just add the appropriate number of terms.

If, however, there is an insulating dielectric between the plates of the capacitor, how can a current flow through it? (Air is also an insulator.) No current actually flows through the capacitor, but as charge is separated, the charge separation acts as a continuous current of electrons from the negative terminal of the battery to the positive terminal. When the battery is first connected, an electron flows from the upper plate of C_1 to the positive terminal of the battery. This electron gives C_1 a net positive charge, and thus an electron from the upper plate of C_2 flows to the lower plate of C_1, leaving C_2 with a net positive charge. An electron then flows from the upper plate of C_3 to the lower plate of C_2. Since C_3 now has a net positive charge, an electron flows from the negative terminal of the battery to the lower plate of C_3, and all three capacitors now have a small charge separation and the polarities shown in Figure 17.6. In this way, all of the capacitors in series will have the same charge, regardless of the size of the individual capacitor. The charge separation "dominoes down" through all of the capacitors in series until the negative terminal of the battery is reached, and charge separation doesn't stop until the voltage across the capacitors in series equals the battery voltage.

The parallel configuration is given in Figure 17.7. If a positive test charge wanted to travel from the positive terminal of the battery to the negative terminal by an external path, it would have three choices. It could go through C_1, C_2, or C_3. These parallel paths mean that capacitors C_1, C_2, and C_3 are in parallel. Notice that the battery is connected directly across each of the three capacitors, so each component in a parallel configuration sees the same voltage. Again the battery doesn't know that there are three capacitors connected to it in a parallel configuration. All it knows is that it can separate a given

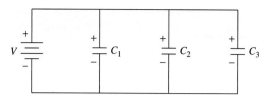

Figure 17.7 Capacitors in Parallel

amount of charge and store energy in the separated charge. What total capacitance does the battery see? A hint is that when capacitors are connected in parallel, the plate area of the connected capacitors is effectively increased, and increased plate area means increased capacitance. Also, the total charge separated is equal to the simple sum of the charge on each capacitor.

$$Q_T = Q_1 + Q_2 + Q_3 \qquad C = \frac{Q}{V} \qquad Q = CV$$

$$C_T V = C_1 V_1 + C_2 V_2 + C_3 V_3 \qquad V_1 = V_2 = V_3 = V$$

$$C_T V = C_1 V + C_2 V + C_3 V \qquad V C_T = V(C_1 + C_2 + C_3)$$

(Equation 17.12)

$$C_T = C_1 + C_2 + C_3 + \cdots + C_n$$

If there are more than three capacitors in parallel, just add the extra capacitors in parallel.

EXAMPLE 17.4

The circuit in Figure 17.8 shows capacitors connected in both a series and a parallel configuration. What is the total capacitance of the circuit?

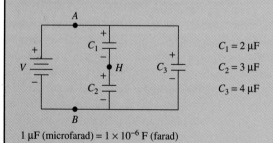

$C_1 = 2\ \mu F$

$C_2 = 3\ \mu F$

$C_3 = 4\ \mu F$

Figure 17.8 Capacitors in Series and Parallel

1 μF (microfarad) = 1×10^{-6} F (farad)

SOLUTION

Capacitors C_1 and C_2 are in series. First find the total capacitance for the series leg, C_{T1}; C_{T1} is then in parallel with C_3 (see Figure 17.9). Add C_{T1} to C_3 to find the total capacitance.

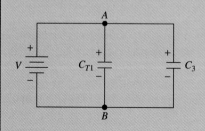

Figure 17.9 Solution to Series-Parallel Capacitor Circuit

$$\frac{1}{C_{T1}} = \frac{1}{C_1} + \frac{1}{C_2} \qquad \frac{1}{C_{T1}} = \frac{C_2 + C_1}{C_1 C_2} \qquad C_{T1} = \frac{C_1 C_2}{C_1 + C_2}$$

$$C_T = C_{T1} + C_3 \qquad\qquad C_T = \frac{C_1 C_2}{C_1 + C_2} + C_3$$

$$C_T = \frac{(2\,\mu\text{F})(3\,\mu\text{F})}{2\,\mu\text{F} + 3\,\mu\text{F}} + 4\,\mu\text{F} \qquad C_T = 1.2\,\mu\text{F} + 4\,\mu\text{F}$$

$$C_T = 5.2\,\mu\text{F}$$

Thus, the total capacitance that the battery sees is 5.2 μF. The battery in this circuit is only a reference point. Notice that we didn't need to know the battery voltage to calculate the total capacitance. The battery is connected between points A and B for the solution. If the battery were connected between points H and B, the answer would be different. Then C_1 and C_3 would be in series, and that series leg would be in parallel with C_2. The total capacitance between these two points would be 4.33 μF. Convince yourself of this fact. It is very helpful to redraw the circuit with a battery connected between the points in question.

17.9   The Cathode-Ray Tube

The cathode-ray tube (CRT) is a device that impinges on our lives every day. Television screens, computer monitors, and oscilloscopes are devices that use a cathode-ray tube to display information on a screen. Modern technology is turning more and more to liquid-crystal displays (LCDs) for information displays, but the CRT will always have an important niche in electronics technology.

Figure 17.10 is a sketch of a cathode-ray tube shown from the side. The CRT essentially accelerates electrons and guides them to a screen where they hit and emit photons of light that our eyes see. The interior of a cathode-ray tube is a vacuum. The electron's journey starts at the filament at the far left of Figure 17.10. The filament is a conductor and is heated so that some electrons have enough energy to leave the metal. The filament is negatively charged so that the electron is repelled from it. The electron feels the electric field of the accelerating anode that is at a high positive voltage relative to the filament, and its attraction causes the electron to be accelerated along the axis of the CRT. The electron then passes through the plates of a pair of parallel-plate capacitors. The electric field inside each capacitor causes the electron to be accelerated toward the positive plate. If the right horizontal deflection plate is positive, the electron moves to the right. If the left one is positive, then the electron moves to the left. The vertical deflection plates work in a similar manner. In this way the place where the electron hits the screen can be controlled very precisely. The back of the screen is coated with a phosphorescent material that emits photons of light every time it is hit by an electron. The phosphorescent glow persists long enough that our eyes see the entire screen as glowing, and the electrons paint a picture on the screen one pixel (picture element) at a time. When the electron gun has covered the screen, it returns to the upper left (or right) corner and paints another picture pixel by pixel.

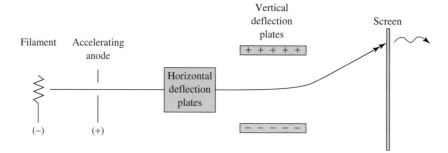

Figure 17.10 Schematic of a
Cathode-Ray Tube

Summary

1. **Voltage** (V) is a measure of energy per unit charge [Equation (17.2)] and is expressed in units of *volts* (V).

2. The equivalent capacitance of a group of capacitors in series is given by

$$\frac{1}{C} = \frac{1}{C_1} + \frac{1}{C_2} + \frac{1}{C_3} + \cdots + \frac{1}{C_n}$$

3. The equivalent capacitance of a group of capacitors in parallel is given by

$$C = C_1 + C_2 + C_3 + \cdots + C_n$$

4. The geometry of a **capacitor** is what dictates its **capacitance** [Equation (17.8)]. Therefore, once a capacitor is constructed, its capacitance will remain constant despite the amount of charge currently being stored within it. One factor may alter this capacitance, however: See *dielectric* below.

Problem Solving Tip 17.1: This fact can be used very often. By utilizing Equation (17.7), we can often find the charge or voltage very easily.

5. A *dielectric* is a material that has a *permitivity* (ε) greater than that of a vacuum (ε_0). The *dielectric constant* (K) is a multiplier of ε_0, and therefore, when placed within a capacitor, it can increase its capacitance.

6. As a charge is built up on the plates of a capacitor, the capacitor will begin to store energy as an increasing electric field. This energy is a function of the capacitance of the capacitor, the voltage across the plates, and the charge currently present. The exact relationship between energy, charge, and voltage is given in Equations (17.4), (17.9), and (17.10).

Problem Solving Tip 17.2: Don't forget that each of these expressions [Equations (17.4), (17.9), and (17.10)] is valid. When choosing which formula to use, simply look at the data given to you! It very often will be a perfect match for one (or more!) of these equations. Also, each of these equations can be used when dealing with problems that use conservation of energy! (See Problem Solving Tip 4.1.)

7. The electric field between a capacitor consisting of two parallel plates remains constant everywhere within the capacitor. The field present is dependent only on the *charge density* (σ) present on the plates [Equation (17.5)].

Problem Solving Tip 17.3: By using Equation (17.3), we can also find the uniform electric field within a parallel-plate capacitor. The field will be equal to the voltage between the plates divided by the plate separation ($E = V/d$).

Key Concepts

CAPACITANCE: The charge stored by a capacitor for a given voltage.

CAPACITOR: An electrical device that stores energy in an electric field produced through the separation of charge.

VOLTAGE: The work done per unit charge; a measure of the PE of a system.

Important Equations

(17.1) $W = \dfrac{1}{4\pi\varepsilon_0}\dfrac{Q_1 Q_2}{r}$ (Work done in separating two charges a distance r)

(17.2) $V = \dfrac{W}{Q}$ (Voltage is the work done per unit charge.)

(17.3) $E = \dfrac{\Delta V}{\Delta r}$ (The electric field equals the spatial change of the voltage.)

(17.4) $U = \dfrac{1}{2}QV$ (Energy storage in a capacitor in terms of charge and voltage)

(17.5) $E = \dfrac{\sigma}{\varepsilon_0}$ (Electric field in terms of plate charge density in a parallel-plate capacitor)

(17.6) $\dfrac{U}{V} = \dfrac{1}{2}\varepsilon_0 E^2$ (Energy density in terms of the electric field)

(17.7) $C = \dfrac{Q}{V}$ (Capacitance in terms of charge and voltage)

(17.8) $C = \dfrac{\varepsilon_0 A}{D}$ (Capacitance in terms of plate area and separation)

(17.9) $U = \dfrac{1}{2}CV^2$ (Energy stored in a capacitor in terms of the capacitance and voltage)

(17.10) $U = \dfrac{Q^2}{2C}$ (Energy storage in terms of charge and capacitance)

(17.11) $\dfrac{1}{C_T} = \dfrac{1}{C_1} + \dfrac{1}{C_2} + \dfrac{1}{C_3} + \cdots + \dfrac{1}{C_n}$ (Total capacitance for capacitors in series)

(17.12) $C_T = C_1 + C_2 + C_3 + \cdots + C_n$ (Total capacitance of capacitors in parallel)

Conceptual Problems

1. Of the following quantities, which one has a vector nature?
   a. Electrical energy   b. Electric field
   c. Charge   d. Voltage

2. In a parallel-plate capacitor, the electric field is:
   a. greatest near the positive plate
   b. greatest near the negative plate
   c. zero
   d. uniform throughout

3. The units for the electric field are the newton per coulomb (N/C), from its definition. The alternative units for the field are:
   a. V/m   b. V-m
   c. V/m^2 d. N/m^2

4. An electron and a proton are accelerated through the same potential difference. Which of the following statements is true?
   a. The electron has the greater speed.
   b. The electron has the greater KE.
   c. The proton has the greater speed.
   d. The proton has the greater KE.

5. The energy stored in a charged capacitor resides in:
   a. the separated charge
   b. the voltage across the capacitor
   c. the electric field
   d. the plates

6. When a dielectric is placed between the plates of a parallel-plate capacitor, the electric field:
   a. increases   b. decreases
   c. remains the same   d. becomes zero

7. The plate area of a parallel-plate capacitor is doubled, and the plate separation is halved. The volume of the capacitor before and after its alteration is the same. The value of the capacitance (C) compared to its original value is now:
   a. $2C$ b. $4C$
   c. C d. $C/4$

8. Installing a dielectric into a parallel-plate capacitor:
   a. increases the electric field
   b. increases the capacitance
   c. decreases the capacitance
   d. increases the voltage

Chapter 17

9. The PE of a system of two charges increases as the charges are brought closer together. What does this tell us about the signs of the two charges?

10. What is the difference between the electric field and the electric potential?

11. The electric field at a spot is zero. Is the electric potential also always zero at that spot? Give an example.

Exercises

Charge in a Uniform Electric Field

*1. Nine hundred volts (900 V) is applied to two parallel conducting plates 80 mm apart. A proton leaves the positive plate, and simultaneously an electron leaves the negative plate. At what distance from the negative plate do the electron and the proton pass each other?

Answer: 79.95 mm

Work Done in Separating Two Charges

2. Two electrons are initially separated by 5 cm. If one electron is then released, what is its speed when the separation reaches 1 m ($m_e = 9.11 \times 10^{-31}$ kg)?

Answer: 98 m/s

3. Assume that we have two charged spheres ($r = 5$ cm), each given 500 μC and separated by 2 m. How much work would be required to force these spheres to touch?

Answer: 21.38 kJ

4. In the absence of gravity, an alpha particle (a helium nucleus with two protons and two neutrons) is shot toward a distant, charged ($+2.5$-μC) target with an initial speed of 1×10^4 m/s. How close to the target will the alpha particle get before being repelled?

Answer: 2.15 cm

*5. A particle has a mass of 3 g and a charge of 8 μC. A second particle has a mass of 6 g and the same charge. The two particles are held apart an initial distance with no initial velocity; then they are released. When the separation between them is 10 cm, the speed of the lighter particle is 125 m/s. What was the initial separation between the particles? (*Hint:* Both energy and momentum are conserved.)

Answer: 1.4 cm

The Volt

*6. The electron gun of a television tube accelerates the electrons with a voltage of 15 kV. The gun has a rated power of 25 W. At what rate do electrons hit the screen?

Answer: 1.04×10^{16} e^-/s

*7. An electron is accelerated horizontally from rest by a potential of 20 kV in a computer monitor. It then passes between two horizontal plates 5 cm long and 1.2 cm apart having a potential difference of 1500 V between them. At what angle from the horizontal will the electron be deflected when it emerges from between the plates?

Answer: 64.4°

8. A flash of lightning transfers 4.5 C of charge at 5 MJ of energy. (a) What was the voltage between cloud and earth when the bolt was generated? (b) How much water could this bolt boil starting from 20°C?

Answer: (a) 2.2×10^6 V
(b) 1.9 kg

Energy Storage in an Electric Field

9. A parallel-plate capacitor is found to have a potential of 5 V and a plate separation of 10 mm. What is the strength of the electric field inside this capacitor?

Answer: 500 V/m

10. The cell membranes within your body are more permeable to positively charged ions (potassium) than to negatively charged ions (sodium). The result is a small potential across the cellular membrane of approximately 0.08 V. If a typical membrane has a thickness of 10×10^{-9} m, what is the electric field associated inside the membrane?

Answer: 8×10^6 N/C

Parallel-Plate Capacitor

11. Two conductive plates are placed 5 mm apart. Surface charge densities of $+1.5$ μC and -1.5 μC per square meter are placed on each of the two plates. What is the voltage across the parallel-plate capacitor?

Answer: 847.5 V

12. The dimensions of a particular parallel-plate capacitor are 1.5 cm $\times$ 2 cm, and the region between the plates is a vacuum. (a) If a field strength of 2000 V/m is desired, what charge must be placed on each plate? (b) If the separation between plates is 15 mm, what is the energy stored in this capacitor?

Answer: (a) 5.3×10^{-12} C
(b) 8×10^{-11} J

13. All three dimensions (length, width, and plate spacing) of a parallel-plate capacitor are doubled. (a) By how much does the capacitance change? (b) If the capacitor keeps the same charge before and after the doubling, what happens to the energy?

Answer: (a) and (b) The capacitance is doubled, and the energy is halved.

*14. A pulsed nitrogen laser has a power supply that consists of a 50-nF capacitor that is charged to 30 kV. If 10% of the maximum stored power in the capacitor is converted to light, and if the pulse is 10 μs long, what is the power output of the laser pulse?

Answer: 225 kW

*15. In a thunderstorm the rain separates charge in the clouds by moving some of it to the ground. The clouds become highly charged, and the potential difference between the clouds and the earth can be as high as 35,000,000 V! The bottoms of the clouds are typically 1500 m above the earth, and the clouds can have areas of around 100 km². (a) What is the capacitance of the earth-cloud system? (b) What is the energy stored in the capacitor? (c) How much charge is stored in the capacitor?

Answer: (a) 590 nF
(b) 351 MJ
(c) 20.65 C

Capacitance

16. It is often dangerous to handle charged capacitors without special precautions. For example, it is possible to buy a capacitor that can store 10 C of charge. (a) If this capacitor were charged to 5 V, how much energy would you receive if you accidentally shorted it out? (b) From what height would a 0.5-kg brick have to be dropped to impart an equal amount of energy?

Answer: (a) 25 J
(b) 5.1 m

17. What is the potential across a 5-pF capacitor with 20 μC of charge?

Answer: 4×10^6 V

*18. An engineer is told to design a portable weapon that can electrically accelerate a 3-kg shell up to a speed of 400 m/s.

She decides to attach a large parallel-plate capacitor to the weapon and discharge it to provide the required energy. If she is limited to only being able to charge the capacitor up to 500 V, and if the efficiency is only 10%, what capacitance must she use to accomplish the job?

Answer: 19.2 F

*19. A parallel-plate capacitor is given a charge Q by connecting it to a battery. The capacitor is then disconnected. How much work is required to pull the plates further apart to twice their original separation?

Answer: The same amount of work as is stored originally in the capacitor.

Dielectric Constant

20. A parallel-plate capacitor measures 2 cm to a side with a separation of 5 mm. The capacitor is found to have a capacitance of 20 pF without a dielectric inserted. (a) If a piece of teflon is placed between the plates, what is the new capacitance? (b) What would the new capacitance be if the empty capacitor were immersed in ethyl alcohol?

Answer: (a) 42 pF
(b) 520 pF

21. An electrician needs to store 2 μJ of energy in a capacitor that will be charged with a 12-V battery. He has an empty capacitor that measures 15 cm $\times$ 20 cm and a plate separation of 1 mm. He realizes that to store the appropriate amount of charge, he will need to insert the proper dielectric material into his capacitor. What value of K will the dielectric constant of this material have?

Answer: K = 105

*22. A 5-μF capacitor is charged with a 9-V battery. (a) What is the potential across the plates of the capacitor? How much charge does the capacitor now hold? (b) If the space between the plates is then filled with water, what would the new potential and charge associated with the capacitor be if the battery were left connected?

Answer: (a) 9 V and 45 μC
(b) 9 V and 3.6 mC

23. The 5-μF capacitor from Exercise 22 is charged with the same 9-V battery and is then disconnected. (a) What are the potential and the charge on the capacitor before and after the disconnection? (b) If the space between the plates is filled with water after the capacitor is disconnected, what would the new potential and charge be?

Answer: (a) 9 V and 45 μC before and after
(b) 9 V and 45 μC before; 0.11 V and 45 μC after

*24. A parallel-plate capacitor has circular plates of 40-cm diameter and a separation of 0.1 cm. The capacitor is charged to 900 V by a voltage source. With the capacitor still connected to the battery, a sheet of mica ($K = 6$) is shoved between the plates, completely filling the space. How much additional charge flows from the battery to the capacitor when the mica sheet is installed?

Answer: 5 μC

Capacitors in Series and in Parallel

*25. What are the equivalent capacitances of the networks shown in Figure 17.11?

Figure 17.11 (Exercise 25)

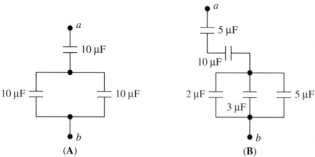

(A)   **(B)**

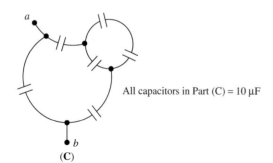

All capacitors in Part (C) = 10 μF

(C)

Answer: (a) 6.67 μF
(b) 2.5 μF
(c) 13.75 μF

*26. A 1-μF capacitor and a 2-μF capacitor are both charged to 600 V. The capacitors are disconnected from the voltage source and then reconnected with a positive-to-positive and negative-to-negative terminal, forming a closed loop. What is the final charge of each capacitor?

Answer: $Q_1 = 6 \times 10^{-4}$ C; $Q_2 = 12 \times 10^{-4}$ C

*27. The two capacitors in Exercise 26 are again charged to 600 V and then disconnected from the voltage source. They are then reconnected, but this time the opposite terminals are connected together: positive to negative and negative to positive. What is the final charge of each capacitor?

Answer: $Q_1 = 2 \times 10^{-4}$ C; $Q_2 = 4 \times 10^{-4}$ C

*28. A 5-μF capacitor is charged to 300 V. The capacitor is then connected to an uncharged 10-μF capacitor. (a) What is the voltage across the pair? (b) What is the energy stored in the pair? (c) What was the energy originally stored in the 5-μF capacitor? (d) Account for any difference.

Answer: (a) 100 V
(b) 0.075 J
(c) 0.225 J

*29. Figure 17.12 shows a parallel-plate capacitor filled with two different dielectric materials. Show that the capacitance of the capacitor is

$$C = \varepsilon_0 A \frac{k_1 + k_2}{2d}$$

where A is the total plate area, and d is the plate separation. Each dielectric material fills half the volume of the capacitor.

Figure 17.12 (Exercise 29)

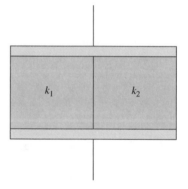

*30. Figure 17.13 shows a parallel-plate capacitor again filled with two dielectric materials. The total plate area is A, the plate separation is d, and each dielectric material fills half the space between the plates. What is the capacitance of the capacitor?

Figure 17.13 (Exercise 30)

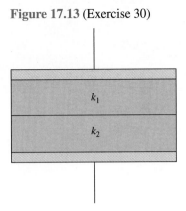

Answer: $C = \dfrac{2\varepsilon_0 A k_1 + k_2}{(k_1 + k_2)d}$

Electric power is generated at a coal-fired power plant as a voltage that can cause a current to flow. This electric power is shipped from the generating station by high-tension transmission lines that ultimately connect a voltage source to your home (the wall sockets) which can deliver current for heat and power for running appliances and machinery.

Electric Current and Resistance

At the end of this chapter, the reader should be able to:

1. Calculate the **current** flowing across a *resistor* in series with a known voltage source.

2. Solve for the *equivalent resistance* of a circuit containing resistors in parallel and resistors in series.

3. Calculate the change in resistance of a given wire during a known change in temperature.

4. Calculate the power dissipated from a resistor possessing a known current.

5. Explain the concept of *impedance matching.*

6. Solve for the charge on a known capacitor at any given point in time after it is connected to a known battery.

7. Write an equation describing a *current loop* using *Kirchhoff's laws.*

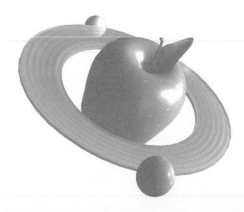

18.1   Electric Current

Current: The flow of electric charge.

The cathode-ray tube (CRT) introduced in Chapter 17 has an electric current flowing through it. That current is the flow of electrons from the hot filament to the screen. *Since each electron has a negative charge, electric **current** is actually a flow of electric charge.* In previous chapters, we encountered flow problems, such as the flow of water. Microscopically, water is made up of countless discrete molecules of H_2O, and *flow rate* was defined as a mass per time, for example, water out of a faucet or into a pan. The flow then becomes mathematically $\Delta m/\Delta t$. However, we perceive water at the macroscopic level as a uniform liquid, and the fact that water is made up of many individual molecules isn't apparent at all. When water flows, we usually perceive it as a continuous process, such as water out of a hose instead of a stream of tiny particles.

The flowing particles in an electric current are electrons, and an electron has both charge and mass. But for an electric current we're interested not in a mass flow but in a charge flow. Since the charge on an electron is small (-1.6×10^{-19} C/electron), and since there are usually many electrons flowing, current is defined as

(Equation 18.1)

$$I = \frac{\Delta q}{\Delta t}$$

Current flow can also be considered a continuous process.

From the point of view of current, materials can be divided into two classes: conductors and insulators. A conductor can carry a flow of current, and an insulator can't. If you look at the cross section of a common extension cord, you'll see both a core of copper conductor that carries the current and a sheath of a plastic or rubber insulator. The insulator protects the two wires in the extension cord both from shorting out against each other and from the environment. The insulator protects us from the high voltage on the wires (typically 120 V_{ac} or 220 V_{ac}). That much voltage could give you a nasty shock if you touched the wire directly.

In copper, the outer electron in each atom is held so loosely that it is free to move about at random throughout the lattice of copper atoms. With billions and billions of copper atoms in a typical wire, this constitutes a "gas" of electrons free to move with the slightest breeze. That "breeze" is an electric field. With no breeze, the movement of electrons is random, and there is no average current. But with a breeze, the gas of electrons drifts in a direction opposite to that of the breeze, and positive charges move in the direction of the breeze.

Up to now we've dealt with positive test charges. Current is also defined in terms of a flow of positive charge. From our macroscopic perspective, it makes no difference that the actual flowing charges are negative. If we take a sheet of paper and plot a vertical y axis and a horizontal x axis and make the origin where the two axes cross, then by convention x coordinates to the left of the origin are negative, and x coordinates to the right of the origin are positive. A velocity along the x axis to the left is defined as negative, and a velocity to the right is defined as positive. If we have a negative charge moving to the left, mathematically we can say that there is a positive charge moving to the right. A negative number times a negative number is a positive number. The derivation that follows should convince you of this point.

Assume that in a copper wire with a circular cross section, there exist n electrons per unit volume that are available to carry a current (see Figure 18.1). An electric field exists down the axis of the wire pointing from left to right. Electrons will tend to move to the left.

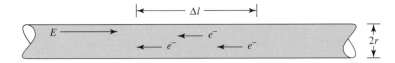

Figure 18.1 Electrons in a Wire Move Because of an Electric Field.

$$I = \frac{\Delta Q}{\Delta t} \qquad \Delta Q = n(-e)\pi r^2\, \Delta l$$

$$\frac{\Delta Q}{\Delta t} = \frac{n(-e)\pi r^2\, \Delta l}{\Delta t} \qquad \frac{\Delta l}{\Delta t} = -V_d$$

where $-V_d$ is the drift velocity.

$$\frac{\Delta Q}{\Delta t} = n(-e)\pi r^2(-V_d)$$

$$I = ne\pi r^2 V_d$$

The total current is a positive number. In other words, positive charges flow to the right. *By convention a normal current is the flow of positive charge, and that is the way most physics and engineering texts define current.* Benjamin Franklin is responsible for this mess! He defined charge convention back in the eighteenth century (early 1700s). It wasn't until the discovery of the electron some 200 years later that scientists found that what usually makes up an electric current is a flow of negative particles.

A steady current means that the drift velocity must be constant. An electric field will accelerate a charged particle, but in the copper crystal lattice, the electron keeps hitting copper atoms and rebounding. The motion of a single electron is quite complex, but overall a constant average drift velocity in a direction opposite the electric field can be given for the conduction electrons in the volume of the copper. The electric field in the conductor is constant from one end of the wire to the other. In previous chapters we discussed that no electric field exists in a conductor, but that situation occurred after equilibrium was reached and all movement of charges had stopped. Here the average motion of the conduction charges is constant. The charges are free to move about in the conductor but won't leave it. The wire stays electrically neutral because even though the charges are in motion, there is no net separation of charge. For every electron leaving an atom, another one moves in to replace it.

18.2 Resistivity and Ohm's Law

A free electron in a current-carrying wire doesn't have an average acceleration because of the obstacles in its path. It's constantly bouncing off something. It does, however, have a net drift in the direction opposite the electric field, which gives rise to the constant drift velocity. These obstacles in the electron's path oppose the motion of the

Resistivity: The opposition to current flow in a given material.

electron's movement. *The opposition to the flow of charge is called **resistivity**.* The greater the resistivity, the smaller the flow of current, and the lesser the resistivity, the greater the flow of current. If there were no obstacles in the path of the conduction charge, it would accelerate. But because of the resistivity of the copper, it has a constant average drift velocity.

Copper is a metal, and metal is a conductor. But there are other metals besides copper. Aluminum, iron, gold, and silver will all carry a current. Each metal has a crystal structure that will have its conduction electrons bumping into other atoms, but the other metals also have different numbers of conduction electrons per unit volume (n). Resistivity is a property of a particular material, and it takes into consideration differences in crystal structure and differing values of n. A bulk property, it gives an indication of the amount of resistance that a material will have to current flow. A perfect insulator will have infinite resistivity, and a perfect conductor will have zero resistivity. If an electric field were put across a rectangular block of copper of height h and cross-sectional area A, its resistivity would be given by

(Equation 18.2)

$$\rho = \frac{E}{J}$$

where ρ is the resistivity, E is the electric field, and J is the current density. The current density is the current per unit of cross-sectional area. The resistivity is constant for most materials, and it is the ratio of the electric field to the current density. For a given material, if the field rises, so does the current density, and vice versa. To place an electric field across the block of copper, we would use a battery, which would give us the same type of field as for the parallel-plate capacitor, only with a conductor between the plates instead of an insulator, as shown in Figure 18.2. Thus,

$$\rho = \frac{E}{J} \qquad E = \frac{V}{h} \qquad J = \frac{I}{A}$$

$$\rho = \frac{V/h}{I/A} \qquad \rho = \frac{V}{I} \frac{A}{h}$$

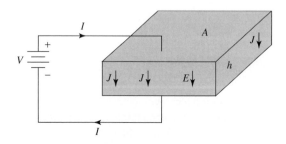

Figure 18.2 Current Density in a Conductor with an Electric Field across It

All that the battery knows is that a given amount of current is being drawn from it for its output voltage. It sees a given *resistance* caused by the block of copper. The resistance is given by

$$R = \frac{V}{I}$$

where R is the resistance in ***ohms*** (Ω), the unit of resistance; V is the battery voltage; and I is the current drawn from the battery. Plugging this definition in the above yields

Ohm (Ω): The unit of resistance (volts per ampere).

$$R = \frac{V}{I} \qquad \rho = R\frac{A}{l}$$

$$R = \frac{\rho l}{A} \qquad \text{(Equation 18.3)}$$

The resistance is defined in terms of the material's resistivity and geometry. The resistance of a material is there whether or not a current is flowing. This definition is somewhat like defining the capacitance of a capacitor in terms of the dielectric and the capacitor's geometry.

Let's take another look at the equation for the definition of resistance above.

$$R = \frac{V}{I}$$

or
$$V = IR \qquad \text{(Equation 18.4)}$$

The above equation, known as *Ohm's law,* establishes the relationship between voltage and current in a resistor. Another way of stating Ohm's law is

$$I = \frac{V}{R}$$

The greater the voltage, the greater the current. The greater the resistance, the smaller the current. This equation implies an electric circuit like the one shown in Figure 18.3, in which a battery is connected by wires to a resistor. A current I flows through the wires. The direction of conventional current flow is from the positive terminal of the battery to its negative terminal. The current magnitude is defined by the battery voltage divided by the resistance in the circuit. The schematic symbol for the resistance is the squiggly line marked R in Figure 18.3. It is often a cylindrical component with wire axial leads emerging from either end, with the cylinder consisting of carbon (a semiconductor). In the circuit shown, the battery and the resistor are discrete components, and all of the resistance in the circuit is contained in the resistor. The wires are supposedly ideal with no resistance at all to an electrical current. There is, however, resistance in the wires, but it is so small in most cases that it can be ignored. If the resistance in a circuit is discrete, it can be a circuit component such as the cylindrical resistor mentioned above, or the filament of a light bulb or the heating coils of an electric stove.

Figure 18.3 Resistive Electrical Circuit

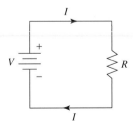

EXAMPLE 18.1

A copper wire 1 mm in diameter carries a current of 1 ampere (1 ampere = 1 coulomb per second, or 1 A = 1 C/s). The free electron density in copper is 8.5×10^{28} electrons per cubic meter. (a) What is the drift velocity of an electron in this wire? (b) How long does it take a given electron to move 1 m along this wire? (c) If the battery voltage is 10 V, what resistance is necessary for a 1-A current to flow? (d) If a cylindrical carbon resistor is 1 cm long, what is its diameter if the resistor has a value of 10 Ω? The resistivity of carbon is 3.5×10^{-5} Ω-m. (e) What is the electric field within the resistor along its length?

SOLUTION

(a) $$I = ne\pi r^2 V_d \qquad V_d = \frac{I}{ne\pi r^2}$$

$$V_d = \frac{1 \text{ C/s}}{(8.5 \times 10^{28}/\text{m}^3)(1.6 \times 10^{-19} \text{ C})\pi(0.0005 \text{ m})^2}$$

$$V_d = 9.36 \times 10^{-5} \text{ m/s}$$

(b) $$V_d = \frac{l}{t} \qquad t = \frac{l}{V_d}$$

$$t = \frac{1 \text{ m}}{9.36 \times 10^{-5} \text{ m/s}} \qquad t = 10,634 \text{ s}$$

$$t = 2.97 \text{ h}$$

The time it takes a single electron to move 1 m along this wire is almost 3 h! So how is it, then, that a communications signal can move along a wire at almost the speed of light (3×10^8 m/s) when a single electron moves so slowly? The answer can be found in Figure 18.4.

Six ping-pong balls are inserted into a cardboard mailing tube, completely filling it, as shown in the figure. A seventh ball, if inserted into the tube, would cause

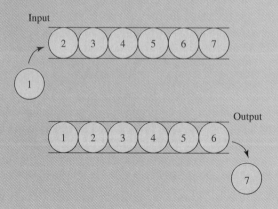

Figure 18.4 Electrical Signal Propagating along a Wire

the end ball of the original six to pop out. If we were interested in the drift velocity of ball 1, we would know that every time a ball was inserted at the input,
ball 1 would move $\frac{1}{6}$ the length of the tube. Therefore, it would take 7 insertion cycles
before ball 1 popped out of the output end. Knowing the length of the tube and the
average time of an insertion cycle, we could come up with the drift velocity of a
given ping-pong ball. However, if we consider inserting ball 1 into the left end of
the tube as a communications signal, then when a ball pops out on the right, that
is a signal that a ball has been inserted in the left side. The communication that a
ball has been inserted in the left end appears almost simultaneously at the right end,
despite the fact that it would take ball 1 quite a while before it popped out at the
right.

(c) $$R = \frac{V}{I} \qquad R = \frac{10\ V}{1\ A}$$

$$R = 10\ \Omega$$

where, as mentioned previously, the ohm (Ω) is the unit of resistance.

(d) $$R = \frac{\rho l}{A} \qquad R = \frac{\rho l}{\pi D^2/4} \qquad R = \frac{4\rho l}{\pi D^2}$$

$$D = \sqrt{\frac{4\rho l}{\pi R}} \qquad D = \sqrt{\frac{4(3.5 \times 10^{-5})\Omega\text{-m}(0.01\ \text{m})}{\pi(10\ \Omega)}}$$

$$D = 2.11 \times 10^{-4}\ \text{m}$$

(e) $$E = \frac{V}{l} \qquad E = \frac{10\ V}{0.01\ \text{m}}$$

$$E = 1000\ \text{N/C}$$

Table 18.1 gives the resistivities for some substances used in electrical applications
and the electronics industry. The range in resistivities is about 10^{24} from conductors to
insulators. Notice that all of the *conductors* are metals and are more tightly grouped than
the other two categories. (Resistivities are within a factor of about 100 of each other.)
Insulators have the widest range of the three groups, from wood (1×10^8 Ω-m) to fused
quartz (75×10^{16} Ω-m). Thus, in the group of insulators the resistivity can vary by a factor of *10 million* and the material can still be considered an insulator. The insulation around
most copper household wiring is a plastic and must be as flexible as the copper of the wires
themselves, so brittle substances such as quartz and glass can't be easily used for this application. But remember that the dielectric in a capacitor must be an insulator, or the capacitor's charge bleeds off. So insulators find use as a dielectric between the plates of a capacitor. To be a good dielectric, a material must have both a large dielectric constant and a
high resistivity.

The *semiconductors* fall intermediately between the ranges of the conductors and the
insulators. We've already discussed carbon and one of its uses in the construction of resistors. Notice that in addition to carbon, silicon and germanium are also listed as semiconductors. Both elements can be doped with impurities which will greatly decrease their
respective resistivities. Both elements are also used in the manufacture of transistors, and

Table 18.1 Resistivity of Some Common Substances

Substance	Resistivity(Ω-m)
Conductors	
Aluminum	2.6×10^{-8}
Constantan (60% Cu, 40% Ni)	49×10^{-8}
Copper	1.7×10^{-8}
Gold	2.44×10^{-8}
Iron	12×10^{-8}
Lead	21×10^{-8}
Manganin (84% Cu, 12% Mn, 4% Ni)	44×10^{-8}
Mercury	98×10^{-8}
Nichrome	112×10^{-8}
Platinum	11×10^{-8}
Silver	1.6×10^{-8}
Semiconductors	
Carbon	3.5×10^{-5}
Germanium	0.5
Silicon	2300
Insulators	
Amber	5×10^{14}
Glass	10^{10}–10^{14}
Lucite	$>10^{13}$
Mica	10^{11}–10^{15}
Quartz (fused)	75×10^{16}
Sulfur	$>10^{13}$
Wood	10^{8}–10^{11}

the ability to make integrated circuits (ICs) with microscopic switching devices (transistors) on a single crystal of silicon launched the explosion in computers and communications that we are now experiencing. For silicon to be used as a current-carrying electronic component, its resistivity must be low enough to efficiently carry a current, but it must also be able to amplify or switch a signal. Silicon has a regular crystal lattice, and the small amount of impurities added to the silicon must fit into the structure just as if the impurities were silicon. As a result, the impurity is forced to give up an electron to the conduction band, or to leave a "hole" at the site of the impurity into which a free electron can drop. These "holes" act just like free positive charges. Silicon doped with impurities that give up free electrons is called *N-type silicon,* and silicon doped with impurities that create holes is called *P-type silicon.* Both types of silicon have resistivities lower than 2300 Ω-m, depending on the strength of the doping. But more important, when an interface is made between P- and N-type silicon, switching characteristics are observed, and the manufacture of miniature electronic switches is possible. Silicon is now the basis for an entire industry. Silicon isn't the only viable semiconductor, but it certainly is the most common and best known.

18.3 Resistors in Series and in Parallel

Resistors can be arranged in series and parallel circuits, just as capacitors were in Chapter 17. In this section we'll derive the formulas for resistors in series and in parallel. A battery doesn't know how many resistors are connected to it or in what configuration. All that it knows is its terminal voltage and the current that's being drawn from it. All that the battery sees is an equivalent resistance. In this section we have a steady current flowing. In Chapter 17, with capacitors in series and in parallel, the charging current started high and trickled down to nothing as the capacitors charged up. In this case the current remains the same as long as the battery is connected to the circuit. With a purely resistive circuit and a battery, the current is said to be dc, or direct current.

In Figure 18.5 there are three resistors in series. Because there is only one path for current flow, the current is the same through all three resistors. According to Ohm's law, if current flows through a resistor, there is a voltage drop across it. You can easily measure this voltage drop by placing a voltmeter across the resistor. A voltage will be shown on the meter as long as current flows through the resistor. The voltage across the resistor will have a definite polarity. The voltmeter should be connected with the red lead to the positive terminal of the resistor and the black lead to the negative terminal of the resistor. If you guess wrong, either the meter will display a negative voltage, or the needle will deflect opposite to the desired direction. (An ideal voltmeter doesn't disturb the circuit that it's measuring, and no current flows through an ideal voltmeter.)

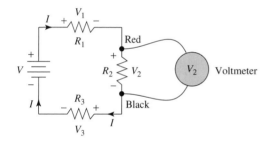

Figure 18.5 Resistors in Series

Just as with capacitors in series, the sum of the voltages across the resistors adds up to the applied battery voltage.

$$V = V_1 + V_2 + V_3 \qquad V = IR_{eq}$$

$$V_1 = IR_1 \qquad V_2 = IR_2 \qquad V_3 = IR_3$$

$$IR_{eq} = IR_1 + IR_2 + IR_3$$
$$IR_{eq} = I(R_1 + R_2 + R_3)$$
$$R_{eq} = R_1 + R_2 + R_3$$

The equivalent resistance that the battery sees or the total resistance for the whole circuit is simply the sum of the resistors in series. If there are more than three resistors, then just add the extra resistor or resistors.

$$R_{eq} = R_1 + R_2 + R_3 + \cdots + R_n \qquad \text{(Equation 18.5)}$$

where n is the total number of resistors. The current through the series circuit is the battery voltage divided by the equivalent resistance. As the equivalent resistance gets larger, the current drops. The more resistors in series, the smaller the current. If there are an infinite number of resistors in series, then no current flows because the equivalent resistance is infinite. This situation is the same as an open circuit, which has a break in the line somewhere in the circuit. If there are no resistors in the circuit ($R_{eq} = 0$), then an infinite current flows, or tries to flow. For this reason, shorting the terminals of an automobile battery is a bad idea. The short-circuit current of an automobile battery is high enough to weld metal, start a fire, or cause the battery to explode!

Next we'll deal with a parallel circuit configuration of resistors. Figure 18.6 shows three resistors in parallel. Each resistor is connected directly across the battery, so the voltage across each resistor is the same (the battery voltage). However, each branch in the circuit draws its own current, so the total current drawn from the battery is the sum of the three branch currents.

$$I = I_1 + I_2 + I_3$$

$$I_1 = \frac{V}{R_1} \qquad I_2 = \frac{V}{R_2} \qquad I_3 = \frac{V}{R_3}$$

$$I = \frac{V}{R_{eq}}$$

$$\frac{V}{R_{eq}} = \frac{V}{R_1} + \frac{V}{R_2} + \frac{V}{R_3}$$

$$V\frac{1}{R_{eq}} = V\left(\frac{1}{R_1} + \frac{1}{R_2} + \frac{1}{R_3}\right)$$

$$\frac{1}{R_{eq}} = \frac{1}{R_1} + \frac{1}{R_2} + \frac{1}{R_3}$$

(Equation 18.6)

$$\frac{1}{R_{eq}} = \frac{1}{R_1} + \frac{1}{R_2} + \frac{1}{R_3} + \cdots + \frac{1}{R_n}$$

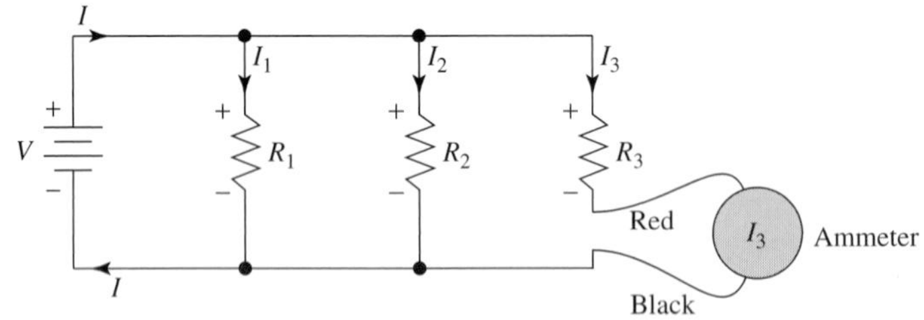

Figure 18.6 Resistors in Parallel

If there are more than three resistors in parallel, then add another term to the equation as shown. For each new resistor added in parallel to the circuit, more current is drawn from the battery. A short circuit across the battery is therefore equivalent to an infinite number of resistors in parallel. For this reason, sometimes when we add too many appliances or loads to an extension cord, a fuse blows or a breaker pops. Every time we add a new load, we effectively add another resistor in parallel, and the necessary current that the source must supply increases. Every time we plug something into a wall socket, we add another load in parallel to the source. The source is 120 V_{ac} in the United States and

220 V_{ac} in other countries. The *ac* stands for "alternating current," which will be covered in a future chapter. The voltage of the source is fixed, and the amount of current drawn from the source depends on the number of loads connected in parallel to it.

The total current is the sum of all of the branch currents. We can demonstrate this fact by connecting an ammeter into each branch as shown in Figure 18.6 to find the branch current and then adding the branch currents to find the total current drawn from the battery. (The total current from the battery can also be measured.) If we wanted to measure I_3, we would first have to cut the wire in branch 3 and then connect it to both sides of the ammeter as shown. For us to measure I_3, *all* of I_3 must flow through the meter.

A good meter won't affect the circuit to which it's attached. It won't change the voltage across a component or the current flowing through a component when connected. The ammeter in Figure 18.6 must have very low resistance, or its resistance must be much lower than the resistance in branch 3. (An ideal ammeter has no resistance.) If the ammeter's resistance were high, it would measure a lower current than without the meter. The meter shouldn't affect the circuit that it's measuring. A voltmeter's resistance, on the other hand, must be very high. A voltmeter is connected *across* a resistor or load, and we don't want any appreciable current to flow through it. If it did, the voltage read across the load would be lower than without the voltmeter, and a good meter doesn't affect the circuit it's measuring! (An ideal voltmeter has an infinite amount of resistance.)

EXAMPLE 18.2

Figure 18.7 shows a series-parallel resistance network connected to a 12-V battery. (a) What current is drawn from the battery? (b) What current runs through resistor R_2?

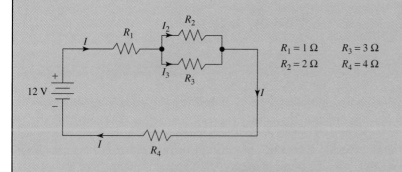

$R_1 = 1\ \Omega \qquad R_3 = 3\ \Omega$
$R_2 = 2\ \Omega \qquad R_4 = 4\ \Omega$

Figure 18.7 Series-Parallel Resistive Circuit

SOLUTION
(a) To find the current drawn from the battery, we must first find the equivalent resistance that the battery sees (see Figure 18.8).

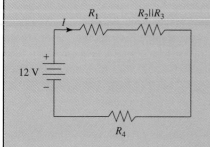

$R_2 \| R_3$ = equivalent resistance
of R_2 in parallel with R_3

$R_{eq} = R_1 + R_2 \| R_3 + R_4$

Figure 18.8 Equivalent Series Circuit of Figure 18.7

$$\frac{1}{R_2 \| R_3} = \frac{1}{R_2} + \frac{1}{R_3} \qquad R_2 \| R_3 = \frac{R_2 R_3}{R_2 + R_3}$$

$$R_{eq} = R_1 + \frac{R_2 R_3}{R_2 + R_3} + R_4 \qquad R_{eq} = 1\,\Omega + \frac{(2\,\Omega)(3\,\Omega)}{2\,\Omega + 3\,\Omega} + 4\,\Omega$$

$$R_{eq} = 6.2\,\Omega$$

$$I = \frac{12\text{ V}}{6.2\,\Omega} = 1.94\text{ A}$$

(b) R_2 and R_3 are in parallel. We must know the voltage drop across the parallel pair before we can find the current in R_2.

$$R_2 \| R_3 = 1.2\,\Omega$$

$$V_{23} = (1.94\text{ A})(1.2\,\Omega) \qquad V_{23} = 2.33\text{ V}$$

$$I_2 = \frac{V_{23}}{R_2} \qquad I_2 = \frac{2.33\text{ V}}{2\,\Omega}$$

$$I_2 = 1.165\text{ A}$$

The current through R_3 is 0.77 A; $I_2 + I_3 = 1.94$ A, which is the total current found in Part (a).

18.4   Temperature Coefficient of Resistance

In Part II on thermodynamics, we learned that the forces between atoms in a molecule or crystal act like tiny springs. The springs are always vibrating with an amplitude that depends on the temperature. The greater the temperature, the greater the amplitude of oscillation. Electrons in a metal drift with a low drift velocity because they are always hitting and rebounding from the atoms in the crystal. If the temperature rises, the interference that the electron sees is greater because of the increased amplitude of vibration of the atoms and molecules. The interference to the conduction electrons is the material's resistivity, so the greater the temperature, the greater the resistivity. A linear formula describes how the resistance of most materials changes with temperature.

(Equation 18.7)

$$\Delta R = \alpha R_o\, \Delta T$$

where ΔR is the change in resistance, R_o is the resistance of the resistor at a given temperature (usually room temperature, 20°C), ΔT is the change in temperature from room temperature, and α is the temperature coefficient of resistance for the given material. The units for α are 1/°C. The temperature coefficient of resistance is a fundamental parameter of a metal or semiconductor.

The values for a number of common conductors and semiconductors are given in Table 18.2. Notice that carbon has a negative coefficient of resistance; that is, its resistance *decreases* with a rising temperature. For most materials the resistance rises with temperature. But as shown in the table, such is not always the case. Carbon does, however, follow the same linear equation as do the listed conductors.

Table 18.2 Temperature Coefficient of Resistance at 20°C

Substance	α (1/°C)
Conductors	
Aluminum	0.0039
Constantan (60% Cu, 40% Ni)	0.000002
Copper	0.0039
Iron	0.0050
Lead	0.0043
Manganin (84% Cu, 12% Mn, 4% Ni)	0.000000
Mercury	0.00088
Nichrome	0.0002
Platinum	0.0036
Silver	0.0038
Semiconductors	
Carbon	−0.0005

EXAMPLE 18.3

A thermistor is a resistor especially designed to take advantage of the change in resistance of a material with temperature. You can use a thermistor as an accurate thermometer by electronically reading the change in resistance from its calibration resistance. A platinum resistor is calibrated at 10 Ω at 20°C. If the resistance becomes 20.1 Ω, what is the temperature of the oven into which it was placed?

SOLUTION

$$\Delta R = \alpha R \, \Delta T \qquad R_f - R_o = \alpha R_o (T_f - T_o)$$

$$R_f - R_o = \alpha R_o T_f - \alpha R_o T_o \qquad \alpha R_o T_f = R_f - R_o + \alpha R_o T_o$$

$$T_f = \frac{R_f - R_o + \alpha R_o T_o}{\alpha R_o}$$

$$T_f = \frac{20.1 \, \Omega - 10 \, \Omega + (0.0036/°C)(10 \, \Omega)(20°C)}{(0.0036/°C)(10 \, \Omega)}$$

$$T_f = 300.6°C$$

In situations such as in Example 18.3, remote sensing would probably be used. The electronics used to read the resistance most likely wouldn't be in the oven along with the platinum resistor. In remote sensing, the leads between the sensor and the instrument are probably long and would also experience a large temperature change. If the reading is to be accurate, you don't want the instrument's leads to change with temperature as the sensor does. Certain conductors are available to use as leads that have a very low or zero temperature coefficient of resistance. Materials such as constantan ($\alpha = 0.000002$) and manganin ($\alpha = 0$) would therefore be useful in extreme sensing environments, provided that the environment doesn't melt the lead!

Electricity is a source of power. Ovens and heaters convert electricity to heat energy. An electric motor converts electricity to mechanical energy. An electronic amplifier might convert electricity into sound energy. Electricity's flexibility and ease of use have made it the most familiar source of power to most people in developed countries. (The next most common source of power is the internal combustion engine.) Of course, electric power must itself be generated by many diverse means such as burning coal and oil; through hydrodynamic and nuclear power plants; and by wind farms, solar collectors, and solar cells. But the bottom line is that the electrical outlet in your wall is in reality a power tap. How does electricity deliver power?

Let's first start with the definition of *power* from Part I on mechanics. Power is the work done in a given amount of time.

$$P = \frac{W}{t}$$

We know that a volt is the work done per unit charge and that current is the rate of flow of charge.

$$V = \frac{W}{q} \qquad W = qV \qquad P = \frac{qV}{t}$$

$$P = \frac{q}{t}V \qquad I = \frac{q}{t}$$

(Equation 18.8)
$$P = IV$$

Electric power is the current times the voltage. The power that an electric device uses is therefore equal to the voltage across it times the current through it.

If you've ever had to pay an electric bill, you're familiar with the term *kilowatt-hour* because the number of kilowatt-hours is the basis for the amount of your bill. A kilowatt-hour is a unit of energy, so you're billed on the basis of your energy use. A kilowatt-hour is a lot of energy: 3.6 *million* joules. To use a kilowatt-hour of energy, you would have to burn ten 100-W light bulbs for an hour. That's enough energy to lift a small car over a thousand feet into the air! If expensive electricity costs 10¢ per kilowatt-hour, it's still cheap at the price.

Up to now we've had to deal with only two electrical components: the capacitor and the resistor. A capacitor stores energy but can't dissipate it. A resistor, on the other hand, stores no energy but dissipates it as heat. Combining Ohm's law with the power formula gives us two equations for power dissipation in a resistor.

$$P = IV \qquad V = IR \qquad P = I(IR)$$

(Equation 18.9)
$$P = I^2R$$

$$P = IV \qquad I = \frac{V}{R} \qquad P = \left(\frac{V}{R}\right)V$$

(Equation 18.10)
$$P = \frac{V^2}{R}$$

Equation (18.9) gives power consumption in terms of the current passing through a resistor, and Equation (18.10) in terms of the voltage drop across it.

EXAMPLE 18.4

An incandescent light bulb rated at 100 W has a tungsten filament. Before the bulb is placed in a socket and switched on, the filament is checked with an ohmmeter and is found to have a resistance of 9.5 Ω. If the temperature coefficient of resistance is $4.5 \times 10^{-3}/°C$, at what temperature is the filament when the bulb is lit?

SOLUTION

$$\Delta R = \alpha R_o \, \Delta T \qquad R_f - R_o = \alpha R_o (T_f - T_i)$$

$$T_i = 20°C$$

Next we must find R_f.

$$P = \frac{V^2}{R} \qquad R_f = \frac{V^2}{P}$$

$$R_f = \frac{(120 \text{ V})^2}{100 \text{ W}} = 144 \ \Omega$$

Now T_f can be calculated.

$$T_f - T_i = \frac{R_f - R_o}{\alpha R_o} \qquad T_f = T_i + \frac{R_f + R_o}{\alpha R_o}$$

$$T_f = 20°C + \frac{144 \ \Omega - 9.5 \ \Omega}{(4.5 \times 10^{-3}/°C)(9.5 \ \Omega)}$$

$$T_f = 3146.2°C$$

18.6   Internal Resistance and Electromotive Force

With nothing connected to it, a battery might have a terminal voltage of 12 V. But when a load is connected between the battery terminals and current starts flowing, the terminal voltage is found to be something less than 12 V. As more current is drawn from the battery, the terminal voltage drops still further because a battery, or any other source of electrical power for that matter, has an internal resistance associated with it, as shown in Figure 18.9.

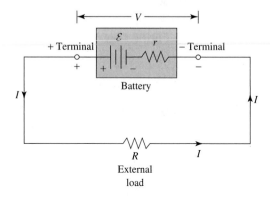

Figure 18.9 Internal Resistance of a Battery

An ideal battery or power source has no internal resistance, but this is a very real world, and losses are inevitably associated with it. All real batteries and power sources have this internal resistance, however slight. This internal resistance is a distributed resistance; that is, it can't be localized as in the discrete resistors used in our examples. It is nevertheless modeled as the series resistor r in Figure 18.6 because all of the current drawn from the battery must pass through it. If no load were connected between the battery terminals, a voltmeter would read the battery's *electromotive force (emf)*. Connecting a load causes current to flow and a voltage drop across r. The terminal voltage of the battery is

Electromotive force (emf):
The no-load (no-current) potential of a voltage source.

$$V = \mathcal{E} - Ir$$

As the voltage drop across r (Ir) becomes bigger, it decreases the voltage across the terminals. The resistor r might be so small that it is hardly noticeable with small battery currents, and that allows us the luxury of treating a battery as ideal in the first approximation, with the terminal voltage approximately the battery's emf. However, with large battery or supply currents, the terminal voltage will drop linearly with the rising current.

The less internal resistance a supply has, the more current it can deliver to a load. An auto battery and nickel-cadmium cells (nicads) have relatively low resistance and are used in high-current situations. Turning an automobile engine over on a cold day might require a momentary current of over 400 A! With a low internal resistance, it is obviously dangerous to short the leads of such a battery. Shorting the leads can melt wiring, provide enough heat to start fires, and cause arcing in explosive environments.

EXAMPLE 18.5

A D cell battery has a terminal voltage of 1.5 V with no load connected. When a lead of 2 Ω is connected across it, its terminal voltage drops to 1.3 V. What is the internal resistance of the battery?

SOLUTION

$$\mathcal{E} = 1.5 \text{ V} \qquad V = IR \qquad I = \frac{V}{R}$$

$$I = \frac{1.3 \text{ V}}{2 \ \Omega}$$

$$I = 0.65 \text{ A}$$

$$V = \mathcal{E} - Ir \qquad Ir = \mathcal{E} - V \qquad r = \frac{\mathcal{E} - V}{I}$$

$$r = \frac{1.5 \text{ V} - 1.3 \text{ V}}{0.65 \text{ A}}$$

$$r = 0.31 \ \Omega$$

18.7   Impedance Matching

Since a battery is a source of electrical energy, then it is used to transfer power to a load. A real battery connected to a resistive load is shown in Figure 18.9. When is power transfer a maximum? Maximum power is transferred when the impedances of the source and the load are matched.

We've already talked about impedance as far back as Chapter 8 in our discussion on waves propagating on a string and collisions between billiard balls. Whenever energy propagates from one medium to another, some energy will be transmitted and some will be reflected. The same is true of batteries and electronic power supplies that supply power to a circuit. In the last section it was shown that all real batteries and electrical power sources have an internal resistance associated with them. This internal resistance is often called the battery's or the supply's *internal impedance*. To match the source and load impedance, the internal resistance of the battery must be equal to that of the load.

To prove this, we must first find an expression for the power dissipated by the circuit in Figure 18.9.

$$P = I^2R \qquad I = \frac{\mathcal{E}}{R + r}$$

$$I^2 = \frac{\mathcal{E}^2}{(R + r)^2} \qquad P = \frac{\mathcal{E}^2R}{(R + r)^2}$$

We now have an equation for the power dissipated in load resistor R. (It is assumed that we can vary the load resistance as needed.) If we modify the above equation somewhat, we can plot the resistance ratio R/r on the horizontal axis and the power dissipated in R relative to maximum power transfer on the vertical axis.

$$P = \frac{(1/r^2)(\mathcal{E}^2R)}{1/r^2(R + r)^2} \qquad P = \frac{\mathcal{E}^2(R/r)}{r(R/r + 1)^2}$$

$$P = \frac{\mathcal{E}^2}{r} \frac{R/r}{(R/r + 1)^2}$$

This ungainly expression is nonlinear and allows the plotting of the graph in Figure 18.10. The coefficient $\mathcal{E}^2/r$ is constant because both the emf and the internal resistance of the battery don't change. As R goes to zero, the power delivered to the load R drops to zero on the left side; and as R gets very large, the power also drops asymptotically to zero on the right side of the graph. The graph has a peak where $R/r = 1$, and this is the point where maximum power is transferred to the load.

When maximum power is transferred to a load, 50% of the total generated power is wasted! Because the load resistance equals the internal resistance, and because the current is the same in both, they both dissipate the same amount of power. Efficient power transfer implies a minimum of wasted power. If we want to deliver power to a load R, we should use a battery or supply that has an internal resistance smaller than R by the biggest margin possible. If $r < R$, then less of the total power dissipated is lost to the internal resistance. If $r = 0.3 \ \Omega$ and $R = 3.0 \ \Omega$, then only about 9% of the total power is wasted.

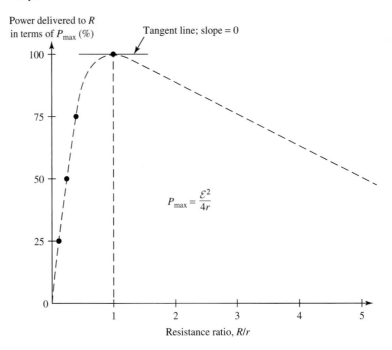

Figure 18.10 Impedance Matching between Source and Load

We've dealt with impedance matching twice before in this book. The first time dealt with elastic collisions and kinetic energy transfer between a moving ball and a stationary ball of different masses. If both balls were the same mass, after the collision the ball with the initial KE is at rest, and the initially stationary ball has all of the original KE. A maximum amount of energy has been transferred. The impedances are matched, the impedance in this case being mass. If the ball with the initial KE has less mass than the stationary ball, then the initial ball will rebound after the collision in the opposite direction as reflected energy.

Our second discussion on impedance involved a pulse traveling along a stretched string. If the pulse hits a junction where two strings of different mass densities are joined, some of the energy in the pulse is transmitted and some of it is reflected back along the way it came. Only if the linear mass densities of the two strings are the same is there no reflection. (Again the reflected pulse can be considered wasted energy.) If the two mass densities are the same, then the string is essentially seamless and homogeneous.

18.8   Capacitive Time Constant

Figure 18.11 Series *RC* Circuit

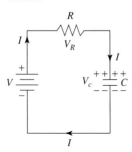

If we charge a capacitor with a battery through a resistor as in Figure 18.11, the current starts high and tapers off exponentially to zero as the capacitor charges. The circuit current doesn't flow through the capacitor as it does through the resistor. Instead it delivers positive charge to the upper plate of the capacitor and takes positive charge from the bottom plate, leaving the lower plate with a net negative charge. Initially the uncharged capacitor acts as a short circuit, and the value of *R* determines the initial current surge. As the capacitor charges, it acquires a voltage with the polarity shown in the figure. This voltage opposes that of the battery, and there is less effective voltage to drive current through the circuit.

$$I = \frac{(V - V_c)}{R}$$

When the capacitor is fully charged, current flow ceases. The voltage across the capacitor equals that of the battery. A fully charged capacitor (charged to the battery voltage) acts as an open circuit to direct current (dc) flow in that no current flows.

The current in the circuit of Figure 18.11 flows exponentially in time, and an equation for the magnitude of the charge on the capacitor at any time t can be derived. We start with the fact that the sum of the voltage drops around the circuit equals the applied (battery) voltage.

$$V = V_R + V_c$$

We now use the appropriate equation for the voltage across each component.

$$V_R = IR \qquad V_c = \frac{Q}{C}$$

$$V = IR + \frac{Q}{C}$$

In the equation above we have both a charge and a current. To solve the equation, we need to put the right side in terms of either all current or all charge. Since current is the rate of charge flow, the right side will be put in terms of charge.

$$I = \frac{\Delta Q}{\Delta t}$$

$$V = R\frac{\Delta Q}{\Delta t} + \frac{Q}{C}$$

We now separate the variables by putting all of the charge terms on the left and all of the time terms on the right.

$$R\frac{\Delta Q}{\Delta t} = V - \frac{Q}{C} \qquad RC\frac{\Delta Q}{\Delta t} = CV - Q$$

$$RC\,\Delta Q = (CV - Q)\Delta t \qquad \frac{\Delta Q}{CV - Q} = \frac{1}{RC}\Delta t$$

What do we do with the above equation? How do we solve it? We look again to the area under a curve.

$$\Sigma\frac{\Delta Q}{CV - Q} = \frac{1}{RC}\Sigma\,\Delta t \qquad \Sigma\frac{\Delta Q}{CV - Q} = \frac{1}{RC}t + K$$

$$-\ln(CV - Q) = \frac{t}{RC} + K \qquad \ln(CV - Q) = \frac{-t}{RC} - K$$

At $t = 0$, $Q = 0$. Thus,

$$\ln(CV) = -K \qquad \ln(CV - Q) = \frac{-t}{RC} + \ln(CV)$$

$$\ln(CV - Q) - \ln(CV) = \frac{-t}{RC} \qquad \ln\left(\frac{CV - Q}{CV}\right) = \frac{-t}{RC}$$

$$\frac{CV - Q}{CV} = e^{-t/RC} \qquad CV - Q = CVe^{-t/RC}$$

$$Q = CV - CVe^{-t/RC}$$

(Equation 18.11)

$$Q = CV(1 - e^{-t/RC})$$

The left side is summed over Q, and the right side is summed over t. The R and C were taken out from behind the summation because they are constant. Remember that the summations mean that we're finding the area under a curve by summing vertical rectangles with a base (Δq or Δt) along the horizontal axis that becomes infinitesimally narrow. We've seen the summation on the right side of the equation before, as long ago as Chapter 1 on uniformly accelerated motion. It sums to t, the total time elapsed. K is a constant that is added to take care of the initial conditions and will be found later. That leaves us with the left side to solve.

If a summation is mathematically the area under a curve, then the curve is $1/(Q - CV)$. Since CV is a constant and only Q is a variable (the charge gets larger as the capacitor charges), the curve is of the generic type $1/x$. The curve $1/x$ is plotted in Figure 18.12A. As x gets larger and larger, $1/x$ goes to zero and is zero only when x is infinitely large. As x goes to zero, the value of $1/x$ goes to infinity and is infinite when x equals zero. So on both sides of the curve we must deal with infinities of one of the variables. How can we find the area under the curve if both sides deal with an infinity?

Luckily the height of the curve goes to zero on the right side as x goes to infinity, and the width of the curve on the left side goes to zero as $1/x$ goes to infinity. Since both the height and the width of the curve go to zero at both extremes, the total area under the curve can be calculated. The area under the curve converges on the value 2.718. (You can find the area under the curve by using gridded graph paper, accurately plotting $1/x$ versus x, and counting the grid squares under the curve. You will find that the area will converge on 2.718 as x goes to infinity in one direction and zero in the other.) This is the value of the natural logarithm e^1, or e. How does this help us? It tells us that the solution to the summation on the left will be either logarithmic or exponential.

The curve $y = \ln x$ is plotted directly below the $1/x$ curve in Figure 18.12B. Notice that if plotted directly above, the slope at all points of the curve $y = \ln x$ becomes the curve $y = 1/x$. On the left of the curve $y = \ln x$, the line becomes almost vertical, and the slope goes to infinity as does the function $1/x$ above. As x goes to infinity, the curve $y = \ln x$ flattens and becomes nearer and nearer to a horizontal line. The slope of a horizontal line is zero, so the function $y = 1/x$ goes to zero as the function $\ln x$ flattens out.

Now let's return to our unsolved summation. From studying the two graphs, we can now say that the summation on the left is a natural logarithm. The slope of the function $\ln x$ at $x = 1$ gives the value of the function $1/x$ at $x = 1$, which is 1. Looking at how the function $\ln x$ changes with respect to x gives us the function $1/x$. But since there is a negative sign in front of the variable Q in the summation, the answer will have a negative sign

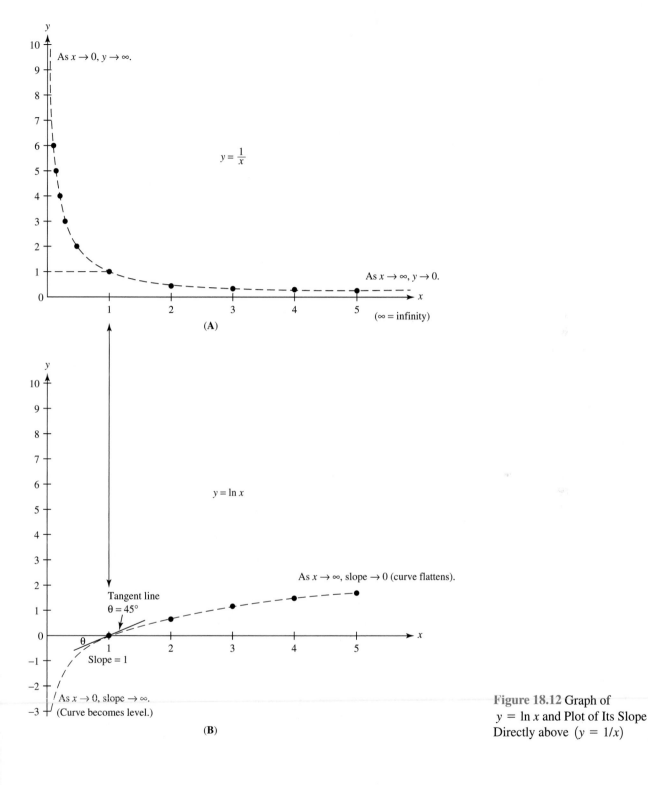

As $x \to 0$, $y \to \infty$.

$y = \dfrac{1}{x}$

As $x \to \infty$, $y \to 0$.

(∞ = infinity)

(A)

$y = \ln x$

As $x \to \infty$, slope $\to 0$ (curve flattens).

Tangent line
$\theta = 45°$

Slope = 1

As $x \to 0$, slope $\to \infty$.
(Curve becomes level.)

(B)

Figure 18.12 Graph of
$y = \ln x$ and Plot of Its Slope
Directly above $(y = 1/x)$

435

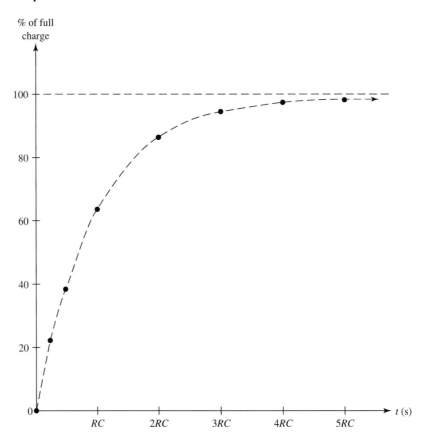

Figure 18.13 Charging Curve of an *RC* Circuit

in front of it as well. The unknown constant K can now be evaluated. We can now combine the natural log terms according to the properties of natural logs. Since we want the time dependence of the charge Q, we have to remove the logarithm terms. The flip side of the equation $y = \ln x$ is $x = e^y$. If we raise e to the power y on the left side of the equation $y = \ln x$, then we must also raise e to the power $\ln x$ on the right side to maintain the equality $e^{\ln x} = x$. (Try the equation $e^{\ln 2} = 2$ on your calculator to convince yourself!) The log terms drop out. Finally! Equation (18.11) gives the charging curve for the *RC* circuit of Figure 18.12. The graph of Figure 18.13 shows the current plotted against time.

We can check the equation at the endpoints. At $t = 0$, the charge Q is zero ($e^0 = 1$). At $t = \infty$, $Q = CV$ and the capacitor is fully charged. The interesting thing about exponential charging is that the capacitor is fully charged only if you wait an infinite amount of time! The full charge upper limit is called an *asymptote*. You can get arbitrarily close to an asymptote by waiting longer and longer periods of time, but you'll *never* reach it. So how long should you wait? After 5 *RC* time constants, the capacitor is over 99% charged, and that is usually considered enough. The dimension of *RC* is time, and the graph's horizontal axis is scaled in terms of the *RC* time constant. To find the time constant, multiply together R and C. Full charge is CV. Any scientific calculator has e^x on the keypad, so the charge can be calculated for any time t in terms of the *RC* time constant.

Before leaving the topic, we should examine the flip side of the coin, or the discharge characteristics of an *RC* circuit such as the one in Figure 18.14. This time we start with the capacitor fully charged and find an equation for how the charge decays with time. All of the energy stored as an electric field in the capacitor is dissipated as heat in the resistor. The discharge current flows through the resistor but not *through* the capacitor. Positive charge flows from the positive upper plate through the resistor and then to the negative lower plate, eventually neutralizing it.

Figure 18.14 Capacitor Discharging through a Resistor

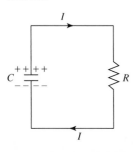

$$V_C = V_R \qquad \frac{Q}{C} = IR \qquad \frac{Q}{C} = R\frac{\Delta Q}{\Delta t}$$

where $\Delta Q/\Delta t$ is the current, and it decreases with time. Therefore, there should be a negative sign in front of it.

$$\frac{Q}{C} = -R\frac{\Delta Q}{\Delta t} \qquad \frac{\Delta Q}{Q} = -\frac{\Delta t}{RC}$$

$$\Sigma \frac{\Delta Q}{Q} = \frac{-1}{RC}\Sigma \Delta t \qquad \ln Q = \frac{-t}{RC} + K$$

At $t = 0$, $Q = Q_0$ or

$$Q_0 = CV$$

$$\ln Q = \frac{-t}{RC} + \ln CV \qquad \ln Q - \ln CV = \frac{-t}{RC}$$

$$\ln\left(\frac{Q}{CV}\right) = \frac{-t}{RC} \qquad \frac{Q}{CV} = e^{-t/RC}$$

$$Q = CVe^{-t/RC} \qquad\qquad \text{(Equation 18.12)}$$

The graph of how the charge decays with time is shown in Figure 18.15. Again we see one of the characteristics of an exponential decay: We must wait an infinite amount of time before the capacitor is fully discharged. Waiting 5 time constants is usually long enough to consider a capacitor fully discharged, though. At $5RC$ the capacitor has only 0.6% left of its original charge.

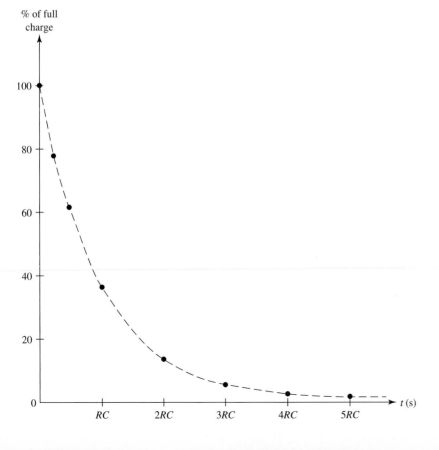

Figure 18.15 Discharge Curve of a Capacitor through a Resistor

EXAMPLE 18.6

A 50-μF capacitor is charged through a 5-kΩ resistor by a 12-V battery as shown in Figure 18.11. (a) What is the voltage across the capacitor after it has been charging for $\frac{3}{4}$ s from an initial charge of zero? (b) What charge is on the capacitor at this voltage? (c) How long does it take the capacitor to discharge from 12 V to 5 V through the 5-kΩ resistor?

See Appendix B, "Calculator Usage," for solving equations with variables in the exponents.

SOLUTION

(a) $$Q = CV(1 - e^{-t/RC}) \qquad V_C = \frac{Q}{C} \qquad V_C = V(1 - e^{-t/RC})$$

$$RC = (5000\ \Omega)(50 \times 10^{-6}\ \text{F}) \qquad RC = 0.25\ \text{s}$$

$$\frac{t}{RC} = \frac{0.75\ \text{s}}{0.25\ \text{s}} = 3$$

$$V_C = (12\ \text{V})(1 - e^{-3})$$

$$V_C = 11.4\ \text{V}$$

(b) $$Q = CV \qquad Q = (50\ \mu\text{F})(11.4\ \text{V})$$

$$Q = 5.7 \times 10^{-4}\ \text{C}$$

(c) $$Q = CVe^{-t/RC} \qquad V_C = \frac{Q}{C} \qquad V_C = Ve^{-t/RC}$$

$$\frac{V_C}{V} = e^{-t/RC} \qquad \ln\left(\frac{V_C}{V}\right) = -\frac{t}{RC}$$

$$t = -RC \ln\left(\frac{V_C}{V}\right) \qquad t = -(5\ \text{k}\Omega)(50\ \mu\text{F}) \ln\left(\frac{5\ \text{V}}{12\ \text{V}}\right)$$

$$t = 0.22\ \text{s}$$

18.9   Kirchhoff's Laws

There are two basic electrical circuit configurations: series and parallel. Unfortunately it's possible to construct circuits that are neither series nor parallel, as in Figure 18.16.

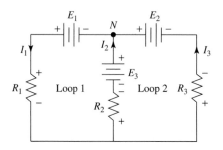

Figure 18.16 A Circuit That Is Neither Series nor Parallel

What do we do when the circuit branch in which a component is located can't be identified as either series or parallel? We can use the fact that the sum of the voltage drops around a closed loop must be equal to zero and solve a number of simultaneous equations. For example, assume that we need to know the current through resistors R_1, R_2, and R_3 in Figure 18.16. First, current direction must be assigned to each of the three branches and labeled. You can't be sure that the direction you assign to each current is correct, but you can make an intelligent guess. It seems logical that current will flow around the outside loop in a counterclockwise direction because batteries E_1 and E_2 are in series and reinforce positive (conventional) current flow in that direction. It also seems logical that batteries E_1 and E_3 will cause current to flow up through R_2. These directions are only an educated guess, but once you've made your guess, stick with it. If you find that you've guessed wrong, don't go back in the middle of your calculations and change the offending direction. You'll get an answer that has the correct magnitude but with a negative sign. The negative sign says that the current is going in the direction opposite to what you guessed. The math will work out even if you guess wrong. Finish your calculation.

Once current direction has been assigned, label the polarities for the resistors. Conventional current flowing into the top of a resistor makes the top positive, and current flowing out the bottom makes the bottom negative. You can readily see this polarity in a real circuit with a voltmeter across the appropriate resistor. A real resistor in a real circuit will exhibit this polarity.

We want to determine currents I_1, I_2, and I_3 in Figure 18.16. To determine these three unknowns, we need three equations. Two of these equations can be voltage loop equations. The sum of the voltages around a closed loop must equal zero. You don't get something for nothing! If we start from node N and go around inner loop 1 in a clockwise direction, we get

$$E_3 - V_2 - V_1 + E_1 = 0 \quad \text{(loop 1)}$$

The polarity signs on the batteries and resistors give the signs for the above equation. As you go around the loop, the sign with which you "enter" each component determines the sign of that voltage in the equation. V_1 and V_2 are the voltages across R_1 and R_2. Going around inner loop 2 in a clockwise manner gives us our second loop equation.

$$E_2 - V_3 + V_2 - E_3 = 0 \quad \text{(loop 2)}$$

We now need a third equation. You might be tempted to use the third outer loop to get another voltage loop equation, but the third loop equation is the simple sum of the two inner loops! (Try it and see.) If you used the outer loop as the third equation in trying to solve for the three currents, you would end up with something like $1 = 1$ after a lot of time spent cranking out calculations—an exercise in futility! The third equation should be a current node equation. The current entering a node equals the current leaving it. For node N, that leads to our third equation.

$$I_2 + I_3 = I_1 \quad \text{(node } N\text{)}$$

Since we're looking for current, we must now put the two loop equations in terms of currents. With Ohm's law ($V = IR$), this can be easily done.

$$I_1R_1 + I_2R_2 = E_1 + E_3 \qquad (\text{loop } 1)$$

$$I_2R_2 - I_3R_3 = E_3 - E_2 \qquad (\text{loop } 2)$$

$$I_1 = I_2 + I_3 \qquad (\text{node } N)$$

Now the electronics is ended, and the pure mathematics begins. All that remains is to solve the three simultaneous equations to find the values for I_1, I_2, and I_3. We solve them in the following example.

EXAMPLE 18.7

With $R_1 = 10\ \Omega$, $R_2 = 20\ \Omega$, $R_3 = 30\ \Omega$, $E_1 = 5$ V, $E_2 = 10$ V, and $E_3 = 15$ V, what are the currents I_1, I_2, and I_3?

SOLUTION

Let's first find I_1. We can eliminate current I_3 by substituting the node equation into the loop 2 equation.

$$I_3 = I_1 - I_2$$

$$I_2R_2 - (I_1 - I_2)R_3 = E_3 - E_2 \qquad I_2R_2 - I_1R_3 + I_2R_3 = E_3 - E_2$$

$$I_1R_3 - I_2(R_2 + R_3) = E_2 - E_3$$

We now combine the equation above in I_1 and I_2 with the loop 1 equation to eliminate I_2.

$$I_1R_1 + I_2R_2 \qquad\qquad = E_1 + E_3$$

$$\underline{I_1R_3 - I_2(R_2 + R_3) = E_2 - E_3}$$

$$I_1R_1(R_2 + R_3) + I_2R_2(R_2 + R_3) = (E_1 + E_3)(R_2 + R_3)$$

$$\underline{I_1R_3R_2 - I_2R_2(R_2 + R_3) = (E_2 - E_3)R_2}$$

$$I_1R_1(R_2 + R_3) + I_1R_2R_3 = (E_1 + E_3)(R_2 + R_3) + (E_2 - E_3)R_2$$

$$I_1R_1R_2 + I_1R_1R_3 + I_1R_2R_3 = E_1R_2 + E_1R_3 + E_3R_2 + E_3R_3 + E_2R_2 - E_3R_2$$

$$I_1(R_1R_2 + R_1R_3 + R_2R_3) = E_1R_2 + E_1R_3 + E_2R_2 + E_3R_3$$

$$I_1 = \frac{E_1(R_2 + R_3) + E_2R_2 + E_3R_3}{R_1R_2 + R_1R_3 + R_2R_3}$$

$$I_1 = \frac{(5\text{ V})(20\ \Omega + 30\ \Omega) + (10\text{ V})(20\ \Omega) + (15\text{ V})(30\ \Omega)}{(10\ \Omega)(20\ \Omega)(10\ \Omega)(30\ \Omega) + (20\ \Omega)(30\ \Omega)}$$

$$I_1 = 818\text{ mA}$$

Be careful with this complicated calculation, because careless math errors (such as reversing a sign) are easy to make.

Having I_1, we can now substitute it into the loop 1 equation to get I_2.

$$I_2 R_2 = E_1 + E_3 - I_1 R_1 \qquad I_2 = \frac{E_1 + E_3 - I_1 R_1}{R_2}$$

$$I_2 = \frac{5 \text{ V} + 15 \text{ V} - (0.818 \text{ A})(10 \text{ }\Omega)}{20 \text{ }\Omega}$$

$$I_2 = 591 \text{ mA}$$

We can now use the node equation to find R_3.

$$I_3 = I_1 - I_2$$

$$I_3 = 818 \text{ mA} - 591 \text{ mA}$$

$$I_3 = 227 \text{ mA}$$

Check your answers by going around a loop or by balancing current flow at a node.

Summary

1. **Current** is defined as a flow of electric charge and is expressed in units of amperes (A). Amperes can also be written as coulombs per second (C/s). The mathematical description of current is given by Equation (18.1).

2. The unit of resistance is the **ohm** (Ω), which can also be expressed as volts per ampere (V/A).

3. *Ohm's law* [Equation (18.4)] calculates the voltage drop across a resistor.

4. As discussed in earlier chapters, electric power is measured in watts (W) and is the electrical energy delivered (or dissipated) per second. There are many equivalent ways to express power, as shown in Equations (18.8), (18.9), and (18.10).

5. When a capacitor is first connected to a circuit containing a battery, it begins to accumulate charge. The rate at which it accumulates this charge drops dramatically in a short amount of time, maximizing at $Q = VC$ [Equation (17.6)].

The amount of charge present at any given time during this charging process can be found by using Equation (18.11). As the capacitor gains charge, the current in the circuit slows, eventually coming to a halt as the capacitor fills completely.

6. When a capacitor is allowed to discharge, the charge remaining in the capacitor at any given time during the discharge process can be calculated using Equation (18.12).

7. **Electromotive force** ($\mathcal{E}$), also called emf, is the potential of a voltage source with no external load. When current is allowed to flow through a voltage source, there is often an internal resistance (r) that must be taken into account. As the current flows through this internal resistance, there is a potential drop. Therefore, the voltage of the source once a load is applied can be expressed as

$$V = \mathcal{E} - Ir$$

Key Concepts

CURRENT: The flow of electric charge.

ELECTROMOTIVE FORCE (EMF): The no-load (no-current) potential of a voltage source.

OHM (Ω): The unit of resistance (volts per ampere).

RESISTIVITY: The opposition to current flow in a given material.

Important Equations

(18.1) $I = \dfrac{\Delta q}{\Delta t}$ (Electric current, or the flow of charge)

(18.2) $\rho = \dfrac{E}{J}$ (Resistivity in terms of electric field and current density)

(18.3) $R = \dfrac{\rho l}{A}$ (Resistance in terms of a material's resistivity, length, and cross-sectional area)

(18.4) $V = IR$ (Ohm's law)

(18.5) $R_{eq} = R_1 + R_2 + R_3 + \cdots + R_n$ (Resistors in series)

(18.6) $\dfrac{1}{R_{eq}} = \dfrac{1}{R_1} + \dfrac{1}{R_2} + \dfrac{1}{R_3} + \cdots + \dfrac{1}{R_n}$ (Resistors in parallel)

(18.7) $\Delta R = \alpha R_o \, \Delta T$ (Change in resistance with a change in temperature)

(18.8) $P = IV$ (Electric power in terms of current and voltage)

(18.9) $P = I^2 R$ (Electric power in terms of current and resistance)

(18.10) $P = \dfrac{V^2}{R}$ (Electric power in terms of voltage and resistance)

(18.11) $Q = CV(1 - e^{-t/RC})$ (Capacitor charging equation)

(18.12) $Q = CVe^{-t/RC}$ (Capacitor discharging equation)

Kirchhoff's law states that the sum of the voltage drops around a closed loop equals zero.

Conceptual Problems

1. Which of the following combinations will give a volume of silver its lowest resistance? (L is length and A is cross-sectional area.)
   a. $2L$ and $\frac{1}{2}A$
   b. $\frac{1}{2}L$ and $2A$
   c. L and A
   d. All are the same because volume is constant.

2. A resistor R_1 dissipates a certain amount of power when connected to a constant voltage source. If another resistor is connected in series with resistor R_1, the amount of power that R_1 dissipates will:
   a. increase
   b. decrease
   c. remain the same
   d. do all of the above, depending on the size of the resistor

3. A resistor dissipates a certain amount of power when connected across a constant voltage source, and another resistor is connected in parallel with the original resistor. The power dissipated by the original resistor will:
   a. increase
   b. decrease
   c. remain the same
   d. do all of the above, depending on the size of the resistor

4. A battery with a given emf and an internal resistance r is connected to a load of equivalent resistance R. If $R = r$, then:
   a. the circuit's current will be a maximum
   b. the power dissipated in the circuit will be a maximum
   c. the circuit's current will be a minimum
   d. the power dissipated in the circuit will be a minimum

5. In setting up a circuit to solve using Kirchhoff's laws, if you assume the wrong direction for a current, the value obtained for the current in that branch will be:
   a. $-I$ b. I
   c. incorrect   d. 0

6. A graph obeying Ohm's law with the vertical axis as the voltage and the horizontal axis as the current:
   a. will be a straight line that tilts down from left to right
   b. will be a straight line that tilts up from left to right
   c. will curve up from left to right
   d. will curve down from left to right

7. When a piece of metal is heated, both its resistivity and its volume increase. Does the increase in volume have anything to do with the increase in resistivity? Why?

8. An electric heater plugs into a 120-V_{ac} wall socket. Will its equivalent resistance be large or small for it to generate the most heat? Why?

9. What is the advantage of arranging batteries in parallel? What is the advantage of arranging them in series?

10. Incandescent lamps that burn out usually do so just as they are turned on. Why is this so?

11. Why is it dangerous to reach into the back of a TV set even after it has been switched off and unplugged?

12. Discuss the advantages and disadvantages of Christmas tree lights strung in series and in parallel.

13. If you have four 60-V lamps and a 120-V source, how could you connect all of them in such a way that you wouldn't burn out any of the lamps?

14. How does the resistance of a room's electrical load change if you have a 60-W bulb burning, and you switch on an additional 100-W bulb?

15. In a series *RC* circuit, current flows from the battery until the voltage across the capacitor equals the battery voltage. Is the total energy given by the battery equal to the energy stored in the capacitor? If not, where did the rest of the energy go?

16. What are the main differences between a voltmeter and an ammeter?

Exercises

Electric Current

1. If a wire carries 4 mA of current, how much charge will pass through a certain point in 2 s?

   Answer: 8 mC

2. An ammeter measures the current onto a capacitor's plate. The capacitor charges at a constant current, the time elapsed was only 2×10^{-8} s, and the ammeter read 2 mA. How much charge does the capacitor gain?

   Answer: 40 pC

3. A cyclotron uses magnetic fields to force charged particles to travel at great speeds along large, circular tracks. Such a cyclotron was set up to push a cluster of 5×10^{10} protons with a speed of 2×10^6 m/s around a circular track 200 m in diameter. What would the equivalent current of this "circuit" be?

   Answer: 25.5 mA

Resistivity and Ohm's Law

4. An electrical technician wishes to build a 12-Ω resistor from a loop of nichrome wire. If the wire has a diameter of 3 mm, how long a piece of the wire must she cut off to produce the desired resistance?

   Answer: 75.7 m

5. How long would an iron rod need to be to have the same resistance as a 10-cm piece of graphite pencil lead of equal diameter?

   Answer: 29.2 m

6. What is the equivalent resistance of a common pencil? The pencil is 17 cm long with a diameter of 7 mm, and the inner cylinder of graphite has a diameter of 1.5 mm. The resistivity of the wooden sheath is 1×10^9 Ω-m, and that of the graphite is 3.5×10^{-5} Ω-m.

   Answer: 3.4 Ω

*7. Aluminum wires are sometimes used instead of copper wires for house wiring. What is the ratio of diameters of the aluminum wire to the copper wire if they both have the same resistance per unit length?

   Answer: 1.24

8. Opposite ends of a long piece of wire are connected to the terminals of a 9-V battery. If the total resistance of the wire is 120 Ω, what is the current running through the circuit?

   Answer: 75 mA

9. The two leads of a 12-V car battery are connected to opposite sides of a 2-cm cube of an unknown material. An ammeter spliced into one of the battery leads reads 0.12 mA. From the table of resistivities (Table 18.1) choose the most likely candidate for this material.

   Answer: Silicon

*10. An electric motor used in a sawmill draws 15 A from a 240-V source 70 m away. What must the minimum diameter of the wire be if the voltage drop along the wires cannot exceed 2%?

   Answer: 2.18 mm

Resistors in Series and in Parallel

11. Three resistors of 100 Ω each are placed in a circuit. What would the equivalent resistance of the circuit be if they were put (a) in series and (b) in parallel?

Answer: (a) 300 Ω
(b) 33.3 Ω

12. A 1.2-V nicad battery is placed in series with two 50-Ω resistors and a 75-Ω resistor. What is the current running through the 75-Ω resistor?

Answer: 6.9 mA

13. What is the equivalent resistance of each of the circuits in Figure 18.17?

Answer: (a) 23.6 Ω
(b) 59.6 Ω
(c) 10.6 Ω

*14. Three identical resistors connected in parallel have a resistance of 10 Ω. What is their resistance if they are connected in series?

Answer: 90 Ω

*15. (a) What is the equivalent resistance of the circuit in Figure 18.18? (b) If 6 V is applied to the circuit, what is the current in the 8-Ω resistor?

Figure 18.18 (Exercise 15)

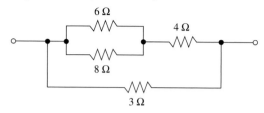

Answer: (a) R_{eq} = 2.14 Ω
(b) 345 mA

16. Fifteen volts is applied to the circuit of Figure 18.19. What is the current in the 6-Ω resistor?

Figure 18.19 (Exercise 16)

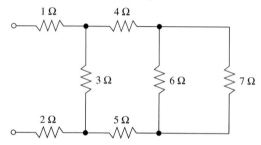

Answer: 305 mA

Figure 18.17 (Exercise 13)

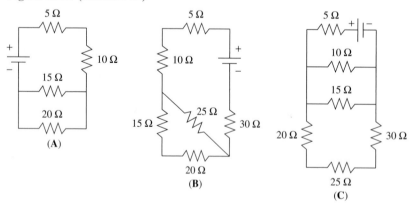

Temperature Coefficient of Resistance

17. A copper wire is heated from 20°C to 80°C. If the wire's resistance is 20 Ω at 20°C, what is the wire's resistance at the higher temperature?

Answer: 24.7 Ω

Electric Power

18. A pair of jumper cables consists of two leads, each of copper wire 2 m long. The wire's diameter is 0.5 cm. If the cables are connected in series with two 12-V car batteries, what is the power initially dissipated by the cables?

Answer: 171 kW

19. How much power is dissipated across each of the resistors in the circuits in Figure 18.20?

Figure 18.20 (Exercise 19)

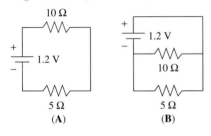

(A)   **(B)**

Answer: (a) 5 Ω, 32 mW; 10 Ω, 64 mW
(b) 10 Ω, 144 mW; 5 Ω, 288 mW

20. A family leaves for vacation and forgets to turn off the lights in the house. If they turned off all appliances and utilities except for ten 60-W bulbs and left the house for the entire month of September, how much would they be billed if electricity costs 7¢ per kilowatt-hour?

Answer: $30.24

21. A clothes dryer is connected to 240 V and draws 15 A of current. If electricity costs 10¢ per kilowatt-hour, how much does it cost to run the dryer for 45 min?

Answer: 27¢

*22. Two 1.5-V batteries are used to power a 3-W flashlight. The batteries are run down after $1\frac{1}{2}$ h of use. If the batteries cost $1.25 apiece, what is the cost of 1 kWh of electricity from these batteries?

Answer: $555.56

*23. A 15-Ω and a 25-Ω resistor are connected in series. The power dissipated in the 25-Ω resistor is 90 W. What is the voltage across the pair?

Answer: 76 V

*24. Each of the resistors in the circuit of Figure 18.21 can dissipate 5 W of power. What is the maximum power that the whole circuit can dissipate?

Figure 18.21 (Exercise 24)

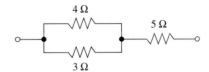

Answer: 6.7 W

Internal Resistance and Electromotive Force

25. Producing $\frac{1}{2}$ A of current, a battery has a terminal voltage of 8.6 V. Its internal resistance is found to be 1.5 Ω. (a) What is the battery's emf? (b) What would the terminal voltage of this battery be, still supplying $\frac{1}{2}$ A, if it were possible to drop the internal resistance to 0.02 Ω as in a nicad battery?

Answer: (a) emf = 9.35 V
(b) 9.34 V

26. A battery has a terminal voltage of 4.0 V when there is no current being drawn from it. If the terminal voltage drops to

3.6 V when 200 mA is drawn from it, what is the battery's internal resistance?

Answer: 2 Ω

*27. A battery has a terminal voltage of 9.6 V when connected to a 200-Ω load. This same battery has a terminal voltage of 10.3 V when connected across a 300-Ω load. What is the battery's emf?

Answer: 12 V

Capacitive Time Constant

28. A 4-μF capacitor is charged with a 5-V power supply as shown in Figure 18.22A. How much charge is left 2 ms after the battery is removed from the circuit and the capacitor discharges through the same 1-kΩ charging resistor as in Figure 18.22B?

Figure 18.22 (Exercise 28)

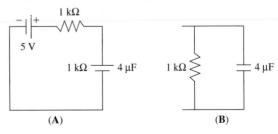

(A)   **(B)**

Answer: 12.1 μC

29. A circuit is constructed from a 9-V battery, a 16-mF capacitor, and a 150-Ω resistor in series. How much time would it take the capacitor to discharge to one-half its original full charge through the 150-Ω resistor?

Answer: 0.83 s

30. A 0.47-μF capacitor is charged to 1500 V. A 100-kΩ resistor is used to discharge it. How much time elapses before the voltage drops to 5 V?

Answer: 0.268 s

31. Four identical batteries have an emf of 1.5 V and an internal resistance of 0.3 Ω. (a) If the four batteries are connected in series to a 3-Ω load, how much current flows through the load? (b) If the four batteries are connected in parallel and then connected to the 3-Ω load, how much current now flows through the load?

Answer: (a) 1.43 A
(b) 1.82 A

32. What are the currents in the three resistors of Figure 18.23?

Figure 18.23 (Exercise 32)

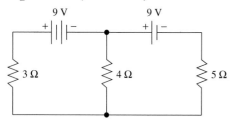

Answer: I_3 = 1.44 A; I_4 = 1.17 A; I_5 = 0.27 A

33. The circuit shown in Figure 18.24 is a battery charging circuit. Both the 9-V and the 6-V batteries are being simultaneously charged by a 12-V power supply. (a) What are the current drawn from the 12-V battery and the charging currents for each battery being charged? (b) What would a voltmeter read if connected between points a and b?

Figure 18.24 (Exercise 33)

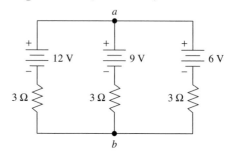

Answer: (a) I_{12} = 1 A; I_9 = 0; I_6 = 1 A
(b) V_{ab} = 9 V

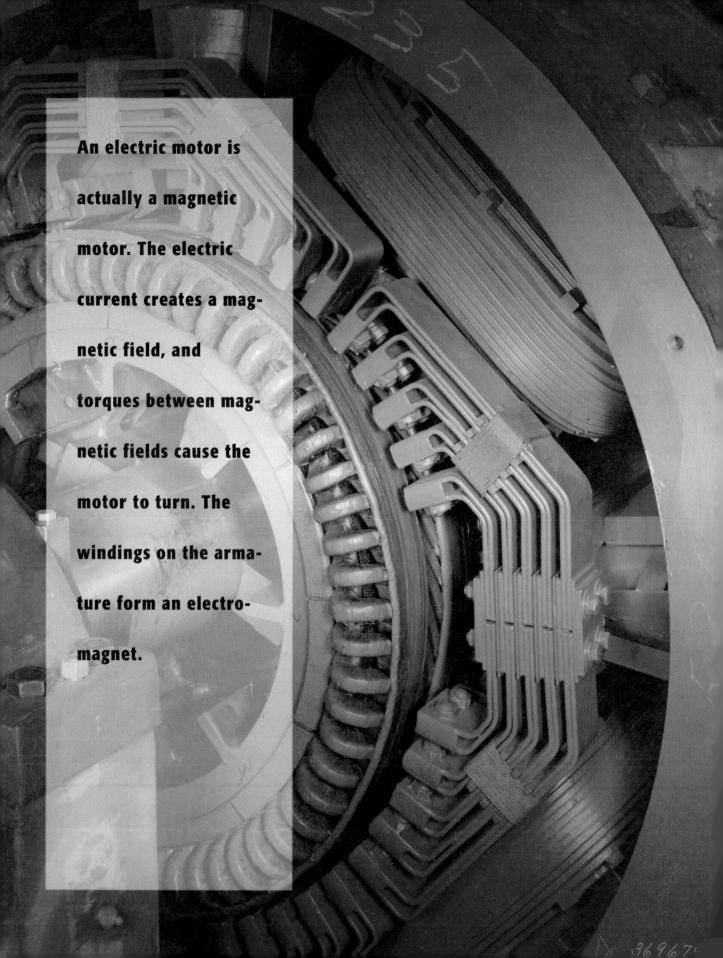

An electric motor is actually a magnetic motor. The electric current creates a magnetic field, and torques between magnetic fields cause the motor to turn. The windings on the armature form an electromagnet.

Magnetism

At the end of this chapter, the reader should be able to:

1. Calculate the strength of a magnetic field a given distance from a wire carrying a known current.

2. Calculate the magnetic field produced by a straight wire, a wire loop, and a solenoid containing a known current.

3. Solve for the speed and direction an electron must be traveling if it is immersed in a known electric or magnetic field.

4. Demonstrate how to use the **right-hand rule** to determine the direction of a magnetic field if the direction of a particle's velocity and its magnetic force are known.

5. Solve for the distance between two parallel wires if their currents are given and the force between them is known.

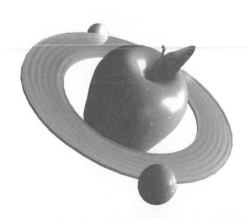

6. Explain why a current-carrying loop in an external magnetic field may experience a torque.

7. Calculate the current needed to impart a given torque on a loop of wire, assuming that the external magnetic field and the geometry of the wire are known.

19.1   Natural Magnets and Compasses

Natural magnets occur in mineral deposits called *magnetite,* a highly concentrated form of iron ore. Iron is a ferromagnetic material; that is, it becomes magnetized very easily. A refrigerator magnet clamps onto the door of a refrigerator because of the iron or steel in the door. A magnet easily picks up objects with a high iron content such as paper clips, pins, or nails.

When magnetite ore is formed in the earth, it forms under high pressure and high temperature. The earth has a magnetic field, and when the magnetite solidified and cooled below a certain temperature, it became a permanent magnet. When it cooled, it literally froze in the prevailing field, which is the earth's.

Probably the earliest use for a permanent magnet was at sea as a compass. The magnetic field lines of the earth run roughly north to south. It was found that if a slender piece of magnetite were suspended from a very thin string or strand of silk, it would align itself with the earth's field (pointing north to south). This was a very useful navigational aid at sea on a cloudy day or night when neither the sun nor the stars could be seen. Permanent magnets are commercially available generally as horseshoe magnets, bar magnets, and the round (or flat) refrigerator magnets, as shown in Figure 19.1. These magnets can affect

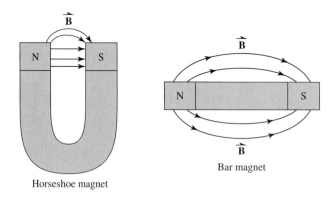

Horseshoe magnet

Bar magnet

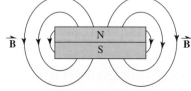

Refrigerator (flat) magnet

Figure 19.1 Various Types of Magnets and Their Field Lines

objects such as nails, pins, and paper clips at a distance with an attractive force. Thus, a force field is associated with a magnet. The field lines are relatively easy to show. For example, place a bar magnet under a thin sheet of transparent plastic, sprinkle iron filings on the plastic, and tap the plastic lightly with your finger. The filings will align themselves with the magnetic field.

If you have two bar magnets and place a south pole next to a north pole, there will be a strong attraction. If you place a north pole near a north pole, or a south pole near a south pole, there will be a strong repulsion. Like poles repel, and unlike poles attract. This law is the same as the one for electric charges. However, positive and negative electric charges can be separated. For magnetic materials, a purely north pole can't be separated from a purely south pole. If we were to cut a bar magnet in half trying to separate the north pole from the south pole, we would end up with two bar magnets, each half the size of the original and both with a north and a south pole. If we cut those two magnets in half, we would have four bar magnets, each a fourth the size of the original. No matter how many times we cut up the original bar magnet, we can't separate a north pole from a south pole. There are no magnetic monopoles! The north pole is intrinsically connected to the south pole.

19.2   The Magnetic Field and the Force on a Moving Charge

Assume that we have a volume in which the magnetic field is constant. Such a volume can be found between the poles of a horseshoe magnet, as shown in Figure 19.1. How would an electric charge react to the magnetic field? If the charge had no velocity relative to the magnetic field, it would just sit there. There would be no force on the charge. But if the charge enters the magnetic field with an initial velocity, it will move in a circle as in Figure 19.2.

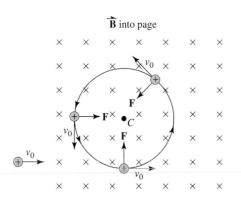

Figure 19.2 Force on a Charge Moving in a Magnetic Field

A magnetic force is exerted on the charge only if it is moving. The force is at right angles to the velocity, causing the charge to move in a circle of constant radius as long as the magnetic field is constant. For a moving charge in a magnetic field, the force law is

$$\mathbf{F} = q\mathbf{v} \times \mathbf{B}$$

The term in the equation above is a cross-product. (Remember torques?) It can be written

(Equation 19.1)

$$\mathbf{F} = \mathbf{v}\mathbf{B}\sin\theta$$

where θ is the angle between $\mathbf{v}$ and $\mathbf{B}$. In the case of Figure 19.2, $\mathbf{v}$ and $\mathbf{B}$ are always at 90°, so $\mathbf{v} \times \mathbf{B}$ will always point to the center of the circle at C. The rule for finding the direction in which the force is pointed is the *first right-hand rule:* The thumb of your right hand should point in the direction of the particle's instantaneous velocity. The fingers of your right hand with your palm opened flat should be in the direction of the magnetic field, $\mathbf{B}$. The open palm of your right hand will point in the direction of the magnetic force on the particle. One reason many people find magnetism difficult to understand is its weird vector nature (it has both magnitude *and* direction). By becoming familiar with the right-hand rule, you will avoid many conceptual problems later.

First right-hand rule: The thumb of your right hand points in the direction of the particle's velocity. The fingers of your right hand with your palm open are in the direction of the magnetic field. Your open palm is in the direction of the force.

If the charged particle in Figure 19.2 is traveling in a circle, there is a centripetal force on it. The centripetal force is caused by, and equal to, the magnetic force.

$$\frac{mv^2}{r} = qvB \qquad \frac{mv}{r} = qB$$

(Equation 19.2)

$$r = \frac{mv}{qB}$$

If the charge on the particle or the magnetic field is increased, the radius of the circle decreases. Increasing the particle's mass or velocity increases the radius of the circle.

$$\frac{mv}{r} = qB \qquad v = \frac{qBr}{m} \qquad v = \frac{2\pi r}{T}$$

where T is the period of 1 revolution.

$$\frac{2\pi r}{T} = \frac{qBr}{m} \qquad \frac{2\pi}{T} = \frac{qB}{m} \qquad \omega = \frac{2\pi}{T}$$

$$\omega = \frac{qB}{m}$$

Uniform circular motion can be characterized by its period or its angular velocity. It's interesting to note that neither the period nor the angular velocity of the spinning particle depends on the radius of the circle. They depend only on the external magnetic field and the ratio of charge to mass (which is often an intrinsic property of subatomic particles).

EXAMPLE **19.1**

An electron initially at rest is accelerated by a potential of 4000 V and is injected into a region with a constant magnetic field of 0.6 T. (The units of the magnetic field are the tesla, abbreviated T, whereas the symbol for the field is usually *B*.) (a) What is the radius of the circle around which the electron spins in the magnetic field? (Assume that the electron enters the field with its velocity at right angles to the magnetic field.) (b) With what frequency does the electron move in the circle?

SOLUTION

(a)
$$V = \frac{W}{q} \qquad W = qV$$

$$e = q \qquad W = eV$$

$$\frac{1}{2}m_e v^2 = eV \qquad v = \sqrt{\frac{2eV}{m_e}}$$

$$v = \sqrt{\frac{2(1.6 \times 10^{-19}\,\text{C})(4000\,\text{V})}{9.11 \times 10^{-31}\,\text{kg}}}$$

$$v = 3.75 \times 10^7\,\text{m/s}$$

$$r = \frac{m_e V}{eB} \qquad r = \frac{(9.11 \times 10^{-31}\,\text{kg})(3.75 \times 10^7\,\text{m/s})}{(1.6 \times 10^{-19}\,\text{C})(0.6\,\text{T})}$$

$$r = 0.356\,\text{mm}$$

(b)
$$\omega = 2\pi f \qquad 2\pi f = \frac{eB}{m}$$

$$f = \frac{eB}{2\pi m_e} \qquad f = \frac{(1.6 \times 10^{-19}\,\text{C})(0.6\,\text{T})}{2\pi(9.11 \times 10^{-31}\,\text{kg})}$$

$$f = 1.68 \times 10^{10}\,\text{rev/s}$$

19.3 Work Done on a Moving Charge by a Magnetic Field

In a magnetic field, the force on a charge and its velocity are always at right angles to each other. The mathematical definition of *work* is $Fx \cos \theta$, where θ is the angle between the force and the displacement caused by the work. If θ is 90°, then the force does no work on the object. If no work is done on the object, its kinetic energy doesn't increase or decrease. Thus, its speed is constant. The speed of the electron in Example 19.1 doesn't change. Therefore, the magnetic field does no work on the electron.

19.4 Magnetic Field of a Current

What gives rise to a magnetic field? A refrigerator magnet has a permanent magnetic field. It doesn't require recharging or a power source to maintain the field, so where does its permanent magnetic field come from? The origin of a permanent magnetic field is at the level of the individual atom. An atom has a positively charged nucleus with negatively charged electrons orbiting it. These orbiting electrons constitute circular currents around the nucleus, and a moving charge generates a magnetic field.

Figure 19.3A shows a view of an atom from above with electrons orbiting the nucleus in a clockwise manner. Figure 19.3B shows a short current segment. (Consider it a short section of a circular tube in which the electrons orbit.) The magnetic field around

A moving charge generates a magnetic field.

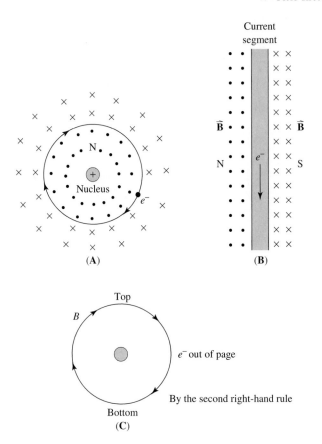

Figure 19.3 Current-
Generated Magnetic Fields

the straight current element is circular. (The wire or current element is the center of the circle.) The current and the magnetic field are proportional. The larger the current, the larger the magnetic field. The magnetic field is also inversely proportional to the distance from the wire or current element. The farther away from the wire, the smaller the field. Putting these two facts together, we have the formula for the magnetic field around a current-carrying wire or current element.

(Equation 19.3)

$$B = \frac{\mu_o I}{2\pi s}$$

where the 2π comes from the circular geometry of the field around the wire. The magnitude of the field is the same at all points around the wire a distance s from the wire. These points describe the circumference of the circle centered on the current segment. μ_o is the magnetic permeability of free space; it is similar to the permeability constant for an electric field ε_o. The greater the permeability of the surrounding space, the larger the value of the magnetic field. The value for the permeability constant μ_o is 1.26×10^{-6} T-m/A, a fundamental physical constant. Notice that it has units of current (amperes, A) in the denominator. The reason is that a magnetic field is generated by a moving charge (current).

A field has a direction, and it will circulate around the wire in either a clockwise or a counterclockwise sense. A **second right-hand rule** is needed to establish the direction of the field around the wire: Your thumb should point in the direction of conventional cur-

Second right-hand rule: The thumb of your right hand points in the direction of the positive current flow. The fingers of your hand naturally curve around the path of the current in the direction of the magnetic field.

rent flow, which is up for Figure 19.3B and counterclockwise for Figure 19.3A. (Remember that an electron has a negative charge.) The fingers of your right hand will naturally curl around the path of current flow in the direction of the magnetic field. Thus, inside the orbit of the electron in Figure 19.3A, the field is coming out of the page, and outside the orbit it is going into the page. In Figure 19.3B, the new right-hand rule has the magnetic field coming out of the page to the left of the current element, and into the page on the right. Below the current element, the field direction is from right to left, and above it from left to right. The field completes a full circle around the current element.

For the atom, the orbiting electrons make the atom look like a very small bar magnet if viewed from the side. The field lines leave a north pole (top) and enter a south pole (bottom). Each atom is called a *magnetic dipole* with a dipole moment. (*Dipole* means that there are two poles, and *moment* means that the poles are separated by a small distance.) In a ferromagnetic material such as iron, all of the atoms in a sample can be aligned with an external magnetic field (such as the earth's), and each individual atom (bar magnet) adds its small moment to the total magnetization of the magnet. In this way a ferromagnetic material can end up having a magnetic field much larger than the original field that was used to initially align the tiny bar magnets.

If all matter is made up of atoms and all atoms have a tiny magnetic field, why isn't all matter magnetic? In an unmagnetized material, the tiny bar magnets all have random orientations. This randomness means that when all of the moments are added, they add to zero. They effectively cancel one another. Most materials can't be effectively lined up like iron and are considered nonmagnetic, when in reality the magnetic nature of the material is masked by randomness.

EXAMPLE 19.2

At what radius from a wire carrying 2.5 A is the field all the way around the wire a constant 1×10^{-5} T?

SOLUTION

$$B = \frac{\mu_o I}{2\pi s} \qquad s = \frac{\mu_o I}{2\pi B}$$

$$s = \frac{(1.26 \times 10^{-6} \text{ T-m/A})(2.5 \text{ A})}{2\pi(1 \times 10^{-5} \text{ T})}$$

$$s = 5 \text{ cm}$$

The magnetic field of the earth is 2×10^{-5} T. A current-carrying wire will swing the needle of a compass. Because of the magnetic field around a current-carrying wire, it is a good idea to keep such wires away from a compass if you want to be sure that the compass will read accurately.

19.5   Magnetic Field at the Center of a Current Loop

Figure 19.3A shows a view of a hydrogen atom from above with its central positively charged nucleus and solitary orbiting electron. The orbiting electron is a current loop. The current loop is just like a wire with a given current flowing through it. If the electron has a given velocity v_o and the orbit a radius r_o, the current can be calculated.

$$I = \frac{\Delta q}{\Delta t} \qquad I = \frac{e^-}{T}$$

where T is the period of one revolution.

$$v_o = \frac{2\pi r_o}{T} \qquad T = \frac{2\pi r_o}{v_o}$$

$$I = e^- \left(\frac{v_o}{2\pi r_o} \right)$$

$$I = \frac{e^- v_o}{2\pi v_o} r$$

Inside the orbit of the electron, or inside the current loop, the magnetic field is surprisingly uniform because of the geometry of the loop. It's this uniformity that makes the current loop act like a bar magnet. At the center of the current loop, the formula for the magnetic field is

$$B = \frac{\mu_o I}{2r}$$

where I is the current and r is the radius of the loop. Since the magnetic field at the center of the loop is caused by a circular wire, we would expect the field formula to be very similar to that of a straight wire. It is, because when the current increases, the field increases, and when the loop's radius increases, the field decreases. The only difference is that the field equation for the loop lacks π in the denominator. Thus, the loop field is a little more than three times as great (π) as the field near a straight wire with the same radius and current. The geometry of the loop and the great symmetry of a circle concentrate the field.

The more loops of wire carrying the same current and with the same radius there are, the greater the field inside the loop. If there are two loops, the field is twice as great; three loops, three times as great; and so on. Taking into consideration the number of loops, the field equation for inside the loop becomes

(Equation 19.4)

$$B = \frac{\mu_o N I}{2r}$$

where N is the number of loops.

19.6   The Solenoid

Figure 19.4 Magnetic Field of a Current Loop Viewed from above

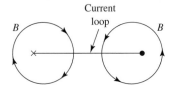

Current loop

If the magnetic field due to a current loop is viewed from above, the field lines will appear as in Figure 19.4. On the left side of the loop looking from above, the current is flowing into the page, and the magnetic field flows clockwise around that portion of the loop. On the right side the current is flowing out of the page, and the magnetic field is counterclockwise around the loop segment. In the plane of the loop inside the loop, the magnetic field lines are all in the same direction and reinforce one another. Outside the plane of the loop, the field lines begin to change direction.

If we want a magnetic field that is constant in both magnitude *and* direction over a fairly large volume, we can take our current loop and wind it as a long tube. If the radius

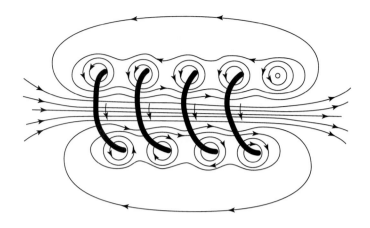

of the tube is small relative to the tube's length, the interior magnetic field is quite homogeneous in both magnitude and direction, as in Figure 19.5. As shown, the magnetic field lines inside the coil are parallel to the axis of the tube. If the length of the coil is much larger than the radius, then the magnitude of the field no longer depends on the radius. For a given number of turns N, the longer the length over which they're stretched (like a coil spring), the smaller the field is. So the field is inversely proportional to the length L. The field equation for the wire-wound tube is

$$B = \mu_o \frac{N}{L} i$$

where N/L is the turns per unit length, written n. *The wire-wound tube with the homogeneous interior magnetic field is called a* **solenoid.** Most solenoids are described in terms of n instead of N/L. The field equation for a solenoid then becomes

Solenoid: A wire-wound tube with a homogeneous magnetic field inside.

$$B = \mu_o n i$$

(Equation 19.5)

The interior field is still too weak, though, to be of much use because μ_o (the permeability constant for free space) is so small. However, a ferromagnetic material such as iron can cause a much larger magnetic field than the small field inside the solenoid with an air core because the small solenoid field can line up all of the atomic dipoles in the iron. As mentioned previously, the sum over all of the aligned dipoles produces a field far in excess of the original field.

Figure 19.6 is a magnetization curve for iron. If a solenoid with an air core has a magnetic field inside of 8×10^{-4} T, the same solenoid with the same previous current will have an internal field of about 1.5 T with an iron core. This is an increase of 1875 times! Without an iron core, to obtain the same field the solenoid's current would have to be increased by a factor of 1875. The reason the magnetization curve levels off at higher solenoid fields is that all of the atomic dipoles in the iron are already lined up, and the iron saturates.

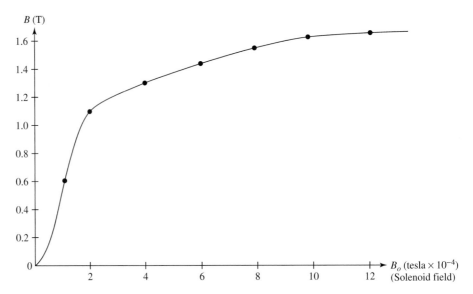

Figure 19.6 Magnetization Curve for Iron

EXAMPLE 19.3

You have a cardboard tube left over from a roll of toilet paper which is 11 cm long and 3.5 cm in diameter. You wind it with one layer of insulated copper wire 0.5 mm in diameter as tightly as possible. (a) If you connect a power supply that has an output of $\frac{1}{2}$ A to each end of your hand-rolled solenoid, what is the field inside? (b) If an iron core were inserted into the solenoid, what would the new field be? (c) Without the iron core, what must the current for a 1.65-T field be?

SOLUTION
(a) The number of turns N possible around the tube is the length of the tube divided by the diameter of the wire.

$$N = \frac{L}{D} \qquad N = \frac{0.11 \text{ m}}{0.5 \times 10^{-3} \text{ m}} \qquad N = 220 \text{ turns}$$

$$B = \mu_o nI \qquad n = \frac{220 \text{ turns}}{0.11 \text{ m}} \qquad n = 2000 \text{ turns/m}$$

$$B_o = (1.26 \times 10^{-6} \text{ T-m/A})(2000/\text{m})(0.5 \text{ A})$$

$$B_o = 1.26 \times 10^{-3} \text{ T}$$

(b) From the graph in Figure 19.6, the field with the iron core would be 1.65 T.

(c) $$\frac{1.65 \text{ T}}{1.26 \times 10^{-3} \text{ T}} = 1309.5 \text{ times increase in field}$$

$$I = (1309.5)(0.5 \text{ A})$$

$$I = 654.75 \text{ A!}$$

Because of the resistance of the copper, the heat generated by the current would melt the wiring long before the 654.75 A were reached.

19.7   Electromagnets and Relays

The solenoid leads to two very practical applications with the electromagnet and the relay. As shown earlier, an iron-core solenoid makes a fairly good magnet that has the advantage of being switched on and off at will. Actually, once the iron core has been magnetized, its magnetization doesn't go all the way back to zero once the current again drops to zero. Rather, it retains a small residual magnetization because of the aligning of the magnetic domains in the iron crystal. This lagging effect is called *hysteresis*. To bring the magnetic moment of the iron back to zero, a small current opposite to that which caused the residual magnetization must be applied. But the permanent magnetization of the iron is small relative to that induced in the iron by the current in the solenoid, and the iron core becomes a switchable magnet whose polarity is also switchable.

An electromagnet is useful for purposes other than lifting metal objects. For example, the speaker cones in a loudspeaker are caused to vibrate by the action of an electromagnet. An electric current with the frequency of the sound being reproduced is fed into the solenoid. The magnetization of the iron core is switched on and off at this frequency, causing an iron plate near the iron core to vibrate. This plate is attached to the speaker cone, and its vibration causes the air in the speaker cone to vibrate, reproducing the original sound.

The iron-core solenoid is also the heart of another electromechanical device, the *relay*. A relay is a current amplifier in that a small current in the solenoid causes the switching of much larger currents, as shown in Figure 19.7.

The relay as an electromechanical switch has a huge number of applications. The modern digital computer stores information as 1s or 0s, or binary digits. Early computers used switching transistors or vacuum tubes as the memory elements, but relays would also serve.

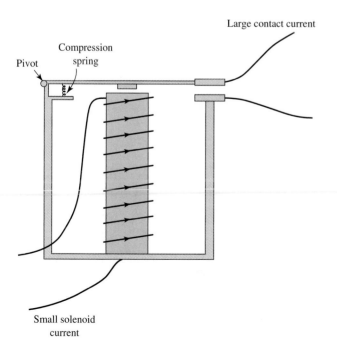

Large contact current

Compression spring

Pivot

Small solenoid current

Figure 19.7 Electromagnetic Relay

A digital computer built entirely of relays was constructed in the late 1930s. A relay has a switching time of milliseconds, so the relay computer was terribly slow by today's standards. But at the time, the very automation of a lengthy calculation was novel and sped such things up considerably. Vacuum tubes could switch in a microsecond, though, and the first electronic computer using vacuum tubes (ENIAC) was on the order of a thousand times faster than a computer using only relays.

A relay can also be used to convert an electromagnetic signal into a linear movement. If the iron core inside the solenoid can move, and if a spring partially withdraws the iron core when the solenoid isn't energized, then when a current is present in the solenoid windings, the core is powerfully retracted into the solenoid. A solenoid used in this manner lies at the heart of a great many electromechanical devices.

19.8   The Force between Two Current-Carrying Wires

If a copper wire is placed between the poles of a horseshoe magnet, as shown in Figure 19.8, when a current flows through the wire, a force on the wire tries to push it either up or down depending on the direction of the current. In this figure, the right-hand rule applies, and the force on the wire is trying to pull the wire down. (The velocity of the positive charges is along the wire in the direction of the arrows.) Common sense allows us to make an educated guess that the force is probably proportional to the current flow. It also is a good guess that the force on the wire is probably proportional to the length of the wire immersed in the magnetic field as well.

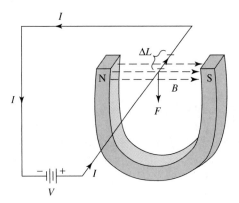

Figure 19.8 Force on a Current-Carrying Wire in a Magnetic Field

If we start from the primal equation for the force on a moving charge, we will be able to derive the force on the wire.

$$\vec{F} = q\vec{v} \times \vec{B} \qquad F = qvB \sin \theta$$

The magnetic field and the velocity of a positive charge are at right angles to one another. Thus,

$$F = qvB$$

We're interested only in the area where the wire is between the poles, and many charges are flowing.

$$F = \Delta q \frac{\Delta L}{\Delta t} B$$

Mathematically it's permissible to change the positions of Δq and ΔL.

$$F = \Delta L \frac{\Delta q}{\Delta t} B \qquad I = \frac{\Delta q}{\Delta t}$$

$$F = I \, \Delta L \, B \qquad\qquad \text{(Equation 19.6)}$$

So in the presence of a magnetic field, there is a force on a current-carrying wire. Thus, there is a force between two current-carrying wires, since both will have magnetic fields around them.

Figure 19.9 depicts the circular magnetic fields around two current-carrying wires. I_1 is out of the page with a counterclockwise magnetic field around the left wire. I_2 is into the page with a clockwise magnetic field around the wire. Between the wires where the two magnetic fields interact, the field lines from both wires point upward. Therefore, both fields in the center have the same polarity and hence repel one another. Adjacent wires with currents in opposite directions repel one another. Prove to yourself that adjacent wires with currents in the same direction will attract one another.

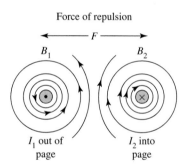

Force of repulsion

B_1 B_2

I_1 out of page I_2 into page

Figure 19.9 Magnetic Fields around Current-Carrying Wires

EXAMPLE 19.4

In Figure 19.9, if the wires are a distance s apart, what is the force of attraction per unit length between them?

SOLUTION

$$F_1 = I_1 \, \Delta L \, B_2 \qquad B_2 = \frac{\mu_o I_2}{2\pi s}$$

$$F_1 = I_1 \, \Delta L \left(\frac{\mu_o I_2}{2\pi s} \right)$$

$$\frac{F_1}{\Delta L} = \frac{\mu_1 I_1 I_2}{2\pi s}$$

It makes no difference which wire you start with, because the force of repulsion is the same for each wire.

19.9   Torque on a Current Loop

If a force is exerted on a current-carrying wire in a magnetic field, then if that wire is formed into a loop that can rotate, a torque acts on the loop. The loop will rotate until it reaches a position where the net torque is zero. Figure 19.10 shows a symmetric loop immersed in an area of constant magnetic field B with the plane of the loop parallel to the field. The loop in the figure is a square of side L with the axis of rotation down the middle. Current is flowing through the wire as shown. Think of the current flow as a collection of individual charges moving in the same direction. In the top view there is no force on the upper and lower wire segments because the charges are moving in the same direction as the magnetic field lines.

$$F = qvB \sin \theta$$

where θ is either $0°$ or $180°$. There is a force on the left and right line segments because current flow is at right angles to the magnetic field lines. From Section 19.8, we know that the force on both the left and the right segments is ILB. This force is constant in

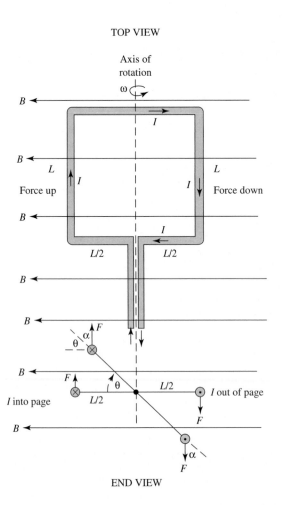

Figure 19.10 Torque on a Current Loop

magnitude and direction because no matter what the orientation of the current loop, the current will cut the magnetic field lines at right angles.

Because the loop will rotate around the axis shown, the forces on the left and right wire segments will tend to make the loop rotate in a clockwise direction as seen from the end view. The total torque on the current loop becomes

$$\vec{\tau} = \vec{r} \times \vec{F} \qquad \tau = \left(\frac{1}{2}L\right)(ILB)\sin\alpha \qquad \tau = \frac{1}{2}IL^2B \sin\alpha$$

where the torque on either wire is the same in both magnitude and direction (clockwise).

$$\theta + \alpha = 90° \qquad \alpha = 90° - \theta$$

$$\tau = \frac{1}{2}IL^2B \sin(90° - \theta)$$

$$\sin(90° - \theta) = \cos\theta$$

$$\tau = \frac{1}{2}IL^2B \cos\theta$$

where θ is the angle through which the loop turns. The total torque is twice the torque on one wire.

$$\tau = 2\left(\frac{1}{2}IL^2B \cos\theta\right)$$

$$\tau = IL^2B \cos\theta$$

where θ is zero when the plane of the loop is parallel to the direction of the magnetic field, as in Figure 19.10.

The above equation for the torque on a current loop can be extended to a much more general configuration. Notice that L^2 is the area inside the current loop. It doesn't matter if the loop is rectangular, octagonal, circular, or an irregular shape. The torque on the loop is proportional to the area of the loop. Notice also that it doesn't specify in the equation where the axis of rotation is; it can split the area in half as in the derivation above, be diagonal across the area, or be along one side. It doesn't matter! So the general form for the equation of the torque on a current loop becomes

$$\tau = IAB \cos\theta \qquad\qquad \text{(Equation 19.7)}$$

As the loop rotates from its initial position where the torque is a maximum, the torque decreases because $\cos\theta$ decreases (θ goes from 0° to 90°). When the loop is vertical, the torque is zero. As θ goes beyond 90°, the torque switches sign and becomes counterclockwise. The loop will find a stable equilibrium position that is vertical in the end view, as in the lower drawing in Figure 19.10. With friction in the system, the loop will eventually come to rest in a vertical position, oscillating for a while around the vertical position with the oscillations eventually damping out.

EXAMPLE 19.5

A drum of radius 3 cm has a cord wound around it with a mass of 150 g hanging from it, as shown in Figure 19.11. The drum is attached to a square wire loop, and a portion of the loop passes through the jaws of a large horseshoe magnet 4 cm thick. (a) What current is required to lift the weight? (b) What must its direction be (into or out of the page)?

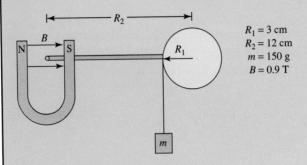

$R_1 = 3$ cm
$R_2 = 12$ cm
$m = 150$ g
$B = 0.9$ T

Figure 19.11 Torque Balancing Act

SOLUTION
(a)
$$\vec{\tau}_{min} = \vec{R}_1 \times m\vec{g} \qquad \tau_{min} = R_1 mg$$

$$\tau_{min} = (0.03 \text{ m})(0.15 \text{ kg})(9.8 \text{ m/s}^2) \qquad \tau_{min} = 0.0441 \text{ N-m}$$

The torque on the wire must equal the minimum torque calculated above to just lift the weight.

$$\vec{\tau} = \vec{R}_2 \times \vec{F} \qquad \tau = R_2 F \sin \theta$$

$$\theta = 90° \qquad \tau = R_2 F$$

$$F = ILB \qquad \tau = R_2 ILB \qquad \tau = \tau_{min}$$

$$R_2 ILB = \tau_{min} \qquad I = \frac{\tau_{min}}{R_2 LB}$$

$$I = \frac{0.0441 \text{ N-m}}{(0.12 \text{ m})(0.04 \text{ m})(0.9 \text{ T})}$$

$$I = 10.2 \text{ A}$$

(b) The force on the segment of wire inside the magnet must be up. According to the right-hand rule, the current must be out of the page.

19.10 Galvanometers and Motors

A galvanometer is a device that uses the torque on a current loop to measure a current (see Figure 19.12). The greater the current in the coil of wire around the soft iron core, the greater the deflection of the needle on the scale. The cylinder of soft iron that forms the core of the windings usually has a coil spring attached to it so that with no current

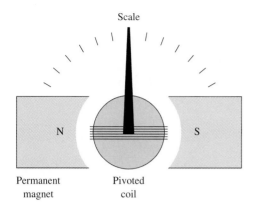

Scale

N   S

Permanent   Pivoted
magnet   coil

Figure 19.12 Galvanometer
Movement

the needle is centered on the scale. If current flows in one direction, the needle might be deflected to the right, and if it flows in the opposite direction, it will then deflect to the left. If many turns of very fine copper wire are used around the iron core, a galvanometer can detect as little current as one-millionth (1×10^{-6}) A, or 1 μA.

A galvanometer that can detect a microampere is far too sensitive to use directly to detect amperes or even milliamperes, because such a current would quickly put the needle in the fully deflected position. However, knowing such parameters of a galvanometer movement as its full-scale deflection current, its internal resistance, and its full-scale deflection voltage, you can use a galvanometer movement to construct a voltmeter, an ammeter, or an ohmmeter for any desired range.

In most cases where an analog electrical measurement is being made by a deflecting needle, a galvanometer movement is at the heart of it. In digital readout meters, an operational amplifier integrated circuit coupled with a liquid-crystal display and an appropriate driver provides the reading.

EXAMPLE 19.6

A galvanometer movement needing $\frac{1}{2}$ mA for full-scale deflection and having 40-Ω internal resistance is to be used to measure a maximum current of 1 A. Design the ammeter using a galvanometer movement.

SOLUTION
We can consider the galvanometer movement as an electronic component of resistance 40 Ω, with the restriction that deflection of the needle is full scale when $\frac{1}{2}$ mA flows through it. Thus, to use the movement in an ammeter, we'll have to construct a shunt in parallel to the movement that allows most of the current to flow around the movement, as shown in Figure 19.13.

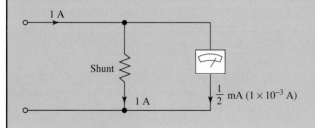

1 A

Shunt

1 A $\frac{1}{2}$ mA $(1 \times 10^{-3}$ A$)$

Figure 19.13 Galvanometer
Movement Used as an
Ammeter

Since the shunt and the galvanometer movement are in parallel, they see the same voltage. At full-scale deflection for the galvanometer, this voltage is the full-scale deflection voltage. For the galvanometer it is

$$V_{fs} = (0.5 \times 10^{-3} \text{ A})(40 \text{ }\Omega) \qquad V_{fs} = 0.020 \text{ V}$$

Knowing the full-scale deflection voltage for the galvanometer, we can calculate the necessary shunt resistance.

$$R_s = \frac{V_{fs}}{I}$$

$$R_s = \frac{0.020 \text{ V}}{1 \text{ A}}$$

$$R_s = 0.020 \text{ }\Omega$$

This shunt resistance is very low and is usually made by cutting a length of wire with known resistance per unit length to the appropriate length. The current through the galvanometer is so much smaller than the current through the shunt that the galvanometer current can be ignored.

A galvanometer's motion is limited to a maximum 180° rotation of the needle. An electric motor must turn through 360°. To make an electric motor out of our galvanometer movement, we must remove the spring that centers the needle on the scale and find a way to keep the loop from being trapped at the equilibrium position. Remember that when we examined the torque on a current loop, if the coil were originally rotating clockwise with a constant current in one direction through the windings, when the windings went through the vertical position, the clockwise torque became counterclockwise. The loop would always be drawn back to the equilibrium position. What we need is to somehow push the coil away from the equilibrium position once it moves through it. If the direction of the current in the windings is changed at the moment it passes through the equilibrium position, then the forces on the loop reverse, and the loop is pushed away from equilibrium (look at Figure 19.10 again). The loop then keeps turning.

The changing of the direction of the current at the proper moment is done by a split ring called a *commutator,* shown in Figure 19.14. The commutator is brazed or soldered to the current loop. The commutator makes electrical contact with the brushes, but the brushes are fixed in place and don't rotate with the commutator. The brushes are called *brushes* because they wipe the commutator as it turns, carrying the current to the current loop. The commutator changes the direction of the current through the loop at the moment the loop passes through the equilibrium position, pushing the loop away from the equilibrium position in the direction it was originally traveling.

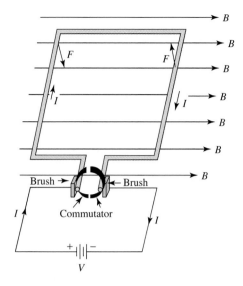

Figure 19.14 Commutator on the Armature of a dc Electric Motor

Thus, an electric motor is actually a magnetic motor because it's the interaction between the magnetic field of the permanent magnets surrounding the current loop and the magnetic field generated by the current flowing in the loop that causes the motor's shaft to turn. Of course, the above-described motor is very simple, but any electric motor will be only a variation on this theme.

Summary

1. A moving charge (or current) will create a *magnetic field,* which may impart a *magnetic force* to any other moving charge (or current).

2. The magnetic field created by a straight current will circle around the wire (according to the **right-hand rule**). As the distance from the wire increases, the magnitude decreases proportionately, as determined by Equation (19.3).

Problem Solving Tip 19.1: When determining the direction of the magnetic field, simply make a "thumbs up" sign with your right hand. If you bend your wrist so that your thumb points in the direction of the current, your fingers will be curled in the appropriate direction (clockwise or counterclockwise) of the magnetic field.

3. When a wire is bent into a loop, the encircling magnetic fields add up in the center of the loop, creating a straight magnetic field. The magnitude of this field is given by Equation (19.4). If there are multiple loops of wire (such as in a **solenoid**), Equation (19.5) can also be used.

Problem Solving Tip 19.2: Similar to Tip 19.1, when determining the direction of the magnetic field at the center of a current-carrying loop, again make the "thumbs up" sign with your right hand. Rotate your wrist so that your fingers curl in the same direction (clockwise or counterclockwise) as the

current in the wire. Your thumb will now be pointed in the correct direction as the magnetic field! Notice the similarities between Tip 19.1 and Tip 19.2. In both cases, your thumb (which is straight) points along with the phenomenon that is straight (whether it be the current or the field), and your fingers curl along with the phenomenon that is curling. And don't forget to use your *right* hand!

4. Whereas an electric force will act in the same direction as the electric field, despite the velocity of a charged particle, a magnetic force acts at right angles to both the magnetic field and the velocity of the charged particle. This direction of the magnetic force can be determined by using the **right-hand rule.** The magnitude of the magnetic force can be determined by using Equation (19.1).

Problem Solving Tip 19.3: Don't forget that this force can still be used in conjunction with Newton's second law. For an example, see Figure 19.2 and its corresponding calculations [solving for Equation (19.2)]. Notice that this calculation uses information from four different chapters! (Velocity = Chapter 1; Newton's second law = Chapter 3; centripetal acceleration = Chapter 7; and magnetic force = Chapter 19.) It may seem difficult to keep track of so many chapters at any given time, but this is what physics is all about: using all of your knowledge to solve problems in the natural world. Keep with it; it gets easier with practice.

5. The magnetic fields around parallel wires carrying current will cause a force to act between the wires. This force may be attractive or repulsive, depending on the directions of the currents within the wires. The direction of the force on either wire can be determined using the **right-hand rule.** The magnitude of the force between the wires can be found by using Equation (19.6).

Problem Solving Tip 19.4: This calculation is often confusing. When performing such an exercise, remember what

Equation (19.6), $F = ILB$, is actually stating: The force on any given wire is equal to *its **own** current* multiplied by its length multiplied by the magnetic field *created by the **other** wire.*

6. A current-carrying loop immersed in an externally created magnetic field may experience a *torque,* causing it to rotate. This is the principle behind electric motors. The torque is dependent on the current in the wire, the total cross-sectional area of the *loop,* the magnetic field that is present, and the angle that the loop makes with the field. This relationship is illustrated by Equation (19.7).

Key Concepts

A moving charge generates a magnetic field.

FIRST RIGHT-HAND RULE: The thumb of your right hand points in the direction of the particle's velocity. The fingers of your right hand with your palm open are in the direction of the magnetic field. Your open palm is in the direction of the force.

SECOND RIGHT-HAND RULE: The thumb of your right hand points in the direction of the positive current flow. The fingers of your hand naturally curve around the path of the current in the direction of the magnetic field.

SOLENOID: A wire-wound tube with a homogeneous magnetic field inside.

Important Equations

(19.1) $\mathbf{F} = \mathbf{v}\mathbf{B} \sin \theta$ (Force on a charged particle moving in a magnetic field)

(19.2) $r = \dfrac{mv}{qB}$ (Radius of the track of a charged particle moving in a magnetic field)

(19.3) $B = \dfrac{\mu_o I}{2\pi s}$ (Magnetic field around a current element)

(19.4) $B = \dfrac{\mu_o NI}{2r}$ (Magnetic field at the center of a current loop)

(19.5) $B = \mu_o ni$ (Magnetic field inside a solenoid)

(19.6) $F = I\,\Delta L\,B$ (Force on a current-carrying wire in a magnetic field)

(19.7) $\tau = IAB \cos \theta$ (Torque on a current loop where θ is the angle of rotation)

Conceptual Problems

1. All magnetic fields have their origin in:
   a. permanent magnets   b. moving electric charges
   c. iron atoms   d. magnetic monopoles

2. A person carrying a meter that detects magnetic and electric fields passes an electron at rest. Her meter reads:
   a. only an electric field
   b. only a magnetic field
   c. no fields
   d. both magnetic and electric fields

3. Magnetic fields have no interaction with:
   a. moving electric charges
   b. stationary electric charges
   c. moving permanent magnets
   d. stationary permanent magnets

4. A positive current is flowing west along a power line. The magnetic field above the power line points:
   a. north   b. south
   c. east   d. west

5. The magnetic field around a long, straight wire carrying a current:

  a. is parallel to the wire

  b. is radially outward from the wire

  c. is radially inward toward the wire

  d. is circular around the wire

6. In the interior of a solenoid, the magnetic field:

  a. increases toward the center

  b. decreases toward the center

  c. is zero

  d. is uniform

7. When a slug of iron is inserted into a solenoid carrying a current:

  a. the magnetic field increases greatly

  b. the magnetic field decreases

  c. the current in the solenoid increases

  d. the current changes direction

8. An electron enters a magnetic field parallel to the field lines. The trajectory of the electron:

  a. is unaffected by the magnetic field

  b. becomes circular

  c. becomes parabolic

  d. slows down and reverses direction

9. Two parallel wires carry currents in opposite directions. Which of the following statements is true?

  a. The wires repel each other.

  b. The wires attract each other.

  c. There is an equal and opposite torque on each wire.

  d. The total current adds to zero.

10. A wire loop carrying a current in a uniform magnetic field tends to rotate:

  a. until the plane of the loop is parallel to the field lines

  b. until the plane of the loop is perpendicular to the field lines

  c. with a constant angular velocity

  d. and generate an opposing electrical field

11. A current is passed through a helical metal spring that can both extend and contract. What happens to the spring, and why?

12. Two iron bars attract each other no matter which ends are placed close together. Are both of the bars magnets? Why?

13. A charged particle is moving with a circular trajectory in a uniform magnetic field. An electric field parallel to the magnetic field is switched on. Describe the new path of the particle.

14. A charged particle enters a region of uniform magnetic field. The KE of the particle:

  a. increases or decreases depending on the direction of the magnetic field

  b. doesn't change

  c. is converted into potential energy

  d. becomes a voltage

Exercises

Magnetic Field and Force on a Moving Charge

1. A charge of 12 μC moving at 4×10^6 m/s suddenly enters a region where it experiences a magnetic force of 9.6×10^{-3} N. What is the magnitude of the magnetic field?

  Answer: 2×10^{-4} T

2. A charged body is moving with a speed of 6×10^4 m/s at an angle of 37°with respect to a magnetic field of strength 7.2×10^{-4} T. It experiences a force of 2×10^{-4} N. What is the magnitude of the charge on the moving body?

  Answer: 7.7 μC

3. Two large magnets are placed one above the other so that a constant magnetic field of 4.8×10^{-2} T runs top to bottom in the small gap between them. If an electron enters this region from the right with a horizontal speed of 6×10^5 m/s, what are the magnitude and the direction of the force exerted on the electron?

  Answer: 4.6×10^{-15} N, into the page

4. A proton is fired through a small aperture into a region with a uniform magnetic field with a velocity of 3.5×10^4 m/s

(see Figure 19.15). If the field strength is 5.6×10^{-3} T, how far from the aperture does the proton strike the plate?

Figure 19.15 (Exercise 4)

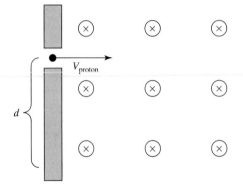

Answer: 13 cm

5. A cyclotron is a device used to accelerate elementary particles such as protons to very high speeds. Two areas of the
cyclotron have uniform magnetic fields for steering the proton. These areas are called "dees" because they are semicir-
cular and look like the letter *D*, as shown in Figure 19.16.
The proton is accelerated by an electric field in a cavity
between the dees. The electric field is generated by putting a
potential difference V_o across the gap. The electric field
must be reversed each time the charged particle travels from
one dee to another so that it is again accelerated when it

Figure 19.16 (Exercise 5)

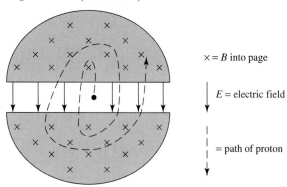

$\times = B$ into page

$E = $ electric field

$= $ path of proton

comes back through the cavity in the opposite direction.
(a) What is the frequency at which the electric field must
change direction? (b) What is the increase in the KE of the
particle for each revolution, assuming that the width of the
cavity is small? (c) If the cyclotron radius is 2 m and the
magnetic field strength is 0.5 T, what is the maximum speed
of a single proton? (d) How many revolutions must the proton make in the cyclotron to attain this speed if the voltage
across the cavity's gap is 30 kV?

Answer: (a) $f = qB/2\pi m$
(b) $KE = 2qV_o$
(c) 9.6×10^7 m/s
(d) 801 rev

6. Show that the period of a charged particle orbiting in a magnetic field is independent of its speed and the radius of its
orbit.

Answer: $T = \dfrac{2\pi m}{qB}$

7. A charged particle follows a circular path when in the presence of a uniform magnetic field. Show that the KE of the
particle is

$$KE = q^2B^2r^2/2m$$

where r is the radius of the circle.

Magnetic Field of Current

8. Power lines often carry currents in excess of 1000 A. (a) If a
particular power line carried 5500 A, and if it were strung
between poles 10 m off the ground, what would the strength
of the magnetic field be at ground level? (b) If you were
1.7 m tall, how strong would the magnetic field be just above
your head if you were standing directly below the power line?

Answer: (a) 1.1×10^{-4} T
(b) 1.3×10^{-4} T

9. If a field strength of 0.0055 T is found 10 cm away from a
wire, how much current is the wire carrying?

Answer: 2750 A

*10. A bullet with a charge of 120 μC was fired with a speed of
300 m/s parallel to a long wire carrying 5 A, as shown in
Figure 19.17. As the bullet emerged from the gun, the separation between it and the wire was 20 cm. (a) What are the
magnitude and the direction of the magnetic force on the
bullet by the wire? (b) If the bullet were held motionless
20 cm from the wire, what would the magnetic force on the
bullet be?

Figure 19.17 (Exercise 10)

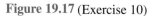

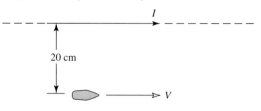

I

20 cm

V

Answer: (a) 1.8×10^{-7} N, toward the wire
(b) 0

11. A technician has a device that cannot work properly in a
magnetic field stronger than 6.8×10^{-3} T. What is the closest that he can place this device to a power line carrying
750 A of current?

Answer: 2.2 cm

12. Two long wires running parallel to each other are separated
by 30 cm, and each has a current of 1.5 A. What is the magnetic field halfway between them if the wires carried their

currents (a) in the same direction and (b) in opposite directions?

Answer: (a) 0

(b) 4×10^{-6} T

*13. Two long parallel wires carry a current I of the same magnitude but in opposite directions. The wires are a distance s apart. What is the magnetic field a distance y away from the plane of the two wires and at an equal distance from each wire?

Answer: $B = \dfrac{4\mu_o Is}{\pi(s^2 + vy^2)}$

Magnetic Field at the Center of a Current Loop

14. A field strength of 1×10^{-5} T is measured at the center of a loop of wire having a diameter of 0.5 cm. What current does the wire carry?

Answer: 0.04 A

15. A wire having a resistance of 100 Ω is bent into a circle with a radius of 10 cm. A 9-V battery is connected to the wire. What is the magnetic field at the center of the circle?

Answer: 5.65×10^{-7} T

The Solenoid

16. A solenoid has 800 turns per meter. (a) If a current of 250 mA were passed through the solenoid, what would the magnetic field be inside? (b) If the solenoid were cut in half, what would the field strength be inside?

Answer: (a) 2.5×10^{-4} T

(b) 2.5×10^{-4} T

17. Solenoids are often rated by their n value, or turns per meter. If an engineer needs a solenoid to produce a field strength of 4×10^{-4} T and have a current of 300 mA, what should the value of n be?

Answer: $n = 1061$ turns/m

18. A solenoid is linked in series with a 12-V battery and a 50-Ω resistor. The solenoid is 10 cm long and has 100 turns, and the wire has a resistance of 10 Ω. What is the magnetic field inside the solenoid?

Answer: 2.5×10^{-4} T

*19. A current of 350 mA passes through a solenoid of $n = 3000$ turns/m. A single wire loop of 36-mm diameter is placed around the solenoid. What current is required in the loop to cancel out the magnetic field at the center of the solenoid directly beneath the coil?

Answer: 37.8 A

20. What is the force per unit length between the two wires in Exercise 12 for both cases? Tell whether the forces are attractive or repulsive.

Answer: 1.5×10^{-6} N/m; attractive same direction; repulsive opposite direction

*21. Three wires are arranged as shown in Figure 19.18. If the force per unit length between wires A and B is the same as the force per unit length between wires A and C, what is the ratio of currents B and C?

Figure 19.18 (Exercise 21)

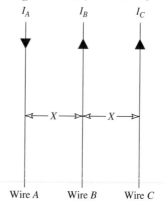

Answer: 0.5

22. The power line in Exercise 8 carries 5500 A. Two such lines run parallel to one another held apart a distance of 3 m. If the poles supporting the lines are 100 m apart, how much force do the lines exert on one another between poles?

Answer: 201.85 N

*23. A rigid rectangular loop of wire is placed near a long wire carrying a current of 2.5 A as shown in Figure 19.19. Does the long wire attract or repel the circuit?

Answer: Repel

Figure 19.19 (Exercise 23)

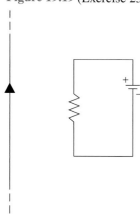

*24. In a lecture demonstration, two horizontal parallel wires, each 1 m long and each having a mass of 15 g, are suspended at each end by 10-cm-long strings attached to an overhead rod. A current I runs through both wires in opposite directions. The wires swing out from each other, and the strings suspending the wires form an angle of 4° with each other. What is the magnitude of the current in the wires?

Answer: 13.4 A

*25. Show that the force on a straight wire perpendicular to a uniform magnetic field is independent of the length of the wire if the potential difference across the wire is held constant.

Answer: $F = \dfrac{VAB}{\rho}$ where A is the cross-sectional area of the wire, and ρ is the wire's resistivity

*26. A rectangular loop of wire sits next to a straight wire as shown in Figure 19.20. Both wires have a current of 3 A in the directions shown. What is the net force on the loop, and what is its direction?

Answer: 5.6×10^{-6} N, repulsive

Figure 19.20 (Exercise 26)

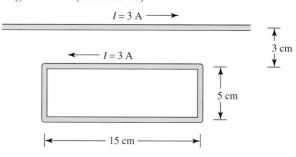

*27. A device to hurl ore from the surface of the moon into lunar orbit is being built. Two conducting horizontal rails are parallel and placed a distance L apart. They are in a region of a uniform magnetic field B directed up. There is a potential difference of V between the rails. When a light metal rod of resistance R is placed across the rails, a current flows, and the rod accelerates. If the coefficient of friction is μ_k between the rails and the rod, what is the relationship between the velocity of the rod along the rails and the distance traveled if the rod starts from rest?

Answer: $v = \sqrt{2\left(\dfrac{VLB}{mR} - \mu_k g\right)s}$

28. Figure 19.21 shows a square loop of wire 20 cm to a side in three different positions with respect to a uniform magnetic field of strength 5×10^{-2} T. What are the corresponding torques on the loop in each position?

Answer: (a) At 0°, 6.6×10^{-3} N-m
(b) At 45°, 4.7×10^{-3} N-m
(c) At 90°, 0

Figure 19.21 (Exercise 28)

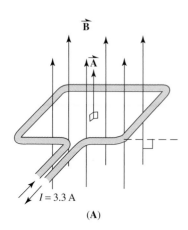

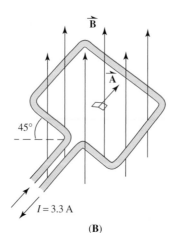

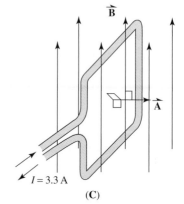

(A)   (B)   (C)

*29. A crane is designed using magnetic principles as shown in Figure 19.22. It consists of a square segment of current-carrying wire immersed in a uniform magnetic field of 2.5×10^{-2} T. Attached as shown is a 5-g mass. What current is required to hold the device in equilibrium at an angle of 45°?

Answer: 19.6 A

Figure 19.22 (Exercise 29)

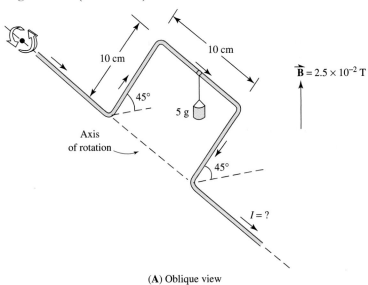

(A) Oblique view

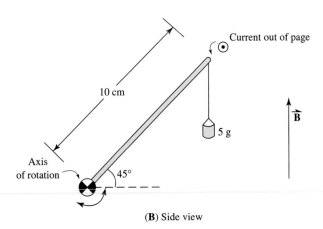

(B) Side view

30. The current loop in Figure 19.23 has a length y and a width x. The axis of rotation is placed as shown. The magnetic field is horizontal and has a direction from left to right. (a) Show that the torque on the loop is $\tau = IAB$. (b) Show that if the loop rotates by an angle θ, the torque on the loop becomes $\tau = IAB \cos \theta$.

Figure 19.23 (Exercise 30)

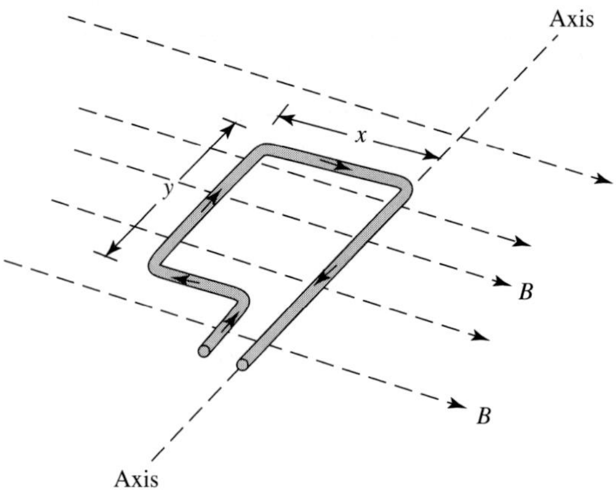

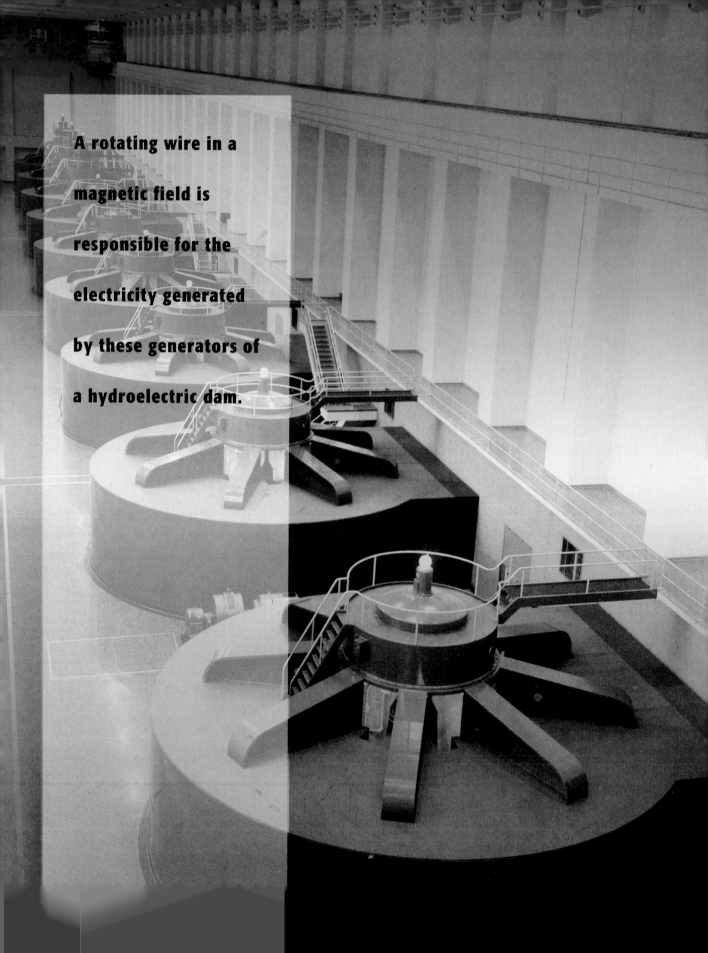

A rotating wire in a magnetic field is responsible for the electricity generated by these generators of a hydroelectric dam.

Induced Voltages and Currents

At the end of this chapter, the reader should be able to:

1. State the difference between **alternating current (ac)** and direct current (dc).

2. Using the principle of magnetic flux, calculate the **induced current** in a wire, given only the geometry of the wire, its resistance, and the rate at which the external magnetic field is changing.

3. Briefly explain how an *electric generator* creates an alternating current.

4. Solve for the current in the secondary coil of a *transformer* if both coils have identical resistance, the current in the primary coil is given, and the number of turns in each coils is known.

5. Describe how alternating current (ac) is converted into direct current (dc).

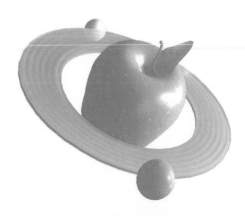

6. Calculate the amount of current flowing through an inductor at a given point in time after a given voltage source is removed from a circuit loop containing only an inductor, a battery, and a resistor.

7. Solve for the *magnetic energy* stored in a known inductor once a set amount of current is allowed to flow freely through it.

20.1   Wire Moving in a Magnetic Field

A common demonstration in a physics lecture is to show that when a wire that is connected to a sensitive galvanometer is passed through a magnetic field, the galvanometer registers a deflection, as shown in Figure 20.1. When manipulating the wire inside the jaws of the horseshoe magnet, you'll notice that when the wire is standing still, there is no deflection. The faster the wire moves in the magnetic field, the greater the deflection. When the wire is moved where there is no magnetic field (outside the jaws), there is no deflection. When the wire is moved in the direction of (with or against) the magnetic field between the jaws, there is no deflection. The strongest current occurs when the wire is moving perpendicularly to the magnetic field lines. Finally, when the direction of the wire's motion is changed, the direction of the current is changed.

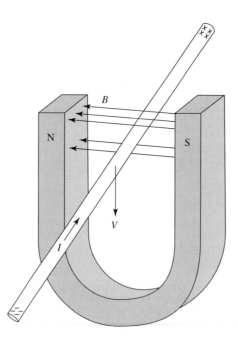

Figure 20.1 Wire Moving in a Magnetic Field

All of these effects can be explained by the equation for the force on a moving charge, $\mathbf{F} = q\mathbf{v} \times \mathbf{B}$. The force is what moves the charge along the wire, and the moving charges are the current that the galvanometer detects when it deflects. This situation is the flip side of the coin of the situation studied in Chapter 19, where we learned that there is a force on a current-carrying wire in a magnetic field. That force tries to move the wire. In this new situation, moving the wire in a magnetic field causes a current to flow. It is said that a current is ***induced*** in the wire.

For a current to flow through the galvanometer, the wire loop must be closed. What if we pass a straight segment of wire through a volume of constant magnetic field, and the segment has nothing attached to either end? Figure 20.1 illustrates this case. Initially a current will flow, and a net charge will begin to build up on either end. Positive charges will collect at one end, and negative charges at the other. From Chapter 17 in the discussion on capacitors, we know that separating charge requires work. These separated charges cause an electric field that is constant along the length of the conductor. The presence of the electric field means that there is a potential difference, or voltage, induced between the ends of the straight wire segment. Current in the open-ended wire stops flowing when the force on an electron due to the electric field equals the force on the electron from the magnetic field, and they cancel one another out.

Again, separating charge requires work, and the work done is the magnetic force pushing the charge along the wire times the distance over which the force is exerted.

$$W = \mathbf{F} \cdot \mathbf{L} \qquad W = FL \cos \theta$$

$$\theta = 0° \qquad W = FL$$

$$\mathbf{F} = q\mathbf{v} \times \mathbf{B} \qquad F = qvB \sin \alpha$$

$$\alpha = 90° \qquad F = qvB$$

$$W = qvBL$$

The angle between **v** and **B** is 90°, and the angle between the force on the charge and the charge's displacement is 0°. In this case, L would be the length of wire between the poles of the horseshoe magnet. The voltage buildup between the two ends is the work done per unit charge.

$$\frac{W}{q} = BLv$$

$$\mathcal{E} = BLv \qquad \text{(Equation 20.1)}$$

The electromagnetic force (emf) induced between the ends of the wire because of the wire's motion in the magnetic field is BLv. The induced emf is used just as the voltage of a battery in determining how much current will flow in a wire if the wire is closed. The only real difference between an emf and a battery's voltage is that the battery's voltage is pretty much localized in the circuit, and an emf is not. It's distributed.

EXAMPLE 20.1

The thickness of the magnet in Figure 20.1 is 4 cm. The wire loop is closed with a length of 21 cm. The wire has a diameter of 0.5 mm and moves through the jaws of the magnet with a speed of 4 m/s. What is the current induced in the wire given that $B = 1.1$ T and $\rho = 1.7 \times 10^{-8}$ Ω-m.

SOLUTION

$$I = \frac{\mathcal{E}}{R}$$

First find the emf induced in the wire.

$$\mathcal{E} = BLv \qquad \mathcal{E} = (1.1 \text{ T})(0.04 \text{ m})(4 \text{ m/s}) \qquad \mathcal{E} = 0.176 \text{ V}$$

Next find the resistance of the wire loop.

$$R = \frac{\rho L}{A}$$

$$A = \pi \left(\frac{0.5 \times 10^{-3} \text{ m}}{2} \right)^2 \qquad A = 1.96 \times 10^{-7} \text{ m}^2$$

$$R = \frac{(1.7 \times 10^{-8} \text{ }\Omega\text{-m})(0.21 \text{ m})}{1.96 \times 10^{-7} \text{ m}^2} \qquad R = 1.82 \times 10^{-2} \text{ }\Omega$$

$$I = \frac{0.176 \text{ V}}{1.82 \times 10^{-2} \text{ }\Omega}$$

$$I = 9.67 \text{ A}$$

20.2   Magnetic Flux and Induced Voltages

In the last section we discovered that a wire moving in a magnetic field has an emf generated across it, and if the wire is closed, a current flows through it. The charges in the wire are unaware if the force pushing them along the wire is caused by an electric field or a magnetic field. So a magnetically induced emf is the same as a voltage caused by an electric field.

Let's look at the same problem as in the last section, but from a new point of view. This new point of view uses the concept of *magnetic flux,* given by the following formula:

Magnetic flux: Magnetic field lines intercepted by the area of a current loop.

(Equation 20.2)

$$\phi = BA \cos \theta$$

The angle θ is due to the fact that magnetic flux is like sunlight falling on a surface. The greatest amount of sunlight falls on the surface when the sun's rays hit the surface at a right angle, or the plane of the surface is perpendicular to the sun's rays. In Figure 20.2, θ is zero. When the plane of the figure is rotated from this maximum flux position, the

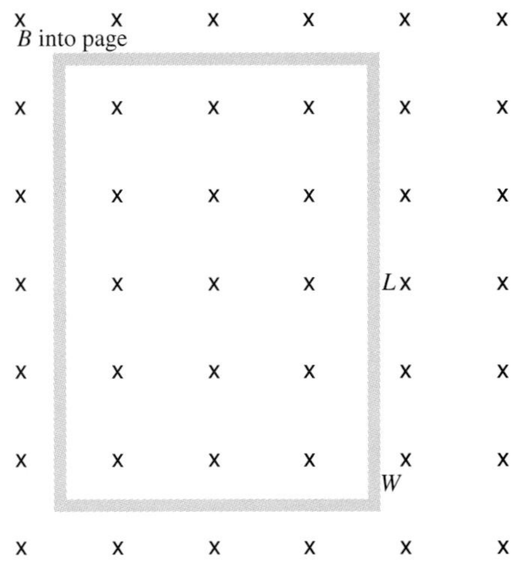

Figure 20.2 Flux of a
Magnetic Field onto a Surface

amount of sunlight intercepted by the surface decreases until it intercepts none when θ is 90°.

In Equation (20.2), B is the magnetic field, A is the area of the loop, and θ is the angle between the direction of the magnetic field and the normal to the loop's area. Area, oddly enough, has a direction, and the direction is that of an arrow pointing perpendicular to the area. In this case, the loop's area would be either into or out of the page. Since the magnetic field in Figure 20.2 is into the page, we can conveniently choose the area's direction as into the page also, and the angle θ is 0°. The total flux for the loop in Figure 20.2 is then

$$\phi = B(Lw)\cos 0° \qquad \phi = BLw$$

If the loop is oriented so that it sticks into and out of the page, then $\theta = 90°$, and the total flux becomes zero. If the loop were to slowly rotate from its initial position in Figure 20.2 until it was perpendicular to the page, its flux would go from maximum to zero ($\cos \theta$ goes from 1 to 0). At maximum flux the loop's area intercepts a maximum number of magnetic field lines, and at minimum (zero in this case) the loop's area intercepts no magnetic field lines. The angle θ defines how much of the area is presented to the flux. As θ goes from 0° to 90°, the area presented to the flux goes from a maximum to zero.

How do we use flux? We first redefine induced emf: The induced emf in a loop is equal to the rate of change of magnetic flux through the loop's enclosed area. Mathematically, this relationship becomes

$$\mathcal{E} = -\frac{\Delta\phi}{\Delta t}$$

(Equation 20.3)

The negative sign comes from the fact that you don't get something for nothing. Nature is conservative and opposes change. An emf in a closed loop causes a current, and the current in the loop causes a magnetic field that opposes the external magnetic field. If this weren't so, then the induced field would add to the external field; the current would get larger, further adding to the external field; and the current would increase without bounds! We would have a way of generating power from nothing. Unfortunately, the world doesn't work that way. You don't get something for nothing, and that negative sign is very important!

Figure 20.3 shows a square wire loop moving with velocity v into a region of magnetic field B. As long as the flux is changing, an emf is induced in the loop.

$$\mathcal{E} = -\frac{\Delta\phi}{\Delta t} \qquad \mathcal{E} = -\frac{\Delta(BA)}{\Delta t} \qquad \theta = 0°$$

The magnetic field is constant, and only the area of the loop in the magnetic field is changing.

$$\mathcal{E} = -B\frac{\Delta A}{\Delta t} \qquad \Delta A = \frac{L\,\Delta L}{\Delta t}$$

$$\frac{\Delta L}{\Delta t} = v \qquad \mathcal{E} = -BLv$$

This formula is exactly the same one derived in Section 20.1. If two ways of looking at the same situation are equally valid, they should predict the same result.

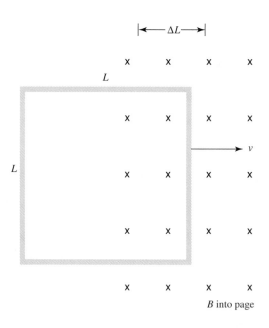

Figure 20.3

What happens when the flux stops changing? In the case above, that could happen when the loop has fully entered the region of the magnetic field without changing its velocity v. There should be no induced emf and therefore no net current in the loop. Look at Figure 20.2 again, and pretend that the loop has a velocity v to the left. Both vertical loop segments will produce currents toward the top of the page of equal magnitude according to the right-hand rule. The one from the left segment will flow clockwise, and the one from the right segment counterclockwise, canceling each other out. No net current means no net emf.

EXAMPLE 20.2

A stationary loop of radius 12 cm is placed in a region where the magnetic field is changing at the rate of 0.6 T/s. If the wire has a resistance of 0.3 Ω/m, what current flows in the wire?

SOLUTION

$$I = \frac{\mathcal{E}}{R} \qquad \mathcal{E} = \frac{\Delta \phi}{\Delta t} \qquad \mathcal{E} = A\frac{\Delta B}{\Delta t}$$

$$A = \pi(0.12 \text{ m})^2 \qquad A = 0.0452 \text{ m}^2$$

$$\mathcal{E} = (0.0452 \text{ m}^2)(0.6 \text{ T/s}) \qquad \mathcal{E} = 0.0271 \text{ V}$$

$$R = 2\pi(0.12 \text{ m})(0.3 \text{ Ω/m}) \qquad R = 0.226 \text{ Ω}$$

$$I = \frac{0.0271 \text{ V}}{0.226 \text{ Ω}}$$

$$I = 0.12 \text{ A}$$

20.3　Generators

A generator is a device that generates electric power. A simple generator is nothing more than a coil of wire rotating in a region where there is a magnetic field. In Section 19.9, we investigated the torque on a current loop. We can use Figure 19.10 to talk about a generator as well as the torque on a current loop. The only difference is that in the generator's case, an external force provides the torque to rotate the loop. We could put a crank on the end of the wire loop and rotate it by hand. Assume that a crank is placed on the end of the loop in Figure 20.2. It can be rotated by hand at a constant angular velocity of ω. Therefore, $\theta = \omega t$.

　　An emf is generated across the ends of the loop by the changing flux.

$$\mathcal{E} = -\frac{\Delta \phi}{\Delta t} \qquad \mathcal{E} = -B\frac{\Delta A}{\Delta t}$$

If the flux is changing but the magnetic field is constant, then the loop's area must be changing in some way. It doesn't change physically, but the area that the loop presents to the magnetic field lines changes with θ. When the angle in Figure 20.2 is 0° (the loop's area is in the plane of the page), the area presented to the magnetic lines is a maximum. When the angle in Figure 20.2 is 90° (the loop is vertical into and out of the page), the area presented to the magnetic field lines is zero. The area must then vary as cos θ.

$$A = L^2 \cos \theta$$

$$\theta = \omega t \qquad A = L^2 \cos \omega t \qquad \frac{\Delta A}{\Delta t} = -\omega L^2 \sin \omega t$$

$$\mathcal{E} = -B\omega L^2 \sin \omega t$$

In Chapter 9, we dealt with sine-cosine relationships (see Figure 9.2), and when we looked at the rate of change of the function with time ($\Delta A/\Delta t$), we were actually looking at the slope of the cosine curve at a time t. Plotting the slope of a cosine curve will give you an inverted sine curve. If you're interested in how a function containing the angle changes with time, remember that since the angle θ is ωt, it will be proportional to ω.

$$\theta = \omega t \qquad \Delta\theta = \omega\,\Delta t \qquad \frac{\Delta\theta}{\Delta t} = \omega$$

where B, ω, and L^2 are all constants. Thus, the emf naturally produced by a loop of wire rotating in a constant magnetic field is a sinusoid. An interesting fact about the emf is that it is greatest where the flux is least, and zero where the flux is greatest. The reason is that it is the rate of change of flux that matters, not the magnitude of the flux.

Alternating current (ac): A current that changes in both amplitude and polarity.

An emf that is sinusoidal is called **alternating current (ac)** *because it alternates in both amplitude and polarity with time.* Figure 20.4 shows an ac sine curve. An oscilloscope is an electronic instrument that can show an electronic signal on a screen. The screen usually has time on the horizontal axis and voltage on the vertical axis. Usually the signal is periodic. The electric power outlets in the United States are 120 V_{ac}, and if an oscilloscope is hooked up to a wall outlet, it will display a voltage signal exactly like that shown in Figure 20.4.

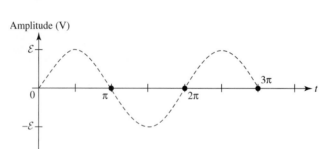

Figure 20.4 An ac Voltage Sine Curve

Figure 19.10 was originally drawn to illustrate the torque on a current loop. Since torque on a current loop is the concept upon which the electric motor is based, it should now be apparent that a motor and a generator are two sides of the same coin. Run a current into the two leads of the loop; the motor turns. Put a crank on the shaft of the loop and start turning; an emf is generated across the ends of the loop.

The sinusoidal form of the voltage coming out of an ac wall socket is there because a sinusoidal emf is the natural output of a coil rotating in a magnetic field. Almost all of the electric power generated comes from a rotating coil in a magnetic field. This electrical power is transmitted as a sinusoidal voltage along the high-voltage transmission lines that link cities to the power stations. It is more efficient to transmit electric power as a high voltage because of the resistive losses in current-carrying lines. For this reason, the voltage on the transmission lines may be 200,000 V or higher! Power is generated at a much lower voltage and boosted to the 200,000-V or higher level to transmit it from the coal-, oil-, or nuclear-fired generating station to where it will be used. At its destination the voltage must be dropped from the very high transmission line level to around 3000 V for local distribution, and finally down to the 120 V_{ac} that appears at our wall sockets. The reason the transmitted power is left as a sinusoid is that the devices used to step up and step down voltages (transformers) operate only with an alternating voltage, and best when that voltage is sinusoidal.

20.4 Transformers

Figure 20.5 shows a wire loop connected to a galvanometer. If a bar magnet is moved toward the loop, the loop sees a changing magnetic field. A current is then induced in the loop in such a direction that the loop's magnetic field will oppose the bar magnet's field. (The magnetic field lines of the loop will be in the opposite direction to those from the bar magnet.) The galvanometer needle registers the current because of the emf induced in the loop. But if the bar magnet slows and stops in the middle of the loop, the galvanometer registers zero. If the bar magnet starts moving again, the galvanometer again registers a current, and the current drops to zero as the magnet moves out and away. An emf is generated in the loop only when it sees a changing magnetic field.

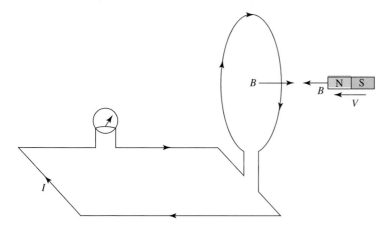

Figure 20.5 Magnet Passing through a Wire Loop

A transformer's primary function is to step up or step down a voltage. It consists of two parts: a primary coil and a secondary coil. The two coils are usually joined with a laminated iron core, as shown in Figure 20.6. An ac voltage is applied to the primary coil. The ac voltage causes an alternating current in the primary coil. The alternating current causes a changing magnetic field in the primary coil. The primary and secondary coils aren't connected electrically. No current is flowing from the primary to the secondary, or vice versa. They are connected magnetically through the shared iron core by magnetic field lines. So the changing magnetic field in the primary is seen by the secondary coil. The changing magnetic field in the secondary induces an emf in the secondary coil, and current can be drawn from it by attaching a load.

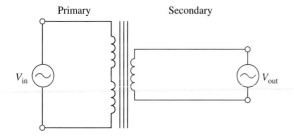

Figure 20.6 Schematic of a Transformer

In Figure 20.6, the number of turns in the primary and the secondary aren't equal. Remember that the magnetic field in a coil of wire is directly proportional to the number of turns of wire, N. The same is true of a transformer, since it's made up of a pair of coils. But since a transformer steps voltages up or down, the side with the greatest number of turns has the greatest voltage. Thus, in Figure 20.6, the primary side has the

A transformer uses coupled changing magnetic fields to step voltages up or down. The voltages appearing on the high-tension transmission lines in the background are far too high to use directly and must be stepped down by a transformer.

greater voltage, and the secondary voltage is less than the primary voltage. The transformer in Figure 20.6 is a step-down transformer. If the primary had fewer turns than the secondary, then the transformer would be a step-up transformer, because the primary voltage would be boosted. By definition the voltage to be transformed (stepped up or stepped down) is connected to the primary side of the transformer, and the load is connected to the secondary. Since voltages are proportional to the number of turns, the equation for the ratio of the primary emf to the secondary emf is a simple one.

(Equation 20.4)

$$\frac{\mathcal{E}_p}{\mathcal{E}_s} = \frac{N_p}{N_s}$$

where N_p is the number of turns on the primary coil, and N_s is the number of turns on the secondary coil.

The power into the primary is equal to the power out of the secondary.

(Equation 20.5)

$$P_p = P_s$$

$$I_p V_p = I_s V_s$$

In an ideal transformer, there is no power loss from primary to secondary. Commercial transformers operate at efficiencies better than 99%, and losses are minimal. Such losses that do occur are usually due to eddy currents in the core of the transformer. If the core is of iron, then the iron inside the windings will have currents induced in it from the changing magnetic field that it sees. These currents will dissipate power as heat because of the resistivity of the iron. Remember also that these induced currents in the core will have a direction that opposes the external magnetic field, which will inevitably lessen the changing flux that the secondary sees.

To lessen these eddy currents, the bulk of the iron core is constructed throughout of thin iron sheets arranged parallel to the magnetic field. The thin sheets are insulated from one another. This kind of construction is called a *laminated core.* The heating in a metal because of the induced currents is called *induction heating,* and induction furnaces are used in foundries to melt metal prior to its being poured in a mold for casting.

A transformer passes an ac voltage from the primary to the secondary, but it blocks dc. If there were a mixed ac/dc signal appearing at the primary like that of Figure 20.7A, then the signal out of the secondary would be that of Figure 20.7B. In Figure 20.7A, a sine wave of 10 V peak-to-peak is riding on a $+5\ V_{dc}$ signal. Since a transformer responds only to an alternating signal, only the sine wave is passed on to the secondary, and the $+5\ V_{dc}$ signal is blocked, because the dc signal generates no changing flux in the secondary. Transformers with a $1:1$ windings ratio are often called *isolation transformers* because they isolate the primary windings from the secondary windings with respect to dc biases.

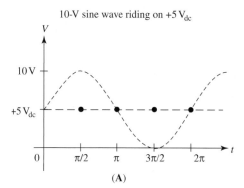

(A)

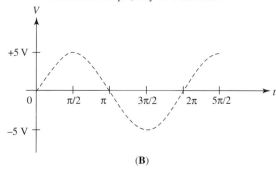

(B)

Figure 20.7 A Transformer Passes ac and Blocks dc.

EXAMPLE 20.3

A load draws a current of 7 A from a transformer secondary that has an output emf of 12 V. The primary is plugged into a wall socket of 120 V_{ac}. (a) What is the turns ratio of the primary to the secondary? (b) What is the current into the primary windings?

SOLUTION

(a)
$$\frac{N_p}{N_s} = \frac{V_p}{V_s} \qquad \frac{N_p}{N_s} = \frac{120\ V}{12\ V} \qquad \frac{N_p}{N_s} = \frac{10}{1}$$

(b)
$$I_p V_p = I_s V_s \qquad I_p = \frac{I_s V_s}{V_p}$$

$$I_p = \frac{(7\ \text{A})(12\ \text{V})}{120\ \text{V}}$$

$$I_p = 0.7\ \text{A} = 700\ \text{mA}$$

20.5   Rectifying ac to dc

Electric power is naturally generated as a sinusoid, and because of the way transformers operate, electric power is transported from place to place as a sinusoid. But what form does this electric power take when we actually have to use it? Incandescent lights, resistance heaters and ovens, and alternating-current motors are designed to operate directly from the 120-V_{ac} signal. But computers, digital logic circuits, TVs, VCRs, and a multitude of other consumer and commercial devices all require dc power at voltages usually well below the line voltage out of the wall socket. The electrical systems for most automobiles are dc, so before that 120-V_{ac} signal from the wall socket can be used in many cases, it must be stepped down and rectified. *Rectification means turning an alternating current, usually sinusoidal voltage, into a direct current voltage like that found in batteries.*

Rectification: Turning an ac signal into a dc signal.

A simple rectifier is the dc motor shown in Figure 19.14. If a crank were placed on the wire loop and turned at a constant angular velocity, we would see the dc signal shown in Figure 20.8B. Remember that the motor's commutator reverses the current in the loop so that it keeps turning in the same direction as it does when it goes past the equilibrium position. In the case of the motor-turned generator,

$$\mathcal{E} = \omega L^2 \sin \omega t \quad \text{and} \quad \omega t = \theta$$

where θ is the angle of rotation referred to in Figure 20.2. As θ goes through each $n\pi$ rad, the sign of $\mathcal{E}$ changes. If we took the output of the generator directly from the ends of the loop, we would have a sinusoidal output as in Figure 20.8A. But with the output taken from the brushes in contact with the commutator, when the sign of the emf changes as the loop swings past equilibrium, the contacts at the end of the loop are simultaneously reversed. This reversal always keeps one of the terminals positive with respect to the other. (Which is positive depends on which way the loop is rotating.)

Figure 20.8B is considered dc because the sign of the emf doesn't change, although its amplitude is continuously changing. To lessen the "lumpiness" of Figure 20.8B, another loop could be added at 90° to the original loop. When the original loop's emf is zero, the second loop will be at its maximum emf. If you looked at the generator's output with an oscilloscope, you would see the superposition signal (solid line). To reduce the lumpiness still further, a third coil could be added with the coils spaced every 60°. The more coils added to the generator, the smoother the output signal.

When the armature (wire loop) of an electric motor rotates, it develops its own emf opposite in direction to the applied emf. This back emf is proportional to the armature's angular speed. Because of this generator effect of electric motors, the applied voltage

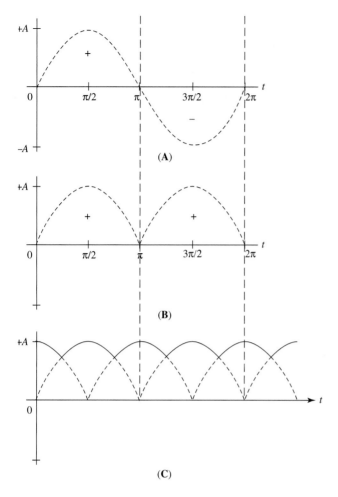

(A)

(B)

(C)

Figure 20.8 Rectified Sine Wave

equals the back emf plus the resistive voltage drop in the wiring of the armature coils and field windings if the motor possesses them. (A permanent-magnet motor doesn't have field windings because the magnets provide the magnetic field with which the armature interacts. But many motors generate the magnetic field in which the armature rotates with an electromagnet, and therefore these motors have field windings as well as armature windings.)

$$V = \mathcal{E}_b + IR$$

(Equation 20.6)

The motor tries to maintain a constant speed for a given applied voltage. If the speed falls, $\mathcal{E}_b$ drops and IR increases. If the motor speed rises, $\mathcal{E}_b$ rises and the current drops. Maximum current occurs when the applied voltage is present and the motor isn't turning (a good way to burn out an electric motor!). A motor spinning under no-load conditions with V applied to the armature requires very little current to keep it spinning, and the current is a minimum. An electric motor is also a generator, as shown by the back emf.

The multiple-armature approach might work for generators and alternators found in automotive applications, but for appliances and electronic equipment that require dc power and that are plugged into a wall socket, the multiple-armature approach to rectification won't work. *Instead, a semiconductor (usually silicon) device that allows current to flow through it in only one direction is used; this component is called a **diode**.*

Diode: An electronic component that allows current to flow in only one direction.

If a diode is connected between a load and an ac source as in Figure 20.9A, then the load sees the emf of Figure 20.9B. The diode allows conduction in the circuit only when the polarity of the source is as shown. When the polarity is reversed, no current flows. The negative half of the sinusoid is blocked.

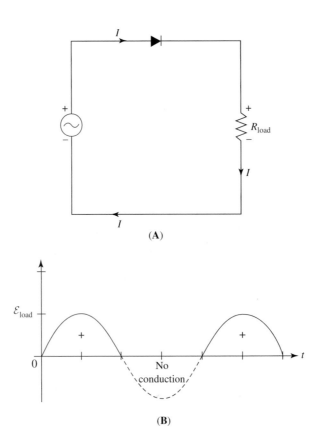

Figure 20.9 Rectification of a Sine Wave by a Single Diode

To improve the efficiency of the rectification, it's desirable to have both the positive and the negative half-cycles of the full sine wave. A diode rectification circuit like the one in Figure 20.10A is used. The load sees the emf of Figure 20.10B. Conventional current flows through the load in only one direction (from bottom to top in the drawing). When the polarity of the source changes from that shown in Figure 20.10A, the diodes direct the current so that it flows through the load in the same direction as in the previous half-cycle. (Trace the current flow for each polarity of the source.) The output looks exactly like that previously shown in Figure 20.8B. A capacitor is often placed across the output to smooth the output. The capacitor charges up to the output voltage, and when the rectified sine wave signal falls below the peak emf, some of the capacitor's charge flows into the load, keeping the voltage relatively constant.

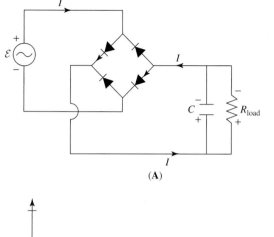

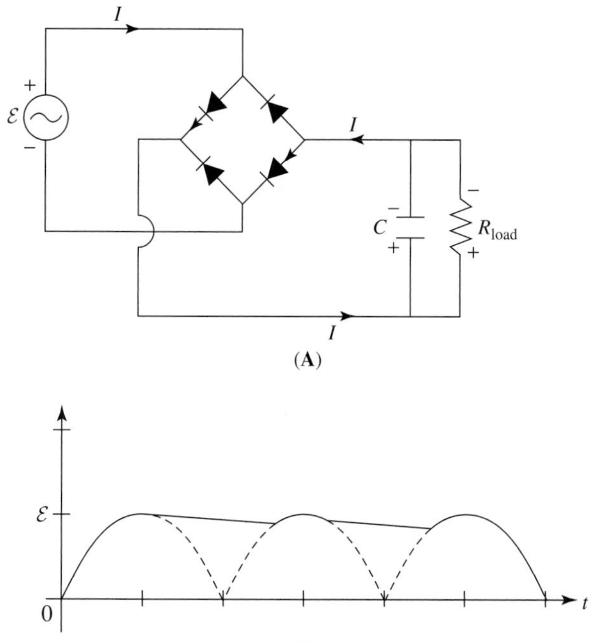

(A)

(B)

Figure 20.10

20.6   Inductors

In the case of a transformer, the changing flux in the primary causes a changing emf at the output of the secondary. The primary and the secondary of a transformer usually share an iron core, but this iron core isn't necessary for the changing flux in the primary to cause an induced emf in the secondary, although the induced emf will be weaker. However, it's not necessary to have two separate coils for an induced emf to appear. If the current is changing in a coil, an induced emf appears in that same coil. This emf is self-induced. Remember that the magnetic field in a coil is caused by the current flowing through it. Changing the current changes the magnetic field which changes the flux which causes the self-induced emf!

A new electronic component called the *inductor* is about to be introduced based on self-inductance discussed above (see Figure 20.11). This component has a characteristic unit of inductance called the *henry* (H) and is rated in units of the microhenry (μH; 1×10^{-6} H) and millihenry (mH; 1×10^{-3} H). More basic units for the henry are

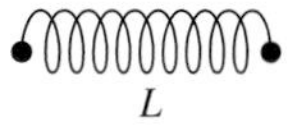

$$1\ \text{H} = 1\ \text{V-s/A}$$

A capacitor stores energy in an electric field by separating charge. We've seen how this energy can be used when a capacitor's charge flows into a load, smoothing out the output voltage of a full-wave rectifier. *An **inductor** is an electronic component that stores energy in a magnetic field.* A capacitor tries to keep the voltage of a circuit constant, and the inductor tries to keep the current flowing in a circuit constant. The definition of inductance is based upon the self-induced emf.

Figure 20.11 Inductor Schematic Symbol

L

Inductor: An electronic device that stores energy in a magnetic field.

$$\mathcal{E} = -\frac{\Delta\phi}{\Delta t} \qquad \phi = NBA$$

where N is the number of turns in the coil, A is the cross-sectional area of the coil, and B is the magnetic field inside the coil. But the magnetic field is directly proportional to the current, and there is an alternative expression for the flux based on the current in the coil.

$$NBA = Li = \text{flux}$$

where L is the inductance of the inductor.

$$\mathcal{E} = -\frac{\Delta(Li)}{\Delta t} \qquad \mathcal{E} = -L\frac{\Delta i}{\Delta t}$$

$$L = \frac{-\mathcal{E}}{\Delta i / \Delta t}$$

(Equation 20.7)

The flux in a coil is proportional to the current through the coil and the inductance L of the coil; L accounts for the coil's area and number of turns. The definition of *inductance* is the ratio of the induced emf across the inductor to the rate of change of current in the inductor. The magnitude of the current has nothing to do with the voltage across the inductor, only with its rate of change.

There is an emf across an inductor proportional to the rate of change of current. The proportionality constant is the inductance. But an inductor possesses its intrinsic inductance whether or not it has current flowing through it, just as a capacitor or a resistor pulled from a bin has a rated capacitance or resistance. Thus, an inductor's inductance is related to how it's constructed and to its geometry. Starting from the original definition of *inductance*, a derivation of the inductance of an inductor follows.

$$L = \frac{-\mathcal{E}}{\Delta i / \Delta t} \qquad \mathcal{E} = \frac{\Delta\phi}{\Delta t}$$

$$L = -\frac{\Delta\phi / \Delta t}{\Delta i / \Delta t} \qquad L = -\frac{\Delta\phi}{\Delta i}$$

$$\phi = NBA$$

where N is the turns in the coil, and A is the coil's cross-sectional area.

$$\Delta\phi = NA\,\Delta B$$

The only thing changing is the magnetic field.

$$L = -\frac{NA\,\Delta B}{\Delta i}$$

where $B = \mu_o ni$ for a solenoid and n is the number of turns per unit length.

$$\Delta B = \mu_o n\,\Delta i$$

$$L = -\frac{NA\mu_o n \,\Delta i}{\Delta i} \qquad L = -NA\mu_o n$$

$$N = nl \qquad L = -(nl)A\mu_o n$$

$$\frac{L}{l} = n^2 \mu_o A \qquad\qquad \text{(Equation 20.8)}$$

The real surprise in the equation for the inductance per unit length is that the inductance is proportional to n^2, and not just to n. The reason is the self-inductance.

EXAMPLE 20.4

An air-core inductor 7 cm long with a diameter of 1 cm is connected to a changing current source where the current changes from 1200 mA to 300 mA in 20 μs. If the inductor is wound with 3500 turns/m, what is the induced emf across the inductor?

SOLUTION

$$\mathcal{E} = L\frac{di}{dt} \qquad L = n^2\mu_o A \qquad \mathcal{E} = n^2\mu_o A\frac{di}{dt}$$

$$A = \pi(0.5 \times 10^{-2}\,\text{m})^2 \qquad A = 7.85 \times 10^{-5}\,\text{m}^2$$

$$\frac{di}{dt} = \frac{1.2\,\text{A} - 0.3\,\text{A}}{20 \times 10^{-6}\,\text{s}} \qquad \frac{di}{dt} = 4.5 \times 10^4\,\text{A/s}$$

$$\mathcal{E} = \left(\frac{3500}{\text{m}}\right)^2\left(\frac{1.26 \times 10^{-6}\,\text{T-m}}{\text{A}}\right)(7.85 \times 10^{-5}\,\text{m}^2)\left(\frac{4.5 \times 10^4\,\text{A}}{\text{s}}\right)$$

$$\mathcal{E} = 54.5\,\text{V}$$

The voltage across the inductor has nothing to do with the nominal amplitude of the voltage to which it's connected; rather, it's related only to the rate of change of current that it sees. Therefore, a highly tuned radio antenna that operates at mega-hertz (MHz) or gigahertz (GHz; 1×10^9 cycles/s) might have hundreds or even thousands of volts across it even though the voltage to which the antenna is connected might be as low as 12 V!

20.7   The Inductive Time Constant

How long does it take the circuit shown in Figure 20.12 to reach its maximum current? The circuit shown in the figure is a series dc circuit. When the switch is closed, current begins to flow, and for a short period of time, a changing current in the inductor causes an induced emf across it. But as the current rises to its maximum with time, the current becomes constant, and the voltage across the inductor drops to zero. An ideal inductor is a short circuit to direct current after the current stops changing. So the maximum current for steady state in Figure 20.12 is $I = V/R$. The inductor has disappeared! There is no voltage drop across it. Contrast this behavior with that of the capacitor, which is an open circuit at steady state in a dc circuit. No current flows through it. Of course, the fine wire windings give the inductor an intrinsic resistance because of the resistivity of the wire.

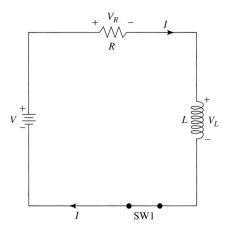

Figure 20.12 Series *RL* Circuit

This resistance is usually so small that it can be ignored. In an ideal inductor this intrinsic resistance is zero.

But how does the current behave in that period of time after the switch has been closed when the current is changing? For a series dc circuit, the voltage drops around the loop must equal the applied voltage.

$$V = V_R + V_L \qquad V_R = iR \qquad V_L = L\frac{\Delta i}{\Delta t}$$

$$V = iR + L\frac{\Delta i}{\Delta t}$$

The equation above is exactly the same mathematically as the one derived when we found the charge and discharge curves for a capacitor in Chapter 18. Using Ohm's law around a closed loop, we arrive at an equation where the variable being solved for is in the denominator of a fraction, or goes as the function $1/x$. The rate of change of the curve $y = \ln x$ is the curve $y = 1/x$. Thus, the summation of $1/x$ finding the area under the curve gives us back $y = \ln x$. Solving for the current takes us from $\ln x$ to e^x, and we arrive at the equation for the current rise in an inductor.

$$L\frac{\Delta i}{\Delta t} = V - iR \qquad \frac{L}{R}\frac{\Delta i}{\Delta t} = \frac{V}{R} - i$$

$$\frac{\Delta i}{V/R - i} = \frac{R}{L}\Delta t \qquad \Sigma \frac{\Delta i}{V/R - i} = \frac{R}{L}\Sigma \Delta t$$

$$-\ln\left(\frac{V}{R} - i\right) = \frac{R}{L}t + K$$

At $t = 0$, $i = 0$.

$$K = -\ln\left(\frac{V}{R}\right)$$

$$-\ln\left(\frac{V}{R} - i\right) = \frac{R}{L}t - \ln\left(\frac{V}{R}\right) \qquad \ln\left(\frac{V}{R} - i\right) = -\frac{R}{L}t + \ln\left(\frac{V}{R}\right)$$

$$\ln\left(\frac{V}{R} - i\right) - \ln\left(\frac{V}{R}\right) = -\frac{R}{L}t \qquad \ln\left(\frac{V/R - i}{V/R}\right) = -\frac{R}{L}t$$

$$\frac{V/R - i}{V/R} = e^{-(R/L)t} \qquad \frac{V}{R} - i = \frac{V}{R}e^{-(R/L)t}$$

$$i = \frac{V}{R} - \frac{V}{R}e^{-(R/L)t}$$

$$i = \frac{V}{R}(1 - e^{-(R/L)t}) \qquad \text{(Equation 20.9)}$$

At time $t = 0$, the exponential is e^0, and anything raised to the zero power is 1. Thus, the current immediately after the switch is closed is zero, and the inductor is an open circuit. At $t = \infty$, the exponential term becomes $1/\infty$, or zero. The maximum current then becomes $I = V/R$, and the inductor acts like a short circuit, as predicted. Between these two extremes, the current in the circuit rises as does the graph in Figure 20.13. After one L/R time constant, the current has risen to 63% of its maximum value. After two time constants it has risen to 86% of its maximum value. After five time constants the current has risen to 99% of its maximum value. But the current reaches its maximum value only after an infinite amount of time! We saw the same behavior with the capacitor in that it never fully charged but got closer and closer to its maximum voltage without actually reaching it. This asymptotic behavior is typical for an exponential. The quantity L/R has dimensions of time, and the exact values for the inductance and the resistance will give the length in seconds of a single time constant.

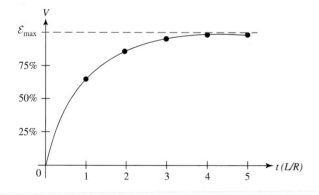

Figure 20.13 Exponential Current Rise in an Inductor

If we had to assign polarities to the resistor and inductor of Figure 20.12 when the current was changing, they would be as shown in the figure. If the inductor weren't in the circuit, the current rise would ideally be instantaneous. There would be no time lag between no current and maximum current. Five time constants is traditionally chosen as the time it takes an inductive circuit to reach maximum current. To account for this time lag, compare the situation to connecting another battery with opposing voltage into the circuit; this battery's voltage would dwindle exponentially.

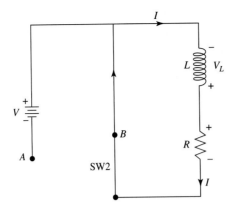

Figure 20.14 Inductor "Discharging" through a Resistor

The flip side of this coin is a situation where an inductor with maximum current flowing through it is switched from position A to B in the circuit of Figure 20.14. How long does it take the current to drop to zero? If the inductor opposes the change, it will want to keep current flowing in its initial direction before the switch is thrown. Since the inductor now takes the place of the applied voltage, the polarity of the induced emf across the inductor must be that shown in Figure 20.14. The sum of the voltages around the loop must equal zero.

$$-V_L + V_R = 0 \qquad V_L = V_R$$

$$L\frac{\Delta i}{\Delta t} = iR \qquad \frac{\Delta i}{\Delta t} = -\frac{R}{L}i$$

The rate of change for current over time is negative because the current decreases as time increases.

$$\frac{\Delta i}{i} = -\frac{R}{L}\Delta t \qquad \Sigma \frac{\Delta i}{i} = -\frac{R}{L}\Sigma \Delta t$$

$$\ln i = -\frac{R}{L}t + C$$

At $t = 0$, $i = i_o$.

$$i_o = \frac{V}{R}$$

It is assumed that the current in the inductor reached its maximum before the switch was thrown from A to B.

$$C = \ln i_o \qquad \ln i = -\frac{R}{L}t + \ln i_o$$

$$\ln i - \ln i_o = -\frac{R}{L}t \qquad \ln\left(\frac{i}{i_o}\right) = -\frac{R}{L}t$$

$$\frac{i}{i_o} = e^{-(R/L)t} \qquad i = i_o e^{-(R/L)t}$$

$$i = \frac{V}{R}(e^{-(R/L)t})$$ (Equation 20.10)

At $t = 0$, $i = V/R$, and as t goes to infinity, the current asymptotically drops to zero. The behavior of the dropping current with time is plotted on the graph of Figure 20.15. Traditionally, after five time constants the current is considered to have dropped to zero.

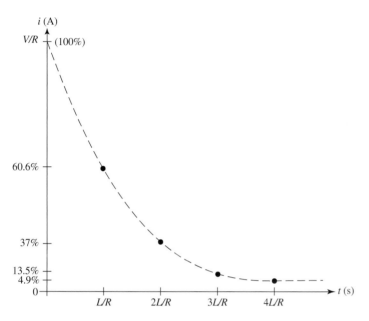

Figure 20.15 Exponentially Decaying Current in an Inductor Connected to a Resistor

EXAMPLE 20.5

A series RL circuit has a 2500-Ω resistor connected to a 30-mH inductor. Both are connected to a 12-V battery through a switch. (a) What is the time constant? (b) Approximately how long does it take the current to reach 75% of its maximum? (Use the graph of Figure 20.13.) (c) What is the maximum current?

SOLUTION

(a)
$$\frac{L}{R} = \frac{30 \text{ mH}}{2500 \ \Omega} \qquad \frac{L}{R} = 12 \ \mu s$$

(b)
$$75\% = 1.5 \text{ time constants} \qquad 75\% = 18 \ \mu s$$

(c)
$$I_{max} = \frac{V}{R} \qquad I_{max} = \frac{12 \text{ V}}{2500 \ \Omega} \qquad I_{max} = 4.8 \text{ mA}$$

20.8   Energy Stored in a Magnetic Field

In Section 20.7 we showed that an inductor acts like a battery when the applied voltage is suddenly switched out, and it provides an exponentially dwindling current to that circuit. This fact implies that energy must be stored in the inductor. Energy is stored in an inductor when the current in the inductor is rising to a maximum. The rate of energy storage in the inductor is the same as the power delivered to the inductor during the time the current is changing. (Remember that when maximum current is reached, no voltage developed is across an ideal inductor, and therefore no power is delivered to or dissipated by it.)

$$P = iV \qquad V = L\frac{\Delta i}{\Delta t}$$

$$P = iL\frac{\Delta i}{\Delta t} \qquad P\,\Delta t = Li\,\Delta i$$

$$P\,\Sigma\,\Delta t = L\,\Sigma\,i\,\Delta i$$

where $P\,\Sigma\,\Delta t$ is the energy stored by the inductor.

$$E = L\,\Sigma\,i\,\Delta i$$

The clock starts at $t = 0$ and begins to tick. The current in the inductor starts at zero and rises to I (maximum current).

(Equation 20.11)

$$E = \frac{1}{2}LI^2$$

In this case, the flux in the inductor changes linearly with current (see Figure 20.16), and it is this changing flux caused by a changing current that gives rise to the energy stored in the inductor.

$$A = \frac{1}{2}(\text{base})(\text{height}) \quad \text{Base} = i$$

$$\text{Height} = \Delta\phi = L\,\Delta i \qquad \Delta i = i - 0$$

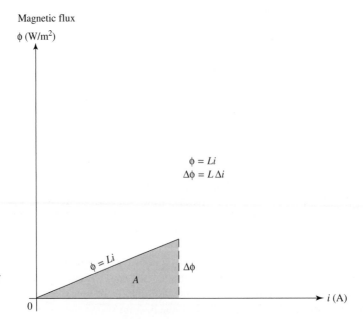

Figure 20.16 The Area under the Curve Is the Energy Stored in the Inductor.

$$\Delta i = i \qquad \text{Height} = Li$$

$$A = \frac{1}{2}(i)(Li) \qquad A = \frac{1}{2}Li^2$$

where A is the energy stored in the magnetic field of the inductor.

Energy stored in a capacitor is stored in the electric field between its plates. The energy stored in an inductor is in the magnetic field inside the inductor.

$$E = \frac{1}{2}Li^2 \qquad Li = NBA = \phi$$

where N is the number of turns in the coil. An alternate definition of *inductance* is the total flux per unit of current.

$$N^2B^2A^2 = L^2i^2 \qquad \frac{N^2B^2A^2}{2L} = \frac{1}{2}Li^2$$

$$E = \frac{N^2B^2A^2}{2L} \qquad N = nl$$

$$L = n^2\mu_o Al$$

where L is the inductance based on the geometry and fabrication of the coil.

$$E = \frac{n^2l^2B^2A^2}{2n^2\mu_o Al} \qquad E = \frac{lB^2A}{2\mu_o}$$

$$Al = V$$

where V is the total volume inside the inductor.

$$E = \frac{VB^2}{2\mu_o}$$

$$\frac{E}{V} = \frac{B^2}{2\mu_o} \qquad \text{(Equation 20.12)}$$

In our electronics parts inventory, we now have five items: the battery (an energy source) and four components (resistor, capacitor, inductor, and diode). One of the components dissipates energy (the resistor), and two others store energy (the capacitor in an electric field and the inductor in a magnetic field). The diode acts as a one-way valve for current flow. The energy-storage devices act contrary to one another upon the application of a voltage: The capacitor is initially a short circuit, and as it charges up, the current ceases to flow and it becomes an open circuit. The inductor is initially an open circuit with no current flowing. As the current through it rises to maximum, the voltage across it drops to zero, and it becomes a short circuit.

20.9   Oscillating Electromagnetic Circuits

In Chapter 9, simple harmonic motion (SHM) was introduced, and both the gravity-driven, simple pendulum and the spring pendulum were used to illustrate cases of SHM. In both, the total energy of the system was constant, and the energy shuttled back and forth between potential energy and kinetic energy. The system oscillated sinusoidally with a characteristic frequency determined by system parameters such as the spring constant and the mass on the end of the spring, and the gravity pendulum's length and the local value of g.

The circuit of Figure 20.17 is also a harmonic oscillator. At time $t = 0$, the circuit is in the configuration shown. The capacitor is charged with a charge q_0 and stores energy in the capacitor's electric field. No current is yet flowing. When switch 1 (SW1) is closed, current begins to flow as the capacitor discharges. The energy that was stored in the electric field is given up as the electric field collapses. As current starts flowing, a magnetic field begins to build in the inductor. Energy is stored in this building magnetic field. When the capacitor is fully discharged (the electric field is zero), the magnetic field in the inductor is at its maximum.

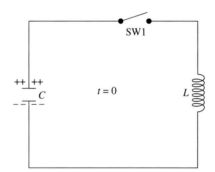

Figure 20.17 An LC Circuit Exhibits Simple Harmonic Motion.

The total energy in the LC circuit is constant, so when the electric field is zero, the magnetic field must be at a maximum. (All of the energy is in the only existing field, the magnetic field.) The magnetic field now in turn begins to collapse, giving up its energy. The collapsing magnetic field causes a current that begins to recharge the capacitor with polarity opposite to its initial charge. The electric field in the capacitor builds while the magnetic field in the inductor collapses. When the magnetic field has totally collapsed, the capacitor has an electric field opposite in direction but with the same magnitude as initially. The electric field begins again to collapse, and the discharging capacitor causes a current that begins to build the magnetic field in the inductor. When the capacitor is totally discharged, the inductor has a maximum magnetic field. This field again begins to collapse, causing a charging current to the capacitor. After the magnetic field has again collapsed, the capacitor has an electric field of both the same magnitude and the same direction as the original. The oscillations in the circuit have completed their first full cycle.

In a physical pendulum, the energy oscillates back and forth between potential and kinetic energy. In this LC circuit, the energy oscillates back and forth between being stored in an electric field and being stored in a magnetic field. The physical pendulum and the electric circuit are very different in technological implementation, but behind their operation lies the same principle: As with the physical pendulum, if energy leaves the system, the oscillations will damp out and eventually die. In the LC

circuit, energy can leave the system if there is a resistance buried somewhere in it. The coils of the inductor are made up of a long length of fine wire, and because of the resistivity of the wire, some resistance is normally associated with an inductor. A capacitor, on the other hand, usually has no resistance associated with it. The oscillations in an LC series circuit will eventually damp out and die because of this resistance to current flow.

What is the frequency of oscillation? The following calculation derives it for a series LC circuit. The total energy E is constant. It can be shared between the energy stored in the capacitor and the energy stored in the inductor.

$$V_L + V_C = 0 \qquad V_C = \frac{Q}{C}$$

$$V_L = L\frac{\Delta i}{\Delta t} \qquad L\frac{\Delta i}{\Delta t} + \frac{Q}{C} = 0$$

$$i = \frac{\Delta Q}{\Delta t} \qquad \frac{\Delta i}{\Delta t} = \frac{\Delta}{\Delta t}\left(\frac{\Delta Q}{\Delta t}\right) \qquad \frac{\Delta i}{\Delta t} = \frac{\Delta^2 Q}{\Delta t^2}$$

The rate of change of current with respect to time is the second rate of change of charge with respect to time.

$$L\frac{\Delta^2 Q}{\Delta t^2} + \frac{Q}{C} = 0$$

$$\frac{\Delta^2 Q}{\Delta t^2} + \frac{1}{LC}Q = 0$$

The above equation is the simple harmonic oscillator equation developed in Chapter 9 for pendulums.

$$\omega = \sqrt{\frac{1}{LC}}$$

$$\omega = \frac{1}{\sqrt{LC}} \qquad\qquad \text{(Equation 20.13)}$$

We started with magnetic fields and induced currents and ended up with simple harmonic motion. Simple harmonic motion in electronic circuits produces oscillators that in turn give rise to radio waves and a whole spectrum of electronic communications devices that are essential to the technological infrastructure of our present civilization. In a thorough study of electronics, these oscillations would be studied in detail in courses on the generation and reception of radio waves through the air and in transmission lines. As it is, the next chapter in this book will look further into the implications of these oscillations.

20.10   Inductors in Series and in Parallel

Two inductors are connected in series to an ac voltage source as in Figure 20.18. What is the total inductance that the voltage source sees?

$$V = V_{L1} + V_{L2}$$

$$V_{L1} = L_1\frac{\Delta i_1}{\Delta t} \qquad V_{L2} = L_2\frac{\Delta i_2}{\Delta t}$$

$$V = L_1\frac{\Delta i_1}{\Delta t} + L_2\frac{\Delta i_2}{\Delta t} \qquad \frac{\Delta i_1}{\Delta t} = \frac{\Delta i_2}{\Delta t}$$

$$V = L_{eq}\frac{\Delta i}{\Delta t}$$

$$L_{eq}\frac{\Delta i}{\Delta t} = L_1\frac{\Delta i}{\Delta t} + L_2\frac{\Delta i}{\Delta t}$$

(Equation 20.14) $$L_{eq} = L_1 + L_2$$

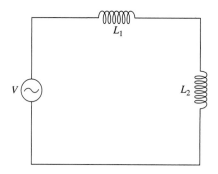

Figure 20.18 Inductors in Series

Two inductors are connected in parallel to an ac voltage source as shown in Figure 20.19. What is the equivalent inductance that the battery sees?

$$i = i_1 + i_2 \qquad \Delta i = \Delta i_1 + \Delta i_2$$

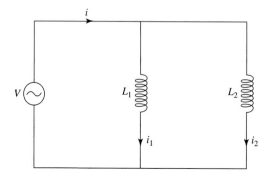

Figure 20.19 Inductors in Parallel

$$\frac{\Delta i}{\Delta t} = \frac{\Delta i_1}{\Delta t} + \frac{\Delta i_2}{\Delta t}$$

$$V_{L1} = L_1 \frac{\Delta i_1}{\Delta t} \qquad V_{L2} = L_2 \frac{\Delta i_2}{\Delta t} \qquad V = L_{eq} \frac{\Delta i}{\Delta t}$$

$$\frac{V}{L_{eq}} = \frac{V_{L1}}{L_1} + \frac{V_{L2}}{L_2} \qquad V = V_{L1} = V_{L2} \qquad \text{(parallel circuit)}$$

$$\frac{V}{L_{eq}} = \frac{V}{L_1} + \frac{V}{L_2}$$

$$\frac{1}{L_{eq}} = \frac{1}{L_1} + \frac{1}{L_2} + \cdots + \frac{1}{L_n} \qquad \text{(Equation 20.15)}$$

Inductors in series and parallel act like resistors in series and parallel, because both have voltages across them involved with the current through them.

Summary

1. As shown in Equation (19.1), a charge moving through a magnetic field will experience a force upon it. This gives rise to an emf ($\mathcal{E}$) that may cause an **induced current** to flow through the wire. The emf experienced can be calculated by using Equation (20.1).

Problem Solving Tip 20.1: This concept of emf can also be used to calculate the induced current in the wire. Since emf acts as a voltage source, Ohm's law can be applied: $\mathcal{E} = IR$, where I is the induced current, and R is the resistance of the wire.

2. **Magnetic flux** can be thought of as the number of magnetic field lines that pass through a given loop of wire. It is determined by the strength of the magnetic field, the dimensions of the loop, and the relative angle that the loop makes with the direction of the magnetic field. This relationship is described by Equation (20.2).

3. If the magnetic flux through a loop of wire changes, it may cause the loop to experience an emf, as shown in Equation (20.3). This emf in turn may also produce an induced current in the loop (see Tip 20.1).

Problem Solving Tip 20.2: There are many ways for magnetic flux to change! If you examine Equation (20.2), you will find three factors that are important when determining flux: (1) the magnetic field, (2) the angle that the loop's normal vector makes with the magnetic field, and (3) the area of the loop. As Equation (20.3) states, $\mathcal{E} = \Delta\phi/\Delta t$. We can replace ϕ with Equation (20.2): $\mathcal{E} = \Delta(BA\cos\theta)/\Delta t$. All we need to do is remove the factors that are constant in our problem from the Δ quantity. Thus, if the area were changing, and if

the angle and the field strength were constant, we could rewrite the above equation as $\mathcal{E} = B\cos\theta(\Delta A/\Delta t)$. After this step, you need only to substitute the relevant values! The tricky part is simply recognizing that the constant terms can be pulled free of the Δ quantity.

4. A *transformer* allows alternating current to be stepped up or down in voltage. If the secondary coil has more turns than the primary, the resulting voltage induced in the secondary will be proportionately higher. If it has fewer turns than the primary, its voltage will be proportionately lower. This relationship is given in Equation (20.4).

Problem Solving Tip 20.3: If needed, you can substitute Ohm's law for the induced voltages in order to solve for the currents in either coil.

5. A coil in a circuit is called an **inductor.** As current changes in a coil (for example, when a battery is first applied to an inductor), it will cause a corresponding change in the coil's magnetic field. This change, as dictated by Equation (20.3), will cause a proportionate amount of induced voltage to be applied in the opposite direction, thus hampering the rate at which a current can change within the coil. The degree of this resistance to change of current is determined by the coil's *inductance*, which in turn is determined by the inductor's physical construction. This quantity can be expressed by both the dimensions of the inductor [Equation (20.8)] and the rate of current flowing through it [Equation (20.7)].

6. When current is initially applied to an inductor, it takes time to reach its maximum value [Equation (20.9)].

7. When a voltage source is removed from a circuit that contains an inductor, the inductor will by its very nature try to impede that drop in current. The current in a *discharging inductor* at any given point in time is given by Equation (20.10).

8. Inductors store energy as magnetic fields. The strength of this magnetic field is determined by the current that is present and the inductance of the coil, as given by Equation (20.11).

Problem Solving Tip 20.4: As before, when you're dealing with moving particles, springs, and capacitors, you can use this energy $\frac{1}{2}LI^2$ in conjunction with *conservation of energy* type of problems. Refer to Tip 4.1 for more information.

9. The unit of inductance is the henry (H).

Key Concepts

ALTERNATING CURRENT (AC): A current that changes in both amplitude and polarity.

DIODE: An electronic component that allows current to flow in only one direction.

INDUCED CURRENT: A current produced by a wire moving in a magnetic field.

INDUCTOR: An electronic device that stores energy in a magnetic field.

MAGNETIC FLUX: Magnetic field lines intercepted by the area of a current loop.

RECTIFICATION: Turning an ac signal into a dc signal.

Important Equations

(20.1)	$\mathcal{E} = BLv$	(emf produced by a wire of length L moving through a magnetic field)
(20.2)	$\phi = BA \cos \theta$	(Magnetic flux)
(20.3)	$\mathcal{E} = -\dfrac{\Delta\phi}{\Delta t}$	(emf produced by a changing flux)
(20.4)	$\dfrac{\mathcal{E}_p}{\mathcal{E}_s} = \dfrac{N_p}{N_s}$	(Transformer primary and secondary voltage ratios in terms of the turns ratios)
(20.5)	$P_p = P_s$	(The power in equals the power out for a transformer.)
(20.6)	$V = \mathcal{E}_b + IR$	(The voltage across a motor equals the back emf plus the voltage drop across the windings.)
(20.7)	$L = \dfrac{-\mathcal{E}}{\Delta i / \Delta t}$	(The inductance equals the emf induced across it over the rate of change of current.)
(20.8)	$\dfrac{L}{l} = n^2 \mu_o A$	(Inductance in terms of an inductor's physical parameters where l is the inductor's length)
(20.9)	$i = \dfrac{V}{R}(1 - e^{-(R/L)t})$	(Exponential rise of the current in an inductor)
(20.10)	$i = \dfrac{V}{R}(e^{-(R/L)t})$	(Exponential decay of the current in an inductor)
(20.11)	$E = \dfrac{1}{2}LI^2$	(Energy stored in an inductor)
(20.12)	$\dfrac{E}{V} = \dfrac{B^2}{2\mu_o}$	(Energy stored per unit volume in a magnetic field)
(20.13)	$\omega = \dfrac{1}{\sqrt{LC}}$	(Angular frequency of oscillation of an LC circuit)
(20.14)	$L_{eq} = L_1 + L_2 + \cdots + L_n$	(Inductors in series)
(20.15)	$\dfrac{1}{L_{eq}} = \dfrac{1}{L_1} + \dfrac{1}{L_2} + \cdots + \dfrac{1}{L_n}$	(Inductors in parallel)

Conceptual Problems

1. A bar magnet passes through a coil of wire. The induced current is greatest when:

   a. the magnet moves slowly so that it's in the coil the longest time

   b. the north pole enters first

   c. the magnet moves quickly so that it's in the coil only a short time

   d. the south pole enters first

2. A wire loop moves parallel to the direction of a magnetic field. The current induced in the loop:

   a. is at a maximum   b. is at a minimum

   c. is zero   d. flows antiparallel to the loop

3. The back emf in an electric motor is at a maximum when:

   a. the motor runs at its maximum speed

   b. the motor is switched on

   c. the motor is increasing in speed.

   d. the motor is decreasing in speed.

4. The secondary voltage in a transformer winding is induced by:

   a. a changing magnetic field in the primary

   b. a dc voltage at the primary

   c. a constant magnetic field in the primary

   d. a changing current in the primary

5. A loop of metal wire is rotated in a magnetic field. There is a force on the loop, and this force always opposes the rotation of the loop. Why?

6. Why is it easier to turn the shaft of a generator when it's unconnected to an outside circuit, but much harder when the generator is connected to a load?

7. Why does an electric motor require more current when it is just switched on than when it is running at full speed?

8. How is the back emf of an electric motor related to the emf of a generator rotating in the same magnetic field at the same speed? What is the difference between the motor and the generator?

9. An inductor and a resistor are wired in series. A battery is connected to the pair, and the current through them rises to its upper limit. As long as the current is changing, there is an emf across the inductor. If the battery were switched out of the circuit, and if the inductor were connected to the resistor, what would happen to the polarity of the emf relative to the emf when the battery was connected and the current changing? Why?

10. A belt-driven generator is being used to charge a 12-V automobile battery. The belt breaks, but the generator continues to turn. Why?

Exercises

Wire Moving in a Magnetic Field

1. A car is traveling on a road where the vertical component of the earth's magnetic field is approximately 3.5×10^{-5} T. If the potential difference between the end of the car's axles (1.8 m long) is 1.6 mV, how fast is the car moving?

   Answer: 58 mph

2. A jet with a wing span of 60 m is flying over the midwestern United States, where the earth's magnetic field is 0.048 mT. The plane travels at 220 m/s perpendicular to the magnetic field. What is the potential difference between the wing tips?

   Answer: 0.63 V

Magnetic Flux and Induced Voltages

3. Consider the circuit shown in Figure 20.20. The wire is ideal, and the resistor has a value of 12 Ω. A 0.35-T magnetic field is directed into the page. At what speed must the bar be moved to produce 50 mA of current in the wires?

   Answer: 2.86 m/s

Figure 20.20 (Exercise 3)

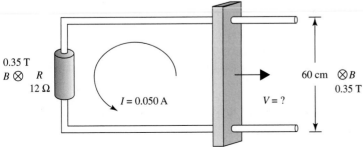

0.35 T
$B \otimes$ R
    12 Ω

$I = 0.050$ A

60 cm $\otimes B$
     0.35 T

$V = ?$

4. A wire loop has a radius of 100 mm and is oriented with its plane perpendicular to a magnetic field. How rapidly should the field change if a potential difference of 1.5 V is to appear across the ends of the loop?

Answer: 47.7 T/s

5. The flux passing through a coil of 200 turns changes uniformly from 0.020 T-m^2 to 0.015 T-m^2 in 1.2 s. What is the magnitude of the emf induced during that interval?

Answer: 0.83 V

6. A uniform magnetic field of 2.5×10^{-3} T passes through a wire loop having an area of 0.05 m^2. (a) If the field makes an angle of 30° with the plane of the loop, what is the magnetic flux through the loop of wire? What would the magnetic flux be if the field were directed (b) perpendicular to the wire loop and (c) parallel to the loop?

Answer: (a) 6.25×10^{-5} T-m^2
(b) 1.25×10^{-4} T-m^2
(c) 0

7. A wire loop originally in the form of a circle of 10-cm radius is in an area where the magnetic field is 0.5 T. The wire in the loop has a resistance of 0.3 Ω. The loop's shape is changed from a circle to a rectangle 30 cm long and 1.4 cm wide. What is the average current in the wire while it is being deformed if it takes 0.15 s to go from the circle to the rectangle?

Answer: 302 mA

*8. A metal rod has a length s, a mass m, and a resistance R. It slides down two parallel rails that make an angle θ with the horizontal. The rails have negligible resistance and friction. The rails are joined at the bottom to form a closed current loop with the sliding rod. The area in which the rod slides down the rails has a magnetic field of B directed vertically downward. What is the terminal velocity of the rod?

Answer: $v = \dfrac{mgR \sin \theta}{B^2 s^2 \cos^2 \theta}$

Transformers

9. A transformer connected to a 240-V_{ac} power line has 150 turns in its primary winding and 30 turns in its secondary winding. When a light bulb is connected to the secondary winding, the current across the secondary is measured at 0.5 A. What is the resistance of the light bulb?

Answer: 96 Ω

10. If the primary winding of a transformer has 150 turns and the secondary 30, what voltage would you expect to find at the secondary if the potential across the primary is 850 V?

Answer: 170 V

11. In a hydroelectric plant the generated power is at 1200 V. If the power put on the transmission lines is at 250,000 V, what is the ratio of the secondary to the primary windings?

Answer: 208:1

12. A transformer with 150 turns in its primary and 1200 turns in its secondary has 120 V_{ac} on its primary and a current of 2.5 A. Assuming 100% efficiency, what are (a) the secondary voltage and (b) the secondary current?

Answer: (a) $i_s = 312.5$ mA
(b) $V_s = 960$ V_{ac}

*13. A transformer has a primary voltage of 3500 V and a primary current of 25 A. The secondary voltage is 120 V with a secondary current of 700 A. If the transformer losses are resistive, what is the temperature of the cooling oil emerging from the transformer if it enters at 20°C? The oil flows through the transformer at a rate of 4 kg/min and has a heat capacity of 3 kJ/kg-°C.

Answer: 37.4°C

Rectifying ac to dc

14. An electric motor designed to be run from 24 V_{dc} has a resistance of 0.5 Ω in the coils on the armature. At its operating speed of 2000 rpm under load, it draws 20 A. (a) What is the efficiency of the motor? (b) What is the back emf produced inside the motor operating under load?

Answer: (a) 58.3%
(b) 14 V

15. If the motor in Exercise 14 were operating at only 1500 rpm, what would its current draw, power output, and efficiency be? (b) What is its starting current?

Answer: (a) 27 A; 115.5 W; 24.1%
(b) 48 A

Inductors

16. The current in a circuit drops from 10 A to 8 A in 0.1 s. If an average emf of 3 V is across an inductor during the current drop, what is the inductance of the circuit?

Answer: 150 mH

17. A solenoid 20 cm long and 10 mm in diameter is labeled as having an inductance of 0.358 mH. How many turns of wire does it contain?

Answer: 852 turns

18. An emf of 450 mV is induced in a 400-turn coil of cross-sectional area 0.05 m^2 at an instant when the current has a value of 1.2 A and is changing at a rate of 0.7 A/s. What is the magnetic flux through the coil?

Answer: 0.77 T-m^2

19. An inductor is made from an iron bar 30 cm long with a diameter of 2 cm. It is wrapped with 900 turns of insulating wire. What is the inductor's inductance? (The iron has a permeability 400 times that of free space.)

Answer: 474 μH

*20. A solenoid with an air core is 15 cm long with a diameter of 30 mm and has 1200 turns of wire. A secondary coil of 400 turns of wire is wrapped around the air-core solenoid. The current in the air-core solenoid changes at a rate of $\frac{1}{2}$ A/s. (a) What is the emf induced in the secondary coil? (b) If an iron core is slid into the solenoid, what is the induced emf? An iron core has a permittivity 400 times that of a vacuum.

Answer: (a) 1.42 mV
(b) 0.568 V

Inductive Time Constant

21. A 12-Ω resistor and a 100-mH inductor are placed in series with a 24-V battery and a switch. What is the current in the circuit 0.015 s after the switch is thrown?

Answer: 1.67 A

22. Show that the unit of the ratio L/R is the second.

23. What is the inductance in a series circuit whose resistance is 20 Ω and whose time constant is 0.15 s?

Answer: 3 H

24. An inductor of 35 mH has a current of 150 mA flowing through it. (a) When a switch connects it to a 400-Ω resistor and the current begins to decay, how long does it take the current to drop to almost zero? (b) What is the current in the inductor 100 μs after the current begins to decay?

Answer: (a) 437.5 μs
(b) 47.8 mA

*25. An iron bar 15 cm long with a diameter of 20 mm is wound tightly with copper wire 1 mm in diameter. What is the time constant of the inductor?

Answer: 232 ms

*26. A potential difference of 9 V is applied suddenly to an inductor of 15 mH and 4-Ω resistance. (a) What are the initial current and the rate of change of current? (b) What current flows through the inductor when the rate of change of current is 200 A/s? (c) What are the final current and the rate of change of current?

Answer: (a) 0; 600 A/s
(b) 1.5 A
(c) 2.25 A; 0 A/s

Energy Stored in a Magnetic Field

27. What is the magnetic energy stored in a 200-turn solenoid in which a current of 1.25 A produces a magnetic flux of 2.3×10^{-4} T-m^2?

Answer: 1.44 $\times$ 10^{-4} J

28. A 10-T magnetic field is very strong and can be found in the sample cavities of instruments such as magnetometers. How much energy would be contained in a $\frac{1}{2}$-liter volume with such a field?

Answer: 20,000 J

Oscillating Electromagnetic Circuits

29. A 2-μF capacitor is charged by connecting it to a 12-V battery. The charged capacitor is then connected to a 7-mH inductor.
(a) What is the frequency of the resulting oscillations?
(b) What is the maximum current flowing through the inductor?

Answer: (a) 1345 Hz
(b) 202.8 mA

*30. A capacitor with charge Q is connected to an inductor L, and oscillations at frequency f occur. How is the maximum current in the circuit related to Q and f?

Answer: $i = 2\pi Q f$

Inductors in Series and in Parallel

31. Two inductors of 200 mH and 330 mH are connected in series with a 700-mA dc constant current source. Calculate the energy stored in the circuit.

Answer: 0.13 J

Radio waves broad-

cast and received from

antennae are sinu-

soidal electromagnetic

radiation.

Sinusoidal Currents and Voltages

At the end of this chapter, the reader should be able to:

1. Calculate **root mean square** (**rms**) current if the resistance of a circuit is known, as well as calculate the maximum peak value of the voltage.

2. Calculate the **reactance** of a known capacitor (or inductor) if the current's frequency passing through it is known.

3. Solve for the total **impedance** and *phase angle* of a circuit of known resistors, capacitors, and inductors.

4. Explain the concept of *resonance*, and calculate the *resonant frequency* of a known circuit.

5. Solve for the *equivalent inductance* of a circuit composed of either inductors in series or inductors in parallel.

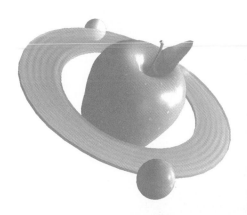

6. Calculate the power dissipated by a circuit composed of resistors, capacitors, and inductors.

7. Explain what causes a *filter* to be *low-pass, high-pass,* or *band-pass.*

21.1   rms Voltages and Currents

Alternating-current (ac) voltages are naturally generated by a coil revolving in a magnetic field. Electric power therefore starts out as a sinusoidal waveform. As already mentioned, it's much more efficient to move electric power from place to place as a high sinusoidal voltage. Transformers step up a low voltage to perhaps hundreds of thousands of volts before putting it on a cross-country transmission line. At the user end, the high voltage is stepped down a few times before being routed into homes and factories. When the electric power emerges from the wall socket, it is still an alternating current, though, and this ac power is directly utilizable by many appliances. Electric heaters, incandescent light bulbs, and ac motors can use the ac power directly. If dc is needed, most appliances or consumer electronic devices have a power supply that rectifies the ac to dc for further use. But many circuits can use the sinusoidally alternating voltage directly. This chapter deals with an alternating voltage and current in an electronic circuit.

Consider, for example, a 100-W light bulb. It's designed to dissipate 100 W of power if it is connected across 120 V. The bulb doesn't care whether this 120 V is ac or dc; in either case it will shine as brightly and dissipate 100 W of power. However, the bulbs in the light fixtures and lamps of homes and apartments are connected directly to the wall socket which supplies 120 V_{ac} in the United States and 220 V_{ac} overseas. But what is meant when a voltage is specified as 120 V_{ac}? The 120 V is the ***root-mean-square (rms) value*** of the sine wave.

Root-mean-square (rms) value: The equivalent dc voltage or current delivered to a load from an ac sinusoidal signal.

A light bulb is connected to an ac line, and the power that it dissipates is an average power over one cycle of the ac sine wave. The simple average of a sine wave is zero. It is negative for as long as it is positive, and the positive and negative half-cycles cancel exactly when simply added together. But power goes as V^2/R or as I^2R, where R is the hot resistance of the bulb's filament. Mathematically, when the ac voltage or current is squared, it is always positive. Much the same operation is performed on the sine wave as when it passes through a full-wave rectifier.

$$P = \frac{V^2}{R} \qquad V = V_0 \sin \omega t$$

$$V^2 = V_0^2 \sin^2 \omega t \qquad P = \frac{V_0^2}{R} \sin^2 \omega t$$

By a trigonometric identity,

$$\sin^2 \omega t = \frac{1}{2}(1 - \cos 2\omega t)$$

$$P = \frac{V_0^2}{R}\left[\frac{1}{2}(1 - \cos 2\omega t)\right]$$

The average power over a full cycle is related to the average voltage over a full cycle. The average of $\cos 2\omega t$ is zero over a full cycle.

$$P_{avg} = \frac{V_0^2}{2R}$$

But

$$P_{avg} = \frac{V_{avg}^2}{R}$$

$$\frac{V_{avg}^2}{R} = \frac{V_0^2}{2R} \qquad V_{avg}^2 = \frac{1}{2}V_0^2$$

$$V_{avg} = \sqrt{\frac{1}{2}}V_0$$

$$V_{avg} = 0.707V_0$$

The average voltage that the bulb sees is $0.707V_0$, where V_0 is the zero-to-peak voltage amplitude of the sine wave. The sine wave in Figure 21.1 is the same picture that an oscilloscope would display if connected to the socket. Note that V_0 is 170 V, or the voltage from zero to peak. The peak-to-peak voltage is 340 V. The root-mean-square (rms) average voltage that the incandescent bulb sees is 120 V. If the bulb sees an rms voltage of 120 V, then the 100-W bulb has an rms current of 833 mA through it. The frequency of the ac voltage here in the United States is 60 cycles/second. Overseas it is 50 cycles/second.

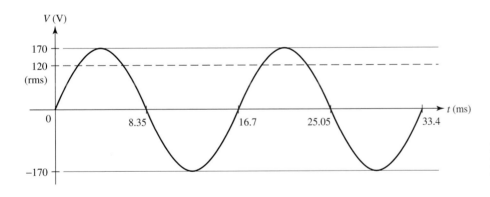

Figure 21.1 Oscilloscope Display of a Wall Socket ac Voltage

21.2 Phase Differences in ac Circuits

The voltage across a resistor and the current through it are governed by Ohm's law, $V = IR$. The current through a resistor is in phase with the voltage across it. Such is not the case for capacitors and inductors with an ac signal applied. For ac voltages across capacitors and inductors, there is either a phase lag of 90° or a phase lead of 90° between the voltage across the capacitor or the inductor and the current through it.

Let's look at the capacitor first (see Figure 21.2). With a phase lag or lead, the current through a capacitor or an inductor doesn't reach a maximum at the same instant

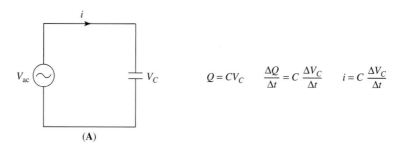

$$Q = CV_C \qquad \frac{\Delta Q}{\Delta t} = C\frac{\Delta V_C}{\Delta t} \qquad i = C\frac{\Delta V_C}{\Delta t}$$

(A)

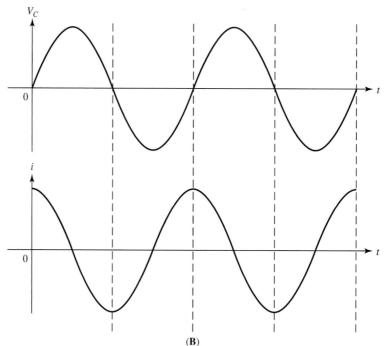

(B)

Figure 21.2 The Current Leads the Voltage by 90° in a Capacitor.

at which the voltage across the component does. The current through the capacitor is proportional to the rate of change of the voltage across the capacitor. Remember that the voltage source is a sinusoid and that these derivations are valid only for a sinusoid. In Chapter 9, we looked at a sine wave and plotted its rate of change with time directly below it. The rate of change of the sine curve at a given point was its tangent (or slope). When the sine's tangent was plotted directly below, we found that the new plot led the old plot by 90° and produced a cosine curve. The current through a capacitor is proportional to the rate of change of the voltage; therefore, the current leads the voltage by 90°. The current maximum occurs before the voltage maximum. You can see this concept by comparing the upper and lower graphs in Figure 21.2B. The current maximum occurs at $t = 0$. The first voltage maximum occurs a quarter-cycle, or 4.18 ms, later.

An inductor connected to an ac voltage source is shown in Figure 21.3A. The voltage appearing across the inductor is proportional to the rate of change of the current with

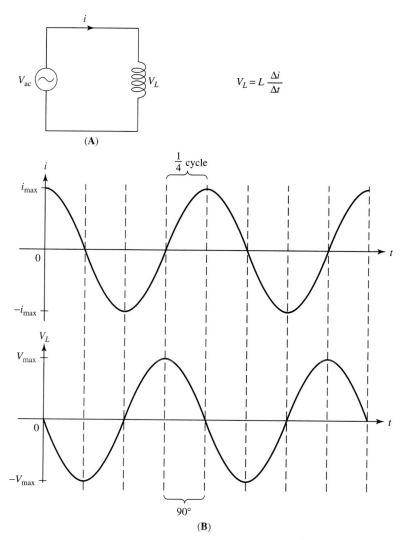

$$V_L = L \frac{\Delta i}{\Delta t}$$

Figure 21.3 The Current Lags the Voltage by 90° in an Inductor.

respect to time passing through the inductor. If the current is a cosine curve, the voltage will be a negative sine curve, according to what we learned in Chapter 9 about sine-cosine relationships. The current lags the voltage by 90° in an inductor. The first voltage maximum in Figure 21.3 occurs a quarter-cycle before the current maximum.

Now we connect a series *LC* circuit, as in Figure 21.4A. It's a series circuit, and the current is everywhere the same, so the alternating current can be used as the baseline (the middle curve in Figure 21.4B). The source current traces out a cosine curve with time. The voltage across the capacitor lags the current by 90°, and the voltage across the inductor leads the current by 90°. Thus, the voltage across the capacitor is 180° out of phase with the voltage across the inductor. If the inductor has a given voltage polarity and amplitude at a given instant, then the capacitor will have the opposite polarity as shown. The implications of this principle will become clear in Section 21.5 dealing with impedance.

Figure 21.4 The ac Voltages across the Capacitor and Inductor Are 180° Out of Phase.

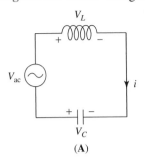

(A)

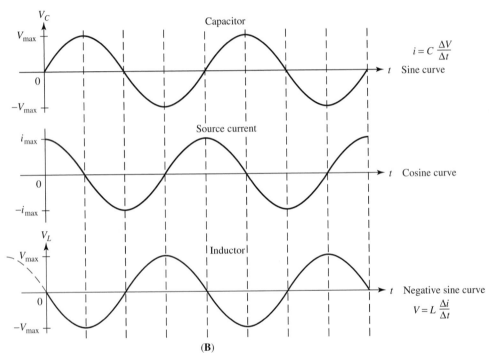

(B)

21.3   Inductive and Capacitive Reactance

The concept of reactance is built directly on the phase differences found in an ac circuit. Up to now, we've treated the frequency in an ac circuit as fixed. If the frequency is zero, it's dc, and for the ac power coming from the wall socket, it's 50 to 60 Hz depending on where you live. But if you ever have the chance to connect a signal generator to an oscilloscope and watch the display, you will find that just by turning a knob, you can vary the frequency over a wide range. So frequency can be a variable.

How do the three electronic components (resistor, inductor, and capacitor) react to a variable frequency? First of all, a resistor is a resistor is a resistor! It doesn't change its character with respect to frequency. However, an inductor and a capacitor do change character with respect to frequency. An inductor is a short circuit at zero frequency (dc) and an open circuit at an infinite frequency. In the dc circuit used to study the inductive time constant, we saw that when the current stopped changing, the inductor disappeared as far as the circuit was concerned. It became a short circuit. Initially when the current was changing at its maximum rate, the current through the circuit was zero, and the inductor acted like an open circuit. The voltage across the inductor is also directly proportional to the rate of change of the current. Thus, the voltage across the inductor is proportional to the frequency. With all of these facts in mind, the formula for the inductive reactance is

$$X_L = \omega L \qquad \omega = 2\pi f$$

$$X_L = 2\pi f L \qquad \text{(Equation 21.1)}$$

Inductive reactance has the same units as resistance: ohms. *Thus, **reactance** opposes current flow in much the same way as resistance.* There is an Ohm's law for reactance.

$$V_L = iX_L \qquad \text{(Equation 21.2)}$$

or
$$i = \frac{V_L}{X_L} \quad \text{or} \quad i = \frac{V_C}{X_C}$$

The greater the reactance, the smaller the current for a given voltage.

A capacitor, on the other hand, acts 180° out of phase with the inductor. Remember that in the *RC* time constant circuit studied earlier, the capacitor was initially a short circuit; when it became fully charged, current stopped flowing, and it acted like an open circuit. In Figure 20.2 we can also see that the ac current through the capacitor is directly proportional to the rate of change of voltage across the capacitor, or proportional to the frequency. Since current increases with frequency, the capacitive reactance should decrease with frequency.

Reactance: Resistance to current flow by an inductor or a capacitor in an ac circuit.

$$X_C = \frac{1}{\omega C}$$

$$X_C = \frac{1}{2\pi f C} \qquad \text{(Equation 21.3)}$$

At zero frequency the reactance is infinite (open circuit), and at an infinite frequency the capacitive reactance is zero (short circuit). For a resistor, Ohm's law is still true regardless of the frequency.

$$V = iR$$

21.4   Resonance

In Chapter 20, we saw that an *LC* circuit will oscillate with simple harmonic motion. The frequency of oscillation was shown to be

$$f = \frac{1}{2\pi\sqrt{LC}}$$

where *f* is the resonant frequency for a circuit with a capacitor and an inductor in it. At this point, the capacitive reactance equals the inductive reactance. Remember that the voltages across the inductor and the capacitor are 180° out of phase. If the reactances are equal, then V_C and V_L cancel each other, and the circuit is purely resistive.

Figure 21.5 displays a graph of the inductive and capacitive reactances on the vertical axis and the frequency on the horizontal axis. The inductive reactance X_L is linear

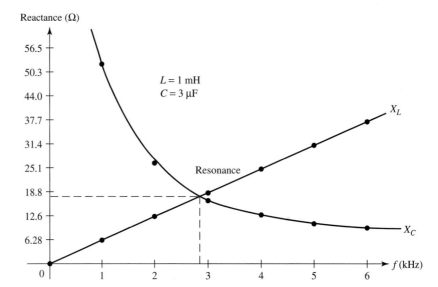

Figure 21.5 Inductive and Capacitive Reactances Plotted against Frequency

and goes through the origin. The capacitive reactance X_C is nonlinear since it goes as the inverse of the frequency and capacitance. The capacitive reactance approaches its limits asymptotically. As the frequency goes to zero, the capacitive reactance approaches infinity. As the frequency approaches infinity, the capacitive reactance approaches zero. But as the frequency is swept higher from zero, at some point the two curves intersect. Where they intersect is the resonance point for the circuit. At this point the capacitive reactance equals the inductive reactance. What is this frequency in terms of circuit parameters such as the inductance and reactance? Set the two expressions for reactance equal and solve for the frequency.

$$X_C = X_L \qquad \frac{1}{2\pi fC} = 2\pi fL$$

$$4\pi^2 f^2 LC = 1 \qquad f = \frac{1}{2\pi \sqrt{LC}}$$

This frequency is the same as that derived in Chapter 20 for the frequency of oscillation for the simple harmonic motion series LC circuit, but we derived it with a lot less trouble. For a capacitance of 3 μF and an inductance of 1 mH, this equation gives a frequency of about 2906 Hz, as shown in Figure 21.5.

21.5   Series Impedance

Assume that we have a series LRC circuit as in Figure 21.6A. For a given frequency, what current is drawn from the source?

To calculate the current, we must know the total impedance of the circuit. *The impedance is the* vector *sum of the reactances and resistances in the circuit.* The sum is a vector sum because of the phase differences among the three components. For a series

Impedance: The total opposition to current seen by an ac source.

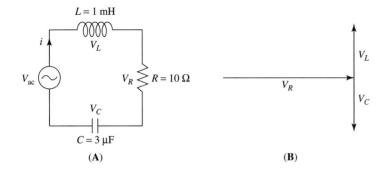

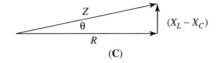

Figure 21.6 Series Impedance

circuit we can start by summing the voltage drops around the circuit. The voltage drops must equal the applied voltage. Figure 20.6B illustrates the relationship among the voltages across the components. Because of the phase differences around the circuit, the voltages are added according to the Pythagorean theorem.

$$\vec{V}_{ac} = \vec{V}_L + \vec{V}_R + \vec{V}_C \qquad V_{ac} = iZ$$

where Z is the symbol for total impedance.

$$V_R = iR \qquad V_L = iX_L$$

$$V_C = iX_C$$

where the current is the same throughout the circuit.

$$V_{ac}^2 = V_R^2 + (V_L - V_C)^2 \qquad V_{ac} = \sqrt{V_R^2 + (V_L - V_C)^2}$$

$$iZ = \sqrt{i^2R^2 + (iX_L - iX_C)^2}$$

$$Z = \sqrt{R^2 + (X_L - X_C)^2} \qquad \text{(Equation 21.4)}$$

where $X_L = 2\pi fL$ and $1/(2\pi fC)$

Because the voltages across the inductor and the capacitor are 180° out of phase, they are subtracted directly from one another. Whatever is left is 90° out of phase with the voltage across the resistor and must be added according to the Pythagorean theorem. Since the circuit is a series circuit, the current drops out on both sides, and we end up with a formula for the total impedance of the circuit. At resonance the inductive and capacitive reactances are equal, and all that is left is the resistance. The inductive and capacitive reactances completely cancel one another, and the inductor and the capacitor essentially drop out of the circuit! The reason is that the inductor and the capacitor juggle the total energy back and forth at resonance, as shown in Chapter 20.

The angle θ is the phase angle for the circuit. If the circuit is purely resistive, the voltage and the current being drawn from the source will be exactly in step (resonance or purely resistive), and θ will be zero. But if the difference between the reactances isn't zero, then there will be a phase difference, and the angle θ will be greater than zero between the source's voltage and current. (The voltage and the current will no longer be in step.) If the inductive reactance is greater than the capacitive reactance, then the current in the circuit will lag the voltage. (The circuit is overall inductive.) If the capacitive reactance is greater than the inductive reactance, then the current will lead the voltage. (The circuit is overall capacitive.) Only at resonance is the circuit wholly resistive.

The angle θ in Figure 21.6C is the phase angle for the source. Because of the reactance in the circuit, there will be a phase difference between the voltage and the current of the source. At resonance the phase difference disappears, and the circuit is totally resistive. If X_L is greater than X_C, then the circuit acts inductively, and the source voltage leads the source current. If X_C is greater than X_L, then the circuit is capacitive in nature, and the source voltage lags the source current. The angle θ is found from trigonometry.

(Equation 21.5)

$$\tan \theta = \frac{|X_L - X_C|}{R}$$

where $|X_L - X_C|$ is the positive magnitude of $X_L - X_C$.

EXAMPLE 21.1

(a) In the circuit of Figure 21.6A, what is X_L at a frequency of 5.0 kHz? (b) What is X_C at this frequency? (c) What is the circuit's impedance at this frequency? (d) What current is drawn from the source if the source's voltage is 14 V_{ac}? (e) What is the phase angle between the source's voltage and current? Does the voltage lead or lag the current? (f) What is the resonant frequency? (g) What is the current drawn from the source at resonance? What is the phase difference between the source's voltage and current?

SOLUTION
(a)

$$X_L = 2\pi f L \qquad X_L = 2\pi(5 \text{ kHz})(1 \text{ mH})$$

$$X_L = 31.4 \ \Omega$$

(b)

$$X_C = \frac{1}{2\pi f C} \qquad X_C = \frac{1}{2\pi(5 \text{ kHz})(3 \ \mu\text{F})}$$

$$X_C = 10.6 \ \Omega$$

(c)

$$Z = \sqrt{(10 \ \Omega)^2 + (31.4 \ \Omega - 10.6 \ \Omega)^2}$$

$$Z = 23.08 \ \Omega$$

(d)
$$i = \frac{V_{ac}}{Z} \qquad i = \frac{14\ V_{ac}}{23.08\ \Omega}$$

$$i = 0.607\ A$$

(e)
$$\tan \theta = \frac{31.4\ \Omega - 10.6\ \Omega}{10\ \Omega}$$

$$\theta = 64.3°$$

Since $X_L > X_C$, the circuit is inductive, and the voltage leads the current.

(f)
$$2\pi f L = \frac{1}{2\pi f C} \qquad 1 = 4\pi^2 f^2 LC$$

$$f = \frac{1}{2\pi \sqrt{LC}} \qquad f = \frac{1}{2\pi \sqrt{(1\ mH)(3\ \mu F)}}$$

$$f = 2906\ Hz$$

(g)
$$i = \frac{14\ V_{ac}}{10\ \Omega}$$

$$i = 1.4\ A$$

At resonance the circuit is resistive, and there is no phase difference between voltage and current.

21.6   Parallel Impedance

The resistor, capacitor, inductor, and ac voltage source from the series circuit of Figure 21.6 have been rearranged to form the parallel circuit of Figure 21.7A. In the series circuit we added the voltage drops around the circuit vectorially. In the parallel circuit the vector sum of the branch currents equals the current drawn from the source. The phase diagram of Figure 21.7B is drawn as the vector sum of the currents. To find the total circuit impedance that the source sees, we must again start with the phase diagram.

$$\vec{i} = \vec{i}_R + \vec{i}_L + \vec{i}_C$$

$$i^2 = i_R^2 + (i_L - i_C)^2$$

$$i = \frac{V}{Z} \qquad i_R = \frac{V}{R} \qquad i_L = \frac{V}{X_L} \qquad i_C = \frac{V}{X_C}$$

$$\frac{V^2}{Z^2} = \frac{V^2}{R^2} + \left(\frac{V}{X_L} - \frac{V}{X_C}\right)^2$$

$$\frac{V^2}{Z^2} = \frac{V^2}{R^2} + V^2\left(\frac{1}{X_L} - \frac{1}{X_C}\right)^2$$

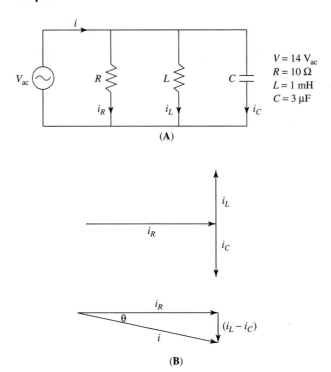

Figure 21.7 Parallel
Impedance

$$\frac{1}{Z^2} = \frac{1}{R^2} + \left(\frac{1}{X_L} - \frac{1}{X_C}\right)^2$$

(Equation 21.6)

$$\frac{1}{Z} = \left[\frac{1}{R^2} + \left(\frac{1}{X_L} - \frac{1}{X_C}\right)^2\right]^{1/2}$$

If the voltages across the inductive and capacitive legs of the circuit are the same, then i_L and i_C will be 180° out of phase. If the currents are 180° out of phase, one can be subtracted from the other, and whatever is left is added by the Pythagorean theorem to the current in the resistive leg. At resonance, $X_L = X_C$ and $Z = R$ as expected.

For a parallel circuit the current determines the circuit's nature. If $X_L > X_C$, then more current will flow through the capacitive leg. As a result, the circuit will be capacitive in nature, and the current will lead the voltage. If $X_C > X_L$, then more current will flow through the inductive leg, the circuit will be inductive in nature, and the current will lag the voltage. This action is opposite to what happens in a series circuit. The phase angle of the source is derived as follows:

$$\tan\theta = \frac{i_L - i_C}{i_R} \qquad \tan\theta = \frac{V/X_L - V/X_C}{V/R}$$

$$\tan\theta = \frac{1/X_L - 1/X_C}{1/R} \qquad \tan\theta = R\left(\frac{1}{X_L} - \frac{1}{X_C}\right)$$

(Equation 21.7)

$$\theta = \arctan\left[R\left(\frac{1}{X_L} - \frac{1}{X_C}\right)\right]$$

EXAMPLE 21.2

Refer to the circuit of Figure 21.7A. (a) What is X_L at 1.5 kHz? (b) What is X_C? (c) What is the circuit impedance that the source sees? (d) What current is drawn from the source? (e) What is the phase angle of the source? (f) Does the source's current lag or lead the voltage? (g) What is the resonant frequency of the circuit? (h) What current is drawn from the source at resonance? (i) Is the current at resonance a maximum or a minimum?

SOLUTION

(a) $$X_L = 2\pi f L \qquad X_L = 2\pi(1.5\ \text{kHz})(1\ \text{mH})$$

$$X_L = 9.4\ \Omega$$

(b) $$X_C = \frac{1}{2\pi f C} \qquad X_C = \frac{1}{2\pi(1.5\ \text{kHz})(3\ \mu\text{F})}$$

$$X_C = 35.4\ \Omega$$

(c) $$\frac{1}{Z} = \left[\frac{1}{R^2} + \left(\frac{1}{X_L} - \frac{1}{X_C}\right)^2\right]^{1/2} \qquad \frac{1}{Z} = \left[\frac{1}{(10\ \Omega)^2} + \left(\frac{1}{9.4\ \Omega} - \frac{1}{35.4\ \Omega}\right)^2\right]^{1/2}$$

$$Z = 7.9\ \Omega$$

(d) $$i = \frac{V}{Z} \qquad i = \frac{14\ \text{V}_{\text{ac}}}{7.9\ \Omega}$$

$$i = 1.8\ \text{A}$$

(e) $$\tan\theta = R\left(\frac{1}{X_L} - \frac{1}{X_C}\right) \qquad \tan\theta = (10\ \Omega)\left(\frac{1}{9.4\ \Omega} - \frac{1}{35.4\ \Omega}\right)$$

$$\theta = 38°$$

(f) The circuit is inductive because more current flows through the inductive leg than the capacitive leg. The current therefore lags the voltage.

(g) $$X_L = X_C$$

$$f = \frac{1}{2\pi\sqrt{LC}} \qquad f = \frac{1}{2\pi\sqrt{(1\ \text{mH})(3\ \mu\text{F})}}$$

$$f = 2906\ \text{Hz}$$

(h) $$i = \frac{V}{R} \qquad i = \frac{14\ \text{V}_{\text{ac}}}{10\ \Omega}$$

$$i = 1.4\ \text{A}$$

(i) Current is at a minimum. For a series circuit at resonance, the current is at a maximum.

21.7   ac Circuit Power

The apparent power in an ac circuit is the current drawn from the source times the voltage of the source, or $P_{app} = iV_{ac}$. If there is a phase difference between the voltage and the current, then the apparent power isn't the power dissipated by the circuit. Remember that inductors and capacitors are energy-storage devices. They store energy, and that stored energy is ready to be used whenever the circuit needs it. If a capacitor charges to a voltage V_{peak}, then when the source voltage falls below that peak voltage, the capacitor starts discharging and tries to keep the voltage at V_{peak}. If the circuit's current is rising, then as the current rises, energy is stored in the inductor in its magnetic field. When the circuit's current drops below that peak current, then the energy stored in the magnetic field is given up when the inductor tries to keep the current at its peak value. The apparent power is measured in volt-amperes (VA), with watts being reserved for the true power dissipated by the circuit. To find the real power dissipated by the circuit, we use the phase angle between the voltage and the current of the source, or we add the power being dissipated in all of the resistors in the circuit. *The cosine of the phase angle is called the* **power factor** *of the circuit.*

Power factor: The cosine of the phase angle.

In a series circuit, $Z \cos \theta$ is the resistance. (Look at the phase diagram for the series circuit in Figure 21.6C.) The real power dissipated by a series circuit is therefore $P_R = i^2 Z \cos \theta$ in terms of the circuit's impedance and current. Since $V_{rms} = iZ$, then $P_R = iV_{rms} \cos \theta$. This holds true for a parallel circuit as well.

Cos θ is the ratio of the resistance to the impedance (R/Z) in a series circuit, and the ratio of the current through the resistor divided by the total current for a parallel circuit. If $\cos \theta = 1$, then the circuit is wholly resistive and at resonance. Thus, the resistance equals the impedance for a series circuit, and the current through the resistor equals the total current for a parallel circuit. (The total circuit impedance for a parallel circuit is also equal to the resistance at resonance.)

EXAMPLE 21.3

Refer to the circuit of Figure 21.6A. (a) What is the apparent power at a frequency of 5 kHz? (b) What is the power dissipated by the resistor? (c) What is the power factor? (d) What is the power factor at resonance?

SOLUTION

(a)
$$P_{app} = iV_{ac} \qquad P_{app} = (0.607 \text{ A})(14 \text{ V})$$
$$P_{app} = 8.5 \text{ VA}$$

(b)
$$P_R = i^2 R \qquad P_R = (0.607 \text{ A})^2 (10 \text{ }\Omega)$$
$$P_R = 3.7 \text{ W}$$

(c)
$$PF = \frac{P_R}{P_{app}} \qquad PF = \frac{3.7 \text{ W}}{8.5 \text{ VA}}$$
$$PF = 0.435$$
$$PF = \cos \theta \qquad 0.435 = \cos \theta \qquad \theta = 64.2°$$

which was the phase angle found in Example 21.1.

(d)
$$PF = 1 \quad \text{at resonance}$$

EXAMPLE 21.4

Refer to the circuit of Figure 21.7A. (a) What is the apparent power at a frequency of 1.5 kHz? (b) What is the power dissipated by the resistor? (c) What is the power factor? (d) What is the power factor at resonance?

SOLUTION

(a)
$$P_{app} = iV_{ac} \qquad P_{app} = (1.8\ A)(14\ V_{ac})$$

$$P_{app} = 25.2\ VA$$

(b)
$$P_R = i_R^2 R \qquad i_R = \frac{14\ V_{ac}}{10\ \Omega}$$

$$i_R = 1.4\ A$$

$$P_R = (1.4\ A)^2(10\ \Omega)$$

$$P_R = 19.6\ W$$

(c)
$$PF = \frac{19.6\ W}{25.2\ VA} \qquad PF = 0.78$$

$$PF = \cos\theta \qquad 0.78 = \cos\theta$$

$$\theta = 38.9°$$

which was the phase angle calculated in Example 21.2.
(d) The circuit is wholly resistive.

$$\theta = 0° \qquad \cos 0° = 1$$

$$PF = 1$$

   Typical ac loads will have a net inductive or capacitive reactance. The net reactance is usually inductive because transformers and electric motors are inductive loads. Lighting loads are often inductive as well. Fluorescent lights have ballasts, which are step-up transformers to supply the high voltage necessary to ionize the gas inside the tube. Incandescent bulbs are also highly inductive because the filament glows more brightly and lasts longer if it is coiled instead of straight. So a net inductive load will have a power factor less than 1, or less than 100%. If the PF is 70%, then for every 700 W of power consumed, 1 kW must be generated. Of course, the extra 300 W is returned to the generator by the load, but the unused power circulating between the generator and the load causes additional resistive losses on the transmission lines, and the extra needed generating capacity is uneconomical. A capacitor is often connected to the power line in such a case to move the PF closer to 100% and so avoid the need for extra generating power and the additional resistive losses in the transmission lines.

21.8   Impedance Matching

If we connect an audio amplifier with an output impedance of 1 kΩ to speakers with an input impedance of 50 Ω, we want maximum power delivered to the speakers with a minimum of distortion. To deliver maximum power, the impedances of amplifier and speaker must be matched. With an ac circuit, mismatched impedances will distort the amplifier's output signal, so an impedance match is desired to minimize distortion as well. For ac circuits, transformers can be used to correct impedance mismatches.

In a transformer, the power delivered to the primary equals the power taken from the secondary.

$$P_p = P_s \qquad i_p V_p = i_s V_s \qquad \frac{V_p}{V_s} = \frac{i_s}{i_p}$$

The turns ratio N_p/N_s equals the ratio of primary voltage to secondary voltage.

$$\frac{N_p}{N_s} = \frac{V_p}{V_s} \qquad \frac{i_s}{i_p} = \frac{N_p}{N_s}$$

$$i_s = \frac{V_s}{Z_s} \qquad i_p = \frac{V_p}{Z_p}$$

$$\frac{V_s/Z_s}{V_p/Z_p} = \frac{N_p}{N_s} \qquad \frac{Z_p}{Z_s} = \frac{N_p}{N_s}\frac{V_p}{V_s}$$

$$\frac{Z_p}{Z_s} = \frac{N_p}{N_s}\frac{N_p}{N_s} \qquad \frac{Z_p}{Z_s} = \frac{N_p^2}{N_s^2}$$

$$\sqrt{\frac{Z_p}{Z_s}} = \frac{N_p}{N_s}$$

(Equation 21.8)

The key to impedance matching is to minimize reflected power. When we studied a wave pulse moving down a string, we noticed that if the pulse met a boundary or a different string with a different linear density, a reflected pulse was generated. The same is true of an electric circuit. Impedance matching doesn't eliminate the difference in impedances but makes the transition gradual instead of sudden, thus mostly eliminating reflected power. The characteristic of a transformer that makes it good for impedance matching is that the power into the primary equals the power out of the secondary.

21.9   Filters

When a radio signal is sent over the air waves, it's often sent as a slow signal modulating a much faster carrier signal at a discrete frequency. If it's a TV signal, it might be a frequency band. When you tune in a television channel, you actually select an electronic filter that passes the desired band and blocks the rest. If we received the entire frequency spectrum, the myriad of signals would interfere with one another, and all we would see or hear would be chaos. A fiber-optic cable multiplexes signals; that is, it simultaneously carries many signals at different frequencies. To disentangle the frequencies from one another, we need a filter.

Filters can be divided into three categories: low-pass, band-pass, and high-pass. A low-pass filter will pass low frequencies and block all of the higher frequencies. Figure 21.8 shows three low-pass filters. Figure 21.8A has an inductor in series with the load; $X_L = 2\pi fL$, and the inductor's reactance is low for low frequencies and high for high frequencies. In series with the load, the inductor passes low frequencies to the load and blocks high frequencies. At high frequencies most of the source voltage appears across the inductor, and very little of it appears across the load. Figure 21.8B is also a low-pass filter but with a capacitor in parallel with the load; $X_C = \frac{1}{2}\pi fC$, and at low frequencies the capacitor's reactance is very high. In a parallel circuit a capacitor with a high reactance acts like an open circuit, and the branch with the capacitor in it disappears. The signal is then across the load. However, the higher the frequency, the lower the reactance, and the capacitor acts like a short circuit, shorting out the load. High-frequency signals are therefore shorted around the load. Figure 21.8C is just Figures 21.8A and B combined; it gives a sharper cutoff than either a single inductor or a single capacitor.

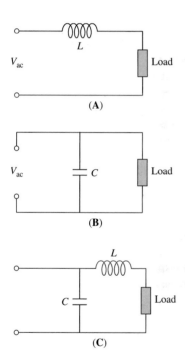

V_{ac} L Load

(A)

V_{ac} C Load

(B)

L C Load

(C)

Figure 21.8 Low-Pass Filters

A high-pass filter passes high frequencies and attenuates (decreases the amplitude of) low frequencies. It is constructed just like a low-pass filter except that the positions of the inductor and capacitor are reversed. In Figure 21.9A, the capacitor blocks low frequencies and passes high frequencies to the load. In Figure 21.9B, the inductor shorts low frequencies around the load and disappears at high frequencies (open circuit), allowing high frequencies across the load. Figure 21.9C both blocks and shorts low frequencies, preventing them from getting to the load. At high frequencies both the inductor and the capacitor virtually disappear (open and short circuit, respectively), passing the high frequencies on to the load.

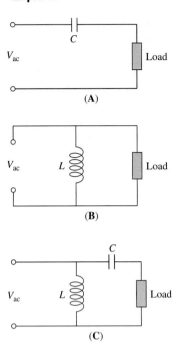

Figure 21.9 High-Pass Filters

A band-pass filter will pass a frequency band on to the load but will reject both the high and the low frequencies. Figure 21.10A has a series *LC* segment, and its impedance at the resonant frequency is zero. So the resonant frequency and a narrow band of frequencies to each side are passed to the load. But off resonance, the low frequencies are blocked by the capacitor, and the high frequencies by the inductor. In Figure 21.10B, a parallel *LC* element is in parallel with the load. At the resonant frequency, the impedance of the parallel *LC* element is virtually infinite, and the resonant frequency and the band of frequencies adjacent to it are passed to the load. Off resonance, the inductor shorts low frequencies around the load, and the capacitor shorts high frequencies around the load.

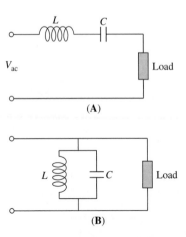

Figure 21.10 Band-Pass Filters

A band-reject filter will reject a narrow band of frequencies around resonance and will pass the rest. Figure 21.11A has a parallel *LC* element in series with the load. At resonance the *LC* element blocks the narrow resonant band. Off resonance the inductor passes the low frequencies, and the capacitor passes the high frequencies on to the load. Figure 21.11B has a series *LC* element in parallel with the load. At resonance the narrow resonant band is shorted around the load. Off resonance the inductor makes the series *LC* element an open circuit for high frequencies and passes them on to the load. The capacitor makes the series *LC* element an open circuit for low frequencies and passes them along to the load.

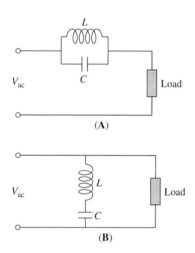

Figure 21.11 Band-Reject Filters

Summary

1. The **root-mean-square (rms)** current or voltage is the average value that is delivered by an alternating source.

$$V_{rms} = 0.707V_{max} \quad \text{and} \quad I_{rms} = 0.707I_{max}$$

Problem Solving Tip 21.1: Most exercises will deal with alternating currents and with rms values rather than maximum peak values.

2. The **reactance** of an inductor or a capacitor is a measure of the "resistance" that the circuit component has to the flow of current. Since capacitors will quickly reach maximum charge with slower current changes, their reactance is inversely proportional to the frequency of the source [see Equation (21.1)]. Inductors tend to resist changes in current but react much less vigorously against gradual changes. Thus, their reactance decreases as frequency decreases [see Equation (21.3)].

Problem Solving Tip 21.2: Ohm's law is still very valid when you're calculating the voltage drop across a capacitor or an inductor. Since reactance is measured in ohms (Ω), simply use the reactance of the component in question in place of

the resistor! Equation (21.2) is an example of this for an inductor. The same can be true for impedance, which is also measured in ohms!

3. The **impedance** of a circuit is the overall resistance that the current experiences in a circuit which may contain resistors, capacitors, and inductors. The impedance of a series circuit is found by using Equation (21.4), and that of a parallel circuit by using Equation (21.6).

4. Since the voltage across an inductor will lead the source current, and since the voltage across a capacitor will lag the source current, the *phase angle* is the measure of how much the resultant voltage of a circuit leads or lags the source current. The phase angle for a series circuit can be determined by using Equation (21.5). See Equation (21.7) for a parallel circuit.

Problem Solving Tip 21.3: Some students find it easier to use a phasor diagram (see Figures 21.6B and C) when solving for whether the circuit's resultant voltage leads or lags the current source. If the reactance of the inductor (X_L) is greater, voltage leads current. If the reactance of the capacitor (X_C) is greater, the opposite is true.

5. When the reactance of the inductors and capacitors of a circuit are equal, the phase angle becomes equal to zero, and the circuit is said to be in *resonance*. When resonance occurs, the impedance of the circuit becomes equal to its equivalent resistance.

6. The equivalent inductance of a group of inductors in series is found by adding them linearly.

$$L = L_1 + L_2 + L_3 + \cdots + L_n$$

The equivalent inductance of a group of inductors in parallel is found by adding them reciprocally.

$$\frac{1}{L} = \frac{1}{L_1} + \frac{1}{L_2} + \frac{1}{L_3} + \cdots + \frac{1}{L_n}$$

Problem Solving Tip 21.4: Note that this rule is the same one that resistors use. If you can remember one of them, you can remember both! Then it is relatively easy to remember that capacitors use the opposite convention.

Key Concepts

IMPEDANCE: The total opposition to current seen by an ac source.

POWER FACTOR: The cosine of the phase angle.

REACTANCE: Resistance to current flow by an inductor or a capacitor in an ac circuit.

ROOT-MEAN-SQUARE (RMS) VALUE: The equivalent dc voltage or current delivered to a load from an ac sinusoidal signal.

Important Equations

(21.1) $X_L = 2\pi f L$ (Inductive reactance)

(21.2) $V_L = iX_L$ (Ohm's law for reactance, both inductive and capacitive)

(21.3) $X_C = \dfrac{1}{2\pi f C}$ (Capacitive reactance)

(21.4) $Z = \sqrt{R^2 + (X_L - X_C)^2}$ (Impedance for a series circuit)

(21.5) $\tan \theta = \dfrac{X_L - X_C}{R}$ (Phase angle for a series circuit)

(21.6) $\dfrac{1}{Z} = \sqrt{\dfrac{1}{R^2} + \left(\dfrac{1}{X_L} - \dfrac{1}{X_C}\right)^2}^{1/2}$ (Impedance for a parallel circuit)

(21.7) $\theta = \arctan\left[R\left(\dfrac{1}{X_L} - \dfrac{1}{X_C}\right)\right]$ (Phase angle for a parallel circuit)

(21.8) $\sqrt{\dfrac{Z_p}{Z_s}} = \dfrac{N_p}{N_s}$ (Turns ratio of an impedance-matching transformer)

Conceptual Problems

1. In an ac circuit, the voltage:
   a. lags the current
   b. leads the current
   c. is in phase with the current
   d. can do/be any of the previous choices

2. The voltage and the current can't be in phase in a circuit that consists of:
   a. only a capacitor
   b. only a resistor
   c. an inductor and a capacitor
   d. an inductor, a capacitor, and a resistor

3. The voltage and the current are in phase in an ac circuit. Which of the following statements is true?
   a. The impedance is zero.
   b. The resistance is zero.

   c. The total reactance is zero.
   d. The phase angle is 90°.

4. The power dissipated by a circuit is a function of:
   a. the capacitive reactance
   b. the inductive reactance
   c. the impedance
   d. the resistance

5. A series circuit contains a resistor, an inductor, and a capacitor. As the frequency increases above the resonant frequency:
   a. the impedance decreases
   b. the impedance increases
   c. the inductance increases
   d. the capacitance decreases

6. A series circuit contains a resistor, an inductor, and a capacitor. At resonance:

a. the current is at a minimum

b. the current is zero

c. the current is at a maximum

d. the impedance is at a maximum

7. An ac circuit has a resistor, a capacitor, and an inductor in parallel with each other. As the frequency increases above resonance:

a. the impedance decreases

b. the impedance increases

c. the inductance increases

d. the capacitance decreases

8. A parallel circuit containing an inductor, a resistor, and a capacitor is at resonance. Which of the following statements is true?

a. The current is at a minimum.

b. The current is at a maximum.

c. The current is zero.

d. The phase angle is 90°.

9. With increasing frequency:

a. capacitive reactance increases

b. inductive reactance decreases

c. resistance remains the same

d. resistance goes to zero

10. What happens to the reactance of an air-core inductor when an iron core is slipped into it, and why?

11. What happens to the reactance of an air-gap, parallel-plate capacitor when a dielectric is slipped between the plates, and why?

12. You have an ammeter, a voltmeter, and an ac power supply that has a variable frequency with a constant amplitude. You also have a black box with two terminals that has a resistor, a capacitor, or an inductor inside. How can you tell what's inside the box without opening it?

13. Audio speakers are often built in two sizes called *woofers* and *tweeters*. A woofer has a large speaker cone and amplifies low frequencies well. A tweeter has a small speaker cone and can reproduce high-frequency sounds more faithfully. The crossover circuit shown in Figure 21.12 divides the output of an audio amplifier into low- and high-frequency components. Which output should be connected to the tweeter, and which to the woofer? Explain why.

Figure 21.12 (Problem 13)

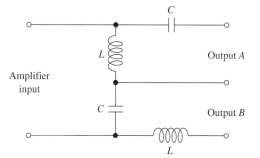

Exercises

rms Voltages and Currents

1. A 2-kΩ resistor is placed in series with an ac generator that produces an rms voltage of 200 V. (a) What is the rms current that passes through the resistor? (b) What is the peak current that passes through the resistor?

> *Answer:* (a) 100 mA
> (b) 141 mA

2. An ammeter in series with an ac circuit reads 20 A, and a voltmeter across the circuit reads 120 V. What are (a) the maximum current and (b) the maximum voltage in this circuit?

> *Answer:* (a) 170 V
> (b) 28.3 A

3. The potential difference across an ac power supply varies sinusoidally with a maximum voltage of 100 V. Find the value of the instantaneous potential difference at (a) one quarter-cycle, (b) one half-cycle, and (c) three-quarters of a full cycle.

> *Answer:* (a) +100 V
> (b) 0 V
> (c) −100 V

4. The voltage that comes out of a wall socket in the United States is 120 V$_{\text{rms}}$ at 60 Hz. What is the voltage at (a) 5 ms, (b) 10 ms, and (c) 15 ms? Assume that the voltage at time $t = 0$ is zero.

> *Answer:* (a) 161.7 V
> (b) −100 V
> (c) −100 V

Inductive and Capacitive Reactance

5. What is the reactance of a 50-mF capacitor at (a) 10 Hz and (b) 10 kHz?

> *Answer:* (a) 0.32 Ω
> (b) 31.8 mΩ

6. What is the reactance of a 50-mH inductor (a) at 10 Hz and (b) at 10 kHz?

> *Answer:* (a) 3.14 Ω
> (b) 3.14 kΩ

7. A 200-mH inductor and a 200-μF capacitor are connected to an ac power supply with a frequency of 60 Hz. What is the reactance of (a) the capacitor and (b) the inductor?

> *Answer:* (a) 13.3 Ω
> (b) 75.4 Ω

8. In order to determine the size of a capacitor, a technician places it in a circuit with a 200-V_{ac} power supply with a frequency of 5000 Hz. He finds that the reactance of the capacitor is 3 kΩ. What is the size of the capacitor?

> *Answer:* 10.6 pF

9. What is the current in a circuit that has a 1000-V_{ac} power supply at 60 Hz connected to (a) a 4-μF capacitor and (b) a 4-mH inductor?

> *Answer:* (a) 1.5 A
> (b) 667 A

10. Show that inductors in series add, just as in the formula for resistors in series using reactances.

$$L_{eq} = L_1 + L_2 + L_3 + \cdots + L_n$$

11. Show that inductors in parallel have the same mathematical formula as resistors in parallel using reactances.

$$\frac{1}{L_{eq}} = \frac{1}{L_1} + \frac{1}{L_2} + \frac{1}{L_3} + \cdots + \frac{1}{L_n}$$

Resonance

12. With what frequency would a circuit containing a 200-mH inductor and a 400-pF capacitor naturally resonate?

> *Answer:* 17.8 kHz

13. A simple radio can be tuned with a capacitor and an inductor in series. An engineer wants to build such a device to receive the transmissions of a radio station broadcasting at 94.7 MHz, and she must use a 0.08-mH inductor. What size capacitor would she need in order to tune in the radio to the desired frequency?

> *Answer:* 0.035 pF

14. In a series ac circuit having a 200-pF capacitor and an unknown inductor, the resonance frequency is found to be 40 kHz. What is the value of the inductor?

> *Answer:* 79 mH

Series Impedance

15. A series ac circuit has an inductive reactance of 20 Ω, a capacitive reactance of 50 Ω, and a resistance of 100 Ω. What is its impedance?

> *Answer:* 104 Ω

16. In Exercise 15, does the voltage lead the current, or does the current lead the voltage?

> *Answer:* The current leads the voltage.

17. A 1-kΩ resistor, a 1-μF capacitor, and a 1-H inductor are connected in series with a 100-V power supply at 60 Hz (see Figure 21.13). (a) What is the impedance of this circuit? (b) What is the phase angle? (c) Does the current lead or lag the source voltage?

> *Answer:* (a) 2.49 kΩ
> (b) 66.3°
> (c) The current leads the voltage.

*18. A capacitor, an inductor, and a resistor are connected in series to an ac power supply. A voltmeter is placed across each component and reads an rms voltage of 30 V, 40 V, and 40 V, respectively. What is the rms voltage of the source?

> *Answer:* 41.2 V

Figure 21.13 (Exercises 17, 23, and 24)

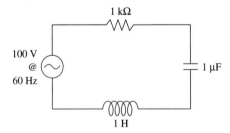

*19. A circuit contains an inductor and a resistor and is connected to a power supply with a variable frequency. At 100 Hz, the circuit's impedance is 50 Ω, and at 500 Hz, its impedance is 100 Ω. What are the values of (a) the inductor and (b) the resistor?

> *Answer:* (a) L = 28.1 mH
> (b) R = 46.8 Ω

Parallel Impedance

20. In Figure 21.14, a 100-Ω resistor, a 300-μF capacitor, and a 500-mH inductor are connected in parallel with each other and a 100-V, 60-Hz power supply. (a) What is the impedance of this circuit? (b) What is the phase angle? (c) Does the current lead or lag the source voltage?

Figure 21.14 (Exercise 20)

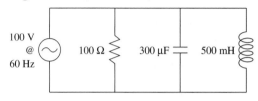

Answer: (a) 9.4 Ω
(b) 88.6°
(c) The current leads the voltage.

21. (a) What is the resonant frequency for the two circuits in Exercises 27 and 28? (b) What is the voltage across the inductor and the capacitor at resonance for the series circuit? (c) What is the current in the series circuit at resonance? (d) What is the current through the inductor and the capacitor at resonance for the parallel circuit? (e) What is the current in the parallel circuit at resonance?

Answer: (a) 91.9 Hz
(b) 294.5 V
(c) 2.55 A
(d) 1.04 A
(e) 2.55 A

ac Circuit Power

22. What is the average power dissipated by a circuit of a 100-V_{rms} power supply and two 400-Ω resistors in series?

Answer: 12.5 W

23. Using Figure 21.13 find (a) the power factor and (b) the power factor at resonance.

Answer: (a) 0.4
(b) 1

24. In Figure 21.13, what is the power dissipated by (a) the resistor, (b) the capacitor, and (c) the inductor?

Answer: (a) 1.6 W
(b) 0 W
(c) 0 W

25. The power dissipated in a series *RLC* circuit is 90 W, and the current is 755 mA. If the circuit is at resonance, determine the voltage of the generator.

Answer: 119.3 V

26. Determine the current drawn by a 1200-W hair dryer plugged into a 120-V_{ac} outlet.

Answer: 10 A

*27. A 15-μF capacitor, a 200-mH inductor, and a 47-Ω resistor are connected in series to a 120-V_{ac}, 60-Hz outlet. (a) What is the current in the circuit? (b) What is the apparent power? (c) What power does the circuit dissipate? (d) What is the phase angle of the circuit? (e) What is the voltage across each component?

Answer: (a) 1.07 A
(b) 128.4 VA
(c) 53.8 W
(d) 65.1°
(e) V_R = 53 V; V_L = 80.7 V; V_C = 189.2 V

*28. A 15-μF capacitor, a 200-mH inductor, and a 47-Ω resistor are connected in parallel across a 120-V_{ac}, 60-Hz outlet. (a) What is the current in the circuit? (b) What is the apparent power? (c) What power does the circuit dissipate? (d) What is the phase angle of the circuit? (e) What is the current through each component?

Answer: (a) 2.7 A
(b) 324 VA
(c) 306.4 W
(d) 19.7°
(e) i_R = 2.55 A; i_L = 1.6 A; i_C = 0.68 A

*29. An electric motor is connected to a 120-V_{ac} source at 60 Hz. It draws 5 A and dissipates 450 W of power. (a) What is the apparent power delivered to the circuit? (b) What is the power factor? (c) What size capacitor is needed in series with the load to bring the power factor to 1? (d) What is the apparent power delivered to the circuit with the capacitor in series?

Answer: (a) 600 VA
(b) 0.75
(c) 167 μF
(d) 450 VA

Impedance Matching

30. An electrician needs to connect an audio amplifier to a number of speakers. The amplifier has an impedance of 20 Ω, and the speakers have a net impedance of 50 Ω. A transformer is used to balance the impedances. What turns ratio is needed?

Answer: 0.63 : 1

Filters

31. A 10-mH inductor is in series with a 50-Ω load, forming a low-pass filter. A 20-mV signal is at the input to the filter. (a) What is the voltage across the load at 1.5 kHz? (b) What is the voltage across the load at 90 kHz?

Answer: (a) 9.4 mV
(b) 0.17 mV

32. A 20-nF capacitor is placed in series with a 50-Ω load, forming a simple high-pass filter. A 20-mV signal is at the input to the filter. (a) What is the voltage across the load at 1.5 kHz? (b) What is the voltage across the load at 90 kHz?

Answer: (a) 0.19 mV
(b) 8.6 mV

33. A 20-nF capacitor and a 10-mH inductor are placed in series with a 50-Ω load, forming a simple band-pass filter. A 20-mV signal is at the input to the filter. (a) What frequency is the filter designed to pass? (b) What is the voltage across the load at resonance? (c) What is the voltage across the load at 1.5 kHz? (d) What is the voltage across the load at 90 kHz?

Answer: (a) 11.25 kHz
(b) 20 mV
(c) 0.19 mV
(d) 0.18 mV

Optics

In this part (Chapters 22–24), we investigate the optical properties of light. It may seem that we have made a complete break with electromagnetism, but such is not the case. Astronomers used to study the heavens using only optical telescopes built with mirrors and lenses. But since World War II, the radio telescope has taken a place in astronomy equal to that of the optical telescope. Radio astronomers are also looking at the "light" flooding in from the heavens, but they are looking at a different part of the spectrum: the part with a longer wavelength. Visible light and radio waves are the same phenomenon, that is, traveling electromagnetic waves, but they have different wavelengths. Optics is thought of as different from electromagnetism because we deal not with electric and magnetic fields, but instead with wave phenomena and laws that require knowledge of geometry more than knowledge of electromagnetic forces. Since light *is* electromagnetic radiation, the study of optics is the study of a special branch of electromagnetism. The connection between optics and electromagnetism will be made clear in the opening sections of Chapter 22, "Light."

PART IV

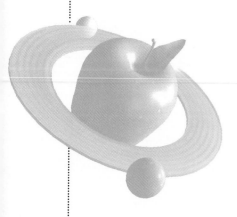

Visible light is electro-magnetic radiation, and here a beam of laser light is being reflected and refracted on an optical bench.

Light

At the end of this chapter, the reader should be able to:

1. Briefly explain how an electromagnetic wave travels through a medium.

2. Describe the differences between **reflection** and **refraction.**

3. Calculate the refracted angle that a light ray makes upon passing through the interface between two media.

4. Solve for the **critical angle** needed to achieve *total internal reflection* from one medium to another.

5. Calculate how far a known light source should be removed to cause a given light *intensity* to fall upon a detector.

22.1   Electromagnetic Waves

In our study of magnetism, we found that an emf is produced by a changing magnetic flux. Since the flux is the magnetic field times the area of the coil immersed in the magnetic field, the emf can be produced by a changing area or a changing magnetic field. An emf is an electromagnetic force, and perhaps this choice of words is unfortunate. We've used an emf in calculations just like a voltage, and a voltage is the potential energy per unit charge. But remember that when a voltage is placed across a wire, a constant electric field is produced inside the wire, causing the free charges to move which in turn gives rise to a current. So a changing magnetic field gives rise to an electric field.

There is a flip side to this coin: It shouldn't be too surprising to know that a changing electric field causes a magnetic field. This magnetic field caused by the changing electric field is usually quite small and could be ignored. The reason why the induced magnetic field is so much smaller than the electric field is that the ratio of the permeability of free space (electric fields) to the permittivity of free space (magnetic fields) is around 142,000 to 1! This ratio implies a correspondingly large ratio between the induced electric and magnetic fields, with the electric field being the larger.

When a capacitor is being charged by a battery through a resistor, positive charge flows from the battery to one plate of the capacitor. As positive charge begins to build up on the positive plate, the electrostatic repulsion on the positive charges on the other plate causes the positive charges to flow away toward the negative terminal of the battery, leaving a net negative charge on the negative plate. The current into the capacitor equals the current out, although no charge flows between the plates. While current is flowing, there is a magnetic field around the wire leading to and away from the capacitor. Despite the fact that the current is discontinuous between the plates of the capacitor, the magnetic field is not. The same magnetic field surrounds the capacitor. Something other than a current causes that magnetic field. Between the capacitor's plates the electric field is growing to a maximum and is therefore changing. The changing electric field between the capacitor's plates causes the changing magnetic field around the capacitor.

A capacitor is an energy-storage device and stores energy in the electric field between its plates. If there is a vacuum between the plates, then the electric field resides in the vacuum. The same can be said of a magnetic field. It can also reside in a vacuum. A solenoid stores energy in a magnetic field, and an air-core solenoid can also have a vacuum in its interior. Then its magnetic field resides in a vacuum. Neither the electric field nor the magnetic field requires a material medium to hold it. Can either field exist alone? We usually spoke of a constant magnetic or electric field residing in either a solenoid or a capacitor. For a capacitor the field between the plates was due to the charges on the plates, and the field was essentially confined to the space between the plates. For an inductor the magnetic field was due to the current in the windings, and the field was confined mostly to the interior of the solenoid. The capacitor and the solenoid allowed us to talk of the two fields as separate entities. If, however, we started a changing magnetic field, it would induce a changing electric field. This changing electric field would re-induce the changing magnetic field, and so on. The two changing fields can bootstrap each other and exist without regard to external charges or currents! Once started, these mutually induced oscillations would continue forever until something drew energy from them and the oscillations decreased. Mutually supporting electromagnetic oscillations can exist in a vacuum for an indefinite period of time.

A stationary electron has an electric field associated with it, and this field is pointing radially inward toward the electron. If the electron starts moving at a constant velocity, then there is a magnetic field encircling the electron's path where the electron happens to be. The electric and magnetic fields are everywhere at right angles. In electromagnetic oscillations in free space, the electric and magnetic fields are also at right angles to one another.

Energy is stored in both electric and magnetic fields. We've seen that a wave pulse transports energy from one point to another. It would seem likely that in free space electromagnetic oscillations would be either transverse or longitudinal traveling waves. If the oscillations took place in a cavity with metallic ends, standing waves could be set up. At the ends would be nodes for the electric field. A standing wave is created when pulses are reflected back down the original path of incidence, and constructive and destructive interference forms the standing wave pattern that seems frozen both in space and in time. All waves move with a characteristic wave speed. The wave speed should be a function of the properties of the "medium" through which it propagates. When we studied transverse waves on a string, we found that the wave speed equaled the square root of the tension in the string divided by the string's linear density. Both the tension and the density were properties of the string. The propagation velocity of an electromagnetic wave equals 1 over the square root of the permeability of free space (ε_o) times the permittivity of free space (μ_o). The two constants are "properties" of a vacuum.

$$c = \frac{1}{\sqrt{\varepsilon_o \mu_o}}$$

(Equation 22.1)

$$c = \frac{1}{\sqrt{(4\pi \times 10^{-7} \text{ Wb/A-m})(8.85 \times 10^{-12} \text{ C}^2/\text{N-m}^2)}}$$

$$c = 3 \times 10^8 \text{ m/s}$$

The wave type is a transverse wave. In three-dimensional space, if the amplitudes are spatially at 90° to one another in the E and B fields (the y and z axes, for example), then the x axis is used for propagation. Recall that for a transverse wave, the direction of propagation is 90° to the direction of its amplitude. Because of the 90° spatial relationship of the amplitudes, a longitudinal wave is ruled out, because the amplitude is in the direction of its propagation, such as a sound wave. Figure 22.1 is a sketch of an electromagnetic transverse wave.

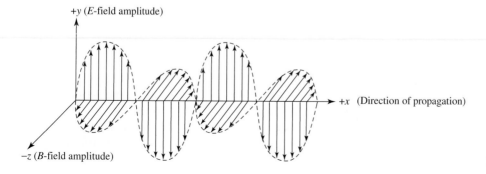

+y (E-field amplitude)

+x (Direction of propagation)

−z (B-field amplitude)

Figure 22.1 Propagation of an Electromagnetic Wave in Free Space

22.2   Generating Electromagnetic Waves

An accelerated charged particle emits electromagnetic radiation.

An electron at rest has an electric field associated with it. Electric lines of force terminate on the electron because of its negative charge. The electric field of a capacitor exists because of the positively and negatively charged plates. A capacitor is a static device (static in the sense that nothing is changing with time), and the charge distributions on the plates are immobile once the capacitor is charged. An electron in motion at a constant speed (such as in a constant current) produces a constant magnetic field. It is the electric current through the windings of an inductor that gives rise to the magnetic field in its interior. Unlike the capacitor, the inductor is a dynamic device because it requires a flow of charge to maintain the magnetic field. But what happens when a charged particle is accelerated? It emits electromagnetic waves. *Any accelerated charged particle will emit electromagnetic waves*. That is, it radiates.

We've looked at three types of acceleration in this book. The first type is a *linear* acceleration in which the accelerated particle changes the magnitude of its speed but not its direction. For example, the electron guns in TV picture tubes and computer monitors must be shielded because of the radio-frequency (RF) electromagnetic radiation generated by the straight-line acceleration of electrons. The second type of acceleration is *centripetal* acceleration in which the accelerated particle changes its direction but not its speed. Electromagnetic radiation generated in this manner is called *synchrotron* radiation. Particle accelerators found in high-energy physics generate synchrotron radiation, and it is usually an unwanted by-product of trying to accelerate charged particles to nearly light speeds. These accelerators are circular, and the resulting centripetal acceleration causes the charged particles to radiate energy away, slowing the particles. An electron or a proton spiraling in a magnetic field is subject to a centripetal acceleration. Simple harmonic motion (SHM) is also accelerated motion. In SHM, the acceleration is sinusoidal, and the magnitude of the acceleration is constantly changing with time. This type of acceleration is most often used to generate electromagnetic waves.

A pebble thrown into a still pond will produce concentric circular wavefronts that move radially out from the center. On the water's surface can be seen wave crests (maximum amplitude) and troughs (minimum amplitude). If we look at any given point through which the wave is moving, we see that the water oscillates up and down sinusoidally with time. From the temporal sinusoidal up-and-down oscillations, we can calculate the period and thus the frequency of the disturbance generating the wave. By measuring the spatial distance from one maximum to another (or from one trough to another), we can measure the wavelength. Using the wave equation ($v = f\lambda$) we can also find the wave speed.

Charges undergoing SHM can also act like a pebble being thrown into a pond. Assume that we have both a positively charged particle and a negatively charged particle. They undergo SHM vertically into and out of the page (see Figure 22.2) and form an electromagnetic disturbance in space much like a pebble being tossed into a still pond. If the charged particles undergo sinusoidal oscillations, then the disturbance propagated through space will also undergo sinusoidal oscillations. The electric field's maximum value will form a wave crest, and its minimum value will form a trough. The magnetic field of the electromagnetic oscillation is at right angles to the electric field. The direction of propagation is everywhere radially outward. For the pebble in the pond, the wavefront travels on the surface of the water and forms a circle. But the wavefront for the electromagnetic oscillations forms the surface of a sphere because the electromagnetic radiation produced radiates throughout all space. The speed of propagation is 3×10^8 m/s as calculated in Section 22.1.

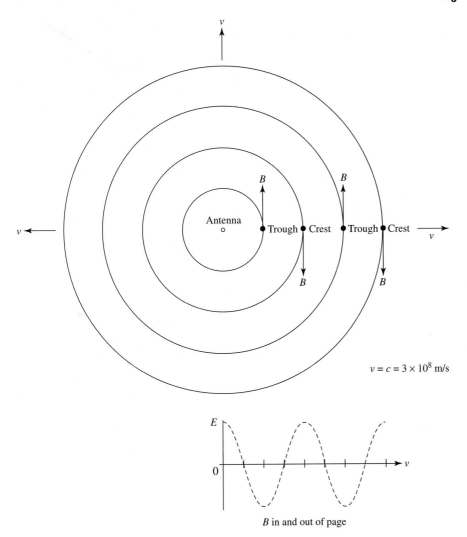

$v = c = 3 \times 10^8$ m/s

Figure 22.2 Antenna Broadcasting Electromagnetic Waves

In a wire the charged particles free to move are negatively charged electrons, and getting them to move with SHM is relatively easy. A sinusoidal voltage is placed across a wire, and the electrons in it will oscillate sinusoidally, producing electromagnetic radiation. This radiating wire is often called an *antenna*. The wave equation applies for electromagnetic radiation, and knowing the frequency, we can easily calculate the wavelength.

These electromagnetic oscillations are picked up by receivers in radio sets, cell phones, and TV tuners. When radio was new, information was sent in the form of Morse code, or a series of dots and dashes. A dot was a short pulse, and a dash a long pulse. The sinusoidal wave that carried the information was turned on at the beginning of a pulse and turned off at the end of a pulse. Later the information was put on the carrier wave by varying the amplitude of the carrier in a process called *amplitude modulation* (AM), as shown in Figure 22.3A. The carrier usually has a much higher frequency than the modulating information.

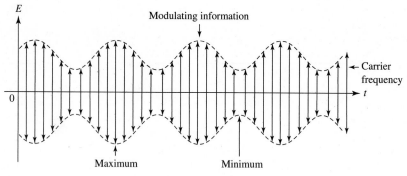

Modulating information

Carrier frequency

Maximum   Minimum

(A) Amplitude modulation

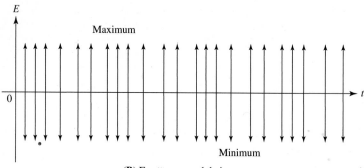

Maximum

Minimum

(B) Frequency modulation

Figure 22.3 AM- and FM-Modulated Carrier Signals

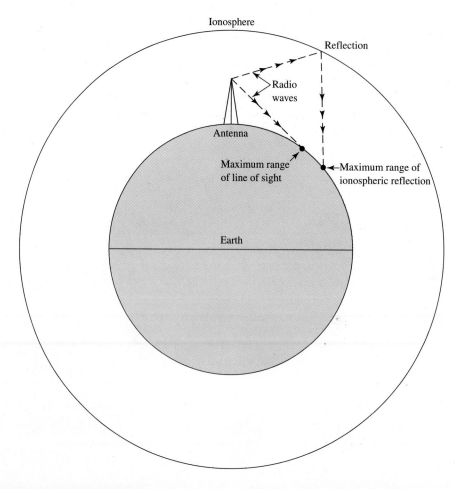

Ionosphere

Reflection

Radio waves

Antenna

Maximum range of line of sight

Maximum range of ionospheric reflection

Earth

Figure 22.4 Radio Waves Reflecting off the Ionosphere

Another form of modulation is *frequency modulation* (FM), as shown in Figure 22.3B. The center frequency is the carrier's base frequency, and the frequency is modulated by the information so that the carrier's frequency will be a little smaller than the base value for a minimum and a little greater for a maximum. If the information being carried is intended for radio, then the modulation frequencies will be in the audio range (1500 to 20,000 Hz). Since radio signals travel in straight lines, it is advantageous to have the transmitting antenna as tall as possible. The direction of propagation of the wavefront in Figure 22.2 is radially outward, or in a straight line outward from the center of the sphere. The signal won't travel through the earth, so the signal is cut off at the horizon, as shown in Figure 22.4. Theoretically, if an antenna were built tall enough, it could cover only half the earth's surface. But the upper regions of the atmosphere are ionized by the sun's radiation, and radio waves can reflect off the ionosphere just as light reflects off a mirror's surface. Thus, the range of an antenna is extended.

22.3   The Electromagnetic Spectrum

Electromagnetic waves propagating through space obey the wave equation.

$$\lambda f = v$$

The velocity of propagation ($c = 3 \times 10^8$ m/s) is fixed. The frequency can vary over a very wide range. We already know from Section 22.2 that radio waves are electromagnetic waves. A typical broadcast frequency for short-wave radio is around 5×10^6 Hz. From the wave equation, that frequency would give a wavelength of 60 m! The 3-meter band of short-wave radio has a broadcast frequency of 100 MHz. The radio-frequency spectrum extends to about 1×10^{12} Hz and includes long-wave (communicating with submerged submarines), short-wave, microwave (microwave ovens), radar, and the millimeter band (see Figure 22.5). X rays are high-energy radiation and are medically useful because of their penetrating ability; they span 5×10^{16} Hz to roughly 5×10^{19} Hz. Following X rays are gamma rays which usually have an extraterrestrial source and have extremely high energies.

As shown in Figure 22.5, microwaves and radar share pretty much the same frequency band; a microwave oven was originally called a *radar range!* A radar set is both a transmitter and a receiver. A short pulse of energy is transmitted in the radar frequency range, and this pulse strikes an object such as an aircraft, a building, or a cloud. The reflected pulse is then received by the radar set, and the short time delay between transmitted and received pulse gives range to the object. The direction in which the antenna is pointed (radar antennas are usually parabolic or dish-shaped) gives the direction to the object. Water efficiently absorbs energy in the microwave frequency band, making weather radar and microwave ovens possible. Food usually has a high water content. When food is placed inside a microwave oven, the microwave radiation is absorbed by the water in the food, and its energy is raised, thereby heating it. When food is cooked in a microwave, it is actually steam-heated from the inside. A bare ceramic dish placed inside a microwave won't get hot because its water content is practically zero. Placing a metallic object inside a microwave oven isn't a good idea because metal objects are excellent reflectors of microwave radiation and can overload the klystron transmitting tube inside the oven and burn it out. A small microwave

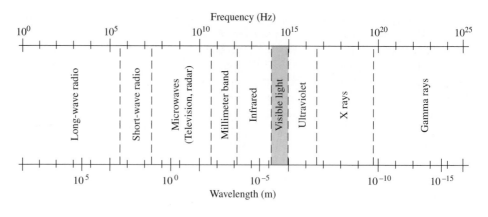

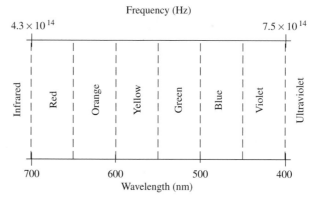

Figure 22.5 Electromagnetic Spectrum

oven can generate 500 W of power, and a 100-W incandescent light bulb is far too hot to touch. Reflecting that much power back into the vitals of the oven will certainly wreak havoc somewhere!

The band of the electromagnetic spectrum that we're most concerned with in this chapter, though, is the frequency range from roughly 4.3×10^{14} Hz to 7.5×10^{14} Hz. This is visible light. Visible light is electromagnetic radiation. Except for the frequency, it is exactly the same as radio waves, X rays, and gamma rays. Optics and electronic communications seem to be two entirely separate disciplines, but they are really one and the same thing. It's just that in optics we traditionally manipulate the narrow band of the electromagnetic spectrum known as *light* with pieces of silvered glass (mirrors) and curved glass lenses instead of electronics. A radio telescope used in astronomy and astrophysics is a telescope just like an optical scope, except that it's receiving its information from a different part of the electromagnetic spectrum instead of the narrow band that comprises visible light.

The visible spectrum consists of somewhat less than 4% of the width of the electromagnetic spectrum shown in Figure 22.5. Our eyes are sensitive to and can see only the wavelengths from about 700 nm (700×10^{-9} m), which is red light, to about 400 nm (400×10^{-9} m), which is violet light. The visible spectrum is bounded on both sides by electromagnetic radiation that can't be seen but is nonetheless there. Infrared can be felt but not seen. The feel of the hot sun on exposed skin on a sunny day is infrared radiation. X rays can't be seen directly, but they can expose a photographic plate and can therefore be seen indirectly.

Radio waves are the same as visible light waves, except with a different wavelength. Astronomers study the sky with optical telescopes as well as radio telescopes pictured here.

White light consists of light that contains all of the wavelengths of the visible spectrum. The sun emits white light, as do incandescent light bulbs. Most sources of illumination are white light sources. If white light is passed through a prism, the multicolored white light spectrum from red to violet results. White light is broken up into its component wavelengths. When white light falls on an object, that object preferentially absorbs and reflects wavelengths in the visible spectrum. A white object, such as the page forming the background to this writing, reflects all wavelengths, and our eyes see white. A black object, such as a letter in the printing on this page, absorbs all wavelengths, and no light is reflected. We see colors because of preferential absorption and reflection of light by an object. The green in grass and leaves is perceived as green because the chlorophyll in green plant matter strongly absorbs the wavelengths around green and reflects wavelengths in the green region of the visible spectrum. A red traffic light lens is red when light passes through it because it passes wavelengths in the red part of the visible light spectrum and absorbs light in the remaining parts of the visible spectrum. If our eyes intercept predominantly red wavelengths coming from an object, we perceive that object as having a red color. Color is the result of a filtered white light spectrum in cases where the illuminating light is white.

EXAMPLE 22.1

A radar set records a time of 20 μs between the transmitted pulse and its return echo. How far away is the object on the radar screen?

SOLUTION

Since the time recorded is for the pulse traveling both to and from the object, dividing the recorded time in half will give the time for a one-way trip.

$$s = \frac{1}{2}ct \qquad s = \frac{1}{2}(3 \times 10^8 \text{ m/s})(20 \times 10^{-6} \text{ s})$$

$$s = 3 \text{ km}$$

EXAMPLE 22.2

A helium-neon laser emits light at 632.8 nm. (a) What color is the light? (b) What is the frequency at which the electromagnetic wave oscillates?

SOLUTION

(a) The light is red.

(b)
$$\lambda f = c \qquad f = \frac{c}{\lambda}$$

$$f = \frac{3 \times 10^8 \text{ m/s}}{632.8 \text{ nm}}$$

$$f = 4.74 \times 10^{14} \text{ Hz}$$

22.4   Huygens's Principle

Light is an electromagnetic wave, and fortunately most optical phenomena can be understood on the basis of wave nature alone without referring to its electromagnetic character. About 300 years ago, the Dutch physicist Christian Huygens devised an approach to the behavior of waves using wavefronts.

Figure 22.2 shows a slice of a spherically propagating wavefront. A similar situation is depicted in Figure 22.6. A line connecting all of the same-order nodes forms a circle around the source. This circle is a wavefront. Successive nodes will form successive wavefronts. A wavefront can also be constructed using a line connecting the same-order maxima, or antinodes, as in Figure 22.2. A wavefront is a surface that joins all of the points where the waves have the same phase (order and magnitude). These wavefronts can be seen when a pebble is dropped into a calm pool. The wavefronts are the circular wave crests (or troughs) that radiate outward.

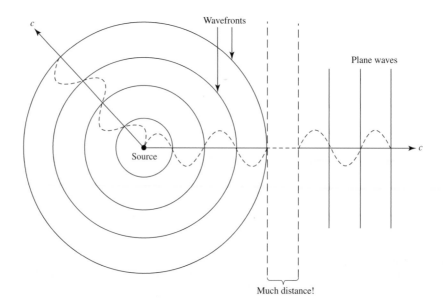

Figure 22.6 Circular Wavefronts Become Plane Waves with Distance.

As a given wavefront moves farther and farther away from the source, its wavefront radius increases. This wavefront radius is like an index for the curvature of the wavefront. The greater the radius, the less the curvature. Wavefronts near the source are highly curved, but wavefronts far away have very little apparent curvature. If the wavefronts are an infinite distance away from the source, they form plane waves as shown in the far right of Figure 22.6. Plane waves are a useful approximation for wavefronts at distances appreciably large from the source.

Huygens's principle states the following: *Every point on a wavefront can be considered as a source of secondary wavelets that spread out in all directions with the wave speed of the medium. The wavefront at any time is the envelope of these wavelets.*

Figure 22.7 shows how Huygens's principle is applied to the motion of a wave through a uniform medium. The primary source is at the center of the concentric wavefronts. The wave spherically propagates from the source. If we choose a wavefront and freeze its motion, any point on that wavefront becomes the source for another spherically propagating wave. These multiple sources (there are an infinite number of them on any given wavefront) create the wavefront of the next instance, interfering in such a way that destructive interference cancels any amplitude inside the circular wavefront and leaves only the infinitesimally short arc length a distance r from the secondary source and at a point on a line joining the primary and secondary sources.

Huygens's principle: Every point on a wavefront can be considered as a source of secondary wavelets that spread out in all directions with the wave speed of the medium.

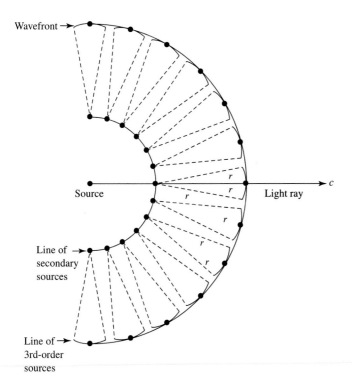

Figure 22.7 Huygens's Principle

*The line joining the primary, secondary, tertiary, and so on, sources is called a light ray. A **light ray** is perpendicular to the wavefronts and points in the direction of propagation.* Light acts in many cases as if it moves in straight lines, and a light ray moves in a straight line radially outward from the primary source.

A light ray is also a consequence of the wave nature of light and Huygens's principle. Although light rays don't explain all optical phenomena, they are extremely useful

Light ray: The straight-line path that light takes radially outward from the primary source.

in describing such optical properties as reflection and refraction. Light rays can be used to model the straight-line motion of light. However, remember that a light ray is a useful theoretical abstraction for explaining certain phenomena, but the underlying reality is the wave nature of light.

22.5 Reflection

Just about every object reflects some of the light that falls upon it. If an object has any color other than black, some of the light is reflected. A white object reflects all of the light that hits it. Both a white board used in classrooms with felt-tip markers (instead of the traditional blackboard and chalk) and a plane mirror are flat, and both reflect all of the visible light that hits them. However, you can see an image of yourself in a plane mirror, but when you look into a white board, all you see is white! What is the difference between them?

Specular reflection: Reflection of light that forms an image.

*Reflection that forms an image is called **specular reflection,** and reflection that doesn't is called* diffused reflection. With diffused reflection, the reflected light rays are reflected randomly in all directions, but with specular reflection they are all reflected in one direction. Specular reflection occurs when a surface is so smooth that any irregularities in it are small relative to the wavelength of the light falling on it. Relative to visible light, a mirror's surface is extremely smooth, whereas a white board's surface is extremely rough. If visible light has a wavelength range of 400 to 700 nm, then the vertical surface irregularities are less than 400 nm to reflect the visible spectrum like a mirror. Radio waves have wavelengths on the order of a million times longer than visible light waves. Thus, a surface needn't be nearly as smooth as a mirror's surface to reflect a radio wave so that it forms an image. For example, a section of chain-link fence can reflect radio waves in the meter region of the spectrum every bit as effectively as a silvered plane mirror can reflect visible light. If you've ever looked at the dish of a parabolic radar receiver or of a radio telescope, you've probably noticed that the surfaces are not smooth and silvery to touch and see, and they might even be formed from lattices with empty space forming part of the surface just as in a chain-link fence. They are literally, however, mirrors to the radio wavelengths concerned and reflect them specularly so that they will form an image or come to a focus.

Earlier it was stated that radio waves are reflected by the ionosphere. Only radio waves in the medium- to high-frequency bands are reflected. Low-frequency (LF) signals are absorbed by the ionosphere, but very-high-frequency (VHF) and ultra-high frequency (UHF) signals pass through the ionosphere. The ionosphere is a layer of positively charged ions in the upper atmosphere, and the charged particles there interact with electromagnetic waves and form a reflecting surface for a band of electromagnetic wavelengths.

Absorbed	{LF	30–300 kHz	
Reflected	{MF	0.3–3.0 MHz	
	{HF	3–30 MHz	
Pass through	{VHF	30–300 MHz	
ionosphere	{UHF	0.3–30 GHz	

The ionosphere in the upper atmosphere is caused by sunlight of nearly ultraviolet and above wavelengths being absorbed by the gaseous molecules in the upper atmosphere. The absorbed energy is enough to drive an electron from the molecule's outer electron cloud and leave it with a net positive charge. Trapped charged particles from the solar wind also contribute to the ionosphere. The sun radiates a "solar wind" as well as visible light. The solar wind is made up of charged particles from the sun's atmosphere flung out into space by interactions with the sun's magnetic field. These charged particles impinging upon the upper atmosphere become part of the ionosphere, and at times of unusually high solar activity (with many sunspots), they can disrupt radio communications around the globe. Near the earth's poles, the ionosphere can actually be seen! The magnetic field lines of the earth's magnetic field are concentrated near the poles. Moving charged particles will spiral around magnetic field lines, and the charged-particle density near the poles is much higher than at the equator. Ionized matter glows with visible light. (A candle flame is a rising column of ionized gases. The light originates when a free electron recombines with the positively charged ion and emits a photon of visible light in the process.) The glowing ionosphere near the poles constitutes the aurora borealis, a transcendental sight if ever there was one!

Figure 22.8 shows the specular reflection of a series of plane wavefronts from the surface of a plane mirror. Plane waves are incident on the mirror from the left. We're interested in tracking plane wave segment AB from its initial contact with the mirror at point A until the top of segment AB hits the mirror at point C. The time it takes is t, and the top of the segment travels a distance ct, where c is the speed of light. When segment AB first touches the mirror, point A becomes a source according to Huygens's principle, and by the time the top of segment AB touches the mirror at point C, the wavefront that started at A has traveled to D. The ray AD points in the direction of the reflected wavefronts. All of the points between A and C where the incident plane wave touches the mirror become successive sources that help define the reflected wavefront CD at time t.

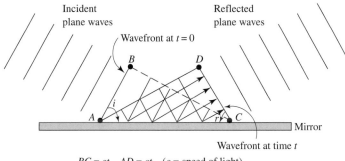

$BC = ct \quad AD = ct \quad (c = \text{speed of light})$

Figure 22.8 Plane Waves Incident on a Mirror

Figure 22.8 was drawn using Huygens's principle, and from it we can determine the relationship between the incident and reflected waveforms using trigonometry. ABC forms a right triangle with the right angle at point B. The angle at which the incident waves hit the mirror is i, and the angle at which the reflected waves leave the mirror is r.

$$\sin i = \frac{BC}{AC} \qquad \sin r = \frac{AD}{AC}$$

$$BC = ct \qquad AD = ct$$

$$\sin i = \frac{ct}{AC} \qquad \sin r = \frac{ct}{AC}$$

$$\sin i = \sin r$$

The angle of incidence equals the angle of reflection for specular reflection relative to the normal line to the surface.

From Figure 22.8 we can deduce that the angle of incidence should equal the angle of reflection. Experimental evidence tells us that this is so.

A mirror forms an image, and using the technique of ray tracing and the fact that the angle of incidence equals the angle of reflection, we can investigate how that image is formed. But first we must discuss the technique of ray tracing. The angle of incidence and the angle of reflection are defined relative to a line normal (or perpendicular) to the surface. In Figure 22.8, *BC* is a light ray associated with the incident light, and *AD* is a ray associated with the reflected light. The rays are pointing either in the direction of propagation of the incident wavefronts or in the direction of the reflected plane wavefronts. If a line is drawn perpendicular to the mirror of Figure 22.8 through the point where *BC* and *AD* intersect, we can show that the angle of incidence relative to this normal equals the angle of reflection. From now on, we will use straight light rays to graphically demonstrate optical properties.

Figure 22.9 shows three light rays radiating from the object on the left. It's as if the object is a point source at the top of the object arrow, and this point source radiates radially with an infinite number of rays pointing equally in all directions. Such radiation is called *isotropic* because it is the same in all directions. A point source can be approximated by the filament in an incandescent flashlight bulb. If a naked flashlight bulb were placed in the middle of a room and switched on, the filament would act like a point source, and the bulb would radiate isotropically and equally illuminate all parts of the room.

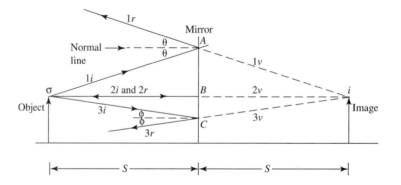

Figure 22.9 Ray Tracing Incident and Reflected Rays at a Mirror's Surface

However, in Figure 22.9, only three rays are chosen from the infinite number of rays. To show the location of image formation, we need a minimum of two rays because the image point is an intersection of rays. The third ray is used to make the image point unambiguous. In Figure 22.9, the first ray travels from the object and is incident on the mirror at point *A* (ray 1*i*). The reflected ray (1*r*) is reflected from the mirror at an angle θ to the normal, which is also the incident angle. A mirror's image appears as if it is behind the mirror, yet the light is reflected from the mirror and never penetrates behind it. Our eyes are fooled into thinking that there is something behind the mirror. It's an optical illusion. When we see an image in a mirror, what our eyes intercept are the reflected rays. Our eyes trace these reflected rays back behind the mirror. If the reflected rays are traced back behind the mirror, they will pass through the image point. These rays, called *virtual rays* because they don't really exist, are represented by dotted lines in the figure. Ray 1*v* is the virtual ray traced back from 1*r*. Ray 2 hits the mirror at an angle of 0° and returns back along itself. Its virtual ray intersects with ray 1*v* at the image point. Ray 3 is incident on the mirror at point C at an angle ϕ to the normal and is reflected on the other side of the normal line also at an angle ϕ. Tracing the reflected ray $3r$ back behind the mirror gives the virtual ray $3v$, and this ray intersects with rays $1v$ and $2v$ at the image point.

Examining the image, we notice that it is erect and the same size as the object. We also notice that it is the same distance behind the mirror that the object is in front of the mirror. When we look at a full-size image of ourselves in a mirror, we notice these same characteristics in our image. The only characteristic not yet mentioned is that if you're left-handed, the mirror image is then right-handed. This left-right reversal is very obvious when you try to read a printed page in a mirror. Confusing? See Figure 22.10!

Figure 22.10 Image Reversal in a Mirror

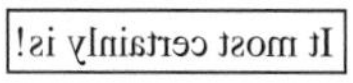

It most certainly is!

22.6 Refraction and Snell's Law

Light travels at different speeds in different media. In a medium such as glass, light interacts with the electrons around each atom. The electrons begin to resonate at the light's frequency, and thus propagation is from electron to electron in a material medium. Because of this intermediary, the speed of propagation is slower than in a vacuum. *The result is **refraction**; that is, light bends from a straight-line path when it hits an interface between two media, as in Figure 22.11.* In medium 1, light is faster than in medium 2 $(c_1 > c_2)$. In a time t, ray BC travels a distance $c_1 t$, and ray AD travels a distance $c_2 t$. Ray AD is shorter than ray BC because light has a lower speed in medium 2. Wherever an incident plane wavefront touches the interface, it becomes a source for a wavefront in medium 2. The incident plane wavefront AB generates the refracted plane wavefront CD in the time interval from 0 to t. The plane wavefront is perpendicular to the ray pointing in the direction of propagation. This fact gives us two right triangles: The one in the faster medium 1 is ABC, and the one in the slower medium 2 is ADC. They both share side AC, the interface, in common. From Figure 22.11, we can derive the relationship between the incident and refracted wavefronts.

Refraction: The bending of a light ray that takes place at an interface when it enters a medium with a different index of refraction.

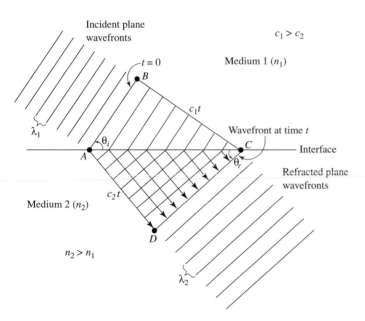

Figure 22.11 Refraction of Light

$$\sin \theta_i = \frac{c_1 t}{AC} \qquad \sin \theta_r = \frac{c_2 t}{AC}$$

$$AC = \frac{c_1 t}{\sin \theta_i} \qquad AC = \frac{c_2 t}{\sin \theta_r}$$

$$AC = AC$$

$$\frac{c_1 t}{\sin \theta_i} = \frac{c_2 t}{\sin \theta_r} \qquad \frac{c_1}{\sin \theta_i} = \frac{c_2}{\sin \theta_r}$$

$$\frac{\sin \theta_i}{\sin \theta_r} = \frac{c_i}{c_r}$$

(Equation 22.2)

Equation (22.2) is known as *Snell's law* after the Dutch astronomer of the seventeenth century who first discovered it.

Figure 22.11 and the subsequent derivation of Snell's law were based upon Huygens's principle, but again we will use ray tracing to work with the optical property of refraction. If Ray AD were slid over to the right so that it started where ray BC ended, and if the incident and refracted angles were defined relative to the normal line as in Figure 22.12, we would still obtain the same relationship known as *Snell's law*.

$$(90° - \theta_i) + \phi_i = 90° \qquad \phi_i = \theta_i$$
$$(90° - \theta_r) + \phi_r = 90° \qquad \phi_r = \theta_r$$

The incident and reflected angles in Snell's law will always be defined from the normal line where the incident ray hits the surface of the medium into which it is refracted.

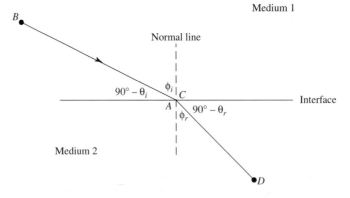

Figure 22.12 Refraction at an Interface Represented Using Ray Tracing

Index of refraction: The ratio of the speed of light in a vacuum to its speed in a given medium.

*The **index of refraction** of a medium is the ratio of the speed of light in a vacuum (c) to its speed in that particular medium (c_m).* Since the speed of light in a vacuum is the fastest, the index of refraction (n) will be greater than or equal to 1. It is a pure number with no units, and if you get an answer to a problem where the index of refraction is less than 1, you undoubtedly have a wrong answer! Indexes of refraction (n) for common substances are given in Table 22.1. As the index of refraction gets larger, the light is bent more and more toward the normal line in the refracting medium. Diamond has the highest refractive index in the list, with zircon (used in jewelry as a pseudo-diamond) not even a close second.

Table 22.1 Indexes of Refraction for Some Common Substances

Substance	n
Air	1.0003
Benzene	1.5
Diamond	2.42
Ethyl alcohol	1.36
Glass	1.55
Ice	1.31
Lucite and plexiglass	1.51
Quartz	1.46
Water	1.33
Zircon	1.92

The wavelength, frequency, and propagation velocity of a wave are governed by the wave equation in both media. The frequency remains the same in both media, but if the propagation velocity changes, so must the wavelength.

EXAMPLE 22.3

A beam of visible light of wavelength 500 nm in air enters a piece of lead glass with index of refraction of 1.6. The beam's incident angle is 20°. (a) What is the angle of refraction? (b) What is the light's speed inside the glass? (c) What is the light's wavelength inside the glass?

SOLUTION

(a)
$$\frac{\sin \theta_i}{\sin \theta_r} = \frac{c_i}{c_r} \qquad \frac{\sin \theta_r}{\sin \theta_i} = \frac{c_r}{c_i} \qquad \sin \theta_r = \frac{c_r}{c_i} \sin \theta_i$$

$$n_{air} = \frac{c}{c_i} \qquad c_i = \frac{c}{n_{air}}$$

$$n_g = \frac{c}{c_r} \qquad c_r = \frac{c}{n_g}$$

$$\sin \theta_r = \frac{c/n_g}{c/n_{air}} \sin \theta_i \qquad \sin \theta_r = \frac{n_{air}}{n_g} \sin \theta_i$$

$$\sin \theta_r = \frac{1}{1.6} \sin 20°$$

$$\theta_r = 12.3°$$

(b)
$$n_g = \frac{c}{c_r} \qquad c_r = \frac{c}{n_g}$$

$$c_r = \frac{3 \times 10^8 \text{ m/s}}{1.6}$$

$$c_r = 1.875 \times 10^8 \text{ m/s}$$

(c)
$$\lambda_g f = c_r \qquad \lambda_g = \frac{c_r}{f}$$

$$f = \frac{c}{\lambda_a} \qquad f = \frac{3 \times 10^8 \text{ m/s}}{500 \text{ nm}} \qquad f = 6 \times 10^{14} \text{ Hz}$$

$$\lambda_g = \frac{1.875 \times 10^8 \text{ m/s}}{6 \times 10^{14} \text{ Hz}}$$

$$\lambda_g = 312.5 \text{ nm}$$

Another form of Snell's law that might be easier to remember is

(Equation 22.3)
$$n_i \sin \theta_i = n_r \sin \theta_r$$

where the subscripts i and r stand for "incident" and "refracted."

22.7   The White Light Spectrum

The index of refraction of a medium isn't constant but depends on the frequency of the incident light. The variation in the index of refraction is very small, and for that reason, we can treat it as a constant as in Table 22.1. Nevertheless, the variation is real and is responsible for that beautiful natural phenomenon, the rainbow. This variable index of refraction based upon frequency is called *dispersion*. For the visible light spectrum, red light will be refracted (bent) less than violet because red light has the smaller frequency. If a thin beam of white light falls upon a prism, when it emerges from the other side, it will have fanned out because of the dispersion. If the light falls upon a white screen, the least refracted light (red) will be at one end of the spectrum, and the most refracted (violet) will be at the other end. In between will be all of the remaining colors of the white light spectrum: orange, yellow, green, and blue.

Since white light can be split up into six primary colors, when these six colors are mixed with one another, the color of the mix should be white. We see the result of such color mixing whenever we watch a TV screen. If you take a magnifying glass and hold it near the surface of a color TV picture tube, you'll notice that the surface consists of many identical elements. Each element is composed of a red, blue, and green phosphor dot. When an electron from the electron gun in the tube hits one of the phosphor dots, it fluoresces for a short time, emitting red, green, or blue light depending on which dot has been hit. By preferentially combining the colors red, green, and blue, all colors of the white light spectrum can be reproduced, including white (all three dots fluoresce) or black (none of the dots in an element are hit by an electron).

You see a rainbow when you are facing falling rain with the sun behind you. Since raindrops are roughly spherical, a rainbow is formed by dispersion as in Figure 22.13. The white light sunlight ray is first refracted as it enters the drop; then it is reflected from the drop's rear surface. The incident ray's refraction initially separates the white light into its constituent spectral colors. The subsequent reflection causes the violet and red rays to cross one another and emerge as shown in Figure 22.13. Between the red and violet rays are the other colors of the spectrum.

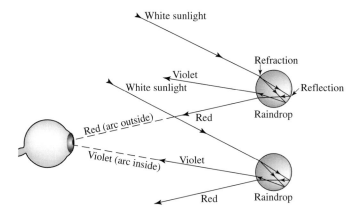

Figure 22.13 Rainbow Formation

In a rain shower there are an almost infinite number of raindrops, all refracting and reflecting the sunlight in the same way. As a result, when the dispersed light leaves the drops, it forms a circle with red on the outside, violet on the inside, and the remaining colors sandwiched in between. Our eyes trace back the reflected rays to perceive the arc of the rainbow. If the violet and red reflected rays are traced back behind the same spherical droplet, then violet will always be below red. We see only an arc (the rainbow's characteristic arc) of this circle if we're standing on the ground, but if we were at a high enough altitude in an aircraft, we would be able to see the entire circle.

If light enters the water droplet at a large enough angle, it can undergo two reflections inside the water droplet. It will form an outer secondary rainbow with the colors in reverse order. The secondary rainbow is fainter than the primary because at each reflection inside the drop, some of the light is lost refracting outside the drop. There is a dark band between the rainbows since no light enters the eye from refraction from the area between the rainbows because of the geometry of the sun, the rain shower, and placement of the observer.

The sparkle of a diamond is caused not only by its high index of refraction but also by the way it is cut. Usually a diamond is cut in such a way that light is reflected twice before leaving it. The optical path length inside the diamond is therefore greater than if the light emerged after only one reflection. A wider separation between the colors of the white light spectrum results.

22.8 Apparent Depth

Refraction is responsible for another phenomenon familiar to anyone who has swum or fished in clear water: An object below the surface of the water appears to be less deep than it really is. If you reach directly for an object on the bottom of a pool, you will generally overshoot the object with your hand if your head is above the water. The reason that the object's apparent depth is less than its real depth is the refraction of the light rays leaving the object as the rays emerge from the water into the air.

As light leaves a medium with a large index of refraction into a medium with a smaller index of refraction, the light bends away from the normal line, as shown in Figure 22.14. Your eye follows the light ray from the object back into the water as if it were a straight line (the dotted virtual ray), and your eye sees the object at a lesser depth than it really is. For this reason, we perceive an oar with only its lower part in the water as being bent.

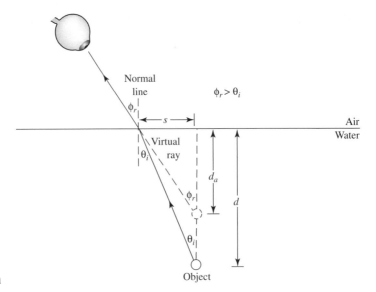

Figure 22.14 Apparent Depth

Spear fishermen must learn to compensate for the refraction, or they'll catch nothing! Only when you are directly above the object in the water is the apparent depth the actual depth, and only then will reaching directly for the object allow you to grasp it.

What is the relationship between apparent depth and actual depth? From Figure 22.14 and some trigonometry, we can derive the mathematical relationship.

$$\tan \phi_r = \frac{s}{d_a} \qquad\qquad \tan \theta_i = \frac{s}{d}$$

$$s = d_a \tan \phi_r \qquad\qquad s = d \tan \theta_i$$

$$s = s \qquad\qquad d_a \tan \phi_r = d \tan \theta_i$$

$$\tan \phi_r \approx \sin \phi_r \qquad\qquad \tan \theta_i \approx \sin \theta_i$$

The above relationship between sine and tangent is true for a small angle not far from the vertical. Try it on your calculator and see. The angles must be in radians, though! By using this relationship, we can use Snell's law.

$$d_a \sin \phi_r = d \sin \theta_i \qquad \frac{d_a}{d} = \frac{\sin \theta_i}{\sin \phi_r}$$

By Snell's law,

$$\frac{\sin \theta_i}{\sin \phi_r} = \frac{n_r}{n_i} \qquad \frac{d_a}{d} = \frac{n_r}{n_i}$$

(Equation 22.4)

$$d_a = \left(\frac{n_r}{n_i}\right) d$$

In the above calculation, the incident ray is from the object, so that n_i is the index of refraction of water and n_r is the index for air. Since $n_r < n_i$, the refracted light ray bends further *away* from the normal line. This small angle approximation is useful as long as the eye is at a small angle from the vertical relative to the object viewed in the other medium. This principle is true under most viewing circumstances. In Section 22.9 on total internal reflection, we will explain another reason that this small angle approximation is made.

EXAMPLE 22.4

A quartz crystal lying at the bottom of a clear creek is found to be $2\frac{1}{2}$ ft below the surface when a yardstick is lowered vertically into the water and touches it. What is the apparent depth of the crystal to someone viewing it from a small angle from above?

SOLUTION

$$d_a = \left(\frac{n_r}{n_i}\right)d \qquad d_a = \left(\frac{1}{1.33}\right)(2.5 \text{ ft})$$

$$d_a = 1.88 \text{ ft}$$

The quartz crystal in Example 22.4 appears a lot closer to the surface than it really is. With the small angle approximation, the apparent depth won't change with viewing angle. With large angles, if the viewing angle is changed, so is the apparent depth.

22.9   Total Internal Reflection

Information, if not transmitted as a radio wave, is usually carried by a copper wire. A single copper wire can carry many messages simultaneously in a process called *multiplexing*. These simultaneous messages are carried on a number of discrete frequencies as electrical signals that are selectively filtered at the receiving end. The propagation speed is near that of light. However, there are limits to the number of discrete frequencies that can be carried by a copper wire, and that maximum number has been reached. A fiber-optic cable, however, can carry much more multiplexed information than a copper wire and on a thinner wire. A fiber-optic cable is a glass cable, and the information carried on it is carried as light.

A fiber-optic cable is actually a light pipe, and light travels down the cable by successive total internal reflections, as shown in Figure 22.15. When a beam of light hits an interface between two media, some of the light is reflected, and some is transmitted and refracted. In the spherical raindrops that disperse light to form a rainbow, at each internal reflection some of the light is lost. That is why the secondary rainbow (if you're lucky enough to see one) is so much fainter than the primary one. But with total internal reflection, *all* of the light is reflected, with none being lost by transmission through the interface.

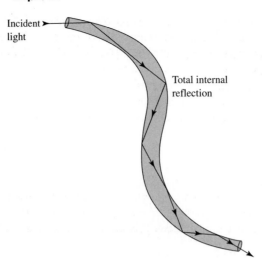

Figure 22.15 Light Propagating in a Fiber-Optic Cable by Total Internal Reflection

Critical angle: The incident angle at which a light ray traveling from a medium with a high index of refraction into a medium with a low index of refraction is refracted at 90°.

    How does total internal reflection come about? Assume that there is a point source of light at the bottom of a pool of water, as shown in Figure 22.16. The light emits rays in all directions, and in the figure, five rays have been singled out. Ray 1 makes an angle of 0° with the normal and travels from the water into the air without bending at the interface. Ray 2 hits the interface at a shallow angle to the normal and is refracted a little as it leaves the surface of the water. Ray 3 hits the interface at a greater angle and is bent further away from the normal. The greater the angle at which the incident ray hits the interface, the farther away from the normal the refracted ray is bent. Eventually the refracted ray is bent at an angle of 90° and doesn't leave the water but travels along the surface as ray 4 does. This angle is called the ***critical angle.*** Any ray that hits the interface at an angle greater than the critical angle will be specularly reflected back into the water and won't be transmitted out of the surface. The formerly transparent surface has become a reflecting one.

    As mentioned earlier in Section 22.7 dealing with the white light spectrum, a glass prism is often used to break white light up into its spectral colors. A prism is also used

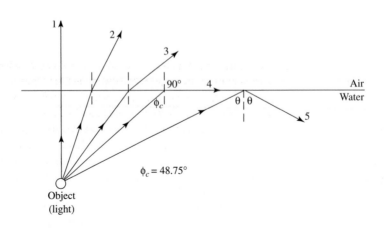

Figure 22.16 Critical Angle and Total Internal Reflection

to manipulate an image in optical instrumentation. Total internal reflection better preserves the brightness and sharpness of an image than does a mirror. Figure 22.17 shows some common image manipulations using a prism.

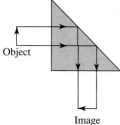

Image
(A) 90° direction change

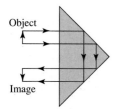

(B) 180° direction change and inversion

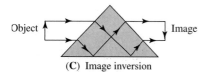

(C) Image inversion

Figure 22.17 Image Manipulation by a Prism

EXAMPLE 22.5

A small pea light is embedded 2 in. below the top surface of a glass block. (a) What is the critical angle for glass ($n = 1.55$)? (b) A circle of light is seen on the top of the block when the light is lit. The light's rays inside the block form a cone. What is the angular width of the cone inside of which all of the rays will leave the surface? (c) What is the diameter of the circle of light on the surface?

SOLUTION
(a)
$$n_i \sin \theta_i = n_r \sin \theta_r \qquad \theta_r = 90°$$

$$n_i = 1.55 \qquad n_r = 1.0$$

$$\sin \theta_i = \frac{n_r}{n_i} \sin \theta_r$$

$$\sin \theta_i = \frac{1}{1.55} \sin 90°$$

$$\theta_i = 40.18°$$

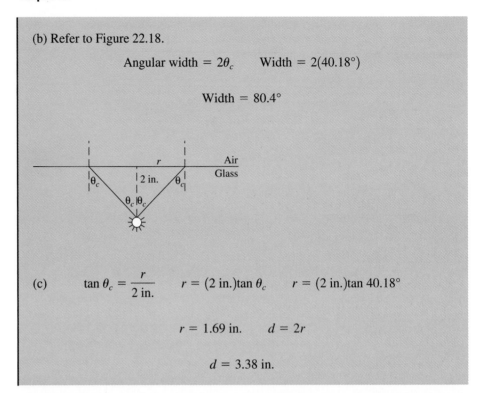

(b) Refer to Figure 22.18.

$$\text{Angular width} = 2\theta_c \qquad \text{Width} = 2(40.18°)$$

$$\text{Width} = 80.4°$$

Figure 22.18 Formation of a Light Cone

(c) $\quad \tan \theta_c = \dfrac{r}{2 \text{ in.}} \qquad r = (2 \text{ in.})\tan \theta_c \qquad r = (2 \text{ in.})\tan 40.18°$

$$r = 1.69 \text{ in.} \qquad d = 2r$$

$$d = 3.38 \text{ in.}$$

22.10 Intensity and Flux

A 100-W light bulb consumes energy at a rate of 100 J/s. The power goes into heating the tungsten filament in the bulb until it glows incandescently. The filament is a resistor, and most of the power applied is given off as heat. Only about 10% of the applied power is converted into light. A fluorescent tube is much more efficient and can emit the same amount of light for a lot less power consumed. The power that an electric light consumes is therefore not a good indication of the illumination that a light provides. What, then, is?

A photocell can give a direct reading of the light intensity falling on a surface. It can give the reading in terms of power, or wattage. *Intensity* is the power per unit area, so knowing the area of the photocell and the conversion efficiency, you can calculate the intensity at the position of the photocell. If a light bulb is placed at the center of a hollow sphere, and if the bulb radiates isotropically, then the sphere will intercept all of the light emitted by the bulb. The light intercepted by the entire inside surface of the sphere is the entire light output of the bulb. The light power emitted by the bulb can be considered a flux because the power flows outward into the surrounding space with spherical symmetry. The intensity of a light source drops off rapidly as you move away from it: It drops off by a factor of $1/r^2$. The inside surface of the sphere intercepting the light equals $4\pi r^2$. The luminous flux emitted by a given light is usually constant, so the flux intercepted by the sphere's inner surface is constant, regardless of the radius of the sphere. But the surface area goes as $4\pi r^2$, so the flux per unit area must drop as $1/r^2$. The $1/r^2$ factor arises because of the spherical symmetry of isotropic radiation. We've already seen this factor twice: in the universal law of gravitation and in the electrical force between two charges. The flux argument isn't new either, since we've used the idea of magnetic flux in induction, and electrical flux is also inherent in Gauss's law.

Intensity is defined as the amount of power per square meter falling on a surface. For an isotropically radiating source, then, the intensity is

$$I = \frac{P}{A}$$

(Equation 22.5)

or
$$I = \frac{\text{total flux}}{4\pi r^2}$$

The units of intensity are watts per square meter (W/m^2). The intensity can be calculated for any distance from the source using Equation (22.5). However, not all sources radiate isotropically, such as a spotlight or a flashlight. A spotlight or a flashlight projects a cone of light, and the area of a cone's radial surface is part of the surface of a sphere with the same curvature as the entire sphere at a radius r. Such areas can be characterized by a solid angle (see Figure 22.19). A solid angle is an angle in three dimensions. The area on the surface of a sphere is $A = 4\pi r^2$. For a complete sphere, the solid angle is 4π. The symbol for a solid angle is Ω (omega). Therefore, the surface area of a sphere is given by $A = \Omega r^2$. If the area on the sphere is less than the entire sphere, then Ω is less than 4π. A solid angle is defined by the equation above. Rearranging, we get

$$\Omega = \frac{A}{r^2}$$

(Equation 22.6)

The area doesn't have to have a circular border on the sphere like that of a cone. It can be "rectangular" as in Figure 22.19. The unit for a solid angle is called the *steradian,* and like the radian or degree it is a dimensionless number.

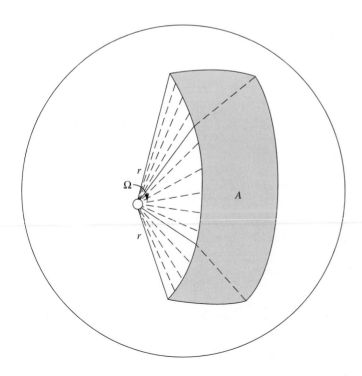

Figure 22.19 Solid Angle

EXAMPLE **22.6**

(a) What is the luminous flux from a 100-W light bulb? (b) What is the light intensity at a distance $\frac{1}{2}$ m from the bulb if it radiates isotropically? (c) How much light power is intercepted by a solar cell at this distance, 2.0 cm on a side? (d) If the conversion efficiency is 25%, how much electrical power is produced by the cell? (e) What is the solid angle subtended by this area relative to the incandescent light bulb?

SOLUTION

(a)
$$F = (0.1)(100 \text{ W}) \qquad F = 10 \text{ W}$$

(b)
$$I = \frac{P}{A} \qquad A = 4\pi r^2$$

$$I = \frac{P}{4\pi r^2} \qquad I = \frac{10 \text{ W}}{4\pi(0.5 \text{ m})^2}$$

$$I = 3.18 \text{ W/m}^2$$

(c)
$$P = IA \qquad P = (3.18 \text{ W/m}^2)(0.02 \text{ m})^2$$

$$P = 1.27 \text{ mW}$$

(d)
$$P = (0.25)(1.27 \text{ mW}) \qquad P = 0.32 \text{ mW}$$

(e)
$$\Omega = \frac{A}{r^2} \qquad \Omega = \frac{(0.02 \text{ m})^2}{(0.5 \text{ m})^2}$$

$$\Omega = 1.6 \times 10^{-3} \text{ steradian}$$

EXAMPLE **22.7**

At the surface of the earth on a clear day, a solar panel 2 m on a side with a 25% conversion efficiency generates 1 kW of electrical power. (a) What is the intensity of the sunlight striking the earth's surface? (b) If 28.6% of the incident sunlight is absorbed by the atmosphere, what is the sunlight's intensity at the top of the atmosphere? (c) If the earth's average orbital radius is 1.5×10^{11} m, what is the solid angle that the solar panel subtends relative to the sun? (d) What is the total luminous flux of the sun?

SOLUTION

(a)
$$P_{\text{out}} = 0.25 P_{\text{in}} \qquad P_{\text{in}} = \frac{P_{\text{out}}}{0.25}$$

$$P_{\text{in}} = \frac{1 \text{ kW}}{0.25} \qquad P_{\text{in}} = 4 \text{ kW}$$

$$P = IA \qquad I = \frac{P}{A}$$

$$I = \frac{4 \text{ kW}}{(2 \text{ m})^2} \qquad I = 1 \text{ kW/m}^2$$

(b) $$100\% - 28.6\% = 71.4\%$$

$$I = \frac{1 \text{ kW/m}^2}{0.714} \qquad I = 1.4 \text{ kW/m}^2$$

(c) $$\Omega = \frac{A}{r^2}$$

$$\Omega = \frac{4 \text{ m}^2}{(1.5 \times 10^{11} \text{ m})^2} \qquad \Omega = 1.78 \times 10^{-22} \text{ steradian}$$

(d) $$P = 4\pi r^2 I \qquad P = 4\pi (1.5 \times 10^{11} \text{ m})^2 (1.4 \text{ kW/m}^2)$$

$$P = 3.96 \times 10^{23} \text{ kW!!}$$

or

$$P = \frac{4\pi}{\Omega} A (1.4 \text{ kW/m}^2) \qquad P = \frac{4\pi}{1.78 \times 10^{-22} \text{ steradian}} (4 \text{ m}^2)(1.4 \text{ kW/m}^2)$$

$$P = 3.96 \times 10^{23} \text{ kW}$$

Summary

1. Light, an electromagnetic wave, is composed of self-propagating electric and magnetic fields. In a vacuum, light travels at a velocity of 3×10^8 m/s. In other media, the speed of light is given by

$$c_{\text{new}} = \frac{c_{\text{vacuum}}}{n}$$

where n is the **index of refraction** of the medium.

2. When light strikes the interface between two media, it can be *reflected, refracted,* or partially reflected *and* refracted.

3. The *reflected angle* of a light ray will be equal to its *angle of incidence.* That is, a light ray will reflect with the same angle that it had when it originally struck the surface.

4. A light ray that refracts through an interface will follow *Snell's law* [Equations (22.2) and (22.3)].

Problem Solving Tip 22.1: Sometimes, more than one application of Snell's law [Equation (22.3) is often more useful] will be required. As a general rule, try to use Snell's law for every interface through which a light ray passes. This procedure often will give you enough equations to solve for the answer in the fewest number of steps.

5. When looking from a medium with a lower index of refraction into a medium with a higher index of refraction, you must take into account the refraction of light when determining the position of objects. An object in the medium with the higher index will appear much closer to the viewer than it is in actuality. This apparent position of a body, called *apparent depth,* can be determined by Equation (22.4).

6. As a point source emits electromagnetic radiation, such as light, the total power delivered must be spread over an ever-growing sphere as the wavefront travels outward. The amount of light energy per second per unit area along this ever-growing sphere is the *intensity* of the light. Intensity is measured in watts per square meter (W/m^2) and can be calculated by Equation (22.5).

7. Whereas there are 2π radians in a circle, there are 4π *steradians* in a sphere. A steradian is the unit that is used for a *solid angle.*

Problem Solving Tip 22.2: Although steradians are seldom used in problems involving circuits, it is still a good idea to be wary of symbols. The unit of the steradian (Ω) does *not* stand for ohms and is not related to resistance in any way!

Key Concepts

An accelerated charged particle emits electromagnetic radiation.

The angle of incidence equals the angle of reflection for specular reflection relative to the normal line to the surface.

CRITICAL ANGLE: The incident angle at which a light ray traveling from a medium with a high index of refraction into a medium with a low index of refraction is refracted at 90°.

HUYGENS'S PRINCIPLE: Every point on a wavefront can be considered as a source of secondary wavelets that spread out in all directions with the wave speed of the medium.

INDEX OF REFRACTION: The ratio of the speed of light in a vacuum to its speed in a given medium.

LIGHT RAY: The straight-line path that light takes radially outward from the primary source.

REFRACTION: The bending of a light ray that takes place at an interface when it enters a medium with a different index of refraction.

SPECULAR REFLECTION: Reflection of light that forms an image.

Important Equations

(22.1) $c = \dfrac{1}{\sqrt{\varepsilon_o \mu_o}}$ (Speed of light in terms of the properties of a vacuum)

(22.2) $\dfrac{\sin \theta_i}{\sin \theta_r} = \dfrac{c_i}{c_r}$ (Snell's law of refraction)

(22.3) $n_i \sin \theta_i = n_r \sin \theta_r$ (Alternate form of Snell's law)

(22.4) $d_a = \left(\dfrac{n_r}{n_i}\right) d$ (Apparent depth)

(22.5) $I = \dfrac{P}{A}$ (Intensity)

(22.6) $\Omega = \dfrac{A}{r^2}$ (Solid angle)

Conceptual Problems

1. A changing electric field generates:
   a. static snow on the screen of a TV
   b. a changing magnetic field
   c. a constant magnetic field
   d. an accelerated charge

2. In an electromagnetic traveling wave, the magnetic field is:
   a. perpendicular to the electric field and the propagation direction
   b. parallel to the electric field and the propagation direction
   c. parallel to the electric field and perpendicular to the propagation direction
   d. perpendicular to the electric field and parallel to the propagation direction

3. The propagation speed of an electromagnetic wave in a vacuum:
   a. is frequency dependent
   b. is wavelength dependent
   c. is a universal constant
   d. depends on the velocity of the source

4. Electromagnetic waves carry:
   a. charge   b. energy
   c. magnetic monopoles   d. current

5. The ionized region of gas in the upper atmosphere is the ionosphere. It's responsible for:
   a. the blue of the sky
   b. the ability of satellites to orbit
   c. long-distance radio communications
   d. our inability to see stars during the day

6. Which color has the longest wavelength?
   a. Violet   b. Yellow
   c. Red   d. Green

7. When an electromagnetic wave interacts with matter, most of the effects are due to the:
   a. magnetic field   b. electric field
   c. speed of propagation   d. transported charge

8. When you see yourself in a plane mirror, you see:
   a. a real image in front of the mirror
   b. a real image behind the mirror
   c. a virtual image behind the mirror
   d. a virtual image in front of the mirror

9. When a ray of light goes from one medium to another, the quantity that remains constant is:
   a. speed   b. frequency
   c. wavelength   d. direction

10. A material's index of refraction:
    a. is always greater than 1
    b. is always less than 1
    c. never exceeds the speed of light
    d. is an index of how much faster the speed of light is in the medium

11. A ray of light goes from one medium into another along the normal line. The refracted angle is:
    a. different from the incident angle
    b. the same as the incident angle
    c. 90°
    d. totally internally reflected

12. How are electromagnetic traveling waves produced?

13. Why can light travel through a vacuum, but sound can't?

14. How does the ionosphere allow for radio communications below the horizon?

15. Why doesn't a pane of window glass cause a spectrum to be produced when a flashlight beam is shone through it?

Exercises

Electromagnetic Waves

1. The average distance between the earth and the moon is approximately 240,000 mi or 386,000 km. If an astronaut on the moon beamed a laser toward the earth, how long would it take to reach us?

   Answer: 1.29 s

Reflection

2. What is the height of the smallest mirror in which a man 1.8 m tall can see himself at full length?

   Answer: 90 cm

*3. A hamster runs perpendicularly toward a mirror at a speed of 1.5 m/s. (a) How fast does the hamster see its reflection approaching it? (b) If the hamster ran toward the mirror at an angle of 30° relative to the normal line of the mirror, how fast would it see its reflection approaching it?

   Answer: (a) 3 m/s
   (b) 2.6 m/s

*4. A mirror is placed flat on the floor. Another mirror is butted up against the flat mirror edge to edge and is tilted away from the mirror on the floor. The second mirror makes an angle of 55° with the horizontal. If a light ray hits the flat mirror at an incident angle of 65°, at what angle does it leave the second mirror relative to the normal?

   Answer: 60°

5. A light beam is initially incident on a plane mirror at an angle of θ. The mirror is then rotated an angle ϕ about an axis in the plane of the mirror and perpendicular to the plane formed by the incident and reflected rays. Show that the reflected ray rotates through an angle of 2ϕ.

Refraction and Snell's Law

6. What is the speed of light in a transparent dielectric that has a dielectric constant of 2.3?

   Answer: 1.98×10^8 m/s

7. A beam of light is incident on a pane of glass with an angle of incidence of 42°. At what angle does the beam emerge from the glass with respect to the normal line?

   Answer: 42°

8. The speed of light in a vacuum is 3×10^8 m/s. What is the speed of light in (a) air and (b) water?

   Answer: (a) 2.99×10^8 m/s
   (b) 2.25×10^8 m/s

*9. A beam of light in air enters a transparent pane of some unknown material at an angle of 40° with respect to the normal line, which is the wall of a container holding a transparent liquid. The beam travels completely through the wall of the container and exits into the liquid at an angle of 30° to the normal. What is the index of refraction of the liquid?

Answer: 1.285

*10. A light in air strikes a pane of some transparent material at an angle of incidence of θ. The light is both refracted and reflected. The reflected and refracted rays are perpendicular to each other. (a) What is the index of refraction of the material? (b) What restriction does the index of refraction place on the angle of incidence?

Answer: (a) $n_g = \tan \theta$
           (b) The angle of incidence can't be less than 45°.

*11. A rectangular transparent block has sides that are parallel. Show that light incident on one side at an incident angle θ emerges from the other side with the same angle. Also show that the displacement of the two beams is proportional to the thickness of the block.

12. A beam of light enters one of the short sides of a 45°–45°–90° prism normal to the surface. Show that any

ray hitting the side normal to the surface will be deflected by 90° when it emerges from the other short side. The index of refraction for the glass is 1.52.

*13. A horizontal ray of white light in air with red light of wavelength 660 nm on one side and violet light of wavelength 410 nm on the other of the visible spectrum strikes a glass prism with an apex angle of 30°. When the prism is sitting on its base, the face that the light ray strikes is vertical, and the apex is the top part of the prism. The index of refraction for the red light is 1.520, and for the violet light it is 1.538. What is the angle of refraction for each ray as it emerges from the opposite side of the prism? The angles should be relative to the normal line of the opposite slanted face.

Answer: $\phi_r = 49.46°$; $\phi_v = 50.26°$

14. A laser beam in the air strikes a glass prism at an angle to the normal of 60°. (a) What is the refracted angle of the beam inside the prism? (b) What is the refracted angle if the prism is immersed in water?

Answer: (a) 34°
          (b) 48°

Apparent Depth

15. A spear fisherman sees a fish in the water in front of him. In order to strike the fish, should he aim directly at the fish or above or below it? Use a diagram to justify your answer.

Answer: Below

16. If a fish appears to be swimming 6 in. below the surface of a pond when seen from behind, how deep is it really?

Answer: 8 in.

17. A student places a plexiglass cube 10 cm to a side on her textbook. When she looks downward into its top face, how far below the top surface of the cube do the letters appear to be?

Answer: 6.6 cm

18. The fish in Exercise 16 is frightened and swims directly downward toward the bottom of the pond. If its actual speed is 3 m/s, how fast does it appear to be diving?

Answer: 2.25 m/s

*19. A beaker has a height of h. The lower half of the beaker is filled with water with an index of refraction n_w. The upper half is filled with oil with an index of refraction n_o. If you're looking down into the beaker from above and slightly to the side, what is the apparent depth of the bottom? Choose a spot on the bottom at the beaker's side ($n_w > n_o > n_a$; $n_a = 1$). Draw a ray diagram, and remember that for small angles the tangent of the angle is approximately equal to the sine of the angle.

Answer: $d = \dfrac{h}{2}\left(\dfrac{1}{n_o} + \dfrac{1}{n_w}\right)$

Total Internal Reflection

*20. What is the maximum angle necessary for an incident ray of light to enter the top face of a block of ice and not exit the side of the block?

Answer: 57.3°

21. A fish below the surface of a pond will see anything above the surface only through a circular "window," the diameter of the window being a function of the depth. What is the diameter of the window for a fish 5 cm below the surface?

Answer: 11.5 m

22. After feeding the fish in a public aquarium, a scuba diver swims to the side of the tank and shines a flashlight into the

glass of a viewing window. What is the minimum incident angle that will result in no light being seen by the spectators looking into the tank?

Answer: 48.9°

*23. A light ray enters one end of a glass cylinder from the air at an angle of incidence of θ. The glass has an index of refraction of n_g. What is the maximum value of θ for which total internal reflection takes place?

Answer: $\sin \theta = \sqrt{n_g^2 - 1}$

Intensity and Flux

24. A moderately large city in Utah consumes 50 GW of power during a hot summer. If the city government decides to convert to solar power using solar cells with an efficiency of 25%, how large an array would it take, assuming that the average intensity of the sunlight striking the earth's surface is 1.5 kW/m²?

Answer: A square array 11.6 km on a side

25. A Boy Scout starting a fire decides to use a magnifying glass and some chips of wood. The magnifying glass has a diameter of 10 cm and focuses the light into a very small point 2 mm in diameter. What is the intensity of light in this point?

Answer: 3.8×10^6 W/m²

26. The moon has a diameter of 3500 km and is approximately 3.86×10^5 km from the earth. The sun has a diameter of just over 1.3×10^6 km and is 1.5×10^8 km from the earth. What is the ratio of the solid angle subtended by the moon as seen from the earth to the solid angle subtended by the sun as seen by the earth?

Answer: 1.1

The refraction of light
through a lens causes
a magnified image of
whatever is behind a
magnifying glass.

Spherical Mirrors and Lenses

At the end of this chapter, the reader should be able to:

1. Draw ray-trace diagrams for both **concave** and **convex** mirrors.

2. Draw a *ray-trace diagram* for a double-lens system involving both a *converging* lens and a *diverging* lens.

3. Calculate the **focal length** of a lens if the *radius of curvature* is known for both of its sides.

4. Describe two different types of image **aberrations.**

5. Explain the difference between a *real* and a *virtual* image, and give examples of both.

6. Solve for the *image magnification* of a given lens and object.

23.1   Image Formation

A lens is generally considered to be a device that focuses visible light. This focused light is considered to be an image. There are two types of images: *real* and *virtual*. A real image can be focused on a screen, such as the one that a movie projector throws on a white screen. A virtual image can't be captured on a screen. Two common examples of virtual images are those from a plane mirror and the object enlargements found in a magnifying glass. Whether real or virtual, images can be formed by spherical lenses or by spherical mirrors. A glass lens forms an image by refracting light, and a curved mirror forms an image through specular reflection.

Images formed by refracted light form from light rays passing through a thin, spherical transparent lens. *Spherical* means that each of the two lens surfaces forms part of the surface of a sphere with a certain radius of curvature. *Thin* means that the thickness of the lens is small compared to the radius of curvature of either lens surface. There are two basic physical profiles for thin lenses: convex and concave. *A **convex lens** curves outward from the center line of the lens, and a **concave lens** curves inward.* The six basic thin lens profiles are made up from convex and concave profiles; these profiles are illustrated in Figure 23.1. A meniscus lens has both a concave and a convex surface.

Convex surface: A surface that curves outward from the lens or mirror.

Concave surface: A surface that curves inward from the lens or mirror.

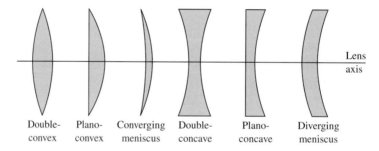

Double-   Plano-   Converging   Double-   Plano-   Diverging
convex   convex   meniscus   concave   concave   meniscus

Figure 23.1 Thin Lens Profiles

How a thin lens uses refraction to form an image is best shown using parallel light rays (plane wavefronts) and a plano-convex lens. As shown in Figure 23.2, the light hitting the rear of the lens passes into the lens unbent because it hits the surface normal line at 0°. A ray on the optical axis passes through the center of the lens unbent because it emerges from the glass at an angle of 0° with the interface's normal. Above and below the optical axis, when the light rays pass out of the lens, they are refracted because they hit the interface at nonzero angles that increase as the distance from the optical axis increases. As the angle increases, the ray is bent farther and farther away from the normal. Because the ray's deviation increases, the rays leaving the lens are no longer parallel. In fact, it can be shown that because of the constant radius of curvature of the convex face of the lens, all of the light rays will pass through the same point on the optical axis. This point is the *focal point* of the lens, and the focal point's distance from the lens is the ***focal length*** of the lens.

Focal length: The length from the center of a lens to the point where parallel light rays entering the lens will intersect on the optical axis.

If you take a large-diameter magnifying glass outside on a sunny day and project the sun's image onto a piece of wood, the wood will start to smoke and char where the sun's image is projected. The image of the sun is broad if the glass is close to the wood since the converging rays aren't well concentrated near the lens. The farther from the wood the lens is withdrawn, the smaller the sun's image is, the more concentrated the rays are, and the greater the intensity of the sun's image becomes. When the lens is withdrawn to a distance from the wood equal to the focal length of the lens, the image is too bright to stare

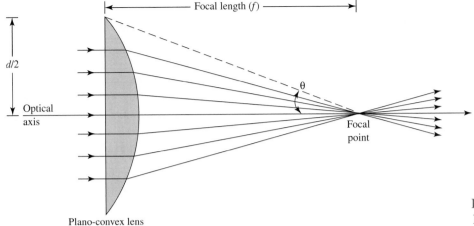

Plano-convex lens

Figure 23.2 Converging Lens
Focusing Sunlight

at, and the wood will smoke, char, and maybe even burst into flame. A lens can concentrate light and the power that hits the lens area. If you look at the area on the wood under the lens, you'll notice that the lens's shadow occupies all of the area under the lens except where the image is. No direct sunlight reaches this shaded area because of the refraction of the light in the lens.

EXAMPLE 23.1

Sunlight hitting the earth's surface has an intensity of 1 kW/m^2. A magnifying glass with a 15-cm diameter and a 12-cm focal length is placed 9 cm from a flat surface. (a) What is the power delivered to the area of the image? (b) What is the intensity in this area?

SOLUTION
(a)
$$P = IA \qquad A = \pi r^2 \qquad P = \pi r^2 I$$

$$P = \pi(0.075 \text{ m})^2(1000 \text{ W/m}^2)$$

$$P = 17.7 \text{ W}$$

The power collected by the glass is the power passing through the image.

(b)
$$I = \frac{P}{A} \qquad A = \pi r^2 \qquad I = \frac{P}{\pi r_i^2}$$

We next must find the image radius.

$$\tan \theta = \frac{d/2}{f} \qquad \tan \theta = \frac{1/2(15 \text{ cm})}{12 \text{ cm}} \qquad \theta = 32°$$

$$\tan \theta = \frac{r_i}{12 \text{ cm} - 9 \text{ cm}} \qquad r_i = (12 \text{ cm} - 9 \text{ cm})\tan 32°$$

$$r_i = 0.0187 \text{ m}$$

$$I = \frac{17.7 \text{ W}}{\pi (0.0187 \text{ m})^2}$$

$$I = 16.1 \text{ kW/m}^2$$

At a distance of 9 cm, the sunlight intensity has been increased 909 times! A lens can concentrate the sun's rays and greatly increase their intensity.

A refracting lens has a well-defined focal point and can concentrate the sun's rays dramatically. A reflecting lens or a spherical mirror also has a focal point and can concentrate the sun's rays. Solar boilers are often spherical mirrors with the liquid to be boiled at the focal point of the mirror. The greater the diameter of the mirror, the greater its collecting area, and the more power available to boil the liquid. Efficient use of the available power requires concentrating the power just where it is needed and nowhere else. A spherical mirror can do this.

In Figure 23.3, the incident parallel light rays are specularly reflected from the inside surface of a concave spherical mirror. The angle of reflection equals the angle of incidence, and by constructing a ray diagram with accurate plotting of angles, we can show that the reflected light will pass through a common point on the optical axis. A spherical mirror implies that the reflecting surface is part of the surface of a sphere with a constant radius of curvature. On the optical axis, the angle of incidence is 0°, and therefore the reflected angle is also 0°. The reflected ray retraces the path of the incident ray. But as the incident parallel rays hit the concave mirror farther and farther from the optical axis, the angle of reflection becomes greater and greater because the angle of incidence becomes greater and greater. As a result, all of the reflected rays intersect at the focal point of the mirror. From calculations similar to those in Example 23.1, we can also show that a multifold increase in the intensity of the incident light can also be obtained.

Although operating on two entirely different physical principles, spherical mirrors and refracting lenses have much in common. This commonality will be stressed in this chapter.

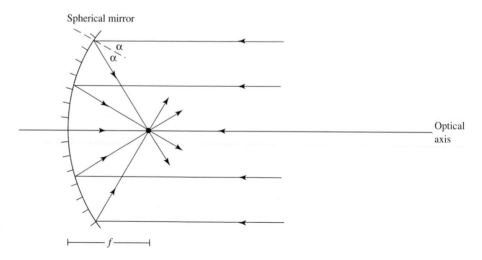

Figure 23.3 Spherical Mirror Focusing Sunlight

23.2 Designing Spherical Lenses and Mirrors

In the designing of lenses and spherical mirrors, usually two parameters are of main interest: light-gathering ability and focal length. Light-gathering ability is a function of area and for round lenses and mirrors is a function of the diameter. The focal length of a lens or a mirror will depend on the radius of curvature of the surfaces for both, and the index of refraction of the lens material for a spherical lens.

To derive the equation necessary to design a spherical lens with a given focal length, we'll start with a plano-convex lens as in Figure 23.4. The radius of curvature of the spherical surface of the lens is R. A line drawn from C (the center of curvature) to the curved surface of the lens will have a constant length R no matter where it contacts the curved surface. The normal line to the spherical surface is an extension of the line of the radius of curvature through the spherical surface.

$$\sin\theta = \frac{r}{R} \qquad\qquad \tan(\phi - \theta) = \frac{r}{f}$$

$$r = R\sin\theta \qquad\qquad r = f\tan(\phi - \theta)$$

$$r = r \qquad\qquad R\sin\theta = f\tan(\phi - \theta)$$

where θ is the incident angle, and ϕ is the refracted angle. We assume small angles, or $R \gg r$ and $f \gg r$.

$$\tan(\phi - \theta) \approx \sin(\phi - \theta)$$

$$\sin(\phi - \theta) = \sin\phi\cos\theta - \cos\phi\sin\theta$$

$$R\sin\theta = f(\sin\phi\cos\theta - \cos\phi\sin\theta)$$

$$R = f\left(\frac{\sin\phi}{\tan\theta} - \cos\phi\right)$$

For a small angle,

$$\tan\theta \approx \sin\theta \qquad \cos\phi \approx 1$$

$$R = f\left(\frac{\sin\phi}{\sin\theta} - 1\right)$$

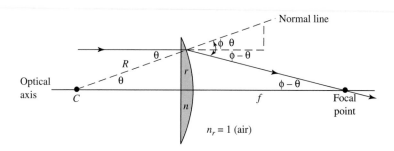

Figure 23.4 Ray Parallel to the Optical Axis Hitting a Converging Lens

From Snell's law,

$$n_i \sin \theta_i = n_r \sin \theta_r \qquad n \sin \theta = (1) \sin \phi$$

$$\frac{\sin \phi}{\sin \theta} = n \qquad R = (n - 1)f$$

$$f = \frac{R}{n - 1}$$

or in the usual form,

$$\frac{1}{f} = (n - 1)\frac{1}{R}$$

The above result is for a lens with one curved surface. A double-convex lens has two curved surfaces. The second curved surface would curve to the left instead of to the right as it does in Figure 23.4. The lensmaker's equation for a double-convex lens could be considered the superposition of two plano-convex lenses back to back with curvatures in opposite directions, and the distance between the lenses being negligible (thin lens approximation). The focal length for a system of two such lenses is

(Equation 23.1)

$$\frac{1}{f} = (n - 1)\left(\frac{1}{R_1} - \frac{1}{R_2}\right)$$

This equation is quite general, working for any of the lens profiles given in Figure 23.1. R_1 is for the first surface upon which the light is incident, and it is positive if the center of curvature is on the outgoing side of the lens. R_2 is for the surface on the far side of the lens and is positive if its radius of curvature is on the outgoing side of the lens. (A radius of curvature on the incoming side of the lens is negative by definition.) The focal length can be positive or negative. A positive focal length means that the lens will cause the light rays to converge as in Figure 23.2. A negative focal length will cause the light rays to diverge or spread out like the beam of a flashlight.

Let's return to the equation for the plano-convex lens in Figure 23.4. R_1 is infinite. (The radius of curvature of a straight line is infinite. This argument is the same one that we made for plane waves. The farther from the source the wavefront travels, the more it approximates a straight line until, when it is at an infinite distance, the wavefront forms a flat plane surface.) Any fraction that has infinity in the denominator equals zero, and the R_1 term drops out. The center of curvature of side 2 is on the incoming side of the lens, and R_2 is negative. Thus, we have the equation that we first derived for the focal length of a plano-convex lens.

$$\frac{1}{f} = (n - 1)\left(\frac{1}{R_1} - \frac{1}{R_2}\right) \qquad R_1 = \infty$$

$$\frac{1}{f} = (n - 1)\left(\frac{1}{\infty} - \frac{1}{-R_2}\right)$$

$$\frac{1}{\infty} = 0 \qquad \frac{1}{f} = (n - 1)\frac{1}{R_2}$$

We now want to come up with an equation to design a spherical mirror for a given focal length. With a spherical mirror we have only the radius of curvature and the law of specular reflection with which to do our calculations (see Figure 23.5).

$$\sin \alpha = \frac{r}{R} \qquad\qquad \tan \beta = \frac{r}{f}$$

$$r = R \sin \alpha \qquad\qquad r = f \tan \beta$$

$$r = r \qquad\qquad R \sin \alpha = f \tan \beta$$

$$\beta = 2\alpha \qquad\qquad R \sin \alpha = f \tan 2\alpha$$

$$R \sin \alpha = f \frac{\sin 2\alpha}{\cos 2\alpha} \qquad R \sin \alpha = f \frac{2 \sin \alpha \cos \alpha}{2 \cos^2 \alpha - 1}$$

$$R = f \frac{2 \cos \alpha}{2 \cos^2 \alpha - 1}$$

We again use the small angle approximation, so $\cos \alpha \approx 1$.

$$R = \frac{2}{2 - 1} f \qquad R = 2f$$

or
$$f = \frac{1}{2} R$$
(Equation 23.2)

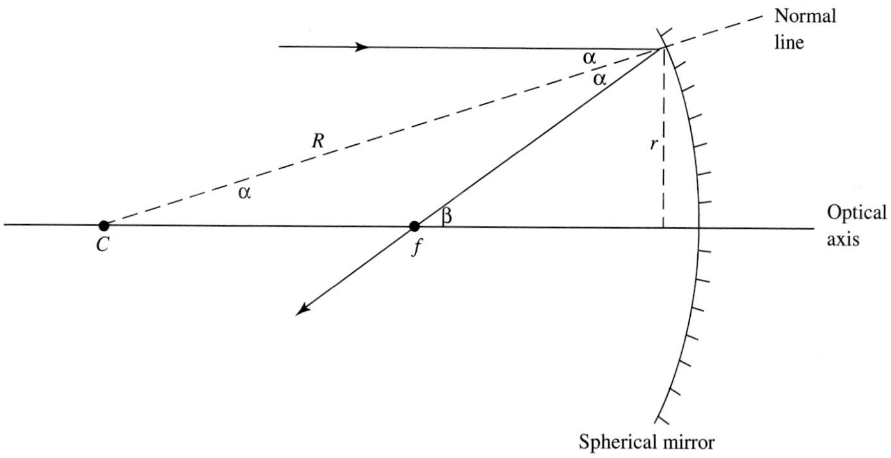

Figure 23.5 Ray Parallel to the Optical Axis Hitting a Spherical Mirror

By using a small angle approximation, we find that for a spherical mirror, the focal length equals one-half the radius of curvature. Note that the reason we were able to use a small angle approximation for the derivation for both the lens and the mirror is that the area of the lens and mirror is small compared to the area of a sphere with the same radius of curvature. Most lenses and mirrors used as lenses have spherical surfaces, and refracting lenses are thin relative to the diameter of the lens. Not all lenses are thin and can use the thin lens, small angle approximation. Sometimes such approximations are unstated, and we assume that they hold for all cases. This isn't true. Later in this chapter we'll examine some cases where the thin lens approximation is invalid.

EXAMPLE 23.2

A double-convex, thin lens has a radius of curvature of 10 cm for side 1 and 15 cm for side 2; $n = 1.5$. (a) What is the focal length of the lens for a ray of light first incident on side 1? (b) What is the focal length of the lens for a ray of light first incident on side 2?

SOLUTION

(a)
$$\frac{1}{f} = (n - 1)\left(\frac{1}{R_1} - \frac{1}{R_2}\right)$$

$$\frac{1}{f_1} = (1.5 - 1)\left(\frac{1}{10\ \text{cm}} - \frac{1}{-15\ \text{cm}}\right)$$

$$f_1 = +12\ \text{cm}$$

(b)
$$\frac{1}{f_2} = (1.5 - 1)\left(\frac{1}{15\ \text{cm}} - \frac{1}{-10\ \text{cm}}\right)$$

$$f_2 = +12\ \text{cm}$$

The two surfaces of a double-convex lens have opposite curvatures, so the two radius-of-curvature terms always add. The lens has the same converging focal length regardless of the side on which the light is first incident.

EXAMPLE 23.3

A meniscus lens has a radius of curvature of 10 cm for side 1 and a radius of curvature of 15 cm for side 2; $n = 1.5$ (see Figure 23.6). (a) What is the focal length of the lens if the light is first incident on side 1? Does the light converge or diverge? (b) What is the focal length of the lens if the light is first incident on side 2? Does the light converge or diverge?

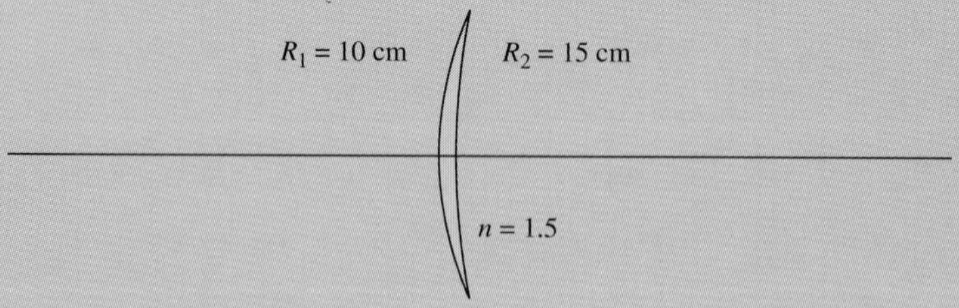

Figure 23.6 Meniscus Lens

SOLUTION

(a)
$$\frac{1}{f_1} = (1.5 - 1)\left(\frac{1}{10\ \text{cm}} - \frac{1}{15\ \text{cm}}\right)$$

$$f_1 = +60\ \text{cm}$$

The light converges.

(b)
$$\frac{1}{f_2} = (1.5 - 1)\left(\frac{1}{-15\text{ cm}} - \frac{1}{-10\text{ cm}}\right)$$

$$f_2 = +60\text{ cm}$$

The light converges.

    The above lens is called a *converging meniscus* because the light converges regardless of whether it is concave or convex to the incident light.

23.3   Ray Tracing and the Lens Equation

We now have equations that will allow us to design a refracting lens or a spherical mirror for a given focal length. Next we need to develop an equation or equations that can predict how a lens will image an object. We can do this with the technique of ray tracing without any further need to invoke Snell's law or the law of specular reflection.

    For refracting lenses we need only three light rays coming from the object to find the image, as shown in Figure 23.7. The three rays will intersect at the image point. The object is usually represented as an arrow on the optical axis, with the three rays originating from the arrow's head (point B). Ray 1 travels parallel to the optical axis. It emerges from the other side and passes through the focal point of the lens. We already know that parallel light hitting a lens will pass through the focal point on the other side. Ray 2 is just the opposite of ray 1. A light ray passing through the focal point will emerge from the other side of the lens parallel to the axis. Notice that there are two focal points in Figure 23.7: one in front of the lens and one behind it. All refracting lenses have these two foci, equally spaced from O, the origin of our ray-tracing coordinate system on the center line of the lens. The reason for both is that any light ray can be reversed. (A light ray originating from the image parallel to the optical axis will emerge on the object side and will pass through the focus on the image side.) The third ray passes through the center of the lens undeviated. The image is also represented as an arrow and is drawn perpendicular to the optical axis to the intersection of the three rays.

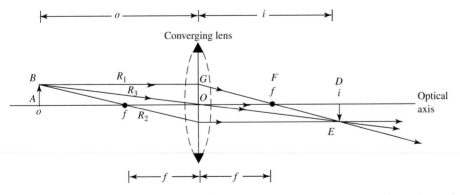

Figure 23.7 Image Formation of a Converging Lens

    An accurately scaled ray-tracing diagram can give us the lens equation using only geometry. We want a relationship among the image, the object, and the focal distances from the center of the lens, O. We can find it using the fact that triangle ABO is similar to triangle DEO in Figure 23.7. With similar triangles, corresponding sides are proportional. Triangle OGF is also similar to triangle DEF. Capital letters are individual points, and two capital letters together symbolize the distance from one point to another. Single lower-case letters will be our standard symbols for image distance (i), object distance (o), and focal length (f).

To obtain a relationship between image distance i and object distance o, we use similar triangles ABO and DEO.

$$\frac{DO}{AO} = \frac{DE}{AB} \qquad DO = i \qquad AO = o$$

$$\frac{DO}{AO} = \frac{DE}{AB} = \frac{i}{o}$$

To obtain a relationship between image distance i and focal length f, we use similar triangles OGF and DEF.

$$\frac{DE}{OG} = \frac{DF}{OF} \qquad DF = i - f \qquad OF = f$$

$$\frac{DE}{OG} = \frac{DF}{OF} = \frac{i - f}{f}$$

Now we must somehow unite both relationships. Since $OG = AB$,

$$\frac{DE}{AB} = \frac{DE}{OG} \qquad \frac{i}{o} = \frac{i - f}{f}$$

$$\frac{i}{o} = \frac{i}{f} - 1$$

Next we divide both sides by the image distance i.

$$\frac{1}{o} = \frac{1}{f} - \frac{1}{i}$$

(Equation 23.3)   or

$$\frac{1}{o} + \frac{1}{i} = \frac{1}{f}$$

Equation (23.3) was derived using a double-convex converging lens, but it is good for all six profiles shown in Figure 23.1. If we look at the image in Figure 23.7, we can immediately see from the ray-tracing diagram that it is real, inverted, and smaller than the object. It's real because all three light rays converging at the image point are real. It's inverted because the image arrow is upside down. The image is "magnified" because the image height and the object height can be different. In the case of Figure 23.7, the image height is less than the object height. The numerical value for the magnification can also be gleaned from Figure 23.7 using similar triangles.

$$m = \frac{DE}{AB} \qquad \frac{DE}{AB} = \frac{i}{o} \qquad m = \frac{i}{o}$$

Since the image is inverted, we add a negative sign.

(Equation 23.4)

$$m = -\frac{i}{o}$$

Equation (23.4) for magnification is also quite general.

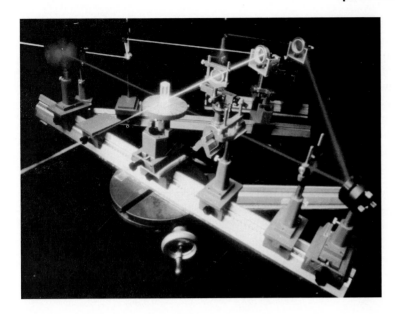

A light ray from a laser is reflected and refracted on this optical bench.

Figure 23.8 shows a diverging lens that will have a virtual image. How do we use the ray-tracing technique with a diverging lens? A diverging lens causes the light rays to spread out instead of focusing. Yet our three standard light rays (parallel to the axis, through the focal point, and through the center of the lens) act in nearly the same way that they did in Figure 23.7. Ray 1 travels from the object parallel to the optical axis. After it encounters the lens, the outgoing ray must pass through the focal point, yet it must also diverge. A line traced back through the lens from outgoing ray 1 will pass through the focal point on the object side of the lens. The real light ray will follow the solid line. The dotted line is a virtual ray. Ray 2 is directed so that if it were undeviated through the lens, it would pass through the focal point on the far side of the lens (the dotted-and-dashed line). When it emerges from the lens, it is traveling parallel to the optical axis. Tracing outgoing ray 2 back through the lens gives us virtual ray 2, and its intersection with virtual ray 1 gives us the image point. Ray 3 passes undeflected through the center of the lens and also passes through the image point. In ray tracing, *all three rays,* virtual and real, must intersect at the image point. If they don't, you've made a mistake with one of the rays. Try again. Any two rays will intersect. Use all three to be sure of the image location.

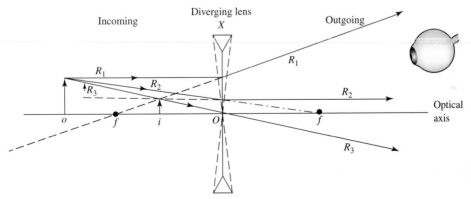

Figure 23.8 Image Formation by a Diverging Lens

A diverging lens always has a virtual image and a negative focal length.

In Figure 23.8, two out of the three rays intersecting at the image point are virtual. Thus, the image is a virtual image. If any one of the three rays intersecting at the image point is virtual, so is the image. *A virtual image has a negative sign. Remember also that the focal length of a diverging lens is also negative.* The image is erect and smaller than the object. How would our eyes perceive such an image? Just as with a plane mirror, they would be fooled into thinking that what they saw was on the object (incoming) side of the lens. The eyes would follow the straight lines formed by the real rays emerging from the lens back through the lens to their virtual intersection at the image point. Your eyes can see the image, but the image couldn't be placed on a screen.

EXAMPLE 23.4

A diverging lens has a focal length of 12 cm. An object 4 cm tall is placed 18 cm in front of the lens. (a) Where is the image? (b) What is the height of the image? (c) Is it erect or inverted?

SOLUTION

(a)
$$\frac{1}{o} + \frac{1}{i} = \frac{1}{f} \qquad f = -12 \text{ cm} \qquad o = 18 \text{ cm}$$

$$\frac{1}{i} = \frac{1}{f} - \frac{1}{o}$$

$$\frac{1}{i} = \frac{1}{-12 \text{ cm}} - \frac{1}{18 \text{ cm}}$$

$$i = -7.2 \text{ cm}$$

The image is negative, so it must be virtual.

(b)
$$m = -\frac{i}{o} \qquad m = -\frac{-7.2 \text{ cm}}{18 \text{ cm}} \qquad m = 0.4$$

$$h = (0.4)(4 \text{ cm})$$

$$h = 1.6 \text{ cm}$$

(c) The sign of the magnification is positive, so the image is erect.

Next we consider mirrors used to image an object. Figure 23.9 shows a concave mirror with the inside of the concave surface mirrored. With the object placed as shown, what kind of image do we have, and where is the image located? To find the image, we again use three rays. A concave mirror is a converging mirror and can concentrate light. Ray 1 travels from the object to the mirror parallel to the optical axis. When it reflects from the mirror, it travels through the focal point of the lens in front of the mirror as shown. Ray 2 travels from the object through the focal point, and when it reflects from the mirror, it returns parallel to the optical axis. This situation is not new; it occurred earlier when we discovered the focal point of a concave mirror. The two rays also act just like the similar rays for refracting lenses. Ray 3 is the ray that passed through the center of the refracting lenses undeviated in the previous two examples. In the case of the concave mirror, after ray 3 reflects from the mirror, it travels back upon itself undeviated. However, in order for a ray reflecting from a mirror to travel back upon itself undeviated, it must pass through the radius of curvature of the sur-

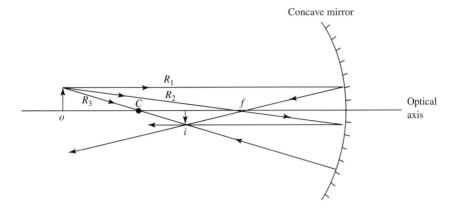

Concave mirror

Figure 23.9 Image Formation by a Concave Mirror

face. Only by passing through the radius of curvature does the ray hit the mirrored surface at an angle of 0°. For a spherical mirror, an undeviated ray passes through the radius of curvature, but for a refracting lens, an undeviated ray passes through the center of the lens.

The image point for our concave mirror is at the intersection of the three rays. All three rays are real, so our image is also real. From Figure 23.9, we can see that the image is inverted and smaller than the object. We should also be able to find a relationship between the object, the image, and the focal distances just as we did for our convex refracting lens. If we do so, we get exactly the same lens equation as for both refracting lenses! We find that the lens equation is general for both spherical lenses and spherical mirrors.

EXAMPLE 23.5

We have a concave mirror with a focal length of 19 cm and an object distance of 52 cm. (a) What is the image distance? (b) What is the magnification? (c) Is the image erect or inverted?

SOLUTION

(a)
$$\frac{1}{o} + \frac{1}{i} = \frac{1}{f} \qquad \frac{1}{i} = \frac{1}{f} - \frac{1}{o}$$

$$\frac{1}{i} = \frac{1}{19 \text{ cm}} - \frac{1}{52 \text{ cm}}$$

$$i = 30 \text{ cm}$$

A concave mirror has a positive focal length, just as a converging lens does. The image is positive and real.

(b)
$$m = -\frac{1}{o} \qquad m = -\frac{30 \text{ cm}}{52 \text{ cm}} \qquad m = -0.58$$

(c) The magnification is negative, so the image is inverted. The image is smaller than the object.

The last stop in our grand tour of single lenses is the convex mirror. A convex mirror causes light to diverge and spreads it out instead of concentrating it (see Figure 23.10). You now know that a diverging lens has a negative focal length, and the image is virtual with a negative image distance. The same is true for a convex mirror. The convex mirror is also governed by the same lens equation and the same magnification equation of the previous three examples.

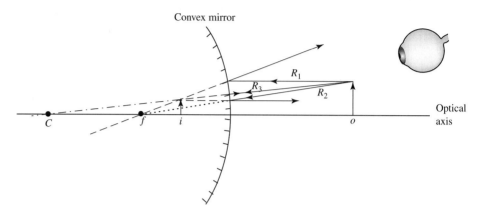

Figure 23.10 Image Formation by a Convex Mirror

In Figure 23.10, ray 1 travels from the object to the mirror parallel to the optical axis. Since this is a diverging mirror, the reflected ray bounces off the mirror in such a way that if a line were traced back through the mirror, it would pass through the focal point behind the mirror (the dashed virtual ray). Ray 2 leaves the object on a path that would pass through the focal point behind the mirror (the dotted line) if the mirror weren't in the way. When ray 2 is specularly reflected from the mirror, it travels parallel to the optical axis. If we trace a line back through the mirror from reflected ray 2 (the dashed virtual ray), it will intersect virtual ray 1 at what should be the image point. To be sure, we use ray 3 which travels from the object on a path that would take it through the center of curvature behind the mirror. When reflected, ray 3 travels back along the path on which it came in. Tracing a line back through the mirror and through the center of curvature (the dashed-and-dotted line), we find that it goes through the intersection of virtual rays 1 and 2. We have found the image location.

From Figure 23.10, we can see that the image is virtual because all of the intersecting rays are virtual. The image is erect and smaller than the object. If our eyes were in front of the mirror on the right, they would be fooled into thinking that the image was behind the mirror. Only real light rays enter the eye, but the eye would follow the straight lines of these light rays back behind the mirror to the virtual image.

EXAMPLE 23.6

An object is placed 9 cm in front of a convex mirror of focal length 6 cm. (a) Where is the image? (b) What is the image magnification? (c) Is the image erect or inverted?

SOLUTION

(a)
$$\frac{1}{o} + \frac{1}{i} = \frac{1}{f} \qquad o = 9 \text{ cm} \qquad f = -6 \text{ cm}$$

$$\frac{1}{i} = \frac{1}{f} - \frac{1}{o}$$

$$\frac{1}{i} = \frac{1}{-6 \text{ cm}} - \frac{1}{9 \text{ cm}}$$

$$i = -3.6 \text{ cm}$$

The image is virtual and behind the mirror.

(b) $\qquad m = -\dfrac{i}{o} \qquad m = -\dfrac{-3.6 \text{ cm}}{9 \text{ cm}} \qquad m = +0.4$

(c) The image is erect.

Note that in our use of the lens equation in the last four examples, there exists a sign convention for focal length, image distance, and even object distance. Table 23.1 lists the sign conventions for spherical lenses and mirrors.

The only object that we've dealt with up to now has been a real object. A virtual object can exist if we have lens systems of two or more lenses in which the image of the first lens becomes the object of the second lens. We'll deal with lens systems in the next section.

Table 23.1 Lens and Mirror Sign Conventions

Quantity	Positive	Negative
Focal length (f)	Converging lens	Diverging lens
	Concave mirror	Convex mirror
Image distance (i)	Real image	Virtual image
Object distance (o)	Real object	Virtual object

23.4 Optical Devices and Lens Systems

Now that we have a pretty good idea of how to use spherical lenses and mirrors, we can examine some of the devices that are based on the principles outlined in the last section. The first such device is the magnifying glass, shown in Figure 23.11. A magnifying glass is usually a double-convex lens with the object placed somewhere between the focal point and the lens. With the object so placed, we can draw two of our standard three rays with no problem (rays 2 and 3). However, ray 1 travels from the object as if it had passed through the focal point on the object side of the mirror, emerging from the other side parallel to the optical axis. The real rays that reach the eye don't converge, so the eye follows the rays back through the lens to the virtual, erect, and *enlarged* image. The lens and magnification equations for the magnifying glass are the same ones previously developed.

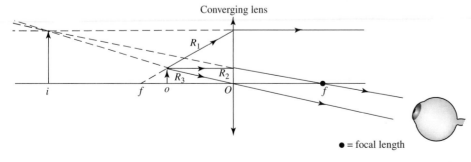

Figure 23.11 Magnifying Glass

A magnifying mirror, such as one used to examine the wrinkles and pits in your face, is ray traced in Figure 23.12. The image is erect and magnified, as expected. You can use a magnifying glass to read the fine print in a contract. Why wouldn't it be a good idea to use a magnifying mirror for the same purpose?

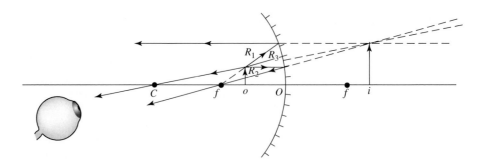

Figure 23.12 Magnifying Mirror

A simple camera is a single-lens device. For objects far away from the camera lens, we can assume that the light rays from the object entering the lens are parallel and that all of the light rays will converge at the focal point of the lens. For this reason, the film plane is slightly beyond the focal point of the lens. The lens of a simple camera is converging, and the image formed on the film plane is real and inverted. The image must be real because the film plane forms a screen, and the light patterns of a real image on the screen cause the chemicals in the film to react and form a negative of the image. However, not all objects are an infinite distance away. Thus, the distance between the lens and the film plane is adjustable so that the image formed by the lens will always fall on the film plane. This situation is best examined using the lens equation.

$$\frac{1}{o} + \frac{1}{i} = \frac{1}{f} \qquad \frac{1}{i} = \frac{1}{f} - \frac{1}{o} \qquad \frac{1}{i} = \frac{o}{of} - \frac{f}{of}$$

$$\frac{1}{i} = \frac{o - f}{of}$$

$$i = \frac{of}{o - f}$$

For the equation above, if the object distance is much greater than the focal length of the lens, the equation for image distance reduces to the focal length. (The focal length of a camera lens is about the distance between a camera's lens and the film plane inside the body of the camera, about the width of your palm. Most objects photographed will be much farther away. The lenses must be ground so that the focal length will fit inside the physical dimensions of the device.) However, if the object distance is within an order of magnitude of the focal length, then if the object distance is known, the lens can be moved to compensate for the image distance. As the object gets closer and closer to the focal point, the image distance gets greater and greater, until if the object is at the focal point, the image is at infinity. Obviously the lens can't adjust for such a case. The object must always be outside the focal length of the lens, or the image distance either is infinite or is negative, implying a virtual image. A virtual image can't be put on a screen, and the camera is then useless! For an object outside the focal length of the camera's lens, the image is upside down. But that is no problem for the light-sensitive film, because when developing the film, the photographer just turns the film right side up!

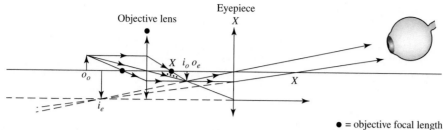

$\bullet$ = objective focal length
X = eyepiece focal length

Figure 23.13 Image
Formation by a Microscope

A microscope enlarges an object that is originally too small to see with the naked eye or that is too small to easily study with the naked eye. The simplest microscope is a two-lens system in which the image of the first lens (the lens closest to the object being examined, the objective lens) becomes the object for the second lens (the eyepiece). Figure 23.13 illustrates a microscope. We're interested in the magnification, and in our analysis we would like to find the lens system's magnification in terms of the focal lengths of the two lenses.

$$m = m_o m_e \qquad m_o = -\frac{i_o}{o_o}$$

$$\frac{1}{o_o} + \frac{1}{i_o} = \frac{1}{f_o} \qquad \frac{1}{o_o} = \frac{1}{f_o} - \frac{1}{i_o}$$

$$\frac{1}{o_o} = \frac{i_o - f_o}{i_o f_o} \qquad o_o = \frac{i_o f_o}{i_e - f_o}$$

$$m_o = -i_o\left(\frac{i_o - f_o}{i_o f_o}\right) \qquad m_o = \frac{f_o - i_o}{f_o} \qquad \text{(where } i_o > f_o)$$

$$m_e = -\frac{i_e}{o_e} \qquad m_e = -\left(\frac{i_e - f_e}{i_e f_e}\right)(i_e)$$

$$m_e = -\frac{i_e - f_e}{f_e} \qquad \text{(where } i_e > f_e)$$

$$m = \left(\frac{f_o - i_o}{f_o}\right)\left(\frac{f_e - i_e}{f_e}\right) \qquad m = \frac{(f_o - i_o)(f_e - i_e)}{f_o f_e}$$

where $i_o \approx f_e$ and $f_e \gg f_o$. Thus,

$$m = \frac{(f_o - f_e)(f_e - i_e)}{f_o f_e} \qquad m \approx \frac{-f_e(f_e - i_e)}{f_o f_e}$$

where $m \approx -(f_e - i_e)/f_o$ and $i_e \gg f_e$.

$$m = \frac{i_e}{f_o} \qquad i_e \approx 25 \text{ cm}$$

$$m = \frac{25 \text{ cm}}{f_o}$$

(Equation 23.5)

The system's magnification is designed to be the product of the magnification of the objective lens times that of the eyepiece. The objective lens has a very short focal length relative to that of the eyepiece. The real inverted image of the objective lens falls inside the focal length of the eyepiece so that it acts like a magnifying glass. If your eye is right up against the eyepiece, then i_1 should be equal to the distance where your eyes can best focus on a near object. This works only if the light rays emerging from the eyepiece are nearly parallel and the object of the eyepiece is just inside the eyepiece's focal length. The object distance for the eyepiece is nearly equal to its focal length. The focal length of the eyepiece is typically about 25 cm, which is the distance where your eyes see best. You can verify this by moving your finger away from your eyes until your eyes can discern the wrinkles on your knuckle without squinting. If you measure the distance from your eye to your knuckle, it will be about 25 cm. The objective lens can have a focal length on the order of a millimeter. This can give a simple microscope a magnification from 200 to 300 times.

The microscope image is also inverted. If you've ever used a microscope and had to manually adjust the stage, you found that it moves opposite to the direction you expect when looking through the eyepiece. A little technique is necessary to avoid losing sight of your specimen altogether. Also, because of the magnification, a very slight nudge of the stage can send the specimen completely out of the viewing area. A very intense light is needed to view an object at large magnification. Because of the small aperture of the objective lens, the lens can't collect enough light to provide anything but a dim view of the specimen. This artificial illumination is often so intense that you can't view the specimen with the naked eye because it's too bright. The turret on a microscope typically contains three lenses. When the magnification of the microscope is changed, the turret is rotated to place the objective lens for the desired magnification under the eyepiece. Changing the focal length of the lens changes the magnification of the microscope as given by Equation (23.5).

A telescope is a device for examining distant objects. Astronomers use an astronomical telescope to examine planets, stars, and galaxies. As shown in Figure 23.14, this telescope consists of two converging lenses, an objective, and an eyepiece, much like a microscope. As with a microscope, the image is inverted, but it makes little difference if the planet Jupiter is upside down in the eyepiece. It causes little loss of orientation. The purpose of a telescope is to bring a faraway object closer to the eye, revealing more detail. This is magnification, but not magnification in the sense of a magnifying glass or microscope. Those two instruments produce an image that is larger than the object. The image that a telescope produces isn't larger than the object; rather, the telescope brings the object closer to the eye by *angular* magnification. The planet Jupiter is already plenty big enough, and it isn't necessary to enlarge it as a microscope does. What we want to do is to bring it closer to the eye so that we can make out details on its surface. The planet Jupiter is 6.28×10^{11} mi away from us at its closest approach, and with a diameter of 1.38×10^8 mi, it subtends an angle of 2.2×10^{-4} rad at our eye. This angle is so small that we see Jupiter as a point of reflected

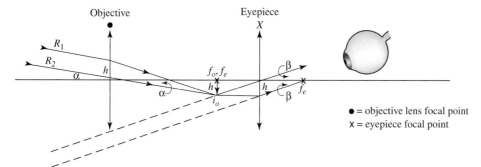

Figure 23.14 Astronomical Telescope

sunlight in the night sky as a star instead of as a circular spot with the naked eye. To be able to discern details on the planet's surface, we must enlarge this angle subtended by the eye.

The angle θ in Figure 23.15 is the angle that is subtended at the eye and that we want to magnify. In the refracting telescope, two converging lenses are arranged so that they are a distance apart equal to the sum of the focal lengths of the objective and the eyepiece. The object is essentially an infinite distance away from the objective lens. Thus, light rays entering the objective lens from the object are parallel. (The distance between the rays is greatly exaggerated for the sake of clarity.) The image of the objective lens falls at the focal point of the objective lens, f_o. If the object is slightly off the optical axis, the parallel rays entering the objective make an angle alpha (α) with the optical axis. The two parallel rays (which could correspond to the upper and lower borders of Jupiter's disc) are a distance h apart when they hit the objective. The image formed by the objective is then h units tall.

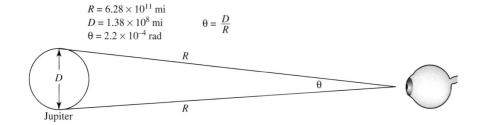

$$R = 6.28 \times 10^{11} \text{ mi}$$
$$D = 1.38 \times 10^{8} \text{ mi} \qquad \theta = \frac{D}{R}$$
$$\theta = 2.2 \times 10^{-4} \text{ rad}$$

Figure 23.15 Angle Subtended by Jupiter When Seen by the Naked Eye

If i_o is at the focal point of the eyepiece, then when the rays from i_o leave the eyepiece, they are parallel and form an angle β with the optical axis of the eyepiece. Because the rays emerging from the eyepiece are parallel, the eye sees the image of Jupiter at infinity. (Follow the virtual rays back.) As the image distance increases, so does its height. But the ratio of its height to its distance remains the same regardless of the distance. The angle subtended at the eye by the image remains constant, and it is this angle that the eye uses to judge the relative closeness of an object. The eyepiece acts like a simple magnifying glass, but the object is at its focal point.

If β is greater than α, then the object appears closer to the eye. The definition of angular magnification is

$$m = \frac{\beta}{\alpha}$$

(The negative sign is ignored.)

$$m = \frac{\beta}{\alpha} \qquad \beta = \frac{h}{f_e} \qquad \alpha = \frac{h}{f_o}$$

$$m = \frac{h/f_e}{h/f_o} \qquad m = \frac{\cancel{h}\, f_o}{f_e\, \cancel{h}}$$

(Equation 23.6)

$$m = \frac{f_o}{f_e}$$

It is advantageous to have the focal length of the objective lens as large as possible, and focal lengths on the order of 20 m are not unheard of. It's also a high priority that the lens collect as much light as possible, so the diameter of the lens must be as large as practicable.

The Andromeda galaxy can be seen above the plane of the Milky Way on a moonless night. But it appears to the naked eye as a fuzzy white patch that is best seen out of the corner of the eye. Why doesn't it appear to the eye as it does in the color posters that you can purchase at science-oriented gift shops? The eye can't collect enough light to discern details, and even a sensitive photographic film using a lens with a diameter dozens of times larger than that of the eye might need an exposure time of around one-half hour to collect enough light to produce a good image. To produce a good image of a distant object, we must both magnify it *and* collect enough light.

A telescope used to view terrestrial objects is disorienting if the image is upside down. For a terrestrial telescope, a third converging lens is placed between the objective and the eyepiece. This third lens takes the upside-down image of the objective lens and inverts it again, making the image right side up. A pair of binoculars also has to have an erect image. But binoculars use a prism both to change the direction of the light and to give the image a double inversion, thereby making it appear erect.

From the refracting telescope, we find that the eye can see an image if the object is at the focal point of a converging lens. Since a telescope's purpose is angular magnification of an object, this leads to the question, What is the magnification of a lens when the object is either at an infinite distance from the lens or at the focal point of the lens?

When $o = \infty$,

$$\frac{1}{o} + \frac{1}{i} = \frac{1}{f} \qquad (\text{where } o \approx \infty)$$

$$\frac{1}{\infty} + \frac{1}{i} = \frac{1}{f} \qquad \left(\text{where } \frac{1}{\infty} \approx 0\right)$$

$$\frac{1}{i} = \frac{1}{f} \qquad i = f$$

$$m = -\frac{i}{o} \qquad m = \frac{-f}{\infty}$$

$$m = 0$$

When $o = f$,

$$\frac{1}{o} + \frac{1}{i} = \frac{1}{f} \qquad \frac{1}{f} + \frac{1}{i} = \frac{1}{f}$$

$$\frac{1}{i} = 0 \qquad (\text{where } i = \infty)$$

$$m = -\frac{i}{o} \qquad m = -\frac{\infty}{f}$$

$$m = -\infty$$

In the first case ($o = \infty$), the object is very far away from the lens, or the object distance is much greater than the focal length. The lens equation coupled with the magnification equation tells us that the magnification is zero, or that the image height is zero which is the same as saying that there is no image present. Yet an image is present, and this can be easily demonstrated using a converging lens and a clear incandescent light bulb with a filament. If the bulb is lit at one end of a room, and if the lens forms an image of the filament on the wall at the other end of the room, the distance between the wall and the lens will be essentially equal to the focal length of the lens ($o \gg f$). The image is inverted, real, and smaller in size than the actual filament of the bulb. That shows the magnification is not zero. Why the discrepancy?

If you look at the rays and hitting the objective lens in Figure 23.15, you'll notice that they're a distance h apart at the lens. The image formed by them at the focal length of the objective lens has a height h also. Yet we can't say that this is a linear magnification of 1 because if we're looking at the planet Jupiter, the image certainly doesn't have a height of the diameter of Jupiter! What we can say, though, is that the angular magnification of the image is 1 from the ray-tracing diagram for the objective lens. The image of the filament on the wall is the same size as what our naked eye would see from the distance of the lens. The farther we move away from some object, the smaller it appears until at an infinite distance the object disappears. The height of the real object doesn't change, but the angle subtended at the eye by the object becomes smaller and smaller until in the limit of infinite distance it becomes zero. An object is never really an infinite distance from a lens, but where the object distance is much greater than the focal length, it's a useful approximation.

For the eyepiece the object is at the focal length of the lens. The rays emerging from the lens are parallel. The eyepiece acts like a simple magnifying glass, and the eye sees a virtual image following the real rays back through the lens. The linear magnification is infinity, and you can see that as the image distance increases, the height of the image must also increase as the eye traces the virtual ray back. But do we see an image of infinite height towering above us? No, because the angle subtended at the eye (β) is well defined and constant. It is the angle β that really determines what the eye interprets as the height of the image, and therefore it is the angular magnification that is important in lens systems.

Large refracting lenses are unsupported over the area of the lens when mounted in a telescope, and very large lenses tend to sag toward the middle. Also, both surfaces of the lens must be precisely ground because light passes through both surfaces. For telescopes requiring very-large-diameter lenses, concave mirrors are used instead of refracting lenses. With the mirror, light is reflected only from the concave surface, so only one surface needs to be ground. Support can also be given to the center portions of the lens to prevent sagging by building a framework under the mirror in contact with the rough, unsilvered side. For these reasons, the great majority of astronomical telescopes use mirrors for the objective lens.

The lens system organization for a reflecting telescope is exactly the same as for a refracting telescope. The objective mirror forms an image at its focal length, and the eyepiece is placed its focal length away from that image. But since the mirror reflects light back the way it came, the eyepiece would be directly in the path of the incoming light. For a large-diameter mirror, this amount of missing light is minuscule. However, since the eyepiece is usually on the optical axis of the mirror, the object being observed must be slightly off axis.

In most cases the eyepiece isn't inside the body of the reflecting telescope. A small plane mirror reflects the light from the objective to the eyepiece placed on the side of the body tube (a Newtonian reflector) or through a small hole in the center of the objective mirror to the eyepiece mounted behind the objective (a Cassegrainian reflector).

23.5   Lens Aberrations

Chromatic aberration: The smearing of an image from a refracting lens caused by differing indexes of refraction for the different wavelengths of light.

Dispersion results from the fact that different wavelengths of light travel at different speeds inside a material medium. It is what gives rise to a white light rainbow spectrum from a prism. The result is that for a given medium, each color of the spectrum has a slightly different index of refraction from all of the other colors. Each color is bent at a slightly different angle as it passes through the lens medium, and instead of a sharp focus for the image, it becomes spread out and smeared chromatically. This smearing is called ***chromatic aberration.*** Chromatic aberration is another reason that most astronomical telescopes use mirrors instead of refracting lenses.

Spherical aberration: For a spherical mirror, the shifting of the focal length toward a mirror's surface as light rays hit the mirror farther and farther from the optical axis.

The shape of a mirror can introduce its own aberrations, though. A spherical mirror forms a sharp focus for parallel incoming light only for those rays close to the optical axis. For rays hitting the mirror well away from the axis, the focus moves closer to the mirror's surface the farther away from the axis the ray hits. For this reason, spherical mirrors usually have a small diameter relative to the radius of curvature of the mirror. This shifting of focal length is called ***spherical aberration.*** To avoid it, the surface of the mirror can be ground to the shape of a parabola. (The equation of a parabola is $y = kx^2$, where k is an arbitrary constant.) From the law of specular reflection for parallel light rays bouncing off a mirror with a parabolic section, it can be shown that all of the light rays go through the focal point no matter how far off the axis they are. Parabolic mirrors are harder to grind, though, and that is why spherical mirrors are used in applications that tolerate a spherical section.

23.6   The Eye

The eye is a lens system and a worthy candidate for examination in this chapter. A profile of the eye is given in Figure 23.16. The eye is roughly spherical in shape with a diameter of about 3 cm. Its shape is maintained by the pressure of the liquid filling the eye, the vitreous humor. There are actually two lenses in the eye: the cornea and the lens. Both are converging lenses, the cornea being a converging meniscus lens and the lens being double-convex. Because the index of refraction of the lens is close to that of the vitreous humor, the lens is a less effective lens than the cornea that interfaces with the air. Therefore, most of the focusing power of the eye is due to the cornea. The shape of the lens can be changed by the ciliary muscle. By relaxing or contracting the ciliary muscle, the eye can focus on objects faraway or close up. When the ciliary muscles relax, the lens focuses objects very far away on the retina (screen) of the inside back surface of the eye-

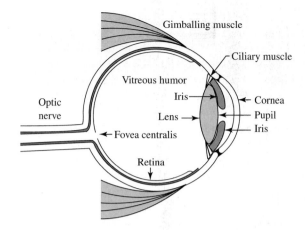

Gimballing muscle

Ciliary muscle

Vitreous humor

Iris

Optic nerve

Cornea

Pupil

Lens

Iris

Fovea centralis

Retina

Figure 23.16 The Eye

ball. The ciliary muscle forms a band around the lens and, when it contracts, forces the lens into a more convex shape that allows it to focus a close-up object on the retina. The focal length of the lens is thereby changed to accommodate close objects. The image distance from the lens to the retina is fixed in the eye. With the camera the situation is reversed. The focal length of the lens is fixed, and the image distance is changed by moving the lens in and out. The eye is very similar to a camera in that it also has a shutter, the iris. The pupil is the front surface of the lens that we see when we look into another's eyes. We say that in bright light the pupil contracts, but what happens is that the iris closes, blocking some of the surface area of the pupil. In dim light the iris opens, allowing light to reach more of the surface area of the lens, in turn allowing the lens to collect more light. (The pupil dilates.)

Like a camera, the image on the focal plane is real and inverted. Our brain makes the image appear erect. The sensors that convert light focused on the retina into nerve impulses sent down the optic nerve to the brain are called *rods* and *cones*. The cones are located in a small region around the optical axis of the eyeball on the retina called the *fovea centralis*. In bright conditions the pupil contracts, and the light admitted to the eye falls principally on the fovea. The cones are responsible for vision in bright light conditions and for our color vision. The rods are on the retina surrounding the fovea. When the light is dim and the pupil dilates, most of this dim light falls on the rods. The rods are sensitive not to color but to degrees of shading. For this reason, colors are difficult to distinguish in dim light. This is also why it's best to view faint objects out of the corner of the eye.

The sensitivity of the eye to light is greatest at a wavelength of 555 nm, or green light. That is right in the middle of the white light spectrum and where the cones are most sensitive. The rods have their greatest sensitivity at 510 nm, or blue light. For this reason, on board a ship when night vision is important, the illumination in the cabin for instrumentation is red, on the other end of the spectrum. Red light affects night vision minimally.

There are no photoreceptors near the optic nerve, and this area is known as the *blind spot*. It's only a very small part of the field of vision, and as in the reflecting telescope it is easily compensated for. The eye compensates in two ways. Our two eyes give overlapping fields of view. The eyes are spread apart in the skull, and the distance between the eyes allows triangulation. Triangulation gives greater depth of field and allows the eyes to image objects in three dimensions. Because of the overlapping fields of view, the blind spot isn't noticed because the blind spots obscure different fields of vision. The eyes are also in constant motion. The eyes are gimballed much like the gimbals that vector the thrust of a rocket motor. Each eyeball has two sets of muscles that allow the eyeball to swivel up and down and left and right. This constant swiveling won't allow the eye's blind spot to block out anything in the field of view.

Those of us who wear glasses are only too well aware of the defects in vision. The two most common defects are farsightedness (hypermetropia) and nearsightedness (myopia). If you are nearsighted, your eyeball is too long. The light coming from an object far away (infinity) focuses in front of the retina. Warping of the lens by the ciliary muscles allows clear vision of nearby objects, but the lens can't accommodate faraway objects. If you are farsighted, your eyeball is too short, and the light from an object at infinity focuses behind the retina. Warping of the lens allows clear vision of objects at a distance, but the lens cannot accommodate objects that are nearby. The answer for both defects is to place a diverging lens in front of each eyeball (a pair of glasses or contacts) to adjust the focus of the cornea so that light from a distant object focuses on the retina when the ciliary muscle is relaxed.

Astigmatism is a structural defect of the eye in which the surface of the cornea departs from a spherical shape and is more sharply curved in one plane than in another. The result is that it's impossible to focus on the horizontal and vertical bars of an object (for example, the letter E) at the same time. The correction is a cylindrical instead of a spherical lens in front of the eyeball.

Summary

1. The *lensmaker's equation* [Equation (23.1)] is used to determine what radius of curvature a lens will require to possess the needed **focal length.**

2. When an object is placed in front of a mirror or a lens, the light rays that are associated with it may be reflected or refracted (respectively) to form an image at another location. Equation (23.2) describes how the *object distance, image distance,* and *lens focal length* are related.

Problem Solving Tip 23.1: When dealing with multiple-lens systems, work with only one lens at a time. Start with the lens closest to your object, and, *while ignoring all other lenses,* find the corresponding image. Use that image as the object for the next lens, ignoring all other lens, *including the lenses already used!* By following this method, you will soon arrive at the image for the final lens (farthest from the object). This image is the resultant image for your multiple-lens system.

    Warning! You must be careful, however. If you ever come across an image that is beyond the next lens, you must still use that image as the next lens's object. This is where positive and negative conventions for focal length become very important. For example, let's assume that we have a system with two converging lenses (both have $+f$). Our object is to the left of the first lens (lens 1). To keep with our method, we will find the image created by our first lens while completely ignoring our second (lens 2). Now, to illustrate the above warning, let's assume that our calculated (or drawn) image falls *to the right* of lens 2. We know that light from our object is actually traveling from left to right (since the object is to the left of our multiple-lens system), so we will call this direction positive. We should use this as our

deciding factor in choosing which focal point to use for lens 2. In our case, we would draw our line to the lens (as usual, although this time it requires us to back up); then we would continue our line to the rightmost focal length (it has a positive focal length, so it must be drawn in the positive direction). Our second line will cross through the center of lens 2 (as usual). The point where these two lines cross is the location of the image. Analytically, in our case, the "new" object distance (for the image from lens 1) will be negative, and light would have to travel a negative distance to reach lens 2 (virtual object).

3. Depending on where an object is positioned, the image may appear bigger or smaller than the original. This *magnification* is the ratio of the image height to the object height. Another way of determining magnification is given by Equation (23.5), which is related to the distances from the lens to both the object and the image.

Problem Solving Tip 23.2: When finding the magnification of a multiple-lens system, simply find the magnification of each lens separately; then multiply them together. A number greater than 1 indicates an image size greater than the object's, and a number smaller than 1 indicates the opposite. A negative magnification indicates an inverted image.

Problem Solving Tip 23.3: An inverted image is upside down *with respect to the object.* So if you start with an upside-down object, an inverted image will be right side up!

4. Images can be found both analytically (using equations) and graphically (using ray-trace diagrams). When solving problems

involving lenses or mirrors with either method, you should keep the following information (from Table 23.1) in mind:

Lens and Mirror Sign Conventions

Quantity	Positive	Negative
Focal length (f)	Converging lens	Diverging lens
	Concave mirror	Convex mirror
Image distance (i)	Real image	Virtual image
Object distance (o)	Real object	Virtual object

Problem Solving Tip 23.4: Another way of discovering whether or not an image is virtual is to draw the ray-trace diagram for your lens system. Examine closely the rays that you have drawn, and determine if they are the actual path of the light ray or the apparent (projected) path of the light ray. If all of the rays that you have drawn that converge on the image are actually the path the light has traveled, the image is real. If any one of them is a projected path, it is virtual.

5. The eye uses a lens to focus light onto the retina. If an eye is too long, the image will focus in front of the retina, which results in nearsightedness. If an eye is too short, the image will focus behind the retina, resulting in farsightedness.

Key Concepts

A diverging lens always has a virtual image and a negative focal length.

CHROMATIC ABERRATION: The smearing of an image from a refracting lens caused by differing indexes of refraction for the different wavelengths of light.

CONCAVE SURFACE: A surface that curves inward from the lens or mirror.

CONVEX SURFACE: A surface that curves outward from the lens or mirror.

FOCAL LENGTH: The length from the center of a lens to the point where parallel light rays entering the lens will intersect on the optical axis.

SPHERICAL ABERRATION: For a spherical mirror, the shifting of the focal length toward a mirror's surface as light rays hit the mirror farther and farther from the optical axis.

Important Equations

(23.1) $\dfrac{1}{f} = (n - 1)\left(\dfrac{1}{R_1} - \dfrac{1}{R_2}\right)$ (Lensmaker's equation)

(23.2) $f = \dfrac{1}{2}R$ (Focal length in terms of the radius of curvature for a mirror)

(23.3) $\dfrac{1}{o} + \dfrac{1}{i} = \dfrac{1}{f}$ (Lens equation)

(23.4) $m = -\dfrac{i}{o}$ (Magnification)

(23.5) $m = \dfrac{-25 \text{ cm}}{f_o}$ (Linear magnification of a simple microscope)

(23.6) $m = \dfrac{f_o}{f_e}$ (Angular magnification of a telescope)

Conceptual Problems

1. A negative focal length comes from a:
   a. convex mirror   b. convex lens
   c. concave mirror   d. virtual image

2. An object is outside the focal point of a converging lens. The image is:
   a. inverted   b. virtual
   c. larger than the object   d. erect

3. An object is inside the focal length of a converging lens. The image is:
   a. inverted   b. virtual
   c. smaller than the object   d. real

4. An object is at the focal length of a converging lens. The light rays leaving the lens:

    a. intersect at the image point

    b. diverge

    c. emerge from the lens parallel to the optical axis

    d. form a virtual image

5. The image of an object produced by a diverging lens is:

    a. always real   b. always virtual

    c. always inverted   d. always erect

6. If the magnification is negative, the image is:

    a. erect   b. inverted

    c. always virtual   d. always real

7. If the image distance is negative, the image is:

    a. real   b. virtual

    c. always erect   d. always inverted

8. If a converging lens is used as a magnifying glass, the object distance must be:

    a. equal to the focal length

    b. greater than the focal length

    c. less than the focal length

    d. equal to the radius of curvature

9. Four lenses can be used as the objective lens of a microscope. Their focal lengths are listed below. Which lens will give the greatest magnification?

    a. 3 mm   b. −3 mm

    c. 5 cm   d. −5 cm

10. A concave mirror produces an erect image when the distance of the object from the lens is:

    a. equal to the focal length

    b. less than the focal length

    c. the same as the focal length

    d. equal to the radius of curvature

11. Under what conditions is a light ray that passes through a lens undeflected?

12. What direction does the incident ray in Figure 23.17 take upon passing through the lens?

13. What direction does the incident ray in Figure 23.18 take upon passing through the lens?

14. What direction does the incident ray in Figure 23.19 take after reflecting from the mirror?

15. What direction does the incident ray in Figure 23.20 take after reflecting from the convex mirror?

16. What's wrong with the ray diagram of Figure 23.21? Where should the image be, and what type of image is it?

17. What's wrong with the ray diagram in Figure 23.22? Where should the image be, and what type of image is it?

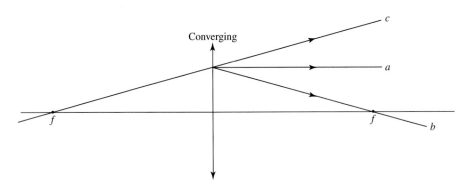

Figure 23.17 (Problem 12)

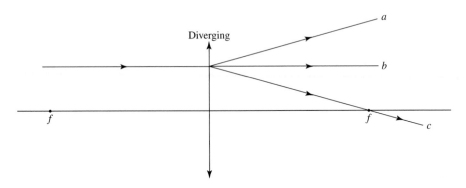

Figure 23.18 (Problem 13)

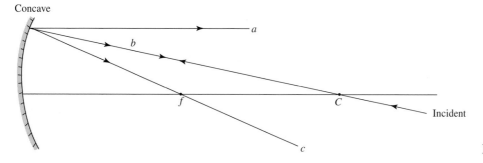

Figure 23.19 (Problem 14)

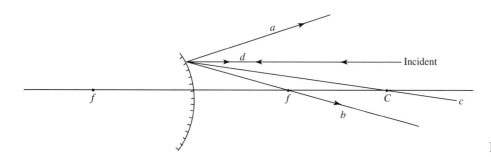

Figure 23.20 (Problem 15)

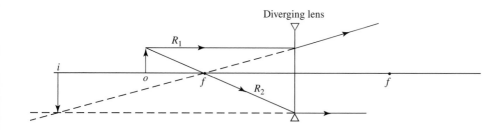

Figure 23.21 (Problem 16)

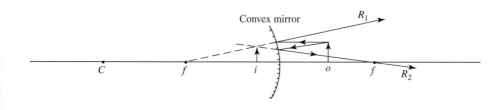

Figure 23.22 (Problem 17)

Exercises

Image Formation

1. A student wishes to use a lens for woodburning. In order to char the wood, she needs an intensity of at least 50 kW/m². If she finds a lens whose diameter is 10 cm with a focal length of 15 cm, what is the minimum distance between the lens and the piece of wood? Assume that the solar intensity is 1 kW/m².

Answer: 12.9 cm

Designing Spherical Lenses and Mirrors

2. A double-convex (converging) thin lens made of crown glass with an index of refraction of 1.5 has the same radius of curvature of 40 cm for both sides. What is the focal length of the lens?

Answer: 40 cm

3. A double-convex lens has radii of curvature 15 cm and 20 cm. The focal length is found to be 10 cm. What is the index of refraction of the material of which the lens is made?

Answer: 1.86

4. An ice sculptor wishes to make a working sculpture of a double-convex (converging) lens for an optometrist's banquet. The required focal length is 2 m, and the diameter of the lens is to be 1.5 m. The radius of curvature of one side of the lens is 4 m. What is the radius of curvature of the other side of the lens?

Answer: −0.79 m

5. A plano-concave lens of focal length 15 cm is to be made from flint glass ($n = 1.55$). What is the required radius of curvature?

Answer: 8.25 cm

*6. A converging lens of glass with an index of refraction in air of 1.6 has a focal length of 15 cm. (a) What is the focal length in water? (b) Is the lens now converging or diverging in water?

Answer: (a) 45 cm
(b) Converging

*7. A diverging lens made of glass with an index of refraction of 1.52 in air and a focal length in air of −65 mm is immersed in carbon disulfide. Carbon disulfide has an index of refraction of 1.63. (a) What is the focal length in the carbon disulfide? (b) Is the lens converging or diverging in carbon disulfide?

Answer: (a) 50.4 cm
(b) Converging

Ray Tracing and the Lens Equation

8. An 11-cm-tall object is placed 30 cm in front of a converging lens with a 20-cm focal length. (a) Where is the image located? (b) What is the image height? (c) Is the image erect or inverted? (d) Is the image real or virtual?

Answer: (a) $i = 60$ cm
(b) $h = 22$ cm
(c) Inverted
(d) Real

9. A 17-cm-tall object is placed 20 cm in front of a converging lens with a focal length of 30 cm. (a) What is the image distance? (b) What is the image height? (c) Will the image be erect or inverted? (d) Will the image be real or inverted?

Answer: (a) $i = -60$ cm
(b) $m = +3$
(c) Erect
(d) Virtual

10. A pencil held 10 cm in front of a lens creates a real image 20 cm beyond the lens. Where does the pencil have to be held in order to produce an image 30 cm beyond the lens?

Answer: 8.57 cm

11. A contractor doesn't wish his clients to understand too well the terms under which his clients have hired him, and he has hidden the terms in the small print. The small print must be magnified by a factor of 4 to be read with normal eyesight. Using a converging lens with a focal length of 5 cm, how far must the lens be held from the page the page for the proper magnification?

Answer: 3.75 cm

12. A magnifying glass is used to view a snowflake 1.1 mm in diameter from a distance of 3 cm. If the focal length of the magnifying glass is 5 cm, how large does the snowflake appear?

Answer: 2.75 mm

13. Cosmetic mirrors are often made from concave mirrors to magnify the image of a face. One such mirror has a radius of curvature of 3 m and is held 35 cm from a person's face. How many times larger will the face appear in the mirror?

Answer: 1.3×

14. The right rear-view mirror on many cars is often a convex mirror. A warning is often printed on such mirrors: "Objects in mirror are closer than they appear." That is, objects will appear smaller, or the magnification is less than 1. If your right rear-view mirror has a radius of curvature of 2.5 m, and if a car is following 3 m behind you, what is the magnification of the following car?

Answer: 0.29

15. A lawn decoration consists of a pedestal topped by a reflective metal sphere having a radius of 15 cm. If a bird 7 cm tall were to land atop the sphere and look down into it, how far beneath the sphere's surface would it see the reflection of its head?

Answer: 3.6 cm

16. To the left of a candle is a concave mirror with a focal length of 10 cm. On the candle's right is a converging lens with a focal length of 15 cm. The separation between the candle and the lens is 40 cm, and the candle is placed at the mirror's radius of curvature. What is the location of the two images formed by this system?

Answer: 60 cm to the right of the candle

17. A concave mirror of focal length 6 cm is placed inside a bird cage. A canary is perched 4 cm from the mirror and, looking into the mirror, becomes very frightened. How much larger does the canary's image appear to the canary?

Answer: 3×

18. A coin 12 mm in diameter is 16 cm in front of a spherical mirror. The image is erect and 4 mm in diameter. (a) Is the mirror concave or convex? (b) What is its radius of curvature?

Answer: (a) Convex
(b) 16 cm

19. An amusement park mirror produces an erect image four times life-size of anyone standing 3 m in front of it. (a) Is the mirror concave or convex? (b) What is its radius of curvature?

Answer: (a) Concave
(b) 8 m

*20. An object 5 cm high can be placed in two positions in front of a concave mirror where the image height is 20 cm. The mirror has a focal length of 50 cm. What are the two object distances? What is the type of image in each position?

Answer: 62.5 cm, real and inverted; 37.5 cm, virtual and erect

*21. An object 2 cm high is placed 10 cm in front of a concave mirror. A virtual image 6 cm high is produced. What is the other object location that will produce a 6-cm image height, and what is the nature of the image?

Answer: 20 cm; real and inverted

*22. A converging lens is held near the bright filament of a flashlight bulb. The image of the filament appears on the wall 1.6 m away. The lens is then moved closer to the wall, and an image is again formed when the lens is 15 cm from the wall. (a) What is the focal length of the lens? (b) How far is the flashlight from the wall?

Answer: (a) 15 cm
(b) 1.77 m

*23. Show that the magnification of an object lying along the optical axis is m^2, where $m = -i/o$. Assume that the distance from the object to the lens is much greater than the length of the object.

Optical Devices and Lens Systems

24. Two converging lenses with focal lengths of 5 cm are separated by a distance of 15 cm. An intense lamp is set up 2 m in front of the first lens. How far behind the second lens must the screen be placed for an observer to see a focused image of the lamp's filament?

Answer: 10.1 cm

25. A diverging lens with a 15-cm focal length is placed 30 cm in front of a converging lens with a focal length of 20 cm. An object is placed 1 m in front of the diverging lens. (a) Where is the image formed? (b) What is the magnification of the image? (c) Is the image erect or inverted? (d) Is the image real or virtual?

Answer: (a) 37.4 cm to the right of the converging lens
(b) $m = 0.11$
(c) Inverted
(d) Real

26. Using the same lens system as in Exercise 25, the object is now placed 10 cm in front of the diverging lens. Where is the image formed?

Answer: 45 cm to the right of the converging lens

*27. Two thin lenses of focal lengths f_1 and f_2 are placed close together so that their separation is negligible compared to the object distance. What is the focal length of the lens system? (*Hint:* The object of the second lens becomes the image of the first.)

Answer: $\dfrac{1}{f} = \dfrac{1}{f_1} + \dfrac{1}{f_2}$

28. A bright lamp filament 30 cm away from a convex lens puts an inverted image on a screen 30 cm away from the lens on the other side. Then a concave lens is placed midway between the filament and the convex lens, and the lamp is moved away from the convex lens 10 cm from its original position, making the filament's image reappear on the screen. What is the focal length of each lens?

Answer: $f_1 = 15$ cm; $f_2 = -37.5$ cm

29. The objective lens of a telescope has a focal length of 75 cm, and the eyepiece a focal length of 2.5 cm. (a) What is the telescope's angular magnification? (b) How far apart are the lenses?

Answer: (a) 30×
(b) 77.5 cm

30. Two converging lenses with focal lengths of 12 cm are placed 60 cm apart. An object is placed 18 cm to the left of the first lens. Where does the image appear relative to the second lens?

Answer: 24 cm to the right

31. A plano-convex lens is ground from crown glass, whose radius of curvature is 15 cm. The index of refraction for red light is 1.517, and for blue light it is 1.523. How far apart are the focal points of the two colors? (This lens defect is called *chromatic aberration.*)

Answer: 3.3 mm

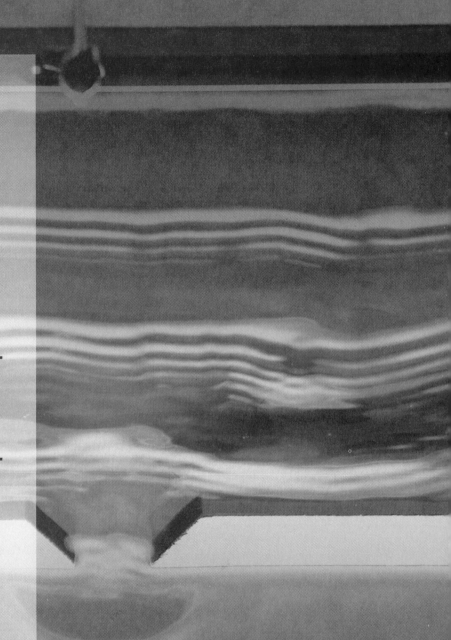

Plane waves hitting a small opening will cause the slit to act like a secondary point source. Two waves from two such secondary sources in phase will interfere in a regular pattern.

The Wave Nature of Light

At the end of this chapter, the reader should be able to:

1. Explain the differences among constructive interference, destructive interference, and **coherence,** and give examples of each.

2. Calculate the minimum thickness that an *antireflection coating* must be to destructively interfere with reflected light of a known wavelength.

3. Solve for a given *minimum* or *maximum* order for a problem involving *diffraction gratings* if the grating, wavelength, and screen distance are known.

4. Calculate the intensity of light that emerges from a *multiple-polarizer* system if the initial *unpolarized* intensity is known.

5. Discuss in detail why the sky is blue.

24.1   Interference and Coherence

Light propagates as a traveling wave, and thus as a wave it should exhibit all of the usual wave phenomena. In Chapter 23, we dealt with light in the form of a light ray that travels in a straight line. The light ray made it easy to develop the theory of lenses. Remember that the concept of a ray was developed from the wave nature of light and Huygens's prin-

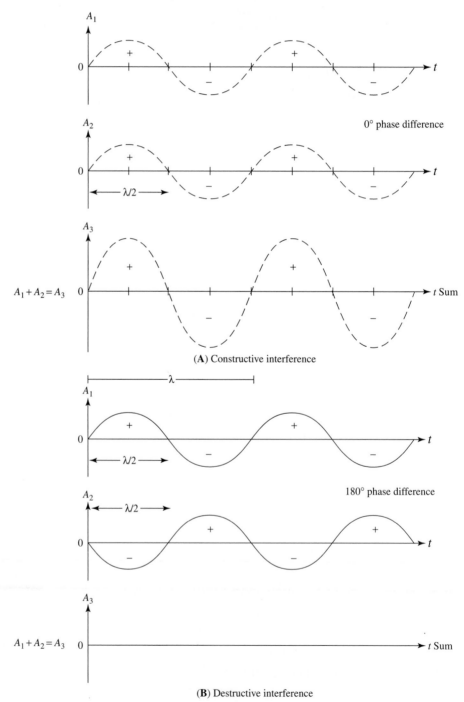

Figure 24.1 Constructive and Destructive Interference

ciple. Since light is indeed a wave, then we should be able to observe light-wave inter-ference patterns. In Chapter 8 on waves, we examined both temporal interference (beats) and spatial interference (standing waves on a string). Beats can be observed with a pair of tuning forks that have slightly different frequencies. There is a slow rising and falling of the sound generated at the tuning fork's frequency with time. Standing waves were observed on a stretched string. An oscillator creates traveling wave pulses that move down a string and are reflected from the other end. The incident and transmitted pulses inter-fere, and at certain well-defined resonant frequencies, standing waves appear on the string. At resonance the string has a transverse amplitude that is sinusoidal. The nodes have no amplitude, while points between the nodes have an amplitude that varies sinusoidally from zero to a maximum and then back to zero.

Any repetitive phenomenon will exhibit a wavelike nature and interfere. Let's refresh our memories as to what interference is and what the requirements are so that sources will interfere in a way that is observable. The two basic types of interference are constructive and destructive as shown in Figure 24.1. Note that the interfering waveforms in Figure 24.1 are coherent. ***Coherence*** *is the property where two waveforms have a fixed phase difference between them.*

Coherence: The state in which two waves have a fixed phase difference.

In constructive interference, such as in Figure 24.1A, the phase difference is 0°. Destructive interference, such as in Figure 24.1B, has a phase difference of 180°. Constructive interference and destructive interference are easily observable at these fixed phase differences. But interference can occur at any phase difference between 0° and 360°. Figure 24.2 shows interference between two waveforms that are 90° out of phase. In some parts of the resultant waveform, the interference is constructive, and in other parts it is destructive.

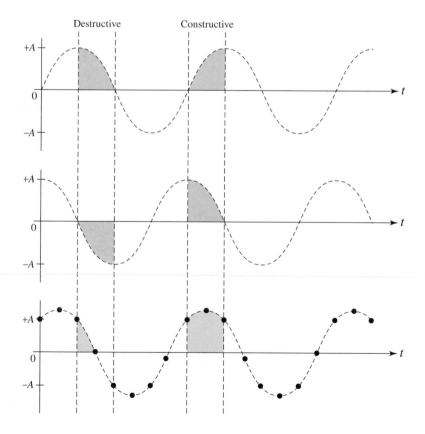

Figure 24.2 Interference between Two Waves 90° Out of Phase

In the filament of an incandescent light bulb, electromagnetic radiation is emitted in the form of light by the oscillating electrons. Each oscillating electron is an independent emitter, and each emitter has a random phase difference that varies from that of the other emitting electrons. The emitters interfere with one another, but because of the randomness of the phase differences, no periodic interference pattern is detectable. We see a fairly constant amplitude over time. Incoherence is the primary reason that the interference of light is so difficult to see in nature.

24.2 Double-Slit Interference

Double-slit interference as observed in the laboratory has a general setup much like that shown in Figure 24.3. A low-power, helium-neon laser that emits red light is often used because laser light is by its very nature coherent. The distance between the double-slit source and the screen is D. The double slits appear as a single source in Figure 24.3 because the spacing between the two slits (d) is very small compared to D ($d \ll D$). The image on the screen is a pattern of light and dark bands. The light bands are intensity maxima where the interference is constructive. The dark bands are where the interference is totally destructive; they lie between the light bands. The light bands are usually used for experimental purposes because they are easier to see and aren't as spread out as the dark bands. The center of a light band is easier to detect than the center of a dark band. The zeroth-order line from the source to the screen can be considered the system's axis, and there is a first-order maximum above and below the axis, a second-order maximum above and below the axis, and so on. There can be many orders of maxima, but only two are shown here for the sake of clarity. The angles that the maxima make with the central axis are θ_1, θ_2, and so on, depending on which order maximum is chosen. The order is determined by the size of the angle that the order makes with the axis,

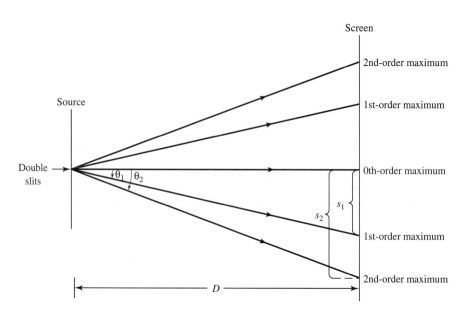

Figure 24.3 Double-Slit Interference

with the smaller angle having the smaller order. The light traveling to the screen is again represented by rays.

So what particular set of circumstances generates the pattern on the screen? For the answer, turn to Figure 24.4, where the spacing between the slits is d. Remember again that $d \ll D$, and thus we need two drawings. The interference pattern on the screen is a result of the path difference between the two rays, $\Delta\lambda$. If we look at the central maximum (zeroth-order), the path difference between the two rays is zero. The upper and lower rays are the same length. As the order above and below the central axis increases, so does the path difference. For an nth-order maximum, where n is the order number, one of the rays will be a little longer than the other. For the nth-order maximum shown in Figure 24.4, the upper ray is a little longer than the lower one. If the length $\Delta\lambda$ were taken out of the longer upper ray, what remained of the upper ray would be the same length as the lower ray. If the path difference $\Delta\lambda$ were equal to one-half wavelength of the light hitting the two slits (a 180° phase difference), then the band on the screen would be a minimum because the two light rays would destructively interfere. If $\Delta\lambda$ were a full wavelength (360° or 0° phase difference), then the band would be a maximum because the two light rays would constructively interfere.

d is greatly enlarged for clarity. Therefore, the upper light ray intersects the screen axis very close to the double slits, and $\tan \theta_n \approx s_n/D$.

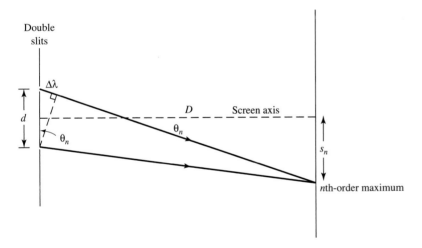

Figure 24.4 Path Difference ($\Delta\lambda$) Creates an Interference Pattern.

In deriving an equation for interference, we would probably know the slit spacing, d; the wavelength of the light, λ; and whether the band we were looking at was a maximum or a minimum as well as its order. Some of this information we could obtain from the screen. We would also have to know the distances D and s_n. Since the interference is a result of the path difference, we need to first concentrate on the small right triangle on the left side of Figure 24.4.

$$\sin \theta_n = \frac{\Delta\lambda}{d} \qquad \Delta\lambda = n\lambda$$

$$\sin \theta_n = \frac{n\lambda}{d}$$

(Equation 24.1)

(Equation 24.2)

$$\tan \theta_n = \frac{s_n}{D}$$

The hypotenuse is d, the slit spacing. $\Delta \lambda$ is the far side of the small right triangle from the order angle, θ_n. Although θ_n can't be measured at the double slits because d is so small, the angle between the screen axis and the light rays to the nth-order maximum is also θ_n because of similar triangles. D and s_n are easily measured, and the tangent of θ_n can be calculated. In Figure 24.4, if $d \ll D$, then the intersection point between the upper ray and the central axis (the dotted line) originates at the double slits instead of between the slits and the screen as in the figure. The result is that the small triangle with d as the hypotenuse and the large triangle with the upper ray as the hypotenuse are similar triangles; θ_n is the same for both triangles.

EXAMPLE 24.1

The distance from a pair of double slits to a screen is 4.8 cm. The distance from the central axis to the second-order maximum is 5.6 cm. The spacing between the slits is 1.67×10^{-6} m. (a) What is the wavelength of the light hitting the double slits that produces the interference pattern? (b) What is the light's color?

SOLUTION

(a) $$\sin \theta_n = \frac{n\lambda}{d} \qquad d \sin \theta_n = n\lambda \qquad \lambda = \frac{d \sin \theta_n}{n}$$

$$\tan \theta_n = \frac{5.6 \text{ cm}}{4.8 \text{ cm}} \qquad \tan \theta_n = 1.17$$

$$\theta_n = 49.4° \qquad \sin \theta_n = 0.76$$

$$\lambda = \frac{(1.67 \times 10^{-6} \text{ m})(0.76)}{2}$$

$$\lambda = 634 \text{ nm}$$

(b) The light is at the red end of the spectrum.

24.3   Diffraction Gratings and X-Ray Diffraction

A diffraction grating is a piece of glass or transparent plastic that has a very large number of parallel lines scratched or etched into the surface. A diffraction grating will give the same kind of interference pattern that a double slit will, with the added benefit that the pattern is more intense and easier to see. With double-slit interference there are only two light sources, and the resulting interference pattern is therefore rather dim. With a diffraction grating it's as if there are many double slits operating simultaneously on the same coherent source and superimposing their maxima for a much brighter line on the screen. The brighter lines on the screen correspond to locations where all of the light that interferes there is in step, or in phase. Thus, the phase differences are 0°, 360°, 720°,

1080°, and so on. Of course, for something periodic such as a light wave, when you go through 360°, it's as if you've returned to your starting point, ready to begin all over again.

If polychromatic (many different wavelengths) white light hits a diffraction grating, a rainbow-like spectrum results. For all orders of maxima, the longest wavelength (red) is deviated most and would be found on the outside (the side away from the screen axis) of the color spread. Blue light with the shortest wavelength is deviated least and would be found on the inside of the color spread. This order of color is opposite the order that results when a white light spectrum is formed by refraction through a prism. Blue light is deviated most when it is refracted, and red light least. Remember that in refraction, the index of refraction in a material medium is the ratio of the speed of light in a vacuum to the speed at which that particular wavelength of light travels in a medium. Blue light travels more slowly than red light in a medium such as glass and is deviated most.

$$\sin \theta = \frac{n\lambda}{d}$$

In the double-slit interference equation, the longer the wavelength, the greater the deviation.

For most materials the index of refraction decreases with increasing wavelength. Blue light has a short wavelength, so it has a higher index of refraction than does red light. A high index of refraction means that the light is bent (deviated) from its path more. In refraction, blue light is deviated more than red light.

EXAMPLE 24.2

The lines on a diffraction grating are ruled 300 lines per millimeter. (a) What is the angular width of the second-order white light spectrum? (b) Does the red end of the second-order spectrum overlap the blue end of the third-order spectrum?

SOLUTION

(a)
$$\sin \theta = \frac{n\lambda}{d} \qquad \text{(where } n = 2)$$

$$\lambda_r = 700 \text{ nm (red)} \qquad \lambda_b = 400 \text{ nm (blue)}$$

$$d = \frac{1}{300/\text{mm}} \qquad d = 3.33 \times 10^{-6} \text{ m}$$

$$\sin \theta_r = \frac{(2)(700 \text{ nm})}{3.33 \times 10^{-6} \text{ m}} \qquad \theta_r = 24.9°$$

$$\sin \theta_b = \frac{(2)(400 \text{ nm})}{3.33 \times 10^{-6} \text{ m}} \qquad \theta_b = 13.9°$$

$$\theta_r - \theta_b = 11°$$

(b)
$$\theta_r = 24.9°$$

$$\sin \theta_b = \frac{(3)(400 \text{ nm})}{3.33 \times 10^{-6} \text{ m}}$$

$$\theta_b = 21.1°$$

where θ_b (third-order) $< \theta_r$ (second-order). The second- and third-order spectra overlap.

Some diffraction gratings are silvered and operate by reflected instead of transmitted light. So when white light hits the surface of an object with many closely spaced lines, such as an old vinyl record or a compact laser disc, a white light spectrum is produced, and we see all of the colors of the rainbow. But any object that has a repetitive structure should produce a diffraction pattern if the wavelength of the light hitting the regularly spaced lines and grooves is short enough.

Much structural information can be gained about molecules that are arranged in regular crystal arrays if the array is used as a diffraction grating. The spacing between atoms in a crystal are on the order of 0.1 nm. In the double-slit equation, $\sin \theta$ can't be greater than 1, and thus the wavelength of the light hitting the crystal can be *at most* 0.1 nm. The shortest wavelength of visible light is 400 nm, or 4000 times too long. If we're going to examine the structure of a crystal lattice using diffraction, the light will have to be in the X-ray region of the electromagnetic spectrum.

Figure 24.5 shows two X rays reflecting off the top two layers of a crystal with a cubic symmetry. (The real crystal would extend into and out of the page.) The bottom incoming ray that penetrates into the first crystal layer travels an extra path length of $2 \Delta\lambda$. If it is to emerge from the crystal in phase with the upper ray, this extra length must equal $n\lambda$. For the two rays to interfere constructively, the path difference must be equal to an integral number of wavelengths.

$$\sin \theta = \frac{\Delta\lambda}{d} \qquad \Delta\lambda = d \sin \theta$$

$$Pd = 2d \sin \theta \qquad Pd = n\lambda$$

$$n\lambda = 2d \sin \theta$$

(Equation 24.3)

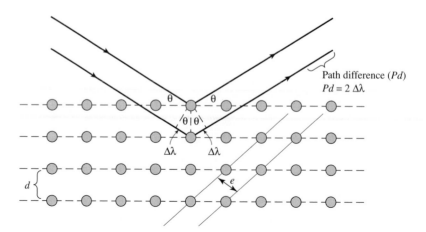

Figure 24.5 X-Ray Diffraction

A detector scans a small crystal sample, moving in a circular arc with the crystal at the center of curvature, to detect the pattern of minima and maxima that gives information on the crystal structure. But the pattern isn't as simple as that found on a screen for a double-slit interference pattern. Even in the simplest crystal structure such as the cubic one shown in Figure 24.5, there is more than one spacing that the detector will see. It can also see a spacing distance e between the diagonal crystal planes as shown in the figure, depending on the orientation of the crystal to the incoming X rays. So the information given by an X-ray diffraction apparatus can be quite a puzzle, and deciphering the structure of a complicated structure requires a lot of experience. The structure of the DNA double-helix crystal was discovered in the early 1950s using X-ray diffraction techniques.

24.4 Thin Films and Antireflection Coatings

Visible light interference patterns can be seen in nature as the rainbow colors that appear in the reflections from thin oil films on water or from soap bubbles. Both the oil film on water and the soap bubble are examples of thin films, as illustrated in Figure 24.6. The rainbow colors are the result of constructive interference, where $n = 0, 1, 2, 3, \ldots$. If two glass plates are very close together but not quite in contact, you'll see a pattern of light and dark bands. The dark bands are the result of destructive interference, where $n = \frac{1}{2}, \frac{3}{2}, \frac{5}{2}, \ldots$. This particular pattern is used as a test for optical flatness.

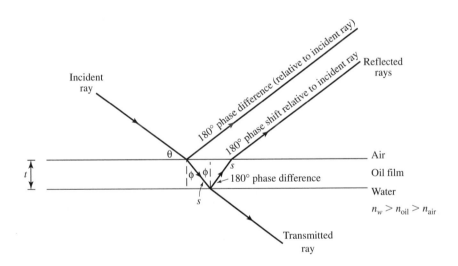

Figure 24.6 Thin-Film Interference

Thin films produce light and dark bands and colorful spectra in the following way: When the incident ray of Figure 24.6 hits the first air-film interface, it is both refracted and reflected. The reflected ray undergoes a 180° phase change, just as a wave pulse does when hitting the fixed end of a string tied to a wall. The refracted ray undergoes no phase change. When the transmitted ray hits the second film-air interface, it is again refracted and reflected. The reflected ray undergoes another 180° phase change, and when it finally reemerges into the air, both reflected rays have undergone a 180° phase change. However, a path difference is caused by the thickness of the film, and this path difference

(*PD*) determines how the reflected light rays will interfere. The thickness of the film must be comparable to the wavelength of the incident light for the interference effects to be obvious.

$$t = \text{film thickness} \qquad PD = \text{path difference}$$

$$PD = 2s \qquad \cos\phi = \frac{t}{s}$$

$$s = \frac{t}{\cos\phi} \qquad PD = \frac{2t}{\cos\phi}$$

$$PD = n\lambda \qquad \text{where } n = 0, 1, 2, 3, \ldots \text{ (constructive interference)}$$

$$n = \tfrac{1}{2}, \tfrac{3}{2}, \tfrac{5}{2}, \ldots \text{ (destructive interference)}$$

$$n\lambda = \frac{2t}{\cos\phi}$$

(Equation 24.4)

Refracting lenses in an optical instrument should be able to transmit as much light as possible. In good optical-quality glass, about 96% of the light striking it is transmitted, and about 4% reflected. That doesn't sound like enough of a loss to worry about, but in a precision instrument there may be quite a number of individual lenses. In such a multiple-lens system, reflective losses can easily exceed 33% of the incident light. Lenses are often coated with an antireflection coating to minimize such losses. For an antireflection coating, Equation (24.4) is used with $n = \tfrac{1}{2}$ for destructive interference. Since light entering a lens is usually parallel to the optical axis, the angle ϕ is 0°, and the result is that the thickness of the coating must be $\tfrac{1}{4}$ the wavelength of the incident light. Remember that the wavelength of the incident light will shorten when it enters a medium with a higher index of refraction. The thickness of the coating must be $\tfrac{1}{4}$ of this shorter wavelength.

A good pair of binoculars will have an antireflection coating on the lens. It is easily noticeable because it makes the lens look purplish. The coating works well only for a particular wavelength of visible light. The wavelength chosen is usually right in the middle of the visible spectrum: green. Thus, what remains of the reflected light from the lens will have only wavelengths from the blue and red ends of the spectrum. Blue and red mix to give purple.

The reflections are reduced to about 1% with a single coating. The single coating will have an index of refraction that is intermediate between that of the air and that of the lens material. If a secondary coating is added, reflection can be reduced still further to about 0.5%. The secondary coating will have an index of refraction intermediate between those of the first coating and the lens material. The more intermediate coatings that are added, the less the reflected light. If there is no reflected light, then there will be an infinite number of coatings, with the index of refraction smoothly rising from 1 to that of the lens material.

An abrupt air-glass interface reflects back some of the light energy, creating an impedance mismatch. The concept of impedance brings to mind elastically colliding spheres. A perfect impedance match occurs when a moving sphere collides elastically with a stationary sphere of the same mass. The originally moving sphere is stationary after the collision, and the originally stationary sphere moves off with all of the initial energy and momentum of the other sphere. If the two spheres were of different masses, then the smaller sphere would rebound back in the direction from which it came. The concept is the same with reflected light. Some of the original energy is lost in the "collision." We saw a very similar occurrence with a traveling wave pulse moving down a stretched string. When the wave pulse encounters a string segment with a linear density higher than the linear density of the string on which it has been traveling, some of the incident energy will be reflected back the way it came. If the pulse encounters a string segment with an effectively infinite linear density (hits a wall), then all of the incident energy will be reflected back 180° out of phase with the original pulse.

The colliding spheres, a traveling wave pulse on a string, and reflected light are three entirely different phenomena. Yet despite their differences, they can all be covered under the single concept of impedance. The concept of simple harmonic motion is also an umbrella concept for seemingly disparate phenomena. An oscillator can be driven by gravitational energy, by stored energy in a spring, or by electromagnetic energy in an inductor-capacitor circuit. But all three are described by the same mathematics!

EXAMPLE 24.3

Magnesium fluoride is used as an antireflection coating for glass lenses. The index of refraction is 1.38. The coating is effective for green light at 550 nm. How thick must the coating be?

SOLUTON

$$n\lambda = \frac{2t}{\cos \phi} \qquad t = \frac{n\lambda}{2} \cos \phi \qquad \lambda = \frac{n_a}{n_c} \lambda_o$$

$$t = \frac{n}{2} \frac{n_a}{n_c} \lambda_o \cos \phi \qquad n = \frac{1}{2} \qquad \text{(Destructive interference)}$$

$$n_a = 1 \qquad \text{(Index of refraction of air)}$$

$$n_c = 1.38 \qquad \text{(Index of refraction of magnesium fluoride)}$$

$$t = \frac{1/2}{2} \frac{1}{1.38}(550 \text{ nm})\cos 0°$$

$$t = 9.96 \times 10^{-8} \text{ m}$$

The film thickness should be about 100 nm.

Impedance matching and a phase shift were mentioned above. *The phase shift occurs in the reflected pulse when the incident pulse meets a string segment with a higher linear density, or in the case of light when the incident ray hits an interface with a higher index of refraction.* In the case of the reflected rays in Figure 24.6, both rays emerge with a shift of 180° relative to the incident ray. When the bottom ray is reflected from the glass, it undergoes a 180° phase shift because glass has a higher index of refraction than the antireflection coating. Since both reflected rays have the same phase relationship, it wasn't necessary to take it into account in calculating the antireflection film thickness.

When a light ray is reflected from an interface with a larger index of refraction than that of the incident medium, the reflected ray undergoes a phase shift of 180°.

24.5   Resolving Power

If monochromatic light shines on a circular aperture where the diameter of the aperture is comparable to the wavelength of the light, a circular diffraction pattern will form. This pattern is called an Airy disk after Sir George Airy, the Astronomer Royal of England who first derived the expression for the intensity pattern. The circular diffraction pattern consists of a central bright circle of light with alternating dark and bright rings around it. The intensity of the rings drops off rapidly, especially if the diameter of the aperture is much greater than the wavelength of the incident light. This circular diffraction pattern limits the resolving power of optical instruments such as microscopes and telescopes. From ray tracing, if parallel light hits the face of a converging lens, it should ideally be focused into a dimensionless point at the focal length of the lens. What we actually get is the circular diffraction pattern known as the Airy disk. The angle subtended by the circle is given by the following equation:

(Equation 24.5)

$$\sin \theta = 1.22 \frac{\lambda}{D}$$

where λ is the wavelength of light and D is the diameter of the circular aperture. There are many orders as in the double-slit interference pattern, but since the intensity falls off rapidly with increasing order, we're concerned only with the first-order bright circle.

Rayleigh's criterion is a widely used convention for the resolution of two point objects. If a car approaches you from a large distance at night, the first thing that you see is a point of light from the approaching car's headlights. But there are two headlights, and when the car is near enough, the single spot becomes two. The eye can resolve two objects that subtend a minimum angle of about 5×10^{-4} rad. At the distance of most distinct vision (25 cm), the eye can resolve two objects about 0.1 mm apart. If the objects are any further away, the two objects merge and appear as one. *Rayleigh's criterion formalizes this concept by stating that two objects are barely resolvable if the center of one Airy diffraction disk from one of the objects coincides with the first minimum of the other.*

Rayleigh's criterion: Two objects are resolvable if the center of one Airy disk coincides with the first minimum of the other.

The result will be an image that looks like two overlapping circles. With that as the definition of *resolvability*, we can use Equation (24.5) for the angular separation of the image centers. The angular separation of the images is the same as that of the objects, and the angle is in radians.

EXAMPLE 24.4

If the two headlights mentioned above were 5 ft apart, what would the furthest distance of the car be for the two headlights to be barely resolvable?

SOLUTION
Refer to Figure 24.7.

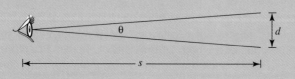

Figure 24.7 Resolving Power of the Eye

$$\theta = \frac{d}{s} \qquad d = 5 \text{ ft} \qquad \theta = 5 \times 10^{-4} \text{ rad}$$

$$s = \frac{d}{\theta} \qquad s = \frac{5 \text{ ft}}{5 \times 10^{-4} \text{ rad}}$$

$$s = 10{,}000 \text{ ft}$$

$$s = 1.9 \text{ mi}$$

24.6   Polarization

Light is a transverse wave with both an electric field and a magnetic field, as shown in Figure 24.8. The light that we see with our eyes is such a wave. The current through a filament in an incandescent light bulb heats up the filament, causing it to glow. Each atom in the filament becomes an emitter of visible light. The atoms emit light not in step but with random phases. This random-phase emission is the reason for the incoherence of most light sources. There is also another result of the atoms emitting randomly: unpolarized light. If you looked at the light wave in Figure 24.8A as if it were coming out of the page directly at you, it would appear as in Figure 24.8B. The electric and magnetic fields of a single wave segment would be at 90° to each other, but the orientation of the E–B axis can be anything from 0° to 360° (θ is 133° in the figure). The E field is oriented in a plane transverse to the direction of propagation, as is the B field. If the direction of the E field is known, so is the direction of the B field. For this reason, the phenomenon of polarization is concerned with the direction of the E field, and unpolarized light has a random orientation for the E field (many out-of-step emitters).

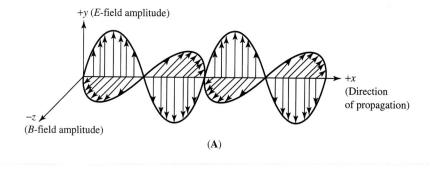

(A)

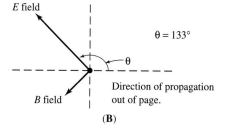

(B)

Figure 24.8 The Electric and Magnetic Fields Are at 90° to Each Other in an Electromagnetic Wave.

Unpolarized light entering the eye with randomly oriented, equal-intensity E fields can be represented as shown in Figure 24.9. (The angle between adjacent electric field vectors is 45°. Only eight electric field vectors are shown for clarity. In reality, there would be a seemingly infinite number with random orientations from 0° to 360°.) The intensity of an electromagnetic wave is proportional to the square of the amplitude, and a negative number squared is positive. Our eyes "see" or record the intensity. For this reason, the vectors in Figure 24.9 don't cancel out to zero. Otherwise, E_3 would cancel E_7, E_4 would cancel E_8, and so on.

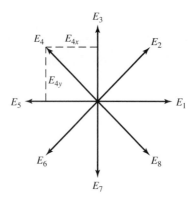

Figure 24.9 Random Electric Field Orientation of Unpolarized Light

If unpolarized light has a completely random orientation for the E field, the E field of the completely plane-polarized light will be oriented in one plane only. If Figure 24.9 represents unpolarized light, then completely plane-polarized light will have only an E field along the E_3–E_7 axis, the E_4–E_8 axis, the E_5–E_1 axis, or the E_6–E_2 axis. In plane-polarized light, all of the E fields are lined up in a single plane. (Thus, all of the B fields are also lined up.)

If most light is unpolarized, what creates polarized light? Many crystals found in nature have an optical axis. When light is incident on the crystal, the only light transmitted has its E field parallel to the optical axis. *If the component perpendicular to the optical axis is absorbed, then the crystal is said to be **dichroic**.* The polarizing filters commonly found in sunglasses are sheets of a dichroic material.

If unpolarized light is incident on a sheet of dichroic polarizing material, the transmitted light is plane-polarized with one-half the intensity of the original incident light. The reason is that the random E fields in Figure 24.9 are vectors, and each vector can be resolved into components parallel to and perpendicular to the optical axis. Because the E fields of the incident light are in random directions, the parallel and perpendicular components will be equal on average. An ideal polarizer will transmit light only with its E field parallel to the optical axis, and therefore it transmits only one-half of the incident intensity. Dichroic polarizing material works well as the lenses of a pair of sunglasses because it halves the intensity of already too intense sunlight.

The eye can't detect polarized light unaided, but a sheet of dichroic material can act as an analyzer to detect partially and completely polarized light. For the analyzer to work, we first need polarized light, which we can get by passing unpolarized light through a piece

Dichroic material: A material with an optical axis such that E-field components parallel to the components are transmitted, and components perpendicular to it are absorbed.

of polarizing material. It will emerge from the polarizing material plane-polarized as shown in Figure 24.10 and as given in the following equations:

$$E_\perp = \text{perpendicular component of } \vec{E}$$

$$E_\parallel = \text{parallel component of } \vec{E}$$

$$\cos\theta = \frac{E_\parallel}{E} \qquad E_\parallel = E\cos\theta \qquad I_{\text{trans}} = E_\parallel^2$$

$$I_{\text{trans}} = E^2\cos^2\theta \qquad E^2 = I_o$$

$$I = I_o\cos^2\theta \qquad\qquad\qquad \text{(Equation 24.6)}$$

Equation (24.6), called *Malus's law,* applies only to plane-polarized light incident on the analyzer. When the optical axis of the analyzer is lined up with the plane of polarization, all of the incident light is transmitted through the polarizer. When the polarizers are crossed ($\theta = 90°$), no light is transmitted.

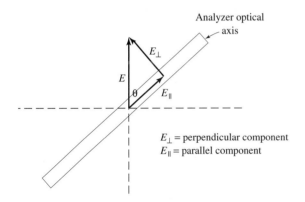

Analyzer optical axis

$E_\perp$ = perpendicular component
$E_\parallel$ = parallel component

Figure 24.10 Malus's Law

Polarized light can be produced by methods other than dichroism. Some crystals, such as calcium carbonate, are birefringent. The index of refraction is different for different directions for the propagation of light through the crystal. Light incident on the crystal is split into two rays that are plane-polarized at 90° to one another. There are two images of an object. If an analyzer is rotated until one of the images disappears, the other image is at its brightest. If the analyzer is then rotated another 90°, the original bright image disappears, and the vanished image reappears. All reflected light is partially polarized and can be completely plane-polarized if the criterion for Brewster's angle is met

> *All reflected light is partially polarized.*

$$\theta + 90° + \phi = 180°$$

$$\theta + \phi = 90° \qquad \text{(Criterion for Brewster's angle)}$$

$$n_a\sin\theta = n_b\sin\phi \qquad \phi = 90° - \theta \qquad n_a\sin\theta = n_b\sin(90° - \theta)$$

$$\sin(90° - \theta) = \cos\theta \qquad n_a\sin\theta = n_b\cos\theta$$

$$\frac{\sin\theta}{\cos\theta} = \frac{n_b}{n_a}$$

(Equation 24.7) $$\tan \theta = \frac{n_b}{n_a} \qquad \text{(Brewster's law)}$$

The criterion for Brewster's angle is that the incident plus refracted angles must be equal to 90°. If that is so, then the reflected light is plane-polarized with the plane of polarization parallel to the interface surface, as in Figure 24.11. The refracted light is partially polarized. The formula for Brewster's angle, given in Equation (24.7), is known as *Brewster's law.*

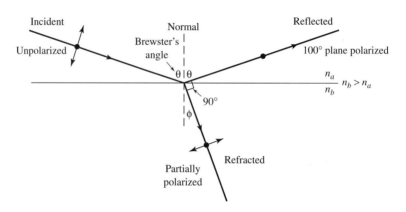

Figure 24.11 Brewster's Angle

Brewster's law is *not* valid for mirrors. There must be a refracted component of the incident light ray. For a mirror all of the incident light is reflected, and there is no refracted ray to carry away the *E*-field component that is perpendicular to the plane of incidence.

Because reflected light is partially or completely polarized, dichroic polarizing material makes good sunglass lenses because it reduces glare. The dichroic material is an analyzer for reflected light and can partially or totally block glare. Polaroid materials also make good sunglasses because they reduce intensity by one-half by polarizing the incident unpolarized light, and they reduce glare significantly by acting as an analyzer for the polarized reflected light.

24.7 Scattering and Blue Skies

The atmosphere scatters sunlight, and for this reason, during the day, light comes from areas of the sky where the sun isn't present. The size of the nitrogen and oxygen molecules predominantly present in the atmosphere are smaller than the wavelengths of visible light, and the molecules scatter light inversely proportionally to the fourth power of the wavelength $(1/\lambda^4)$. The shorter the wavelength, the greater the proportion of light scattered. If a light ray coming from the sun is partially scattered, the scattered light will have a direction perpendicular to the original ray. The scattered light will have an intensity much reduced from that of the original ray, and the scattering centers will be all over the sky. The light entering the eye will be the sum of the light from all of these individual scattering centers, and that is why we see a uniform background glow from all over the sky. Blue light has a shorter wavelength than red light, so it is scattered more than red light. That's why the color of the sky is blue.

The blue light from the sky is also partially polarized. The oscillating electric field of a traveling light wave will cause the electrons in the molecules that scatter the light to oscillate in a plane perpendicular to the direction of propagation. These accelerated charges in turn reradiate in a direction perpendicular to the original direction of propagation. Thus, scattered light is polarized. If you take a pair of Polaroid sunglasses, hold a lens to the blue sky, and then rotate it, you'll see the blue darken and then lighten. The lens acts as an analyzer. The blue isn't completely polarized because it comes from all over the sky and not just from one source.

Sunsets are red because much of the blue in white sunlight has been scattered out of it, leaving it reddish. When the sun is on the horizon at sunset or sunrise, you can look directly at it because the sunlight has to travel through a much greater thickness of atmosphere than when it is overhead. The atmosphere absorbs some of the light, and that fact, added to the scattering, greatly reduces the intensity. Without an atmosphere the sky would appear black, as it does at night or in space, because there would be no scattering.

Summary

1. When two light waves experience *superposition,* they may experience *constructive interference,* where their amplitudes add; *destructive interference,* where their amplitudes subtract; or a combination of both.

2. If two parallel light waves have the same phase, they are said to be **coherent.**

3. When a plane wave strikes an opaque surface containing a double slit, some light will *diffract* through the openings and will cause an *interference pattern* on a far screen. Equation (24.1) describes the pattern that will be produced, where n is the *order number* desired, λ is the wavelength of the light, and d is the separation between slits.

4. An antireflection coating is a *thin film* on a surface that is designed to destructively interfere with any reflected light of an undesired wavelength of light [Equation (24.4)].

5. Light will also diffract when shining through a circular hole. The angle of diffraction is dependent on the diameter of the hole and the wavelength of light involved, as shown in Equation (24.5).

6. When unpolarized light enters a *polarizer,* one-half of the light intensity is lost. When linearly polarized light enters a polarizer, the amount of intensity that is lost is dependent on the angle between the polarization of the polarizers. This relationship can be seen with *Malus's law* [Equation (24.6)].

Problem Solving Tip 24.1: When dealing with polarized light passing through multiple polarizers, remember that the angle in Malus's law is the angle difference between two adjacent polarizers. Thus, if we had vertically polarized light passing through a polarizer angled 10° to the vertical, θ would be 10°. If it then passed through another polarizer angled 30° with the vertical, the next θ would be 20° (there is a 20° difference between the first and second polarizer), and so on. Don't forget to drop the intensity by one-half if the original light was unpolarized.

Key Concepts

All reflected light is partially polarized.

When a light is reflected from an interface with a larger index of refraction than that of the incident light medium, the reflected ray undergoes a phase shift of 90°.

COHERENCE: The state in which two waves have a fixed phase difference.

DICHROIC MATERIAL: A material with an optical axis such that E-field components parallel to the components are transmitted, and components perpendicular to it are absorbed.

RAYLEIGH'S CRITERION: Two objects are resolvable if the center of one Airy disk coincides with the first minimum of the other.

Important Equations

(24.1) $\sin \theta_n = \dfrac{n\lambda}{d}$ (Double-slit diffraction)

(24.2) $\tan \theta_n = \dfrac{s_n}{D}$ (Formula for finding the angle in double-slit diffraction)

(24.3) $n\lambda = 2d \sin \theta$ (Crystal diffraction)

(24.4) $n\lambda = \dfrac{2t}{\cos \phi}$ (Thin film)

(24.5) $\sin \theta = 1.22 \dfrac{\lambda}{D}$ (Resolving power)

(24.6) $I = I_o \cos^2\theta$ (Malus's law for polarization)

(24.7) $\tan \theta = \dfrac{n_b}{n_a}$ (Brewster's law)

Conceptual Problems

1. To produce an interference pattern, the waves involved must be:
   a. in a narrow beam   b. electromagnetic
   c. coherent   d. the same wavelength

2. When a spectrum is produced by a diffraction grating, the light deviated least is:
   a. red   b. blue
   c. green   d. dependent upon the slit spacing

3. In a double-slit interference pattern, the path difference for the first-order maxima on both sides of the central maximum is:
   a. one wavelength
   b. two wavelengths
   c. one-quarter wavelength
   d. one-half wavelength

4. The greater the number of lines ruled onto a grating per unit width:
   a. the narrower the spectrum produced
   b. the shorter the wavelengths that can be diffracted
   c. the broader the spectrum produced
   d. the longer the wavelengths that can be diffracted

5. X rays are used in crystal diffraction because:
   a. their wavelengths are much shorter than the atomic spacings
   b. their wavelengths are comparable to atomic spacings
   c. their wavelengths are much longer than atomic spacings
   d. they are easy to polarize

6. It's impossible to polarize:
   a. white light   b. radio waves
   c. sound waves   d. X rays

7. The sky is blue because:
   a. the lens of the eye reflects long wavelengths of light
   b. air molecules are blue
   c. light scattering is more efficient for long wavelengths of light
   d. light scattering is more efficient for short wavelengths of light

8. Sunsets are red because:
   a. the blue light is scattered out by the thickness of the atmosphere when the sun is low
   b. the short-wavelength light is refracted more by the thickness of the atmosphere
   c. the long-wavelength light is refracted more by the thickness of the atmosphere
   d. the thick atmosphere absorbs the short wavelengths

9. When light hits a thin slit comparable to its wavelength:
   a. it can pass only straight through to a screen on the other side
   b. the slit acts as a second source radiating light in a hemisphere from the slit
   c. it can pass through only if there is light from a second slit with which to interfere
   d. it is blocked because the slit is too narrow to pass through

10. For reflected and refracted rays of light:
    a. there is a 180° phase change when light reflects from a medium with a higher index of refraction
    b. there is a 180° phase change when light reflects from a medium with a lower index of refraction
    c. there is a 180° phase change when light refracts into a medium with a higher index of refraction
    d. there is no phase change when light reflects from a medium with a higher index of refraction

11. For polarized light:
    a. the electric vector of electromagnetic radiation is absorbed by dichroic material
    b. the magnetic vector of electromagnetic radiation is absorbed by dichroic material

c. the polarizer blocks anything not lined up with its optical axis

d. a reflected light ray is partially polarized perpendicular to the surface from which it was reflected

Exercises

Interference and Coherence

1. Draw the wave that results from the superposition of waves *A* and *B* in Figure 24.12.

2. Draw the wave that results from the superposition of waves *A* and *C* in Figure 24.12.

3. Draw the wave that results from the superposition of waves *B* and *C* in Figure 24.12.

Figure 24.12 (Exercises 1, 2, and 3)

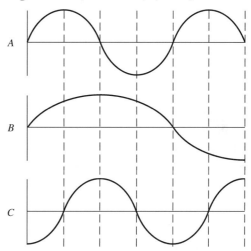

Double-Slit Interference

4. An infrared light source is directed onto a wall with two narrow slits in it 25 mm apart. Five meters beyond the wall is a heat-sensitive screen with an interference pattern on it. The distance from the central peak to the first-order maximum is 6 cm. What is the source's wavelength?

    Answer: 0.3 mm

5. Two narrow slits 2 mm apart are cut into a thin wooden plank. The plank is placed 3 m from a wall, and a red laser with a wavelength of 652 nm illuminates the slits. (a) What is the distance between the central and the first-order maxima? (b) What is the distance if the light is blue and of wavelength 400 nm?

    Answer: (a) 0.98 mm
    (b) 0.6 mm

6. A lab technician is instructed to measure the slit separation for each unmarked double slit in a collection used for a student interference lab. The slits are much too small to measure the separation, and a microscope isn't available, so the technician decides to determine the separation indirectly. Using a helium-neon gas laser of wavelength 652 nm, he sets up an interference pattern on the wall of his office 2 m away. The separation between the central maximum and the first-order maximum is 15 cm. What is the slit separation?

    Answer: 8.69 μm

7. The lines on a diffraction grating are ruled 400 lines per millimeter on a screen 50 cm distant. (a) How far from the central maximum is the third-order maximum for green light of wavelength 550 nm? (b) Which is closer to the central axis, the third-order green peak or the third-order blue (400-nm) peak?

    Answer: (a) 33 cm
    (b) Third-order blue peak

8. A blue light of wavelength 410 nm and a green light of wavelength 550 nm are shone into a diffraction grating ruled 250 lines per millimeter. (a) How many blue peaks are visible on a screen placed 2 m away? (b) How many green peaks are visible on the same screen?

    Answer: (a) 9 peaks
    (b) 7 peaks

9. With the same equipment used in Exercise 8, is the fourth-order green or the third-order blue maximum closer to the central axis of the screen?

Answer: Third-order blue

10. White light with wavelengths within the range 400 to 700 nm is shone through a diffraction grating with 3000 lines per centimeter. If the light shines on a screen 3 m away from the grating, what is the width of the first-order spectra?

Answer: 27 cm

11. A beam of blue light of wavelength 440 nm falls on a double slit and forms an interference pattern on a screen across the room. If the light source is changed, what wavelength of light will produce a *minimum* at the position of the first-order maximum of the blue light?

Answer: Infrared light of 880 nm

*12. Laser light of wavelength 633 nm is normally incident on a thin double slit of separation 0.1 mm cut into a piece of opaque plastic floating on the surface of a pool. What is the width of the first-order maximum at the bottom of the pool 4.5 m below?

Answer: 4.3 cm

*13. Show that for white light hitting a diffraction grating, the violet (400-nm) portion of the third order overlaps the yellow-orange (600-nm) portion of the second order regardless of the grating's spacing.

*14. Light of two different wavelengths is used in a double-slit experiment. The location of the third-order maximum for yellow light of wavelength 600 nm coincides with the location of the fourth-order maximum of the other wavelength of light. What is the wavelength of the other light?

Answer: 450 nm

Diffraction Gratings and X-Ray Diffraction

15. A radio transmitting complex operating at 16 MHz has a row of vertical antennas spaced 60 m apart along a north-south line. How many intensity maxima are there?

Answer: 14

16. The radio transmitter operating in Exercise 15 now changes its broadcasting frequency to 5 MHz. How many intensity maxima are there?

Answer: 4

Thin Films and Antireflection Coatings

17. A thin film 110 nm thick covers a window pane 4 mm thick. The film is found to be antireflective for cyan light of 490 nm. What is the film's index of refraction?

Answer: 1.11

18. An engineer wishes to apply a thin coating to a glass lens that will reflect any red light that hits it but that will pass all other wavelengths. The coating has an index of refraction of 1.38, and the red light has a wavelength of 700 nm. What is the thickness of the coating?

Answer: 254 nm

19. Soap with an index of refraction of 1.4 forms a thin film approximately 100 nm thick on the surface of water. If a flashlight is shone directly down into the water, what color does the film appear in the light?

Answer: Blue-violet

*20. A soap bubble floating in the air appears to be violet or blue-red under white light, indicating that green light of around 550-nm wavelength is not being reflected from the film. The soap film has an index of refraction of 1.25. What is the minimum thickness of the soap film?

Answer: 220 nm

Resolving Power

21. A jet has running lights under each wing tip. How high would the jet have to be if the wing span were 70 ft (about 21 m) and an observer couldn't resolve the two lights from the ground?

Answer: 26.25 mi

22. An aircraft is sent to investigate a UFO hovering over Albuquerque, New Mexico. At a range of 6.5 km, what was thought to be one bright light is actually discovered to be two. How far apart are the lights?

Answer: 3.25 m

23. A radar can resolve objects 30 ft apart at a range of 1 mi. If the radar operates at 15,000 MHz, what is the minimum width of the radar's antenna?

Answer: 4.3 m

24. The Hale telescope at Mount Palomar in California has a concave mirror with a diameter of 5 m. How far apart must two objects on the moon be before the telescope can resolve them? The distance to the moon is 3.8×10^5 km. Use 550 nm as the light's wavelength.

Answer: 50.9 m

Polarization

25. An unpolarized beam of light is incident on a polarizer with its optical axis vertical. (a) What percentage of the light passes through the polarizer? (b) On the other side of the polarizer, an analyzer is set up with its optical axis horizontal. What percentage of the light passed by the original polarizer passes through to the other side of the analyzer? (c) A second analyzer with its optical axis oriented 45° to the vertical is placed between the polarizer and the initial analyzer. How much of the original unpolarized light passes through the polarizer and the two analyzers to be emitted from the second analyzer?

Answer: (a) 50%
(b) 0%
(c) 12.5%

26. Unpolarized light shines into a polarizer/analyzer apparatus as shown in Figure 24.13. If the light's original intensity is

Figure 24.13 (Exercise 26)

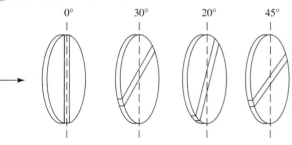

500 W/m^2, what is the intensity of the light as it emerges from the far side of the apparatus?

Answer: 15 W/m^2

27. At what angle must an incident ray of sunlight strike the surface of a calm pond for the reflected light to be completely polarized?

Answer: 36.9°

28. Vertically polarized light is incident on dichroic material with its optical axis at an angle of 20° to the vertical. At what angle should the optical axis of another piece of polarizing material be placed relative to the first piece so that the light passing through the second polarizer is 10% of the original intensity?

Answer: 70.3°

*29. What is Brewster's angle for a piece of glass of index of refraction 1.51 submerged in water?

Answer: 41.4°

*30. An intense ray of sunlight is incident on a glass auto windshield of index of refraction 1.58, but the reflected glare can be completely blocked by a pair of polarized sunglasses. What is the angle of refraction for the beam?

Answer: 32.3°

Modern Physics

These last three chapters (Chapters 25–27) are a short (and necessarily incomplete) overview of some of the topics found in textbooks on modern physics. Modern physics encompasses such disciplines as relativity, quantum mechanics, nuclear physics, cosmology, and subatomic particles. All of the physics studied up to now in this text involves concepts and principles that can be directly experienced by the five senses. However, the worlds described by modern physics are beyond the direct experiences of the senses and form the far boundaries of our local reality. The realities of modern physics are often mathematical realities. Thus, the concepts of modern physics are often difficult to express in terms of common sense! From special relativity we find that time isn't experienced at the same rate to observers moving relative to one another near the speed of light, and this idea spells the death of simultaneity. At the subatomic level, matter becomes a wave function, and electrons become standing waves in the presence of a nucleus. Space itself is bent by a gravitational field. Gravitational fields so strong that not even light can escape are generated by massive stars called *black holes,* whose mass forever collapses inward.

However, this short overview of modern physics will tie some of these concepts to current uses that have been found for them (both good and bad) and will open a small window on the very strange workings of our universe.

PART V

Matter waves from speeding electrons make possible the electron microscope.

Waves, Particles, and Special Relativity

At the end of this chapter, the reader should be able to:

1. Solve for the frequency of a **photon** with known energy.

2. Calculate the kinetic energy that an electron would have if it were stripped off a known metal by an incident photon of given frequency in the photoelectric effect.

3. Solve for the *time dilation* that you would observe if you viewed a clock receding away from yourself at a known *relativistic* speed.

4. Calculate the *de Broglie wavelength* of a particle traveling with a given momentum.

5. Solve for the uncertainty in a particle's position if its velocity is known to lie within a given range of values.

25.1   The Photoelectric Effect

Photoelectric effect: Effect produced when light shining on certain metals liberates electrons from the surface.

Work function: The minimum energy needed for an electron to leave the surface of a metal.

Double-slit and thin-film interference had proved to physicists in the late 1800s that light was definitely a wave. The theoretical work of James Clerk Maxwell on electromagnetic radiation in the 1860s had also predicted radio waves and had shown that electromagnetism and light were one and the same wave phenomenon, with the only difference being the wavelength. Physicists thought that all electromagnetic phenomena could be explained by wave theory, and these scientists had a very difficult time explaining the *photoelectric effect* because the explanation couldn't be put in terms of wave theory. Since light was known to be a wave, physicists were stumped!

It was known that when light shone on certain metals, such as sodium, the metals would emit electrons, and these free electrons could be detected as a weak current by a galvanometer, as shown in Figure 25.1. Note that the sodium metal is enclosed by a glass tube with a vacuum inside. When light shines on the sodium, electrons are liberated from the surface of the sodium metal by the energy that the light waves carry. These electrons travel to another metal plate known as a *collector.* At the collector, the electrons enter a wire and are detected as a weak electric current by the galvanometer. It takes a certain amount of energy to get the electrons to leave the surface of the sodium. This minimum energy is known as the *work function.* Electrons that absorb only this minimum energy are able to leave the sodium, but they don't have any kinetic energy to travel to the collector on their own. They must be helped along by a positive voltage applied to the collector. If the electrons have energy in excess of the work function, the energy will show up as kinetic energy, and you can measure the kinetic energy of the electrons by putting a negative charge on the collector as in Figure 25.1. The negative charge repels the electrons, and only those electrons with sufficient kinetic energy to overcome the repulsion will make it to the collector.

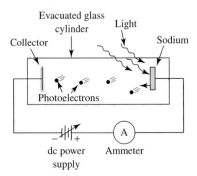

Figure 25.1 Photoelectric Effect

In studying the photoelectric effect, when light with varying intensity and wavelength was shone on the sodium metal, the results illustrated in Figure 25.2 were observed. Note in the graph that the power supply voltage is varied. It can be varied in both magnitude and polarity. With light of a given wavelength and intensity shining on the sodium, the power supply voltage is varied, and the photocurrent is plotted as a function of the voltage as shown in Figure 25.2A. If the collector voltage is positive and high enough, all of the available photoelectrons hit it, and the photocurrent saturates. As the collector voltage becomes more and more negative, fewer electrons have the necessary kinetic energy to reach it. When the collector current drops

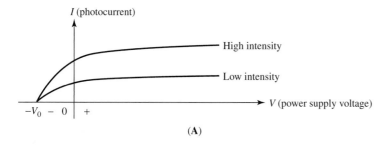

(A)

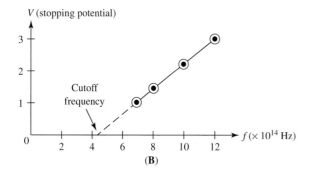

(B)

Figure 25.2 Characteristics of the Photoelectric Effect

to zero, the photoelectrons with the highest kinetic energy no longer have enough energy to reach it. *The negative voltage where the photocurrent is zero is called the stopping potential.* If the light intensity increases, the saturation photocurrent increases. When the intensity decreases, the saturation photocurrent decreases. However, the stopping potential remains the same for a given wavelength of incident light regardless of the intensity. In Figure 25.2B, the frequency of the incident light is varied, and the stopping potential is plotted as a function of the frequency. As the frequency decreases, the stopping potential decreases, until at the ***cutoff frequency*** the photoelectric effect ceases entirely.

The above results contradict the wave theory of light in the following ways:

1. The intensity of light is measured in power per unit area. The kinetic energy imparted to photoelectrons should be a function of the intensity. The greater the intensity, the more kinetic energy the photoelectrons should have. But the kinetic energy of the photoelectrons is not a function of the intensity as shown in Figure 25.2A. The stopping potential $(-V_0)$ is a measure of the electron's KE, and it is the same for two very different intensities.
2. According to the wave theory of light, the photoelectric effect should occur for any wavelength. If the intensity was low, then there should be a time lag before the photoelectrons are emitted while the electrons in the metal surface soak up enough energy to escape the surface. Figure 25.2B shows that there is a characteristic cutoff frequency below which the photoelectric effect ceases entirely. The cutoff frequency varies for different metals, but all metals that exhibit the photoelectric effect have a cutoff frequency.
3. If the incident light is feeble (low intensity), then there should be a time lag before the photoelectrons are emitted. No time lag has ever been observed.

Albert Einstein pondered over the problem of the photoelectric effect and came to the conclusion that light travels through space in concentrated bundles, or wave packets.

Stopping potential: The negative collector voltage when the photocurrent becomes zero.

Cutoff frequency: The frequency of light below which the photoelectric effect ceases entirely.

Photon: A discrete wave packet (particle) of light that is always moving at the speed of light.

(Equation 25.1)

These wave packets were named ***photons,*** and each photon's energy is a function of the frequency of the light.

$$E = h\nu$$

where ν (the Greek letter nu) is the frequency of the light. If light were emitted as wave packets instead of as a wave, then there would be no time lag and no reason for contradiction (3) listed above. A wave travels spread out in space, and it would take some time before an electron would have enough energy to break out of the surface of the metal regardless of the intensity. A high intensity would only shorten the lag time, not eliminate it. If a photon of light acts like a particle, then as soon as the photon strikes the surface of the metal, a photoelectron can be emitted. The amount of energy needed for an electron to leave the surface of a given metal is called the metal's *work function.* If light acts like a particle in striking the metal's surface, then if the particle doesn't carry enough energy, no electrons will be emitted, no matter how many particles hit the surface. When the particle's energy is high enough, then electrons will leave the surface. Above the cutoff frequency, the more wave packets that hit the surface, the more photoelectrons will be emitted, and the greater the photocurrent will be, explaining Figure 25.2A. If a wave packet has more energy than that needed to knock an electron out of the surface, the excess energy will show up as the kinetic energy of the electron. If an electron is knocked out of the metal from below the surface, it takes more energy to free it, and its KE will be less. Thus, there will be a range of kinetic energies for a given wavelength of incident light, but the maximum KE will be from an electron knocked from the top surface of the metal, and that is what the stopping potential measures. Taking this into account, Einstein was able to write a simple linear equation describing the photoelectric effect.

(Equation 25.2)

$$h\nu = E_o + \text{KE}_{\text{max}}$$

where E_o is the work function of the metal. When the kinetic energy of the emitted electrons is zero, then the frequency is the cutoff frequency below which the photoelectric effect ceases for that particular metal.

The interpretation of the photoelectric effect by Einstein showed that light had the nature of a particle as well as the nature of a wave. A particle and a wave are poles apart from each other. Light and dark are polar, as well as up and down and north and south. How can anything be both light and dark at the same time? There is a deep contradiction here that physicists found hard to accept at the time. This contradiction still causes problems long after the particle nature of light has been resignedly accepted. Experiments show either the particle nature of light or the wave nature of light. But no experiment has as yet shown both natures simultaneously. Light is the Dr. Jekyll and Mr. Hyde of physics!

The slope of Figure 25.2B can give us the value of h if the electron's charge is known; h was found to be equal to 6.57×10^{-34} J-s and is called *Planck's constant.* The reason it wasn't called Einstein's constant is that it had been seen before. At the turn of the century (from the 1800s to the 1900s), Max Planck in Germany was working on what was then known as the *ultraviolet catastrophe* (a good name for a rock band!). Classical theory couldn't predict the way that a blackbody radiated electromagnetic energy. Planck ran into a problem similar to Einstein's with the photoelectric effect, and he could only theoretically predict the known radiation characteristics of a blackbody by "fudging" the theoretical equations and assuming that the oscillating electrons in the body's surface emitted discontinuously (like a particle) and with multiples of his constant. With the constant h, theory matched reality quite well, but Planck was quite uneasy about his fudged constant. Planck's discovery was, however, the first hint of the quantum nature of light. Einstein stumbled onto Planck's constant, and the photoelectric effect and blackbody radiation explanations compelled physicists to accept the quantum (particle) nature of light.

EXAMPLE 25.1

Light of frequency 5×10^{14} Hz falls on sodium metal whose cutoff frequency is 4.39×10^{14} Hz. (a) If an *electron-volt (eV)* is the work done on an electron by a potential difference of 1 V, what is the work function of sodium in electron-volts? (b) What is the maximum speed of the ejected photoelectron from the surface of the sodium? (c) What is the stopping potential for this electron?

Electron-volt (eV): The work done on an electron by a potential of 1 V ($1 \text{ eV} = 1.6 \times 10^{-19}$ J).

SOLUTION

(a)
$$E_o = h\lambda_o$$

$$E_o = (6.57 \times 10^{-34} \text{ J-s})\left(\frac{4.39 \times 10^{14}}{\text{s}}\right)$$

$$E_o = 2.88 \times 10^{-19} \text{ J}$$

$$1 \text{ eV} = 1.6 \times 10^{-19} \text{ J}$$

$$E = (2.88 \times 10^{-19} \text{ J})\left(\frac{1 \text{eV}}{1.6 \times 10^{-19} \text{ J}}\right)$$

$$E_o = 1.8 \text{ eV}$$

(b)
$$E = h\nu$$

$$E = (6.57 \times 10^{-34} \text{ J-s})(5 \times 10^{14}/\text{s})$$

$$E = 3.29 \times 10^{-19} \text{ J}$$

$$h\nu = E_o + \text{KE}_{\text{max}} \qquad \text{KE}_{\text{max}} = h\nu - E_o$$

$$\text{KE}_{\text{max}} = 3.29 \times 10^{-19} \text{ J} - 2.88 \times 10^{-19} \text{ J}$$

$$\text{KE}_{\text{max}} = 0.41 \times 10^{-19} \text{ J}$$

$$\text{KE} = \frac{1}{2}m\nu^2 \qquad \nu = \sqrt{\frac{2\text{KE}}{m}}$$

$$\nu = \sqrt{\frac{2(0.41 \times 10^{-19} \text{ J})}{9.11 \times 10^{-31} \text{ kg}}}$$

$$\nu = 3 \times 10^5 \text{ m/s}$$

(c)
$$\text{KE} = 4.1 \times 10^{-20} \text{ J} \qquad \text{KE} = eV$$

$$V = \frac{KE}{e}$$

where e is the charge of the electron and V is the voltage.

$$V = \frac{4.1 \times 10^{-20} \text{ J}}{1.6 \times 10^{-19} \text{ C}}$$

$$V = 0.26 \text{ V}$$

25.2   Special Relativity

With the photoelectric effect we found that light is both a wave *and* a particle. The wave nature of light was (and still is) firmly established, but light as a wave has become not the only interpretation of the nature of light. At the turn of the century, physicists received another shock when a prediction of the wave theory of light couldn't be verified experimentally. According to wave theory, a wave travels through a medium. Water waves travel on the surface of a pond or a pool through the water, sound waves propagate through the air, and traveling waves on a string cause the string to move transversely to the direction of propagation. The energy that a wave carries is stored in the displacement of the medium. Scientists felt that light should move through a medium of some sort, but such movement had never been detected. The American physicists Albert Michelson and Edward Morley designed an experiment in the 1880s that attempted to detect the medium, called the *ether,* through which light moved; it was assumed to permeate all space.

An interferometer was constructed that split a monochromatic beam of light into two paths at right angles to each other. The two beams bounced off mirrors at the ends of their respective paths and recombined. If the path lengths were equal, no interference effects between the beams were observed, but if they were unequal, interference effects were observed. The interferometer used the fact that the apparent speed of a wave through a medium changes with the velocity of an object moving through the medium. A good example is the bow of a boat. It disturbs the surface of calm water much as a pebble dropped into a still pool does. If the boat is at rest relative to the water, then ripples leave the bow (if it is gently rocking up and down) with the wave speed that is characteristic of water. If the boat has a forward speed less than the wave speed, then the ripples move forward with a speed that is the difference between the wave speed and the boat's speed. If the boat is traveling at a speed greater than the wave speed, then no ripples propagate forward, and the boat leaves the characteristic V-shaped bow wave, with the tip of the V being the boat's bow. The wave can't precede the boat because the boat's speed is in excess of the wave's speed. This change in wave speed was thought to occur in the ether also, and the speed of the earth in orbit about the sun was used like the speed of the boat on the water. The physicists theorized that if one of the arms of the interferometer were aligned in the direction of the earth's movement through the ether, then the speed of light in that arm would be slightly less than the speed in the arm at right angles to it. The arm at right angles would have a wave speed analogous to that of the boat at rest in the water. This slight difference in the speed of light would cause interference effects in the interferometer, and the ether could be detected. No interference effects were ever detected!

Physicists knew that the viscosity of the ether must be vanishingly small for it to have avoided detection and for the speed of light to be so high. They were so wedded to the idea of an ether, though, that most of them weren't prepared to accept the fact that there was no ether, since that would strongly violate an important part of wave theory. Albert Einstein, however, was prepared to accept that fact. He postulated that light was the fastest thing in the universe and that nothing was faster. Moreover, he also postulated that the speed of light was the same for all observers regardless of the speed of the source. A missile launched from a moving aircraft will have the speed of the aircraft plus the missile's speed according to an observer on the ground. With light, such is not the case. If the landing lights of an aircraft were turned on, the speed of the photons emitted from them would be the same to an observer on the ground regardless of whether or not the aircraft was moving.

Einstein took these two postulates and redefined the space-time background against which all of physics occurs. He came up with some very bizarre conclusions. To illustrate the most bizarre, assume that a clock is constructed that uses a short light pulse to mark time. The readout for this clock is a chart recorder that registers a spike whenever a detector sees a light pulse. The first short pulse puts a spike on the recorder and travels to a mirror from which it is reflected. The reflected pulse then returns to the detector, and another spike is registered on the paper of the chart recorder. The detector triggers another pulse whenever it detects a reflected pulse. The basic time unit for our clock is the time between the emitted pulse and the reflected pulse. Suppose that one such clock is put aboard a spaceship traveling at a constant velocity relative to the earth. If the distance between the detector and the mirror is D, then a crew member on board the spaceship will detect the basic time interval of the clock, which is Δt_o.

$$\Delta t_o = \frac{2D}{c}$$

where the path traveled by the light to and from the mirror is $2D$. But what time would an observer on the earth's surface record? As shown in Figure 25.3, to an observer on the earth's surface, the light would travel a distance of $2s$ because of the velocity of the spaceship.

$$s = \sqrt{D^2 + x^2} \qquad 2s = 2\sqrt{D^2 + x^2}$$

$$2s = c\,\Delta t$$

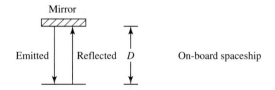

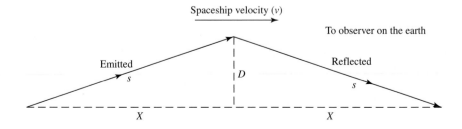

Figure 25.3 Light Pulse as Seen by a Stationary Observer and a Moving Observer

where $2s$ is the distance that the light travels in the earth observer's time Δt.

$$x = \frac{1}{2}v\,\Delta t \qquad c\,\Delta t = 2\left(D^2 + \left[\frac{v\,\Delta t}{2}\right]^2\right)^{1/2} \qquad c\,\Delta t = 2\sqrt{D^2 + \frac{v^2\,\Delta t^2}{4}}$$

$$c^2\,\Delta t^2 = 4\left(D^2 + \frac{v^2\,\Delta t^2}{4}\right) \qquad c^2\,\Delta t^2 = 4D^2 + v^2\,\Delta t^2 \qquad c^2\,\Delta t^2 - v^2\,\Delta t^2 = 4D^2$$

$$(c^2 - v^2)\Delta t^2 = 4D^2 \qquad \Delta t^2 = \frac{4D^2}{c^2 - v^2}$$

$$\Delta t = \frac{2D}{\sqrt{c^2 - v^2}}$$

$$\Delta t = \frac{2D}{c}\frac{1}{\sqrt{1 - (v^2/c^2)}}$$

But $2D/c$ is the time that the spaceship crew member observes, Δt_o. Thus,

(Equation 25.3)

$$\Delta t = \frac{\Delta t_o}{\sqrt{1 - (v^2/c^2)}}$$

The numerator is always less than 1, so Δt will be greater than Δt_o. The observer on the earth records a longer time for the same event than does the spaceship crew member! This is called *time dilation* and spells out the death of simultaneity. We assume that two events that occur in two reference frames moving with respect to one another are simultaneous if they occur at the same time. But if there are two different times for an observer in each reference frame, how can they possibly be simultaneous?

Time dilation isn't the only thing that gets strange when the relative speed gets close to that of light. Close to light speed, an object contracts in the direction of its motion. An object gets shorter according to a stationary observer measuring its length by the factor

$$\sqrt{1 - \frac{v^2}{c^2}}$$

or

(Equation 25.4)

$$L = L_o\sqrt{1 - \frac{v^2}{c^2}}$$

where L_o is the uncontracted length of the object, and L is the object's length according to an observer who sees it speed by with a velocity v relative to himself. Moreover, the object's mass increases by the same factor.

(Equation 25.5)

$$m = \frac{m_o}{\sqrt{1 - (v^2/c^2)}}$$

Equation (25.5) means that at the speed of light, the relative mass of an object is infinite! Thus, if you read between the lines, you'll realize that the speed of an object can asymptotically approach the speed of light but can never reach it.

All of the above effects have been verified many times in experiments designed to test the special theory of relativity. Time dilation particularly has been verified using particles created in the upper atmosphere by cosmic rays and by atomic clocks. A *muon* is a particle that has a well-known lifetime at rest. It will decay in a period of 2.2×10^{-6} s when at rest to an observer. But when a muon is created by a collision of a cosmic ray with an atom in the upper reaches of the earth's atmosphere, the resulting muon streaks earthward with a speed over 99% that of light. The time dilation of the speeding muon relative to an observer on the earth gives the muon a decay time of 35×10^{-6} s. The longer time allows the muon to be detected at the earth's surface, while the shorter, nonrelativistic decay time would allow the muon to move only less than 1 km before disintegrating. Muons would be detected only in the upper atmosphere and not at the earth's surface.

The detection of muons at the earth's surface is one proof of the existence of relativistic time dilation. Another proof uses two identical atomic clocks. An atomic clock can measure an incredibly small amount of time with precision. In an experiment, two such clocks were synchronized, and one clock remained stationary while the other clock was flown around the earth in a jet plane against the earth's rotation. When the flying clock returned, the time measured by the stationary clock was compared to the time recorded by the moving clock. The difference in time between the two clocks was the difference predicted by the special theory of relativity.

EXAMPLE 25.2

The crew of a starship travels to a distant star 10 light-years from the earth. (A *light-year* is the distance that light travels in 1 year.) The trip is accomplished at 99.9% the speed of light with no acceleration or deceleration. The crew on the starship age 10 years on the outbound journey. By how much will a family member left behind on the earth have aged when the starship arrives at its destination?

SOLUTION

$$\Delta t = \frac{\Delta t_o}{\sqrt{1 - (v^2/c^2)}} \qquad \frac{v}{c} = 0.999$$

$$\frac{v^2}{c^2} = 0.998 \qquad \sqrt{1 - \frac{v^2}{c^2}} = 0.0447$$

$$\Delta t = \frac{10 \text{ yr}}{0.0447}$$

$$\Delta t = 223.6 \text{ yr}$$

Since a light-year is the distance that light travels in 1 year, if the nearest stars are on the order of a dozen light-years away, the fastest that a starship can reach them is 12 years. Since light is the limiting speed of any material object in the universe, special relativity serves to make the universe even more immense than mere distances can indicate. With classical physics, if we want to get to a distant object more quickly, all we have to do is to increase our speed. Special relativity puts a cap on the maximum speed attainable. But Example 25.2 makes the universe seem even lonelier. The crew of the starship ages 10 years, which is a time span familiar to any adult. But for a crew member who left family behind, not only would all of her immediate family have died, but the world would probably have changed beyond recognition in the 224 years that would have elapsed for the people left behind. The starship crew members would have left not only loved ones behind, but also their histories. If the ship were to return, almost 500 years would have elapsed: a time span comparable to the time between the present and the time Columbus "discovered" America.

25.3   X Rays

X rays are high-energy electromagnetic waves. *High-energy* means that the waves have short wavelengths in the approximate interval from 0.01 nm to about 10 nm. (Recall that visible light runs from 400 nm to 700 nm.) These highly energetic waves, discovered in 1895 by Wilhelm Roentgen, were called X rays (a generic designation) because it was about 10 years before their true nature was known. It was eventually found that X rays could be polarized and would interfere just like light rays, so their nature as electromagnetic waves was established.

X rays are produced by a reverse photoelectric effect, as illustrated in Figure 25.4. An electric current runs through a filament which grows hot as a result of resistive heating. (The filament glows red.) The filament is hot enough that electrons have enough kinetic energy to leave the surface. The filament is at one end of an evacuated glass tube. A metal target is at the other end and has a positive potential on it of thousands of volts. The negatively charged electron is attracted to the positively charged target and smashes into it at a high rate of speed. (The tube is evacuated so that the high-energy electrons don't lose any of their kinetic energy through collisions with other atoms or molecules.) The electrons that hit the target have an energy of their charge times the voltage, or thousands of electron-volts. A charged particle that is accelerated radiates electromagnetic radiation. When the electron slams into the target, it undergoes a violent deceleration, and high-energy photons, or X rays, are produced. The acceleration of the electrons through the potential difference is much lower than the deceleration at the target, and thus no X rays are produced. Most of the electrons hitting the target undergo multiple collisions with the metal and also decelerate too gradually to emit X rays. Their energy goes into heating the target. For this reason, the target is usually made of a metal with a high melting point and is water cooled. Some of the electrons lose most or all of their energy through a collision with a single atom. These electrons are the ones that emit X rays.

The work function of a metal is usually only a few electron-volts and is negligible with respect to an electron with an energy of tens of thousands of electron-volts. If an X-ray tube is the photoelectric effect in reverse, then the equation governing the frequency of the generated X rays is

(Equation 25.6)

$$hf = eV$$

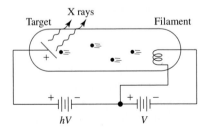

Figure 25.4 Generation of X Rays

EXAMPLE 25.3

(a) What is the speed of an electron that passes through a potential difference of 41 kV? (b) What is the wavelength of the X rays emitted by such electrons?

SOLUTION

(a)
$$eV = \frac{1}{2}mv^2 \qquad v = \sqrt{\frac{2eV}{m}}$$

$$v = \sqrt{\frac{2(1.6 \times 10^{-19}\text{ C})(41{,}000\text{ J/C})}{9.11 \times 10^{-31}\text{ kg}}}$$

$$v = 1.2 \times 10^8 \text{ m/s}$$

The above speed is only about 40% of the speed of light, and relativistic effects are too small to worry about.

(b)
$$hf = eV \qquad f = \frac{eV}{h}$$

$$f = \frac{(1.6 \times 10^{-19}\text{ C})(41{,}000\text{ J/C})}{6.63 \times 10^{-34}\text{ J-s}}$$

$$f = 9.89 \times 10^{18}\text{ Hz}$$

$$\lambda f = c \qquad \lambda = \frac{c}{f}$$

$$\lambda = \frac{3 \times 10^8 \text{ m/s}}{9.89 \times 10^{18}\text{ Hz}}$$

$$\lambda = 0.03 \text{ nm}$$

25.4   Matter Waves

When it was found that what was formerly considered a wave (light) also had the characteristics of a particle, some physicists began wondering if the reverse could also be true. Could a particle also have the characteristics of a wave? In 1923, Louis de Broglie suggested that the wavelength of a particle with momentum p is

$$\lambda = \frac{h}{p}$$

(Equation 25.7)

Equation (25.7) was confirmed several years later when the physicists Davisson and Germer directed a beam of electrons onto crystals of nickel. The electrons exhibited a diffraction behavior like X-ray diffraction. Recall that much information can be generated about crystal structure by X-ray diffraction. The crystal structure is a regularly spaced lattice, and the diffraction patterns from diffracted light can tell much about how the crystal lattice is constructed. However, the interatomic spacing is much smaller than that of visible light, so electromagnetic radiation with a much shorter wavelength must be used. X rays have a short enough wavelength so that beams of X rays that are shone on a crystal will exhibit diffraction patterns and the crystal structure can be figured out. Davisson and Germer directed a beam of electrons with the right speed onto a nickel crystal, and there was a diffraction pattern similar to that produced by X rays. Particles also have the properties of waves!

EXAMPLE 25.4

(a) What is the de Broglie wavelength for a baseball traveling at 90 mph (40.2 m/s)? A baseball has a mass of about 0.15 kg. (b) What is the de Broglie wavelength for an electron traveling at a speed of 5×10^6 m/s?

SOLUTION

(a)
$$\lambda = \frac{h}{p} \qquad p = mv \qquad \lambda = \frac{h}{mv}$$

$$\lambda = \frac{6.63 \times 10^{-34} \text{ J-s}}{(0.15 \text{ kg})(40.2 \text{ m/s})}$$

$$\lambda = 1.1 \times 10^{-34} \text{ m}$$

(b)
$$\lambda = \frac{h}{mv}$$

$$\lambda = \frac{6.63 \times 10^{-34} \text{ J-s}}{(9.1 \times 10^{-31} \text{ kg})(5 \times 10^6 \text{ m/s})}$$

$$\lambda = 1.46 \times 10^{-10} \text{ m}$$

The reason that we don't see baseballs exhibiting wavelike properties when they are thrown is that the wavelength is so much shorter than the diameter of the ball. The electron, on the other hand, is truly a point particle. Notice that no diameter is given in data tables for an electron. In such a case, an electron can much more easily behave as a wave; it readily exhibits wave properties even though it was first discovered as a particle, and it acts like one in most circumstances. The electron microscope takes advantage of this characteristic. Because the shortest wavelength of visible light is around 400 nm (blue light), a light microscope has an upper limit to the magnification that it can provide. Details inside a cell or microbes and viruses are often much smaller than the shortest wavelength of visible light that can be shone on them. Because of its pointlike character and its ability to act like a wave with a much shorter wavelength than visible light, the electron is a good candidate for constructing a microscope to magnify very small structures. The electron microscope is the result. Most people have never used one or even seen one, but most have seen the images that such a microscope can produce in magazines or books. The lenses of an electron microscope are magnetic and electric fields. They can bend and focus a beam of electrons by interacting with its negative charge. The de Broglie wavelength of the resulting beam can also be tailored to the size of the object under examination. The higher the energy (momentum), the shorter the wavelength. Recall that the baseball moving at 90 mph had a very short wavelength of about 1×10^{-34} m, and this short wavelength would work very well to illuminate a very small object. But the physical size of a baseball rules out its use in such a microscope, and a stream of baseballs hitting a bacterium (if that were the object under examination) would very probably scramble it to where observation was pointless anyway!

25.5   The Heisenberg Uncertainty Principle

Since waves are particles and particles are waves, there should be some way of reconciling the two, although they seem mutually exclusive. There is a way, and to understand it, we must first recall the concept of beats. If two tuning forks have slightly different frequencies, and if both are struck with a hammer, then we will hear the slowly rising and falling amplitude of a tone very near the two original frequencies. What we hear are the two sine waves interfering both destructively and constructively in time, as shown in Figure 25.5. The two sine waves are superimposed upon one another, and the sum of the two is what we hear. The two original sine waves have a constant amplitude, but the sum of the two has a slowly varying amplitude envelope. The amplitude envelope has chopped up the originally continuous waveforms into a number of wave packets. These wave packets can be isolated by adding more and more sine waves that differ in frequency by infinitesimal amounts. If an infinite number of sine waves with infinitesimally differing frequencies are added, then we end up with a single wave packet whose width either temporally or spatially goes to zero. This is our particle.

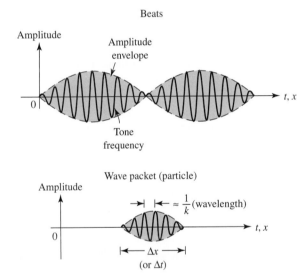

Figure 25.5 Wave Packet

The beat phenomenon is caused by temporal interference of sound waves whose amplitudes vary sinusoidally with time. A sine wave has both a temporal and a spatial component.

$$A = A_o \sin 2\pi(kx - ft)$$

If a sine wave can interfere temporally, it can also interfere spatially. We get the same wave packet phenomenon if the sine waves differ infinitesimally in wave number (k, or inverse wavelength). The wave packet is formed because all of the individual sine waves that are added are in phase in the middle of the packet and add constructively to give a large amplitude. As you move farther away from the center of the packet in either direction, the sine waves shift out of phase, and beyond a certain point they are completely out of phase, never to get back into phase again. The larger the range of frequencies or wavelengths that are added together, the shorter the final isolated wave packet will be. It can be shown that Δx (the width of the wave packet) is inversely proportional to the range of the wave numbers, and from wave theory this proportionality is about equal to 1 over 2π.

(Equation 25.8)

$$\Delta x \, \Delta k = \frac{1}{2\pi}$$

Equation (25.8) is true for all wave motion. The wave number, k, is the inverse of the wavelength. Using the formula for the de Broglie wavelength, we can derive the Heisenberg uncertainty principle.

$$\Delta x \, \Delta k = \frac{1}{2\pi} \qquad k = \frac{1}{\lambda}$$

$$\lambda = \frac{h}{p} \qquad k = \frac{p}{h}$$

$$\Delta k = \frac{1}{h} \Delta p \qquad \Delta x \left(\frac{1}{h} \Delta p \right) = \frac{1}{2\pi}$$

(Equation 25.9)

$$\Delta x \, \Delta p = \frac{h}{2\pi}$$

The Heisenberg uncertainty principle was first derived by Werner Heisenberg in 1927. It puts a limit on the precision with which we can simultaneously know both the position and the momentum of a particle. The better we know the position, the more uncertainty there is in its momentum. Because of the small value of Planck's constant, the uncertainty principle isn't important in classical physics. But it is extremely important over small distances and momenta in examining atomic and nuclear domains. The implications of the uncertainty principle are profound. A good experimental measurement is supposed to leave the system that it's measuring undisturbed. However, at the atomic level, every measurement by its very nature disturbs the system being examined. Science is objective, and an experiment is usually considered invalid if the act of performing an experiment influences its results. The uncertainty principle creates in science a basic subjectivity that biases the investigation of a system simply by the very act of trying to measure the system!

25.6   The Schrödinger Equation

An electron is a point mass, yet at the level of the electron, is the electron a hard, dense sphere with a vanishingly small diameter, or is it a very large set of mutually interfering waves? If the number of waves added together is large enough, then the waves become a wave packet with a vanishingly small width. An electron possesses a property that points to the wave description as the one closer to reality: An electron can tunnel. If a bowling ball is set rolling and it encounters a hump in an otherwise smooth floor, it can roll over the hump only if its kinetic energy is greater than the gravitational potential that it will have at the top of the hump. If the kinetic energy is too small, it will roll only partway up the hump and be unable to get to the other side. If the kinetic energy is too small, it is impossible classically for the bowling ball to clear the hump. But an electron, acting classically like a particle in most cases, can "tunnel" through such a barrier without having the necessary kinetic energy. A commercial electronic device called a *tunnel diode* operates on this principle. Tunneling can be explained only by viewing the electron as a wave packet.

Much of physics deals with how a particle moves under the influence of forces. At the atomic level we need an equation that will predict how a wave packet will propagate under such conditions. (Recall that an electron can be moved around by electric and magnetic fields.) The equation derived to predict how such wave packets propagate is the Schrödinger equation developed by Austrian physicist Erwin Schrödinger in 1925. Schrödinger started with the relationship between wavelength and energy $(E = hf)$ discovered by Einstein in the photoelectric effect and the de Broglie wavelength $(\lambda = h/p)$. He hung these two equations on the principle of the conservation of energy, $E = KE + PE$, and he used a sine wave to represent a typical wave.

$$\Psi = \sin(kx - \omega t)$$

The total energy can be put in terms of frequency using Equation (25.2) describing the photoelectric effect, and the kinetic energy can be put in terms of momentum and therefore wavelength (wave number k) from the de Broglie wavelength equation [Equation (25.7)]. The equation for the conservation of energy using wave parameters that results is

$$\frac{\hbar^2}{2m} K^2 + V_o = \hbar\omega$$

where V_o is the potential energy. If the wave function is represented by the sine function psi (Ψ), then we can find K^2 term by looking at how Ψ changes with respect to distance twice, and the ω term by looking at how Ψ changes with respect to time once. (We did so earlier in this book when we developed the concept of the sine wave in Chapter 8 on waves.) Scientists wanted the resulting wave equation to be linear in Ψ, and therefore all terms in the above energy equation were multiplied by either Ψ or one of its derivatives with respect to time or distance. Linearity also required that Ψ be a function of sines *and* cosines, and then an equation was formulated that could finally be solved. The result was the famous Schrödinger equation, given below and illustrated in Figure 25.6.

$$-\frac{\hbar^2}{2m} \frac{\delta^2 \Psi}{\delta x^2} + V\Psi = i\hbar\frac{\delta \Psi}{\delta t}$$

(Equation 25.10)

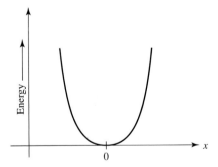

Figure 25.6 Harmonic Oscillator

Equation (25.10) predicts how a wave packet will propagate in space and time. It predicted the tunneling diode. If the potential energy is known (usually as a function of distance, x), much useful information about a system can be learned by its solution. An atom in a solid, for instance, is often held in place by electromagnetic forces that increase linearly with distance and where the potential energy goes as the square of the displacement from equilibrium. This is the case for a harmonic oscillator, and the potential is called a *harmonic oscillator potential.* Like standing waves on a stretched string with a finite length, only certain, well-defined discrete energy levels are allowed. The allowed energies of vibration become quantized. The solution of the Schrödinger equation for such a potential can find the discrete, quantized energy levels. When an atom jumps between these discrete levels, it therefore naturally emits or absorbs quantized radiation, which is what Planck had to introduce in the first place to start the chain of events that led to the Schrödinger equation!

The i in the Schrödinger equation is the square root of negative 1.

$$i = \sqrt{-1}$$

In classical physics this number isn't allowed and is considered impossible. However, in the derivation of the Schrödinger equation, it is necessary for a solution. When the Schrödinger equation is solved, the math takes us into an imaginary space where all of the mathematical manipulation is done, and it brings us back to reality when we arrive at an answer. Mathematics requires only an internal consistency and need not have any link with reality. There is a whole branch of mathematics built around the square root of negative 1, but it is extremely useful in obtaining real answers at the atomic level in quantum physics. Reality at the level of the atom is very far removed from what we experience with our senses!

Summary

1. Light exhibits both wave and particle characteristics. A particle of light, called a **photon,** has an energy that is solely dependent on its frequency, as shown in Equation (25.1).

2. Particles also exhibit both wave and particle characteristics, although in most cases the particle nature predominates. The wave nature can be considered by calculating the associated *de Broglie wavelength,* which is dependent on an object's momentum [see Equation (25.7)].

3. The **photoelectric effect** involves photons striking a metal surface to dislodge electrons. Before an electron can be removed from its parent atom, the photon must possess enough energy to overcome the metal's characteristic **work function** (E_o). Any energy left over is used to impart the freed electron with kinetic energy. The relationship between the energy of the incident photon, the work function of the metal, and the resultant kinetic energy of the electron is described by Equation (25.2).

4. Light appears to travel at the same velocity to all observers, despite what speed the observers may possess at the time. This quality causes space and time to distort. Although imperceptible at slow speeds, the effects become very apparent at velocities approaching the speed of light. This theory is called the *special theory of relativity.* At very high

velocities, time and length appear to shorten, or *dilate,* and mass increases. The amount that these values change can be calculated using Equations (25.3), (25.4), and (25.5).

Problem Solving Tip 25.1: When using these equations, remember that the Δt_o term is the time that passes in the moving frame of reference, whereas Δt is the time that a "stationary" observer records for the same event. Both the L_o and the M_o terms are the quantities that would be measured if the object were at rest, where the L and the M terms are the apparent quantities once the object reaches relativistic speeds.

5. When you are examining matter in very small quantities, a useful unit of energy is the **electron-volt (eV),** which is the work done on an electron by a potential difference of 1 V: $1 \text{ eV} = 1.6 \times 10^{-19} \text{ J}$.

6. There is a fundamental limit on how accurately a particle can be measured. As the position of a particle becomes more certain, the less we know of its velocity. Likewise, if the value of a particle's velocity becomes well known, its exact position becomes less certain. This balance between knowing the exact position and velocity of a particle at any given time is known as the *Heisenberg uncertainty principle* [see Equation (25.9)].

Key Concepts

Cutoff frequency: The frequency of light below which the photoelectric effect ceases entirely.

Electron-volt (eV): The work done on an electron by a potential of 1 V ($1 \text{ eV} = 1.6 \times 10^{-19} \text{ J}$).

Photoelectric effect: Effect produced when light shining on certain metals liberates electrons from the surface.

Photon: A discrete wave packet (particle) of light that is always moving at the speed of light.

Stopping potential: The negative collector voltage when the photocurrent becomes zero.

Work function: The minimum energy needed for an electron to leave the surface of a metal.

Important Equations

(25.1) $E = h\nu$ (Energy carried by a photon is proportional to its frequency.)

(25.2) $h\nu = E_o + KE_{max}$ (Photoelectric effect where E_o is the work function for metal)

(25.3) $\Delta t = \dfrac{\Delta t_o}{\sqrt{1 - (v^2/c^2)}}$ (Time dilation where Δt_o is the time measured in the object traveling at speed v)

(25.4) $L = L_o \sqrt{1 - \dfrac{v^2}{c^2}}$ (Length dilation where L_o is the length measured in the object traveling at speed v)

(25.5) $m = \dfrac{m_o}{\sqrt{1 - (v^2/c^2)}}$ (The mass of an object traveling at a speed v increases.)

(25.6) $hf = eV$ (The energy of an X-ray photon equals the KE of an electron that is accelerated through a potential V.)

(25.7) $\lambda = \dfrac{h}{p}$ (The wavelength of a matter wave is inversely proportional to its momentum.)

(25.8) $\Delta x\, \Delta k = \dfrac{1}{2\pi}$ (The length of a wave packet is inversely proportional to the range of the wave numbers.)

(25.9) $\Delta x\, \Delta p = \dfrac{h}{2\pi}$ (Heisenberg uncertainty principle)

(25.10) $-\dfrac{\hbar^2}{2m}\dfrac{\delta^2\Psi}{\delta x^2} + V\Psi = i\hbar\dfrac{\delta\Psi}{\delta t}$ (Schrödinger equation)

Conceptual Problems

1. A metal surface emits photoelectrons only when the light shining on it exceeds:
   a. a certain intensity
   b. a certain wavelength
   c. a certain frequency
   d. a certain photon speed

2. Increasing the brightness of light hitting the metal surface of a photocell:
   a. increases the photoelectron's energy
   b. increases the photoelectron's speed
   c. increases the number of photoelectrons
   d. increases the stopping potential of the photocell

3. The photoelectric effect can best be understood on the basis of:
   a. the principle of superposition of waves
   b. the electromagnetic theory of light
   c. the special theory of relativity
   d. light as particles called *photons*

4. In a vacuum, all photons have the same:
   a. speed
   b. frequency
   c. wavelength
   d. energy

5. The rest mass of a photon:
   a. is the same as that of a neutron
   b. is the same as that of an electron
   c. depends on the photon's charge
   d. is zero

6. Increasing the voltage across an X-ray tube:
   a. increases the frequency of the X-ray photons
   b. increases the wavelength of the X-ray photons
   c. increases the number of X-ray photons emitted
   d. increases the speed of the X-ray photons emitted

7. Matter waves are caused by:
   a. the superposition of matter fields
   b. the masking of the charges that make up all atoms
   c. the relative movement of a massive object
   d. the conservation of momentum

8. What is the key principle behind the production of X rays from an X-ray tube?

9. Increasing the voltage across an X-ray tube increases the output energy of the X rays. What is the chain of events leading to the increase in energy?

10. A number of waves interfere constructively and destructively to form a wave packet. As the range of wavelengths that superimpose decreases, what happens to the wave packet?

11. The basis for the predictions of special relativity come from:
    a. the speed of a photon being equal to the speed of light
    b. the difference of the speed of light in the ether being dependent on the source's speed
    c. the speed of light being the same regardless of the speed of the source
    d. $E = mc^2$

12. If the momentum of a particle is known with increasing accuracy:
    a. its position is known with decreasing accuracy
    b. its energy is known with decreasing accuracy

    c. it's impossible to locate the particle
    d. the particle is easier to locate because it is slower

Exercises

The Photoelectric Effect

1. What is the energy of a photon having a wavelength of (a) 300 nm and (b) 700 nm?

   Answer: (a) 6.63×10^{-19} J
   (b) 2.85×10^{-19} J

2. Photons having 3.1×10^{-19} J of energy strike a metal plate. Electrons are released with a maximum kinetic energy of 4.5×10^{-20} J. What is the work function of the metal?

   Answer: 2.65×10^{-19} J

3. With what maximum speed will an electron escape a plate of sodium metal if green light of wavelength 550 nm shines on the plate?

   Answer: 6.96×10^5 ms

4. What wavelength of light is required to strike a plate of sodium with a work function of 2.88×10^{-19} J so that it is released with a maximum speed of 4.2×10^5 m/s?

   Answer: 540 nm

5. Green light of wavelength 546.1 nm ejects photoelectrons from the surface of a metal with a maximum kinetic energy of 0.13 eV. What is the work function of the metal?

   Answer: 2.15 eV

*6. A silver ball with a work function of 4.7 eV is suspended in a vacuum chamber by a string. (a) If ultraviolet light of wavelength 190 nm shines on the ball, what potential will the ball acquire after the light has shone on the ball for a while? (b) What charge does the ball have if the ball has a diameter of 1 cm?

   Answer: (a) 1.84 V
   (b) 1.02×10^{-12} C

*7. Blue light of wavelength 430 nm falls on the surface of a cesium plate with a work function of 1.9 eV. The light falls on the plate at a rate of 6 mW. If one photoelectron is emitted from the plate's surface for every 10,000 photons that are absorbed by the plate, what photocurrent does the photocell generate?

   Answer: 208 nA

Special Relativity

8. An observer on the earth is watching a starship recede from the earth with a speed of $0.87c$. (a) How much time has passed on the starship after 1 min has passed on the earth? (b) To the crew of the starship, how many seconds have passed on the earth after 1 min has passed on the starship?

   Answer: (a) 30 s
   (b) 30 s

9. (a) Using Newton's second law, calculate the force necessary to accelerate a 10-kg mass by 2 m/s². (b) Now calculate the force necessary to accelerate the mass traveling at $0.98c$ by 2 m/s². (c) Again calculate the force if the mass is traveling at $0.998c$.

   Answer: (a) 20 N
   (b) 100.5 N
   (c) 316.4 N

10. A meter stick is moving at $0.88c$. How long is it to someone standing next to its path?

    Answer: 48 cm

11. A proton in an accelerator is accelerated to the speed of 99.999% the speed of light. What is its mass to the physicist operating the accelerator?

    Answer: 3.73×10^{-25} kg

X Rays

12. An electron is accelerated through a potential difference of 25 kV and strikes a metallic surface. What is the shortest X-ray wavelength produced?

    Answer: 0.050 nm

13. In TV tubes and computer monitors, electrons are accelerated through a potential difference of about 10 kV before they hit the screen. In what part of the electromagnetic spectrum will the most energetic radiation from the screen lie?

    Answer: X rays

Matter Waves

14. What is the de Broglie wavelength of a ping-pong ball of mass 8 g during a fast serve (20 m/s)?

Answer: 4.1×10^{-33} m

15. How fast would a small automobile of mass 1600 kg have to travel to have the same de Broglie wavelength as the ping-pong ball in Exercise 14?

Answer: 0.1 mm/s

16. Through what potential must an electron be accelerated to have a wavelength of 0.01 nm?

Answer: 15 kV

17. A slowly moving electron can be localized to within 0.5 mm. What is the uncertainty in its velocity?

Answer: 0.23 m/s

*18. Two wavelengths are beat together spatially to form wave packets. The two wavelengths are nearly the same, so that the product of the two equals λ_o^2, and $\Delta\lambda = \lambda_2 - \lambda_1$. What is the length of the wave packet?

Answer: $\Delta x = \dfrac{\lambda_o^2}{2\pi\,\Delta\lambda}$

The energy source of our sun is atomic fusion, where hydrogen nuclei slam into one another at very high speeds, fusing to form a nucleus of helium and liberating vast amounts of energy in the process. The energy is vast because a little matter is transformed to energy by the relationship $E = mc^2$ for each fusion event.

Physics inside the Atom

At the end of this chapter, the reader should be able to:

1. Describe briefly the particles that make up an atom.

2. Calculate the energy required to ionize a known element.

3. Explain the concept of a *half-life*.

4. Solve for how much of a known radioactive material will remain after a given amount of time.

5. Explain in detail the difference between **fission** and *fusion*.

6. Calculate the *latent energy* in a given amount of mass.

26.1   The Atom and the Atomic Nucleus

All material objects are composed of countless billions of basic building blocks called *atoms*. There are 103 different types of building blocks commonly used in the construction of the material world around us. The blocks themselves are called *atoms,* and the name of each different type of block is the name of an element on the periodic table (see Appendix E). How atoms fit together and are held together to form the many varieties of matter with which we come into contact is the concern of the next chapter. In this chapter we're interested in the details of the building blocks and in what makes them different from one another.

Proton: A nuclear particle with a charge of $+1.6 \times 10^{-19}$ C and a mass of 1.67×10^{-27} kg.

Neutron: A nuclear particle with no charge and about the same mass as the proton.

An atom is composed of a solid, dense core called a *nucleus,* as shown in Figure 26.1. Two types of particles are found in the nucleus: protons and neutrons. ***Protons** are positively charged and have the same magnitude of charge as does the electron,* 1.6×10^{-19} C. A proton has a rest mass of 1.67×10^{-27} kg (over 1800 times more massive than an electron). The ***neutron** also has about the same mass, but no charge. The protons and neutrons are tightly bound together in the nucleus and carry most of the mass of the atom. Surrounding the nucleus is an orbiting cloud of electrons. There is one electron for every proton in the nucleus. A hydrogen atom, the lightest element in the periodic table, normally has only a single proton in its nucleus. A helium atom has two protons in its nucleus; a lithium atom, three; beryllium, four; and so on. *The number of protons in the nucleus determines what element that particular atom is.* Atoms are usually electrically neutral. Since there is an electron for every positively charged proton, and since the magnitudes of the charges are equal, the electrons and protons cancel out or neutralize each other.

The number of protons in the nucleus determines the elemental identity of an atom.

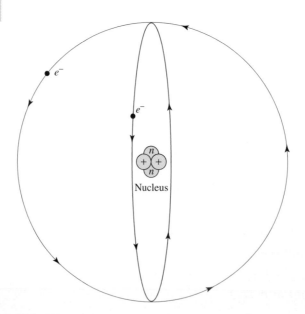

Figure 26.1 Schematic Drawing of a Helium Atom

The nucleus is small and dense and at the center of the atom. The orbiting cloud of electrons is relatively remote from the nucleus. Originally the atom was thought to be much like a rice-and-raisin pudding. The protons and electrons were thought to be all mixed up together in the atom. The British physicist Ernest Rutherford discovered the nucleus in an experiment in which a beam of protons was directed on a thin, gold foil. The protons were deflected by the like positive charge on the nuclei of the gold atoms and were scattered. However, the scattering angle for a very small percentage of the incoming protons was 180°. This angle indicated that the proton had hit something solid and had rebounded back along its original path. The very small percentage of protons scattering back through 180° meant that whatever the protons were hitting was very small compared to the volume of the atom. It was later deduced that what the protons were hitting was the nucleus and that an atom is mostly empty space! The electrons orbit the nucleus at a distance.

The discovery brought new problems to physicists interested in the problem of the atom. Protons are positively charged particles, and thus they repel one another. Even with the small charge on a proton, the separation distances are so small inside the atom that the repulsive forces present between protons would be enormous. The neutrons in the nucleus serve to keep the protons in the nucleus apart, and that keeps the repulsive forces from becoming infinitely high. But there still had to be a force stronger than the electromagnetic force operating over very short interatomic distances that kept the nucleus from flying apart. That force, now called the *strong force,* is one of the four forces known in nature. The two others studied so far in this book are the *gravitational* and *electromagnetic forces.*

The orbiting electrons pose another problem for classical physics. An accelerated charge radiates energy. The orbiting electrons are accelerated. An acceleration is caused either by a change in the magnitude of a velocity or by a change in the velocity's direction. The circling electrons are constantly changing their direction. They are accelerated centripetally. As the electrons orbit, they should radiate away their orbital kinetic energy, and the orbits should decay. The electrons should spiral into the nucleus, and the atom should be impossible classically. But the atom does exist, and the electrons don't spiral into the nucleus. Thus, somehow the electrons orbiting the nucleus don't radiate. Why? X rays are caused by suddenly decelerating electrons, and a radio antenna radiates an electromagnetic wave because the electrons in it are oscillating with simple harmonic motion and are constantly accelerating and decelerating sinusoidally. The reason electrons don't radiate is this: When near a proton, an electron acts like a wave. We've already seen an electron act like a wave with the de Broglie wavelength, and that's what makes the electron microscope possible. The electrons orbiting a nucleus act like standing waves on a string. Only an integral number of wavelengths are allowed on a stretched string of finite length, and each set of standing waves is a resonance for that string. The string might begin and end on a node, for example. But this stretched string is closed upon itself, as in the orbit of an electron around a nucleus. If a nonintegral number of wavelengths were allowed on the string, there would be destructive interference at the string's endpoint/origin. This destructive interference would cause the electron's wave function to decay, and the atom couldn't exist. Since nonintegral wavelength states are forbidden to the electron in its orbit around the nucleus, when the electron gains or loses energy, it must do so in a quantized manner. The energy level of the electron must jump over these forbidden states.

EXAMPLE 26.1

(a) If it takes 13.6 eV to remove the electron from a hydrogen atom, what is the atomic radius of the hydrogen atom? (The hydrogen atom is ionized when this happens.) (b) If the electron were treated as a particle, what would its orbital speed be? (c) What is the de Broglie wavelength of the electron? (d) How does this wavelength compare to the length of the electron's path around the nucleus?

SOLUTION

(a) Classically, the force of attraction of the electric field equals the centripetal force.

$$\frac{1}{4\pi\varepsilon_o}\frac{e^2}{r^2} = \frac{mv^2}{r} \qquad \frac{e^2}{4\pi\varepsilon_o r} = mv^2 \qquad r = \frac{e^2}{4\pi\varepsilon_o mv^2}$$

$$E = 13.6 \text{ eV} \qquad \frac{1}{2}mv^2 = E \qquad mv^2 = 2E$$

$$r = \frac{e^2}{4\pi\varepsilon_o(2E)} \qquad r = \frac{e^2}{8\pi\varepsilon_o E}$$

$$E = (13.6 \text{ eV})\left(\frac{1.6 \times 10^{-19} \text{ J}}{\text{eV}}\right)$$

$$E = 2.18 \times 10^{-18} \text{ J}$$

$$r = \frac{(1.6 \times 10^{-19}\,\text{C})^2}{8\pi(8.85 \times 10^{-12}\,\text{C}^2/\text{N-m}^2)(2.18 \times 10^{-18}\,\text{J})}$$

$$r = 5.3 \times 10^{-11} \text{ m}$$

(b)
$$E = \frac{1}{2}mv^2 \qquad v = \sqrt{\frac{2E}{m}}$$

$$v = \sqrt{\frac{2(2.18 \times 10^{-18}\,\text{J})}{9.11 \times 10^{-31}\,\text{kg}}}$$

$$v = 2.2 \times 10^6 \text{ m/s}$$

(c)
$$\lambda = \frac{h}{mv}$$

$$\lambda = \frac{6.63 \times 10^{-34}\,\text{J-s}}{(9.11 \times 10^{-31}\,\text{kg})(2.2 \times 10^6\,\text{m/s})}$$

$$\lambda = 3.3 \times 10^{-10} \text{ m}$$

(d)
$$L = 2\pi r$$

$$L = 2\pi(5.3 \times 10^{-11} \text{ m})$$

$$L = 3.3 \times 10^{-10} \text{ m}$$

The de Broglie wavelength exactly equals the electron's path around the nucleus! This is a very strong indication that the electron is actually a standing wave around the nucleus instead of a classical particle. And yet we treated the electron as a particle in calculating the radius of the electron's orbit and the electron's speed around the nucleus. The electron is also, like light, both a particle and a wave. Near an atom nucleus, it acts like a wave.

26.2   Energy Levels and Atomic Spectra

As mentioned in Section 25.6 on Schrödinger's equation and its solution for a simple harmonic oscillator potential, the energy levels for the electrons orbiting the nucleus of an atom are quantized. The quantization of the energy levels can be arrived at almost classically with a simple quantum assumption. We first start with the total energy of the electron. The total energy of the electron in orbit around the nucleus of a hydrogen atom is equal to the sum of the kinetic energy and the potential energy. The potential energy is negative because the electron has been captured by the nucleus.

$$E = \text{KE} + \text{PE} \qquad \text{KE} = \frac{1}{2}mv^2 \qquad \text{PE} = \frac{-1}{4\pi\varepsilon_o}\frac{e^2}{r}$$

$$E = \frac{mv^2}{2} - \frac{e^2}{4\pi\varepsilon_o r}$$

The centripetal force must equal the force of attraction between the nucleus and the electron.

$$\frac{mv^2}{r} = \frac{1}{4\pi\varepsilon_o}\frac{e^2}{r^2} \qquad mv^2 = \frac{1}{4\pi\varepsilon_o}\frac{e^2}{r}$$

$$E = \frac{e^2}{8\pi\varepsilon_o r} - \frac{e^2}{4\pi\varepsilon_o r}$$

$$E = \frac{-e^2}{8\pi\varepsilon_o r} \qquad\qquad \text{(Equation 26.1)}$$

The total energy of the electron is also negative because the electron has been captured by the nucleus. It resides in an energy well. This energy is the ionization energy for the electron. Doing this much work on the electron will remove it from the atom. The energy is in terms of the radius of the orbit, and you must know the orbital radius to know what the energy is.

Now comes the quantum assumption. It was made by Danish physicist Niels Bohr in the early days of quantum theory. He assumed that the electron's angular momentum was quantized. Planck's constant equals 6.63×10^{-34} J-s. The units of Planck's constant are those for angular momentum, which is no coincidence. What Planck saw when he was studying the thermal radiation of a blackbody was related to the energies associated with electrons in orbit around the nucleus. Such an electron has angular momentum.

$$mvr = n\frac{h}{2\pi} \qquad v = \frac{nh}{2\pi mr}$$

where $n = 1, 2, 3, 4, \ldots$.

$$\frac{mv^2}{r} = \frac{1}{4\pi\varepsilon_o}\frac{e^2}{r^2} \qquad \frac{m}{r}\left(\frac{nh}{2\pi mr}\right)^2 = \frac{1}{4\pi\varepsilon_o}\frac{e^2}{r^2}$$

$$\frac{n^2 h^2 m}{4\pi^2 m^2 r^3} = \frac{e^2}{4\pi \varepsilon_o r^2}$$

$$\frac{n^2 h^2}{\pi m r} = \frac{e^2}{\varepsilon_o} \qquad n^2 h^2 \varepsilon_o = \pi m r e^2$$

(Equation 26.2)

$$r = \frac{\varepsilon_o h^2}{\pi m e^2} n^2$$

$$E = \frac{-e^2}{8\pi \varepsilon_o r} \qquad E = \frac{-e^2}{8\pi \varepsilon_o}\left(\frac{\pi m e^2}{\varepsilon_o h^2 n^2}\right)$$

(Equation 26.3)

$$E = \frac{-m e^4}{8 \varepsilon_o^2 h^2} \frac{1}{n^2}$$

$$\frac{m e^4}{8 \varepsilon_o^2 h^2} = 13.6 \text{ eV!!} \qquad \text{(The ionization energy of a hydrogen atom)}$$

$$E = -\frac{1}{n^2}(13.6 \text{ eV}) \qquad \text{(For a hydrogen atom)}$$

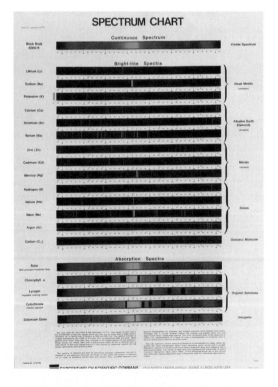

The discrete wavelengths from atomic spectra are caused by electrons jumping between allowed energy levels while in orbit around an atomic nucleus.

For $n = 1$, the calculated radius for the hydrogen atom is 5.3×10^{-11} m, or the radius found earlier from the ionization energy.

When a substance is heated to incandescence, it begins to glow. When hydrogen is heated to incandescence and the light is passed through a prism, the light is broken up into its various wavelength components. In the visible spectrum three lines are seen: a red line at a wavelength of about 690 nm, a green line at about 620 nm, and a blue line at about 450 nm. The fact that we see three discrete lines instead of a continuous spectrum such as a white light spectrum is proof that the energy levels of the electrons orbiting the nucleus of the hydrogen atom are quantized. As the hydrogen atom absorbs energy, the electron in the $n = 1$ state jumps up to the orbital with $n = 2$. The electron is unstable in this higher-energy state, and it drops back down to the lower-energy state, emitting a photon in the process. The energy of the photon is the difference in energy between the two electron orbital energies. If the photon's energy is in the visible spectrum, it can be seen.

EXAMPLE 26.2

What are the two quantum numbers for the red line in the visible spectrum of the hydrogen atom?

SOLUTION
First we must calculate the energy found in the red photon emitted when the electron in the hydrogen atom jumps from the higher-energy orbital to the lower-energy orbital.

$$\Delta E = hf \qquad \lambda f = c$$

$$f = \frac{c}{\lambda} \qquad \Delta E = \frac{hc}{\lambda}$$

$$\Delta E = \frac{(6.63 \times 10^{-34} \text{ J-s})(3 \times 10^8 \text{ m/s})}{690 \text{ nm}}$$

$$\Delta E = 2.88 \times 10^{-19} \text{ J}$$

$$\Delta E = (2.88 \times 10^{-19} \text{ J})\left(\frac{1 \text{ eV}}{1.6 \times 10^{-19} \text{ J}}\right)$$

$$\Delta E = 1.8 \text{ eV}$$

Next we must calculate the two quantum numbers responsible for the red line. We can initially assume that the quantum numbers are adjacent, or separated by 1.

$$E = (-13.6 \text{ eV})\frac{1}{n^2} \qquad \Delta E = (-13.6 \text{ eV})\left(\frac{1}{n_f^2} - \frac{1}{n_i^2}\right)$$

$$n_f - n_i = 1 \qquad n_f = 1 + n_i$$

$$1.8 \text{ eV} = (-13.6 \text{ eV})\left(\frac{1}{(1 + n_i)^2} - \frac{1}{n_i^2}\right) \qquad -0.132 = \frac{1}{(1 + n_i)^2} - \frac{1}{n_i^2}$$

$$-(0.132)(1 + n_i)^2 n_i^2 = n_i^2 - (1 + n_i)^2$$

$$-0.132(1 + 2n_i + n_i^2)n_i^2 = n_i^2 - (1 + 2n_i + n_i^2)$$

$$(0.132)(n_i^2 + 2n_i + 1)n_i^2 = 1 + 2n_i + n_i^2 - n_i^2$$

$$(0.132)(n_i^2 + 2n_i + 1)n_i^2 = 1 + 2n_i$$

$$0.132n_i^4 + 0.264n_i^3 + 0.132n_i^2 = 2n_i + 1$$

$$0.132n_i^4 + 0.264n_i^3 + 0.132n_i^2 - 2n_i - 1 = 0$$

Knowing that n_i is an integer, we can start substituting $n_i = 1, 2, 3, \ldots$ into the above equation to find out which integer solves it. Luckily, we can stop at $n_i = 2$, and thus the final quantum number is 3. So the orbiting electron jumps from allowed orbit number 3 to allowed orbit number 2 when a red photon is emitted. The initial quantum state for the visible spectrum isn't the ground state, or $n = 1$. The spectral series starting with $n = 1$ is in the ultraviolet region and can't be seen.

As the quantum number gets larger, the energy spacing between allowed orbits gets smaller. For very large quantum numbers, the energy spectrum is very small and appears to be virtually continuous, as classical theory predicts. For very large quantum numbers, a system acts classically. If the electron orbiting the hydrogen nucleus is given 13.6 eV, it is no longer bound to the nucleus. The energy is now zero, and the free electron can have any positive energy that you care to give it. The electron now acts like a classical particle.

The electrons in the outer orbits of atoms are loosely bound, and it takes only a few electron-volts to raise them to a higher orbital from which they can drop back down and emit a photon. The visible spectrum has a range of wavelengths from roughly 700 nm to 400 nm, or energies from 1.78 eV to 4.98 eV. Our eyes are sensitive to photon energies in this range. Seeing is a direct result of electrons jumping between orbitals around the nuclei of the elemental atoms in the periodic table (see Appendix E).

26.3   The Pauli Exclusion Principle

There is more than the one element hydrogen, and there is more than one quantum number, n. What kind of element a particular atom is depends on the number of protons in its nucleus. If the atom has one proton, it's hydrogen; two protons, it's helium; three protons, it's lithium; and so on up the periodic table. For each proton in the nucleus there is a corresponding electron in orbit around the nucleus. *But the chemical properties of the atom are determined not by the protons in the nucleus but by the electron cloud around the nucleus.* It's the outer electrons in an atom's cloud that interact with the outer electrons of another atom's cloud. If all of the electrons in an atom's cloud crowded into the same orbit, all atoms would appear to be the same regardless of the number of protons in the nucleus. The Pauli exclusion principle sorts the electrons in the cloud into the various shells that are successively filled to create the periodic table and the distinct elements in it. *The **Pauli exclusion principle** forbids any two electrons in the same atom from having the same quantum numbers.* The different quantum numbers for each electron create the different properties of the elements in the periodic table.

Pauli exclusion principle: No two electrons in orbit around a single atom can have exactly the same quantum numbers.

The principal quantum number, n, has already been discussed. There is a second quantum number associated with the orbital angular momentum of the electron, ℓ, and it has integer values for each principal quantum number *(n)* from zero up to $\ell = (n - 1)$. If $n = 3$, then ℓ can have possible values 0, 1, and 2. The third quantum number is the magnetic quantum number, m_ℓ, and is associated with the orientation of the angular momentum vector in space. The electron in orbit is a current around the nucleus, and there is a magnetic field around a current element. The direction of this magnetic field is opposite the direction of the electron's angular momentum vector because the electron has a negative charge. The atom has a magnetic field like that of a small bar magnet. Its possible values are determined by the orbital angular momentum quantum number, ℓ. Note that m_ℓ has both positive and negative integer values for each value of ℓ. If $\ell = 3$, for instance, then m_ℓ has seven possible values: $-3, -2, -1, 0, 1, 2$, and 3, or $m_\ell = (2\ell + 1)$. The fourth and last quantum number is the electron spin quantum number, m_s. If the electron were spinning around its own axis, its angular momentum could be oriented "up" or "down." The electron spin is defined to be $+\frac{1}{2}$ or $-\frac{1}{2}$. The four quantum numbers are summarized in Table 26.1.

A look at the periodic table in Appendix E will show how the four quantum numbers order the rows of the table. If $n = 1$, then $\ell = 0$ and $m_\ell = 0$. Only the two states of the spin magnetic number are allowed. The first row of the periodic table has only two elements in it: hydrogen and helium. If $n = 2$, then $\ell = 1$, $m_\ell = 3$ and $m_s = \pm\frac{1}{2}$. There are now eight allowed states for electrons in the electron cloud. The second and third rows of the periodic table both have eight elements in them. But for $n = 3$, the rows in the periodic table jump to 18 members for the next three rows. If $n = 3$, then there are three values for the orbital angular momentum, and for each value of ℓ there is a subshell with $2\ell + 1$ orbitals for the orbital magnetic quantum number. Thus, there are nine possible subshells for the electrons to fill up. Multiply this by 2 for the spin quantum number, and there is a total of 18 possible orbits for the electrons in the electron cloud. The electrons fill the shells from the one with the lowest energies first.

So in conclusion the number of protons in the nucleus determines how many electrons are in the cloud around the nucleus. But it is the set of quantum numbers *for the electrons* that determines the elemental properties of the atom. Almost all of the interactions that an atom has with other atoms take place in the outer part of the atom's electron cloud. The discipline of chemistry is primarily concerned with this fact. Very rarely does an interaction take place between the nuclei of the interacting atoms. These *internuclear interactions* are discussed in the next two sections.

Table 26.1 Quantum Numbers

Quantum Number		Allowed Values	Number of Allowed Values
Principal	n	1, 2, 3, ...	Infinite
Orbital angular momentum	ℓ	0, 1, 2, 3, ..., $(n - 1)$	n (for each n)
Orbital magnetic	m_ℓ	0, ± 1, ± 2, ± 3, ..., $\pm\ell$	$2\ell + 1$ (for each ℓ)
Spin magnetic	m_s	$\pm\frac{1}{2}$	2 (for each m_ℓ)

26.4   Radioactivity and Nuclear Fission

The strong force binds the nucleus together against the mutual repulsion of the positively charged protons. But occasionally an otherwise stable nucleus will spontaneously fly apart, emitting radiation as it does so. Such nuclei are called *radioactive* and were first studied in the early 1900s by Henri Becquerel, Marie and Pierre Curie, and others. The tendency for a nucleus to decay radioactively increases with the number of protons because the mutual repulsive force increases. If Z is the number of protons in the nucleus, the force of repulsion goes as Z^2. Remember that the neutrons act to keep the protons from getting too close and from thereby allowing the repulsive force to get too large. For a low number of nuclear protons, the ratio of neutrons to protons is around 1:1. But as the number of protons increases, so does the neutron-to-proton ratio. For the element uranium, the ratio is around 2.5 : 1.

The shorthand for specifying an element is as follows:

$$^A_Z X$$

where X is the chemical symbol for the element; Z is the atomic number of the element, or the number of protons in the nucleus; and A is the mass number of the nucleus. The common helium atom is written as shown in the shorthand defined above:

$$^4_2 He$$

where $Z = 2$ for two protons in the nucleus, and 4 is the mass number. Thus, helium has a total of two neutrons in the nucleus $(4 - 2 = 2)$. (To find the neutrons present in a given nucleus, just subtract the atomic number from the mass number.) A given element can have only one atomic number but more than one mass number. *Atoms with these different mass numbers for an element are* **isotopes** *of that element.* The most common version of carbon has a mass number of about 12 with 6 protons and 6 neutrons in the nucleus. But there is another version of carbon, carbon-14, with 8 neutrons in the nucleus. Both versions must have 6 protons to be carbon, but the mass number can vary without the element losing its identity. In a sample of carbon, carbon-12 will be the largest component of the sample, making up approximately 99.9% with carbon-14 making up the rest of it. Chemically the two versions of carbon are identical, the only difference being that carbon-14 is slightly heavier than carbon-12. The mass number of an element is measured in atomic mass units, or amu. One amu is defined as 1.66×10^{-27} kg, or about the mass of either a neutron or a proton. A very well known pair of isotopes are U-235 and U-238.

Isotopes: Atoms of an element with the same atomic number but different mass numbers.

$$\text{U-235} = {}^{235}_{92}U \qquad \text{U-238} = {}^{238}_{92}U$$

These two isotopes of uranium are very much associated with radioactivity and the atomic bomb. The isotope U-238 isn't radioactive and makes up approximately 99.3% of a sample of uranium refined from ore. The other 0.7% is U-235; it is radioactive and will spontaneously decay into lighter nuclei, emitting radiation when it does so. It is this rare isotope of uranium that causes all of the trouble and is responsible for nuclear fission.

Why do heavy nuclei tend to be more unstable than lighter ones? The graph in Figure 26.2 plots the mass number of a nucleus horizontally and the binding energy per nucleon or particle in the nucleus vertically. Note that the most stable nucleus in the periodic table is the one of Fe-56. This isotope of iron has the most tightly bound nucleons and isn't very likely to decay radioactively. The nuclei to the right or to the left of Fe-56 have less binding energy per nucleon. U-235 is at the right end of the graph. Notice that U-235 has fewer neutrons in the nucleus than U-238. Since more neutrons mean more stability for a heavy nucleus, then U-235 should be more unstable than U-238, and it is. When a nucleus of U-235 decays, the decay fragments try to attain a state more stable than that of the parent nucleus. They climb the binding energy curve toward Fe-56. When a radioactive nucleus decays, it gives off a considerable amount of energy. This shedding of energy allows the nucleons in the decay particles to be bound more tightly together.

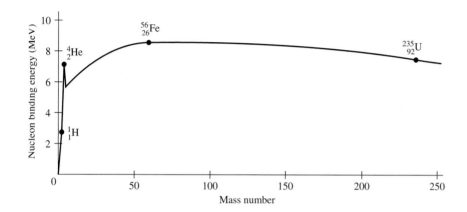

Figure 26.2 Nuclear Stability Curve

The amount of energy given off by a radioactively decaying uranium nucleus is indeed considerable. If a U-235 nucleus splits into two roughly equal parts, the binding energy goes from 7.6 MeV/nucleon to that of a much lighter element with a mass number of about 118 amu. The protons and neutrons in the fragment nuclei are now bound much more tightly and more stably, with a binding energy of about 8.4 MeV/nucleon. Each nucleon had to give up 0.8 MeV to attain the more stable nuclear binding. Since 235 nucleons took part in the reaction, the total energy liberated when the uranium nucleus split was 188 MeV. Remember that 1 MeV is *one million* electron-volts. A chemical reaction takes place when the outer electrons in an electron cloud interact with those in another atom. The energy is only on the order of 1 or 2 electron-volts. A nuclear reaction can deliver 100 million times more energy than a chemical reaction! A "small" nuclear device can deliver a punch of 20 to 30 kilotons of TNT, or 20,000 *tons* of dynamite. Where do nuclear reactions find such energy?

Earlier we studied Einstein's special theory of relativity. Einstein was interested in what happened when a material object traveled at speeds close to that of light. There was time dilation, and a shortening of the length in the direction of motion and an increase in mass as the object got closer and closer to the speed of light. Another result of Einstein's theoretical studies was the famous equation

$$E = mc^2$$

(Equation 26.4)

The energy latent in a mass is equal to the mass of the object times the square of the speed of light. Since the speed of light is 3×10^8 m/s (already a very large number), its square is much larger still. Einstein's equation equates matter and energy. When a nuclear reaction takes place, matter is converted to energy. Thus, the fragment nuclei weigh a little less than the parent uranium nucleus. That's one of the reasons that the mass numbers shown in the periodic table aren't integers. Whenever a nucleus is formed or split, a small fraction of its mass is created or destroyed. Another reason is that the neutron and the proton don't have the same mass. The neutron has a bit more mass than the proton. The atomic number is the number of protons in the nucleus and must be an integer. But the mass number must give the total mass of all of the nucleons in the nucleus. It isn't an integer because the neutron has a little bit more mass than the proton, and because a little mass is either created or destroyed when the nucleus is either formed or split.

*The splitting of an atomic nucleus into smaller fragments is called **fission.*** The most widely known form of nuclear fission is the splitting of uranium or plutonium nuclei. A rapid splitting leads to an explosion such as that produced by an atomic bomb. A slow, controlled splitting allows the construction and use of nuclear reactors for the production of power. But in both cases a nuclear chain reaction takes place in which many nuclei are split in a continuous fashion. How can a continuous splitting of uranium nuclei be set up? When a U-235 nucleus undergoes spontaneous fission, it splits into two fission fragments, but in doing so, it also liberates two fast neutrons. Because a neutron has no charge, it can collide with another uranium nucleus and cause it to undergo fission as well. This induced fission produces another pair of neutrons that can (ideally) cause two more U-235 nuclei to split, and so on. With each fissioning, the uranium nuclei release the energy stored in the nucleus, and this energy is unimaginably large because a little matter is converted into energy with each fissioning.

The key to a chain reaction is the two neutrons that are liberated when a nucleus splits. Under ideal conditions the first split yields two neutrons, which causes 2 more nuclei to split, which can split 4 more, which can split 8 more, which can split 16 more, leading to 32 more, 64 more, 128 more, 256 more, and so on, in a geometric progression that rapidly grows into an uncontrolled explosion. Remember, though, that only the U-235 isotope is radioactive and that in any sample of uranium processed from ore, 99.3% of the sample is U-238 and only 0.7% is U-235 which readily undergoes fission. The naturally processed uranium ore has too few U-235 nuclei to allow the fissioning to grow critical, or to sustain itself, let alone to explode. To allow a chain reaction, the component of U-235 must be enriched to around 3% so that the uranium fuel can be "burned" in a reactor. At this concentration it can't explode, but under the proper conditions in a reactor, the nuclear reaction will continue to sustain itself. Remember that two isotopes of the same

Fission: The splitting of a heavy atomic nucleus into smaller fragments to produce energy.

The explosion of a thermonuclear device occurs using both fission and fusion. The fission reaction creates the temperatures necessary for fusion to take place, which accounts for most of the energy released by the bomb.

element are exactly the same chemically. The only difference between the isotopes is their masses. One way in which uranium is enriched is that it is first converted into uranium hexafluoride gas and then centrifuged. A centrifuge spins the gas at a high rate of speed, and the heavier component (the uranium hexafluoride gas with the U-238) finds itself on the outside of the centrifuge, and the lighter component (the gas containing the U-235) on the inside. Of course, complete separation is never achieved, but if the gas on the inside of the centrifuge is processed through many stages, then the uranium gas can be enriched, or the U-235 component increased, over what is now found in nature many times. Weapons-grade uranium can have a U-235 component of over 90%.

In a fission bomb the uranium is mostly U-235, and, simply enough, if enough of it is brought together in one big lump, it will spontaneously go critical and explode. Making a bomb out of U-235 is easy; the trick is to get enough of it. A nuclear reactor is used to generate power so that a controlled reaction is what is desired, not an explosion. For this purpose the nuclear fuel is uranium enriched to only about 3% U-235. But without help it still doesn't have enough U-235 to sustain an ongoing fission reaction. The neutrons released by a fissioning uranium nucleus are given energy from the conversion of matter to energy and are highly energetic. Since the energy is in the form of kinetic energy, the neutrons move very fast, too fast to interact with other uranium nuclei. Before they can efficiently promote other fissions, they must be slowed down. One way to slow a fast neutron down is to have it collide with another particle about as massive as it is. Remember collisions and the conservation of energy and momentum? A ball will lose most energy in a collision if it collides with another ball with the same mass. If the collision is head-on, then the original ball stops, and all of the energy of the collision is passed on to the other ball, much as when the cue ball in billiards hits another ball solidly. As a result, hydrogen is a good moderator of fast neutrons. Hydrogen has a single proton in the nucleus and has about the same mass as a neutron. So hydrogen in the form of water (H_2O) is used in most reactors as a moderator to slow the fast neutrons down. The reactor core is built so that the neutrons that attempt to leave the core are reflected back into it by the walls of the containment vessel. Other materials are used to absorb the neutrons in the core so that the chain reaction can be slowed down or stopped if necessary. These materials can be partially or fully lowered into a reactor core to "throttle" the reaction. A reactor so built will "burn" evenly and will produce a steady output of thermal energy. The thermal energy is used just like the thermal energy from coal-burning and oil-burning power plants: to boil water that turns a turbine that generates electricity. So a nuclear reactor is just a fancy way to boil water. Even on a nuclear submarine the reactor heat boils water that turns a turbine that turns a propeller shaft. The reactor in a submarine must be small, so in order to obtain the necessary power output, the uranium might be weapons grade, or better than 90% enriched.

When a U-235 nucleus undergoes fission, the fragment nuclei may start a long-lived chain of radioactive decay in search of a stable nuclear configuration. It starts ascending the binding energy curve toward Fe-56. The fragment nuclei usually decay by emitting three types of radiation: an alpha particle, a beta particle, or a gamma ray. An alpha particle is $_2He^4$ or the nucleus of the garden variety of helium with two protons and two neutrons. It emerges from the decaying nucleus with quite a bit of energy and is ionized and has a positive charge. The beta particle is an electron with a negative charge and also carries away a lot of kinetic energy from the decaying nucleus. The electron wasn't originally in the nucleus, but one of the neutrons in the unstable parent nucleus decays into a proton and an electron, and the electron is ejected from the nucleus. In this way a nucleus with too many neutrons with respect to its protons can reduce the neutron count by 1 and simultaneously increase the proton count by 1. In both alpha and beta decay, the parent nucleus becomes a different element because the atomic number is changed. For alpha decay the atomic number drops by 2, but for beta decay it rises by 1. A gamma ray, however, is a highly energetic proton, and the atomic number stays the same, or there is no transmutation of the element.

Another way that a nucleus can decay is by *positron emission.* A positron is a positive electron. It's actually a bit of antimatter. By emitting a positron, an unstable nucleus with too many protons relative to the neutrons can decrease the proton count by 1. An interesting aside is that when a positron meets an electron, they annihilate one another. All of the mass of the two particles (a positron has the same mass as an electron) is converted to energy.

Plutonium is the other element infamous for its part in nuclear weaponry. It had virtually disappeared on the earth by the time man had attained nuclear capabilities. All plutonium is fissionable, and just about all of it had decayed away long before man had even appeared on the planet. There wasn't any plutonium ore to mine as there was uranium ore. So where did the plutonium that now exists come from? Plutonium has its place in the periodic table but was extremely rare and could be found only in trace amounts. It had to be manufactured by man in nuclear reactors. The U-238 is inert and doesn't take part in the chain reaction that produces power or an explosion. But it can capture a fast neutron and become $_{92}U^{239}$, which is another unstable isotope of uranium. This particular isotope can decay twice by beta (electron) decay, and the atomic number of the nucleus is raised by 2. The element with an atomic number of 94 is plutonium. A breeder reactor is a reactor that takes advantage of this nuclear reaction. All plutonium is fissionable, so if a reactor that produced power could also produce plutonium as a by-product, the reactor could actually produce more fuel than it consumes by transmuting the inert U-238 into plutonium. The plutonium is easier to separate out from the uranium because it's another element and can be separated chemically. But because all plutonium is fissionable, it is weapons grade. It is very difficult to turn raw uranium into a bomb because of the need to highly enrich it, but that's not the case with plutonium. At the end of WWII when the two atomic bombs were used on Japan, the first bomb used on Hiroshima was an enriched uranium bomb, but the second one used on Nagasaki was a plutonium bomb. We had learned how to make plutonium very early in the nuclear era.

26.5   Half-Life

One of the problems with radioactive decay is that too much exposure to radiation can cause tissue damage and can cause cancer long after the initial exposure. Radiation can also cause genetic mutations. If radioactive nuclei get out into the environment, they are taken up by plants and animals and are concentrated the farther they migrate in the food chain. In this way, an initially small dose of radiation can become dangerously concentrated. If the radiation associated with a splitting uranium or plutonium nucleus happened only when the nucleus underwent fission and the fragment nuclei became stable soon after, then the problem of radioactivity would be much more manageable. But the problem is that the fragment nuclei are also radioactive and can persist in their radioactivity for many years afterward, in fact, for hundreds and even thousands of years. Eventually the fragment nuclei would decay to stable, nonradioactive nuclei, but for the hundreds and thousands of years that the radioactive waste continues to emit radiation, it must be sealed away from the environment and humans.

*The time it takes a sample of radioactive isotope to decay until only one-half of it is left is called its **half-life.*** The decay is an exponential decay, which we have seen before. (For example, the charge bleeding off a capacitor through a resistor is an exponential decay.) The rate at which a radioactive sample decays is proportional to the amount of

Half-life: The time it takes a radioactive isotope to decay to one-half of its original amount.

the sample present, and this rate is negative because the sample decreases with time. If N is the number of radioactive nuclei present in a sample at any time t, then

$$\frac{\Delta N}{\Delta t} = -kN \qquad \frac{\Delta N}{N} = -k\,\Delta t \qquad \Sigma\frac{\Delta N}{N} = -k\Sigma\,\Delta t$$

where k is an empirically determined constant called the *decay constant,* and the summation on the left is the natural log of N (as in RC time constants).

$$\ln N = -kt + C$$

At $t = 0$, $N = N_o$ (the original concentration of N). Thus,

$$\ln N_o = C \qquad \ln N = -kt + \ln N_o$$

$$\ln N - \ln N_o = -kt \qquad \ln\left(\frac{N}{N_o}\right) = -kt$$

$$e^{\ln(N/N_o)} = e^{-kt} \qquad \frac{N}{N_o} = e^{-kt}$$

$$N = N_o e^{-kt}$$

The sample decays exponentially, but we want to put k in terms of the half-life. At $t = t_{1/2}$, $N = \frac{1}{2}N_o$.

$$\frac{N_o}{2} = N_o e^{-kt_{1/2}} \qquad \frac{1}{2} = e^{-kt_{1/2}}$$

$$2^{-1} = e^{-kt_{1/2}} \qquad \ln(2)^{-1} = -kt_{1/2}$$

$$-\ln 2 = -kt_{1/2} \qquad \ln 2 = kt_{1/2}$$

$$k = \frac{\ln 2}{t_{1/2}} \qquad N = N_o e^{-(\ln 2)t/t_{1/2}}$$

$$N = N_o e^{-0.693t/t_{1/2}} \tag{Equation 26.5}$$

Equation (26.5) looks complicated but can be solved using most scientific calculators. However, after one half-life only one-half of the original sample is left. Each further half-life reduces the sample by another half.

> After 1 half-life: $\frac{1}{2}$ of the original sample is left.
> After 2 half-lives: only $\frac{1}{4}$ of the original sample is left.
> After 3 half-lives: only $\frac{1}{8}$ of the original sample is left.
> After 4 half-lives: only $\frac{1}{16}$ of the original sample is left.
> After 5 half-lives: only $\frac{1}{32}$ of the original sample is left.
> And so on.

After n half-lives (where n is an integer), only $\frac{1}{2}^n$th of the original sample is left.

Table 26.2 gives a list of some radioactive elements, including their primary decay modes and their half-lives. Half-lives can range from times unimaginably short (for exam-

Table 26.2 Primary Decay Modes and Half-Lives of Some Radioactive Elements

Element	Decay Mode	Half-Life
Beryllium-8 ($^{8}_{4}$Be)	Alpha	1×10^{-16} s
Polonium-213 ($^{213}_{84}$Po)	Alpha	4×10^{-16} s
Oxygen-16 ($^{19}_{8}$O)	Beta	27 s
Fluorine-17 ($^{17}_{9}$F)	Positron, electron capture	66 s
Polonium-218 ($^{218}_{84}$Po)	Alpha, beta	3.05 min
Technetium-104 ($^{104}_{43}$Te)	Beta	18 min
Krypton-76 ($^{76}_{36}$Kr)	Electron capture	14.8 h
Magnesium-28 ($^{28}_{12}$Mg)	Beta	21 h
Iodine-123 ($^{123}_{53}$I)	Electron capture	13.3 h
Radon-222 ($^{222}_{86}$Rn)	Alpha	3.82 days
Cobalt-60 ($^{60}_{27}$Co)	Beta	5.3 yr
Strontium-90 ($^{90}_{38}$Sr)	Beta	28 yr
Radium-226 ($^{226}_{88}$Ra)	Alpha	1600 yr
Carbon-14 ($^{14}_{6}$C)	Beta	5730 yr
Plutonium-239 ($^{239}_{94}$Pu)	Alpha	2.4×10^{4} yr
Uranium-238 ($^{238}_{92}$U)	Alpha	4.5×10^{9} yr
Rubidium-87 ($^{87}_{37}$Rb)	Beta	4.7×10^{10} yr

ple, beryllium-8) to times comparable to the age of our planet (uranium-238 and rubidium-87). U-238 is a stable nucleus! It's the isotope U-235 that causes all of the trouble. Plutonium is readily fissionable, yet it has a fairly long half-life of 24,000 years.

Archaeologists have turned the half-life of the radioactive isotope carbon-14 into a dating tool. All biological organisms take carbon up from their immediate environment. The ratio of carbon-12 to carbon-14 in the environment is constant and known. When an organism dies, the carbon-14 begins to decay radioactively, and the ratio of carbon-12 to carbon-14 starts increasing. After 5760 years the dead organism has only one-half the original C-14 left; after 11,520 years, one-fourth as much; and so on. By determining the ratio of C-12 to C-14, archaeologists can date artifacts such as wooden implements, charcoal from fires, and cloth and leather as far back as 70,000 years. Geologists can date rocks back much further using radioactive isotopes with much longer half-lives. Rocks 3.8 billion years old have been found in Greenland using radioactive dating.

The chief reason that radioactivity is so dangerous to biological tissues is that it ionizes it. An ionized atom or molecule has a net electrical charge and is very reactive. The alpha and beta particles are fast ions, and when they hit something, they usually leave an ionized path of material in their wake. It is this ionization that allows us to detect the radiation. A Geiger counter is nothing more than a tube with a positive electrode down its center with the inside surface negatively charged. An ion entering the tube is attracted to either the center electrode or the inside surface. When the ion hits either electrode, a current pulse is detected, and either an audible click is heard or the pulse registers a count in an adder circuit to be displayed as a rate. Large accelerators use cloud chambers to examine the products of collisions between subatomic particles. The results of such a collision are very energetic and ionized and are passed through a chamber that is filled with water vapor with a strong magnetic field applied. The ionized particles begin to spiral in the magnetic field and leave an ionization trail in the vapor. This evidence can be interpreted by physicists to yield information about the collisions in terms of identification of reaction products and energies.

26.6   Nuclear Fusion

Nuclear fusion is the opposite of nuclear fission. *In fusion the nuclei of light atoms join together to yield nuclear energy.* If you look at the part of the nuclear binding energy curve to the left of Fe-56 in Figure 26.2, you'll notice an upward spike. The tip of the spike is the binding energy of He-4, the second most abundant element in the universe. To the left of that spike is the hydrogen nucleus, or a bare proton. Hydrogen is the most abundant element in the universe. (The universe is about 90% hydrogen, 9% helium, and the remaining percent the other elements in the periodic table.) The difference in binding energies between hydrogen and helium (which is more stable) is quite a bit greater per nucleon than the difference in energy liberated when uranium or plutonium undergoes fission. If a reactor could be designed that would fuse hydrogen into helium, then much more energy would be liberated than from a fission reactor, and you would never run out of fuel. You would have to burn up most of the universe to run out! Hydrogen is abundant on the earth, but it is locked up in chemical compounds and molecules because it is very reactive in its pure gaseous state. It wouldn't last long as a gas in our atmosphere because it is too light for our gravitational field to hold it, and it would slowly leak off into space. Much hydrogen is locked into water (H_2O), however, and there is an abundant supply in the earth's oceans. What's more, both the beginning points and the endpoints are stable, non-radioactive nuclei. Fusing hydrogen into helium would also eliminate the radioactive waste problems that have plagued fission reactors. This solution sounds like an almost perfect remedy to the energy crisis, so why aren't we harnessing fusion for our energy needs now or planning to do so in the very near future?

The problem is that to initiate a fusion reaction, you must slam hydrogen nuclei together with enormous velocities. The kinetic energies of the nuclei must be high enough to overcome the electrical repulsion and penetrate the nucleus of the target helium atom. The required enormous velocities translate to extremely high temperatures, in the range of millions of degrees. (Remember that temperature is the average kinetic energy of a collection of atoms and molecules.) Achieving such temperatures in a controlled way in a fusion reactor so that more energy is produced than it takes to make the nuclei fuse is a very difficult engineering feat, one that has not yet been achieved despite decades of trying. This goal is one of the Holy Grails of physics.

Fusion is routinely achieved in the universe at large, though. Every point of light you see in the night sky (except the planets, moon, and comets) is a fusion fire. Stars radiate a vast amount of energy and are driven by fusion at their cores. Our sun is a star about which all of the planets orbit, so it too is driven by fusion. In the early universe not long after the Big Bang, nucleosynthesis took place, and the light elements formed. These light elements were mostly hydrogen and a little helium. Later, when the force of gravity began

Fusion: The joining of the nuclei of light atoms to yield nuclear energy.

to exert its rather feeble force on the expanding matter of the universe, large clouds of hydrogen gas began to aggregate and to collapse in upon themselves. As the clouds collapsed, the temperature and pressure at the core increased. When they became high enough, fusion started and the star ignited. The star began converting its hydrogen into helium and giving off light and heat. Initially a star burns $_1H^1$ in the following sequence: two $_1H^1$ nuclei collide to form deuterium, or $_1H^2$. A $_1H^1$ nucleus then collides with this deuterium nucleus to form an isotope of helium, $_2He^3$. Two of these helium nuclei then collide to form the stable helium nucleus $_2He^4$ and another $_1H^1$ nucleus. From beginning to end of the proton cycle, a total of 18.9 MeV of energy is liberated. When a star initially starts burning its hydrogen, the outward radiation pressure equals the inward gravitational pressure, and the inward collapse is halted. But when a star burns up a good percentage of the hydrogen in its core, the radiation pressure decreases and the inward collapse starts again. The temperatures and pressures then climb still higher in the core until the star starts fusing its helium into carbon. The fusing of helium to carbon liberates more energy than the burning of hydrogen, or the star couldn't resist the far greater gravitational pressures trying to crush it. When the helium at a star's core runs out, the collapse of the core begins yet again, continuing until the carbon begins to fuse, yielding in turn even more energy than the fusing of helium and temporarily halting the core's collapse. The end product of fusing carbon is Fe-56, the nucleus with the greatest binding energy per nucleon and the end of the road as far as fusion reactions are concerned.

So where does a star go from here? The collapse starts again, and at this point the star can go nova, or blow off its outer layers of still unfused hydrogen, helium, carbon, and iron. In the process of going nova, elements heavier than iron are formed and are flung out into space, surrounding the star with an expanding nebula of gas seeded with the heavier elements. Our planet earth has an iron core, and our sun is a second-generation star. The iron in the earth's core came from a star that preceded our sun!

So conditions at the core of a star are favorable for thermonuclear fusion, but these conditions are hard to create from here on earth. Fusion has occurred here on the earth, but the first fusion reactions here were produced by bombs. A hydrogen bomb is more powerful than a fission bomb, but it requires a fission bomb to trigger it. This fact should give you an idea as to how difficult fusion is to achieve terrestrially. Two different approaches are being tried to build a viable fusion reactor to supply the energy needs of societies here on the earth. The first is a reactor called a *tokamak*. In this reactor a plasma of hydrogen ions is heated to fusion temperatures. (A plasma is an ionized gas.) The hydrogen gas in the tokamak reactor is introduced at a very low pressure and is ionized. When the hydrogen is ionized, electric and magnetic fields can pump the temperature of the plasma to fusion temperatures. Although fusion has been achieved in such a reactor, it still takes more energy to start the fusion process than the reactor produces when fusion occurs.

In another approach, called *inertial fusion,* a pellet of a hydrogen compound is placed into a cavity. A very powerful laser illuminates the hydrogen-containing pellet and causes it to implode, or explode inward. As the pellet contracts under the implosion, the pressure and temperature at its core approach conditions like those at the core of a star. The pellet's core fuses and releases a burst of energy. Then another hydrogen-containing pellet is introduced into the cavity, and the process starts all over again. But lasers are still very inefficient, and the efficiencies must be improved by orders of magnitude before a viable reactor can be built. A variation on the inertial theme is to bombard the pellet from all sides with a particle beam. The beam would be a beam of ions or electrons that are accelerated by electric and magnetic fields to high velocities and directed onto the pellet from all sides. As a result, the pellet would collapse inward and the core would fuse, releasing a burst of energy. Although this approach, too, has its problems and has yet to produce a viable fusion reactor, the technology of a viable fusion reactor is within reach, and there is no theoretical reason why it can't be achieved by one or both approaches.

Hydrogen exists in three isotopes: (1) regular hydrogen with a single proton in the nucleus, (2) deuterium or "heavy water" with a proton and a neutron in the nucleus, and (3) tritium with two neutrons in the nucleus. The isotope that burns most easily is tritium, followed by deuterium and then normal hydrogen. The isotopes can be separated by centrifuging the hydrogen gas. Deuterium has a mass twice that of normal hydrogen, and tritium three times that of normal hydrogen. The job of centrifuging the hydrogen to separate the isotopes is made easier than the job of enriching uranium because the mass difference for uranium between the two isotopes is only a little greater than 1%, whereas deuterium and tritium are, respectively, 200% and 300% heavier than common hydrogen.

26.7　The Weak Force

There are four forces in nature, and this book has dealt rather extensively with two of them: *gravitation* and the *electromagnetic force.* The third one, the *strong force* present in the nucleus that holds it together (not always successfully) against the electromagnetic forces trying to tear it apart, has been discussed in the preceding sections of this chapter. The fourth force, known as the *weak force,* was discovered when it was found that a neutron alone isn't a stable particle: It decays into a proton and an electron. Unlike a nucleus with more than one proton that is straining to come apart because of the mutual repulsion of the protons, there was no good reason why a neutron should decay in such a manner, so a fourth fundamental force was postulated to account for it: the weak force. It is called *weak* because it is weaker than even the gravitational force. The weak force is responsible for beta decay, and therefore it plays a part in radioactive decays because many unstable isotopes decay by beta decay, in which the nucleus throws out an electron, causing a neutron to change into a proton.

A shadow particle called a *neutrino* is also often produced during beta decay. Physicists observed that the beta particle couldn't always account for all of the kinetic energy produced when a nucleus decayed by emitting an electron. A particle was postulated to account for the "lost" kinetic energy, the neutrino. The neutrino has no electrical charge and is practically massless. It therefore doesn't interact very strongly with normal matter and has proved to be very elusive. Physicists still don't know whether or not it contains a small amount of mass or whether it is massless and travels at the speed of light as a photon does.

Summary

1. *Atoms* are composed of a cloud of electrons that surround a dense nucleus of **protons** and **neutrons.** Protons have a positive charge equal to the magnitude of the electron $(1.6 \times 10^{-19} \text{C})$, whereas neutrons are electrically neutral. Both protons and neutrons have nearly the same mass of 1.67×10^{-27} kg.

2. Electrons that orbit nuclei are allowed to have only certain energies. Electrons that wish to gain or lose energy must do so in finite "jumps," called energy *quanta.* It is the transi-

tion between a higher-energy state to a lower-energy state that causes the energy difference to escape from the atom in the form of a photon. The energy of a given state can be calculated by Equations (26.1) and (26.3), where r is the radius of the electron orbit, and n is the quantum number of the electron.

3. All radioactive substances possess a **half-life,** which is the time required for one-half of a material to decay into another form. This exponential decay is dependent on the

amount of material originally present, the half-life of the material, and the time elapsed [see Equation (26.4)].

4. **Fission** is the splitting of heavy atomic nuclei, such as uranium and plutonium, into smaller fragments to produce energy. **Fusion** is the joining of very light nuclei, such as hydrogen and helium, to form heavier elements to produce energy. Fusion produces a far greater amount of energy than

does fission, and it avoids many of fusion's nuclear waste problems. However, fusion is extremely difficult to achieve.

5. There are four forces of nature: *gravitational force, electromagnetic force, strong nuclear force,* and *weak nuclear force.* The strong nuclear force is responsible for holding protons together in atomic nuclei, whereas the weak nuclear force is responsible for beta decay within the nucleus.

Key Concepts

The number of protons in the nucleus determines the elemental identity of an atom.

FISSION: The splitting of a heavy atomic nucleus into smaller fragments to produce energy.

FUSION: The joining of the nuclei of light atoms to yield nuclear energy.

HALF-LIFE: The time it takes a radioactive isotope to decay to one-half of its original amount.

ISOTOPES: Atoms of an element with the same atomic number but different mass numbers.

NEUTRON: A nuclear particle with no charge and about the same mass as the proton.

PAULI EXCLUSION PRINCIPLE: No two electrons in orbit around a single atom can have exactly the same quantum numbers.

PROTON: A nuclear particle with a charge of $+1.6 \times 10^{-19}$ C and a mass of 1.67×10^{-27} kg.

Important Equations

(26.1) $E = \dfrac{-e^2}{8\pi\varepsilon_o r}$ (Total energy of an electron in orbit around a hydrogen nucleus)

(26.2) $r = \dfrac{\varepsilon_o h^2}{\pi m e^2} n^2$ (Quantized electron orbital radius around a single proton or hydrogen nucleus)

(26.3) $E = \dfrac{-me^4}{8\varepsilon_o^2 h^2} \dfrac{1}{n^2}$ (Quantized electron orbital energy around a hydrogen nucleus)

(26.4) $E = mc^2$ (Equivalence between matter and energy)

(26.5) $N = N_o e^{(-0.693t/t_{1/2})}$ (Half-life)

Conceptual Problems

1. Why is the total energy of an electron in orbit around a nucleus considered negative?

2. What classical quantity is quantized in the orbit of an electron around a nucleus?

3. What kind of behavior does an electron in orbit around a nucleus exhibit?

4. As an electron's principal quantum number gets larger, what happens to the adjacent energy levels, and what change takes place in the electron's behavior?

5. Why are the spectral lines from the light emitted by an element such as helium discrete?

6. What identifies an atom as a particular element?

7. What determines how an atom binds chemically?

8. Why are nuclear fission and fusion so much more powerful than chemical reactions?

9. All isotopes of a given element have the same:
   a. binding energy   b. number of nucleons
   c. number of protons   d. number of neutrons

10. The atomic number of an atom equals:
    a. the number of protons
    b. the number of neutrons
    c. the number of protons and neutrons
    d. the number of electrons

11. The mass number of an atom equals:
    a. the number of protons
    b. the number of neutrons
    c. the number of protons and neutrons
    d. the number of electrons

12. Carbon has 6 protons in the nucleus. The number of neutrons in the nucleus of the carbon-14 isotope is:
    a. 8   b. 20
    c. 6   d. 14

13. If a nucleus undergoes alpha decay:
    a. the atomic number of the nucleus drops by 2
    b. the atomic number of the nucleus rises by 1

c. the atomic number of the nucleus stays the same

d. the nucleus splits into 2 roughly equal fragments

14. If a nucleus undergoes beta decay:

a. the atomic number of the nucleus drops by 2

b. the atomic number of the nucleus rises by 1

c. the atomic number of the nucleus stays the same

d. the nucleus splits into 2 roughly equal fragments

15. In a nuclear fission reactor, what keeps the chain reaction going?

16. In a nuclear fission reactor, what keeps the chain reaction from becoming an explosion?

17. The sun's energy is produced by:

a. fusion   b. fission

c. radioactivity   d. the conversion of helium to hydrogen

Exercises

The Atom and the Atomic Nucleus

1. Classically, what is the orbital speed of the electron in the $n = 2$ orbit around the hydrogen nucleus?

Answer: 1.09×10^6 m/s

2. (a) What is the de Broglie wavelength of the electron in Exercise 1? (b) What is the orbital path length for an electron circling a hydrogen nucleus in the $n = 2$ state? (c) If the electron "wave" in the $n = 2$ orbital is a standing wave, how many wavelengths fit exactly into the orbital circumference?

Answer: (a) 6.68×10^{-10} m

(b) 13×10^{-10} m

(c) 2

3. Assuming that protons are spherical, what is the repulsive force between two protons that are touching one another? The nucleus of the hydrogen atom is 1×10^{-15} m across.

Answer: 506 lb!

Energy Levels and Atomic Spectra

4. How much energy does it take to ionize a hydrogen atom if the electron is in the $n = 2$ state?

Answer: 3.39 eV

5. Instead of leaving the atom entirely, the electron in Exercise 4 drops down to the ground state. (a) What wavelength of light does the photon emitted have? (b) What type of light is it?

Answer: (a) 122 nm

(b) Ultraviolet

6. What is the orbital radius of the electron in the $n = 2$ state around the nucleus of the hydrogen atom?

Answer: 2.12×10^{-10} m

Radioactivity and Nuclear Fission

7. One kilogram of gasoline yields 47.3 MJ of energy when burned. How many times the energy of burning is the latent energy of the same mass of gasoline?

Answer: 1.9×10^9

Half-Life

8. Radium's half-life is 1600 years. How long will it take the original sample to decay until only $\frac{1}{16}$ is left?

Answer: 6400 yr

9. Only 12.5% of a sample of thorium-227 is left after 54 days. What is the half-life of the thorium isotope?

Answer: 18 days

10. Radioactive isotopes are used to power deep space probes to other planets in orbits beyond Mars where sunlight is too weak to use effectively with solar cells. A power source using polonium-210 has an output of 150 W when it is built. What is its output five years later? Polonium-210 has a half-life of 138 days.

Answer: 15.7 mW

Nuclear Fusion

11. Fusion is the process that powers the sun. Four hydrogen nuclei of mass 1.673×10^{-27} kg join to form a helium nucleus of mass 6.646×10^{-27} kg. (a) How much energy is liberated each time four hydrogen nuclei fuse into one helium nucleus? (b) How much mass is converted into energy when 1 kg of hydrogen is "burned" in such a fashion? (c) How much energy is liberated when 1 kg of hydrogen is "burned" in such a fashion? (d) How many kilograms of gasoline must be burned to produce the above energy?

> *Answer:* (a) 4.14×10^{-12} J
> (b) 6.87 g
> (c) 6.18×10^{14} J
> (d) 13 million

The chemical bonding of semiconductors allows electronic switches to be fabricated from them. The ability to etch tiny conductive patterns on the surfaces of semiconductors has enabled production of the integrated circuits that form the backbone of the communications and computer industries.

Chemical Bonding among and between Atoms

At the end of this chapter, the reader should be able to:

1. Discuss how a **laser** operates.
2. Explain the difference between a **covalent bond** and an **ionic bond.**
3. Calculate the temperature of a gas using **Van der Waals equation.**
4. Briefly describe the basic operation of diodes and transistors.

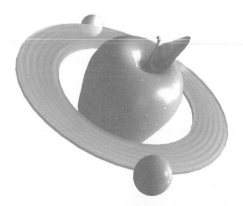

Laser: Light amplification by stimulated emission of radiation.

*The word **laser** is an acronym for "light amplification by stimulated emission of radiation."* The light is caused by electrons jumping among the discrete allowed energy levels for orbiting electrons. Laser light has some remarkable properties:

1. It is monochromatic, with only one wavelength.
2. It is coherent, with all of the waves in phase with one another.
3. The beam is well collimated, with little divergence over distance.
4. The beam is very intense.

These properties make a laser very useful. For example, an intense beam that can be precisely positioned has medical uses, and some surgical procedures have been revolutionized using the laser. The beam's intensity can be potentially useful in driving inertial fusion as discussed in Chapter 26. The fact that the beam is monochromatic is exploited in communications. In a process called *multiplexing,* solid-state lasers are connected to a fiber-optic cable, and many messages can be placed on a single cable simultaneously if the separate messages are carried by light of different frequencies. Fiber-optic cables can have an information density that is far greater than that of a copper wire, and fiber optics might totally replace copper wire in the future for communications. This revolution in networking is made possible by totally internal reflection and the glass fiber, combined with the solid-state, monochromatic tunable laser.

So why and how does a laser work? Remember that we see because electrons in the outer shells of an atom jump from a higher allowed energy state to a lower allowed energy state, emitting a photon of light in the process. In situation (1) of Figure 27.1, we see electrons boosted to an excited state E_3 above the ground state E_1 by absorbing

Metastable state: An allowed energy state for orbiting electrons above the ground state where they may linger for a while before dropping back to a lower-energy state.

photons of energy hf'. The electrons then drop down to a **metastable state** E_2 where they are able to linger for a while. When an electron is in an allowed energy state above the ground state, it usually drops down to a lower-energy level very quickly if it can. But in a metastable energy state, the electron can remain there for relatively a long time before dropping down to a lower-energy state, emitting a photon in the process. The electrons in the metastable state E_2 are then stimulated by a photon of energy hf, and they simultaneously drop down to the ground state E_1, emitting photons with the laser frequency in the process.

Spontaneous emission: An event in which an electron drops down from an excited allowed energy state, emitting a photon, without any external stimulation.

The metastable state is the key to how a laser works. From situation (1) to situation (2), **spontaneous emission** occurs. It occurs on the order of 1×10^{-8} s after the electrons have been boosted to energy level E_3 and the electrons have dropped down to the metastable state E_2. Photons of energy hf'' are emitted, but their energy is small. The electrons in the metastable state can linger there for around 1×10^{-3} s, or about 1 ms. Although that doesn't sound long to us, it's quite a long time for the electrons in the metastable state. It's on the order of 100,000 times longer than they remain in the excited state E_3. Because of the long time that the electrons spend in the metastable state, many electrons are able to gather at energy level E_2. Soon most of the electrons in the optical cavity are in the metastable state. This condition is called a **population inversion** because most of the electrons are in a high-energy state, in contrast to the normal population when most of the electrons are in the ground state. Because of the short time it takes for spontaneous emission to occur, no large fraction of the available population of electrons gets a chance to gather at the excited energy level E_3. Metastable isn't very stable, though. All of the electrons gathered in energy level E_2 are waiting for just the right trigger to simultaneously drop down to energy level E_1. That trigger is a photon of energy hf. It triggers an avalanche of electrons falling from E_2 to E_1. All of

Population inversion: The state in which most of a population of electrons are in a metastable state instead of in the ground state. The population is considered inverted because most electrons are normally found in the ground state.

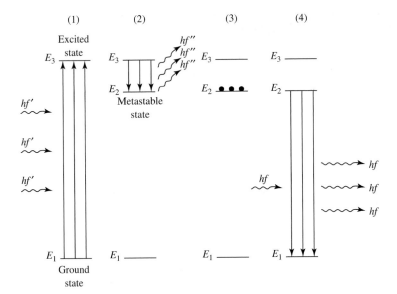

Figure 27.1 Electron Energy
States in a Laser

the emitted photons have the same frequency (and therefore wavelength) and the same phase because they were all emitted at the same time.

This laser action of the electrons takes place in what is called an *optical cavity.* Each end of the cavity has a mirror in it. The mirrors are parallel to each other, and one of them is half-silvered. The photons in the cavity bounce back and forth between the mirrors until the avalanche occurs. Then they pass through the half-silvered mirror and emerge as a coherent, monochromatic, intense, and collimated beam. Because the mirrors are parallel, the emerging beam is highly collimated; that is, it won't spread with distance as does a flashlight beam. The length of the optical cavity is usually an integral number of wavelengths of the characteristic wavelength of the laser. In this way a standing wave is set up inside the cavity. The induced emissions that form the avalanche are stimulated by this standing wave, and the avalanche occurs because it causes all of the emitted photons to be in step. The intensity is due to the fact that all of the inverted population undergoes induced emission essentially simultaneously. The laser is usually pulsed, and the high intensity lasts only for a very short time.

The first lasers had a ruby rod as the optical cavity. Ruby is a crystal of Al_2O_3 or aluminum oxide. The aluminum atoms have a charge of $+3$ in the molecule, and some of them are replaced with chromium (Cr) atoms and are forced to take on the same charge of $+3$ by the crystal array. The chromium atoms are what give the ruby its red color. The chromium ions in the ruby crystal provide the metastable state necessary for laser action to take place. The population inversion is caused by optical pumping. A xenon flash tube is wound around the ruby rod. (The xenon flash tube is often used as a flash for cameras.) When the tube flashes, many photons of the right energy are emitted, so that the chromium ions form a population inversion ready for laser action.

The helium-neon laser uses another mechanism to achieve its population inversion. The optical cavity is filled with a mixture of helium and neon gas. A high-frequency electric discharge is passed through the cavity, and the He and Ne atoms are excited. This process is the same one by which a neon sign glows. Some of the helium atoms give up their energy to ground-state neon atoms, and a population inversion of neon atoms is created. Because the collisions that produce the population inversion occur all of the time, a helium-neon laser operates continuously, unlike the ruby laser that is pulsed.

Optical cavity: The volume within which laser action takes place.

27.2   Bonding and the Periodic Table

Individual atoms are rarely found in nature. They bind together to form molecules and crystals. These molecules and crystals then form solids, liquids, and gases, or the stuff of the physical world. The bonds between and among atoms are formed by the electrons in the outer electron cloud of each atom. Atoms are normally electrically neutral, so there is no electromagnetic force drawing the atoms together, and when they do approach one another, the outer electrons in the electron cloud surrounding each atom repel one another. So how do atoms bond with one another? The answer lies in how the various electron orbital shells surrounding the nucleus are filled. A completely filled shell is a stable shell. The noble gases in the last column of the periodic table have completely filled shells and are stable. They are so stable that the noble gases won't form compounds with the other elements or with one another. They stand aloof. They can be found naturally as isolated atoms because they refuse to bind with other elements, and for this reason they are gases.

The periodic table of the elements is given in Appendix E.

Helium gas is lighter than air and is used to fill balloons and blimps to give them buoyancy in the atmosphere. It is favored over hydrogen gas because hydrogen gas is flammable. When hydrogen burns in the presence of oxygen, it combines with the oxygen to form water vapor. Helium can't burn because it won't form a compound with anything. The next noble gas down the column is neon, the gas found in neon signs. A large electric field ionizes the neon in a glass tube filled with neon gas, and the neon glows with the characteristic red color of the sign. Neon is safe to use in such an environment because it is inert and thus ionizing it won't cause a possible explosion or create danger from a toxic gas if the glass tube breaks.

The outer electron shell of each element wants to attain the stability of a closed noble gas shell. This is the key to much of the bonding between and among non-noble gas elements. Highly reactive elements are in columns near the noble gas column. The elements in the Group I column (hydrogen, lithium, sodium, potassium, and so on) will readily give up an electron in bonding to achieve the electron shell configuration of a noble gas. The elements in the Group VII column (fluorine, chlorine, bromine, iodine, and so on) will readily gain an electron in bonding to attain a noble gas electron shell. The farther away an element is from the noble gases column, the more electrons it must give up or gain to have a filled noble gas electron shell. Some elements, though, are ambivalent as to whether they gain or lose electrons to achieve the noble gas shell. Carbon and silicon are such elements. Both carbon and silicon can either gain or lose four electrons to obtain the stability of the noble gas shell. As a result, these elements have a tremendous amount of flexibility as to the types of chemical structures that can be built with them.

The key to most chemical bonding is that the outer electrons of an atom wish to have the stability of a noble gas shell.

27.3   Molecules and Covalent Bonds

Molecule: A group of atoms bonded together strongly enough to act as a single particle.

*A **molecule** is a group of atoms that are bonded together strongly enough to act as a single particle.* For example, hydrogen is usually found as the hydrogen molecule, H_2. Hydrogen needs one electron to achieve the stability of a noble gas electron shell. When a pair of hydrogen atoms bond, each atom in the pair contributes its single electron to the shared molecular bond. In this way, each atom appears to have the noble gas configuration. Since the noble gas configuration is the most stable, the molecule as a whole has less energy than a single hydrogen atom alone, and for this reason chemical bonding occurs. The molecule has 4.5 eV less energy than the atoms alone; in other words, it takes

4.5 eV of energy to break up a hydrogen molecule. When the two hydrogen atoms bond, the electron orbitals are distorted in such a way that the electrons spend most of their time between the two protons, and the net negative charge there is enough to counterbalance the repulsion that the two protons feel for each other.

The major two components of the air that we breathe are nitrogen (nearly 80%) and oxygen (nearly 20%). Both nitrogen and oxygen exist as molecules. Oxygen needs two electrons to complete the noble gas shell, and both oxygen atoms contribute two electrons. Nitrogen requires three electrons, and in the molecule each nitrogen atom contributes three electrons. This sharing of electrons is called **covalent bonding.** The shared electron doesn't belong to either atom but is shared between or among them for molecules with more than two atoms.

Figure 27.2 illustrates the bonding for some molecules and compounds. In the figure, the dashes (−) represent electrons in the electron cloud surrounding each atom. The number of electrons within an oval around an atom are the electrons that the atom sees in its cloud because of the bonding between or among atoms. The overlapping of the ovals are the electrons shared between or among two or more atoms because of the bonding. In the hydrogen molecule (H_2), each hydrogen atom has one electron but needs another to reach the first noble gas shell which has two electrons in the closed orbital. Because of the molecular bonding, each hydrogen atom sees a closed noble gas shell, and the two electrons are shared between the two protons. Oxygen must gain 2 electrons to have a filled noble gas shell, which is that of neon. Thus, 10 electrons are in orbit around the oxygen atom. An oxygen atom has a total of 8 electrons, so each oxygen atom shares 2 of its electrons in bonding. Four electrons are participating in the molecular bond, so each oxygen atom sees a closed noble gas shell of 10 electrons. Nitrogen has 7 electrons, so an N_2 molecule has a total of 14 electrons. Nitrogen needs 10 electrons around it for a closed shell, so each nitrogen atom contributes 3 electrons to the molecular bond. Notice that the bonding electrons (in the overlap of the ovals) occur in pairs. Each electron pair is a single bond. Thus, H_2 has a single bond, O_2 has a double bond, and N_2 has a triple bond. The more bonds between atoms in a molecule, the more strongly the atoms are bonded together. Nitrogen is an element that is far from inert, because it bonds with many elements and is a chief ingredient in explosives. However, the nitrogen molecule is practically inert. Its stability comes from the fact that to break the triple bond requires so much energy and is so difficult that very violent conditions are required to do so. In the hydrogen molecule both electrons are

Covalent bond: A bond in which atoms share electrons between or among themselves, the shared electrons belonging to no single atom.

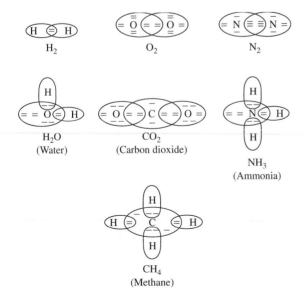

Figure 27.2 Covalent Bonding among Atoms to Form Molecules

shared. But in the oxygen molecule only 2 of the 8 electrons in each atom are shared. The other 6 don't participate in the bond. In the nitrogen molecule, 4 out of the 7 electrons from each atom don't participate in the bonding but remain associated only with their specific nitrogen atom.

Water is a molecule with more than 2 atoms in it. The oxygen atom participates in 2 separate bonds with 2 hydrogen atoms. The 2 hydrogen atoms see the closed noble gas shell of the helium atom, and the oxygen atom sees 10 electrons around it giving it the closed shell of neon. The electrons in the water molecule spend most of their time around the oxygen atom and very little time around the hydrogen atoms, creating a net negative charge on the oxygen atom and a net positive charge on each hydrogen atom. The molecule is polarized. This condition is noticeable at the surface of a body of water. The water molecules at the surface align themselves with the help of this polarization so that an oxygen atom from an adjacent molecule is near the 2 hydrogen atoms of another molecule. This polar alignment is the cause of surface tension, which acts like a rubber membrane at the water's surface. The electric field between adjacent molecules acts like tiny springs and gives the surface tension its elasticity. Steel doesn't float in water, but a steel needle can be supported by the water's surface tension and can appear to float. If a detergent is introduced into the water, it breaks down the surface tension, and the needle immediately sinks! This polarization of water is responsible for its ability to act as a solvent. Water molecules are also lighter than the gas molecules of O_2 and N_2 of the air. The reason that water is a liquid and not a gas at room temperature is the polarization of the water molecule. Neither the oxygen molecule nor the nitrogen molecule is polarized; they form a gas at room temperature. Many molecules are polarized by molecular bonding, and much chemistry is based on it.

Covalent bonding in carbon is so versatile that thousands of carbon compounds are known. Carbon's versatility is due to the fact that it can either gain or lose 4 electrons to achieve the noble gas shell of either neon or helium, respectively. Molecules of carbon atoms can contain hundreds of thousands of molecules. It is no accident that life on our planet is carbon-based. The DNA molecule, which contains the genetic blueprints for life, is a helix with a carbon backbone.

27.4   Ionic Bonds

Covalent bonding is the sharing of electrons between atoms in a molecule. The electron doesn't belong to either atom. But in ionic bonding, one or more electrons are transferred from one atom to another. A positive ion and a negative ion are produced, and they attract one another. The elements closest to the noble gas column in the periodic table are those most likely to participate in ionic bonding. The elements in the Group I column will gladly give up an electron to achieve the closed noble gas shell. The elements in the Group VII column will gladly take another electron to get the coveted stability of a noble gas shell. Table salt (NaCl) is a compound that bonds ionically. The sodium ion (Na^+) has a positive charge, and the chlorine ion (Cl^-) has a negative charge. The ions act like single point charges and attract ions of opposite charge. The sodium and chlorine atoms bond together ionically, forming a regular crystal structure. The ionic bonding is also the reason that salt dissolves so readily in water. The positive sodium atom is pulled from the salt crystal by the negatively polarized oxygen atom in water, and the negatively charged chlorine atom is pulled from the crystal by the positively polarized hydrogen atoms in the water molecule. Pure distilled water doesn't conduct an electric current very well, but if salt is added to the water, its conduction increases tremendously because of the Na^+ and Cl^- ions dissolved in the water. The ions are charges and readily carry a current. Strong acids have an ionic nature. Hydrochloric acid (HCl) and hydrofluoric acid (HF) are ionic liquids and decompose into H^+ and Cl^- and into H^+ and F^-, respectively. Other acids such as battery acid (sulfuric acid, H_2SO_4) decompose into 2 ions: H^+ and HSO_4^-. The positive hydrogen ion is very reactive and readily interacts with many compounds and molecules. Acidity is judged by the concentration of hydrogen ions that a solution contains. The opposite of an acid is a base. Bases also decompose into ions. Sodium hydroxide is a base that decomposes into a positive sodium ion (Na^+) and a negative hydroxyl ion (OH^-). The hydroxyl ion is also very reactive. Interestingly, water is both an acid and a base. We normally wouldn't think of drinking either an acid or a base. Drinking battery acid or Drano® (a base used for unclogging plumbing) is not a good idea! But we don't consider water as dangerous to drink. A glass of water exists mostly as H_2O molecules, but a very small concentration of the water molecules decomposes into H^+ and OH^- ions. The H^+ ions are acidic, and the OH^- ions are basic. This small concentration is maintained by the constant decomposition of H_2O into H^+ and OH^- ions and by the recombination of the ions back into a water molecule.

27.5 Van der Waals Forces

The ideal gas law ($PV = nRT$) works fairly well in calculating pressures, volumes, and temperatures for gases. But *ideal* is an approximation: No gas is ideal. The molecules and atoms in a real gas occupy space, and the ideal gas law assumes that a gas is made up of point particles. As was pointed out in Section 27.3, many molecules are polarized with regions of positive and negative charge. When the molecules approach one another closely, these repulsive and attractive electromagnetic forces cause the particles to interact and cause a weak attractive force called the *Van der Waals force.*

    The ideal gas law assumes that the particles are noninteracting except through collisions. Since the Van der Waals force is weakly attractive, the molecules must do work to overcome this attraction. As a result, the molecules in a gas are slowed down somewhat compared to molecules that don't interact. Thus, the average kinetic energy of the gas is less, and therefore the temperature of a real gas is less than that predicted by the ideal gas law.

    The Van der Waals equation that takes into account the nonideal nature of real gases is

Van der Waals forces: The weak attraction between atoms and molecules in a gas caused by polarized molecules.

$$\left(P + \frac{n^2a}{V^2}\right)(V - nb) = nRT$$

(Equation 27.1)

where a and b are constants that depend on the particular gas. If a and b are zero, then Equation (27.1) reduces to the ideal gas equation. For a rarefied gas where the particle density (n/V) is negligible, the ideal gas law works well. In most cases the ideal gas law works well in predicting pressures, temperatures, and volumes for gases. The differences between the result from the Van der Waals equation and that from the ideal gas law are small. Only under very unusual conditions of pressure and density is the Van der Waals equation necessary. The sun, for instance, is a big ball of hot hydrogen and helium gas. At the center the pressures and temperatures are truly astronomical, with a piece of plasma the size of a golf ball at the sun's center weighing tons. Yet even under these extreme conditions, the hydrogen plasma is a gas, and to calculate interactions under these conditions the Van der Waals equation is necessary. Yet for most applications on the earth, the ideal gas law works fine.

27.6   Silicon Bonding and Semiconductors

Silicon has an important place in the electronics industry. Modern electronics is largely based on silicon, just as life on this planet is based on carbon chemistry. Oddly enough, both silicon and carbon are in the same column of the periodic table. In fact, silicon is directly under carbon. The reason carbon forms the basis for all life on this planet is that carbon has tremendous flexibility in the configurations that it can take because of its relationship with the noble gas column. It can gain 4 electrons to have the noble gas configuration of neon, or it can lose 4 electrons to have the noble gas configuration of helium. The ability to share electrons so flexibly allows carbon to have literally thousands of known compounds. There are 6 electrons in orbit around a carbon atom, and 4 of them are available for bonding. Since silicon is in the same column, it has similar bonding characteristics. Silicon can lose 4 electrons to have the noble gas configuration of neon, or it can gain 4 electrons to have the noble gas configuration of argon. The closed noble gas shell below silicon in the periodic table is neon, and when silicon has a full noble gas shell, it has 18 electrons in orbit around it. Silicon also shares electrons and so bonds covalently.

Silicon is known as a *semiconductor*. It will conduct an electric current, but not very well, having a significant resistivity. Carbon also conducts a current and has a place in the electronics industry. Many resistors are manufactured from carbon because through careful control of the diameter, length, and composition of a sintered carbon powder cylinder, the resistor can be tailored to whatever resistance is desired to close tolerances. However, silicon has a resistivity that is much larger than that of carbon. Carbon has a resistivity of 3.5×10^{-5} Ω-m, and silicon 550 Ω-m. Silicon is actually classified as a glass, and glass isn't known as a good conductor. So how did silicon come to occupy such a central place in the electronics industry if it's such a poor conductor?

First of all, it is worth looking at a good conductor such as the metal copper from the point of view of bonding. Remember that the difference between energy levels in a shell goes as $1/n^2$, where n is the prime quantum number of the electron in orbit. The greater n is, the smaller is the difference in energy. In an amorphous crystal of copper, the outermost electrons in each atom are very loosely held, and it is very easy for an electron in a given cooper atom to get enough energy (either thermally or from an electric field) to jump into the conduction band of the crystal. Because of this ease, there is literally a gas of electrons free to travel anywhere in the copper wire, and that gas of electrons is what allows copper and most other metals to conduct both electricity and heat so effectively. So if there is any movement of free electrons in a material, there must be electrons in the conduction band of the material.

Silicon has 8 electron positions available for bonding in the unfilled outer electron shell. It has only 4 of its own electrons in that shell and so must share with other atoms to attain the filled shell of argon. Silicon will bond with up to 4 other atoms. If pure silicon is grown as an ordered crystal, then it will have 4 bonding partners arranged around it. Most semiconductors use crystalline silicon with this unvarying crystal structure. If trace amounts of any other element are introduced into the crystal, then they will be forced into the bonding pattern of the crystal. They will be forced to bond with the four adjacent silicon atoms as in Figure 27.3. As shown, each silicon atom bonds with 4 other silicon atoms with covalent bonds. The crystal extends left and right and up and down the page ad infinitum with the same bonding pattern. But in two locations there are foreign atoms. These atoms are forced to bond with the same structure of the silicon crystal. Phosphorus (P) is inserted into the crystal pattern in the lower right. Phosphorus has 5 electrons in its outer electron cloud and can use only 4 of them in its bonding with the silicon crystal. The extra electron is forced into the conduction band of the crystal. Remember that the conduction electrons form a gas of electrons in a metal and are free to migrate anywhere in the crys-

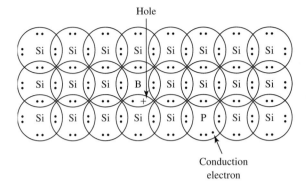

Conduction electron

Figure 27.3 Doping of a Silicon Crystal

tal. This accounts for the low resistivity of metals. A silicon crystal *doped* with an element such as phosphorus that has an excess of electrons is called N-type silicon because the free charges are negative.

Boron (B) is another element with which a silicon crystal can be doped. Boron has only 3 electrons to use for bonding in its outer electron shell. When forced into the bonding pattern of the silicon crystal, it creates a hole. *The hole is an electron bond that goes unused.* One of the silicon atoms around the boron atom has an unpaired electron. The hole, once created, is not necessarily tied to the boron atom. If an electron from a neighboring bond drops into the hole near the boron atom, then another hole is created where the electron came from. In this way the hole has the ability to migrate anywhere in the crystal just as a conduction electron does. It acts like a positive charge in the conduction band of the crystal and in this way significantly reduces the resistivity of the crystal. Silicon crystals doped with elements such as boron are called P-type silicon because the free charges are positive.

The resistivities of P-type and N-type silicon crystals are much lower than the resistivity of glassy pure silicon by orders of magnitude. The reduction of the resistivity is accomplished by adding only a trace amount of dopant. The dopant might be one part per million or less of the silicon crystal. The resistivity of the silicon is still well above that of a metal, though, but the fact that there is a single material with both positive and negative free charges allows the construction of small, solid-state electronic switches and amplifiers on the same silicon crystal. Such a device is called an *integrated circuit,* or a chip. This chip has been the device upon which the computer revolution has been constructed. The simplest device that can be built from the P- and N-type silicon is a diode. A diode is used to rectify an ac signal, as discussed in Chapter 20 on ac signals.

When a P-type and an N-type silicon crystal are joined together as in Figure 27.4, a depletion zone forms between them. Negative charges from the N-type silicon diffuse into the holes in the P-type material, and the holes from the P-type material diffuse into the N-type material. This diffusion results in an area near the junction depleted of free charges. The diffusion continues until an electric field large enough to inhibit further diffusion across the junction is built up. The electric field is characterized by a voltage drop across the junction of about 0.7 V with the polarity shown in the figure. This device will allow current to flow in only one direction and will block its flow in the other direction. With the positive terminal of a battery on the lead with the P-type material, electrons are encouraged to jump the junction voltage and flow through the diode. The electrons in the P-type depletion zone are pulled out of it toward the positive pole of the battery, and the holes migrate toward the negative terminal of the battery, creating *forward bias.* Free conduction occurs limited only by the resistance of the diode itself. However, for conduction to take place at all, the battery voltage must exceed the 0.7-V junction voltage. If the

Doping: The addition of small concentrations of impurities to a crystal to alter the electrical characteristics of the crystal.

Hole: An unused electron bond in a doped crystal that acts like a free positive charge.

Integrated circuit: Many tiny, solid-state components etched and interconnected on a single crystal of semiconductor material.

Forward bias: External voltage that is applied across a PN junction and that allows conduction to occur.

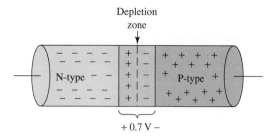

Depletion zone

Figure 27.4 Semiconductor Silicon Diode

N-type   P-type

+ 0.7 V −

Reverse bias: External voltage that occurs across a PN junction, widening the depletion zone and blocking conduction.

negative terminal of the battery is connected to the P-type material, and the positive to the N-type, no conduction will take place. All this does is to widen the depletion zone until enough electrons and holes have been separated in the depletion zone to equal the battery voltage, creating **reverse bias.**

The transistor is an electronic signal amplifier. It turns a small electronic signal into a large one. The transistor is essentially two diodes joined together as in Figure 27.5. The transistor is a three-lead device, and the leads are commonly labeled the *collector,* the *base,* and the *emitter.* A dc voltage is usually connected between the collector and the emitter. With no voltage on the base lead, the lower diode is reverse biased, and no conduction takes place. However, when a positive voltage in excess of 0.7 V greater than the emitter voltage is placed on the base, the lower diode becomes forward biased. Since the upper diode is already forward biased, the transistor as a whole conducts, with current flowing from the collector to the emitter. But the current necessary to forward-bias the base-emitter junction is very small compared to the current that flows from the collector to the emitter. A transistor is a current amplifier and is used to amplify small signals. Small signals are usually smaller than the 0.7-V junction voltage, so dc biasing circuits are built around the transistor so that the base-emitter junction is already conducting. The larger collector-emitter current will then accurately follow the smaller base-emitter current. Without the biasing circuitry there would be a 0.7-V drop-out from the base signal as seen in the collector-emitter signal.

Amplifiers are usually built in stages. The stages are separated by capacitors because direct currents can't flow through them, and ac signals pass through. In this way the ac signal can pass from one stage to another, and the dc biasing circuits of each stage remain separated. The transistor shown in Figure 27.5 is an NPN transistor. A positive voltage on the base will forward-bias the transistor. Another type of transistor is the PNP transistor. It takes a negative signal on the base to forward-bias the transistor.

Another type of transistor operates on a very different principle than do the two discussed above. Called a *field-effect transistor* (FET), it consists of a channel of conducting material, such as N-type silicon, surrounded by P-type material. When a negative voltage is applied to the P-type material, the depletion zone between the N-type and the P-type material increases, and the conduction channel narrows, allowing less current to flow. If a positive voltage is applied to the P-type material, the depletion zone decreases, and the channel for current flow is wider. The P-type material is where the signal to be amplified is applied and is called the *gate.* Very little current flows in the gate circuit, and it is essentially an electric field that narrows or widens the conduction channel. With no signal on the gate circuit, the transistor is conducting. Field-effect transistors require very little current and make possible photocell-powered calculators.

Figure 27.5 The Transistor

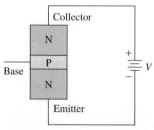

Collector

N

P   + — V

Base

N

Emitter

Summary

1. The term **laser** is an acronym for "*l*ight *a*mplification by *s*timulated *e*mission of *r*adiation." When in operation, a laser produces an intense, *monochromatic,* coherent beam of light that diverges very little over distance. Lasers are used commonly in commercial, industrial, and medical applications.

2. A **molecule** is a group of atoms that are bonded together strongly enough to act as a single particle. This bonding can take two forms: **covalent** and *ionic.* Ionic bonds occur in cases where one atom gives all of its excess electrons to another atom. When both atoms share their electrons, covalent bonding occurs. In both types of bonding, each atom's outermost *electron shell* gains the stability similar to that of a *noble gas.*

3. In certain extreme conditions, such as regions of very high temperatures or pressures, the ideal gas law is not sufficient to describe the behavior of a gas. In such circumstances, Van der Waals equation [Equation (27.1)] is used.

Key Concepts

The key to most chemical bonding is that the outer electrons of an atom wish to have the stability of a noble gas shell.

COVALENT BOND: A bond in which atoms share electrons between or among themselves, the shared electrons belonging to no single atom.

DOPING: The addition of small concentrations of impurities to a crystal to alter the electrical characteristics of the crystal.

FORWARD BIAS: External voltage that is applied across a PN junction and that allows conduction to occur.

HOLE: An unused electron bond in a doped crystal that acts like a free positive charge.

INTEGRATED CIRCUIT: Many tiny, solid-state components etched and interconnected on a single crystal of semiconductor material.

LASER: Light amplification by stimulated emission of radiation.

METASTABLE STATE: An allowed energy state for orbiting electrons above the ground state where they may linger for a while before dropping back to a lower-energy state.

MOLECULE: A group of atoms bonded together strongly enough to act as a single particle.

OPTICAL CAVITY: The volume within which laser action takes place.

POPULATION INVERSION: The state in which most of a population of electrons are in a metastable state instead of in the ground state. The population is considered inverted because most electrons are normally found in the ground state.

REVERSE BIAS: External voltage that occurs across a PN junction, widening the depletion zone and blocking conduction.

SPONTANEOUS EMISSION: An event in which an electron drops down from an excited allowed energy state, emitting a photon, without any external stimulation.

VAN DER WAALS FORCES: The weak attraction between atoms and molecules in a gas caused by polarized molecules.

Important Equation

(27.1) $\left(P + \dfrac{n^2 a}{V^2}\right)(V - nb) = nRT$ (Van der Waals equation)

Conceptual Problems

1. The light from a given laser:
   a. contains many wavelengths of light
   b. is coherent
   c. spreads considerably with distance
   d. contains photons of many different energies

2. A metastable state:
   a. allows a population inversion in a laser cavity
   b. robs a laser of intensity because atoms get hung up in this state
   c. is an energy level that has an unusually quick transition to a lower-energy level
   d. is the chief reason that lasers have low efficiencies

3. A population inversion:
   a. has most of the electrons in a laser cavity in a metastable state
   b. has most of the electrons in a laser cavity in the ground state
   c. is caused by spontaneous emission when an electron goes from a low- to high-energy level
   d. is caused by stimulated emission when an electron goes from a low- to a high-energy level

4. Stimulated emission occurs when:

   a. a photon jumps from a lower- to a higher-energy level

   b. an electron spontaneously jumps from a higher- to a lower-energy level

   c. an electron goes to a metastable state

   d. other photons in the laser cavity trigger an avalanche of photons from a metastable state

5. Most chemical bonding occurs because:

   a. the outer shell electrons want the electron configuration of the Group I elements

   b. the outer shell electrons want the electron configuration of the Group VII elements

   c. the outer shell electrons want the electron configuration of the noble gases

   d. the outer shell electrons can't be trapped in the unreactive electron configuration of the noble gases

6. In a covalent bond:

   a. the electrons in the inner shells of an atom are shared with electrons in the outer shells

   b. outer shell electrons are shared between and among atoms

   c. outer shell electrons are completely given away to another atom

   d. the molecules formed are usually polar

7. In an ionic bond:

   a. the electrons in the inner shells of an atom are given up to the outer shells

   b. outer shell electrons are completely given away to another atom

   c. outer shell electrons are shared between and among atoms

   d. the molecules formed are usually nonpolar

8. Which of the following statements is false?

   a. Water is a liquid at room temperature because of its polar nature.

   b. Water has a high surface tension because of the water molecule's polar nature.

   c. Water evaporates quickly in air because of its polar nature.

   d. Water is a good solvent because of its polar nature.

9. In a water molecule:

   a. the hydrogen atoms are slightly positive whereas the oxygen atom is slightly negative

   b. the hydrogen atoms are slightly negative whereas the oxygen atom is slightly positive

   c. the hydrogen and oxygen atoms alternate between being positive and negative

   d. the molecule has a net charge that causes its polar character

10. A water molecule:

   a. is an acid

   b. is a base

   c. is both an acid and a base

   d. is neither an acid nor a base

11. Which of the following statements is false?

   a. Table salt dissolves readily because of its ionic nature.

   b. Table salt dissolved in water has positive and negative ions in solution.

   c. Table salt melts easily because of the ionic nature of the crystal bonds.

   d. The ions in solution in salt water are the reason that salt water has a higher conductivity than pure water.

12. Which of the following statements is false? The Van der Waals equation:

   a. corrects for the fact that real gas molecules have volume and attract each other

   b. is used when dealing with hot, dense plasmas

   c. assumes point particles whose only interactions are through collisions

   d. is the reason why temperatures in real gases are slightly lower than those in an ideal gas

13. Silicon as used in most semiconductors:

   a. bonds covalently into a three-dimensional crystal

   b. bonds ionically into a three-dimensional crystal

   c. is a glass and therefore can't form a long-range, three-dimensional crystal

   d. is in a different row of the periodic table and therefore has bonding characteristics very different from those of carbon

14. Which of the following statements is false? A small amount of another element introduced into a crystal of silicon:

   a. destroys the periodicity of the crystal lattice

   b. is forced to bond like the silicon in the crystal

   c. alters the electrical characteristics of the silicon to a great degree

   d. bonds covalently with its four nearest neighbors

15. A small proportion of phosphorus added to a silicon crystal:

   a. puts an electron into the conduction band because of the forced covalent bonding with the surrounding silicon

   b. leaves a mobile hole in the conduction band because of the forced covalent bonding with the surrounding silicon

   c. is able to migrate throughout the crystal as an ion, increasing the crystal's conductivity

   d. decreases the crystal's conductivity because it blocks the flow of current

16. Which of the following statements is false? The depletion zone at a PN junction:

  a. is caused by the migration of electrons and holes across the junction until an electric field is built up that opposes further migration

  b. is a volume at the PN junction devoid of mobile charge carriers

  c. is what makes a semiconductor diode possible

  d. has a voltage drop across it with the positive polarity on the P-type silicon side

17. A reverse bias across a PN junction:

  a. allows current flow if the biasing voltage is greater than approximately 0.7 V

  b. increases the size of the depletion zone, inhibiting current flow

  c. has the positive pole of the external voltage on the N-type silicon

  d. pulls all of the charge carriers out of the silicon, stopping current flow

Exercises

Van der Waals Forces

1. Starting from the ideal gas law, derive a new gas law for pressure that takes into account the small but finite volume excluded from interactions by the size of the molecules and atoms in a sample of gas n moles large. Assume that the excluded volume is proportional to the number of moles of a gas.

$$\textit{Answer: } P = \frac{nRT}{V - nb}$$

2. Because of the small attractive forces between atoms and molecules in a gas, the pressure is greater than in the ideal gas law. Assuming that the increase in pressure is proportional to the square of the particle density (n/V), finish the derivation of the Van der Waals equation started in Exercise 1.

3. One mole of nitrogen gas is at 1 atm of pressure and has a volume of 22.4 liters. (a) What is the temperature of this gas using the ideal gas law? (b) What is the temperature of this gas using Van der Waals gas law? ($a = 1.39$ liter²-atm/mole²; $b = 0.039$ liter/mole.)

$$\textit{Answer: } \text{(a) 273.2 K}$$
$$\text{(b) 273.45 K}$$

Mathematics

An equation in the form $ax^2 + bx + c = 0$ is called a *quadratic equation,* where x is a variable and a, b, and c are constant coefficients. A general solution to the problem is

$$x = \frac{-b \pm \sqrt{b^2 - 4ac}}{2a}$$

There are two solutions because the square of either a positive *or* a negative number will always be positive. The term with the square root sign in the numerator is called the *discriminator.* To find the first solution, add the discriminator in the numerator. To find the second solution, subtract the discriminator in the numerator.

EXAMPLE A.1

Solve the following equation:

$$x^2 + x - 2 = 0$$

SOLUTION
In the above equation, $a = 1$, $b = 1$, and $c = -2$. To solve for x, substitute the values for a, b, and c into the general solution given above.

$$x = \frac{-1 \pm \sqrt{1^2 - 4(1)(-2)}}{2(1)}$$

$$x = \frac{-1 \pm \sqrt{9}}{2}$$

$$x = \frac{-1 \pm 3}{2}$$

679

$$x = \frac{-1 + 3}{2} \quad \text{and} \quad x = \frac{-1 - 3}{2}$$

$$x = 1 \quad \text{and} \quad x = -2$$

The solutions to $x^2 + x - 2 = 0$ are 1 and -2.

The number under the square root sign in the discriminator cannot, for the purposes of this book, be negative. If this number is negative in the solution to a problem, something is wrong with the solution.

A.2 Simultaneous Equations

In solving some physics problems, you might derive an equation to solve for the desired variable but find that there are actually *two* unknowns in the equation and only one equation! Such situations arise when, for instance, you solve for the tension in a cable in a problem with more than one cable, or when you want to find the current in a branch of an electrical circuit that has multiple branches. What you must do first is study the system and derive as many equations describing the system as there are unknown variables in the system. This system of equations is known as a system of *simultaneous* equations because they all describe the same system operating at the same time.

Once you have derived the system of equations, you must solve them. A pair of such simultaneous equations is given in Example A.2, as well as the technique for solving them. There are only two equations here, but the technique will work for more than two equations.

EXAMPLE A.2

Solve the following system of equations:

$$2x + y = 4 \qquad x - 2y = 0$$

SOLUTION
First reduce the two equations in two variables to one equation in one variable. A numerical solution for x will be sought first. Thus, y must be eliminated. We solve the equation $2x + y = 4$ for y.

$$y = 4 - 2x$$

Next substitute this value for y into the equation $x - 2y = 0$.

$$x - 2(4 - 2x) = 0$$

$$x - 8 + 4x = 0 \qquad 5x = 8 \qquad x = \frac{8}{5}$$

Knowing the numerical value for x, we can now obtain a numerical value for y by substituting $x = \frac{8}{5}$ into either of the two initial equations.

$$y = 4 - 2\left(\frac{8}{5}\right) \qquad y = 4 - \frac{16}{5}$$

$$y = \frac{20}{5} - \frac{16}{5} \qquad y = \frac{4}{5}$$

The solution to the two equations is $x = \frac{8}{5}$ and $y = \frac{4}{5}$.

To solve for more than two equations, for example, three equations, first reduce the three equations in three variables to two equations in two variables. If the variables are x, y, and z, then eliminate z by solving one of the equations for z and substituting that value into the other two equations. This method will give two equations in only two variables: x and y. Then solve the two equations in two unknowns as shown above. Once you have a numerical value for one of the initial variables, then substitute that value into the intermediate equations for two variables, and find a numerical value for a second variable. Then substitute the values for the first two variables into one of the original simultaneous equations, and all three unknown variables will be known. Be sure to substitute the numerical solutions in one of the original simultaneous equations to check that the solutions are correct.

A.3   Trigonometric Identities

A rotating object or a cyclic phenomenon such as a simple pendulum can be described using angles and trigonometric functions of the angles, as shown in Figures A.1, A.2, and A.3 and as given in the followng equations.

First, the *trigonometric identities* are as follows (see Figure A.1):

$$\sin \theta = \frac{\text{opposite}}{\text{hypotenuse}} \qquad \cos \theta = \frac{\text{adjacent}}{\text{hypotenuse}} \qquad \tan \theta = \frac{\text{opposite}}{\text{adjacent}}$$

$$\sin^2\theta + \cos^2\theta = 1$$

$$\sin(\theta \pm \phi) = \sin \theta \cos \phi \pm \cos \theta \sin \phi \qquad \sin 2\theta = 2 \sin \theta \cos \theta$$

$$\cos(\theta \pm \phi) = \cos \theta \cos \phi \pm \sin \theta \sin \phi \qquad \cos 2\theta = \cos^2\theta - \sin^2\theta$$

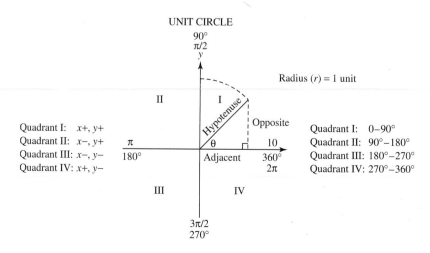

Figure A.1 Trigonometric Identities

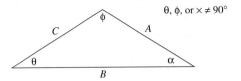

$\theta, \phi, \text{ or } \times \neq 90°$

Figure A.2 Law of Cosines

Second, the *law of cosines* is (see Figure A.2)

$$A^2 = B^2 + C^2 - 2BC \cos \theta$$

The above equation reduces to the Pythagorean theorem if $\theta = 90°$.

The sum of the interior angles of any triangle drawn on a flat surface is 180°. For the law of cosines triangle in Figure A.2, $\theta + \phi + \alpha = 180°$.

$$\sin(90° + \theta) = \cos \theta \qquad \sin(\theta - 90°) = -\cos \theta$$

$$\cos(\theta + 90°) = -\sin \theta \qquad \cos(\theta - 90°) = \sin \theta$$

Last, *radians* are defined as follows (see Figure A.3):

$$2\pi \text{ radians} = 360° \qquad 1 \text{ radian} = 57.3°$$

For small angles (in radians),

$$\sin \theta \approx \theta \qquad \cos \theta = 1 \qquad \tan \theta \approx \theta$$

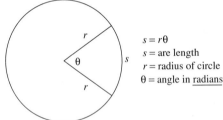

$s = r\theta$
s = are length
r = radius of circle
θ = angle in <u>radians</u>

Figure A.3 Radians

A.4 Scientific Notation and Exponents

Physics often deals with very large and very small numbers. These numbers are compactly expressed with scientific notation, or using powers of 10.

$$1000 = 10 \times 10 \times 10 = 10^3$$

$$100 = 10 \times 10 = 10^2$$

$$10 = 10^1$$

$$1 = 10^0 \qquad \text{(Any number to the zeroth power is 1.)}$$

$$\frac{1}{1000} = \frac{1}{10 \times 10 \times 10} = \frac{1}{10^3} = 10^{-3} = 0.001$$

$$\frac{1}{100} = \frac{1}{10 \times 10} = \frac{1}{10^2} = 10^{-2} = 0.01$$

$$\frac{1}{10} = \frac{1}{10^1} = 10^{-1} = 0.1$$

Speed of light $= c = 3 \times 10^8$ m/s $= 300,000,000$ m/s (eight zeros)

Gravitational constant $= G = 6.67 \times 10^{-11}$ N-m^2/kg^2
$$= 0.0000000000667 \text{ N-m}^2/\text{kg}^2$$
(The first significant figure occupies the eleventh position behind the decimal point.)

For n and m any positive or negative number,

$$\frac{1}{10^n} = 10^{-n} \qquad 10^n \times 10^m = 10^{m+n} \qquad \frac{10^n}{10^m} = 10^{n-m}$$

The roots of a number such as a square root or a cube can also be expressed as exponents.

$$\sqrt{x} = x^{1/2} \qquad \text{(Square root of } x\text{)}$$
$$\sqrt[3]{x} = x^{1/3} \qquad \text{(Cube root of } x\text{)}$$

A.5 Natural Logarithms

The natural logarithm (e) has a base of 2.718 and is found in such phenomena as the half-life of radioactive decay, the charging and discharging of capacitors, and the rise and fall of current through an inductor.

$$e^0 = 1 \qquad\qquad e^1 = 2.7182818$$
$$e^x = y \qquad\qquad \ln(y) = x$$
$$\ln(AB) = \ln A + \ln B \qquad \ln(A/B) = \ln A - \ln B$$
$$\ln A^x = x \ln A$$

A.6 The Straight Line

The generic equation of a straight line is

$$y = mx + b$$

where y is the vertical coordinate, x the horizontal coordinate, m the slope or tilt of the line, and b the y intercept. A typical straight line in the appropriate coordinate system is shown in Figure A.4.

$$\text{Slope} = m = \frac{\text{rise}}{\text{run}} = \frac{\Delta y}{\Delta x}$$
$$\Delta y = y_f - y_i \qquad \Delta x = x_f - x_i$$
$$\text{Slope} = \frac{y_f - y_i}{x_f - x_i}$$

Any straight line can be put into the form of the above equation.

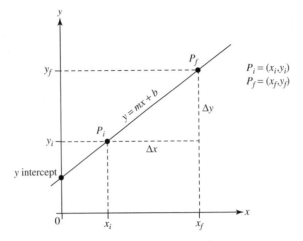

Figure A.4 Straight Line

The y intercept is the point where the line crosses the vertical y axis ($x = 0$). If the line goes through the origin, the y intercept is zero, and the line's equation is usually written $y = mx$. (Dropping a zero doesn't mean that there is no y intercept!) The slope of the line is the tilt of the line, usually expressed as the ratio of the rise over the run. The greater the slope, the more tilt the line has. A vertical line has a slope of infinity because there is no run. The slope equation must divide by zero, and if you divide by zero, you get an infinite result. A horizontal line has a slope of zero because the rise is zero. To calculate the slope of a given line, choose a pair of any two points on the line: P_i (the initial point) and P_f (the final point). Each point will have two coordinates, $P_i = (x_i, y_i)$ and $P_f = (x_f, y_f)$, as shown in Figure A.4. The slope can be either positive or negative in sign. If the line rises from the left to right as in Figure A.4, the slope is positive. If the line falls from left to right, the line has a negative slope.

If you plot a straight line on an accurately scaled set of axes, you can derive the equation of that line using the information above. But if you are given the equation of a straight line, you can plot it by generating a data table. Two variables can be changed: x and y. The x variable is known as the *independent* variable because its values can be chosen arbitrarily. The y variable is known as the *dependent* variable because its value depends on the value previously chosen for the x variable. To generate a data table, create two columns next to each other. The left column should be for the x variable, and the right column for the y variable. Choose a range of x values, and then calculate the y values for the adjacent column from the x values using the equation of the line. The table is now a column of points on the line. If you initially start with five x values, you will generate five points on the line. The points must be plotted. Next determine the correct size for each axis. If the x axis has a range of 10, then making the x axis 5 in. long will scale the x axis at 2 per inch. If the y axis has a range of 40, and if you make the y axis 5 in. tall, the y-axis scaling will be 8 per inch. To plot a point, draw a light vertical line through the x-coordinate position on the x axis. Draw a horizontal line through the y-coordinate position on the y-axis. Where these two lines meet is the position of the plotted point. Plot all of the points generated by the data table, and then connect the points with a straight line.

Calculator Usage

A calculator will considerably simplify solving the equations that you manipulate and derive in solving the exercises in this book. Thus, along with acquiring a certain mathematical fluency in studying this book, you will also begin to know your way around a calculator.

When buying a calculator, you will find a bewildering array of brands and degrees of complexity from which to choose. However, a simple scientific calculator is all that you need. *Simple* means that the calculator does not need the ability to plot functions or store programs. It is only necessary that it have the following features:

- a memory register
- trigonometric functions
- the ability to work with angles in degrees or radians
- natural logarithm functions
- the ability to square a number and take a square root
- the ability to do powers and roots
- an inverse button
- a button that changes the sign of a number, positive or negative
- the ability to move back and forth between a floating-point (decimal) display of a number and an exponential display for very large or very small numbers

Such a calculator can usually be purchased for under $20.00. The simple scientific calculator is easy to use, quite small, and portable.

Often a seemingly complex calculation can be solved in one pass with proper calculator usage. Solving the quadratic equation is one such application. A quadratic equation is an equation that has a number in it raised to the second power (squared) (see Appendix A). The generic form of any quadratic equation is

$$ax^2 + bx + c = 0$$

where a, b, and c are constant coefficients with x the variable to be determined. Its solution is

$$x = \frac{-b \pm \sqrt{b^2 - 4ac}}{2a}$$

Notice the plus-or-minus sign in front of the square root. This sign means that there are always two possible solutions to the quadratic equation. If you need both of them, then you must solve the equation twice.

To solve the equation in one pass instead of going through a number of intermediate stages, you must expand the expression from the inside out. Starting with the term $b^2 - 4ac$ under the square root sign:

- Enter the number for b.
- Square it; then hit the subtract button.
- Enter 4; then hit the multiplication button.
- Enter the number for coefficient a; then hit the multiplication button again.
- Enter the number for the coefficient c; then hit the equal button.

The number displayed is the result of evaluating the expression $b^2 - 4ac$ under the square root sign. Next hit the button with the square root sign on it. If the number under the square root sign is negative, most calculators will give you an error display. If the number isn't negative, the number displayed is the result of the calculations performed in evaluating the expression under the square root sign. Hitting the equal button before you hit the square root sign is important to obtain the correct answer. If you don't hit the equal button, the calculator will take the square root of only c instead of $b^2 - 4ac$.

If you want the negative solution to the quadratic, hit the $\boxed{+/-}$ button, changing the sign of the evaluated square root to a negative number. Then hit the addition $\boxed{+}$ button; enter the number for coefficient b; again hit the $\boxed{+/-}$ button to make the number entered for b negative; and hit the $\boxed{=}$ button. These steps should give you the numerator for the solution of the quadratic equation. Next hit the divide $\boxed{\div}$ button; enter 2; hit the divide button again; and enter the number for the coefficient a. Hitting the $\boxed{=}$ button one final time will give the solution to the quadratic equation.

All of the above can be done in one pass to solve the quadratic equation without the need for intermediate solutions or for use of one of the memory registers to store an intermediate result. However, if there were a square root sign in the denominator with an expression in it requiring addition and subtraction, then you would have to store the number obtained for the numerator in a memory register, evaluate the denominator, hit the inverse $\boxed{1/x}$ button, and multiply that by the number stored in the memory register.

Many scientific calculators have buttons that have dual or triple functions. The prime function is written directly on the button. Pushing the button performs the prime function. The same button may also perform a secondary function. This secondary function is usually printed directly above the button. To perform the secondary function, press the calculator function button labeled $\boxed{\text{2ndF}}$ before pressing the button that actually performs the function. For example, many calculators have a button labeled $\boxed{x^2}$ with a square root sign printed directly over the button. To square a number, press the $\boxed{x^2}$ button; but if you want the square root of a number, first push the $\boxed{\text{2ndF}}$ button; then press the $\boxed{x^2}$ button. The trigonometric function buttons are often set up this way. Printed above the sin button might be $\boxed{\sin^{-1}}$, which stands for the inverse sine. In using this function, enter the sine of an angle into the calculator, and the calculator will return the value of the angle. If you push the $\boxed{\sin}$ button first, you obtain the sine of the angle entered. But if you push $\boxed{\text{2ndF}}$ first and then push the $\boxed{\sin}$ button, the angle is obtained for the entered sine.

Because the number pi ($3.14159 \ldots$, or π) is used often, it may have a special key. Another special key is the $\boxed{F <> E}$ button. It will convert a decimal number into an exponential display, and vice versa. To enter a number as an exponent, you must use the $\boxed{\text{EE}}$ or $\boxed{\text{EXP}}$ button. For example, to enter the number 0.0042 as an exponent, enter 4; then enter (.); and then enter 2. Next hit the $\boxed{\text{EXP}}$ key; enter 3; and hit the $\boxed{+/-}$ button. The number displayed will be 4.2–03, which is 4.2×10^{-3}.

There are many tricks to learn in using calculators, and most new calculators will provide a user's manual with the purchase. The manual will give detailed information as to the function keys and will lead you through complicated computations. Keep the manual for future reference, because even the simple scientific calculators have many functions that will rarely be used but that need explanation when the rare use does arise. Most simple scientific calculators will do some statistics and summations, for instance, and many will perform logic functions and Boolean algebra and will use base two (digital) and hexadecimal numbering systems.

Dimensional Analysis

All symbols in **bold** type are vector quantities. The notations for the basic dimensions given in the third column in Table C.1 (see page 690) are

$$M = \text{mass} \qquad L = \text{length} \qquad T = \text{time} \qquad Q = \text{charge}$$

The names of the derived units are given in the fourth column. There may be more than one name for a given derived unit. A given symbol may also be used for more than one physical quantity. T, for example can represent temperature, tension, or period.

Table C.1 Physical Quantities: Common Symbols, Dimensions, and Units

Physical Quantity	Common Symbol	Dimensions	Units (Derived)
Acceleration	$\mathbf{a}$	L/T^2	m/s² (N/kg)
Angular acceleration	α	$1/T^2$	rad/s²
Angular displacement	θ	—	rad
Angular momentum	$\mathbf{L}$	ML^2/T	kg-m²/s
Angular velocity	ω	$1/T$	rad/s
Area	A	L^2	m²
Capacitance	C	Q^2T^2/ML^2	farad (F)
Charge	q	Q	coulomb (C)
Current	I	Q/T	ampere (A)
Current density	$\mathbf{i}$	Q/L^2T	A/m²
Electric field	$\mathbf{E}$	ML/T^2Q	N/C (V/m)
Electric potential	V	ML^2/T^2Q	volt (V)
Electromotive force	$\mathcal{E}$	ML^2/T^2Q	V
Displacement	$\mathbf{x, y, r, s, d}$	L	m
Energy, total	E	ML^2/T^2	kg-m²/s² (joule, J)
Kinetic	KE	ML^2/T^2	kg-m²/s² (J)
Potential	PE	ML^2/T^2	kg-m²/s² (J)
Force	$\mathbf{F}$	ML/T^2	kg-m/s² (newtons, N)
Frequency	f, ν	$1/T$	hertz (Hz); cycles/s
Heat	Q	ML^2/T^2	kg-m²/s² (J)
Inductance	L	ML^2/Q^2	henry (H)
Magnetic field strength	$\mathbf{H}$	LQ/T	A-m
Magnetic flux	ϕ	ML^2/QT	V-s; weber (Wb)
Magnetic induction	$\mathbf{B}$	M/QT	tesla; Wb/m²
Magnetization	$\mathbf{M}$	Q/LT	A/m
Mass density	ρ	M/L^3	kg/m³
Momentum	$\mathbf{p}$	ML/T	kg-m/s
Period	T	T	s
Permeability (magnetic)	μ	ML/Q^2	H/m
Permittivity (electric)	ε	Q^2T^2/ML^3	F/m
Power	P	ML^2/T^3	watts (W)
Pressure	p	M/LT^2	N/m²
Resistance	R	ML^2/Q^2T	ohm (Ω)
Resistivity	ρ	ML^3/Q^2T	ohm-meter (Ω-m)
Rotational inertia	I	ML^2	kg-m²
Temperature	T	—	degree [Celsius, °C; Fahrenheit, °F; kelvin, K (absolute)]
Time	t	T	s
Torque	τ	ML^2/T^2	kg-m²/s² (N-m)
Velocity	v	L/T	m/s
Voltage	V	ML^2/QT^2	V
Wavelength	λ	L	m
Work	W	ML^2/T^2	kg-m²/s² (J)

Conversion Factors

D.1 Length

$$1 \text{ in.} = 2.54 \text{ cm} \qquad 1 \text{ ft} = 0.305 \text{ m}$$
$$1 \text{ mi} = 5280 \text{ ft} = 1.61 \text{ km} \qquad 1 \text{ m} = 3.28 \text{ ft}$$
$$1 \text{ m} = 39.4 \text{ in.}$$

D.2 Area

$$1 \text{ m}^2 = 10^4 \text{ cm}^2 = 10.76 \text{ ft}^2 \qquad 1 \text{ ft}^2 = 144 \text{ in.}^2 = 9.29 \times 10^{-2} \text{ m}^2$$
$$1 \text{ hectare} = 10^4 \text{ m}^2 = 2.47 \text{ acres} \qquad 1 \text{ acre} = 43{,}560 \text{ ft}^2 = 0.405 \text{ hectare}$$
$$1 \text{ in.}^2 = 6.45 \text{ cm}^2 = 6.45 \times 10^{-4} \text{ m}^2$$

D.3 Volume

$$1 \text{ cm}^3 = 10^{-6} \text{ m}^3 = 3.53 \times 10^{-5} \text{ ft}^3 = 6.1 \times 10^{-2} \text{ in.}^3$$
$$1 \text{ m}^3 = 10^6 \text{ cm}^3 = 10^3 \text{ liters} = 35.3 \text{ ft}^3 = 6.1 \times 10^4 \text{ in.}^3 = 264 \text{ gal}$$
$$1 \text{ liter} = 10^3 \text{ cm}^3 = 10^{-3} \text{ m}^3 = 1.06 \text{ qt} = 0.264 \text{ gal}$$
$$1 \text{ in.}^3 = 5.79 \times 10^{-4} \text{ ft}^3 = 16.4 \text{ cm}^3 = 1.64 \times 10^{-5} \text{ m}^3$$
$$1 \text{ ft}^3 = 1728 \text{ in.}^3 = 7.48 \text{ gal} = 0.0283 \text{ m}^3 = 28.3 \text{ liters}$$
$$1 \text{ qt} = 2 \text{ pints} = 946 \text{ cm}^3 = 0.946 \text{ liter}$$
$$1 \text{ gal} = 4 \text{ qt} = 231 \text{ in.}^3 = 0.134 \text{ ft}^3 = 3.785 \text{ liters}$$

D.4 Time

$$1 \text{ h} = 60 \text{ min} = 3600 \text{ s} \qquad\qquad 1 \text{ day} = 24 \text{ h} = 1440 \text{ min} = 8.64 \times 10^4 \text{ s}$$

$$1 \text{ yr} = 365 \text{ days} = 8.76 \times 10^3 \text{ h}$$

$$= 5.26 \times 10^5 \text{ min} = 3.16 \times 10^7 \text{ s}$$

D.5 Mass

$$1 \text{ g} = 10^{-3} \text{ kg} = 6.85 \times 10^{-5} \text{ slug} \qquad 1 \text{ kg} = 10^3 \text{ g} = 6.85 \times 10^{-2} \text{ slug}$$

$$1 \text{ slug} = 1.46 \times 10^4 \text{ g} = 14.6 \text{ kg} \qquad\qquad 1 \text{ kg} = 2.2 \text{ lb} \qquad \text{(scales conversion)}$$

$$1 \text{ ton} = 2000 \text{ lb}$$

D.6 Angle

$$1 \text{ rad} = 57.3° \qquad\qquad\qquad 1° = 0.0175 \text{ rad}$$

$$90° = \frac{\pi}{2} \text{ rad}$$

$$1 \text{ rpm} = 0.105 \text{ rad/s} = 6°/\text{s} \qquad 1 \text{ rad/s} = 9.55 \text{ rpm}$$

D.7 Speed

$$1 \text{ m/s} = 3.28 \text{ ft/s} = 3.6 \text{ kph} = 2.24 \text{ mph}$$

$$1 \text{ ft/s} = 0.305 \text{ m/s} = 0.682 \text{ mph} = 1.1 \text{ kph}$$

$$1 \text{ kph} = 0.278 \text{ m/s} = 0.621 \text{ mph} = 0.911 \text{ ft/s}$$

$$1 \text{ mph} = 1.47 \text{ ft/s} = 1.61 \text{ kph} = 0.45 \text{ m/s}$$

$$60 \text{ mph} = 88 \text{ ft/s}$$

$$100 \text{ kph} = 62.5 \text{ mph}$$

D.8 Force

$$1 \text{ N} = 0.225 \text{ lb} = 10^5 \text{ dynes} \qquad 1 \text{ lb} = 4.45 \text{ N} = 4.45 \times 10^5 \text{ dynes}$$

$$1 \text{ dyne} = 10^{-5} \text{ N} = 2.25 \times 10^{-6} \text{ lb}$$

The equivalent weight of a 1-kg (9.8-N) mass = 2.2 lb.

D.9 Work and Energy

$$1 \text{ J} = 0.738 \text{ ft-lb} = 0.239 \text{ cal} = 9.48 \times 10^{-4} \text{ Btu} = 6.24 \times 10^{18} \text{ eV}$$

$$1 \text{ kcal} = 4186 \text{ J} = 3.968 \text{ Btu} \qquad 1 \text{ cal} = 4.186 \text{ J} = 3.97 \times 10^{-3} \text{ Btu}$$

$$1 \text{ Btu} = 1055 \text{ J} = 778 \text{ ft-lb} = 0.252 \text{ kcal}$$

$$1 \text{ ft-lb} = 1.36 \text{ J} = 1.29 \times 10^{-3} \text{ Btu} \qquad 1 \text{ eV} = 1.6 \times 10^{-19} \text{ J}$$

$$1 \text{ kWh} = 3.6 \times 10^6 \text{ J} = 860.4 \text{ kcal}$$

D.10 Power

$$1 \text{ W} = 0.738 \text{ ft-lb/s} = 1.34 \times 10^{-3} \text{ hp} = 3.41 \text{ Btu/h}$$

$$1 \text{ hp} = 745.7 \text{ W} = 550 \text{ ft-lb/s} = 2545 \text{ Btu/h}$$

$$1 \text{ kW} = 1.34 \text{ hp}$$

D.11 Pressure

$$1 \text{ pascal (N/m}^2) = 1.45 \times 10^{-4} \text{ lb/in.}^2 = 7.5 \times 10^{-3} \text{ torr (mm Hg)}$$

$$1 \text{ torr (mm Hg)} = 133 \text{ Pa} = 0.02 \text{ lb/in.}^2$$

$$1 \text{ atm} = 14.7 \text{ lb/in.}^2 = 1.01 \times 10^5 \text{ N/m}^2 = 30 \text{ in. Hg} = 76 \text{ cm Hg}$$

$$1 \text{ bar} = 10^5 \text{ Pa} = 1 \text{ atm}$$

D.12 Temperature

$$T_{\text{F}} = \frac{9}{5}T_{\text{C}} + 32$$

$$T_{\text{C}} = \frac{5}{9}(T_{\text{F}} - 32)$$

$$T_{\text{K}} = T_{\text{C}} + 273.2$$

Periodic Table of the Elements

I A	II A	III B	IV B	V B	VI B	VII B	VIII			I B	II B	III A	IV A	V A	VI A	VII A	INERT GASES	
1 **H** 1.00797																1 **H** 1.00797	2 **He** 4.0026	2
3 **Li** 6.939	4 **Be** 9.0122											5 **B** 10.811	6 **C** 12.01115	7 **N** 14.0067	8 **O** 15.9994	9 **F** 18.9984	10 **Ne** 20.183	2 8
11 **Na** 22.9898	12 **Mg** 24.312											13 **Al** 26.9815	14 **Si** 28.086	15 **P** 30.9738	16 **S** 32.064	17 **Cl** 35.453	18 **Ar** 39.948	2 8 8
19 **K** 39.102	20 **Ca** 40.08	21 **Sc** 44.956	22 **Ti** 47.90	23 **V** 50.942	24 **Cr** 51.996	25 **Mn** 54.9380	26 **Fe** 55.847	27 **Co** 58.9332	28 **Ni** 58.71	29 **Cu** 63.54	30 **Zn** 65.37	31 **Ga** 69.72	32 **Ge** 72.59	33 **As** 74.9216	34 **Se** 78.96	35 **Br** 79.909	36 **Kr** 83.80	2 8 18 8
37 **Rb** 85.47	38 **Sr** 87.62	39 **Y** 88.905	40 **Zr** 91.22	41 **Nb** 92.906	42 **Mo** 95.94	43 **Tc** (99)	44 **Ru** 101.07	45 **Rh** 102.905	46 **Pd** 106.4	47 **Ag** 107.870	48 **Cd** 112.40	49 **In** 114.82	50 **Sn** 118.69	51 **Sb** 121.75	52 **Te** 127.60	53 **I** 126.9044	54 **Xe** 131.30	2 8 18 18 8
55 **Cs** 132.905	56 **Ba** 137.34	57 ***La** 138.91	72 **Hf** 178.49	73 **Ta** 180.948	74 **W** 183.85	75 **Re** 186.2	76 **Os** 190.2	77 **Ir** 192.2	78 **Pt** 195.09	79 **Au** 196.967	80 **Hg** 200.59	81 **Ti** 204.37	82 **Pb** 207.19	83 **Bi** 208.980	84 **Po** (210)	85 **At** (210)	86 **Rn** (222)	2 8 18 32 18 8
87 **Fr** (223)	88 **Ra** (226)	89 †**Ac** (227)																

( ) Numbers in parentheses are mass numbers of most stable or most common isotope.

Atomic weights corrected to conform to the 1963 values of the Commission of Atomic Weights.

*Lanthanum Series

58 **Ce** 140.12	59 **Pr** 140.907	60 **Nd** 144.24	61 **Pm** (147)	62 **Sm** 150.35	63 **Eu** 151.96	64 **Gd** 157.25	65 **Tb** 158.924	66 **Dy** 162.50	67 **Ho** 164.930	68 **Er** 167.26	69 **Tm** 168.934	70 **Yb** 173.04	71 **Lu** 174.97	2 8 18 32 9 2

†Actinium Series

90 **Th** 232.038	91 **Pa** (231)	92 **U** 238.03	93 **Np** (237)	94 **Pu** (242)	95 **Am** (243)	96 **Cm** (247)	97 **Bk** (247)	98 **Cf** (249)	99 **Es** (254)	100 **Fm** (253)	101 **Md** (256)	102 **No** (256)	103 **Lr** (257)	2 8 18 32 9 2

Glossary

ABSOLUTE PRESSURE The total pressure on an object, or the gauge pressure plus the atmospheric pressure.

ACCELERATION The rate of change of speed over time.

ADIABATIC PROCESS A process in which there is no heat transfer from inside a system to outside the system.

ALTERNATING CURRENT (AC) A current that changes in both amplitude and polarity.

ANGULAR MOMENTUM The product of an object's moment of inertia and its angular speed.

AVERAGE VELOCITY The total distance traveled over the time a trip takes.

BROWNIAN MOTION The erratic motion of smoke particles caused by the particles being unevenly hit on the sides by the molecules of the air in which they are suspended.

BUOYANT FORCE The force acting up on an object immersed in a fluid that is equal to the weight of the displaced fluid.

CAPACITANCE The charge stored by a capacitor for a given voltage.

CAPACITOR An electrical device that stores energy in an electric field produced through the separation of charge.

CELSIUS OR CENTIGRADE TEMPERATURE SCALE The temperature scale tied to the phase changes of water.

CENTER OF MASS (CM) The point at which all of a body's mass seems to be concentrated.

CENTRIFUGAL FORCE The reaction (from Newton's third law) to the centripetal force. We experience it as the normal force from the object in which we're riding when the object moves in a curved path.

CENTRIPETAL FORCE The force generated from the change in direction of a velocity directed toward the center of the circle in which the object moves.

CHROMATIC ABERRATION The smearing of an image from a refracting lens caused by differing indexes of refraction for the different wavelengths of light.

COEFFICIENT OF RESTITUTION (*e*) The ratio of the speed of an object just after a collision with an immobile surface to the speed just before the collision.

COHERENCE The state in which two waves have a fixed phase difference.

COMPRESSION A force pair applied to a material that causes the material to shorten.

CONCAVE SURFACE A surface that curves inward from the lens or mirror.

CONSERVATION OF ENERGY Energy conversion from one form to another without losses.

CONVECTION Heat flow in a fluid caused by mixing through density gradients.

CONVEX SURFACE A surface that curves outward from the lens or mirror.

COVALENT BOND A bond in which atoms share electrons between or among themselves, the shared electrons belonging to no single atom.

CRITICAL ANGLE The incident angle at which a light ray traveling from a medium with a high index of refraction into a medium with a low index of refraction is refracted at 90°.

CROSS PRODUCT The vector product between two vector quantities and the sine of the angle between them.

CURRENT The flow of electric charge.

CUTOFF FREQUENCY The frequency of light below which the photoelectric effect ceases entirely.

DENSITY The ratio of a mass to the space occupied by that mass. The space is usually a volume, but it can be an area or a linear distance.

DICHROIC MATERIAL A material with an optical axis such that E-field components parallel to the components are transmitted, and components perpendicular to it are absorbed.

DIFFERENTIAL A small change in some quantity, such as distance or time.

DIODE An electronic component that allows current to flow in only one direction.

DOPING The addition of small concentrations of impurities to a crystal to alter the electrical characteristics of the crystal.

DOPPLER EFFECT An apparent frequency shift of a wave due to the relative movement of the source toward or away from you.

DUCTILE MATERIAL A material that can be deformed plastically under tension into a new shape.

ELASTIC COLLISION A collision in which kinetic energy is conserved.

ELECTROMOTIVE FORCE (EMF) The no-load (no-current) potential of a voltage source.

ELECTRON A negatively charged particle that orbits the nucleus and that is much lighter than the proton or the neutron.

ELECTRON-VOLT (eV) The work done on an electron by a potential of 1 V ($1 \text{ eV} = 1.6 \times 10^{-19}$ J).

EQUIPARTITION OF ENERGY The principle stating that the motion of molecules and atoms in each of the three spatial dimensions add equally to give the internal energy of a sample of gas.

FIRST LAW OF THERMODYNAMICS Law stating that energy can be neither created nor destroyed but can change forms.

FIRST RIGHT-HAND RULE The thumb of your right hand points in the direction of the particle's velocity. The fingers of your right hand with your palm open are in the direction of the magnetic field. Your open palm is in the direction of the force.

FISSION The splitting of a heavy atomic nucleus into smaller fragments to produce energy.

FLOW The rate at which mass passes a given point, or mass per unit time.

FOCAL LENGTH The length from the center of a lens to the point where parallel light rays entering the lens will intersect on the optical axis.

FORWARD BIAS External voltage that is applied across a PN junction and that allows conduction to occur.

FUSION The joining of the nuclei of light atoms to yield nuclear energy.

GAUGE PRESSURE The differential pressure between the inside and the outside of a volume.

HALF-LIFE The time it takes a radioactive isotope to decay to one-half of its original amount.

HEAT CAPACITY The ability of a material to store heat.

HOLE An unused electron bond in a doped crystal that acts like a free positive charge.

HOOKE'S LAW The greater the extension or compression of a spring from its equilibrium position, the greater the force on it: $\mathbf{F} = k\mathbf{x}$.

HUYGENS'S PRINCIPLE Every point on a wavefront can be considered as a source of secondary wavelets that spread out in all directions with the wave speed of the medium.

IMPEDANCE An abrupt change from one medium to another involving energy transfer that can cause loss of energy through a rebound or a reflection. Also, the total opposition to current seen by an ac source.

IMPULSE A large force acting on a body over a very short period of time.

INDEX OF REFRACTION The ratio of the speed of light in a vacuum to its speed in a given medium.

INDUCED CURRENT A current produced by a wire moving in a magnetic field.

INDUCTOR An electronic device that stores energy in a magnetic field.

INELASTIC COLLISION A collision in which kinetic energy is not conserved.

INTEGRATED CIRCUIT Many tiny, solid-state components etched and interconnected on a single crystal of semiconductor material.

INTENSITY The power received per unit area from a distant power source.

ISOBARIC PROCESS A process in which there is constant pressure.

ISOTHERMAL PROCESS A process in which there is constant temperature.

ISOTOPES Atoms of an element with the same atomic number but different mass numbers.

KINETIC ENERGY Energy in the motion of an object.

KINETIC FRICTION The force of contact between two bodies in motion that opposes that motion.

LAMINAR FLOW Fluid flow without turbulence.

LASER Light amplification by stimulated emission of radiation.

LIGHT RAY The straight-line path that light takes radially outward from the primary source.

LONGITUDINAL WAVE A wave in which the amplitude variation is in the direction of the propagation.

MAGNETIC FLUX Magnetic field lines intercepted by the area of a current loop.

MALLEABLE MATERIAL A material that can be deformed into a new shape under compression.

MASS The property of a body that opposes a change in its motion. Resistance to motion is also called *inertia*.

METASTABLE STATE An allowed energy state for orbiting electrons above the ground state where they may linger for a while before dropping back to a lower-energy state.

MOLECULE A group of atoms bonded together strongly enough to act as a single particle.

MOMENT OF INERTIA The resistance to the rotation of a given body about a given axis.

MOMENTUM The product of the mass and velocity of a body. If the momentum of a body changes, a force has acted on that body.

NEUTRON A neutral particle in the nucleus of an atom that is used to separate the protons and that is about the same mass as the proton. Also, a nuclear particle with no charge and about the same mass as the proton.

NEWTON'S FIRST LAW Every body persists in its state of rest or of uniform motion in a straight line unless it is compelled to change that state by forces impressed upon it.

NEWTON'S SECOND LAW A net force applied to a massive object causes the object to accelerate ($\mathbf{F} = m\mathbf{a}$).

NEWTON'S THIRD LAW For every action there is an equal and opposite reaction.

OHM (Ω) The unit of resistance (volts per ampere).

OPTICAL CAVITY The volume within which laser action takes place.

PATH INDEPENDENCE The work done by conservative forces is independent of the path taken in doing the work.

PAULI EXCLUSION PRINCIPLE No two electrons in orbit around a single atom can have exactly the same quantum numbers.

PHOTOELECTRIC EFFECT Effect produced when light shining on certain metals liberates electrons from the surface.

PHOTON A discrete wave packet (particle) of light that is always moving at the speed of light

PLASTIC DEFORMATION The response of some materials such that when they are placed under enough tension, they will lengthen and not return to their original equilibrium position.

POPULATION INVERSION The state in which most of a population of electrons are in a metastable state instead of in the ground state. The population is considered inverted because most electrons are normally found in the ground state.

POTENTIAL ENERGY Energy stored in the position of an object.

POWER The rate at which work is done.

POWER FACTOR The cosine of the phase angle.

PROTON A positively charged particle in the nucleus of an atom. The number of protons in the nucleus defines the atom's elemental identity. Also, a nuclear particle with a charge of $+1.6 \times 10^{-19}$ C and a mass of 1.67×10^{-27} kg.

RADIAN An angular measure for the circle based on pi (π), where 2π radians = 360°.

RADIATION Heat flow caused by the emission of photons.

RAYLEIGH'S CRITERION Two objects are resolvable if the center of one Airy disk coincides with the first minimum of the other.

REACTANCE Resistance to current flow by an inductor or a capacitor in an ac circuit.

RECTIFICATION Turning an ac signal into a dc signal.

REFRACTION The bending of a light ray that takes place at an interface when it enters a medium with a different index of refraction.

RESISTIVITY The opposition to current flow in a given material.

REVERSE BIAS External voltage that occurs across a PN junction, widening the depletion zone and blocking conduction.

RIGHT-HAND RULE 1 The right-hand rule used to find the direction of a torque or an angular momentum.

RIGHT-HAND RULE 2 The right-hand rule used to find the direction of an angular velocity.

ROOT-MEAN-SQUARE (RMS) VALUE The equivalent dc voltage or current delivered to a load from an ac sinusoidal signal.

ROTATIONAL EQUILIBRIUM The condition that occurs when the sum of the torques acting on a body equals zero.

SCALAR A quantity that has only magnitude.

SECOND LAW OF THERMODYNAMICS Law stating that natural processes tend to move in the direction of greater disorder (entropy).

SECOND RIGHT-HAND RULE The thumb of your right hand points in the direction of the positive current flow. The fingers of your hand naturally curve around the path of the current in the direction of the magnetic field.

SHEAR Two opposite and equal forces acting on a body, but *not* in the same line.

SIMPLE HARMONIC MOTION (SHM) A periodic phenomenon in which a system naturally oscillates sinusoidally.

SOLENOID A wire-wound tube with a homogeneous magnetic field inside.

SPECULAR REFLECTION Reflection of light that forms an image.

SPEED The rate of change of distance over time.

SPHERICAL ABERRATION For a spherical mirror, the shifting of the focal length toward a mirror's surface as light rays hit the mirror farther and farther from the optical axis.

SPONTANEOUS EMISSION An event in which an electron drops down from an excited allowed energy state, emitting a photon, without any external stimulation.

STATIC FRICTION The force of contact between two bodies at rest that opposes motion.

STOPPING POTENTIAL The negative collector voltage when the photocurrent becomes zero.

STRAIN The change in length of an object undergoing stress.

STRESS The application of equal and opposite forces to a body.

SUPERPOSITION Constructive or destructive interference of one wave with another. The waves occupy the same space at the same time as they move through one another.

TEMPERATURE The average kinetic energy of the molecules of a gas.

TENSILE STRESS An equal and opposite force pair applied to a body, causing it to lengthen.

THERMAL CONDUCTIVITY The ability of heat to flow through a material.

THERMAL RESISTANCE The ability of a material to block heat flow, or to act as an insulator.

THERMOMETER An instrument used to measure temperature.

TRANSLATIONAL EQUILIBRIUM The condition that occurs when the sum of the forces acting on a body equals zero.

TRANSVERSE WAVE A wave in which the direction of propagation is 90° to the amplitude.

ULTIMATE STRENGTH The tensile stress that causes a material to break.

VAN DER WAALS FORCES The weak attraction between atoms and molecules in a gas caused by polarized molecules.

VECTOR A quantity that has both magnitude *and* direction.

VELOCITY Speed with the direction of motion given.

VOLTAGE The work done per unit charge; a measure of the PE of a system.

WORK The work done by a force **F** acting on an object undergoing a displacement **x** is equal to the magnitude of the force component in the direction of the displacement multiplied by the magnitude of the displacement.

WORK FUNCTION The minimum energy needed for an electron to leave the surface of a metal.

Index

Vertical plane, motion in a, 38–40, 46
Very-high frequency (VHF) signals, 546
Virtual images, 568, 577, 580–81
Virtual rays, 548–49, 577–78
Viscosity, 287–88
Visible light, 542. *See also* Light
Voltage, 396, 410. *See also* Electric *listings*; Induced voltages and currents; Sinusoidal currents and voltages
Volume expansion, 309

W

Water
    as an acid and a base, 671
    boiling point, 335
    conductivity, electrical, 670
    covalent bonding, 670
    density, 267
    dielectric constant, 402–3
    fusion, heat of, 335
    heat capacity, 331, 332
    index of refraction, 551
    ionic bonds, 670
    melting point, 335
    thermal conductivity, 337
    triple point of, 343–44
    vaporization and fusion curves for, 342–43
    vaporization, heat of, 335
    volume expansion, 309
Watts, 86–87
Wave motion
    beats, 200–202
    concepts, key, 203
    Doppler effect, 191–92
    electromagnetic waves, 536–41
    equations, important, 203
    equation, wave, 541

    exercises, 204–6
    intensity, 196–97
    matter waves, 620
    microwaves, 541–42
    problems, conceptual, 203–4
    pulses on a string, 178–80, 182–83
    radar and sonar, 197–200
    radiation, 342
    radio waves, 539–41, 543, 546
    reflections at a boundary, 180–81
    sonic boom, 192–95
    sound waves, 189–91
    standing waves on a string, 184–89
    summary, chapter, 202
    superposition principle, 181–82
    traveling wave, 210–12
Wave nature of light, 195–96
    concepts, key, 613
    diffraction gratings and X-ray diffraction, 602–5
    double-slit interference, 600–602
    equations, important, 614
    exercises, 615–17
    interference and coherence, 598–600
    particle nature of light contrasted with, 624, 631–34
    polarization, 609–12
    problems, conceptual, 614–15
    resolving power, 608–9
    scattering and blue skies, 612–13
    summary, chapter, 613
    thin films and antireflection coatings, 605–7
Waves/particles and special relativity
    concepts, key, 636
    equations, important, 637
    exercises, 638–39

Heisenberg uncertainty principle, 633–34
    matter waves, 631–32
    photoelectric effect, 622–26
    problems, conceptual, 637–38
    relativity, special, 316–17, 626–30, 638, 651
    Schrödinger equation, 634–36
    summary, chapter, 636
    X-rays, 630–31
Waxed paper and dielectric constant, 403
Weak force, 659
Weight and mass, relationship between, 52
White light spectrum, 543, 552–53, 603
Wind as a vector, 30
Wires and magnetism, 460–61, 478–80
Wood
    construction methods, 276–77
    density, 267
    heat capacity, 331
    resistance, electrical, 422
    thermal conductivity, 337
Work. *See* Energy and work
Work function, 622, 624, 630

X

X-ray diffraction, 604–5
X rays, 630–31, 638, 643

Y

Young's modulus, 273, 277, 280

Z

Zircon and index of refraction, 551